Air Pollution Modeling and Its Application XIII

Previous Volumes in this Mini-Series

Volumes I–XII were included in the NATO Challenges of Modern Society Series.

AIR POLLUTION MODELING AND ITS APPLICATION I
Edited by C. De Wispelaere

AIR POLLUTION MODELING AND ITS APPLICATION II
Edited by C. De Wispelaere

AIR POLLUTION MODELING AND ITS APPLICATION III
Edited by C. De Wispelaere

AIR POLLUTION MODELING AND ITS APPLICATION IV
Edited by C. De Wispelaere

AIR POLLUTION MODELING AND ITS APPLICATION V
Edited by C. De Wispelaere, Francis A. Schiermeier, and Noor V. Gillani

AIR POLLUTION MODELING AND ITS APPLICATION VI
Edited by Han van Dop

AIR POLLUTION MODELING AND ITS APPLICATION VII
Edited by Han van Dop

AIR POLLUTION MODELING AND ITS APPLICATION VIII
Edited by Han van Dop and Douw G. Steyn

AIR POLLUTION MODELING AND ITS APPLICATION IX
Edited by Han van Dop and George Kallos

AIR POLLUTION MODELING AND ITS APPLICATION X
Edited by Sven-Erik Gryning and Millán M. Millán

AIR POLLUTION MODELING AND ITS APPLICATION XI
Edited by Sven-Erik Gryning and Francis A. Schiermeier

AIR POLLUTION MODELING AND ITS APPLICATION XII
Edited by Sven-Erik Gryning and Nadine Chaumerliac

Air Pollution Modeling and Its Application XIII

Edited by

Sven-Erik Gryning
Risø National Laboratory
Roskilde, Denmark

and

Ekaterina Batchvarova
National Institute of Meteorology and Hydrology
Sofia, Bulgaria

Kluwer Academic / Plenum Publishers
New York, Boston, Dordrecht, London, Moscow

Library of Congress Cataloging-in-Publication Data

Air pollution modeling and its application XIII / Sven-Erik Gryning and Ekaterina Batchvarova.
p. cm.
Includes bibliographical references and index.
ISBN 0-306-46188-9
1. Air--Pollution--Mathematical models Congresses. 2. Atmospheric diffusion--Mathematical models Congresses. I. Gryning, Sven-Erik. II. Batchvarova, Ekaterina. III. NATO/CCMS International Technical Meeting on Air Pollution Modeling and Its Application (23rd : 1998 : Varna, Bulgaria)
TD881.A47524 1999
628.5'3'015118--dc21 99-37360
CIP

FRONT COVER

Nessebar is one of the oldest towns in Bulgaria, famous for the ancient ruins, medieval and renaissance architecture and contemporary artistic spirit. Two thousand years B.C. Thracian people established the town of Messembria. In 510 B.C. the town became an ancient Greek colony. Starting from first century A.D., the town belonged to the Roman and later to the Byzantine empire. The medieval time is connected with the history of the Bulgarian state.

The town in now included in the list of the world cultural heritage by UNESCO and has the status of a town-museum in Bulgaria. Tourists walk along the narrow streets of Nessebar, looking to the blue sea and the stands of the artists. The artist Emanuil Thaliovsky has drawn one of the picturesque small streets of Nessebar for the 23rd NATO/CCMS ITM.

Nessebar is situated on a small peninsula, 35 kilometers north of Burgass and 100 kilometers south of Varna. The excellent sand beaches with natural dunes attract many Bulgarian and foreign tourists.

Proceedings of the Twenty-Third NATO/CCMS International Technical Meeting on Air Pollution Modeling and Its Application, held September 28–October 2, 1998, in Riviera near Varna, Bulgaria

ISBN: 0-306-46188-9

233 Spring Street, New York, New York 10013

http://www.wkap.nl/

10 9 8 7 6 5 4 3 2 1

A C.I.P. record for this book is available from the Library of Congress

Printed in the United States of America

PREFACE

The ITM conference series has always had a strong spirit of cooperation under the NATO/CCMS umbrella, and with the considerable interest from Partner countries to participate in the ITM conferences, it provides an excellent opportunity to create ties between scientists. Whereas all previous ITM conferences have taken place in NATO countries, the 23rd ITM takes place in a Cooperative Partner country, Bulgaria, and is hosted by the National Institute of Meteorology and Hydrology, Bulgarian Academy of Sciences. This fact reflects a general wish for a closer connection and collaboration among scientists from Partner and NATO countries.

This volume contains the papers from the 23rd NATO/CCMS International Technical Meetings on Air Pollution Modelling and Its Application, being held September 28 – October 2, 1998, at Riviera Holiday Club, Varna, Bulgaria. It was attended by 120 participants from 30 countries. Thanks are due to all who made it possible to plan, carry through, and follow up the meeting, and to the participants who made the conference so successful. Special thanks are due to the sponsoring institutions:

NATO/CCMS EURASAP RISØ

NIMH BAS

3M Prestige CIM

NATO/CCMS - Committee on the Challenges of Modern Society
EURASAP - European Association for the Science of Air Pollution
RISØ - Risø National Laboratory, Denmark
NIMH - National Institute of Meteorology and Hydrology, Bulgaria
BAS - Bulgarian Academy of Sciences
3M Representation office, Bulgaria
Prestige Business Ltd., Bulgaria

The excellent collaboration with CIM (Company for International Meetings Ltd., Bulgaria) is specially acknowledged. The next conference in this series, the Millennium ITM, will be held in the Western USA.

Sven-Erik Gryning
Chairman Scientific Committee

Ekaterina Batchvarova and Dimiter Syrakov
Local hosts, Bulgaria

THE SCIENTIFIC COMMITTEE OF THE 23rd NATO/CCMS INTERNATIONAL TECHNICAL MEETINGS ON AIR POLLUTION MODELLING AND ITS APPLICATION

G. Schayes, Belgium
D. Syrakov, Bulgaria
D. Steyn, Canada
S. E. Gryning (chairman), Denmark
H. R. Olesen, Denmark
W. Klug, Germany
N. Chaumerliac, France
G. Kallos, Greece
D. Anfossi, Italy
H. van Dop, Holland
T. Iversen, Norway
C. Borrego, Portugal
R. Salvador, Spain
B. Fisher, United Kingdom
F. A. Schiermeier, USA
Y. Schiffmann, USA

HISTORY OF THE NATO/CCMS AIR POLLUTION PILOT STUDIES
Pilot Study on Air Pollution: International Technical Meetings (ITM) on Air Pollution Modelling and Its Application

Dates of Completed Pilot Studies:

1969 - 1974 Air Pollution Pilot Study (United States Pilot Country)
1975 - 1979 Air Pollution Assessment Methodologies and Modelling (Germany)
1980 - 1984 Air Pollution Control Strategies and Impact Modelling (Germany)

Dates and Locations of Pilot Study Follow-Up Meetings:

Pilot Country - United States (R.A. McCormick, L.E. Niemeyer)

February 1971 - Eindhoven, The Netherlands, First Conference on Low Pollution Power Systems Development
July 1971 - Paris, France, Second Meeting of the Expert Panel on Air Pollution Modelling

---

All following meetings were entitled NATO/CCMS International Technical Meetings (ITM) on Air Pollution Modelling and Its Application

---

October 1972 - Paris, France - Third ITM
May 1973 - Oberursel, Federal Republic of Germany - Fourth ITM
June 1974 - Roskilde, Denmark - Fifth ITM

Pilot Country - Germany (Erich Weber)

September 1975 - Frankfurt, Federal Republic of Germany - Sixth ITM
September 1976 - Airlie House, Virginia, USA - Seventh ITM
September 1977 - Louvain-La-Neuve, Belgium - Eighth ITM
August 1978 - Toronto, Ontario, Canada - Ninth ITM
October 1979 - Rome, Italy - Tenth ITM

Pilot Country - Belgium (C. De Wispelaere)

November 1980 - Amsterdam, The Netherlands - Eleventh ITM
August 1981 - Menlo Park, California, USA - Twelfth ITM
September 1982 - Ile des Embiez, France - Thirteenth ITM
September 1983 - Copenhagen, Denmark - Fourteenth ITM
April 1985 - St. Louis, Missouri, USA - Fifteenth ITM

Pilot Country - The Netherlands (Han van Dop)

April 1987 - Lindau, Federal Republic of Germany - Sixteenth ITM
September 1988 - Cambridge, United Kingdom - Seventeenth ITM
May 1990 - Vancouver, British Columbia, Canada - Eighteenth ITM
September 1991 - Ierapetra, Crete, Greece - Nineteenth ITM

Pilot Country - Denmark (Sven-Erik Gryning)

November 1993 - Valencia, Spain - Twentieth ITM
November 1995 - Baltimore, Maryland, USA - Twenty-first ITM
June 1997 - Clermont-Ferrand, France - Twenty-second ITM
September 1998 - Varna, Bulgaria - Twenty-third ITM

CONTENTS

INTEGRATED REGIONAL MODELLING

GLOBAL AND LONG-RANGE TRANSPORT

ETEX SESSION

NEW DEVELOPMENTS

ACCIDENTAL RELEASES

MODEL ASSESSMENT AND VERIFICATION

POSTER SESSION

INTEGRATED REGIONAL MODELLING

chairmen:	D. Anfossi
	C. Borrego
	S. E. Gryning
	M. Rotach
	F. Schiermeier
rapporteurs:	M. Kaasik
	M. Kolarova
	J. Wilkinson
	M. Tayanc
	S. Rafailidis

AN INTEGRATED MODELING AND OBSERVATIONAL APPROACH FOR DESIGNING OZONE CONTROL STRATEGIES FOR THE EASTERN U. S.

S. Trivikrama Rao[1,2], Christian Hogrefe[1], Gopal Sistla[2], Shiang-Yuh Wu[2], Winston Hao[2], Eric Zalewsky[2], Igor G. Zurbenko[1], P. Steven Porter[3], George Kallos[4], Kostas Lagouvardos[4], Vassiliki Kotroni[4], D. Alan Hansen[5], and Martin Smith[6]

[1] State University of New York - Albany, Albany, NY 12222
[2] New York State Department of Environmental Conservation, Albany, NY 12233
[3] University of Idaho, Idaho Falls, Idaho 83405
[4] University of Athens, Athens 10680, Greece
[5] Electric Power Research Institute, Palo Alto, CA 94303
[6] Niagara Mohawk Power Corporation, Syracuse, NY 13202

INTRODUCTION

Despite vigorous attempts to control the ozone problem during the past three decades, ozone levels in many areas over the Eastern United States continue to exceed the National Ambient Air Quality Standards. Until recently, photochemical models were applied to simulate historical ozone episodic events to examine the future ozone non-attainment problem. When the model's ability to reproduce the observed ozone air quality was deemed acceptable, control measures needed to meet and maintain the ozone standards were evaluated using projected emissions inventories and historical episodic meteorological conditions. Since the episodic meteorological events under which the model has performed best may never occur in the future, there is an inherent uncertainty in the controls identified as required to comply with the ozone standards.

Because of the recent revisions to the ozone standards in the United States, there is an interest in performing regional-scale ozone modeling on a seasonal basis. Bastrup-Birk et al (1997) applied a photochemical model for a 7-year period over Europe to study cumulative ozone exposures. Such seasonal ozone modeling would allow a more thorough examination of model reproduction of the observed spatial and temporal scales and patterns in ozone concentrations. In addition, seasonal modeling would instill greater confidence in emission control strategies.

In this paper, we evaluate the modeling system's performance for the Eastern United States during the summer of 1995 using UAM-V (version 1.24), a photochemical model developed by Systems Applications Incorporated (SAI, 1995), and RAMS, a meteorological model discussed by Pielke and Uliasz (1998). UAM-V is a regional-scale ozone air quality model, which contains features such as multi-scale modeling capability, grid-nesting, plume-in-grid treatment for elevated sources, etc.

Model performance is determined via comparison of observed and model-predicted ozone concentrations at various time and space scales. Time scales include intra-day, diurnal, synoptic, and longer-term variations. Space scales are described in terms of the decay of the process correlation in space (decay of the correlations between different stations as a function of distance from a given location). Techniques for comparing the modeled and observed ozone time series data on different time scales are presented. Finally, we examine changes in ozone at various time scales as emissions are changed and discuss the implications of these modeling results to air quality management practices. An approach to using the models in a regulatory setting which incorporates the concept of time and space scales is discussed.

Weather-related (synoptic-scale) variations in ambient ozone data over the Eastern United States decay exponentially downwind with an e-folding distance of about 600 km. The e-folding times, reflecting the synoptic-scale forcings in the data, range from 1-3 days, depending upon the location. The large spatial extent of ozone footprints are further evidence that the ozone problem in the Eastern United States needs to be addressed from a regional perspective.

The results of this study reveal that model-predicted ozone concentrations do not closely resemble ambient data: correlations (r) between the model-predicted and measured hourly ozone concentrations in the Eastern U.S averaged about 0.5 at the 496 grid points having monitoring sites. The correlations are also different among different temporal components of modeled and observed concentrations. The model simulates the diurnal component reasonably well, with spatially-averaged correlations of 0.7; this is not surprising since both observations and modeled values respond to day and night differences. For synoptic and baseline (seasonal plus longer-term) components, the spatially-averaged correlations are 0.3 and 0.4, respectively. Intra-day variations resemble a white noise process, and do not contain information either in time or in space. At most locations, the model underestimates the variances of shorter-term (intra-day and diurnal) components and overestimates the variances of longer-term (synoptic and baseline) components when compared with those in the observed data. As expected, the level of the ozone baseline is reduced when emissions are reduced.

METHODS

Observation data base

Hourly ozone concentrations from various monitoring stations in the Eastern United States were obtained from the U.S. Environmental Protection Agency's (EPA) AIRS data base for the 1995 summer (4 June -31 August).

Photochemical Modeling

Model Set Up

The UAM-V (version 1.24) was designed for application to the Eastern United States with a fine (12 km) grid extending from 92^0 W to 69.5^0 W (long.) and from 32^0 N to 44^0 N (lat.), and a coarse grid (36 km) extending from 99^0 W to 67^0 W and from 26^0 N to 47^0 N. There are 14 vertical layers, extending from the surface to 4 km. The meteorological fields for the summer of 1995 were derived by applying the RAMS3b (RAMS-version 3b) with nested grids at a horizontal grid resolution of 12, 36, and 108 km, where the outer grid covered most of North America. Hourly meteorological data for 1 June to 31 August, 1995 were simulated with RAMS3b using 4-dimensional data assimilation (4-DDA) techniques (Kallos et al., 1997). The anthropogenic emissions inventory for the summer months of 1995 was updated using all available source-specific information. Natural emissions were estimated from EPA's BEIS2 model and temperature-adjusted based on the RAMS3b temperature fields. Similarly, day-specific mobile source emissions were calculated from EPA's MOBILE5a model and temperature-corrected using RAMS3b data. Additional details on UAM-V and its previous applications to the Eastern United States can be found in OTAG (1996). The ramp-up days for this study are June 1-3.

Modeling scenarios

Several UAM-V simulations were performed for all three summer months of 1995 even though there were only three episodic events with high ozone levels in the Eastern United States. In addition to the base case, NOx/VOC emission reduction scenarios, reflecting an across-the-board reduction in elevated and ground-level emissions of magnitudes 25%/25%, 25%/50%, 50%/25%, and 75%/25%, respectively, were simulated. These reduction scenarios permit examination of the relative efficacies of VOC-focused versus NOx-focused emission controls over the modeling domain.

Data analysis

Extraction of time scales embedded in modeled and measured ozone

It is well known that meteorological and air quality variables contain intra-day, diurnal, synoptic, seasonal, and longer-term variations, each attributable to different dynamical processes. In ozone, intra-day variations are attributable to the photochemical production and meteorological processes, diurnal variation to night and day differences, synoptic-scale variations to the weather and short-term fluctuations in precursor emissions, seasonal scale to the variation in the solar angle, and long-term scale to changes in climate, policy, and/or economics (Rao et al., 1998). Thus, the time series of hourly ozone data can be represented by:

$$O(t) = T(t) + I(t) + D(t) + W(t) + S(t) + e(t) \quad \textbf{(1)}$$

where O(t) is the natural logarithm of the original ozone time series, T(t) is turbulence-related changes, I(t) is intra-day variation, D(t) is diurnal variation, W(t) is weather-induced variation, S(t) is seasonal change, e(t) is long-term (trend) component, and t is time. The measured and modeled ozone concentrations were log-transformed to homogenize the variance. Since we deal with hourly ozone, T(t) is not considered further. The additive decomposition of log-transformed ozone time series in equation (1) exemplifies the aspect that ozone concentrations are interpreted as a baseline caused by polluting sources and multiplied by factors which are dependent on meteorological conditions.

In this study, we applied the Kolmogorov-Zurbenko (KZ) iterative moving average filter (Zurbenko, 1991) to separate the different physical processes (components) described by (1). Properties of the filter as well as applications to ozone and meteorological variables have been described in Rao et al. (1996), Porter et al. (1996), and Milanchus et al. (1998). The performance of other methods in separating the scales of motion has been discussed by Eskridge et al.(1997). Sample estimates of individual components were obtained from the 89 days of hourly data as follows.

$$\text{estimate of } D(t) = KZ_{3,3} - KZ_{13,5} \tag{2}$$

$$\text{estimate of } W(t) = KZ_{13,5} - KZ_{103,5} \tag{3}$$

$$\text{estimate of } S(t) + e(t) = KZ_{103,5} \tag{4}$$

$$\text{estimate of } I(t) = O(t) - KZ_{3,3} \tag{5}$$

where $KZ_{m,k}$ refers to k iterations of a simple moving average filter of width m hours. The 50% cutoff frequencies (see Rao et al., 1997) for the $KZ_{3,3}$, $KZ_{13,5}$, and $KZ_{103,5}$ filters are 1/11, 1/64, and 1/505 ($hour^{-1}$). Since this modeling study spans only three months in 1995, equation (5) contains only a small amount of long-term information. Although we have chosen the KZ filter here, we should note that any spectral decomposition technique (e.g., wavelet transform) capable of cleanly separating the different components will achieve the goal of providing a better understanding of the underlying physical processes that affect ozone concentrations.

The different time scales imbedded in 10 years of ozone time series data at the Greenbelt, MD monitoring site when separated by the KZ filter reveal that whereas the intra-day component contributes approximately 4%, the diurnal and synoptic components contribute about 46% and 32%, respectively, to the total variance in the ozone data. During the July 1995 episode, hourly ozone concentrations at this site exceeded the 124 ppb level on July 15 (Figure 1). The decomposition of time series data in the log-scale provides multiplicative effect in the ppb-scale. The σ values noted in Figure 1 provide approximate percentage contribution above the level of the ozone baseline. Whether it is an episode or a non-episode, the intra-day component reflects stochastic variation at a fixed level. The variation in the diurnal component during episodes and non-episodes is similar. However, the level of the ozone baseline on July 15, the episode day, is about 15% higher than that on July 12, the non-episode day. Also, the synoptic forcing on July 15 is about 25% higher than that on July 12. It is evident in Figure 1 that ozone baseline and synoptic forcing work together to produce ozone exceedances. Hence, ozone management efforts should focus on reducing the ozone baseline; it is through the baseline that it is possible to address the peak ozone levels in a non-attainment area.

Spatial smoothing

Since observations are not available at each model grid point, we have reconstituted data at grid locations that do not have a monitor with a simple data restoration algorithm that

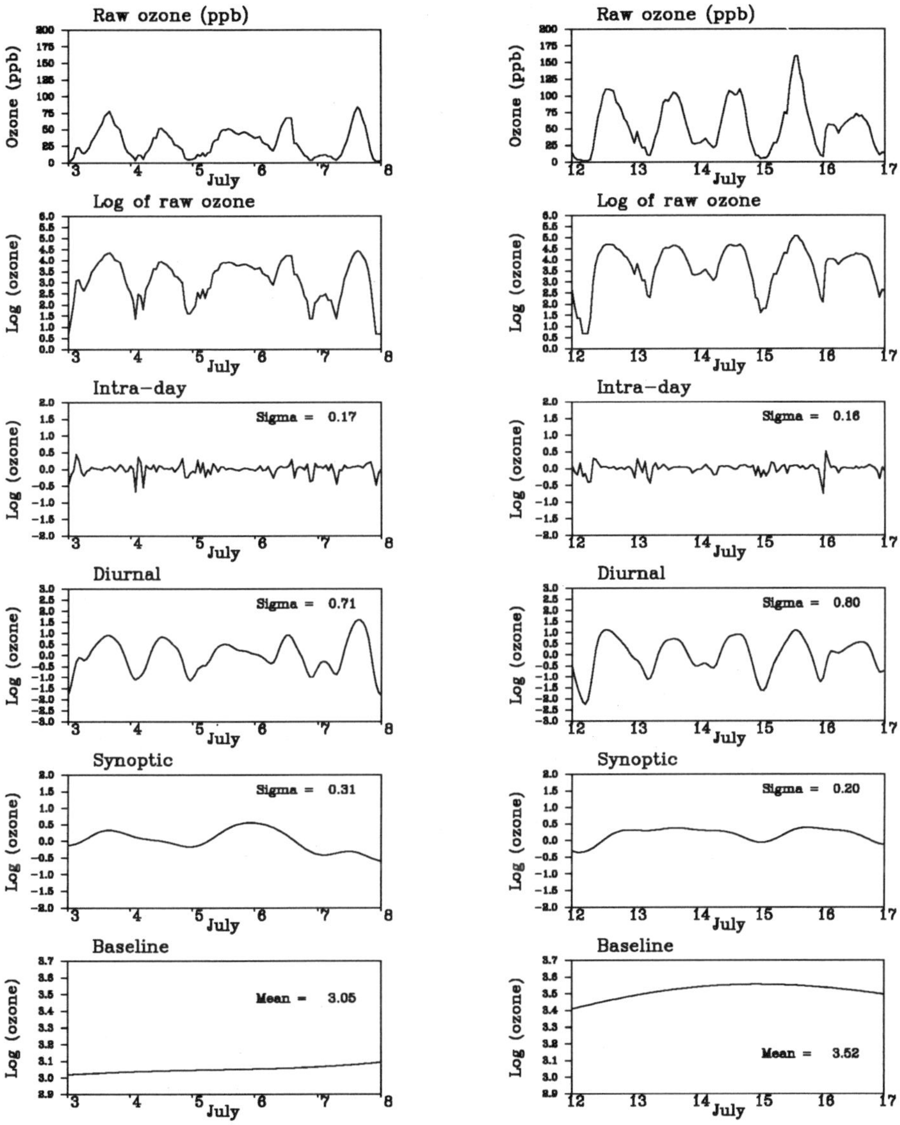

Figure 1. Time series data of ozone in ppb-scale, log-scale, and the intra-day, diurnal, synoptic and baseline components during non-episodic and episodic days in July 1995

Note: Ozone in the ppb-scale is represented as:
O(t) = exp(baseline) * exp(synoptic) * exp(diurnal) * exp(intra-day)

For example on July 15, the peak ozone concentration is about 165 ppb and it can be calculated as:
exp(3.53) * exp(0.5) * exp(0.9) * exp(0.16) ~ 165 ppb
For the intra-day component, the calculated sigma value was used. This value represents the percentage contribution of this stochastic component to the total concentration at the one-sigma confidence level.

averages the daily maximum ozone over a 5x5 grid (as in the model grid) nearest to the observed location. The averaging process is iterated until all grid points are filled. At each iteration and in the final smoothed surface, actual observations are used at grid points that have monitors. Thus, this spatial kernel smoothing procedure leaves the actual observed values untouched while extrapolating the observations to the grids that have no data.

Traditional statistical methods for model performance evaluation

It is a common practice among the air pollution modeling community to compare raw data of modeled and observed concentration values employing traditional statistical methods such as mean bias, normalized error, peak prediction accuracy, etc. (Morris et al., 1990). Wilmott (1991) suggested the use of measures such as index of agreement, systematic and unsystematic errors, etc., in evaluating the model performance. The index of agreement is a measure of the degree to which the observed variate is accurately estimated by the simulated variate. When the mean square error is entirely systematic, further refinement in the model is needed since the mean square error should be minimized for the model to predict at its maximum possible accuracy. If the mean square error is composed entirely of unsystematic error, perhaps the model is as good as it can be and may not need major modification.

RESULTS

Assessing model performance

The results from traditional performance evaluation measures for the modeling system are presented in Table 1. There is good agreement between the observed and predicted maxima over the domain. The index of agreement measure suggests that the model's performance is slightly better for the Northeast than the Midwest. Correlation analyses between observed and modeled daily maxima of 1-hr and 8-hr values at each grid point indicates that although the correlations are not very high (r values less than 0.8), the model performance is much better along the eastern seaboard than inland (Figure 2).

Over the 1995 summer season, the spatial pattern of exceedances of 1-hr, 124 ppb and 8-hr, 84 ppb ozone thresholds are quite different (Figure 3). Exceedances of the 1-hr ozone threshold value of 124 ppb are confined to the northeastern urban corridor, Atlanta and Lake Michigan areas. When 8-hr average ozone concentrations are considered, however, the non-attainment region covers a large portion of the Midwest. Thus, the ozone non-attainment problem relating to the new standard covers most of the Eastern United States.

Model Assessment on Different Time Scales

In analyzing the performance of the model on different scales, it should be noted that we did not use the spatially-smoothed observations, but worked with the observations at each individual monitoring location and modeled values at that location. The 4th highest ozone concentration during the summer of 1995 at each monitoring station based upon observations and modeled values over the modeling domain, presented in Figure 5, reveal the model's overprediction at many locations. Power spectra of measured and model-predicted ozone concentrations at Greenbelt, MD grid location indicate that the model performs differently on different scales. It underestimates the variances of intra-day and diurnal components and overestimates the variances of synoptic and longer-term components (Figure 4a). Both in the

Table 1. Traditional statistics on the observed and predicted peak ozone concentrations for domain-wide seasonal and each episode basis, on all episodic days, and on the Northeast (NE), Southeast (SE), and Midwest (MW) domains of the regional model.

Metric	Seasonal	June18-20	July13-15	July31-Aug2	All Episodes	NE	SE	MW
Sample Size	47097	1589	1558	1597	4744	15826	9712	20427
Obs. Station-wide Max.	198	163	184	172	184	184	198	178
Pred. Max. Dom. wide	232	155	232	181	232	185	205	232.69
Pred. Max. Station-wide	196	142	176	151	176	176	196	163.45
Std. Dev. Obs.	21.46	22.29	23.37	26.82	25.25	23.57	21.47	19.72
Std. Dev. Pred.	21.32	17.88	23.96	23.62	23.08	23.59	20.74	19.76
Unpaired Acc.	-0.01	-0.12	-0.04	-0.12	-0.04	-0.04	-0.01	-0.08
Normalized Bias	0.12	-0.04	0.03	0.15	0.05	0.11	0.18	0.1
Avg. Abs. Gross Error	14.86	14.52	15.77	17.42	15.9	14.96	14.3	15.11
Root Mean Sq. Error	19.29	18.46	20.71	22.15	20.5	19.65	18.55	19.45
Index of Agreement	0.77	0.77	0.79	0.78	0.8	0.81	0.79	0.72
Systematic Error (%)	52	47	50	53	50	51	53	51
Unsystematic Error (%)	48	53	50	47	50	49	47	49

Table 2. Percent reduction in the σ of each component from the base case for different control runs for the Greenbelt, MD grid cell. The spatially-averaged reduction is shown in the shaded area. (For the baseline, the value represents the percent reduction in the mean ozone concentration.)

Component (at one location and over the whole model domain)	25%NO_x 25%VOC	25%NO_x 50%VOC	50%NO_x 25%VOC	75%NO_x 25%VOC
Intra-day Greenbelt, MD	13%	13%	23%	34%
Intra-day Spatially-Averaged	11%	12%	23%	34%
Diurnal Greenbelt, MD	9%	11%	20%	36%
Diurnal Spatially-Averaged	9%	9%	19%	30%
Synoptic Greenbelt, MD	15%	14%	30%	48%
Synoptic Spatially-Averaged	11%	11%	22%	35%
Baseline Greenbelt, MD	1.0%	2.4%	4.3%	15.7%
Baseline Spatially-Averaged	1.2%	2.7%	3.9%	13.8%

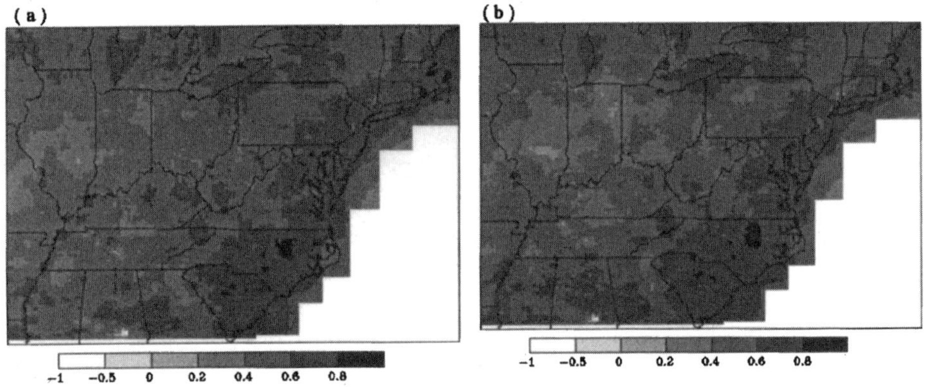

Fig 2. Correlation between the daily maxima of observed and predicted ozone concentrations over the summer of 1995; (a) 1-hr ozone; (2) 8-hr ozone

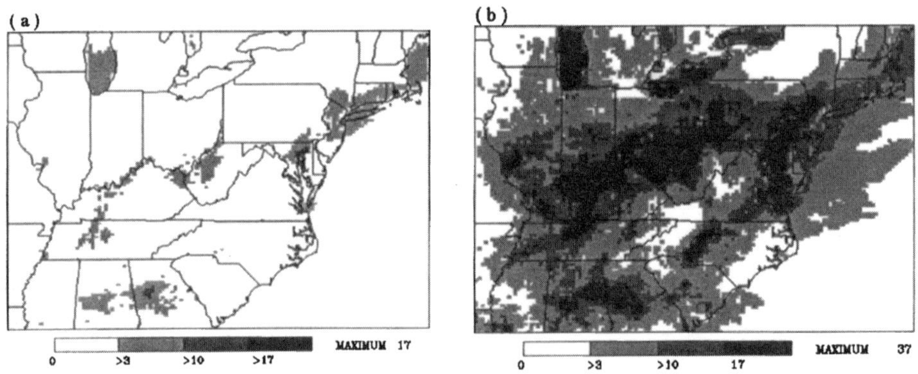

Fig. 3. Number of ozone exceedances over the summer of 1995; (a) 1-hr exceedences of 124 ppb; (b) 8-hr exceedences of 84 ppb

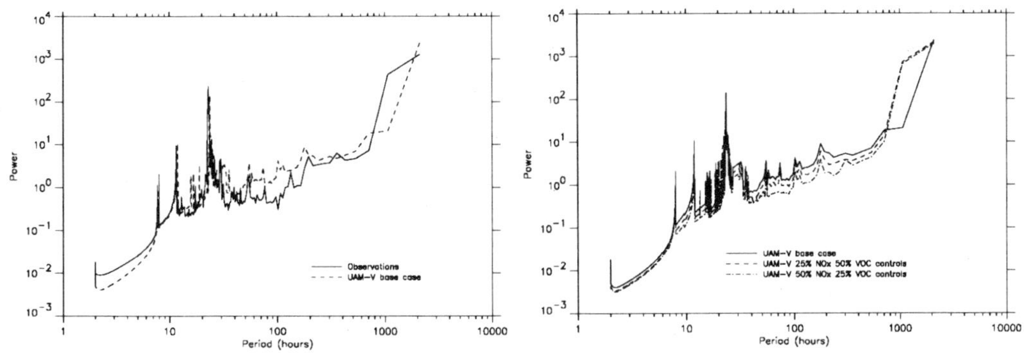

Figure 4. (a) power spectrum of observed and predicted hourly ozone concentrations at Greenbelt, MD for the summer of 1995 (b) same as (a) except for the model-predicted ozone under different emission reduction scenarios.

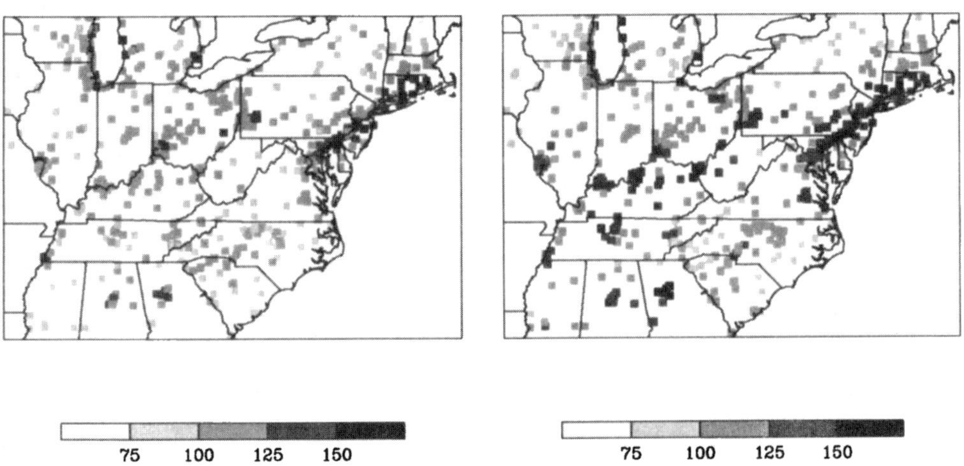

Figure 5. Spatial pattern of the 4th highest 1-hr ozone concentrations for the summer of 1995: (a) observed data (b) model predicted values

observations and modeled values, about 60% of the energy is in the diurnal oscillation and about 35% in the synoptic and longer-term components.

Simulated emission reductions lead to variance reductions on all temporal scales (Figure 4b and Table 2). Since one can compute the magnitude of the ozone concentration for each hour of the day as the sum of the baseline concentration on that day and percent contribution from each component to the baseline value, these variance reductions need to be considered when the results from emission reduction scenarios are applied to the observations.

Intra-Day Scale

Agreement between the intra-day components of modeled and observed values is poor with r values less than 0.35 at all monitoring stations in the Eastern United States. In space, the correlation among the intra-day components of observed ozone is near-zero even for the closest monitoring stations. However, the spatial correlations among intra-day components in modeled values are higher than those in the observations and do not decrease with distance as rapidly as those of the observations. Also, the spatial correlation among the modeled intra-day components decreases up to 200 km and then begins to increase slightly to a nearly constant value of 0.45 over the entire modeling domain. The autocorrelogram of the intra-day component in observations and modeled ozone decreases rapidly to near-zero values in 2 hours.

Diurnal Scale

The correlation between the observed and modeled diurnal components is about 0.7 over the modeling domain; the spatial variability of this correlation is less than that for the synoptic and baseline components. The spatial correlations among diurnal components in the model are similar to those in the observations. The rate of decay of correlations among the diurnal components with downwind distance in the predicted and observed data is also quite similar.

Synoptic Scale

The correlations (r) between the synoptic components of modeled values and observations over the Eastern United Stations exhibit some spatial variability with r values ranging from 0.6 in the Southeast and Northeast to less than 0.4 in the Midwest. Synoptic ozone ($O^S(t,x,y)$) contains a very precise spatial law related to the directional decay of their spatial correlations (see Rao, et al., 1997); that is, there is a strong relationship between exponential decay in correlations between two monitors m_o, m' and their distance of separation, d, along the direction ϕ:

$$\log \mathrm{Cor}\left(O^S(t,m_o),\ O^S(t,m')\right) = -a(m_o,\phi)d + b \tag{6}$$

where 'a' depends on reference monitor location (x_o ,y_o) and spatial direction ϕ. In other words, there is a Markov relationship between short-term (synoptic) components in space. The Markov property for a Gaussian process yields:

$$O^S(t,m') = \exp\{-ad + b\}\, O^S(t,m_o) + O^{loc}(t,m') \tag{7}$$

where $O^{loc}(t,m')$ is a local contribution portion at the location m'. Therefore, expression (7) separates the transport portion of $O^S(t,m')$ from m_o to m' and local production. The exponential term in (7) is related to the synoptic transport of the pollution, weather conditions, and emissions which created ozone.

The spatial extent of the correlation structure among synoptic ozone components was extracted for the Greenbelt, MD monitoring location. The orientation and spatial structure in the ozone "footprint" is dictated by weather patterns. Although there is a good agreement between the orientations of the ozone footprint in modeled and measured values, the spatial correlations among the synoptic components in the model are higher in the near field and lower in the far field than those in the observations. Whereas the correlation coefficient decreases by about 27% in the model, it decreases only by about 12% per 100 km distance downwind in the observations. The e-folding distances (scales) from Greenbelt, MD is about 600 km for observed synoptic-scale ozone while it is approximately 300 km for the modeled ozone. These results imply, in other words, that the model overstates the importance of local influences and underestimates the synoptic-scale transport of pollution.

Baseline (seasonal plus longer-term)

The correlations (r) between baseline components of modeled and observed values is about 0.8 for the Southeast and Northeast, and about 0.4 for the Midwest. The spatial extent of the ozone footprint in baseline ozone is larger in the model than in the observations.

Analysis of Emission Reduction Scenarios

As expected, the model-predicted ozone levels decrease when emissions are reduced. The index of improvement measure [(Base-control)/Base] suggests that NOx-focused controls are far more effective in reducing ozone concentrations regionally than VOC-focused controls. For urban areas such as the New York Metropolitan area and Lake Michigan region, VOC-focused controls provide a greater improvement in ozone levels than NOx-focused controls. Over the Ohio River Valley, NOx reductions are more effective in reducing ozone exceedances than are VOC reductions. Given a 75% reduction in NOx emissions, the model did not predict exceedances of the 1-hr threshold of 124 ppb anywhere in the domain, but predicted more than 3 exceedances of the 8-hr threshold of 84 ppb in the New York, Atlanta and Chicago areas for the summer of 1995. These results demonstrate the need to pursue VOC control strategies for major urban areas while reducing regional ozone levels with NOx controls.

Using photochemical models in the regulatory framework

It has been a common practice to use the absolute ozone concentrations predicted by the modeling system for a few episodes in designing emission control strategies. Since neither the baseline nor synoptic forcing can be calculated from the information provided by few days of episodic modeling, it is impossible to explain the behavior of ozone in the daily ppb-scale by ignoring the synoptic and baseline scales; we need to model for longer time periods to properly explain the ozone exceedance events. However, because of the poor agreement between the modeled and observed ozone values on the various time scales, use of model-predicted ozone concentrations in an absolute sense to assess ozone compliance doesn't appear to be wise. Instead, since the model appears to capture the key dynamical processes, it might be more reasonable to use model results in a relative rather than in an absolute sense for regulatory purposes.

We suggest here three alternative approaches that entail the use of both modeled and measured data for evaluating the efficacy of emission control strategies within the regulatory framework.

1. For a given scenario, adjust the observed hourly ozone concentrations in proportion to that predicted by the model (modeled emission control strategy/model base case). Compute the number of exceedances (say, the 4th maximum) from the adjusted observations.

2. Carry out Approach 1 for each temporal component (intra-day, diurnal, synoptic, and longer-term components). In other words, for a given emission control case, adjust each temporal component of the observed ozone hourly concentrations in proportion to that predicted by the model. Again, compute the number of exceedances (say, the 4th maximum) from the adjusted observations.

3. Carry out Approach 2 but use the unadjusted intra-day component of the observations. This approach acknowledges that there is little relationship between the observations and the model base case for this component.

When observations are adjusted according to the model-predicted changes from Approach 1, the so-called NOx disbenefits are evident for the NOx-focused control case (Figure 6) over the VOC-limited areas. When ozone observations are adjusted according to Approach 3, the 50% NOx and 25% VOC control scenario produces a spatial pattern with much lower values of 4th maxima than the 25% NOx and 50% VOC control case (Figure 6). Given an across-the-board emission reduction of 50% NOx and 25% VOC, only two regions, the Lake Michigan and the New York-New Jersey-Connecticut areas, have 4th maxima in excess of 124 ppb.

SUMMARY

To understand the behavior of models, it is important to consider how well they reproduce the various time scales embedded in the data. Application of techniques like those described here permit assessment of the model's ability to simulate the spatial and temporal features found in ambient ozone data. For example, along the direction of the prevailing wind, the spatial information extracted from the synoptic-scale forcing in the observations decays more slowly than that derived from the model, suggesting that the model tends to underestimate pollutant transport.

How one would use the modeling results in a regulatory setting is a topic of considerable debate. The results presented here provide insight into the model's behavior on different time scales and how this information should be taken into consideration in designing ozone control strategies. Clearly, it is erroneous to use model results in an absolute sense in the process of attainment demonstration. Since the ozone exceedances are dictated by the level of the ozone baseline and synoptic forcing, ozone management efforts should focus on reducing the baseline ozone level because the synoptic forcing cannot be controlled.

Among the useful results of these analyses, unavailable in a meaningful form from raw ozone data, is a description of the spatial and temporal information in ozone data, needed for ozone management efforts. The techniques discussed here would be useful for evaluating the model's ability to simulate the spatial and temporal features found in the ozone observations. Also, it is necessary to integrate observations and modeling results to increase our confidence in models used in a regulatory framework.

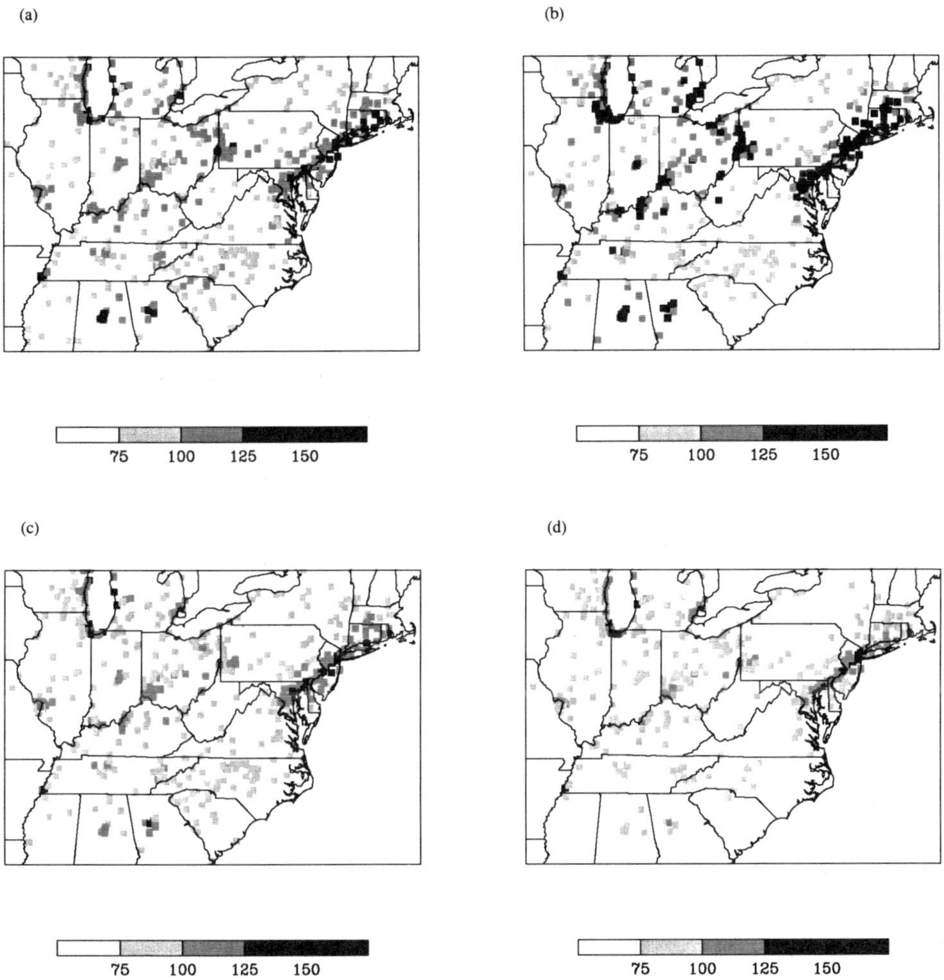

Figure 6. Spatial pattern of the 4th highest 1-hr ozone concentrations for adjusted observations for the summer of 1995: (a) 25% NOx and 50% VOC reduction - Approach 1 (b) 50% NOx and 25% VOC reduction - Approach 1 (c) same as (a) except for Approach 3 (d) same as (b) except for Approach 3

ACKNOWLEDGMENTS

This research is supported by the U.S. Environmental Protection Agency under grant #R825260-01-0, Electric Power Research Institute under contract #WO3189-12, Niagara Mohawk Power Corporation under contract #LC66765AML, and New York State Department of Environmental Conservation.

REFERENCES

Bastrup-Birk, A., Brandt, J. And Z. Zlatev. Studying cumulative ozone exposures in Europe during a 7-year period. *J. Geophy. Res.*, Vol. 102, No. D 20, October 1997, pp 23917-23935.

Eskridge, R.E., Ku, J.Y., Rao, S.T., Porter, P.S. and I.G. Zurbenko. Separating different scales of motion in time series of meteorological variables. *Bull. Amer. Meteor. Soc.*, Vol. 78, No. 10. October 1977, pp 1473-1483.

Kallos, G., Lagouvardos, and V. Kotrani. Atmospheric modeling simulations over the Eastern United States with the RAMS3b model for the summer of 1995. Final Report to EPRI, December 1997.

Lurmann, F.W. and Kumar, N. Evaluation of the UAM-V model performance in OTAG simulations Phase I: Summary of performance against surface observations. Draft Final Report #STI-996120-1605-DFR, Sonoma Technology, Inc., September 30, 1996.

Milanchus, M.L., Rao, S.T., and I.G. Zurbenko. Evaluating the effectiveness of ozone management efforts in the presence of meteorological variability. *J. Air & Waste Manage. Assoc.*, Vol. 48, March 1998, pp 201-215.

Pielke, R.A. and M. Uliasz. Use of meteorological models as input to regional and mesoscale air quality models - Limitations and strengths. Atmos. Environ., Vol. 32, No. 8, 1998. pp 1255-1466.

Systems Applications International. Users Guide to the Variable Grid Urban Airshed Model (UAM-V), San Rafael, CA, 1995.

Ozone Transport Assessment Group (OTAG) Modeling Report, Volume 1.1, February 12, 1997. See OTAG Website (http://www.iceis.mcnc.org.80/OTAGDC/index.html).

Porter, P.S., Rao, S.T., Zurbenko, I.G., Zalewsky, E., Henry, R.F. and J.Y. Ku. Statistical characteristics of spectrally-decomposed ambient ozone time series data. 1996, OTAG Web Site (http://capita.wustl.edu/otag/reports).

Rao, S.T., Zurbenko, I.G., Porter, P.S., Ku, J.Y and R.F. Henry. Dealing with the ozone non-attainment problem in the Eastern United States, *Environ. Manage.*,1996, pp17-31.

Rao, S.T., Zurbenko, I.G., Neagu, R., Porter, P.S., Ku, J.Y and R.F. Henry. Space and time scales in ambient ozone data. *Bull. Amer. Meteor. Soc.*, Vol. 78, No. 10, October 1997, pp 2153-2166.

Rao, S.T., Porter, P.S., Zurbenko, I.G., Dunker, A.M and G.T. Wolff. Ozone air quality over North America: A critical review of trend detection techniques and assessments. Submitted to *Atmospheric Environment*, May 1998. Also available on the NARSTO Website (http://narsto@odyesseus.owt.com/Narsto/assess_activities.html).

Wilmott, C.J. On the validation of models. *Phys. Geogrh.*, Vol. 2, No. 2, 1991, pp 168-194.

Zurbenko, I.G. *The Spectral Analysis of Time Series*, North Holland, 1986.

DISCUSSION

B. FISHER: For episodes of high ozone, the main factors are synoptic and baseline components. One cannot do anything about the synoptic component, nor the natural part of the baseline. By how much can we realistically influence the rest of the baseline component?

S. T. RAO: Let me reiterate that the baseline component in our paper should not be misunderstood as the natural background; it contains fluctuations in ozone with periods longer than 21 days. You are correct in stating that there is an irreducible part in ozone because of the natural background and that we can only attempt to reduce the impact from man-made pollution. We assessed this problem by removing all anthropogenic emissions and modeled the ozone season as before. From this analysis, it appears that the reducible part is at least 50% of the ozone baseline.

R. YAMARTINO: The ozone power spectrum shows some structure. Have you looked at co-spectra, quad-spectra, or coherence measures to probe into dominant causal factors?

S. T. RAO: No, not yet.

Z. ZLATEV: The spatial domain you showed us in your first slide was the Eastern part of USA. After that you showed results for Los Angeles. Have you extended the spatial domain to cover the whole USA?

S. T. RAO: No, the modeling domain covers only the Eastern United States. I showed ambient data from Los Angeles, California and Cliffside Park, New Jersey for two different time periods during which time there were changes in anthropogenic emissions due to controls implemented in the real-world as a way to illustrate how the energies in the intra-day, diurnal, synoptic and baseline components have changed and to examine the reliability of the model predicted changes on each time due to an emission reduction scenario.

Z. ZLATEV: The processes involved in comprehensive air pollution modeling are very complicated; some of them (the chemical reactions) are non-linear. Therefore, while I do believe that you can obtain some qualitative estimates about the components of the pollutant due to different physical processes, I should like to know how reliable are your quantitative estimates.

S. T. RAO: Yes. I concur with you that atmospheric processes affecting ozone pollution levels are very complicated; that is exactly what the models are trying to predict. I have illustrated how well the model is reproducing the dynamical processes influencing the spatial and temporal features embedded in the observations. From my presentation, you can see that there is a very good agreement between the variances of modeled and observed values over the modeling domain. That this is fortuitous is evident from a more detailed examination of the modeling results; the model is underestimating the variance in the shorter time scales (intra-day and diurnal) and overestimating at the longer time scales (synoptic and baseline components). Also, I showed results which reveal that the model overstates the effects of local emissions in the near-field and understates the effects of synoptic-scale transport in the far-field. This is an example of a case where the model has compensating errors. We need to evaluate the different dynamical processes (scales) to properly assess how well the model is reproducing the observed spatial and temporal features found in the observations. Only through this type of qualitative and quantitative analysis that we can readily identify the strengths and limitations of models. Traditional model evaluation methods are not very powerful in revealing to us why the model has performed the way it did.

A REGIONAL MODELLING STUDY OF TRANSPORT AFFECTING THE CANADIAN SOUTHERN ATLANTIC REGION OXIDANTS

Wanmin Gong[1], Xiude Lin[2], Sylvain Ménard[3], Pierre Pellerin[4], and Robert Benoit[4]

[1]Atmospheric Environment Service, Downsview, Ontario, Canada
[2]Ontario Hydro, Toronto, Ontario, Canada
[3]40 Lucien Thériault, Notre-Dame-de-l'Ile Perrot, Québec, Canada
[4]RPN, Atmospheric Environment Service, Dorval, Québec, Canada

INTRODUCTION

Elevated ozone levels in the Canadian southern Atlantic region (SAR) are usually associated with regional ozone episodes over the northeastern US and southern Ontario/Quebec, typically under the synoptic settings of a high pressure system to the east over the Atlantic Ocean and a low pressure system to the northwest. Under this condition the southwesterly flow at the backside of the high pressure cell favours the transport of ozone and its precursors from the US northeastern seaboard to the SAR. Observations during the 1993 North Atlantic Regional Experiment field campaign indicate that high ozone observed over the Gulf and over the coast of Nova Scotia is generally related to moderate to fast transport from the Boston-New York corridor, typically in the warm sector ahead of an approaching cold front. However, due to the complicated vertical layering structure that dominates the lower atmosphere over the Gulf of Maine, especially towards the Nova Scotia side, high ozone concentrations are often observed well above the surface unaccompanied by surface ozone events.

Angevine et al. (1996) described two idealized scenarios of transport over the Gulf of Maine: 1). warm polluted continental air advecting onto a cooler water surface being thermally modified at lower levels, in which case the marine boundary layer over the Gulf would be just as polluted, and would have a direct impact on the ground-level ozone in the SAR; and 2). advecting warm polluted continental air encountering and over-riding the sea-breeze front resulting in a situation where the marine boundary layer would remain relatively clean, and the overriding polluted air would have an impact at ground level only through entrainment at the top of the internal boundary layer or through the mechanism of coastal fumigation. In this paper we will explore the roles and mechanisms of transport and transformation affecting ground-level ozone in the SAR based on a model simulation of an

Air Pollution Modeling and Its Application XIII, edited by S.-E. Gryning and E. Batchvarova, Kluwer Academic/Plenum Publishers, 2000

episode in summer 1988. This is a part of a more extensive paper being submitted for journal publication (Gong et al., 1998).

STUDY PERIOD AND NUMERICAL SIMULATION

The simulation period for the present study is the first week of August 1988. This week coincides with a high ozone episode in eastern Canada during the first Canadian hyper-intensive campaign (July 25 - August 5, 1988) of the Eulerian Model Evaluation Field Study (EMEFS-1). EMEFS was a joint two-year effort between US and Canada to collect extensive chemistry data in Eastern United States and Canada for evaluating regional Eulerian acid deposition models (Hensen et al. 1991). During this one week period, eastern North America experienced a regional episode of high ground-level ozone. From August 1 to 3 the northeastern US and southern Ontario were in the warm sector behind a surface warm front with an upper level ridge residing overhead. This synoptic setting resulted in generally light southwesterly winds and high day time temperatures for the region, a favourable meteorological condition for surface ozone level to build up. On August 4, with the approach of an upper-level trough from the west, a surface cold front was moving into the region resulting in much stronger southwesterly wind along the US northeastern seaboard which was responsible for transporting the New York - Boston plume to the SAR later in the day. The cold front swept through the entire region on August 5 ending the ozone episode.

The simulation was carried out with a 3-D comprehensive regional oxidant modelling system, ADOM/MC2. The modelling system and some preliminary results, as well as some of the sensitivity tests were presented in previous CCMS ITMs (Lin et al., 1996, Gong et al., 1998). The actual model runs were performed for an eight day period (00 Z July 30 - 23 Z August 6, 1988). The first two days are to allow ADOM to "spin-up". The spin-up time for the meteorological model is much shorter, e.g. less than 6 hours at 21 km resolution. Both the meteorological model, MC2, and the oxidant model, ADOM, were run on two grids, a coarse grid with 127 km resolution in horizontal and a nested fine grid with 21.17 km resolution. The coarse ADOM domain covers most of eastern North America, while the nested ADOM fine mesh domain, consisting of 108 by 60 horizontal grid cells, was designed to mostly focus on the SAR, with its left boundary extending west of Ohio to include some of the major source regions in the north-eastern US, the Detroit/Windsor-Quebec City Corridor and most of the industrial Ohio Valley. There were 12 vertical layers in ADOM, for both the coarse and the fine resolution runs, logarithmically distributed between 1 and 10000 m. Eight layers were located below 2000 m.

The CMC global objective analysis provided the initial and temporal boundary conditions for the MC2 coarse resolution run, which in turn provided the boundary conditions for the fine resolution run. As for the oxidant modelling, the concentration fields were initialized with horizontally uniform but vertically varying background values. The hourly output of concentrations of 29 advected species from the ADOM coarse mesh run was used to construct the time-dependant boundary conditions for the nested ADOM fine mesh runs. In the ADOM nested fine mesh run, the original 30-minute timestep for advection and diffusion calculations was reduced to 5 minutes to ensure that the Courant number remained less than one at each grid during the integration.

Emission (anthropogenic and biogenic) data available for the simulation were compiled by ORTECH Environmental Consulting Company based on a 1985 inventory (coordinated by the US National Acidic Precipitation Assessment Programme). All emissions except biogenic emissions are seasonal averages with three categories (weekday, Saturday and Sunday) and with seasonal averaged diurnal variation. Biogenic emissions were generated specifically for the EMEFS-1 period.

RESULTS AND DISCUSSION

Figure 1 illustrates the simulated ozone episode during the study period. As seen from Figure 1a, the simulated ozone during the beginning of the week reflects a stagnated, light wind situation where elevated ozone levels are predicted over southern Ontario and the US Northeast regions with localized centres especially on the east coast. In contrast, the simulated ozone for 20Z August 4 shown in Figure 1b exemplifies the situation during the second part of the study period. Due to stronger synoptic flow, the New York - Boston plume, though considerably weaker as compared to the first half of the week, extends much further downstream. This particular plume resulted in a high ozone event later during the day over the SAR.

(a)

(b)

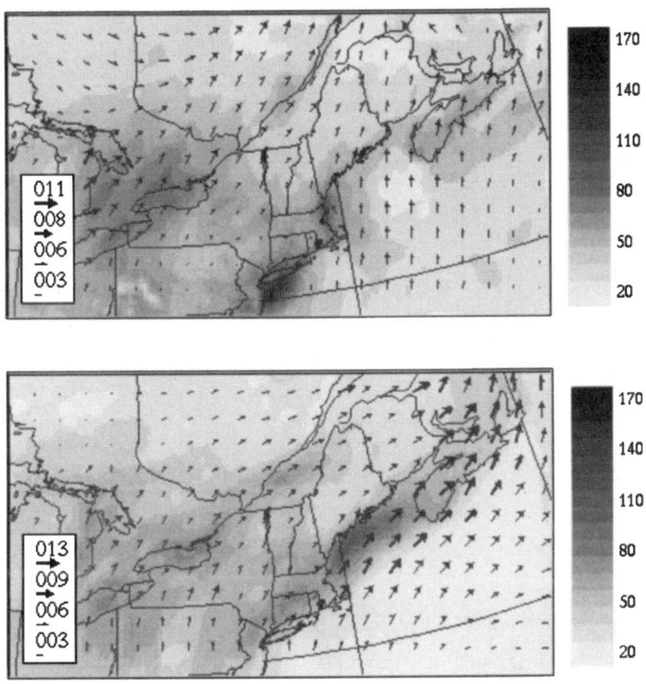

Figure 1. Modelled ozone (ppb) and wind vectors at the first model level for a) 20Z August 1, 1988 and b) 20Z August 4, 1988.

Comparison with observations

As a model evaluation procedure, the simulated surface ozone is compared with observations from two sources: the EMEFS database and the Canadian National Air Pollutant Surveillance (NAPS) database. The comparisons for a selected sites shown in Figure 2 are presented here. More extensive and detailed evaluation can be found in Multistakeholder (1997) and Gong et al. (1998).

Time series of the modelled and the observed ozone for 4 sites are shown in Figure 3. The first two sites (Madison and Newburyport) are fairly close to source regions and, as seen from both the simulation and the observation, the time series at these two sites are dominated by strong diurnal signals. In contrast, the other two sites (Winterport and Kejimkujik) are farther downwind from source regions and the influence of transport is

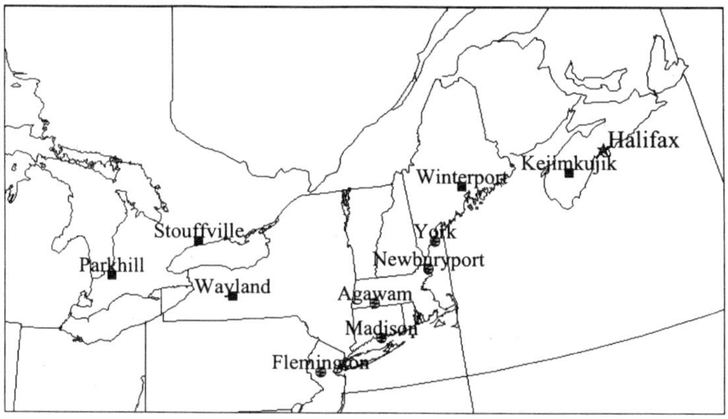

Figure 2. Location of sites for the present model-observation comparison. Squares denote the sites from the EMEFS database and circles denote the sites from the NAPS database; the location of Halifax is also given.

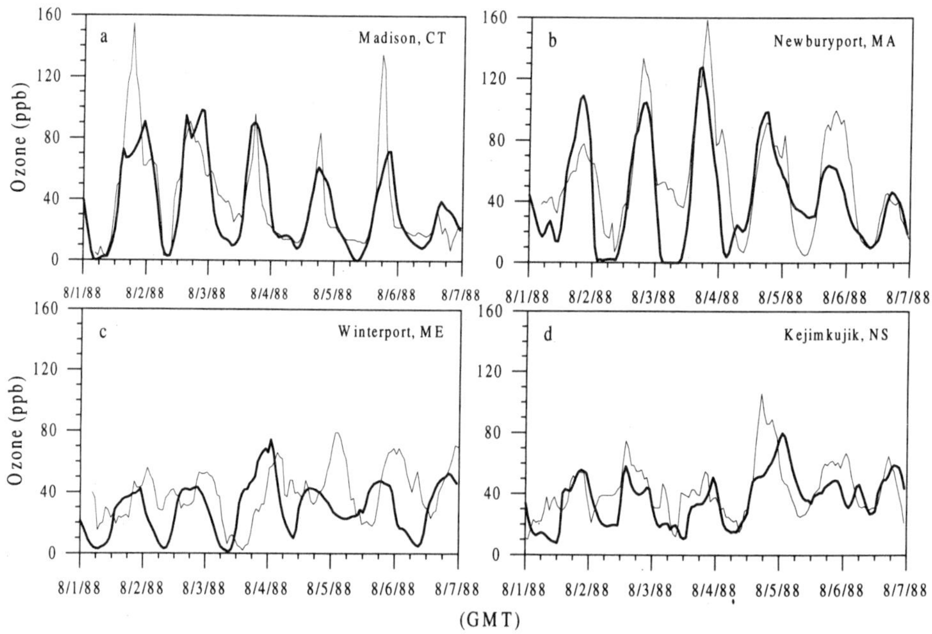

Figure 3. Time series of observed (thin lines) and modelled (thick lines) surface ozone concentration at a) Madison, CT; b) Newburyport, MA; c) Winterport, ME; and d) Kejimkujik, NS.

clearly seen in the time series at these two sites. A comparison of some of the statistics calculated from the modelled and the observed 6-day time series, for the 10 sites shown in Figure 2, is given in Table 1. This includes the mean and the standard deviation. A time lagged linear correlation between the observed and the modelled ozone time series is calculated for these 10 sites. The maximum correlation coefficients and the associated time shifts are also included in Table 1. A positive time shift indicates that the modelled time series is generally lagged behind the observed time series, and a negative time shift indicates otherwise. As seen the correlation between the simulated and the observed ozone time series is quite good with correlation coefficients in the range of 0.7 and higher. It is interesting to note that although the model seems to systematically under-predict the mean ozone concentration for the sites in consideration, the observed standard deviation of ozone concentration (the range of variation) is reasonably well simulated in most cases.

Table 1. Statistics of the modelled and the observed ozone concentration for the 10 Sites

Station	Mean (ppb)		Standard deviation		model-obs. Correlation	
	observed	modelled	observed	modelled	r	Δt
Flemington, NJ	40.7	28.2	36.9	28.7	0.93	0
Madison, CT	39.3	36.1	31.4	28.3	0.80	2
Agawam, MA	31.2	35.9	25.9	27.9	0.70	0
Newburyport, MA	53.2	43.5	34.0	33.0	0.79	0
York, ME	59.4	43.9	30.1	24.3	0.67	0
Winterport, ME	40.2	29.5	17.8	16.9	0.69	-1
Kejimkujik, NS	42.8	36.6	17.0	16.2	0.69	2
Stouffville, ON	56.9	39.2	27.1	26.0	0.73	0
Parkhill, ON	52.3	39.5	24.8	18.6	0.77	-1
Wayland, NY	45.9	38.5	23.0	14.0	0.76	0

SAR ozone and transport over the Gulf of Maine

It is shown, from both the observation and the model simulation, that the high ozone event (exceeding 80 ppb) in the SAR during this regional ozone episode over eastern North America occurred towards the end, when the synoptic setting changed from a more stagnated condition to a more transient condition with strong southwesterly flow due to an advancing frontal system. In this study the modelled concentrations of various species and their changes due to different processes are analysed along the back-trajectories arriving at the model grid containing the Kejimkujik site at various time and heights to investigate the role of transport affecting the SAR ground-level ozone. This receptor-oriented analysis is shown to be quite effective in explaining the SAR ozone event during the study period.

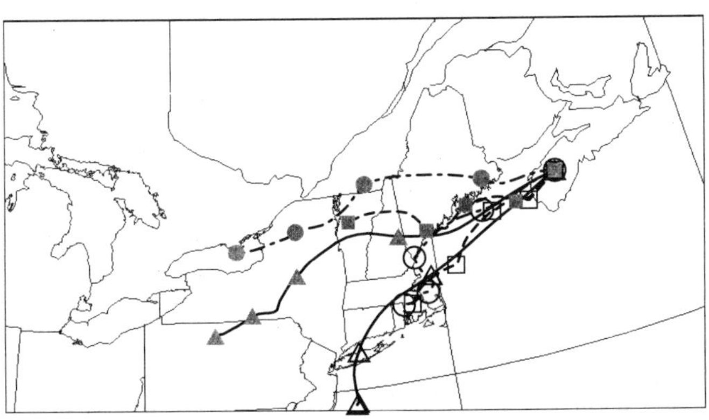

Figure 4. Back-trajectories arriving at 00Z August 3 (circles), 4 (squares) and 5 (triangles). Open symbols denote trajectories arriving at the 2nd model layer, and filled symbols denote trajectories arriving at the 5th model layer.

Figure 4 shows the trajectories arriving at the Kejimkujik grid at the second (centred at 96 m) and the fifth (centred at 536 m) model layers for several arrival times (00Z, August 3, 4 and 5). These arrival times either coincide with or are close to the peak ozone concentration as seen in Figure 3d. In general the trajectories arriving at the higher level are from the west and southwest, while the trajectories arriving at the lower level are from the southwest. Interestingly, all the lower level trajectories shown in Figure 4 pass over the Boston area at one time before arriving at Kejimkujik. However only one of these trajectories (arriving at 00Z August 5) is associated with the SAR high "ground-level" ozone event (modelled) during this one-week period. Therefore, the simple fact that the air parcel comes from the source region does not always lead to high ground-level ozone in the SAR. What also matters is what happens to the air parcel as it travels to the SAR.

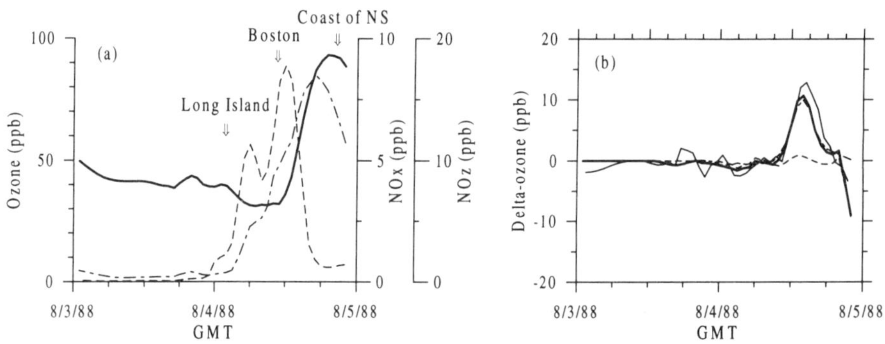

Figure 5. a) Time series of ozone (solid line), NOx (dashed line) and NOz (dash-dot line) concentrations along the lower trajectory arriving at 00Z August 5; b) Tendency of ozone (thin solid line) along the trajectory and individual contributions from vertical diffusion (dashed line) and chemistry (dash-dot line) components, as well as the net contribution to the tendency combining the chemistry and vertical diffusion components (thick solid line).

To look at the event on late August 4 (or early August 5, GMT) in more details, the time series of modelled ozone, NOx ($NO + NO_2$) and NOz (HNO_3 + PAN) along the trajectory arriving at the lower level are plotted in Figure 5a.. The concentrations are taken from the nearest grid. In the meantime the total time tendency of ozone along the trajectory as well as the contributions from the individual terms, i.e. vertical mixing, chemistry and dry deposition (if the air parcel is in the first model layer) are shown in Figure 5b. Note that the horizontal mixing, which includes, in the present case, both physical and numerical dispersion, will also contribute to the concentration changes in the air parcel. Note also that although the analysis attempted here is to follow an "air parcel" along a trajectory in a Lagrangian framework, the model is formulated in an Eulerian framework.

As seen from Figure 4, the air parcel along the trajectory was located over Long Island at 00 Z August 4 and was over Boston at about 10 Z (06 EDT). Under the strong southwesterly flow, the air parcel travelled across the Gulf in only 14 hours to reach the Kejimkujik grid. As shown in Figure 5a, NOx in the air parcel increased from below 1 ppb to about 10 ppb when the it was travelling from Long Island to the Boston coast during the night of August 3rd and the early morning of August 4th, followed by a rapid decrease during the day which coincided with the increase in both ozone and NOz. Figure 5b clearly indicates that the increase of the ozone concentration in the air parcel was almost exclusively due to photochemical production over the ocean during the day with the precursors picked up earlier over the emission regions. Vertical mixing was greatly suppressed during the night and during the day over the ocean, and it only came into effect when the air parcel was over the Nova Scotia peninsula. The vertical mixing was responsible for the decrease of ozone in the air parcel over the last few hours before arrival, driven by dry deposition in the air below. The ozone level in the air parcel reached a maximum of about 90 ppb just before reaching the Nova Scotia coast and was still over 80 ppb as it arrived at the Kejimkujik grid in the evening.

The process analysis along the trajectory shown above in Figure 5b does not identify the "mixing-down-from-above" process as a significant mechanism for this ground level ozone event. The vertical cross sections of modelled temperature and ozone concentration at 20Z August 4 taken through Agawam-Newburyport-Kejimkujik-Halifax, more or less aligned with the lower trajectory path, are given in Figure 6. The air parcel location at the time is indicated in the cross section. It is seen that the plume that later reached the Kejimkujik grid and led to the high ground-level ozone event was trapped within the stable marine boundary layer, i.e. a transport situation more in line with condition #1 described in the introduction. This is in contrast to the situation described in Merrill and Moody (1996) and Moody et al. (1996). Based on a trajectory analysis focused on two ozone events during NARE-93 (August 25 - 28, 1993), they suggested that the incidences of high ground-level ozone observed in this period over southern Nova Scotia were a result of weakened stability at low levels to allow the mixing down of the ozone from aloft, a case of condition #2 in the introduction.

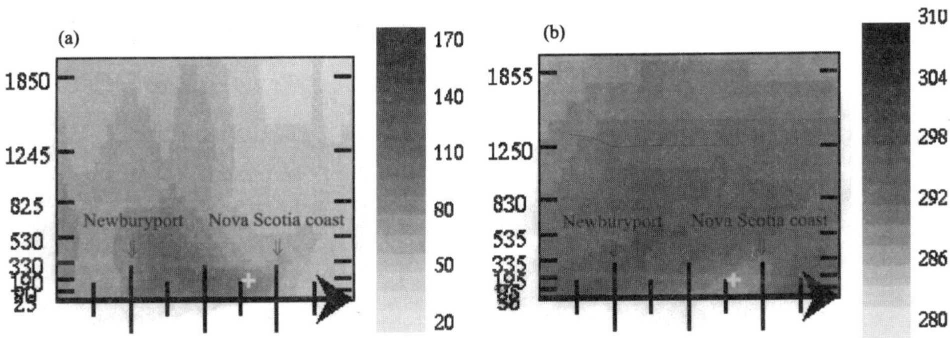

... sections of a) ozone (ppb) and b) temperature (K) at 20Z August 4 through Agawam-Newburyport-Kejimkujik-Halifax. The air parcel's position in the cross section is indicated by a "plus". The positions of Newburyport and the Nova Scotia coast are also indicated.

The other two lower level trajectories included in Figure 4, in contrast, had all taken more than 24 hours to track across the Gulf under weak flow situations. For example, the air parcel arriving at 00Z August 3 had spent most of the daytime on August 2 travelling along the east coast between Boston and the New Hampshire coast (just north of Newburyport). The ozone concentration along the trajectory reached over 90 ppb during this day, primarily due to photochemical production consuming most of NOx picked up earlier by the air parcel over the Boston area. Some fresh injection of NOx later on that day over the New Hampshire coastal area resulted in the reduction of ozone overnight by titration. During the following day (August 3), with a much lower NOx level in the air parcel at the start of the day and no further emission over the water, the ozone production was much reduced compared with the previous day. The ozone concentration in the air parcel peaked at about 60 ppb for this day just west of the Nova Scotia coast, but decreased to about 45 ppb when arriving at Kejimkujik.. The reduction was mainly driven by dry deposition over land.

In comparing the above cases, it seems that a fast transport situation over the Gulf of Maine (with the right timing for photochemical production over the Gulf) is more favourable for the occurrence of high ground-level ozone events in the SAR than a weak flow, slow transport situation for the present study period.

SUMMARY AND CONCLUSIONS

In this paper, we have presented a modelling study of oxidants in the Canadian SAR based on one regional ozone episode case (EMEFS-1 intensive period). The comparisons with the available observations show that the model has reasonable skill to reproduce the general features of the regional ozone episode over eastern Canada and the north eastern United States. Investigation into the roles of transport and other related processes on the SAR oxidants was carried out. It was found that the elevated ozone at the Kejimkujik site during the study period occurred under fast transport driven by a strong southwesterly flow over the Gulf of Maine, and is likely due to the transport at low levels rather than a case of downward mixing from upper levels. It was also shown that the high ozone brought into southern Nova Scotia during this event was mostly photochemically produced over the water with the precursors picked up earlier over the emission regions on the US east coast. The analysis suggests that a multi-day transport over the Gulf of Maine, a weak transport situation, is unfavourable to the occurrence of high ground-level ozone in the SAR under the low level transport scenario.

REFERENCES

Angevine, W.M., M. Trainer, S.A.McKeen and C.M. Berkowitz, 1996, Meteorology of the New England coast, Gulf of Maine, and Nova Scotia: Overview, J. Geophys. Res., Vol. 101, No. D22, 28,893-28,901.

Gong, W., X. Lin, S. Menard, P. Pellerin, and R. Benoit, 1998: Modelling the Canadian Southern Atlantic Region oxidants - A study of a Canadian EMEFS-1 hyper-intensive period. Submitted to J.Geophys. Res.

Gong, W., S. Menard and X. Lin, 1998, Uncertainties in Regional Oxidant Modelling - Experience with a Comprehensive Regional Model, in "Air Pollution Modelling and Its Applications, XII", edit.S.E. Gryning and N. Chaumerliac,, 559-566, Plenum Publishing.

Hansen, D.A., K.J. Puckett, J.J. Jansen, M. Lusis and J.S. Vickery, 1991, The Eulerian model evaluation field study (EMEFS), Proc. 7th Joint Conference on Applications of Air Pollution Meteorology with AWMA, AMS, 58-62.

Lin, X., W. Gong and S. Ménard, 1996, A New MC2/ADOM Modelling System and Its Application to the Study of Oxidants in the Canadian Southern Atlantic Region, in "Air Pollution Modelling and Its Applications, XI", edit. S. Gryning, and F.A. Schiermeier, 275-282, Plenum Publishing.

Merrill, J.T. and J.L. Moody, 1996, Synoptic meteorology and transport during the North Atlantic Experiment (NARE) intensive: Overview, J. Geophy. Res., Vol. 101, 28,903-28,921.

Moody, J.L., J.C. Davenport, J.T. Merrill, S.J. Oltmans, D.D. Parrish, J.S. Holloway, H. Levy II, G.L. Forbes, M. Trainer and M. Buhr, 1996, Meteorological mechanism for transporting O3 over the western North Atlantic Ocean: a case study for August 24-29, 1993, J. Geophys. Res., Vol. 101, No. D22, 29,213-29,227.

Multi-stakeholder, 1997: Modelling of ground-level ozone in the Lower Fraser Valley. Report of Canadian 1996 NOx/VOC Science Assessment, Sept., ISBN 1-896997-08-2, 157 pp. [Available from APAC, Atmospheric Environment Service, Downsview, Ontario.]

DISCUSSION

R. YAMARTINO: Over-water transport is very sensitive to the assumed $K_z(z)$ profile over water. How did you modify the ADOM diffusivity algorithm to handle this?

W. GONG We use a prognostic mesoscale meteorological model (MC2) to drive ADOM runs. Vertical diffusivities (3D) are taken from the meteorological model along with other meteorological fields required by ADOM. MC2 has relatively complete and detailed physics and uses a one-and-half order turbulence closure by solving a prognostic turbulent kinetic energy equation. With a horizontal resolution of 21 km and 14 vertical levels below 2000 m, the model produces reasonable diffusivity profiles over the Gulf of Maine.

PROGRAMME OF CONTROL CONCEPTS AND MEASURES FOR OZONE ("SUMMER SMOG")

Arno Graff

Federal Environmental Agency
Bismarckplatz 1
D-14193 Berlin
Germany

INTRODUCTION

In mid-summer episodes, highly elevated concentrations of ground-level ozone are frequently measured in various regions of Germany. The related risks for human health and for eco-systems increasingly concerned environmental policy at national and local levels in the early 90s. In public discussions, the call for ozone abatement strategies was raised. However, knowledge about the efficiency of control strategies and measures under discussion was widely lacking. Surveys analysing the effects of possible measures to reduce the emissions of ozone precursor substances NO_x and VOCs (such as speed limits, driving restrictions or specific sector-related or species-related measures at local or national levels) on the emission levels and, subsequently, on peak ozone concentrations had not been conducted thus far. Against this background, the research objective of the "Programme of Control Concepts and Measures for Ozone (summer smog)" as commissioned by the German Federal Environmental Agency (Umweltbundesamt)[1] was defined as follows: **Determination and evaluation of the effectiveness of regional scale and of local scale emission control measures on elevated ground-level ozone concentrations in mid-summer episodes.** Thus, the research project was designed to contribute to scientific conclusions as regards the effectiveness of ozone abatement strategies. Furthermore, taking into account the ongoing political discussions to advance legislation for ozone abatement at federal and at state levels, the findings of this project were to contribute to improving the basis for decision-making. To this end, the survey programme is split into two phases:

Phase A: Analysis of the efficiency of measures within the regional scale (nation-wide) domain, applying regional scale photochemical dispersion modelling;

Phase B: Analysis of the efficiency of measures at local level in three selected

[1] German Federal Ministry of the Environment - Environment Research Plan, UFOPLAN No. 104 02 812/01

regions[2], based on photochemical dispersion modelling with high spatial resolution. In comparison to studies performed thus far, this research programme exhibits a number of significant features:

The photochemical model simulations are preceded by a comprehensive analysis of the efficiency of a wide spectrum of potential measures to reduce emissions of ozone precursors NO_x and VOCs, in order to assess the reduction rates of realistic and policy-oriented control strategies (such as speed limits, driving restrictions, etc.). The photochemical simulations are therefore not based on global reduction rates but on rates for various defined measures and for sets of measures (scenarios), each of them highly differentiated in terms of sector, time, substance and location.

REGIONAL SCALE (NATIONAL) ANALYSIS OF EFFECTS OF CONTROL MEASURES (PHASE A)

Emission-related effects of regional scale (national) temporary and permanent control measures

Current as well as future reduction potentials of all emitter groups of ozone precursor substances NO_x and VOC are analysed and grouped in scenarios for the photochemical simulation runs. For this purpose, model based estimations for various control measures in different emitter sectors are performed, as well as evaluations of prior studies and conclusions derived from current research work of the German Federal Environmental Agency. Within the framework of Phase A, two permanent and six temporary reduction scenarios at the regional scale level have been analysed. The main characteristics of these scenarios and the emission-related analysis of effects of measures are summarized in Table 1.

Effectiveness of regional scale emission reduction measures on peak ozone concentrations in Germany

The effects of regional scale emission reduction measures on ground-level ozone concentrations according to the scenarios in table 1 are analysed by performing simulations with two different photochemical modelling systems: EURAD and REM3[3]. Both are complex three-dimensional meso scale Eulerian dispersion models simulating the formation, transport and removal of photo-oxidants in the troposphere resp. in the planetary boundary layer. The spatial resolution (modelling grid) used is 36km x 36km for EURAD and approximately 30km x 30km for REM3 in the horizontal and 15 respectively 3 layers in the vertical.

The simulations were conducted for two ozone episodes in 1990 and in 1994, each of a fortnight's duration. These two episodes, exhibited quite different patterns of meteorological conditions. Ground-level peak ozone concentrations of 250 to 300 $\mu g/m^3$ (1-h-values) were observed in the afternoon. The EURAD and REM3 model calculations are based on a uniform emission data base. In all, 22 calculation runs were performed, including the base case runs for the simulation of the episodic patterns and for the testing of the performance of the models in replicating observed ozone concentrations. For each

[2] Regions of Rhine-Main-Neckar (Frankfurt), Berlin-Brandenburg and Dresden

[3] **EURAD** - **EUR**opean **A**ir Pollution **D**ispersion Model , Rheinisches Institut für Umweltforschung, Universität Köln; **REM3** - **R**egional **E**ulerian **M**odel with **3** chemistry schemes (R. Stern), Institut für Meteorologie, Freie Universität Berlin.

modelling run of the 14-day episodes, hourly values for O_3, NO, NO_2 and other substances are computed at every grid point in the modelling domain. Both models arrive basically at the same findings as regards the effectiveness of the reduction measures on the (afternoon) peak ozone concentrations.

Table 1. Regional scale (national) emission control scenarios

D-1	**Trend-2005**	Reductions of emissions of ozone precursor substances, which will be fully achieved by the year 2005 by emission control measures already implemented (EC Directives, national environmental legislation, etc.) -37% NO_x; -42% VOC
D-2	**Reduction-2005**	Trend-2005 with the addition of measures for the exploitation of known further reduction potential in the fields of traffic, industry and solvent emissions -64% NO_x; -72% VOC
T-1	**Speed limits at federal level**	Speed limits (nation-wide; 80 km/h for cars and 60 km/h for trucks on highways and 60 km/h on non-urban roads; 80% compliance) – implementation the day after observed ozone concentrations have exceeded 240µg/m^3 (1h value) -15% NO_x; -1% VOC
T-2	**Driving bans in agglomerations**	Driving bans for cars without a regulated 3-way catalytic convertor on inner urban roads in agglomerations of more than 250.000 inhabitants – implementation as with T-1 -2% NO_x; -5% VOC
T-3	**T-1 plus T-2**	Combination of the temporary measures T-1 and T-2 – Implementation as with T-1 -16% NO_x; -5% VOC
T-4	**T-3 plus voluntary reductions in case of stationary sources and solvents**	T-3 plus voluntary reductions at stationary sources and furthermore, appeals to the public concerning the use of solvents– implementation as with T-1 -19% NO_x; -12% VOC
T-5	**Driving bans for non-low emission cars**	Nation-wide driving bans for cars without a regulated 3-way catalytic convertor as well as for non-low emission diesel cars – implementation as with T-1 -29% NO_x; -32% VOC
T-6	**T-5 plus voluntary reductions in case of stationary sources and solvents**	T5 plus voluntary reductions at stationary sources and furthermore, appeals to the public concerning the use of solvents– implementation as with T-1 -32% NO_x; -39% VOC

Permanent reduction measures. In the case of permanent reduction measures according to the scenario "Trend-2005" (D-1), decreases of the afternoon's ozone peak concentrations ranging from 15% to 25% were computed in large parts of the modelling domain. Peak values of 300 µg/m^3, for instance, would thus be lowered by 60µg/m^3 on an average. The calculated number of grid hours at ground-level, during which in the base case run threshold values of 180 or 240 μg/m^3 are exceeded, would be reduced by 70 to 80% in the scenario "Trend-2005".[4]

In the case of additional permanent reduction measures according to the scenario "Reduction 2005", the calculated afternoon's peak concentrations are 30 to 40% lower than in the base case run. The calculated frequency of number of grid hours exceeding threshold values of 180 or 240 μg/m^3 respectively is reduced by approximately 90%.

Temporary reduction measures. In the case of a "severe" nation-wide speed limit (T-1), the model simulation shows a decrease in the calculated frequency of grid hours exceeding the threshold of ground-level O_3 concentrations of 180 μg/m^3 by some approx. 14%. The domain-related rates of reductions of peak ozone concentrations in the afternoon hours range from 2% to 6%.

In the case of local driving bans in all larger agglomerations (T-2), for which only minimal emission-related reductions at the federal level were computed, only marginal reductions of peak ozone concentrations (2% to 5%) are simulated.

In the case of a nation-wide driving ban for passenger cars without 3-way catalytic convertor (T-5), the simulation shows a 29% decrease of the calculated number of grid hours with ground-level O_3 concentrations above 180 μg/m^3. The domain-related rates of reduction of peak ozone concentrations in the afternoon hours range from 5% to 10%.

LOCAL SCALE ANALYSIS OF EFFECTS OF CONTROL MEASURES (PHASE B)

Data bases for the selected modelling regions Rhine-Main-Neckar, Dresden and Berlin-Brandenburg

The local scale analysis of the effectiveness of control measures were performed for three selected modelling regions: Rhine-Main-Neckar (Frankfurt), Dresden and Berlin-Brandenburg. In all three regions, ozone peak concentrations were significantly above 200 μg/m^3 (1h value) over several days in the episode studied (23.7-8.8.1994).

The surface covered by the modelling regions ranges from 7'500 km^2 to 11'000 km^2. The socio-economical and the transport-related structures as well as the topography of the three regions are rather different, and so are the area source emission profiles.

The NO_x emission intensity in the Rhine-Main-Neckar region is 4.2 times that of Dresden and 1.8 times that of Berlin-Brandenburg. The VOC emission density related to anthropogenic sources is roughly the same in the cases of Rhine-Main-Neckar and Berlin-Brandenburg, but more than halved in the case of Dresden. The contribution of VOC emissions from biogenic sources (forests) to the total VOCs in the (mid-summer) episode amounts to approximately 51% in the case of Dresden, against 27% and 33% in the cases of Rhine-Main-Neckar and Berlin-Brandenburg, respectively. The VOC/NO_x emission ratio is almost balanced with 1.1 in Rhine-Main-Neckar. The region of Dresden (with 2.9) and the area of Berlin-Brandenburg (with 1.9) show highly VOC-marked emission profiles.

[4] The number of grid hours during which a given O_3 threshold has been exceeded indicates the number of hours during which this concentration threshold value is exceeded in the model grid cells of the surface layer (i.e., in the first model layer) during the episode modelled.

Effects of local emission reduction measures

Corresponding to the procedure adopted in the research Phase A, several temporary local measures were assessed and then grouped into selected reduction scenarios. Prevailing criteria for the selection of the measures and scenarios were the links to the scenarios of Phase A, the consideration of specific local starting points for control strategies as well as the temporal and spatial dimension of the measures. In total, 11 sets of measures (reduction scenarios) have been defined. They are presented in Table 2. In the case of the maximum realization of regional reduction potential the corresponding scenarios are predominantly marked by high NO_x reduction rates. It must also be pointed out that the effects of the measures vary in intensity from one region to another depending on the domain's emissions profile.

Effects of local emission control measures on peak ozone concentrations in the modelling regions

Like in Phase A, again two different photochemical modelling systems were applied. The analysis for the modelling region of Dresden is performed with the EURAD modelling system. In the case of the areas of Rhine-Main-Neckar and Berlin-Brandenburg the CALGRID modelling system[5] is used. The EURAD modelling approach is based on an integrated two-way nesting. For CALGRID, the initial and boundary inputs were provided by the meso-scale model REM3. In order to assess the quality of model performance, base case simulations of local ozone concentrations and patterns were extensively compared to the data available from measuring networks and show well matching results.

Permanently effective, local scale emission controls (S-10) result in a decrease of the calculated peak ozone concentrations within the region by 30% to 50%. At afternoon peak values of 240-280 $\mu g/m^3$, this would be equivalent to a drop of peak loads below the 200 $\mu g/m^3$ threshold. The effectiveness of permanent local scale control measures is significantly higher than with temporary measures, although the emission-related reduction effects are "only" in the magnitude of -30% to -40%. The main reason for higher efficiency of permanent control measures is to be seen in the decrease of precursor emissions at the national (European) level mentioned previously for the scenario S-10 (Trend-2005). Thus, the regional scale background concentration of ozone, of ozone precursors and of other reactive components is reduced.

Local control strategies, temporarily implemented, are able to achieve moderate decreases in the afternoon O_3 peak values in that domain exposed to the measures when there are conditions of very little exchange of air masses. Even by exhausting all available local control potential (and hence by applying the most stringent actions), the effects on peak ozone levels cannot be placed on a par with those of permanent regional (nation-wide and less intensive) emission control.

Temporary single measures which are solely region-related lead to relatively small emission reductions (up to 25%) and exert only minimal effects on peak ozone concentrations (S-1, local speed limits and S-2, local driving bans for non-low emission diesel cars). That local measures solely have an impact on the local ozone production accounts for the low efficiency of this type of measures. Thus, background concentrations of ozone and precursors remain unaffected.

---

[5] CALGRID, 3-dimensional photochemical transport model system (containing moduls for transport, diffusion, dry deposition and photochemical reactions); Yamartino, R. et al. (1989), A mesoscale photochemical grid model, model formulation, Atm. Env. 26A, pp. 1493-1512

Table 2. Emission control scenarios for the three modelling domains

No.	Scenario	NO_x effect in % Frankfurt/Dresden/Berlin	VOC effect in % Frankfurt/Dresden/Berlin
S-1	**Regio-speed limit;** severe speed limit in the region	-14% / -11% / -10%	-1% / -1% / -1%
S-2	**Regio-car traffic;** driving restrictions for cars in the region	-25% / -22% / -19%	-27% / -25% / -28%
S-4a	**Regio-road transport;** severe speed limit plus driving restrictions for cars and trucks in the region (S-1 plus S-2)	-53% / n.c. / -51% n.c.: no calculation	-32% / n.c. / -31%
S-4b	**Combi-road transport-max;** S-4a plus driving restrictions for cars in adjacent states (Bundesländer)	-53% / -43% / -51% plus effects in adj. states	-32% / -27% / -31% plus effects in adj. states
S-6	**Regio-focus NO_x;** speed limits, driving restrictions, temporary closure of larger NO_x emission sources, public appeals	-74% / n.c. / -66%	-33% / n.c. / -33%
S-7	**Regio-focus;** driving restrictions for cars temporary closure of larger VOC emission sources, public appeals to business and households	-25% / -22% / -19%	-36% / -30% / -36%
S-8a	**Regio-max NO_x/VOC;** S-6 plus S-7	-74% / -74% / -66%	-36% / -33% / -40%
S-8b	**Combi-max NO_x/VOC;** S-8a plus driving restrictions in adjacent states (Bundesländer)	-74% / -74% / -66% plus effects in adj. states	-36% / -33% / -40% plus effects in adj. states
S-9	**Reformulated fuels;** Reformulated fuels for Otto and Diesel engines	n.c. / -3% / -3% plus effects in adj. states	n.c. / -6% / -7% plus effects in adj. states
S-10	**Trend 2005;** Full effect of all measures enacted in Germany and Europe	-30% / -22% / -21% plus effects in Germany/ Europe	-31% / -27% / -30% plus effects in Germany/ Europe

The enlargement of the area subject to control measures leads to a reduction of pre-loads in the air masses transported into the domain under consideration from (upwind) adjacent regions. As a consequence, the efficiency of local control measures is markedly higher due to a lower background concentration. The effectiveness of local control strategies increases with the intensity of the control measures.To a large extent, the efficiency analysis of the reduction measures shows coinciding results in the three different modelling regions. But they also put in evidence deviations that can be ascribed to the specific local emission profiles and intensities as well as to the specific local meteorological conditions in the episode.

CONCLUSIONS

The long-term and European-wide formulated emission reduction scenarios "Trend 2005" and especially its more ambitious variation, "Reduction 2005", are by far the most effective measures to reduce high ozone concentrations during the summer months. If fully implemented (by the year 2005), it can be expected that peak ozone concentrations will decrease by as much as 40%. The warning threshold ozone value of 180 μg/m3 would still be exceeded; however, the number of exceedances would be clearly lower than at present.

Temporary and locally restrictive measures, such as a speed limit, a ban on non-cat motor vehicles, or the burning of reformulated fuels, are not very effective in decreasing high ozone levels, because the attainable emission reduction is small. For each single measure, the decrease of peak ozone concentrations would be at most 5%.

Only the combination of several such locally restrictive measures can lead to a distinct reduction of the precursor emissions (50% or more) and therefore also to a distinct reduction of peak ozone concentrations. However, even if all possible local measures would be implemented, the resulting decrease of peak ozone concentrations would be at most 20%.

The expansion of an emission control domain increases the efficiency of ozone reduction measures. A speed limit or a traffic ban for non-cat cars put into force in several states or even in all of Germany would be much more effective than local implementation of the same measure.

Temporary, locally restrictive measures are not very effective, if the emissions reduction potential is below 50% .

Control measures which are permanent and implemented on a large control domain (national or even international) are even with lower overall percentage emissions reduction more effective in reducing high ozone concentrations than temporary and locally restrictive measures.

REFERENCES

The final documentation of this R&D project No. 104 02 812/01 of the German Federal Environmental Agency takes into account that each section of the study program is to be considered as a study of its own, and to be documented as such. Therefore, the documentation will encompass 15 separate reports as follows:

Hartmann, A., Motz, G., 1997, Inputdatenbasis 1994 der Modellregion Rhein-Main-Neckar - kleinräumige Emissionen nach Emittentengruppen, Emissionsminderungs-Szenarien, Landnutzung und Orographie, Ergebnisbericht zum Teilvorhaben B-1-1, Prognos AG, Basel

Hartmann, A., Motz, G., 1997, Inputdatenbasis 1994 der Modellregion Dresden - kleinräumige Emissionen nach Emittentengruppen, Emissionsminderungs-Szenarien, Landnutzung und Orographie, Ergebnisbericht zum Teilvorhaben B-1-2, Prognos AG, Basel

Hartmann, A., Motz, G., 1996, Inputdatenbasis 1994 der Modellregion Berlin-Brandenburg - kleinräumige Emissionen nach Emittentengruppen, Emissionsminderungs-Szenarien, Landnutzung und Orographie, Ergebnisbericht zum Teilvorhaben B-1-1, Prognos AG, Basel

Meeuwsen H. J. A. et al,. 1994, Emissions- und Landnutzungsbasis 1990 für das PHOXA-Untersuchungsgebiet, TNO, Apeldorn

Memmesheimer, M., Jakobs, H. J.,Piekorz, G., Ebel, A., Hass, H., 1995, Großräumige Ozon-Modellrechnungen mit verschiedenen Emissionsminderungsszenarien mit Hilfe des Modellsystems EURAD, Abschlußbericht zum Teilvorhaben A-5 "Großräumige Ozon-Ausbreitungsrechnungen", Förderverein des Rheinischen Instituts für Umweltforschung an der Universität zu Köln e.V.

Motz, G., Hartmann, A., 1997, Ermittlung und Bewertung der Effekte von lokalen, regionalen und überregionalen Emissionsminderungs-Strategien auf bodennahe Ozonkonzentrationen in hochsommerlichen Episoden, **Gesamtbericht** zur Untersuchung, Prognos AG, Basel

Motz, G., Hartmann, A., 1997, Ermittlung und Bewertung der Effekte von lokalen, regionalen und überregionalen Emissionsminderungs-Strategien auf bodennahe Ozonkonzentrationen in hochsommerlichen Episoden, **Kurzfassung,** Prognos AG, Basel

Motz, G., Hartmann, A., 1997, Determination and evaluation of effects of local, regional and larger-scale (national) emission control strategies on ground level peak ozone concentrations in summer episodes by means of emission analysis and photochemical modeling, **Summary** , Prognos AG, Basel

Motz, G., Hartmann, A., Plinke, E, 1996, Ermittlung und Bewertung der Effekte von großräumigen kurz- und langfristigen Emissionsminderungs-Strategien auf bodennahe Ozonkonzentrationen in hochsommerlichen Episoden, Ergebnisbericht zur **Stufe A**, Prognos AG, Basel.

Plinke, E., Hartmann, A. und Motz, G., 1996, Emissionsseitige Wirkungen von langfristigen und von temporären großräumigen Emissionsminderungs-Maßnahmen und Szenarien, Ergebnisbericht zum Teilvorhaben A-2, Prognos AG, Basel

Stern, R., Fath, J. ,Reimer, E., 1995, Bewertung der Effektivität von Ozonminderungsstrategien mit Hilfe des photochemischen Ausbreitungsmodells REM3, Abschlußbericht zum Teilvorhaben A-5 "Großräumige Ozon-Ausbreitungsrechnungen", Freie Universität Berlin, Institut für Meteorologie

Stern, R., Scherer, B., Fath, J., 1997, Ermittlung und Bewertung der Wirkung von Ozonminderungsmaßnahmen im Rhein-Main-Gebiet mit Hilfe des photochemischen Ausbreitungsmodells CALGRID, Abschlußbericht zum Teilvorhaben B-3 "Kleinräumige Ozon-Ausbreitungsrechnungen", Freie Universität Berlin, Institut für Meteorologie

Stern, R., Scherer, B., Fath, J. Reimer, E., 1997, Ermittlung und Bewertung der Wirkung von Ozonminderungsmaßnahmen im Großraum Berlin-Brandenburg mit Hilfe des photochemischen Ausbreitungsmodells CALGRID, Abschlußbericht zum Teilvorhaben B-3 "Kleinräumige Ozon-Ausbreitungsrechnungen", Freie Universität Berlin, Institut für Meteorologie

Stern, R., 1997, Untersuchung des Einflußes des Emissionswochengangs auf die Ozonkonzentrationen mit Hilfe des photochemischen Ausbreitungsmodells REM3, Abschlußbericht zum Teilvorhaben A-5 "Großräumige Ozon-Ausbreitungsrechnungen", Freie Universität Berlin, Institut für Meteorologie

Tippke, J., Memmesheimer, M., Jakobs, H., Piekorz, G., Ihrig, M., Kessler, C., Roeben, B., Ebel, A., 1997, Ermittlung und Bewertung der Wirkungen von Ozonminderungsmaßnahmen im Großraum Dresden mittels Simulationsrechnungen mit dem EURAD-Modellsystem für die Episode Juli/August 1994. Abschlußbericht zum Teilvorhaben B-3 "Modellregion Dresden", Förderverein des Rheinischen Instituts für Umweltforschung an der Universität zu Köln e.V.

These reports are available on CD-ROM, thus enabling anyone interested to view the reports on-screen and to make use of the data according to individual needs. On the CD-ROM all reports will be in PDF format with accessing software. Furthermore, there is an access to the results of this research project presented via internet to address a broader public: http://www.umweltbundesamt.de/ozon-e. Additionally it is possible to download the final report in German as well as the summary in German and English from the internet.

DISCUSSION

D. STEYN: Are the politicians and public aware that even after the great effort needed to achieve the emissions reductions (TREND 2005), there will still be occasional exceedances of the warning level (90 ppb)? Furthermore, do they understand the massive adjustments needed to reduce these exceedances to zero?

A. GRAFF: At least the policy makers know about these problems and they are aware that only an Europeanwide strategy is appropriate to tackle high ground level ozone concentrations in summertime episodes. Furthermore awareness increases to look at the hemispheric background concentrations of ozone precursors like CO and CH_4 as well in parallel. The German government's concept is thus to focus on long term permanent emissions reductions measures at first and secondly, combining those with temporary measures in case of episodes with extremely high ground level ozone concentrations to strengthen the effect on ozone concentrations reductions.

The aim to reduce exceedances of 90 ppb ground level ozone concentrations to zero has to be set against the principle of proportionality (with respect to emissions reductions measures).

IMPACT OF SPECIAL FEATURES OF NUMERICALLY PREDICTED AND ANALYSED METEOROLOGICAL DATA ON THE RESULTS OF OZONE FORECAST BY A PBL-CTM

J. Flemming, E. Reimer, R. Stern

Freie Universität Berlin
Institut für Meteorologie
Carl-Heinrich-Becker-Weg 6-10
12165 Berlin
Germany

INTRODUCTION

The Freie Universität Berlin and IVU [1]develop a complex ozone forecast system, which combines statistical and fuzzy methods with regional Eulerian transport modelling. Each method is an independent forecast tool focussing on different spatial and temporal scales. The statistically and fuzzy based forecast is locally applied, whereas the Eulerian transport model predicts the large scale development of ozone patterns. The results of the transport modelling will be used as an additional parameter for the statistical and fuzzy methods in a framework of a neuro fuzzy module. Ideally, the combination of both approaches shall improve the forecast performance (Reimer and Dlabka, 1998).

In the preliminary step, the performance of the model REM3 as a stand alone tool for the ozone forecast is tested. The test focusses on the impact of the meteorological input data on the model performance. Therefore, REM3 results gained by employing a numerically predicted data base will be compared to results gained by an analysed data base for the summer 1997.

The forecast error of the prognostic data base will influence the model simulation. However, the different statistical properties of the two data sets (temporal and spatial variability, resolution etc.) causes substantial differences in the calculated ozone fields too. The cloud cover is chosen as an example to illustrate the differences of the two data sets and their impact on the model calculation.

[1] Gesellschaft für Informatik, Verkehr und Umwelt

MODEL DESCRIPTION AND FORECAST SETUP

REM3 (= Regional Eulerian Model) is a photochemical transport model of the planetary boundary layer in a regional scale (Stern 1994). The current model domain is Central Europe with an horizontal resolution of about 25*25 km^2. The vertical resolution of the model is based on 3 dynamically changing layers. A terrain following layer with a fixed depth of 50m is assumed at the ground level. The model calculation for this layer are considered being corresponding to ground based ozone measurements. The height of the interface between the 2nd and 3rd layer is controlled by the external calculated mixing layer height. Consequently the 2nd layer describes the mixing layer and the 3rd layer works as an reservoir layer. The emission data are taken from the PHOXA data base which was adapted to the situation in 1994. The chemical mechanism CBM4 is used in this application.

Model forecast run starts in the afternoon (15.00 UTC) when the PBL is well mixed. The run time is 72 h. The calculated initial ozone fields for the next day are adapted to available ozone measurements (currently only for Germany) by an optimum interpolation procedure. Both ground and mixing layer fields are changed in the area of representativity of the ozone measurements. As the PBL is well mixed, the representivity of the ground measurements is extended to the 2nd model layer. The calculated initial concentration fields of NO and NO2 are adjusted to the new initial ozone concentration field assuming photo stationary equilibrium

METEOROLOGICAL INPUT DATA

The model is not accompanied by its own meteorological model. Instead meteorological raw data were analysed and interpolated to the REM3 coordinate system. The required PBL parameters such as mixing layer height, MO-length, u-star were calculated by a 1-D PBL bulk model (Van Ulden and Holtslag 1985).

In the episodical and climatologial application of REM3 only observed meterological raw data have been utilized to prepare the required input data. For the real time application of REM3 the diagnostic data had to be replaced by operationally numerical predicted data. The processing for obtaining the REM3 data input had to be adapted to this data typ but remained unchanged in its philosophy. In order to test the model perfomance with the prognostic data the simulation was repeated with the common diagnostic data base.

The diagnostic data base is derived routinely at the Freie-Universität-Berlin from synoptic meteorological surface data, upper air data and climatological information (Reimer and Scherer 1992). By means of a statistical interpolation scheme the data are analysed on isentropic surfaces with a horizontal grid resolution of about 25 x 25 km^2.

The prognostic data base is supplied by a numerical weather prediction model of the German Weather Service (EUROPA-Modell). It consists of 3-hourly meteorological data fields on σ - layers with horizontal grid resolution of about 60 x 60 km^2.

Differences of prognostic and diagnostic meteorological data

The predicted data contain an unknown forecast error, which influences the ozone forecast. Moreover, the two data sets differ in their statistical or climatological properties, i.e. differences in the temporal and spatial variability. This fact has to be taken into account, when the differences of the ozone calculations are analysed. The cloud cover is a good example to demonstrate both features. The simulated ozone concentrations depend strongly on the cloud cover because it controlls the photolysis rates as well as the development of the mixing height.

Forecast error. The forecast error is variable, depends on the meteorological situation and is likely to increase with increasing forecast time. It is different for different variables. The forecast of sub - scale depended parameters, such as cloud cover, tends to be more erroneous than the one of large scale parameters such as pressure fields. The forecast error can be estimated by comparing the forecasted field with the analysed initial field of the numerical weather prediction (figure 1).

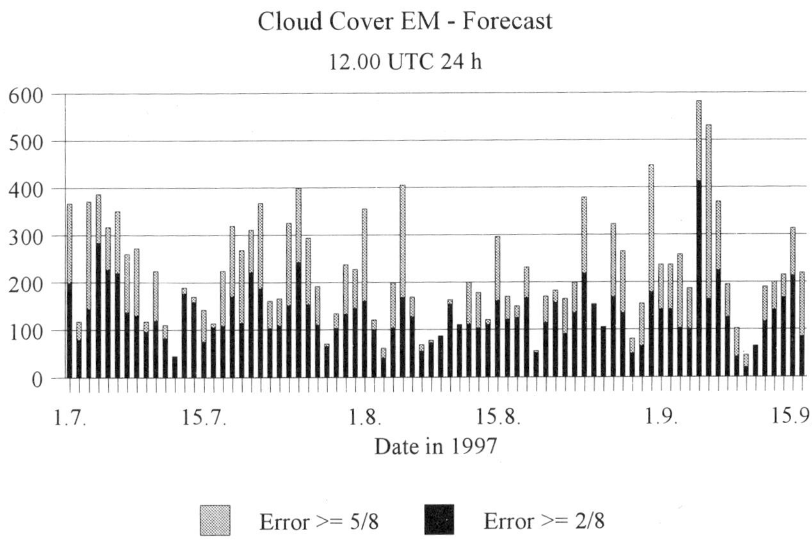

Figure 1. Graph of the cloud cover forecast error of the EUROPA-model: Percentage of grid points over Germany predicted erroneously after 24 hours over Germany. The error is determined by comparing the predicted cloud cover data with the corresponding EUROPA Modell analysis.

Variability difference. It is far more difficult to judge the differences due to the different origin of the two data types. However, the different variability of the data, beside the forecast error, may cause substantial differences in the calculation of the concentration fields by REM3. In general, the prognostic data are smoother because the horizontal resolution of this data base is coarser. Especially the cloud cover fields differ in patterns and variability: EUROPA model cloud cover results from parameterisation of large scale humidity fields, whereas diagnostic data were interpolated and averaged from local and subjective observations (figure 2).

COMPARISON OF REM3 OZONE FORECAST

From 1st of July to 15th of September 1997, REM3 performed daily a 72-h ozone forecast by using prognostic meteorological data. The procedure has been repeated by employing diagnostic data. The comparison with measurements is used to evaluate the general ozone forecast capability of the model. The intercomparison of the model results will depict the influences of the different data types on the model performance.

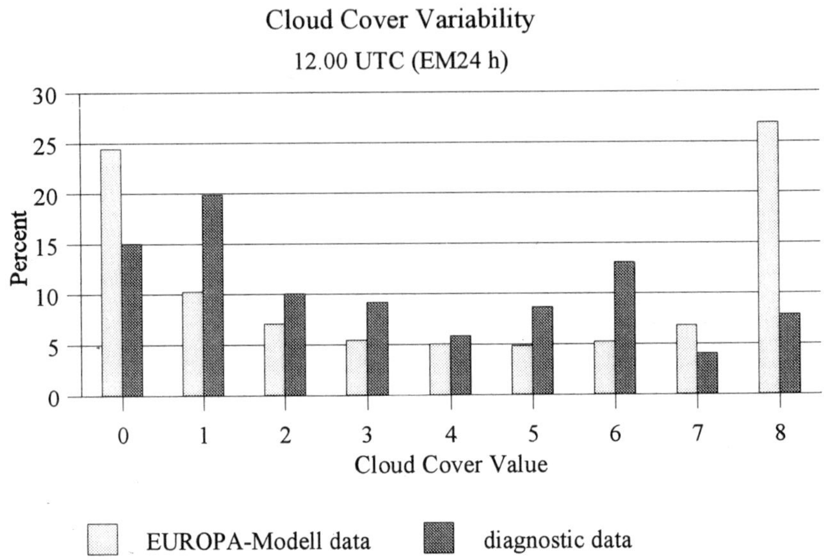

Figure 2. The occurrence of cloud cover values (0 - 8) in the forecast period in Germany. The EUROPA Modell favours no or total cloud coverage.

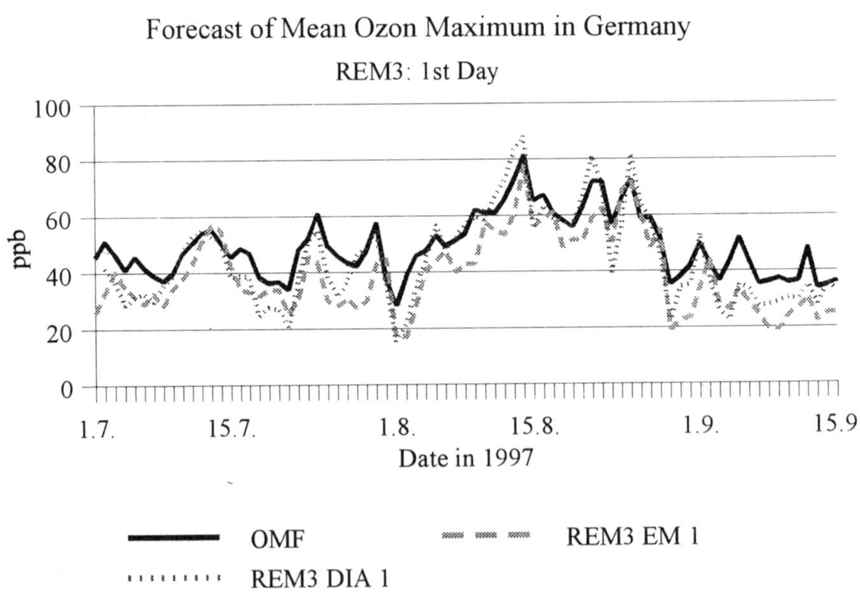

Figure 3. Time series of the mean ozone maximum (DMMA) in Germany calculated by REM3 and that of measurements (OMF). The REM3 fields were calculated using prognostic EM-data (REM3 EM) or diagnostic meteorological data (REM3 DIA), 24-hours forecast

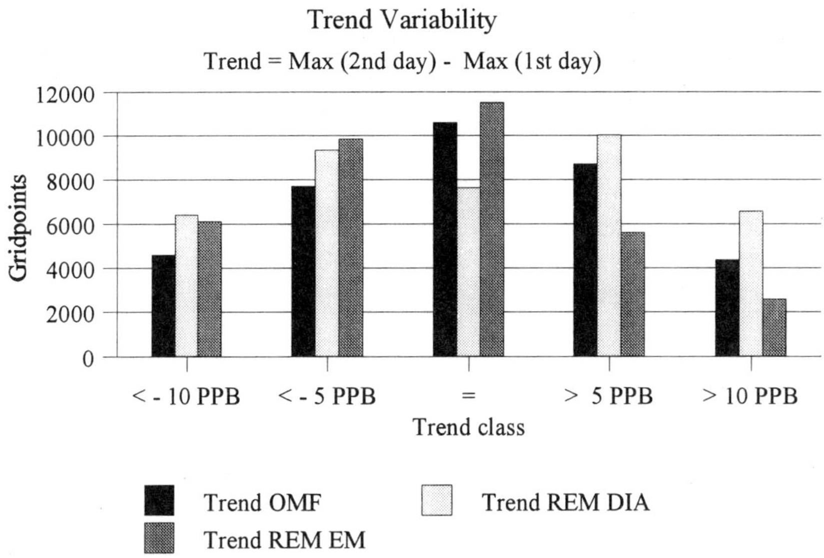

Figure 4. Comparison of the temporal variability of REM3 ozone fields and measured ozone fields (OMF). The accumulative occurrence of "no trend" (=) is the lowest for the REM3 fields using diagnostic data, meaning that the temporal variability is high.

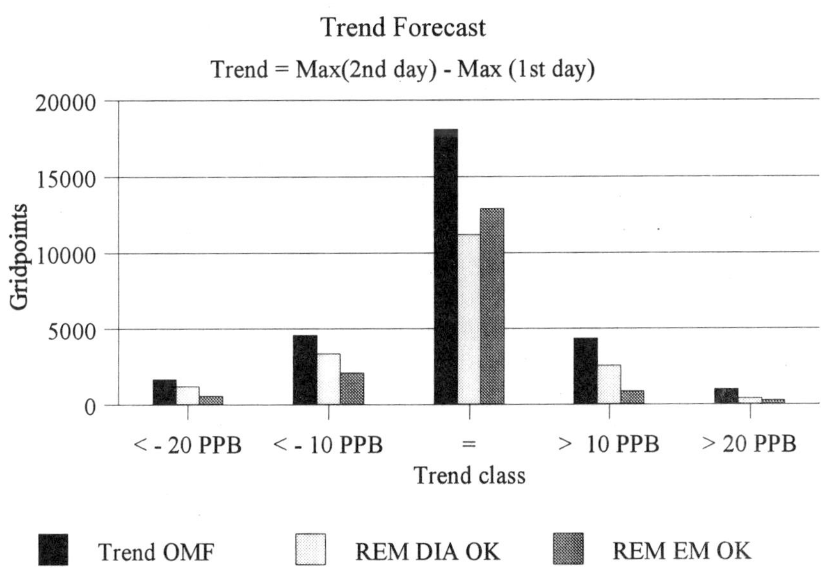

Figure 5. Capability of REM 3 in forecasting trends of the daily ozone maximum. The accumulative number of a correct trend forecast (REM DIA OK / REM EM OK) at the grid point is compared to the occurrence of measurement fields trends. (trend OMF).

Methodology

With respect to the intended application, only the forecast of the daily ozone maximum is considered. Beside the absolute values, the temporal and spatial variability of the ozone concentration is taken into account.

The ozone maximum at the 1st, 2nd and 3rd day of the model run is compared to measurements of the German ozone monitoring network. The model performance is tested not by direct comparison with station measurements but by point wise comparison with a gridded field of the measurements (OMF). This field is produced by employing an optimum interpolation technique. The calculated and the interpolated field of the observations are comparable within Germany. This is an attempt to close the conceptual gap between the site related measurements and the calculated grid box averages of the model. Additionally, the use of the OMF avoids problems with the daily fluctuating number of available station measurements.

Forecast of the ozone maximum

The daily mean (DMMA) of the observed maximum field (OMF) over all grid points in Germany gives an idea of the German wide ozone situation at every day in the period (figure 3). Although there is some bias, the model is capable to reproduce DMMA in its temporal change rather well. The DMMA of the diagnostic run tends to be higher then that of the prognostic run. This fact may have many reasons. One of them could be the more frequent occurrence of total cloud cover 8/8 in the prognostic data (figure 2). In general the model performs better when the measured ozone concentration tends to be medium or high. The underestimation of low ozone values is not easy to explain but does not trouble too much the intentions of the operational ozone forecast.

Forecast of ozone concentration trends

In regard of the application, the forecast of the expected change in the extreme values (trend) is of crucial importance. The change from day to day at every grid point in Germany is expressed in five trend classes (no trend, negative trend, stronger negative trend, positive trend, stronger positive trend). The accumulative occurrence of the trend classes shows the temporal variability of the REM3 calculations and of the interpolated observations (figure 4). Obviously diagnostic REM3 results are less persistent then that of the measurements and the prognostic run.

The occurrence of daily trends indicates only the climatological behaviour of the measurements and of the simulations. In order to evaluate the forecast applicability of the model, the local correspondence of the simulated and the observed trends has to be considered (figure 5). Even if the model fails to forecast the correct absolute value the forecasted trends might be correct. The correct forecasted trends will supply important information and might be a useful additional parameter for the statistical methods within the complex ozone forecast system.

Acknowledgements

The project is funded by the environmental agency of Germany (UBA F&E Vorhaben 10402817)).

References

DWD, *Quarterly Report of the Operational NWP-Models of the Deutsche Wetterdienst,* No. 12, Offenbach (1997).

Flemming, J., *Ozonprognose mit dem photochemischen Ausbreitungsmodell REM3*, Zwischenbericht 2/1998 zum F&E Vorhaben 10402817 des UBA, Berlin (1998).

Reimer, E. and B. Scherer, An operationel meteorological diagnostic system for regional air pollution analysis and long-tem modelling, in: *Air Pollution Modelling and its Application IX* ,Plenum Press, New York (1992).

Reimer, E. and M. Dlabka, Local forecast of low ozone by use of cluster methods and fuzzy-models in: *Air Pollution Modelling and its Application XIII*, Plenum Press, New York (1998).

Stern, R.., *Entwicklung und Anwendung eines dreidimensionalen photochemischen Ausbreitungsmodells mit verschiedenen chemischen Mechanismen*, Meteorologische Abhandlungen des Instituts für Meteorologie der FU-Berlin. Serie A, Band 8, Berlin (1994).

Van Ulden, A.P. and Holtslag, Estimation of atmospheric boundary layer parameters for diffusion application, *J. Climate and A. Met.*, 24:11, (1985).

DISCUSSION

P. BUILTJES: In the future you want to improve the cloud cover description in the meteorological data. Can you give an idea how you are going to do that?

J. FLEMMING: We intend to use satellite information to improve the diagnostic cloud cover data. For the prognostic applications, we are thinking of making corrections of the predicted cloud cover by statistical methods. Moreover, a model run without any clouds should give us an idea of the potential ozone production in certain critical situations.

MODELLING PHOTOCHEMICAL SMOG EPISODES IN LOMBARDY REGION

A. De Martini[1], F. Pasi[2], G. Maffeis[1], M. G. Longoni[1], M. Tamponi[3], G. Graziani[2], G. Lanzani[4]

[1] Environmental consultants at Regione Lombardia, Via Fara 26, Milano, Italy
[2] Joint Research Centre, Environment Institute, I-21020 Ispra, Italy
[3] Asl 7, P.M.I.P. U.O. FISICA & T.A., Via Longoni 13, I-22048 Oggiono, Italy
[4] Amministrazione Provinciale di Como, Via Borgovico 148, Como

INTRODUCTION

Lombardy is the most industrialized and populated region in Italy. As a matter of fact it has more than 9 millions inhabitants (16% of the Italian population) over a territory of about 24000 km^2 (8% of Italy). Because of its complex orography the distribution of the population is not homogeneous, for example, there are 55 compared to 1440 inhabitants per km^2 in the province of Sondrio (the most northern and mountainous) to the province of Milan. The area centered in Milan city and extending north and north-west for about 30 km, can be considered a single metropolitan area, with 3,700,000 inhabitants, with 1,550,000 workers (data taken from the National Statistical Institute) and 2,500,000 vehicles running. It has been estimated that the total emissions in this area in 1990 were about 94,000 tons/year of NO_X, 186,000 tons/year VOC and 529,000 tons/year of CO (ENEA, 1995).

These high emissions and the meteorological situation, characterized by frequent conditions of weak winds and high solar radiation, are responsible for several ozone (and other secondary pollutants) exceedances, measured by the air quality network (187 monitoring sites through out the region). The particular topography of the territory (mountains in the North and flat in the South) creates a valley breeze that transports pollutants, worsening the pollution situation north, where there are not so big emission sources.

In this paper the results of two model systems, CALMET-CALGRID and TVM-LCC, are compared with the ground measurements of the Lombardy air quality network. The main goal is to study the photochemical episode occurred from 20th till 23rd July 1994, chosen because it's a typical exceedance episode happening several times in summer. After briefly describing the model systems, the simulation domain and the input used, the results of the comparison between the estimated and measured ground level ozone concentrations at seven monitoring stations (chosen so that are just partially or not influenced by traffic emissions) are presented. A brief statistical analysis of the outputs of the two photochemical models has been done.

Air Pollution Modeling and Its Application XIII, edited by S.-E. Gryning and E. Batchvarova, Kluwer Academic/Plenum Publishers, 2000

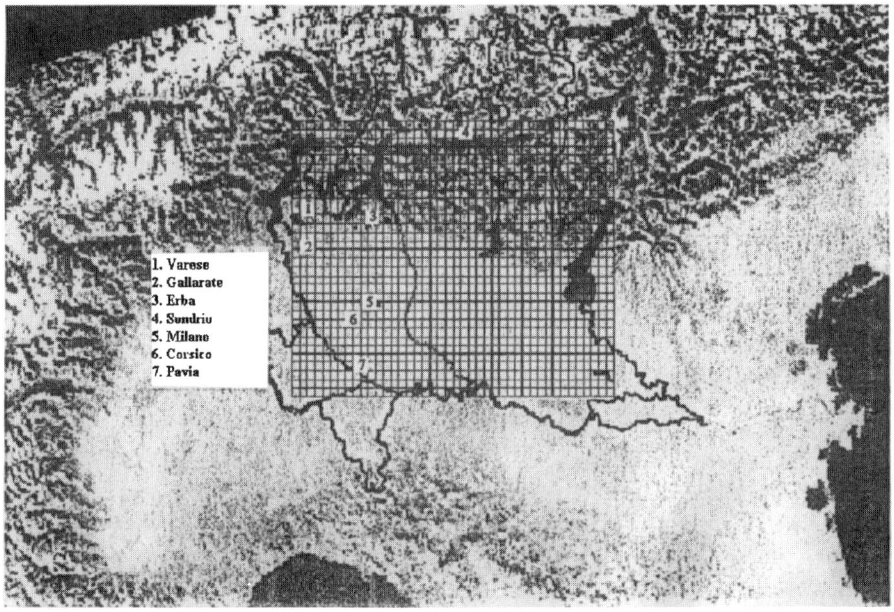

Figure 1. Domain of CALMET/CALGRID and TVM/LCC simulations. The table besides includes the air quality monitoring stations considered in the study.

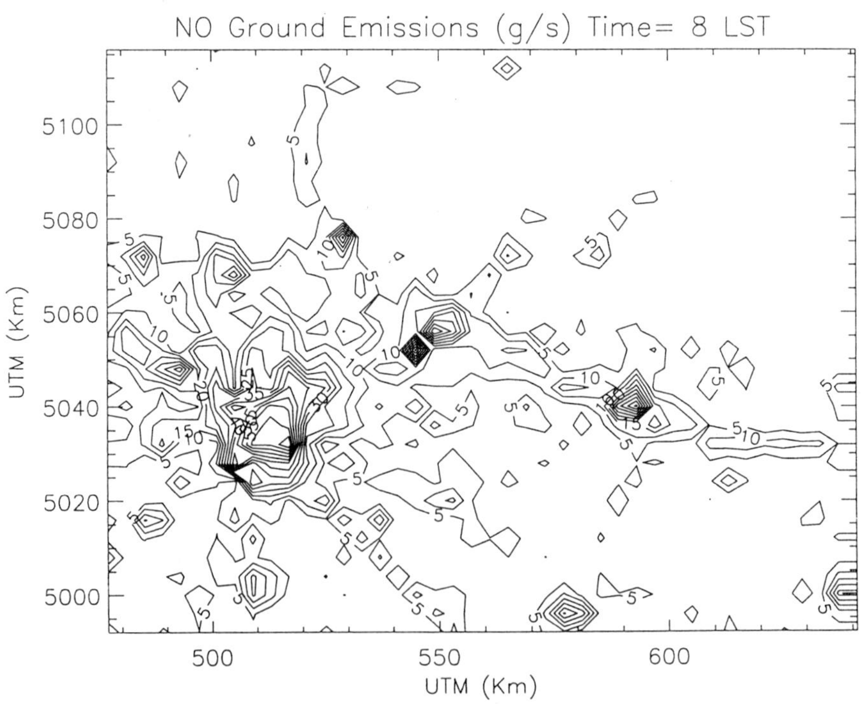

Figure 2. NO (g/s) emitted at 8 a.m. in the domain of the simulations.

SIMULATION DESCRIPTION

Models

Two model systems have been used to study the photochemical episode occurred in July 1994. The first one includes the mesoscale diagnostic meteorological model CALMET (Scire, 1997) and the photochemical model CALGRID (Yamartino, 1992). CALGRID is an Eulerian model with modules that simulate the transport, the chemical activity and the dry deposition. The chemical mechanism included is the SAPRC-90 chemical scheme (Carter, 1990). The other model system is the mesoscale meteorological prognostic model TVM (Topographic Vorticity Mesoscale Model, Thunis, 1995) with the chemical solver of CIT (McRae et al., 1983) and the LCC mechanism (Lurmann et al., 1987).

Domain

The horizontal domain consists of 42 x 32 grid points with a resolution of 4 x 4 km^2, covering most of the Lombardy region and southern part of the Canton Ticino (Switzerland). The morphology of the territory is very different through out the region, passing from the Alps (about 4000 m asl) to the flat Pianura Padana, through the Prealps and hills. The vertical grid is different for the two model systems: CALMET-CALGRID uses 10 layers up to 3100 m, while TVM-LCC respectively uses 25 layers up to 8500 m and 15 layers up to 3500 m. Both systems, starting from the lowest layer at 20 m above the ground, apply incremental vertical spacing with a maximum of 500 m.

The CALMET wind field is produced by an objective analysis of the regional meteorological monitoring stations data and the vertical sounding of Milan Linate Airport. TVM takes into account the synoptic conditions through a one-way nesting technique with the global circulation model ECMWF (Reading). Both meteorological models reproduce correctly the local circulation showing similar behavior, which, in summertime high pressure conditions, is characterized by a valley breeze circulation, featuring southerly winds in the afternoon (with a typical intensity of 2-3 m/s), shifting to weaker northerly one during the night. The wind fields obtained are then used as input to the coupled photochemical models that are responsible for the transport of the pollutants. CALGRID needs 3D initial and boundary conditions. In this simulation the outputs of the regional photochemical model LOTOS (Builtjes, 1992) have been interpolated in order to construct these inputs. LCC considers fixed boundary condition values, extrapolated from the measurements.

The hourly areal emissions estimates in the domain of the simulations have been done by using the bottom up approach, i.e. by considering local data and information, and the specific emission factors. The inventory includes areal emissions, urban and extra-urban traffic, productive activities, biogenic, and 13 point sources emissions. Both models have used this inventory (figure 2).

STATISTICAL EVALUATION

To evaluate the models performance the mean gross error and the average station peak estimation accuracy have been used. The mean gross error is defined as:

$$E_{ng}=\frac{1}{N}\sum_{i=1}^{N} \left|(c_e(x_i,t)-c_o(x_i,t))\right| \tag{1}$$

where N is the number of monitoring stations, and $c_e(x_i,t)$ and $c_o(x_i,t)$ are the estimated and observed hourly concentrations at the same location x_i and time t. It indicates the mean absolute discrepancy between hourly estimates and observations concentrations. The average station peak estimation accuracy is defined as:

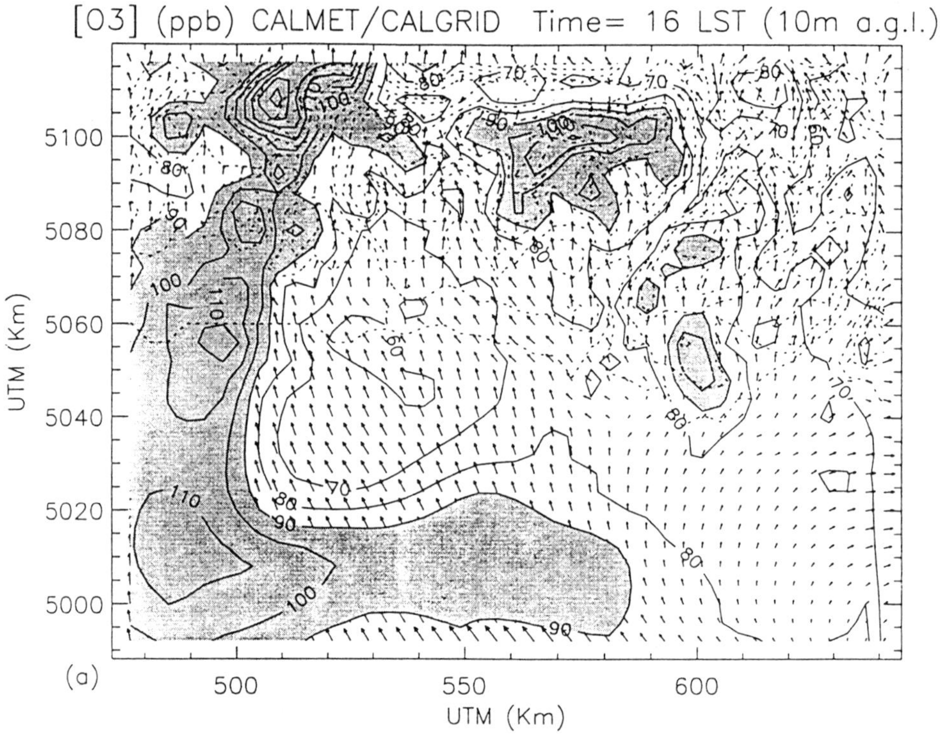

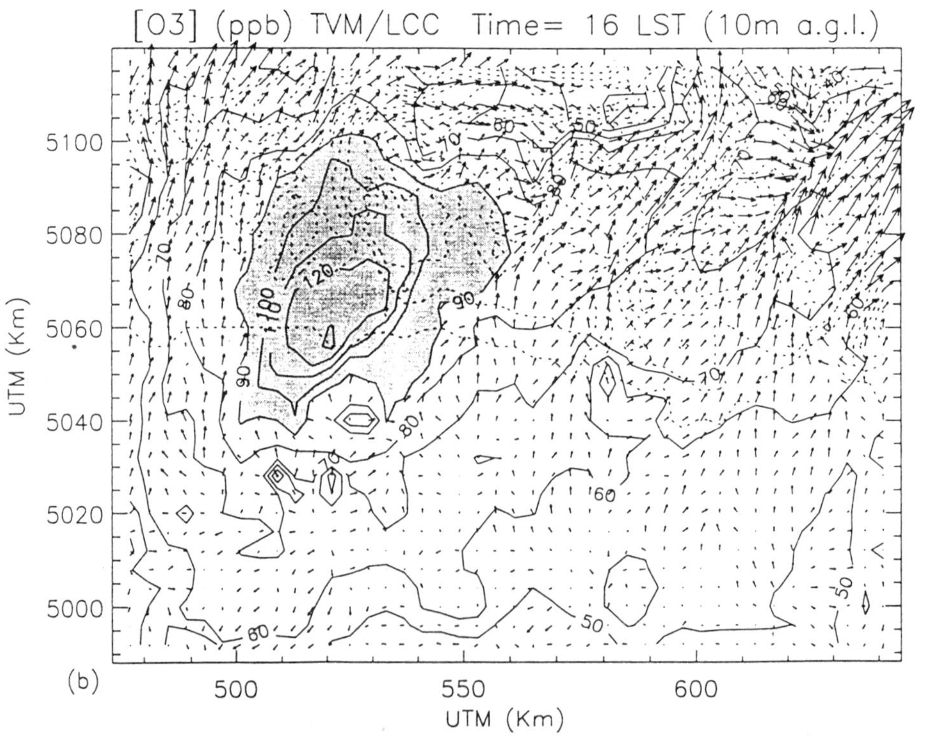

Figure 3a-3b. Ground level hourly ozone concentration isopleth with wind field overplotted on 22 July 1994 at 4 p.m.. Shaded areas represent concentration above 90 ppb.

$$A = \frac{1}{N} \sum_{i=1}^{N} |A_{si}| \tag{2}$$

where

$$A_{si} = \frac{c_e(\bar{x}_i, t) - c_o(\bar{x}_i, \bar{t})}{c_o(\bar{x}_i, \bar{t})} * 100\% \tag{3}$$

A_{si} is the spatially paired peak estimation accuracy for each monitoring station, $c_o(\bar{x}_i, \bar{t})$ is the peak concentration at the location x_i and $c_e(\bar{x}_i, \bar{t})$ is the highest estimated concentration at the same location, within a specified time interval centered on the peak hour, three hours in these simulations. The average station peak estimation accuracy describes how well the maximum concentration at each station is reproduced (Tesche, 1990).

RESULTS

In this paragraph results of the simulations of July 22 1994 are presented. The choice of the day is mainly due to chemical (gas phase reactions predominance) and computational reason (need of prerun days, July 20-21, to reach steady state). The regional air quality monitoring network measured ozone exceedances particularly in the northern part of the domain. In fact it's typical to find high concentrations of secondary pollutants (such as ozone) at 30-40 km from the main source (Milan metropolitan area). The afternoon transport of the plume of Milan area towards north is represented in figures 3a-3b, which show the concentrations isopleth (every 10 ppb) at 4 p.m. at 10 m above the ground, overplotted to the wind fields. The two estimated concentrations fields agree with the measurements showing higher concentrations as one moves north, even though the location of the peaks are different, due to the different wind fields: TVM computes a weaker wind towards north, while CALMET describes a stronger north-west one. Analyzing CALGRID results (figure 3a), one can notice the presence of more peaks: from the upper left corner of the domain the first maximum is probably due to the plume coming from the Milan area, the second is the result of the channeling of pollutants into the Valtellina Valley and from the city of Bergamo, while the third is due to the local contribution of the city of Brescia (see emission figure 2). CALGRID simulation shows a strong dependancy on the boundary conditions on the south-west corner of the domain, probably due to the nesting with LOTOS. After several testing simulations, this seems not to influence the Milan and its downwind area. This problem is under study anyway. LCC (figure 3b) shows one extended maximum north of Milan. This maximum is located a lot closer to Milan compared to the CALGRID ones: this is probably due to the weaker transport, which also doesn't allow the plume to channel into the Valtellina Valley (see also Sondrio station on figure 4b).

The two model systems ground estimates have been then compared to the air quality measurements of six monitoring stations (see figure 1). The models predictions, used in the comparison, are the result of a weighted average over the four cells closer to the station. Varese and Erba stations clearly show the presence of the plume of Milan, measuring peaks of concentrations up to nearly 130 ppb. Both models underestimate the concentrations in Varese (CALGRID 100 ppb, LCC 70 ppb). The strong LCC underestimate is probably caused by a wrong transport direction of the plume due to the wind field. Erba station concentration is well represented by LCC, while CALGRID underestimates the maximum. Gallarate station, which is still under the influence of the Milan plume but closer to the emission source, measures lower ozone concentrations (110 ppb). In this case CALGRID seems to be more reasonable, because LCC plume is not transported so west. The station in Milan (Verziere) is well reproduced by the models.

The models results have also been compared to the measurements done in Corsico, south of Milan (figure 5a-5b), at a typical traffic station. As figure 5a shows, the models

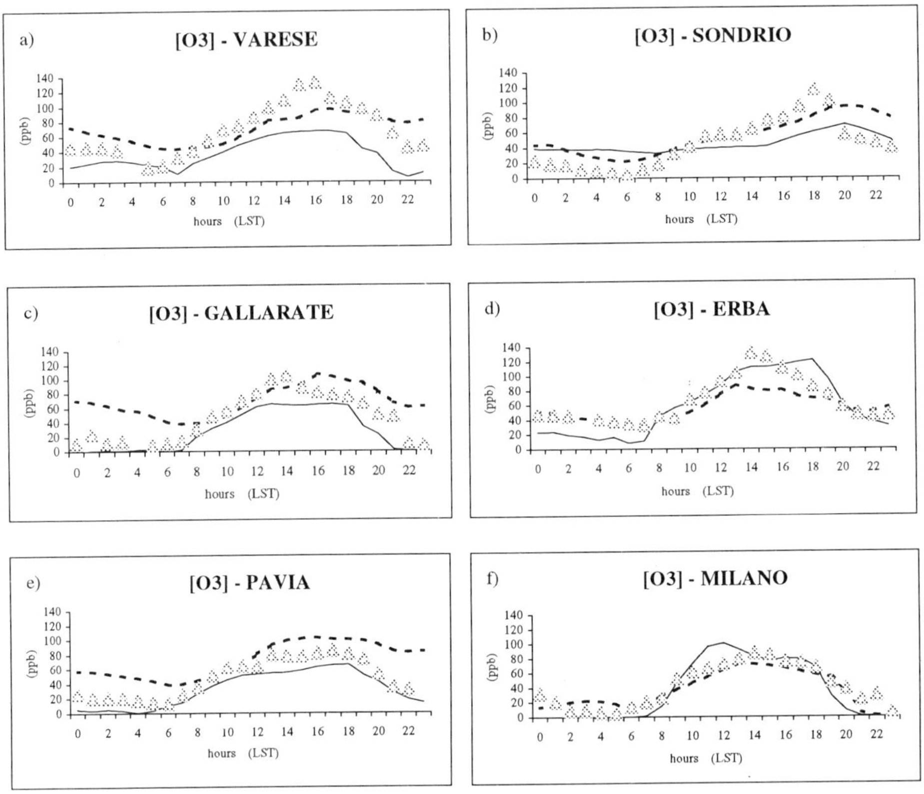

Δ : measurements, ----- : CALGRID, ——: TVM

Figure 4. Comparison of measured and estimated $[O_3]$ time series at six regional stations.

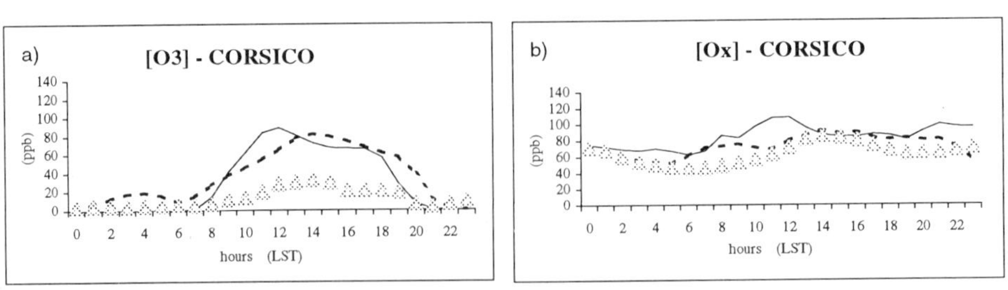

Δ : measurements, ----- : CALGRID, —— : TVM

Figure 5. Comparison of measured and estimated $[O_3]$ and $[O_x]$ at the Corsico monitoring station.

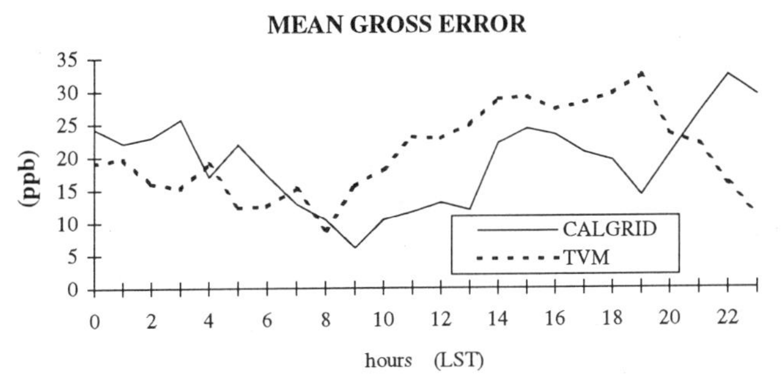

Figure 6. Hourly gross error averaged over the six monitoring stations considered in the analysis.

hardly reproduce the ozone concentrations where traffic (high NOx emissions) is intense. On the other hand the quantity O_x (= O_3 + NO_2) seems to be more representative, because of its stability. In this case (figure 5b) the results are matching better the measurements.

Finally a brief statistical analysis has been performed to evaluate the general behavior of the two photochemical systems. Figure 6, which shows the mean gross error over the six monitoring stations, clearly points out that CALGRID is not able to reproduce well the strong decrease of ozone at night (see also figure 4a), probably due to a strong vertical mixing, and that TVM underestimates the day concentrations. This is also evident in table 1 with the mean gross error calculated for the day and night separately and the average station peak estimation accuracy, which seems to be higher for CALGRID.

CONCLUSIONS

Two model systems have been used to describe an ozone episode occurred in Lombardy on July 1994. The two simulations show different ozone concentration isopleth maps. Their comparison with measurements of the air quality monitoring network gives good results in terms of value and time of the peaks, as also the average station peak estimation accuracy indicates.

In the domain under study, the influence of the wind is particularly important to describe the plume. In fact the wind field is the only different input, variable in the domain, which can explain, at least partially, the ozone isopleth maps.

In certain cases, such as at station close to a very traffic road, O_3 doesn't seem to be a good indicator especially at night, both for CALGRID and LCC. When using instead Ox, the two models work better.

Table 1. Average gross error calculated over 24 hours, during the day and night, and average station peak estimation accuracy

		TVM	CALGRID
Average Station Peak Estimation Accuracy		36.9%	21.5%
Mean Gross Error	24 hrs	20	19
Mean Gross Error	6 a.m. - 7 p.m.	23	16
Mean Gross Error	0-5 a.m. and 8-23 p.m.	17	24

ACKNOWLEDGEMENTS

We would like to thank Peter Builtjes and his team, for providing us the LOTOS data, Bob Yamartino and Joe Scire for their precious help.

REFERENCES

Builtjes P.H., 1992, *The LOTOS - Long Term Ozone Simulation Project*, Summary Report, TNO Report, IMW-R92 \ 240,Delft, The Netherland.

Carter W., 1990, *A Detailed Mechanism for the Gas Phase Atmospheric Reactions of Organic Compounds*, Atm. Env. 24A.

ENEA-Dip. Ambiente, 1995, *CORINAIR PROJECT: Air Pollutant Emission Inventory in Italy for the year 1990*, Final Report

Lurman F. W., Carter W. and Coyner L. A., *A Surrogate Species Chemical Reaction Mechanism for Urban-Scale Air Quality Simulation Models, I: Adaptation of the Mechanism*, Epa Report N, 600/3-87-014A.

McRae G. J., Goodin W. R. and Seinfeld J. H., 1983, *Mathematical Modelling of Photochemical Air Pollution,* EQL Report no.18, Environmental Quality Laboratory, CALTECH, Pasadena, CA.

Scire J. et al., 1997, *A User's Guide for the Calmet Metereological Model*, Earth Tech.

Tesche T. W., Georgopoulos P.,Seinfeld J. H., Cass G., Lurmann F.L. and Roth P. M., 1990, *Improvement of Proceduree for Evaluating Photochemical Models*, Report to the California Air Resources Board, contract A832-103, Radian Corp., Sacramento, Ca.

Thunis P., 1995, *Formulation and Evaluation of a Non-Hydrostatic Vorticity-Mode Mesoscale Model*, PhD Dissertation, Universite' Catholic de Louvain-La-Neuve, Belgium.

Yamartino B. et al., 1992, *The Calgrid Mesoscale Photochemical Grid Model-I. Model Formulation*, Atm. Env. 26A.

DISCUSSION

D. STEYN

Referring to figures 3a and 3b: clearly both figures cannot be correct. Intuitively the TVM/LCC appears correct. The only way to establish which is correct is to perform a detailed chemical analysis at selected points in the domain. I suggest examining terms in the O_3 tendency equation as well as calculating the Sillmann (1994) ratios.

A. DE MARTINI:

We still are working on this episode, in particular on the wind fields, which present some differences. These are not the final results. There is clearly a difference in the O_3 concentration fields. It's hard to say which model performs better, since for example CALMET/CALGRID seems to simulate the peaks better than TVM/LCC. Both models seem to perform in a similar way according to the gross error. Not even if we distinguish the gross error for the night and the day, the results are very different. Also the comparisons between the concentrations measured and calculated, don't show clearly who is predicting better. We'll certainly try to examine more chemical species and to calculate the Sillmann ratios.

R. YAMARTINO:

Comment: There are continuos efforts at Earth Tech to improve the nighttime K_y profiles and the detailed treatment of the canopy layers, as the ability to predict O_3 at night is closely tied to the ability to model O_3 deposition removal and mixing at night. Also note that recent applications in Europe (e.g. Stern at al.) and in the U.S., have shown improved nighttime O_3 performance particularly over the rural stations.

LOCAL FORECAST OF LOW LEVEL OZONE BY USE OF CLUSTER METHODS AND FUZZY-MODELS

E. Reimer[1] and M. Dlabka[2]

[1] Freie Universität Berlin
Institut für Meteorologie
Carl-Heinrich-Becker-Weg 6-10
12165 Berlin
Germany

[2] Fachhochschule für
Technik und Wirtschaft Berlin

INTRODUCTION

At the Free Universität Berlin a complex ozone forecast system is being developed, which is sponsored by the German Environmental Agency. The main objectiv of this scientific project is the development of a tested and applicable model system consiting of different procedures to forecast:

- daily maxima of 1h ozone concentrations over Germany (threshold values 180 and 240 μg)
- up to 3 day forecast of ozone trend over Middle Europe

Investigations have been started to develop, evaluate and compare the different procedures to get an improved forecast system. The basic idea is that a weighted combination of specific approaches might be more successful than one scheme alone. The approaches are:

- statistical analysis of local time series by combining hierarchical cluster analysis and regression methods. For these examinations the meteorological and concentration time series and a special procedure for determining series of objective Großwetterlagen is taken in consideration.

- fuzzy and neurofuzzy models to get a different description of the local and highly nonlinear processes of ozone generation and the errorneous time series of observations. Such systems can deal with poorly defined parameter distributions to some extent.
  Different concepts are developed to get rulebases by use of special cluster algorithms.

- the Eulerian chemical transport model REM3 with a grid size of about 30km. The model area covers West and Central Europe and uses the Phoxa emission database. The meteorological forecast is supplied by the Europe-Model from the German Weather Service.

Each method can be used as an independent forecast tool focussing on different spatial and temporal scales. The statistically based forecast will be locally applied, whereas the Eulerian transport model will predict the large scale development of ozone patterns. Ideally, the combination of deterministic transport modelling with statistical or fuzzy methods shall improve the forecast performance. Hence, in the framework of a neurofuzzy module, the large scale ozone forecast of the photochemical transport model will be used as an additional parameter for the site related statistical models.

The separate tools will be combined into a neurofuzzy system in order to find a best combination for each region and episode. Neurofuzzy systems combine automatic learning procedures with fuzzy systems.
In this paper the Fuzzy and Neurofuzzy modelling is discussed.

DATA BASE

As shown in table 1 the meteorological time series of data are received from German Weather Service.

- 1990 to 1997:
  time series of hourly surface observations for special WMO stations.

- 1990 to 1998:
  time series of 3-hourly surface observations for all German WMO stations (ca 300), received in realtime at Institute for Meteorology from FU Berlin.

- 1990 to 1998:
  radiosonde measurements from Germany.

The prognostic data sets from DWD are

- Since summer 1997 up to 1998:
  point forecasts from „Deutschlandmodell" for special surface locations corrected by Kalman-filter for 72 hours and time steps of 3 hours.
  Forecast starts at 00 and 12 UTC.

Table 1. Conception of the ozone forecast system

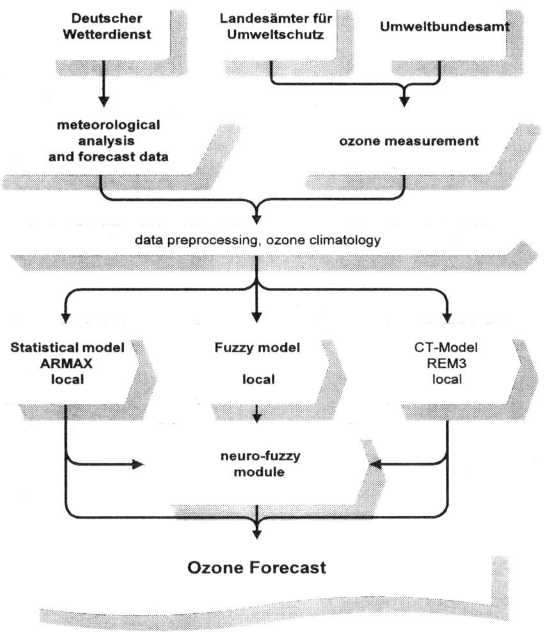

- Since summer 1997 up to 1998:
  gridded field forecasts from „Europamodell" for an middle European area. Forecasts are prepared up to 72 hours and for time steps of 3 hours starting at 00 and 12 UTC.

The concentration data time series and real time data are prepared by Umweltbundesamt (UBA) and environmental services. All time series were checked and a special interpolation procedure was used to fill missing data.

For this part of the project commonly observed meteorological parameters were used:

surface: total coverage by clouds, coverage with low clouds, cloudtype, horizontal visibility, actual weather, air pressure, air pressure tendency, temperature, dewpoint, windvector.

troposphere: geopotential, temperature, humidity, windvector.

Some sets of derived parameters were prepared from gridded forecasts and analyses from German Weather Service:

Großwetterlagen derived by use of Cluster methods, mixing height,

chemical parameter:

hourly conzentrations of ozone

Table 2 . Parameters for fuzzy-model example.
(t+1) - forecast day, (t) - last observation, (t-1) - the day before,
Delta - 24 hour difference between forecast and las observation.

(t-1)	(t)	(t+1)	Delta	Parameter
OZA1	**OZA**	**OZ**	**DOZ**	**Ozone (daily max. concentration)**
	NNA	**NN**	**DNN**	**cloud coverage (09+12+15z)**
	NLA	**NL**	**DNL**	**low cloud coverage (09+12+15z)**
	W6			**synopt. weather class (09+12+15z)**
	VVA	**VV**	**DVV**	**visibility (09+12+15z)**
	TTA	**TT**	**DTT**	**temperature (daily maximum)**
	RHA	**RH**	**DRH**	**humidity (09+12+15z)**
PPA1	**PPA**	**PP**	**DPP**	**air pressure (12z)**
		DPT	**DDPT**	**pressure tendency (12z)**
	DIA	**DI**	**DDI**	**winddirection (12z)**
SPA1	**SPA**	**SP**		**windspeed (12z)**
		GWL		**Großwetterlagen**

Different parts of these time series are used for tests with Neurofuzzy, however for Fuzzy modelling special time differences and observations were chosen. Table 2 displays these parameters.

FUZZY AND NEUROFUZZY MODELLING

The generation and loss of ozone is a highly nonlinear process, depending on meteorological conditions and the variable occurrence of precursors like VOC's or NOx.

A large number of physical/chemical models have been developed to handle these processes from a deterministic point of view in combination to a meteorological model (Flemming et.al., 1998). These models are very complex so that the evaluation and application cause several problems, because of uncertainties in the initial model conditions and errors in emissions. Moreover, insufficient knowledge of the relevant processes produces corresponding errors in the determination of the ozone budget (Motz et al. , 1996).

Therefore many other methods have been introduced like regression equations or other statistical approaches, neural networks or fuzzy systems to develop empirical models for that problem. Such procedures utilize local observations and therefore have the advantage to be closer to observed local reality than the physical models. However, only a relatively small number of observing stations for meteorological and chemical parameters exist relatively in space and the parameters shown above cannot explain the ozone formation at all, but mostly accompany these processes. Nevertheless, correlations and rules can be found to develop practical forecast models.

Fuzzy systems are well known for controlling technical processes (Kosko, 1992). The system of rules is interpretable and the membership functions, which can be determined by cluster methods, are open for interpretation and manipulation by an expert. However, the tuning of fuzzy systems brings about problems.
Therefore different procedures have been developed to fit the rule system to a special parameter presentation of a process:

- the fuzzy pattern classification (Bocklisch, 1987 and Burmeister, 1995), where by fuzzy clustering and a special algorithm for the determination of membership functions a system of rules is adapted to empirical data sets,

- the Sugeno fuzzy system (Hellendoorn, 1997), which combines the learning procedure of a neural network and the rule system of a fuzzy system.

Both procedures are tested within this project and will be compared within the ozone forecast system in the next time.

The fuzzy rules have the simple principle form:

IF (fuzzy condition) THEN (fuzzy conclusion) OR
IF(................

However, the fitting of rules to empirical data is a complex procedure.
Within the Sugeno fuzzy the conclusion is not a fuzzy membership function, but might be a system of algebraic equations, e.g. ARMAX models (Dlabka, 1996).

IF $(x^{(a)} = X_1^{(a)})$ AND $(x^{(b)} = X_1^{(b)})$ THEN y_{11} OR
IF $(x^{(a)} = X_2^{(a)})$ AND $(x^{(b)} = X_1^{(b)})$ THEN y_{21} OR
IF $(x^{(a)} = X_1^{(a)})$ AND $(x^{(b)} = X_2^{(b)})$ THEN y_{12} OR
IF $(x^{(a)} = X_2^{(a)})$ AND $(x^{(b)} = X_2^{(b)})$ THEN y_{22}

with a local model: $y_{ij} = a_1^{ij} x^{(a)} + a_2^{ij} x^{(b)} + b^{ij}$

and the rule matrix:

	$X_1^{(a)}$	$X_2^{(a)}$
$X_1^{(b)}$	y_{11}	y_{21}
$X_2^{(b)}$	y_{12}	y_{22}

x input values
a,b parameters
X_k fuzzy classes
i,j clusters and variables
The result is a weighted sum of e.g. ARMAX models yi,j

The membership functions and their representation is determined by fuzzy cluster methods with an adaptation procedure which minimizes the RMS of predictions and observations (Klawon, 1996 and Gronostay, 1998).

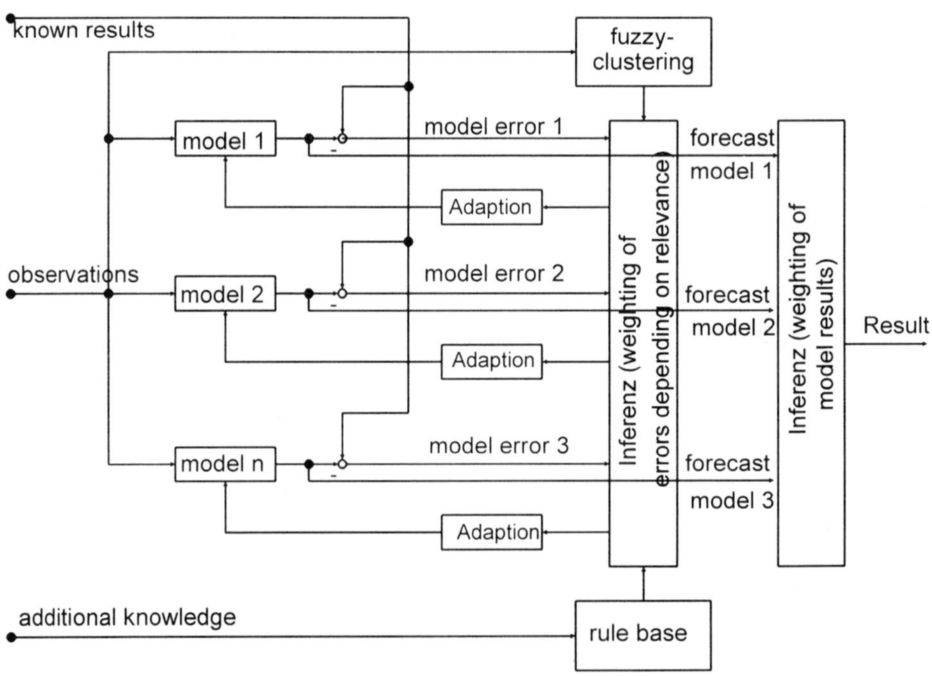

Figure 1. Principle of Neuro-Fuzzy System.

Both systems, fuzzy pattern classification and Sugeno fuzzy system were trained with a number of time series as shown above for the period of 1990 to 1996.

The forecast tests were done with the data sets from 1997 as a perfect prognosis in relation to meteorological data. Only ozone concentration are unknown.

The next step is the use of forecasted meteorological parameters from Europa-model for 1998, where the meteorological forecast errors must be considered in addition.

EXAMPLE

In this chapter a short example is given for the tests with the fuzzy pattern classification. For the ozone time series of the station Waldhof from Umweltbundesamt different parameter groups from Table 2 were tested, to get some information about the most efficient combination. At the start all data were ordered to Großwetterlagen 1 to 10, which were determined by a special cluster procedure using of geopotential height of 1000 and 500 hPa, the relative topography 500/1000 and surface ozone concentrations (Enke, 1997). In Table 3 these different groups of parameters are shown for the training phase., where the cluster procedure were applied to grouped data sets for 1990 to 1996. The bias and RMS is shown.

Table 3. training phase for Waldhof 1990-96 (April - September)
Mean errors of RMS differences between observations and model forecast of daily maxima of ozone concentrations for different cluster combinations and Großwetterlagen, dimension: hourly means [μg].

DD	winddirection 12UTC
DOZ	daily change of maximum of ozone concentrations
NN	total cloud coverage (9 - 15UTC)
GWL	Grosswetterlagen
OZA	daily ozone maximum yesterday
OZ	daily ozone maximum today
TT	daily temperature maximum

	Großwetterlagen									
	01	02	03	04	05	06	07	08	09	10
bias										
GWL/OZA/TT/OZ	5	2	2	1	-3	-2	5	-4	-2	9
GWL/OZA/DD/OZ	5	2	-1	0	0	-1	5	-1	-4	8
GWL/OZA/TT/DOZ	-1	0	1	3	-1	-1	0	0	3	2
GWL/OZA/DD/DOZ	-6	0	1	1	1	-1	3	-1	1	6
GWL/OZA/NN/DOZ	-7	2	1	0	2	1	3	0	1	1
GWL/OZA/TT/DD/DOZ	-1	1	0	-1	1	2	3	0	-1	3
RMS										
GWL/OZA/TT/OZ	24	18	14	14	19	13	24	18	19	28
GWL/OZA/DD/OZ	28	21	15	15	17	10	24	20	23	28
GWL/OZA/TT/DOZ	29	18	17	17	16	13	21	18	20	27
GWL/OZA/DD/DOZ	29	19	15	14	16	14	24	21	21	28
GWL/OZA/NN/DOZ	35	20	18	16	17	12	24	20	23	28
GWL/OZA/TT/DD/DOZ	29	20	15	17	19	15	25	20	22	28
persistence (RMS)										
	36	27	22	24	20	17	31	25	31	31
number of observations										
	172	99	158	53	69	65	129	47	97	308

Table 4 displays the forecast test for the same station Waldhof for the year 1997. by using the rule system developed with the training data set.

Table 4. Test forecast 1997 (April - September)
Mean errors of RMS differences between observations and model forecast of daily maxima of ozone concentrations for different cluster combinations and Großwetterlagen, dimension: hourly means [μg].

	Großwetterlagen 01	02	03	04	05	06	07	08	09	10
bias										
GWL/OZA/TT/OZ	4	3	-5	-2	-14	-18	11	-15	14	8
GWL/OZA/DD/OZ	12	3	-8	-7	-14	-16	12	-12	14	6
GWL/OZA/TT/DOZ	-3	0	-5	-4	-21	-12	7	-3	19	3
GWL/OZA/DD/DOZ	-3	-2	-8	6	-8	-6	10	-2	17	2
GWL/OZA/NN/DOZ	-3	-4	-3	-	-1	-12	13	-4	10	0
GWL/OZA/TT/DD/DOZ	2	-1	-9	4	-15	-4	12	1	16	2
RMS										
GWL/OZA/TT/OZ	23	14	16	6	21	28	28	24	25	25
GWL/OZA/DD/OZ	34	15	17	12	16	25	29	25	23	26
GWL/OZA/TT/DOZ	27	15	15	6	25	19	22	17	28	21
GWL/OZA/DD/DOZ	33	16	18	8	17	20	27	19	27	24
GWL/OZA/NN/DOZ	34	14	16	-	13	21	27	24	24	24
GWL/OZA/TT/DD/DOZ	30	16	18	8	19	16	28	18	28	23
persistence (RMS)	34	16	17	7	19	16	37	21	23	25
number of observations	28	18	19	3	4	14	13	8	17	51

It is obvious that for each Großwetterlage a different group of parameters is statistically optimal. Partly that might be a realistic effect, because different Großwetterlagen, which represent different transport pattern over Europe, show a specific influence of clouds, wind direction or other parameters on the ozone pattern.
The mean absolute errors for these example are less then 10% as the best.

Table 5 shows the optimal parameter combinations for each Großwetterlage for this example together with the climatological percentile values. It is obvious, that only special Großwetterlagen are of interest in relation to the formation of high ozone concentrations. On the other hand the list of parameters used here might not be optimal for ozone forecast. Therefore new parameters will be introduced, like photolysis frequencies or mixing height, to get a better description of ozone formation in relation to meteorological weather situations.

Next the statistical, CTM and Fuzzy procedures will be compared with the data sets from 1998 with the aim, to get an optimal combination of these forecast tools.

For 1999 the forecast system is planned to be installed for real time application.

Table 5 . Concentration observation station Waldhof
„optimal" parameter combination depending on Großwetterlagen.

Großwetterlage			**Training data (1990 - 96)**	**Test data (1997)**
	90%	**98%**		
Hoch - Mitteleuropa **1**	150	205	GWL/OZA/TT/OZ	GWL/OZA/TT/OZ
Tief - Frankreich **2**	132	158	GWL/OZA/TT/DOZ	GWL/OZA/NN/DOZ
Hoch - Südwesteuropa **3**	112	130	GWL/OZA/TT/OZ	GWL/OZA/TT/DOZ
Tief - Genua **4**	153	177	GWL/OZA/TT/OZ	GWL/OZA/TT/OZ
Tief - Mitteleuropa **5**	160	171	GWL/OZA/TT/DOZ	GWL/OZA/NN/DOZ
Hoch - Britische Inseln **6**	138	123	GWL/OZA/DD/OZ	GWL/OZA/TT/DD/DOZ
Tief - Britische Inseln **7**	147	172	GWL/OZA/TT/DOZ	GWL/OZA/TT/DOZ
Hoch - Skandinavien **8**	164	162	GWL/OZA/TT/OZ	GWL/OZA/TT/DOZ
Suedlage **9**	178	198	GWL/OZA/TT/OZ	GWL/OZA/TT/OZ
Westlage **10**	137	187	GWL/OZA/TT/DOZ	GWL/OZA/TT/DOZ

ACKNOWLEDGEMENTS

The project is funded by the environmental agency of Germany (UBA F&E Vorhaben 10402817).

REFERENCES

S.Bocklisch S. Prozeßanalyse mit unscharfen Verfahren
VEB Verlag Technik, Berlin 1987

Dlabka M., Anhang 2 zum Skript 'Intelligente Systeme': Lösung von linearen Gleichungssystemen mit der Methode der kleinsten Quadrate auf der Basis der Givens-Rotation. *FH-Telekom*, Berlin 1996

DWD, *Quarterly Report of the Operational NWP-Models of the Deutsche Wetterdienst*, No. 12, Offenbach (1997).

Flemming J., E. Reimer, R. Stern. Impact of special Features of numerically predicted and analysed meteorological data on the result of ozone forecast by a PBL-CTM, in: *Air Pollution Modelling and its Application XIII*, Plenum Press, New York (1998).

Fuzzy Classification Toolbox Software (FCT). *TAT Ingenieurbürobüro Dr. Jörn Burmeister*. Benutzerhandbuch, Version 2.0. Berlin/Chemnitz, 1995.

Gronostay R., Einsatz von Clusterverfahren bei der Vorhersage von meteorologischen Prozessen. *Diplomarbeit at FHTW Berlin and at FU Berlin*, 1998

Hellendoorn H., D.Driankov (Eds.). Fuzzy Modell Identifikation.
Springer Verlag, Berlin 1997

Klawon Th. Empirische Modellbildung mit Sugeno-Fuzzy-Modellen
Diplomarbeit an der FH-Telekom, Berlin und an der FU, Berlin, 1996

Kosko, B. *Neuronal Networks and Fuzzy Systems*. Prentice Hall, 1992.

Motz, B. und A. Hartmann und E. Plinke. *Aktionsprogramm und Maßnahmenplan Ozon (Sommersmog) - Großräumige Wirkungsanalysen*. F&E-Vorhaben 10402812 des Umweltbundesamtes, Berlin/Basel, 1996.

Reimer, E. and B. Scherer, An operationel meteorological diagnostic system forregional air pollution analysis and long-tem modelling, in: *Air Pollution Modelling and its Application IX* ,Plenum Press, New York (1992).

IMPROVEMENTS TO HORIZONTAL TRANSPORT IN GRID MODELS

R.J. Yamartino

Atmospheric Studies Group
EarthTech, Inc.
196 Baker Avenue
Concord, MA 01742 USA

INTRODUCTION

The last decade has witnessed significant improvements in the quality of advection schemes in terms of reduced amounts of numerical dispersion and diffusion. The levels of numerical diffusion are now down to the point where one has to be concerned about adding back amounts of diffusion that are appropriate to the atmospheric conditions being modeled. Determination of the appropriate horizontal diffusion coefficients that should be used in a model requires that one compute them as the total desired diffusivity minus the numerical diffusivity already accompanying the advection scheme; that is, $K_{HM} = K_{HT} - K_{HN}$. Even the numerical diffusivity, K_{HN}, known to vanish at Courant (CFL) numbers, $\varepsilon_x = u{\cdot}\Delta t/\Delta x$, of 0, -1, and +1 with flux formulation algorithms and peak at CFL of +1/2 and -1/2, must first be modeled if K_{HM} is to be properly characterized as a function of CFL. It also is important to realize that the "total" diffusivity may involve a number of different processes, such as temporal wind field variations (e.g., sub-grid-scale turbulence and grid-scale wind meander) that are unresolved or dissipated by the wind field model. In addition, there are other transport errors which begin to become evident with these higher quality transport schemes; specifically, operator splitting errors and shear flow errors, which were heretofore buried by the numerical diffusion. Some of the effects of wind shear, for example, are genuinely transportive rather than diffusive, but may be lumped into the diffusivity for expediency. The mathematical characterization of these diffusivity and error terms will be detailed in the paper along with proposed corrections.

OPERATOR SPLITTING CORRECTIONS

Nearly all of the mesoscale and regional scale oxidant and acid rain models in existence today employ operator splitting (Yanenko, 1971) to solve the advection-diffusion partial differential equation (PDE) including linear and non-linear (i.e., 2nd order chemistry) source and sink terms. The primary motivation for using operator splitting is that it enables the optimal numerical methods to be employed for specific elements of the overall time-development operator. Numerous authors, including McRae et al. (1982) and Yamartino et al. (1992), discuss the validity, limitations and ramifications of invoking the symmetric operator splitting approximation, but the most important ramification from the point of view of choosing the horizontal advection scheme is that the truncation error of a model based on operator splitting plus operator reversal is second-order in time (Marchuk, 1975) even for suboperators having first-order in time truncation errors. Consequently, there is no clear motivation to use time-marching schemes in the advection operator above first-order in time and similarly no need to store or manipulate more than one time level of species concentrations.

Nevertheless, the second-order accuracy resulting from operator splitting can create some genuine problems, particularly in shear flows. These problems can be seen when advecting a uniform concentration distribution (i.e., C = 1) in a flow with net outflows $\Delta\varepsilon_x$ and $\Delta\varepsilon_y$ (i.e., expressed as the difference of cell face Courant numbers, $\varepsilon_x = u_f \Delta x/\Delta t$) in the x and y directions respectively. After both operators are applied, the mass remaining in the cell is correctly given as

$$M_r = (1 - \Delta\varepsilon_x)(1 - \Delta\varepsilon_y) \tag{1}$$

rather than as the incorrect result

$$M_r = (1 - \Delta\varepsilon_x - \Delta\varepsilon_y) \tag{2}$$

given by the single 2-d operator. Given that Eq.(2) resembles more closely the operators emerging from a differential analysis, previous works have concerned themselves with the opposite problem: assuming that Eq. (2) was correct and that Eq. (1) was an approximation in need of correction. For example, this problem was outlined by Flatoy (1993) and his solution approach involved an iterative adjustment to the winds to ensure a constant mixing ratio. However, differential analysis usually ignores quantities that are of higher order in the differential, whereas a finite time step analysis shows that it is Eq. (2) which ignores the higher-order terms which Eq. (1) captures. One is convinced of this if one thinks of cutting slices of size ε_x and ε_y away from a unit square of material and asking how much material (i.e., area) is left. Even if Eq.(1) gives the correct amount of material remaining, the masses exiting the x and y faces are not in proper proportion due to the fact that the first operator took away the entire 'corner' of area, $a=\varepsilon_x\varepsilon_y$, when it should have only taken half the shared area or a/2. Given the operator order X then Y, one can show that the x Courant numbers, ε_x must be corrected by the multiplicative factor

$$(1 - \Delta\varepsilon_y/2) \tag{3a}$$

whereas the y (i.e., the 2nd operation) Courant numbers, ε_y must be corrected by the multiplicative factor

$$(1 - \Delta\varepsilon_x/2)/[1 - \Delta\varepsilon_x(1-\Delta\varepsilon_y/2)] \quad , \tag{3b}$$

where the denominator of this factor effectively adjusts concentrations for the depleting effect of the first operator.

Clappier (1998) has recently proposed a comprehensive set of corrections to be made to advective fluxes (i.e., rather than the velocities) for 2-d and 3-d split-operator transport. His corrections completely eliminate the errors for the wind fields he has considered, and may constitute a complete solution even in very complex flow situations.

Though the perfect solution to this problem may presently be somewhat elusive, such Courant number corrections should be inserted into all operator split models subsequent to the 'last molecule' and other wind-shear specific corrections described below.

WIND SHEAR CORRECTIONS

The Along-Wind Gradient Transport Correction

Most Eulerian numerical transport schemes are routinely tested under flows (e.g., uniform, rotational) that contain no shear. Under moderate shears, Eulerian formulations have not performed as well as Lagrangian or semi-Lagrangian formulations which include shear corrections to their computation of the back trajectories from each grid point. With most (or all?) split-operator (i.e., separate 1-d operators for x and y) Eulerian advection algorithms, mixing ratio ripples will develop and grow with successive time steps for a time-independent, deformational flow field. For example, using the vortex (deformational) wind field suggested by Bott (1989), Bott's transport scheme, and a non-dimensional time step of 0.5, tests indicate that it took about 25 time steps for this ripple amplitude to reach 10% of the initial concentration and continued to grow to a maximum of about 20%. Fortunately, this problem can be substantially reduced using solutions that were originally introduced by Staniforth and Pudykiewicz (1984) for semi-Lagrangian schemes and more recently by Emde (1992) and Easter (1993). Similar corrections can now be made to Eulerian schemes.

The proposed Lagrangian corrections are basically equivalent to defining a corrected flux-face Courant number, $\varepsilon' = \varepsilon / (1.0 + \beta/2)$, where $\beta = \Delta u^* \Delta t/\Delta x$ and Δu is the difference in velocities between the flux face and its upwind counterpart. The basic principal of this approach was to find the point, $\varepsilon' \cdot \Delta x$, within the cell where the last molecule originated that passed through the flux face during the advection time step Δt. This correction is approximate and valid only for $\beta \ll 1$; however, if one integrates the expression $dt = dx / u(x)$ over the time step, Δt, assuming a linear variation of the wind in space, the exact result,

$$\varepsilon' = \varepsilon \cdot (1.0 - \exp(-\beta)) / \beta \ , \tag{4}$$

where $\beta = \Delta u \cdot \Delta t/\Delta x$ and Δu is the difference in velocities between the flux face and its upwind counterpart, is obtained. This correction is valid for all Courant numbers and β, though the usual stability criteria, $|\varepsilon| < 1$, and the supplementary criteria, $|\beta| < 1$, should be considered.

Tests conducted using this formulation showed that the ripple 'instability' was confined to about 1% amplitude even after 200 time steps and, thus, should no longer be considered a serious problem relative to the various other approximations associated with an Eulerian photochemical grid model. In addition, a realistic simulation involving time dependent winds and diffusion operators would act to diffuse any such ripples. It might also be mentioned that nearly all of this minor instability arises from the failure to correct effective Courant numbers at cell faces with the gradient in wind speed and is not an intrinsic limitation associated with operator splitting. The latter interpretation gained credibility because the instability is made more severe if one neglects to do operator reversal.

Flatoy (1993) attributed most of the above instability to errors intrinsically associated with operator splitting and he utilized an iterative adjustment to the winds to ensure a constant mixing ratio. We now see that two separate corrections: one for wind shear in the transport direction and one for operator splitting are required.

The Cross-wind Gradient Transport Correction

A final correction relates to advective fluxes due to distortion or stress in the horizontal wind field that are most often approximated as diffusive terms. For example, the CALGRID model (with cell dimensions $\Delta x = \Delta y$) used the moderately simple Smagorinsky (1963) formulation

$$K_S = \alpha_o^2 \, |D| \, (\Delta x)^2 \tag{5a}$$

where $\alpha_o \simeq 0.28$ and

$$|D| = \left[\left(\frac{\partial v}{\partial x} + \frac{\partial u}{\partial y} \right)^2 + \left(\frac{\partial u}{\partial x} - \frac{\partial v}{\partial y} \right)^2 \right]^{1/2} , \tag{5b}$$

to characterize the impact of stress in the wind field.

In his review of horizontal diffusion processes and their representation, Hanna (1994) points out the problem that 'horizontal diffusion follows a linear growth rate with time for travel times out to a day or more' whereas K-theory diffusion leads to $t^{1/2}$ plume growth. Even if true for only a few hours, this fact stresses the importance of capturing as much of the wind field shear in the advective portion of the transport flux vector, F, given as

$$F = VC - K\rho\nabla(C/\rho) \quad ; \tag{6}$$

unfortunately, this cannot be done for the scales that are smaller than Δx. For these smaller scales we first consider the problem of u (i.e., x) transport through the face of a single grid cell. We then may write the local flux as either an advective flux,

$$F_a = \left[u + \left(\frac{\partial u}{\partial y} \right) y + \left(\frac{\partial u}{\partial z} \right) z \right] \left[C + \left(\frac{\partial C}{\partial y} \right) y + \left(\frac{\partial C}{\partial z} \right) z \right] \tag{7a}$$

or as a diffusive flux,

$$F_d = \rho \left[K_{xx} \left(\frac{\partial (C/\rho)}{\partial x} \right) + K_{xy} \left(\frac{\partial (C/\rho)}{\partial y} \right) + K_{xz} \left(\frac{\partial (C/\rho)}{\partial z} \right) \right] \quad . \tag{7b}$$

Integrating Eqs.(6-7) over the facial area $\Delta y \cdot \Delta z$ of the cell and noting that the $u \cdot C$ term is the purely advective term already handled by our advection scheme and the term containing K_{xx} is the traditional diffusion term, the matching of terms enables one to extract expressions for the K tensor elements, such as:

$$K_{xy} = \frac{(\Delta\ y)^2}{12} \left[\left(\frac{\partial u}{\partial y} \right) \left(\frac{\partial C}{\partial y} \right) \right] / \left[\left(\frac{\partial C}{\partial y} \right) - \frac{C}{\rho} \left(\frac{\partial \rho}{\partial y} \right) \right] \tag{8a}$$

and

$$K_{xz} = \frac{(\Delta\ z)^2}{12} \left[\left(\frac{\partial u}{\partial z} \right) \left(\frac{\partial C}{\partial z} \right) \right] / \left[\left(\frac{\partial C}{\partial z} \right) - \frac{C}{\rho} \left(\frac{\partial \rho}{\partial z} \right) \right] \tag{8b}$$

which in the case of negligible density variation become

$$K_{xy} = \frac{(\Delta\ y)^2}{12} \left(\frac{\partial u}{\partial y} \right) \tag{9a}$$

and

$$K_{xz} = \frac{(\Delta\ z)^2}{12} \left(\frac{\partial u}{\partial z} \right) \quad . \tag{9b}$$

Extension to the y flux-related tensor components K_{yx}, and K_{yz} is straightforward and the similarity between Eq(9) and Smagorinsky's Eq(5) becomes more striking when one realizes that $\alpha_o^2 = (0.28)^2 = 1/12.8$ or $\alpha_o^2 \approx 1/12$. Further, what this comparative analysis tells us is that rather than use the directionally blind K_S for wind shear related lateral 'diffusion' in both directions x and y, one may now selectively employ the direction-specific advective exchange fluxes, such as:

$$F_a = \left[u\ C + \left(\frac{\partial u}{\partial y} \right) \left(\frac{\partial C}{\partial y} \right) \frac{\Delta y^2}{12} + \left(\frac{\partial u}{\partial z} \right) \left(\frac{\partial C}{\partial z} \right) \frac{\Delta z^2}{12} \right] \quad . \tag{9c}$$

What is significant about Eq.(9c) is that the sign or direction of the latter two, x-direction, mass flow terms have no connection with the sign of the concentration x gradient, dC/dx. Such 'counter-gradient' flows are not normally equated with a diffusive process, but such is the nature of the oft-ignored 'off-diagonal' terms of the diffusivity tensor. Du and Venkatram (1998) demonstrate the practical importance of the K_{xz} term even in the absense of a velocity gradient, $\partial u/\partial z$; thus, the right hand side of Eq.(9b) probably represents a portion of the off-diagonal K_{xz}.

OTHER WIND FIELD CORRECTIONS

Photochemical grid models are frequently driven by the output of diagnostic wind field generators. In such models, temperature, pressure and hence, density fields, are often computed by methods that are decoupled from the wind field computations. In addition, there is rarely any connection between the density fields of time step n with those of time step n+1. This means that while mass (e.g., grams of mass on the grid) might be completely conserved during and between both time step, ripples in the mixing ratio may appear in time step n+1 that were not present after time step n simply because the new density field (at step n+1) is unrelated to the densities at step n. The only way to eliminate these problems is to constrain the system to satisfy the full divergence relation, $\nabla\cdot(\rho V) = -\partial\rho/\partial t$ as is done in prognostic meteorological models. While use of the correct conservation equation should be included as part of the meteorological model's divergence reduction procedure, corrections to achieve this could be approximated 'on the fly' within the dispersion model.

ACKNOWLEDGMENTS

This work was supported in part by a contract from the Research Division of the California Air Resources Board.

REFERENCES

Bott, A., 1993: The monotone area-preserving flux-form advection algorithm: Reducing the time-splitting error in two dimensional flow fields. *Mon. Wea. Rev.*, **121**, 2637-2641.

Bott, A., 1992: Monotone flux limitation in the area-preserving flux-form advection algorithm. *Mon. Wea. Rev.*, **120**, 2592-2602.

Bott, A., 1989: A positive definite advection scheme obtained by nonlinear renormalization of the advection fluxes. *Mon. Wea. Rev.*, **117**, 1006-1015.

Clappier, A., 1989: A correction method for use in multidimensional time-splitting advection algorithms: Applications to two- and three-dimensional transport. *Mon. Wea. Rev.*, **126**, 232-242.

Du, S. and A. Venkatram, 1998: The effect of streamwise diffusion on ground-level concentrations. *Atmos. Environ.*, **32**, 1955-1961.

Easter, R.C., 1993: Two modified versions of Bott's positive definite numerical advection scheme. *Mon. Wea. Rev.*, **126**, 232-242.

Emde, K.V.D., 1992: Solving conservation laws with parabolic and cubic splines. *Mon. Wea. Rev.*, **120**, 482-492.

Flatoy, 1993: Balanced wind in advanced advection schemes when species with long lifetimes are transported. *Atmos. Environ.*, **27A**, 1809-1819

Marchuk, G.I., 1975: *Methods of Numerical Mathematics*. New York, Springer-Verlag. 316 pp.

McRae, G.J., W.R. Goodin and J.H. Seinfeld, 1982: Numerical solution of the atmospheric diffusion equation for chemically reacting flows. *J. Comp. Phys.*, **45**, 1-42.

Yamartino, R.J., J.S. Scire, G.R. Carmichael and Y.S. Chang, 1992: The CALGRID mesoscale photochemical grid model--I. Model formulation. *Atmos. Environ.*, **26A**, 1493-1512.

Yanenko, N.N., 1971: *The Method of Fractional Steps*. Springer, New York.

DISCUSSION

E. GENIKHOVICH: Is my understanding correct that the problems discussed in your presentation are mainly due to splitting and that they could be eliminated automatically if one wouldn't use this technique?

R. YAMARTINO: It is true that some of the issues in the paper are solely related to operator splitting and could be eliminated via a correctly formulated multidimensional operator, but I would question whether the fluxes are correctly formulated in most current multidimensional operators. For example, I would guess that the traditional 2-d operator yields a remaining cell mass, $M_r = (1 - \Delta\varepsilon_x - \Delta\varepsilon_y)$, rather than include the 2nd-order cross term to yield $M_r = (1 - \Delta\varepsilon_x)(1 - \Delta\varepsilon_y)$.

Z. ZLATEV: Many of your results rely heavily on the classical concept of "Courant numbers". This concept is related to simple PDEs with a constant wind velocity (i.e., $\partial c/\partial t = -u\ \partial c/\partial x$), but you are using your results for much more complicated 2-d and 3-d PDEs. Is this generalization based on an assumption that all these properties do not change too much when you extend the results from the equation $\partial c/\partial t = -u\ \partial c/\partial x$ to more complicated cases?

B. YAMARTINO: Despite the simplicity of some of my notation (e.g., u_f for face velocity), subsequent usage hopefully makes it clear that this u_f is defined at the right face of cell i and that, for example, $\Delta\varepsilon_x = (u_{f,i} - u_{f,i-1})\Delta t/\Delta x$. Thus, the results are valid for an arbitrary wind field and are generally valid except where noted (e.g., some diffusivity results hold only in the limit of small density variation). I usually err in the direction of excessive complexity and regret that the oversimplification this time suggested a lack of generality.

REDISTRIBUTION OF CARBON MONOXIDE IN CENTRAL AFRICA DURING BIOMASS BURNING EVENTS (EXPRESSO EXPERIMENT)

D. Poulet and S. Cautenet

LaMP/OPGC, CNRS and Université Blaise Pascal, 24 Av. des Landais
63177 Aubière, France
Tel. 33-04-73-40-73-59, Fax 33-04-73
e-mail : poulet@opgc.univ-bpclermont.fr, cautenet@opgc.univ-bpclermont.fr

INTRODUCTION

Tropical biome exhibits some of the most dynamic, yet most poorly understood on the earth. The half of the world's savanna biomass is located in Africa (Andreae,1991). The savanna fires are estimated to contribute about 50 – 67% of tropical biomass fire emissions (Hao and Liu, 1994). The CO emissions from savanna fires are about 30% of the release from industrial sources (WMO, 1995).

Since the last years, a few campaigns are conducted to confirm the importance of the tropics on the regional and global composition and chemistry of the atmosphere. We can cite ABLE 2A and 2B in Amazonia (1985 and 1987), DECAFE in Congo and Ivory Coast (Africa) (1988), then STARE/TRACE in South America (1992) and TRACE/SAFARI in South Africa (1992). More recently, in 1996 the EXPRESSO experiment in Central Africa could investigate processes controlling chemical composition of the tropical troposphere and consider the impact on the global atmosphere.

This work shows a mesoscale modelling over Central Africa. The advantage of this type of simulation is that the large scale fields are taken into account but also the local circulations (valley and mountains or sea breezes). In this case, in dry season, the ITCZ (Inter Tropical Zone of Convergence) is located about 7°N-8°N. This zone separates the flows of Harmattan from North East and Monsoon from South West. In Central Africa, during this period, near ITCZ the winds are weak (~ $2 ms^{-1}$) and convergence zones alternate with divergence zones. The redistribution of pollutants emitted by biomass burning in savanna regions, is difficult to understand and to simulate because there are recirculations of compounds.

During dry season linked to biomass burning occurrence, aircraft measurements were achieved over Central Africa (EXPRESSO experiment). on November and December 1996. A remote sensing has been improved with NOAA-AVHRR satellite pictures which allow to known the localizations of fires day by day (Gregoire and al, 1998, Pereira and al, 1998). A 60m walkup tower, providing access to the canopy was installed at a nearly undisturbed tropical forest site in the northern Congo forest (Vierling and al, 1998).

In this study, we examine using a meso-scale model, RAMS (Regional Atmospheric Modeling System: see Pielke and al, 1992) in its nonhydrostatic version coupled with CO emisson module, the metorological fields over Central Africa and the redistribution of CO in harmattan and monsoon flows. We don't consider the microphysical processes which will take into account in a future work. The RAMS model has been initialized with ECMWF data. For the three dimensionnal modelling we considered two nested grids in order to take into account the dynamics which are associated with ITCZ and with three different vegetal covers: savanna, rain forest, mixed savanna forest.

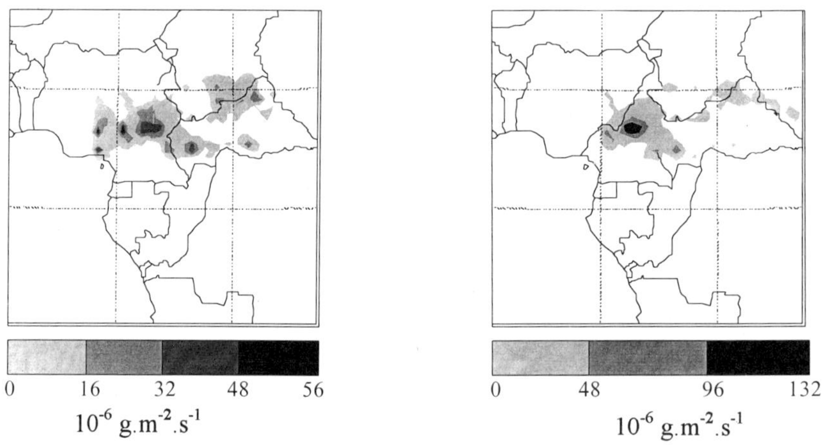

Figure 1. Carbon monoxide emission at the ground for the days: 24th November and 25th November 1996.

EXPRESSO CAMPAIGN

EXPRESSO is an international and interdisciplinary experiment (Delmas and al, 1998). Three main research tools were deployed: the French research aircraft (ARAT, Fokker 27), two satellites receivers of NOAA-AVHRR-HRTP and tower installed at a tropical forest site.

Eleven flights are performed above Central African Republic from 21st November to 2nd December 1996 (during the dry season). The observations include dynamical and thermodynamical mesureaments, radiative, microphysical and chemical measurements. Four flights are on the savanna and five on the forest zone. The two others flights are transects flights.

CALCULATION OF CO EMISSION

The actives fires and the burnt areas are provided by images AVHRR for the domain 4°N-12°N, 8°E-28°E (see figure 1).

The CO emission is given from the number of detected fires and from the biomass density, by:

$$CO\ emitted = Sf*d*CE*EF(CO) \quad (1)$$

Where Sf is the area of the fires, d is the biomass density, EF is the emission factor for the carbon monoxide and CE is the combustion efficiency. The image AVHRR give Sf. d, CE and EF are provided by ground based experiments done in Central African Republic. CE is equal to 0.8 and EF to 65 g/kg dry matter. Between 31 October and 2 December 1996, 514000 km^2 have been burnt. Then the formula (1) give 16 Mt of CO emitted, let 313 $kg.ha^{-1}$ of burnt area. These values are more important than measurements obtained in SAFARI experiment where about 200 $kg.ha^{-1}$ was found.

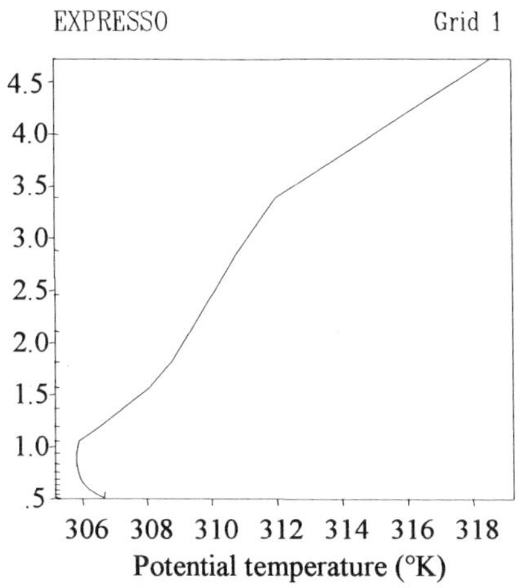

Figure 2. Potential temperature profile simulated near Bangui for the 24th November.

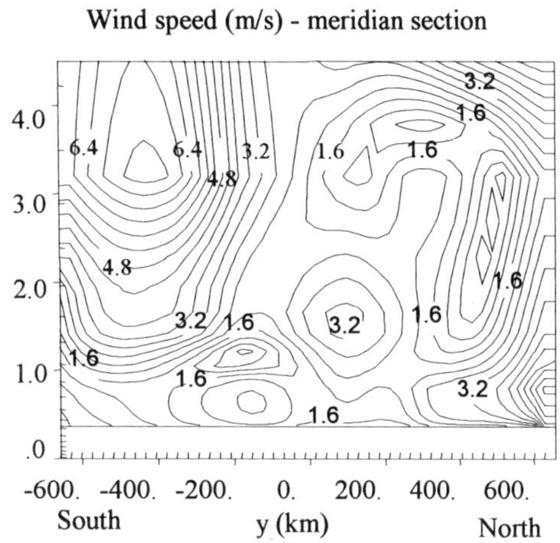

Figure 3. Meridian section between 1°N and 13°N of wind speed.

THE EXPRESSO MODELLING

For this study, we use the RAMS model of the Colorado State University, in its non-hydrostatic version. The initialization is made from synoptic data and assimilation of soundings and surface data every 24 hours This model includes a soil and vegetation module. Then the simulated domain is displayed by three classes: evergreen forest, Guinean savanna and Sahelian savanna. A nudging is prescribed every 24 hours. A detailed description of the model, including the basic equations, is given by Pielke et al (1992). We simulate four days, from the 23rd November to 26th November 1996. During these days, three flights are performed: one in the savanna zone (24th November) and two over the rain forest (25th and 26th November).

We have achieved two different simulations: two-dimensional and three-dimensional. The microphysical processes like convection are not taken into account.

TWO DIMENSIONAL MODELLING OF THE REDISTRIBUTION OF CO OVER CENTRAL AFRICA DURING BIOMASS BURNING EVENT

We have carried out a two dimensional simulation along a domain oriented south-north between 1° North and 13° North where exists the vegetation contrast. For this run, the

CO emissions calculated by (1) are averaged along longitude. For the four days simulated, near Bangui, we note two main features (see figure 2 and 3): Firstly, the boundary layer is not very thick (1000 m above the ground). In this layer, the winds are weak (2 m.s^{-1}). Secondly, above this layer, the troposphere is very stable which lead to a weak vertical motion (some cm.s^{-1}). Nevertheless the meridian and the zonal winds are significant (10 m.s^{-1}). These observations are in good agreement with the data measured during the EXPRESSO campaign.

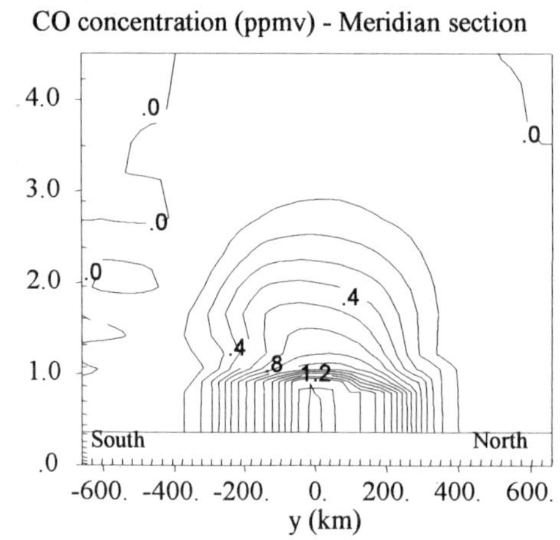

Figure 4a. Meridian section between 1°N and 13° N of CO emission simulated for the 24th November at 12H (36 h of modeling). The CO only begins to reach the free troposphere.

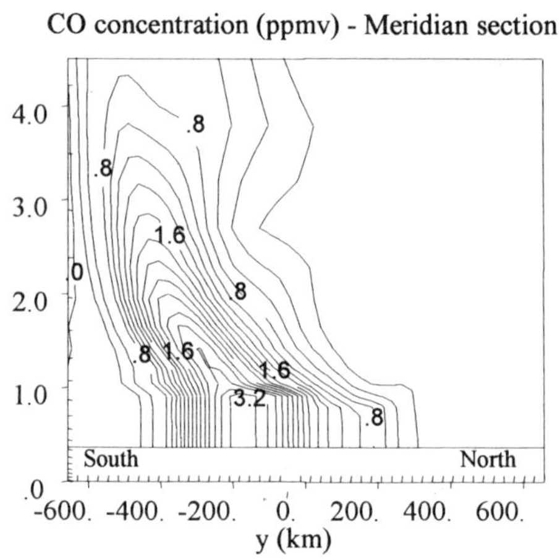

Figure 4b. Meridian section between 1°N and 13° N of CO emission simulated for the 26th November at 12H (84 h of modeling): Accumulation of CO in the boundary layer and transport in the free troposphere.

The redistribution of the CO is largely determined by these characteristics (see figure 4a and 4b): because of the stability of the atmosphere above 1500 meters, the exchanges between boundary layer and free troposphere are very difficult. Moreover the emissions due to bush fires are locally very important. The modeling shows that these emissions start to reach the free troposphere only after one day. Then the CO is accumulated in this layer and can reach very high values (3 ppmv). So, in some regions and for days unless convection, we can be afraid of a very high level of pollution. On the contrary, in the free troposphere, the meridian winds higher and the African Easterly Jet can effectively redistribute the chemical compounds. This results in an important zonal transport (toward west) and in the moving toward south of the CO above the rain forest (see figure 3b).

The vertical motion is principally lead by the convergence zone of the wind in the boundary layer, and can locally transport the chemical species above 4000 meters. The clouds can't cause the vertical transport since the microphysical processes are not taken into account in this modeling.

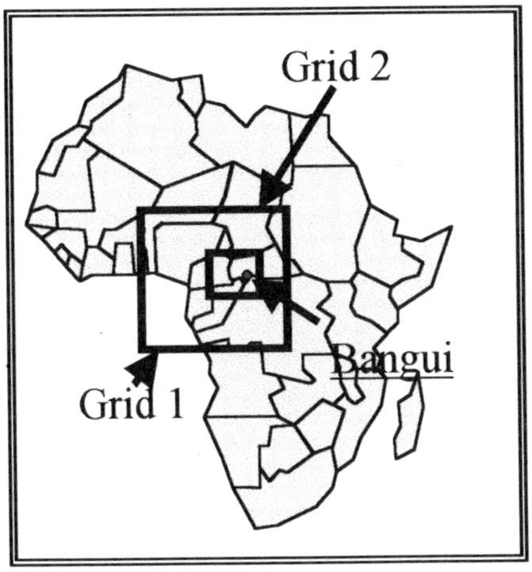

Figure 5. Location of the two nested grid in the RAMS model.

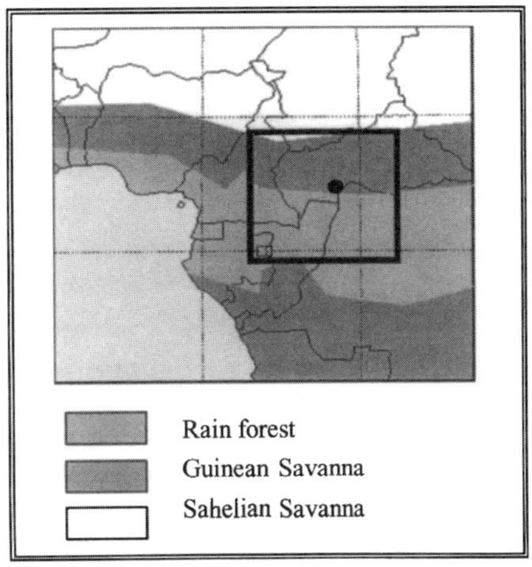

Figure 6. Vegetal cover used in the 3D modeling.

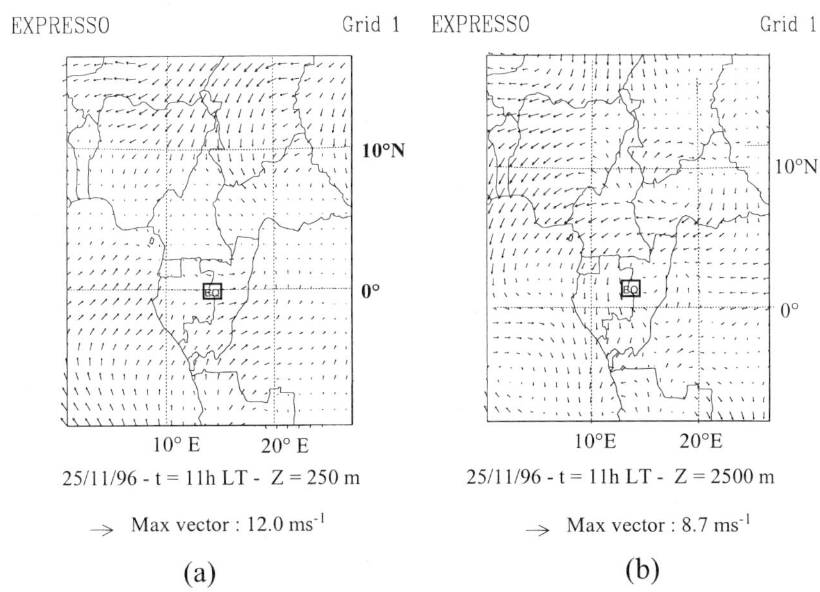

Figures 7a and 7b. Horizontal section at 250 m and 2500 m of wind simulated in the grid 1.

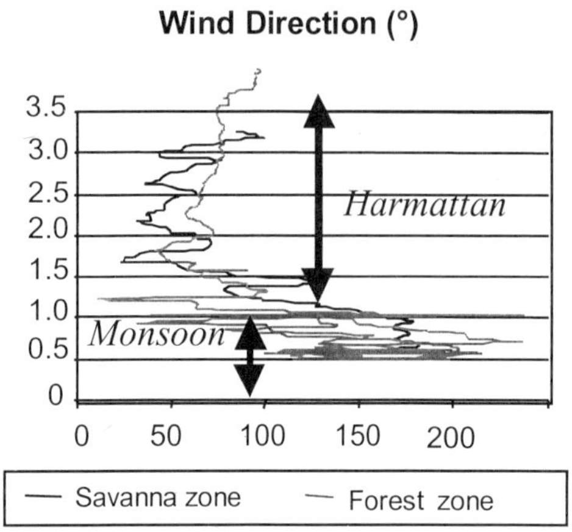

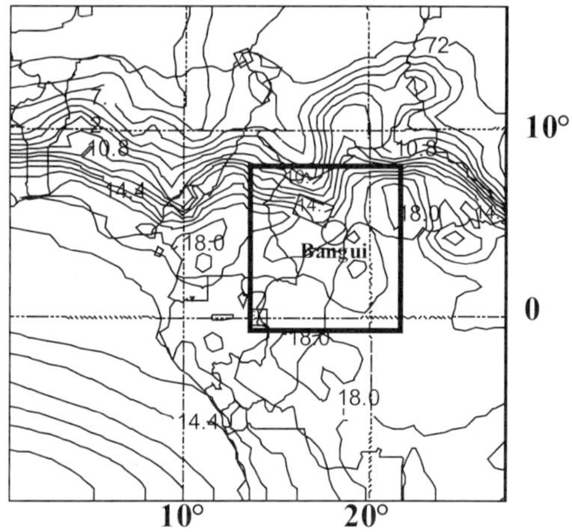

Figure 8. Profiles of wind direction measured during the EXPRESSO campaign.

Figure 9. Horizontal section at 250 m of total mixing ratio (g/kg) simulated for the 24th November at 12 H.

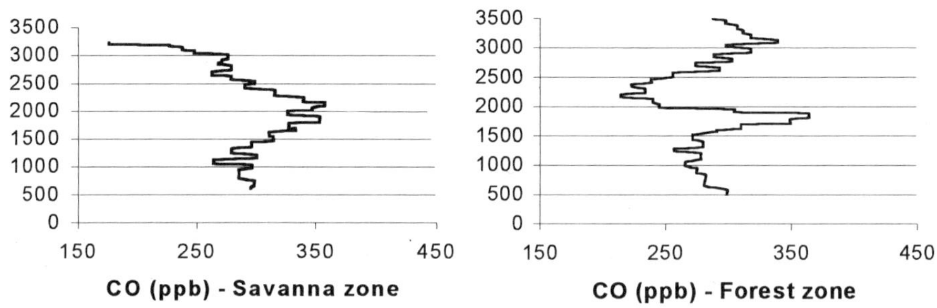

Figure 10a and 10b. CO profiles measured during the EXPRESSO campaign in savanna zone and in forest zone during the flight 45 (24th November 1996) and 46 (25th November 1996).

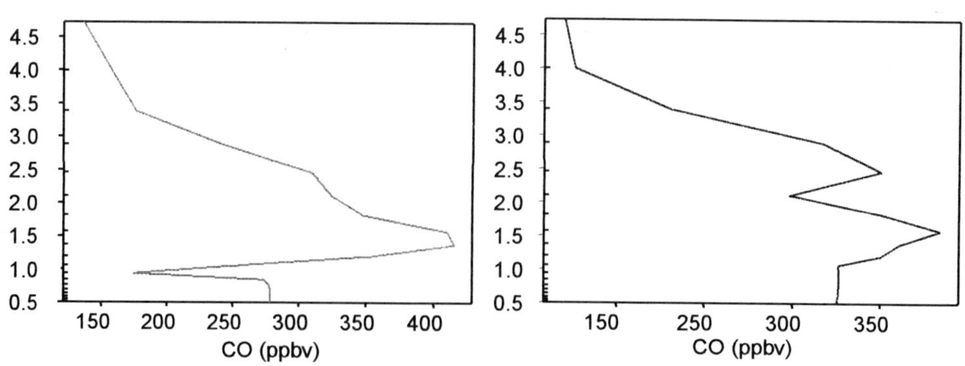

Figure 11a and 11b. CO profiles simulated for the days 24th November 1996 and 25th November 1996 in savanna

THREE DIMENSIONAL SIMULATION OF THE REDISTRIBUTION OF CO OVER CENTRAL AFRICA DURING BIOMASS BURNING EVENT

In order to evaluate the redistribution of CO over Central Africa, we have achieved the simulation of an event of 4 days (23rd to 26th November 1996). The localization of the CO sources determined by the AVHRR image is adjusted to the modeling for the two grids. On each mesh and at each time step, the simulation introduces a CO concentration squaring with the fire activity detected on this location and for this day. The CO concentration calculated by the formula (1) is distributed on twelve hours between 7 and 19 H UT. In order to take into account the dynamics which are associated with ITCZ, we consider two nesting grids: the first on the zone 10°S-20°N, 0°E-30°E, the second over the zone of the EXPRESSO campaign (see figure 5). The figure 6 shows the location in the model of the three different vegetal covers: rain forest, Sahelian savanna and Guinean savanna.

Meteorological results

The synoptic meteorology of Central Africa is greatly influenced by the seasonal migration of the Inter tropical Convergence Zone (ITCZ) and the associated African easterly jet (AEJ) located at an altitude of approximately 3.5 km. We remark (Figure 7a) that on the horizontal cross section at 250 m of altitude, the Harmattan flow from North East is located above 7°N and that the Monsoon flow comes from South West. In Bangui region (4°N, 18°W), the wind is very weak. At 3000m of altitude, on the Figure 7b, the wind has the direction east: it is the AEJ. These results are in good agreement with the aircraft observations (Figure 8).

The ITCZ associated with those two flow regimes give rise to a strong gradient of moisture between the dry air advected by the Harmattan and the oceanic air of the monsoon flow. We retrieve the humidity field and the correct position of ITCZ: 7 – 8° North (Figure 9).

Redistribution of carbon monoxide over Central Africa

We retrieve some characteristics of the two dimensional modeling: the vertical transport is generally very weak except for the convergence zone. The horizontal transport plays a very important role, in particular because of the African easterly Jet. The figures 10a and 10b shows two CO profiles extracted from the flights number 45 and 46 (24th and 25th November 1996) of the EXPRESSO campaign: the first one for the savanna zone and the second one for the forest zone. The CO profiles simulated for the same location are very similar (see figure 11a and 11b). For the forest zone, we have abnormally high CO concentrations, which show that the complex dynamical situation can lead to the transport of the pyrogenic emissions above the forest. Then many phenomenon like acid rain, can affect the tropical forest. The modeling shows that the two maximums of CO concentration measured at 1500 m and 3000 m are due to the motion of the compounds emitted by bush fires located for the first one in Central African Republic and for the second one in Cameroon where the active fires are very important. On the profiles measured in savanna zone, we note a maximum of CO concentration at 2000 meters and not near the ground. But this location is far from the fires for these days and the high CO concentration is due to an horizontal transport from the bush fires located on the north-west.

CONCLUSION

We can say that this work has allowed a good restitution of the meteorological conditions measured during the EXPRESSO campaign. The localization of the bush fires by the AVHRR images and the evaluation of the features of theses fires (biomass density, combustion efficiency and emission factor) have seemed supply a good estimation of the carbon monoxide emission. The two and three dimensional modeling have shown that, for days with no convective motion, the important stability of the free troposphere and the weakness of the wind in the boundary layer obstruct the redistribution of the CO. This can result in very high values of chemical species concentrations near regions affected by an important biomass burning. Nevertheless, an effective transport is possible when the pollutants reach the free troposphere where they can cover a very long distance. Moreover, the complex meteorological situation can transport the compounds emitted during biomass burning event over the rain forest and then affect the vegetal life (acid rain).

Our future work will consider the impact of the convection. This phenomenon plays a very important role in the redistribution of the bush fires emissions because it allows an intense vertical transport and tropical zones are regions of very active and deep convection. Then, the model will simulate the main processes of redistribution and the chemical module of RAMS will allow taking into account many non-inert chemical species.

REFERENCES

Delmas R. A. et al. : Experiment for Regional Sources and Sinks of Oxidants (EXPRESSO) : an overview. *J. Geophys. Res.*, submitted 1998.

Gregoire, J. M., S. Pinnock., Satellite monitoring of vegetation fires for EXPRESSO : Outline of the activity and relative importance of the study area in the global picture of biomass burning, *J. Geosphys. Res.*, submitted 1998.

Pereira, J. M., B. S. Pereira, P. Barbosa, D. Stroppiana, Quantitative assessment of the overall fire activity in the EXPRESSO study area during the dry season experiment : active fires, burnt areas and biomass density, *J. Geophys. Res.*, submitted 1998.

Pielke R.A., W.R. Cotton, R.L. Walko, C.J. Tremback, M.E. Nicholls, M.D. Moran, D.A. Wesley, T.J. Lee, and J.H. Copeland, A comprehensive meteorogical modeling system - RAMS, *Meteor. Atmos. Phys.*, 49, 69-91 (1992).

Vierling, L., B., Baker, J. Greenberg, A. Guenther, P. Harley, L. Klinger, D. Serca, C. Bouka-Biona, Ecosystem-atmosphere exhange of isoprene and CO2 at the EXPRESSO African tropical rainforest site, *J. Geophys. Res.*, submitted 1998.

DISCUSSION

G. KALLOS: Your simulation is for several days. Even though CO is slowly reacting your estimated concentrations cannot be considered as very accurate if this reactivity is not taken into the account. This is especially true for cases or areas with strong transport component?

D. POULET: In this work, we just want to estimated the transport of the CO emission. We think that the effect of the CO reactivity is negligible compared with accurate of the evaluation of the CO emission and of the modeling even on four days. However, the reactions of photochemistry of ozone and others gases like CO will be taken into account in our future work that will allowed an estimation of the effect of the CO reactivity.

R. YAMARTINO: Do you consider the heat flux from the fires in the RAMS modeling or in plume rise computations for the CO?

D. POULET: At the present time, we have just increased the temperature of the firsts levels above areas where bush fires exist. But the effect of the heat from the fires need complementary studies (better estimation of the temperature of the fires and of the mechanisms of heat transport between fire and atmosphere...). So, in this work, the heat flux from the fires have not taken into account.

D. ANFOSSI: Have you any observation of the average plume rise of the fires you are studying?

D. POULET: We don't have observation for the bush fires during the EXPRESSO experiment, because the flights of this campaign were located relatively far from the fires. However, we have several observations from others campaigns. In the SAFARI campaign, the averaged observation of plume rise was ranged to 500m up to 2000 m.

THE ROLE OF THE BLACK SEA ON THE LONG-RANGE TRANSPORT FROM SOUTHEASTERN EUROPE TOWARDS MIDDLE EAST DURING SUMMER

G. Kallos, V. Kotroni, and K. Lagouvardos

University of Athens, Department of Applied Physics, Meteorology Laboratory, Panepistimioupolis, PHYS-5, 15784 Athens, Greece

INTRODUCTION

The Black Sea is closed from all sides by land and communicates with the Aegean Sea and Mediterranean through the Vosporous strait (Fig. 1). It is surrounded by land with elevation varying from a few meters to a few kilometers. The most important orographic features are located at the eastern and south-eastern sides. More specifically, there is the Caucasus Peninsula in the East and the Pontine Mountains across the northern Turkish coast. Towards west, there is the opening of the Danube Valley with the Carpathian mountains to the North of it and the Balkan Peninsula to the South in a considerable distance from the coast. The northern part is adjacent to almost flat land with relatively complicated coastline. At a larger distance, the Balkan and Asia Minor Peninsulas together with the Aegean Sea between them and the Mediterranean Region in general are forming a unique pattern for the development of atmospheric circulations of a wide spectrum ranging from the synoptic to meso-gamma scale with significant implications to the air quality of the region.

Air pollution sources located in the area around the Black Sea are of both industrial (power plants, refineries, cement plants etc.) and urban type. There are urban conglomerates with population from a few hundred thousands to more than ten million (Istanbul). In addition to that, the emissions from the commercial shipping is considerable especially in the vicinity of the Vosporus strait. It is also important to note that all this industrial activity lacks any emissions control at any stage of the processes; thus the emissions of sulphur dioxide and particulate matter are very high. Tuncel (personal communication), measured extremely high concentrations of the industrial type of air pollutants in some locations of the Turkish coast, under certain meteorological conditions, mainly during stagnation.

Most of the studies in the past were emphasising in the behaviour of the urban plumes for relatively small spatial and temporal scales of episodic conditions usually associated with phenomena exhibiting strong diurnal variations (e.g. sea/land-breezes, upslope/downslope and drainage flows, orographic effects, heat islands etc.). During the last few years, the influence of regional scale forcing on the formation of specific air quality conditions was found to be important. Kallos et al. (1993) reported that the regional scale phenomena should contribute significantly in the formation of specific air quality conditions in the Greater Athens Area (GAA). Luria et al. (1996) showed that significant degradation of the air quality in some areas should be attributed to regional scale transport phenomena. Kallos et al. (1997b, 1998) identified the major paths of transport of air pollutants from Europe to the SE Mediterranean Region based on model simulations. The importance of the emissions from the sources around the Black Sea to the sulfate and nitrate concentrations in the East Mediterranean was pointed out in Luria et al., (1996) and Kallos et al., (1997a,b).

Air Pollution Modeling and Its Application XIII, edited by S.-E. Gryning and E. Batchvarova, Kluwer Academic/Plenum Publishers, 2000

In this presentation, the dispersion of air pollutants from various sources around the Black Sea is investigated with the aid of the Regional Atmospheric Modeling System (RAMS) and the Hybrid Particle and Concentration Transport Package (HYPACT). Emphasis is given to the multiscale transport under stagnant conditions during summer. The residence time of the polluted air masses over the Black Sea is considered as important for the particulate formation which in general is considered as very high especially during the warm period of the year.

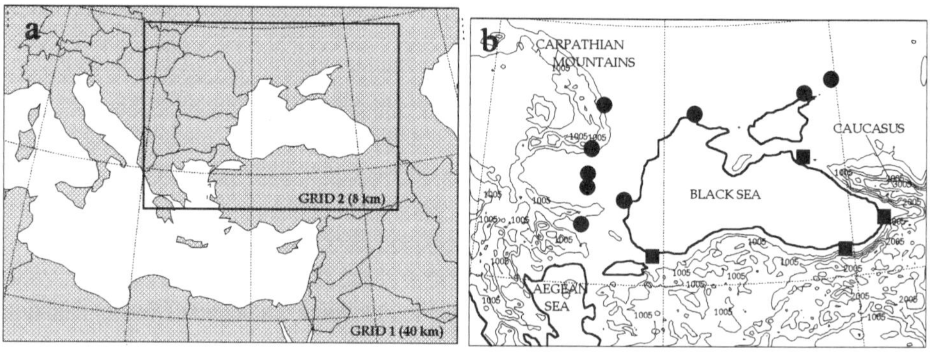

Figure 1. (a) Horizontal extension of RAMS two nested grids (Grid 1, Grid 2). For each grid the horizontal grid increment is also given. (b) Topography inside the second grid of RAMS (at 500 m interval). The bold line denotes the coastline. Full circles denote the power plant locations used in the dispersion model simulations. Full squares denote urban sources of Istanbul, Trabzon, Poti and Novorossiysk.

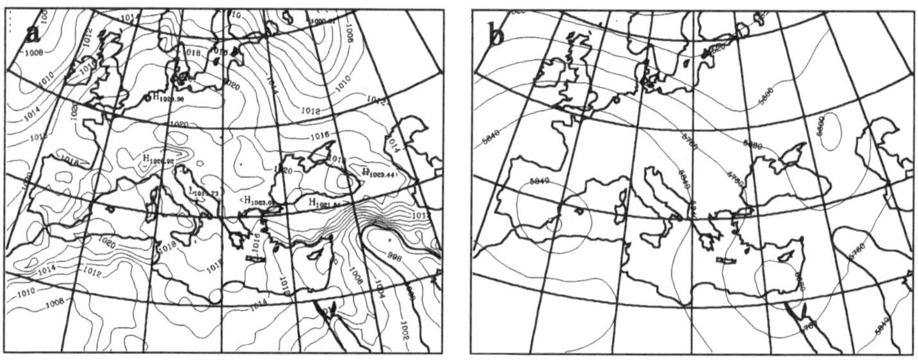

Figure 2. ECMWF analysis of (a) mean sea-level pressure (at 2 hPa intervals), (b) 500 hPa geopotential height (at 40 m intervals), valid at 0000 UTC 28 June 1994.

GENERAL CLIMATIC PATTERNS OF THE REGION

The climatic conditions in the Black Sea Region exhibit some similar characteristics with the Mediterranean Region but there are some considerable differences.

During the warm period of the year, because of the differential heating between the land of North Africa and South Europe the resulting flow is from North. A characteristic phenomenon resulted from this general flow pattern is the etesians along the Aegean Sea (Kallos et al., 1993, 1997a,b and references therein). In general, the anticyclonic system of the Mediterranean and the thermal low over the Anatolian Plateau are always in critical balance. This balance defines either the persistence of the Etesians or the development of strong thermal circulations. The Black Sea Region is strongly affected by this formation. More specifically, during the days where the anticyclonic system over the Mediterranean extends eastward, the Black Sea Region is characterized by local thermal circulations and especially sea/land breeze formations. In several places, stagnation is observed especially during days where the land is relatively wet or the vegetation is rich. Relatively low wind speeds are observed over the sea. When this anticyclonic system weakens, the thermal low over Asia Minor extends towards NW and forms a significant pressure gradient forcing over the Black Sea especially over its SW part. This mechanism favours the transport of the stagnated air masses of the previous days towards the East Mediterranean. Such kind of transport was investigated by Kallos et al. (1996) at the framework of the T-TRAPEM project. An important climatic difference between the East Mediterranean and the Black Sea Region is that the latter is affected by the passage of low pressure systems and therefore it rains more often.

During the transient seasons the weather type changes between summer and winter-type. This change occurs rapidly during spring and slower during autumn.

MODEL DESCRIPTION AND CONFIGURATION

For this study, the Regional Atmospheric Modelling System (RAMS) was used. A general description of the model and its capacities is given in Pielke et al (1992). One of its unique capabilities is the two-way interactive nesting which is very useful for the purpose of this study where the multiscale flow interaction defines the resulting flow features and therefore dispersion.

For dispersion calculations, the Hybrid Particle and Concentration Transport Package (HYPACT) was used. It is a combination of a Lagrangian particle model and an Eulerian concentration transport model. It was developed at the Colorado State University and ASTeR, Inc (Tremback et al., 1993). The Lagrangian model in HYPACT is driven by the velocity and turbulence fields simulated by RAMS,

In order to simulate the summer-type atmospheric circulations on an accurate way, RAMS and HYPACT were configured on different ways. For the final run the following configuration was defined: An outer grid covering an area 3600 km x 2600 km, with 40 km horizontal grid increment centered 28° E and 40° N (90x65x23 grid points). An inner grid was defined over an area 2016 km x 1416 km, with 8 km horizontal grid increment centered 32° E and 43.8° N (252x177x23 grid points). The two model grids and the topography used for the inner one are illustrated in Figs. 1a, b. For test cases, finer resolution grids were nested over the Western Coast of the Black Sea in order to examine the accuracy on the description of local circulations. As it was found, the two-grid configuration was adequate for describing most of the features on a satisfactory way and at a reasonable computer time. These simulations are computer expensive and this is the main reason for not performing such runs for other selected cases.

The model was initialized by using 0.5°x0.5° ECMWF analysis fields available every 6 hours at the standard pressure levels. In addition to this data set the SST fields from ECMWF were used.

The sources used for the simulations are of two types: industrial and urban. The industrial sources used for the simulations are power plants in Bulgaria (Maritza III, Varna, Rasgrad, Ruse), Former Soviet Union Countries (Nowoczerkassak, Starobeszewskaya, Moldawskaya) and Romania (Bicaz and Doicesti) as they are referred to the global data base for mercury emissions prepared by Pacyna (personal communication). Since emission data for the surrounding the Black Sea are not considered as accurate, qualitative-type dispersion simulations were performed. For the urban-type of sources the cities of Istanbul and Trabzon from Turkey, Poti from Georgia and Novorossiysk from Former Soviet Union were used for illustrative purposes covering various places around the Black Sea. In addition to that, two other sources in Yugoslavia were used. The position of these sources is given in Fig. 1b.

MODEL RESULTS

The selected period for simulation is from 0000 UTC of June 28 to 0000 UTC of July 1, 1994. As it is shown in Fig.2a a weak pressure field covers the area around the Black Sea where the pressure drops from W to E Mediterranean for about 15 hPa. In the upper-air, the Atlantic Anticyclone has been shifted over the Iberian Peninsula while in the vicinity of Cyprus there is relatively deep trough (Fig. 2b). This pattern was almost unchanged during the next three days. Because of the weak pressure gradients over the Black Sea Region, thermal circulations start to form. During the night-hours the flow over the western and southwestern coast is from West, at the northern part from N to NW while over the Turkish and Georgian Coast is variable with small speeds (Fig. 3a). An important feature is the deflection of the flow around the flat area N of Caucasus Mountain. This feature is evident during day and night-hours. Over the main body of the Black Sea the flow is from W to E during night-hours while during day when the sea-breeze mechanisms are strong significant variations near the coastline are observed (Fig. 3b). The sea-breeze is even strong enough over the Turkish coast during the day, despite the fact that the topography in the area is relatively high with important slopes facing North and significant vegetation cover. This is partially due to the heating of the bare land of Central Turkey. Since there are significant sources of air pollutants located near the western coast and because of the important diurnal variation of the flow, the flow conditions will be studied in more detail.

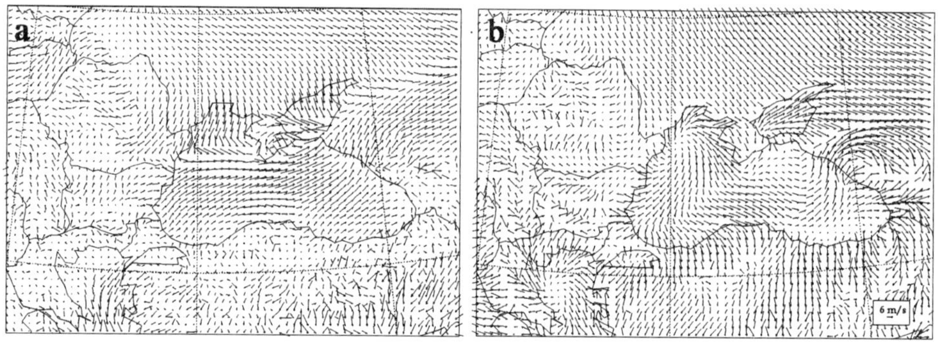

Figure 3. Wind field at the first model level inside the inner grid of RAMS at (a) 0000 UTC, (b) 1200 UTC, 29 June 1994. Wind vectors are plotted every fourth grid point. Scaling vector is given at the bottom Fig. 3b.

As it shown in Fig. 4, the sea-breeze is strong enough to penetrate several tens of kilometers inland. This penetration is supported by the upslopes developed over the East facing slopes of the mountains. In addition to that, a clock-wise rotation of the flow is evident over the Bulgarian and Romanian coast and the vicinity around. This clock-wise rotation is similar to what observed over the East Coast of the Attica Peninsula by Steyn and Kallos (1992). As it was discussed by these authors, the clock or anti-clock rotation of the sea-breezes depends on the shape and orientation of the coastline and the position of significant orography with respect to the coastline. In the case of the western coast of the Black Sea both the orientation of the coastline and the position of some important orographic features are similar with the eastern part of the Attica Peninsula.

The lack of important orographic features over the northern coast and the orientation of the flow from the synoptic scale does not permit the penetration of the sea-breeze deep inland (Fig. 4). The rotation of the flow during the diurnal cycle might create a complicated dispersion pattern where recirculation of pollutants can occur.

The depth of the marine boundary layer was found to be similar to what was found over the East Mediterranean (around 300 m) while over the land varies significantly from a few hundred meters during the nigh to about 1.5 during the first afternoon hours with even higher over areas with significant topographic features.

The model results were compared with the observations provided by the surface synoptic network (Fig. 5). The comparison revealed that the model resolved the flow near the ground very well.

Several dispersion calculations were performed in order to form the best possible generalized picture. In order to examine the effects of the clockwise rotation of the flow over the western coast of the Black Sea only one point source was used: the power plant near Varna. The plume is directed over the Black Sea during the night hours (Fig. 6a), then starts to rotate with the wind towards South (Fig. 6b), starts moving over land with the onset of the sea-breeze (Figs. 6c, d) and repeats the same rotation during the next day (Figs. 6e, f). The "aged particles" are spread around the Vosporous strait and directed towards the SE Mediterranean mainly along the Aegean Sea. The particles transported initially over the sea remain within a

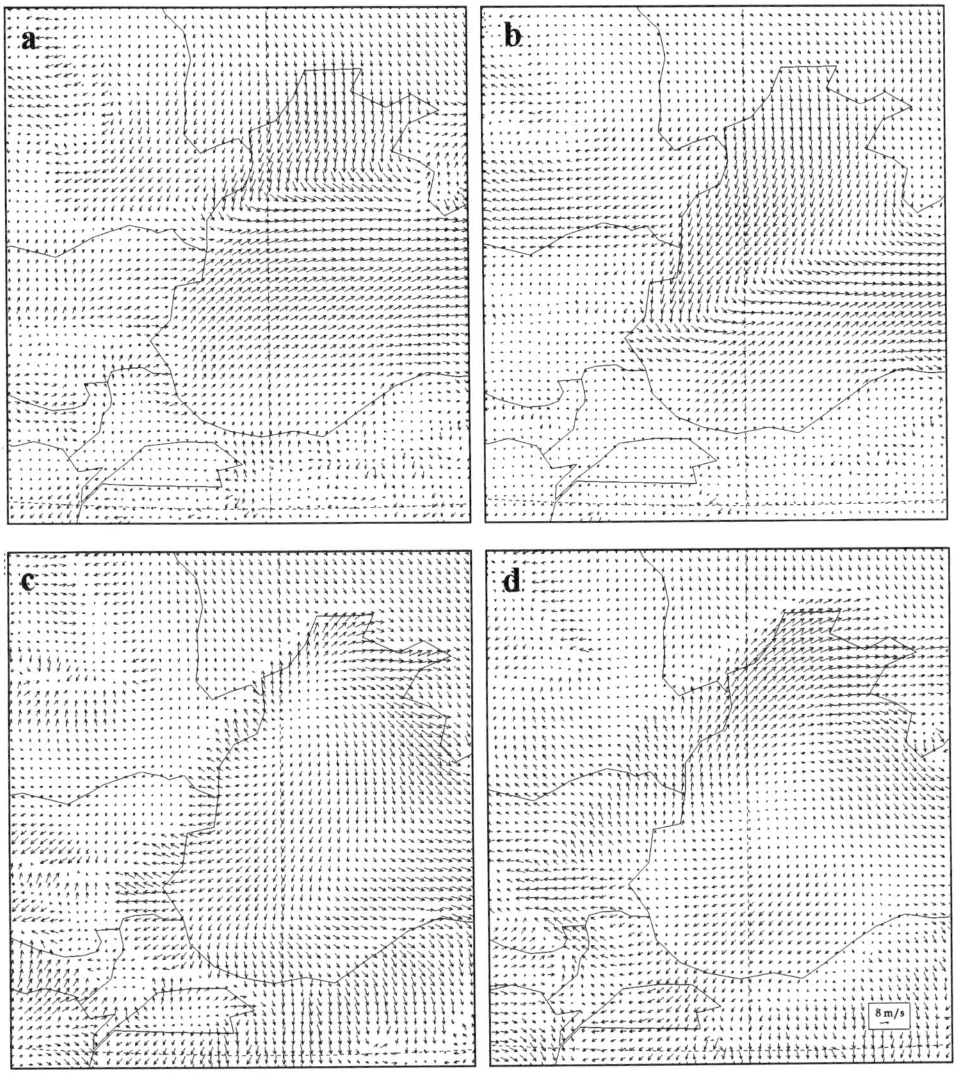

Figure 4. Wind field at at the first model level (zoom inside the inner grid of RAMS) at (a) 0000 UTC, (b) 0600 UTC, (c) 1200 UTC, (d) 1800 UTC, 29 June 1994. Wind vectors are plotted every second grid point. Scaling vector is given at the bottom of Fig. 4d.

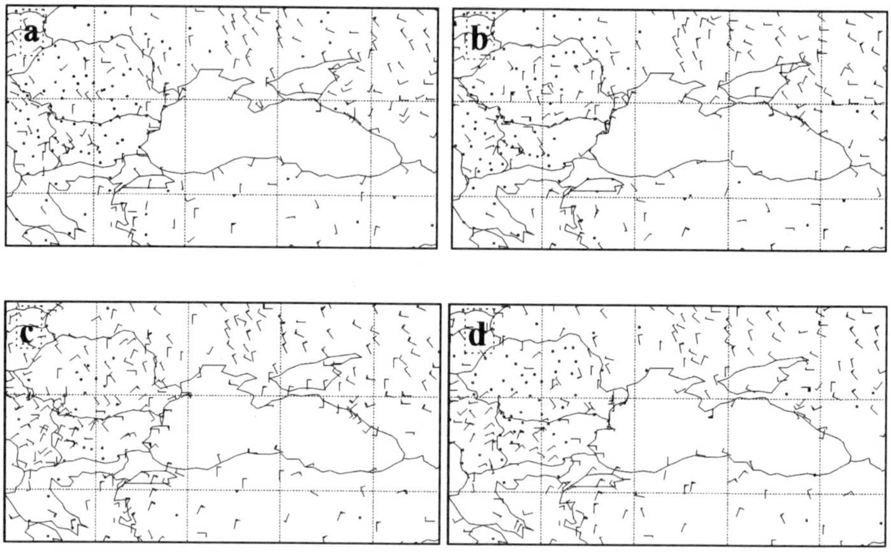

Figure 5. Winds at the anemometric level (z=12 m) reported from the surface synoptic network at (a) 0000 UTC, (b) 0600 UTC, (c) 1200 UTC, (d) 1800 UTC, 29 June 1994. Open circles denote calms. One barb equals 4 ms^{-1}, one half barb $2ms^{-1}$.

relatively shallow layer over the water until they reach the land during the day-hours where the mixing layer is deeper. When the plume enters within the strong updrafts at the cliffs of the mountains the particles are injected up to 2-3 km. Similar patterns are observed for the plumes from the other point sources near the west coast. The plumes from the industrial sources near the northern coasts are directed initially over the sea and then transported towards SE either over the sea or over the flat land N of Caucasus or trapped over the eastern coast of the Black Sea.

The dispersion pattern of some coastal urban plumes (Istanbul, Trabzon, Poti and Novorossyisk) is shown in Fig. 7 for two different hours, day and night. As it is seen, the plume of Istanbul disperses slowly towards SW and continues southward mainly along the Aegean Sea. The prevailing winds over the eastern part of the Black sea are usually from the West with significant variations mainly in strength during the day-hours. This flow favours the transport of the plumes of Trabzon and Poti towards SE, while the plumes of Novorossyisk (or other urban conglomerates in the region) are transported always over the land North of Caucasus Mountain.

An important feature in the Eastern European part is the flow pattern along the Danube and the Bulgarian Valleys. As it was documented in similar areas, under stagnant conditions, the air masses are sloshing in an out without the development of a thermal circulation complete cell. Because of the grid resolution and the size of the Danube Valley is was difficult to observe such phenomenon. In order to examine the possibility of transport of air pollutants from western to the eastern Danube Valley we performed particle releases from two point sources located near Belgrade and Novi Sad in Yugoslavia. As it shown in Fig. 8a the particles disperse towards West during all hours. Despite this evidence the transport along the Danube Valley needs further investigation.

In order to examine the path of air pollutants from the aforementioned sources for longer time periods continuous releases were performed for three days. The general dispersion pattern is shown in Fig. 8b. As it is shown, the followed paths of the plumes located near the coast of the western half of the region are directed over the Balkan Region, the Aegean Sea and even the Ionian Sea on their way towards the SE Mediterranean Region. The plumes from sources located in the coastal area of the eastern half of the Black Sea prefer paths either over Turkey or towards East around the Caucasus Peninsula. The residence time of the pollutants over the Black sea is at the order of a day which is considered as relatively small in order to have significant amounts of SO_2 and NO_x converted in sulfate and nitrate particles respectively.

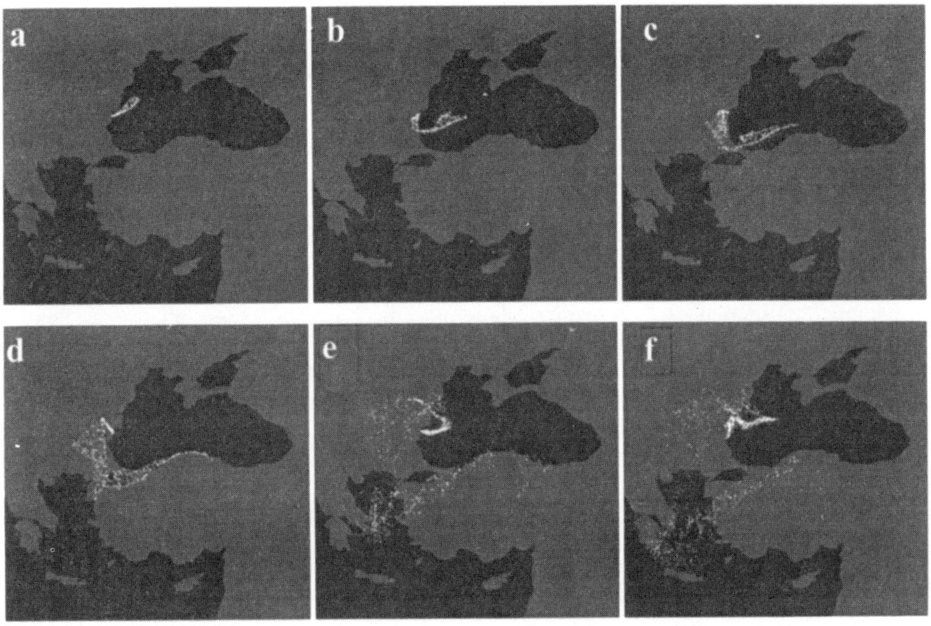

Figure 6. Particle projection from HYPACT dispersion model, at (a) 0000 UTC, 29 June 1994, (b) 0600 UTC, 29 June, (c) 1200 UTC 29 June, (d) 1800 UTC 29 June (e) 0600 UTC 30 June and (f) 1200 UTC 30 June.

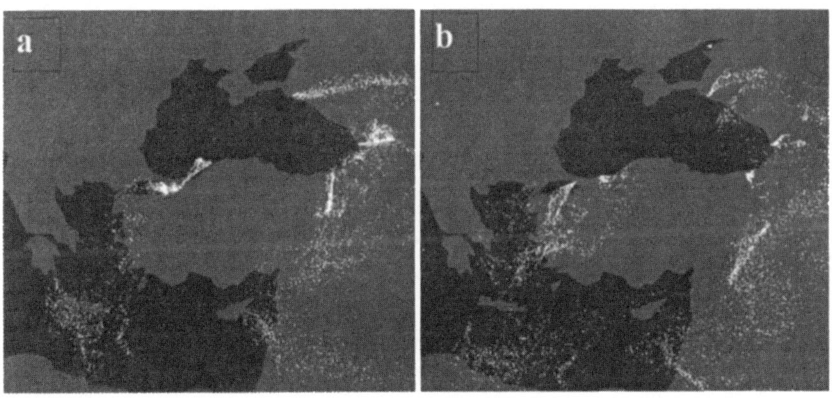

Figure 7. Particle projection from HYPACT dispersion model, at (a) 1200 UTC, 29 June 1994 (after 36 hours of particle release), (b) 0000 UTC, 30 June 1994 (after 48 hours of particle release) from urban sources.

SOME CONCLUDING REMARKS

In this presentation, an attempt was made to identify the dispersion patterns over the Black Sea Region during summer with weak synoptic forcing. Based on these model simulations some useful conclusions are drawn:

- The flow pattern in the Black Sea Region exhibits significant temporal and spatial variations owing to the multiscale flow interaction and the distribution of the physiographic characteristics. The anchor systems which define most of the local flow characteristics during summer are the high pressure system over the Mediterranean and the thermal low over the Asia Minor and Middle East. The strength of the pressure gradient between them in the vicinity of Vosporus defines the appearance (or not) of a series of local circulations.
- Clockwise rotation of the flow field over the western part of the Black Sea Coast where important air pollution sources exist was documented. This implies complicated dispersion patterns in this region.

- Stagnation usually occurs in the SW and SE part of the Black Sea in areas where major urban conglomerates exist. The dispersion over remote areas of pollutants trapped within these stagnation or rotation zones is delayed for about one day. This delay is expected to be longer during the cool and transition seasons. The transformation of air pollutants under such conditions, occurs on a very complicated matter. Mixing between plumes from different origins and ages makes the physico-chemical processes even more complicated.
- For long range transport, three major paths followed by the air pollutants were identified. One path is from the SW part of the Black Sea and Vosporus towards the Aegean Sea and the Balkan Region and further to the East Mediterranean. The time scale for such transport is about three days. Another path was identified over the SE Black Sea Coast. This path results from the geomorphological characteristics of the region and the synoptic-scale flow pattern. Transport through it leads to the eastern part of Turkey and other areas South of Caucasus. The third path is from West to East over the flat area North of the Caucasus Peninsula. Because of the relatively strong winds the transport of polluted air masses mainly from the northern part of the Black Sea Coast towards East, is quite efficient.
- The movement of the air masses along the Danube Valley is very complicated and needs farther investigation because of its importance for the air quality in the region.

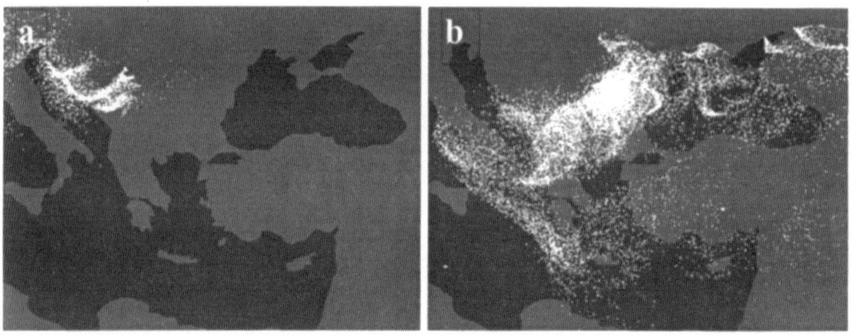

Figure 8. Particle projection from HYPACT dispersion model, at 0000 UTC, 01 July 1994 (after 72 hours of release) from (a) Belgrade and Novisad (b) the power plant installations around the Black Sea.

Acknowledgements: This research was supported by the research projects T-TRAPEM (EV*-CT92-0005) and MAMCS (ENV4-CT97-0593) of the DG-XII of EU. Acknowledgement is also made to NCAR, which is sponsored by the National Science Foundation, for some of the data used in this research.

REFERENCES

Kallos, G., P. Kassomenos, and R.A. Pielke, 1993: Synoptic and mesoscale weather conditions during air pollution episodes in Athens, Greece. *Boundary-Layer Meteorol.* 62, 163-184.

Kallos, G., Kotroni V., Lagouvardos K., and Varinou M., M. Luria, M. Peleg, G. Sharf, V. Matveev, D. Alper-SimanTov, A. Vanger, G Tuncel, S. Tuncel, N. Aras, G. Gullu, M. Idrees, F. Al-Momani, 1997a: Transport and Transformation of Air Pollutants from Europe to the East Mediterranean (T-TRAPEM). Final Report for the DGXII of EU project AVI*-CT92-0005. ISBN 960-8468-07-8. Pp 298.

Kallos, G., V. Kotroni, K. Lagouvardos, A. Papadopoulos, M. Varinou, O. Kakaliagou, M. Luria, M. Peleg, A. Wanger, and M. Uliasz, 1997b: Temporal and spatial scales for transport and transformation processes in the Mediterranean. Proc. of the 22nd NATO/CCMS ITM on Air Pollution Modelling and Its Application, 2-6 June 1997, Clermont Ferrand, France. Edited by S-E. Gryning and N. Chaumerliac, Plenum Press, New York, Vol 20, pp.8.

Kallos, G., V. Kotroni, K. Lagouvardos, and A. Papadopoulos, 1998: On the transport of air pollutants from Europe to North Africa. *Geophys.Res.Let.* 25, No 5, 619-622.

Luria M., M. Peleg, G. Sharf, D. Siman Tov-Alper, N. Schpitz, Y. Ben Ami, Z. Gawi, B. Lifschitz, A. Yitzchaki, and I. Seter, 1996: Atmospheric Sulphur over the East Mediterranean region. *J. Geophys. Res.,* 101, 25917-25930.

Pielke, R.A., W.R. Cotton, R.L. Walko, C.J. Tremback, W.A. Lyons, L.D. Grasso, M.E. Nicholls, M.D. Moran, D.A.Wesley,T.J. Lee, and J.H. Copeland, 1992: A comprehensive meteorological modelling system - RAMS. *Meteorol. Atmos. Phys.* 49, 69-91.

Steyn, D., and G. Kallos, 1992: "A study of the dynamics of hodograph rotation in the sea breezes of Attica, Greece". *Boundary-Layer Meteorol.* 58, pp. 215-228.

Tremback, C.J., W.A. Lyons, W.P. Thorson, and R.L. Walko, 1993: An emergency response and local weather forecasting software system. Proceedings of the 20th ITM on Air Pollution and its Application, X, 423-429.

DISCUSSION

M. BALDI: What is the mean SST of the Black Sea compared to the SST in the Aegean Sea?

G. KALLOS: For the simulations we have used variable SST fields available from ECMWF. During summer the SST of the Black Sea is comparable to the SST in the Aegean Sea and normally is around 22-25 degrees. In general, in both Seas the SST is lower by 2-3 degrees than the SST of the Eastern Mediterranean.

M. SOFIEV: Just for clarification- you have mentioned mercury emission inventory for the Black Sea area. Have you included any types of chemistry transformation in the model?

G. KALLOS: I mentioned mercury emission inventory because we used it to identify the power plants and other major industrial installations in the Region. We have not done any modeling of mercury processes for this presentation. We do mercury simulations for the scope of the MAMCS project. The Lagrangian releases we have shown can be valid for any pollutant behaving as ideal gas. We did these simulations just to show the spatial-temporal behaviour of dispersion in the Region.

PRODUCTION AND TRANSPORT OF OZONE IN LOCAL FLOWS OVER CENTRAL JAPAN–COMPARISON OF NUMERICAL CALCULATION WITH AIRBORNE OBSERVATION–

Toshihiro Kitada, Kiyoshi Okamura, Hideki Nakanishi,and Hiroaki Mori

Dept. of Ecological Engineering, Toyohashi University of Technology
Tempaku-cho, Toyohashi, Japan 441-8580

INTRODUCTION

The Central Japan has notable topographical features such that high mountains of the Japanese Alps exist at its center, large plains extend in its coastal area, and it is bounded with complex coast lines at the south and north sides facing to the Pacific Ocean and the Sea of Japan, respectively. Thus various types of local flows develop in summer season. In this situation, medium range transport of air pollutants from highly developed urban area to mountainous rural area has attracted concern of many people in applied meteorology and atmospheric environment in Japan, since it occurs in an interesting combination of several types of local winds such as land and sea breezes, mountain and valley winds, and plain-plateau winds, and the transport repeated in summer season probably causes severe damage to eco-systems in the mountainous area. Kurita et al. (1985) pointed out that the pollutants discharged in Tokyo Bay area migrate to the mountains in central Japan, i.e. the travel distance of about 150 km within one day. The distance far exceeded what was expected in simple sea breeze situation in mid-latitudes, and thus the phenomena suggested importance of combined effects of sea breeze, valley wind, and plain to plateau wind for the transport. Later Kondo (1990) numerically verified using meso-scale meteorological model that the presence of the high mountains in the central Japan, which are taller than 2000 m, is a key for the formation of the flow resulting in the medium range transport of pollutants. Kitada et al. (1998) also discussed dominant role of the high mountains of the central Japan in temporal variation of sea breezes over coastal plain area in the Pacific Ocean side, i.e., the Nohbi Plain. It was made clear more than 10 years ago, as noted above, that the pollutants released from huge urban areas in the coastal Japan such as Tokyo can be transported more than 150 km within 12 hours mostly during the daytime and reach to the mountains. However, the synoptic scale weather with light gradient wind under high pressure system usually continues for several days, and the behaviors of the primary and secondary pollutants during those days, once brought in the mountainous area, are not well understood. In this paper, to make these phenomena clear we have investigated three days behavior of the pollutants emitted in the coastal urban areas using a combined model of mesoscale meteorology (Kitada et al., 1998) and transport/chemistry/deposition of air pollutants (e.g., Kitada et al., 1993) together with airborne observation of pollutants over the Nohbi Plain and the Ise Bay area in the Central Japan. The period for the investigation was 26 to 28 July, 1995.

The simulation results showed that the high ozone once produced and transported to mountainous area is returned back, on the subsequent day, into the upper layer of 1 to 1.5 km high over the coastal region, and thus tends to accumulate in the air mass over the coastal plain area. These computed results were well compared with the vertical profiles of ozone obtained with aircraft observations over the coastal Nohbi Plain and the Ise Bay; the aircraft observations were performed on the occasion of environmental assessment for a new international airport planned over the Ise Bay. Furthermore, the simulation results showed very high concentrations of $SO_4^=$ and $NO_3^-(HNO_3)$, as well as O_3, at heights between 1 and 1.5 km above mean sea level in central mountainous region during midnight to early morning. This suggests the vegetation and eco-system in the mountains of the Central Japan are being affected by the pollutants discharged in the coastal huge urban areas quite frequently in warm season; our previous study showed that the so-called "extended sea breeze" occurs about 80 days in a year.

NUMERICAL MODELS

Transport/Chemistry/Deposition Model

The model deals with processes of advection, diffusion, chemical reactions and deposition of trace chemical species, and the basic frame of the model is described in Kitada et al. (1984), Carmichael et al. (1986) and Kitada et al. (1993). Equations are written for advected species,

$$\rho\frac{\partial C_i}{\partial t} + \rho U_j\frac{\partial C_i}{\partial x_j} = \frac{\partial}{\partial x_j}\left(\rho K_{jj}\frac{\partial C_i}{\partial x_j}\right) + R_i + S_i, \qquad i = 1, 2, \cdots, IL \tag{1}$$

where C_i denotes dimensionless concentration of species i, and R_i and S_i for chemical reaction rate and emission rate due to point source for the species i, respectively, ρ is air density, x_j spatial coordinates (i.e., j=1, 2, and 3 for x, y, and z, respectively), K_{11} $(= K_{22} = K_H)$ and K_{33} $(= K_V)$ for eddy diffusivity in horizontal and vertical directions, respectively, and U_j for wind velocity in x_j direction. Meteorological fields such as wind velocity, temperature, and eddy diffusivity were all calculated using mesoscale model described in Kitada et al. (1998). Equation (1) was applied for 45 advected species (i.e., IL=45) such as NO, NO_2, HNO_3, PAN, NH_3, NH_4NO_3, SO_2, SO_4^{2-}, O_3, H_2O_2, various anthropogenic and biogenic hydrocarbons, and others. The reaction mechanism used in the model is based on that of Lurmann et al. (1986), and additionally includes α-pinene chemistry of Lloyd et al. (1983), resulting in a system of 159 chemical reactions. The model also applies some of the revised reaction rate coefficients from Grey et al. (1989) and DeMore et al. (1997).

Horizontal boundary conditions for east-west (x_1) direction, for example, were expressed as follows. For inflow case,

$$-\rho K_H\frac{\partial C_i}{\partial x_1} + \rho U_1 C_i = \rho U_1 C_{B,i} \tag{2}$$

where $C_{B,i}$ stands for concentration at outside of the boundary and can be renewed, and for outflow case,

$$\frac{\partial C_i}{\partial x_1} = 0 \tag{3}$$

Vertical boundary condition was given as follows. At the earth surface boundary,

$$-\rho K_V\frac{\partial C_i}{\partial x_3} = Q_i - \rho V_{gi} C_i \tag{4}$$

where Q_i denotes the area emission source, and $V_{g,i}$ the dry deposition velocity for species i, which was evaluated by the method calculating resistances to mass transfer from air to surface (Wesely and Hicks, 1977, Sheih et al., 1979, Kitada et al., 1986 etc.). The dry deposition process was applied to the chemical species of NO_2, HNO_3, NH_3, NO_3^-, SO_2, SO_4^{2-}, O_3, PAN, and H_2O_2. At top boundary, basically the same formulations as (2) and (3) were applied. For 36 free radical species such as OH and HO_2, the steady state assumption was applied.

A conventional terrain following coordinate transformation was applied to the equations with the fixed top boundary above MSL at 6.5 km.

Mesoscale Meteorological Model

Input meteorological fields to the simulation of air pollution transport, such as flow, temperature, moisture, and eddy diffusivity, were computed with a mesoscale model with $k - \varepsilon$ turbulence module. The basis of the model is described elsewhere (Kitada et al., 1998). Here we explain only what was added into the model in this application. The model is hydrostatic, and the equations of momentum transport for horizontal directions are expressed as follows:

$$\begin{aligned}\frac{DU_j}{Dt} &= -\frac{1}{\rho}\frac{\partial p}{\partial x_j} + \frac{\partial}{\partial x_k}(-\overline{u_j u_k}) + f_1\left[\delta_{j1}(U_2 - U_{g_2}) - \delta_{j2}(U_1 - U_{g_1})\right] - f_2\delta_{j1}U_3 \\ &\quad + G_{U_j}\left(U_{obs_j} - U_j\right), \qquad j = 1, 2 \end{aligned} \tag{5}$$

where U_j and u_j denote the mean wind velocity to be predicted and the turbulent fluctuation, U_{g_j} stands for the geostrophic wind component estimated from the observed vertical profiles of the atmospheric pressure at 9 aerological observation sites over Japan, and U_{obs_j} the observed winds, where the suffix j is for coordinate direction with j=3 for vertical direction. Moreover, p denotes the mesoscale pressure perturbation from the synoptic scale pressure field, the Coriolis parameters $f_1 = 2\Omega\cos\varphi$ and $f_2 = 2\Omega\sin\varphi$, the nudging coefficient G_{U_i} (= $10^{-3}s^{-1}$ was used after some numerical experiments), δ_{j1} and δ_{j2} the Kronecker's delta, and $D/Dt = \partial/\partial t + U_j\partial/\partial x_j$ the substantial derivative.

The equations of heat transport:

$$\frac{D\Theta}{Dt} = \frac{\partial}{\partial x_j}\left(-\overline{u_j\theta}\right) + \frac{\Theta_0 - \Theta}{\tau_0} \tag{6}$$

where Θ, Θ_0 and θ stand for the mean potential temperature to be predicted, its synoptic scale component to be estimated from aerological data at seven observation sites in central Japan, and the turbulent fluctuation component, respectively, and τ_0 is a relaxation time scale which was set at $10^4 s$ and determined after some numerical experiments. The turbulent fluxes in (5) and (6) were evaluated using eddy diffusivity and local gradient of corresponding dependent variables, where the eddy diffusivity was calculated applying $k - \varepsilon$ turbulence model.

METEOROLOGICAL INPUTS, EMISSION SOURCES AND SIMULATION CASES

Meteorological Input Data

Numerical simulations were performed for two days from 06 JST (Japan Standard Time) 26 July to 06 JST 28 July 1995. During these days the Central Japan had been under the influence of high pressure system with sunny sky and light gradient wind, as seen in weather chart in Fig. 1. Thus these were typical days suitable for the development of local winds such as land and sea breeze, mountain and valley wind, and plain-to-plateau/plateau-to-plain wind. Photochemical smog is also known to likely occur in this situation.

To obtain meteorological fields for the simulations of air pollution transport and chemistry, the meteorological model was numerically integrated in the region shown in Fig. 2a. The size of the region is 582.9 km × 480.5 km × 6.5 km for east-west(x), north-south(y), and vertical(z) directions, respectively. Variable grid sizes were applied with its minimum values of 2.28 km, 1.85 km, and about 3 m for x, y, and z directions, respectively, and the maximum of 11.43 km, 9.24 km, and 200 m also for x, y, and z directions. The smallest grid size was used for the Nohbi Plain region in Fig. 2b, which is our primarily concerned area. Observational data were taken into the computation through U_{g_1}, U_{g_2}, and U_{obs_j} in (5), and Θ_0 in (6). These observation includes: (i) aerological data at 9 sites over Japan with 12 hours interval, (ii) data by 6 pilot balloons and 2 rawinsondes up to 2 km in the Nohbi Plain area with an hour and a half interval during daytime, and (iii) wind data at every one hour at about 360 surface meteorological observation sites operated over the whole calculation domain either by Meteorological Agency of Japan or by local governments. To get grid-based pressure field representing its characteristics in synoptic scale, the kriging method with spherical distribution model was applied for the sparse aerological data.

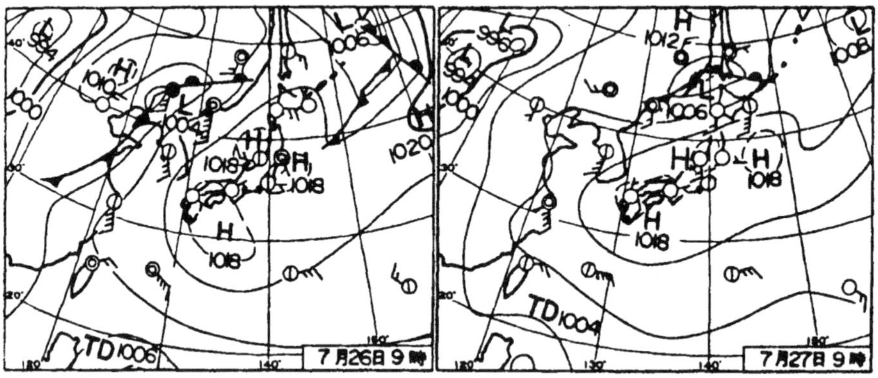

Fig. 1. Weather charts in Japan area at 09 JST 26 and 27 July 1995.

Emission Sources and Simulation Cases

Emission sources of NO_x, SO_x, and various anthropogenic non-methane-hydrocarbons (hereafter abbreviated to ant-NMHC) in and near the Nohbi Plain (see calculation domain in Fig. 2b) were

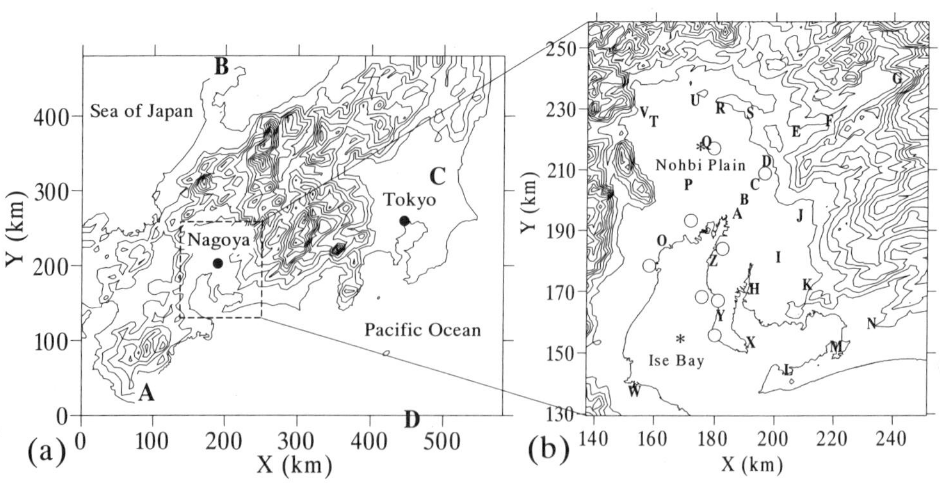

Fig. 2. Calculation domain : (a) whole area, and (b) the magnified for the Nohbi Plain and Ise Bay. A,B,C, and D in Fig. 2a represent aerological observation sites. A through Z in Fig. 2b are for observation sites at "surface" level, open circles for pilot balloon or rawinzonde sites, and * for the locations where spiral flights were carried out, i.e., one over Ichinomiya ("Q") and another over Ise Bay.

estimated from data compiled by the Aichi Prefectural Office (1995). NO_x and SO_x emission distributions were explicitly given in the data; they were computed on the basis of consumption of various fuels in energy sectors. The ant-NMHC were grouped into six categories of C_2H_6, C_3H_8, ALKA, C_2H_4, ALKE, and AROM as in the original chemistry model (Lurmann et al., 1986). Then their emission values were evaluated as ratios of those to NO_x emission by making reference to data in Aichi Prefectural Office (1995).

Two biogenic hydrocarbons of isoprene and α-pinene are considered in the model. Emission sources of isoprene and α-pinene were estimated using the vegetation data file that covers whole Japan area with about 1 km x 1 km grid cell (Research Center for Natural Environment, 1993) and the BEIS (Biogenic Emissions Inventory System) described in Pierce et al. (1990). Table 1 summarizes the emission source strength for the domain in Fig. 2b.

Table 1. Summary of emission sources for the region in Fig. 2b

Chemical species.	Emission rate (kmol day^{-1})	Ratio[3]
NO_x	7.90×10^3	1
Ant – NMHC[1]	7.72×10^3	0.978
C_2H_6	0.47×10^3	(0.059)
$C_3H_8^*$	0.81×10^3	(0.103)
ALKA$^\#$	2.90×10^3	(0.367)
C_2H_4	0.52×10^3	(0.066)
ALKE$^+$	0.70×10^3	(0.089)
AROM$^=$	2.32×10^3	(0.294)
Bio – NMHC[2]	2.10×10^3	
ISOP	1.64×10^3	
PINENE	0.46×10^3	

1, Ant-NMHC = C_2H_6 + C_3H_8 + ALKA + C_2H_4 + ALKE + AROM.
2, Bio-NMHC = ISOP + PINENE. 3, Ratio of emissions of various anthropogenic hydrocarbons to that of NO_x. $*$, C_3H_8 stands for propane and benzene.
#, ALKA for $\geq C_4$ alkanes. +, ALKE for $\geq C_3$ alkenes. =, AROM for alkylbenzenes.

Emission sources in the whole domain (Fig. 2a) were estimated, except for the Nohbi Plain region (Fig. 2b), by using $1°\times1°$ Asian distribution of NO_x and SO_x (Akimoto and Narita, 1994). Emissions of ant-NMHCs were computed by multiplying the NO_x emission distribution by the ratios listed in Table 1.

Two simulations were performed: case "241rxvv" was base case, and case "242rxvv" turned off all the emission sources after 00JST on 27 July (see Table 2). By comparing results of cases 241rxvv and 242rxvv, for example, effects of emissions in the previous day on ozone production can be estimated.

Table 2. Simulation cases

Case No.	Description
241rxvv (base)	normal
242rxvv	off*

$*$, All emission sources were turned off after 00 JST, i.e. midnight, on 27 July.

RESULTS AND DISCUSSION

Computed Flow Fields

Mesoscale meteorological model was integrated for four days from 00 JST 25 July to 00 JST 29 July 1995. Then the outputs were used for two days' air pollution transport simulation from 06 JST 26 July to 06 JST 28 July.

In the simulation, winds in upper layer acquired by pilot balloon and rawinsonde were used for 4 dimensional data assimilation. Figure 3 illustrates an example how the assimilation process worked: Figs. 3a-c show computed wind fields at 00 JST 27 July, i.e., Fig. 3a for surface wind at about 10 m above ground level(AGL) in whole calculation domain, and Figs. 3b,c for those at 500 m and 1000 m

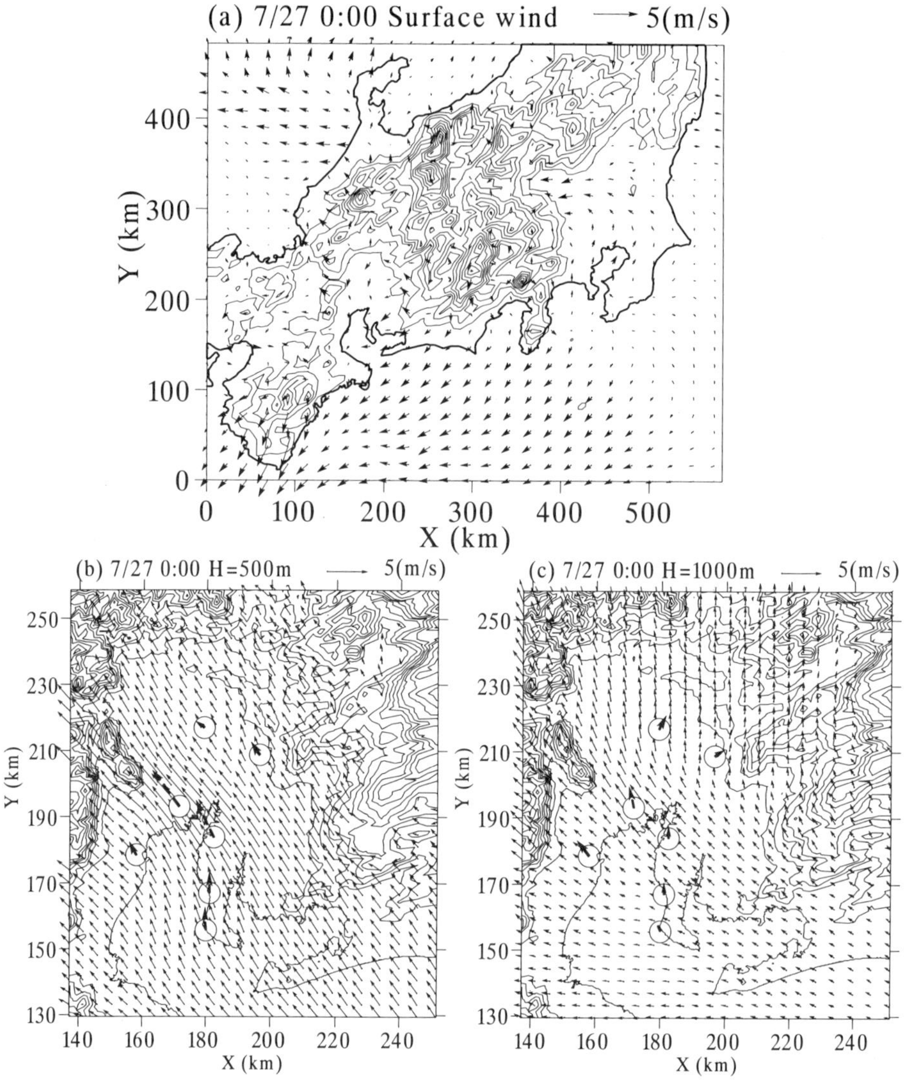

Fig. 3. Computed wind fields : (a) surface wind (at 10m above ground), (b) and (c) upper winds at 500 m and 1000 m high, respectively, at 00 JST 27 July.

above mean sea level(AMSL), where the observed winds are also shown with thick dotted arrow. The results show reasonable modifications of the computed winds, and thus indicate that the nudging coefficient, $G_{Uj} = 10^{-3}\ s^{-1}$, the weighted average of observed winds with a weight of the reciprocal of the squared distance between grid point and observation site, and the radius of influence of 100 km below 2 km AGL all worked well. The computed winds also well reproduced characteristic diurnal pattern of 3D wind field noted in Mori et al.(1998).

Dynamical Behavior of Ozone in Local Winds

As shown in Figs. 2a,b, the Nohbi Plain-Ise Bay area is bounded by mountains on the north, west, and east. This topographic feature suggests that accumulation of both anthropogenic and natural trace chemical species occur in the basin when high pressure system is prevailing with stable stratification in upper layer. Multiple days' simulation indicates this accumulation can take place in the area. We will show this with ozone as a typical photochemical product; Fig. 4 illustrates the computed two-days' behavior of ozone in the north-south vertical cross section, which passes over the city of Ichinomiya marked with symbol "Q" in Fig. 2b, in BASE case (Table 2); Figures 4a-d show the O_3 behavior on 26 July (i.e., 14LST to midnight) and Figs. 4e-h that on 27 July (i.e., 10LST to midnight). In the afternoon O_3 is produced in the pollutant rich urban air mass, and moves inland with a combined flow of sea breeze, valley wind, and plain-to-plateau wind (Figs. 4a-c, and 4f,g). Around midnight the

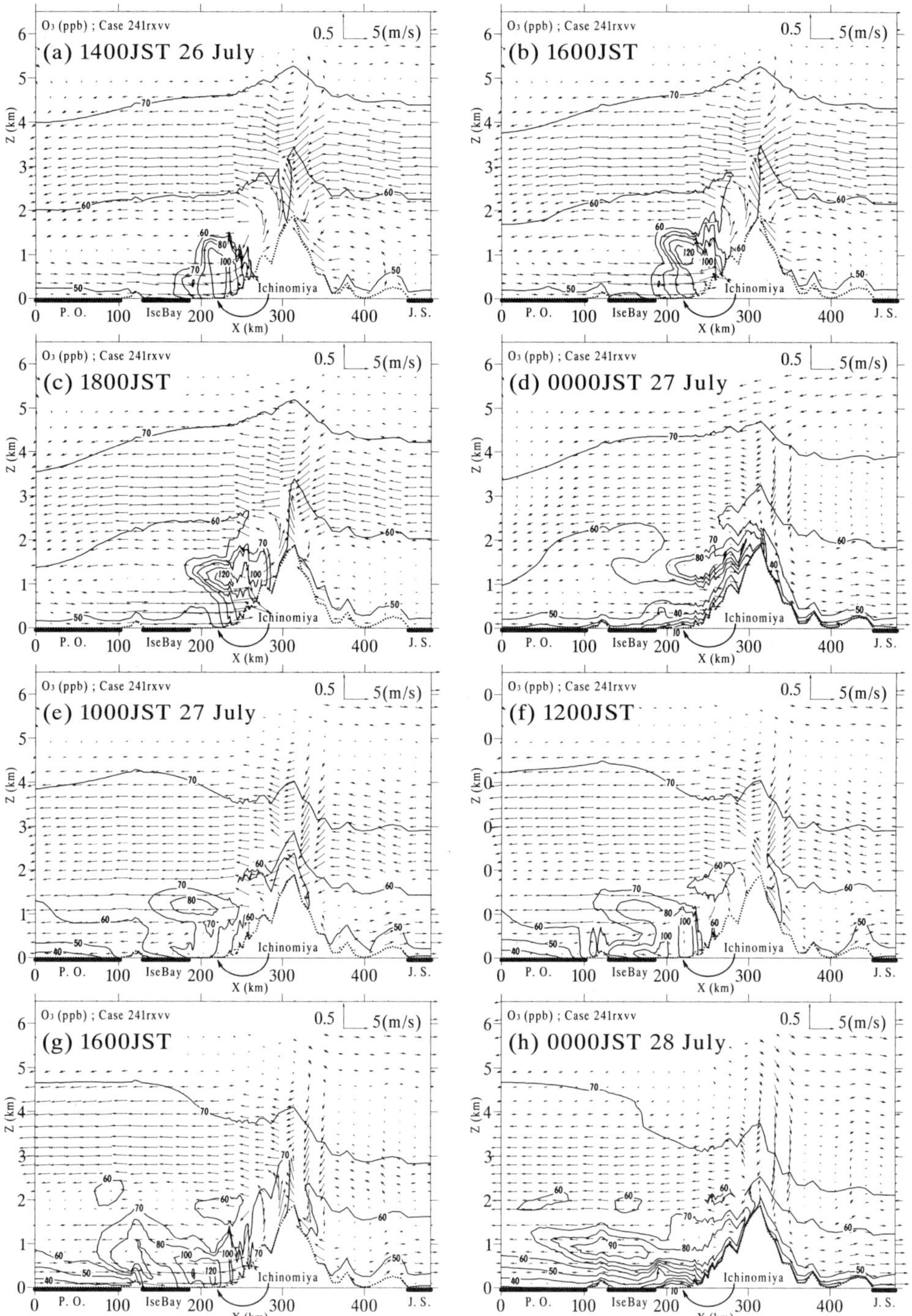

Fig. 4. Vertical cross sections of the computed O_3 (BASE; see Table 2) at x = 196 km (symbol "Q" in Fig. 2b) at (a) 14, (b) 16, (c) 18 JST 26 July, (d) 00, (e) 10, (f) 12, (g) 16 JST 27 July, and (h) 00 JST 28 July.

front of the polluted air mass reaches to the ridge area which divides the Central Japan to the Pacific Ocean side and the Sea of Japan side(Figs. 4d,h), and stagnates there.

Additionally, Figs. 4a-d indicate a high O_3 concentration zone is formed at around 1 km high and y =220 km (see Fig. 2b for the location) over plain area and continues on the next day (Fig. 4d on 27 July); formation of the high O_3 zone is due to the vertical transport of O_3 and its precursors into the upper weak wind layer, which locates above the mixed layer and between sea breeze and its return flow, by both convective motion in the mixed layer and updrafts associated with sea breeze front and upslope/valley wind (e.g., Kitada and Kitagawa, 1990).

At midnight (Fig. 4d), O_3 is very low near the surface in the Nohbi Plain (Fig. 2b) due to NO_x emissions from the urban area. As noted, however, higher O_3 layer (i.e., 80 ppb) remains at around 1.3 to 1.5 km high and extends over 100 km in north-south direction in plain and mountain areas. The O_3-rich air mass in upper layer starts to move to the south over the Nohbi Plain-Ise Bay area in prevailing northerly wind in the morning on 27 July (Figs. 4e,f). Then Fig. 4f shows that at noon the developing mixed layer over the plain area entrains O_3 and other chemical species in the upper layer, which have, as noted earlier, origin in the previous day's emissions. Thus the background O_3 concentration in the Nohbi Plain shows higher values on 27 July in Figs. 4g,h than those on 26 July in Figs. 4b,d, demonstrating accumulation of the pollutants in the air mass extending over bay, plain, and mountain areas, when high pressure system with light gradient wind continues for several days.

Vertical Profiles of O_3 —Observation and Calculation—

Airborne observation of O_3, NO_x etc. was performed three times a day during 26 and 27 July. In each flight, vertical profiles of the chemical species were obtained at Ise Bay and at Ichinomiya in the Nohbi Plain; these places are marked with asterisk in Fig. 2b. Figure 5a shows O_3 profiles over Ise Bay at around 09JST 27 July; thick solid line stands for observed and thin solid line with marks for computed in BASE case, i.e., square, circle, and triangle are at 08, 10, and 12JST, respectively. The observed O_3 (thick solid line) has characteristic structure of clear double maxima, i.e., one at lower altitude of 200 m and the other at 1200 m high. The computed O_3 profiles (e.g., triangle in Fig. 5a) also show the same double maxima structure and thus indicate the simulation captures the characteristic behavior of O_3, though there is some time delay and difference in concentration level between the observed and the computed profiles. Computed O_3 cross sections in Figs. 4d-f suggest how the double maxima structure is developed; one maximum at 1200 m is mainly due to the O_3 that was transported to the height over plain and mountain areas on the previous day, as noted earlier in this paper, and was carried over Ise Bay by the northerly or northwesterly wind in upper layer in the

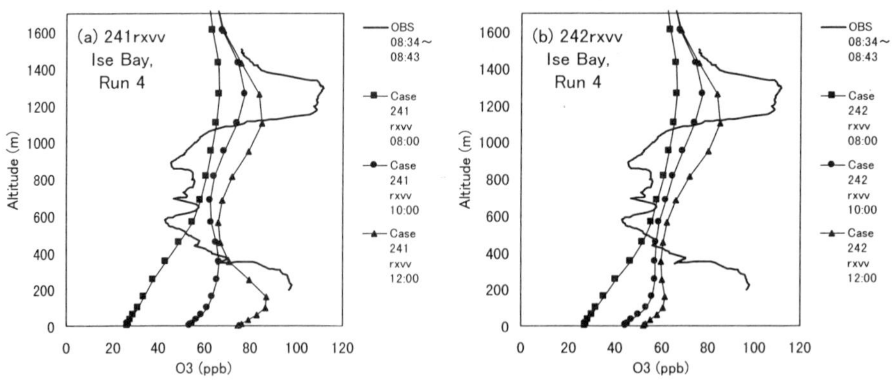

Fig. 5. Vertical profiles of O_3 observed and calculated over Ise Bay: (a) case 241rxvv and (b) case 242rxvv. The location of spiral flight over Ise Bay is marked with "*" in Fig. 2b. The observations were carried out from 0730 to 0900 JST 27 July.

morning on 27 July. On the other hand the O_3 maximum at 200 m could be largely produced from the precursors released in the rim area of Ise Bay after 00JST 27 July and transported to the center of the Bay by northerly and westerly land breeze and mountain wind in the early morning. This can be further verified by comparing case 241rxvv (BASE) with case 242rxvv (Table 2), where all the emission sources were turned off after 00 JST 27 July. The O_3 profiles in case 242rxvv (Fig. 5b) show substantially smaller maximum in the lower layer compared with those in Fig. 5a, while the maximum in the upper layer at 1200 m is almost same in Figs. 5a,b, thus supporting the noted explanation.

Calculated High Concentrations of Photochemical Reaction Products in Mountainous Area during Nighttime

As can be expected from the topography in the domain (Fig. 2a), local flows such as sea breeze, valley wind, and plain-to-plateau wind tend to converge to the central mountainous area during daytime. Thus various photochemical products may form high concentration zone at high altitudes. To see this, horizontal distributions of computed O_3, SO_4^{2-}, and $HNO_3 + NO_3^-$ at 1500 m high at 00JST 27 July (BASE case) are illustrated in Figs. 6a-c, respectively. In the central Japan area, two remarkable high concentration zones can be seen: one is to the west of Tokyo, and covers the Mountains of Chichibu and Tanzawa, and the Izu Peninsula. The other zone is to the north of Nagoya, and extends over the Hida and Takayama areas. These pollutants are the products of photochemical smog reactions, and the present simulation clearly suggests that they are originated from the huge urban areas of Tokyo

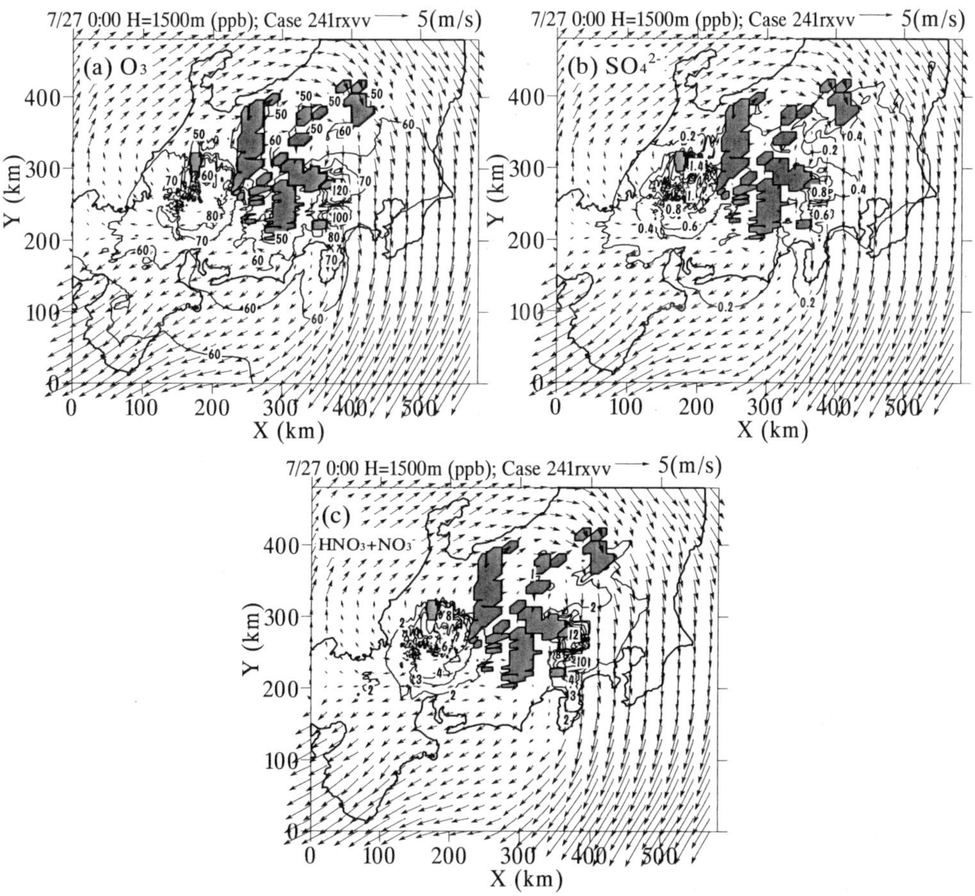

Fig. 6. Computed horizontal cross sections of (a) O_3, (b) SO_4^{2-}, (c) $HNO_3 + NO_3^-$ at 1500 m high on 00 JST 27 July (case 241rxvv, i.e. BASE).

and Nagoya. The mountain side of Chichibu and Tanzawa near Tokyo has already be known because of the severe damage to its vegetation due to acidic fog (Igawa et al., 1998).

Although in the Hida area, such a damage has not yet been clarified, Fig. 6 strongly suggests the same thing as Chichibu and Tanzawa may be taking place. A statistical study indicates that the similar meteorological condition occurs quite frequently in the Central Japan area, i.e., about 80 days during warm season from April to October (Mori et al., 1994).

SUMMARY AND CONCLUSIONS

To see the multiple days' behavior of air pollutants released in coastal urban areas in the Central Japan, two days simulations of meso-scale flow and air pollution transport were performed from 06 JST 26 July to 06 JST 28 July 1995. During the simulated time, high pressure system, with light gradient wind and sunny sky, was dominant over the region. Thus various types of local flows developed in the domain. In the Nohbi Plain, observations using aircraft, pilot balloon, and rawinsonde were performed for the same time, and these observed data were utilized in simulations as inputs for nudging of the computed wind, and for evaluation of computed concentration profiles.

The simulations made clear two days' behavior of O_3 in developed local wind system. The O_3, sulfate, nitrate etc. produced in polluted air mass from coastal urban area, i.e., the Nohbi Plain, reached to the central mountainous region at midnight on the first day. They formed high O_3 concentration zone at altitudes between 1 and 1.5 km; the zone extended more than 100 km from the mountain to the plain. On the following day, the high O_3 in the upper layer moved back to the coastal area, and notable double maxima structure was generated in vertical profile of O_3 over Ise Bay in the morning, with one maximum at lower layer due to fresh O_3 produced on the same day and the other maximum at upper layer due to the O_3 formed on the previous day and transported back from mountainous area. This double maxima structure over Ise Bay in the morning was also observed by aircraft.

That O_3 and other photochemical products accumulate in the air mass over mountain, plain, and coastal areas under several day's dominance of high pressure system was well explained through numerical simulations.

That O_3 and other acidic products gather in the central mountainous area in the nighttime and make high concentrations layer between altitudes of about 1 and 1.5 km was predicted. Two particularly polluted zones appear at the Mountains of Chichibu and Tanzawa, and the Izu Peninsula, to the west of Tokyo, and at the Hida area and Mt. Hakusan, to the north of Nagoya; Tanzawa area has already been known for damage to its vegetation. Statistical study shows that the Central Japan has about 80 days of the local-flow-developing synoptic weather in a year, indicating quite frequent air pollution loading from Japanese coastal mega-cities to the ecosystem in the Japanese Alps.

ACKNOWLEDGMENT

This work was supported in part by the Ministry of Education, Culture, and Science, Japan through grant No. 08650645 and 10650538. Authors are grateful to Mr. Nishizawa, Ph. D. candidate, and Miss Fumiyo Tani, MS student at Toyohashi Univ. of Technology for their assistance in preparing this paper.

REFERENCES

Aichi Prefectural Office, Japan, 1995, Data compiled by Division of Environment.

Akimoto, H., and Narita, H., 1994, Distribution of SO_2, NO_x and CO_2 emissions from fuel combustion and industrial activities in Asia with $1^\circ \times 1^\circ$ resolution. *Atmos. Environ.*, 28:213.

Carmichael, G.R., Peters, L.K., and Kitada, T., 1986, A second generation model for regional-scale transport/chemistry/deposition. *Atmos. Environ.*, 20:173.

DeMore, W.B., Spander, S.P., Hampson, R.F., Kurylo, M.J., Howard, C.J., Kolb, C.E., and Molina, M.J., 1997, *Chemical Kinetics and Photochemical Data for Use in Stratospheric Modeling -Evaluation Number 12*, JPL Publication 97-4, NASA/Jet Propulsio Laboratory.

Gery, M. W., Whitten, G. Z., Killus, J. P., and Dodge, M. C., 1989, A photochemical kinetics mechanism for urban and regional scale computer modeling. *J. Geophys. Res.*, 94:12925.

Igawa, M., Tsutsumi, Y., Mori, T., and Okochi, H., 1998, Fogwater chemistry at a mountainside forest and the estimation of the air pollutant deposition via fog droplets based on th atmospheric quality at the mountain base. *Environ. Sci. Technol.*, 32:1566.

Kitada, T., Carmichael, G.R. and Peters, L.K., 1984, Numerical simulation of the transport of chemically reactive species under land- and sea-breeze circulations. *J. Climate Appl. Meteorol.*, 23, 1153.

Kitada, T., Carmichael, G.R. and Peters, L.K., 1986, Effects of dry deposition on the concentration-distributions of atmospheric pollutants within land- and sea-breeze circulations. *Atmos. Environ.*, 20:10, 1999-2010.

Kitada, T., and Kitagawa, E., 1990, Numerical analysis of the role of sea breeze fronts on air quality in coastal and inland polluted areas. *Atmos. Environ.*, 24A:1545.

Kitada, T., Lee, P.C.-S., and Ueda, H., 1993, Numerical modeling of long-range transport of acidic species in association with meso- β -convective clouds across the Japan sea resulting in acid snow over coastal Japan -I. Model description and qualitative verifications. *Atmos. Environ.*, 27A:1061.

Kitada, T., Okamura, K., and Tanaka, S., 1998, Effects of topography and urbanization on local winds and thermal environment in Nohbi Plain, coastal region of central Japan -A numerical analysis by meso-scale meteorological model with k- ε turbulence model. *J. Appl. Meteor.*, 37:1026.

Kondo, H., A numerical experiment on the interaction between sea breeze and valley wind to generate the so-called "Extended Sea Breeze". *J. Meteor. Soc. Japan*, 68:435, 1990

Kurita, H., Sasaki, K., Muroga, H., Ueda, H., and Wakamatsu, S., 1985, Long-range transport of air pollution under light gradient wind conditions. *J. Climate Appl. Meteorol.*, 24:133.

Lloyd, A.C., Atkinson, R., Lurmann, F.W., and Nitta, B., 1983, Modeling potential ozone impacts from natural hydrocarbons - I. Development and testing of a chemical mechanism for the NO_x-air photooxidations of isoprene and α-pinene under ambient conditions. *Atmos. Environ.*, 17:1931-1950.

Lurmann, F.W., Lloyd, A.C., and Atkinson, R., 1986, A chemical mechanism for use in long-range transport/acid deposition computer modeling. *J. Geophys. Res.*, 91:10905.

Mori, H., Kitada, T., and Iyoda, K., 1998, Vertical structure of land and sea breezes in the Nohbi plain during a passage of an anticyclone over the central Japan. *Tenki*, Meteorol. Soc. Japan, 45:515 (in Japanese).

Mori, H., Ogawa, H., and Kitada, T., 1994, Characteristics of land and sea breezes in the Nohbi plain, and the conditions of occurrence of the "extended sea breeze". *Tenki*, Meteorol. Soc. Japan, 41:379 (in Japanese).

Pierce, T.E., Lamb, B.K, and van Meter, A.R., 1990, Development of a biogenic emissions inventory system for regional scale air pollution models. *Proc. AWMA Annu. Meet., 83rd*, 16pp.

Research Center for Natural Environment, 1993, *Vegetation Data File for Japan Area*, Tokyo.

Shawn, J. R., Thomas, E. P., and Kenneth, L. S., 1991, The sensitivity of regional ozone modeling to biogenic hydrocarbons. *J. Geophys. Res.*, 96:7371.

Sheih, C.M., Wesely, M.L., and Hicks, B.B., 1979, Estimated dry deposition velocities of sulfur over the eastern United States and surrounding regions. *Atmos. Environ.*, 12, 2055-2087.

Wesely, M.L., and Hicks, B.B., 1977, Some factors that affect the deposition rates of sulfur dioxide and similar gases on vegetation. *J. Geophys. Res.*, 100:11,447.

DISCUSSION

X. LIN: What is the effect if you include the emissions from China in the study on a regional scale under different meteorological conditions other than the "sea breeze" condition?

T. KITADA: In summer time, an anticyclone, with its center in the west Pacific Ocean, is usually dominant in Japan area. Under this condition synoptic scale gradient wind is weak and we have often combined local flows of sea breeze, valley wind and plain-to-plateau wind, which converge in the central mountainous area. In this situation we do hardly see effect of emissions in China on air quality in the central Japan. We observe effects of China emissions most in winter and early spring. I have also experience of numerical simulation of air pollution transport from China for these seasons.

ON THE ACCOUNTING FOR SOME SUB-SCALE EFFECTS IN THE LONG-RANGE AIR POLLUTION MODELLING OVER A COMPLEX TERRAIN

Evgeni Syrakov[1] and Kostadin Ganev[2]

[1]University of Sofia, Faculty of Physics
J. Baucher str. 5, Sofia 1126, Bulgaria
[2]Institute of Geophysics, Bulgarian Academy of Sciences
Akad.G. Bonchev str., block 3, Sofia 1113, Bulgaria

INTRODUCTION

The present work aims at a study of some sub-scale (concerning the long-range transport) pollution effects. The accounting for them is especially important for powerful source, located in or close to regions with complex topography and for interpretation of data from mountain background stations[1,2] . In both the cases these effects can lead to significant corrections of the standard long-range pollution characteristics in regions close around the sources, as well as for the integral characteristics of the corresponding cell from the EMEP grid. It should be underlined that these effect concern scales (intermediate) between the long- and local- transport scales (see Fig. 1.). That is why they can form environmental loads, which are smaller then the local regulatory standards, but at the same time are significant fluctuations of the long-range estimations.

MODELS

The IMSM (Integrated Multi-Scale Model) [1,2,3,4,5,6,7,8,9,10], introduced in the present paper, aims at producing an embracing and complex evaluation of the air pollution, accounting for the specific characteristics of the different transport scales (Fig.1.).

IMSM incorporates several coordinated air pollution transport models:

- a 3-D Eulerian pollution transport model;
- air pollution mass-balance relations for each of the pollutants in an arbitrary sub-region D_1 of the integration domain D, and for a time period $[\,0-T\,]$:

$$P = Q + F + P_0 + P_{in} - (P_d + P_{g+} + P_{out} + P_{wet}), \tag{1}$$

Air Pollution Modeling and Its Application XIII, edited by S.-E. Gryning
and E. Batchvarova, Kluwer Academic/Plenum Publishers, 2000

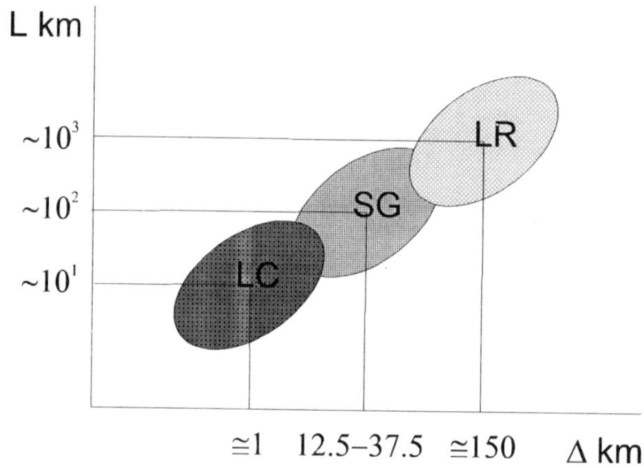

Fig.1. Range of appliance of the IMSM in dependence of the scale of horizontal transport (L) and the corresponding space discretisation Δ used in the present study.

where Q is the emission, P - the total quantity in the moment T, P_0 is the initial quantity, P_d is the dry deposition, P_g is the pollution quantity which had passed the upper boundary of D_1, P_{wet} is the wet deposition (washed-out pollution quantity), P_{out} are the wastes of the given pollutant due to chemical transformations, P_{in} are the influx of the given pollutant due to chemical transformations, F is the pollution quantity which had passed the side boundaries of D_1 (F>0 means inflow and F<0 means outflow of the pollutant).

- functions of influence problem, concerning ecologically important pollution characteristics of a given "protected" region. The most popular of them are perhaps K_1 - the time averaged pollution quantity, $K_2 \equiv P$, $K_3 \equiv P_d$ and $K_4 \equiv F$.
- "Puff" model, coordinated with (1) and a trajectory model (basic and back-wind trajectories).

The IMSM accounts for to 10 interacting Sulfur, Nitrogen and Ammonia compounds, using the standard EMEP chemical transformation scheme[11]

In dependence of the transport scale specifics an appropriate combination of dynamics and transport models is chosen and so, according to Fig.1. three basic operational regimes of the IMSM exist - $IMSM_{LC}$, $IMSM_{SG}$, $IMSM_{LR}$:

- $IMSM_{LC}$ describes the pollution transport in the local domain around a given source LC (<20-25 km). It contains a splited-up 3D Eulerian model, which treats the vertical diffusion numerically, and applies a Gausian distribution in horizontal direction. It utilizes a PBL model (stationary or evolutionary) parameterized either in accordance with the resistance laws, accounting for inversions, baroclynicy, terrain slope, or in accordance with the Pasquill-Turner stability classes;
- $IMSM_{LR}$ describes the pollution transport in the LR area (~10^3km). It treats the above stated set of models numerically, using the standard EMEP discretization ($\Delta = 150km$) and emission inventory [11]. A PBL model parameterized by external aerologo-synoptical parameters is used for the purpose;
- $IMSM_{SG}$ describes intermediate (in respect with the LR and LC scales) sub-grid (SG) effects of pollution in a grid nested (in the present with Δ=37.5km) in the standard EMEP one, with a more detailed emission inventory and accounting of topography effects. A mass-balance diagnostic model (see [12,1,2,5]) is used for the purpose, with a

procedure for coordinated surface and aerological data interpolation, especially designed in correspondence of the physical-geographic specifics of the country.

NUMERICAL SIMULATIONS AND DISCUSSION OF THE RESULTS

As it was underlined above, the major objective of the present work is the study of the SG pollution effects. They are of special importance in the cases of powerful sources. That is why the present work is focused on studying the pollution from the Thermal Power Plants (TPP) from the Maritza-East complex (sources M-1,-2,-3, and-4 from Table 1, located in the (33,18) cell of the EMEP grid). For the sake of completeness the other major Bulgarian TPP (Bobov Dol-BD, Varna-VN and Ruse-RS from Table 1., correspondingly located in sells (32,17), (33,20) and (32,19) of the EMEP grid) are also considered. The emission data for these sources is taken from [11]. It should be noticed that the emissions from the last TPP are much smaller, and so their SG effects are not so well manifested.

The high and powerful sources from the Maritza-East complex cause significant air pollution, which can be followed in the three transport scales. Detailed pollution assessment in the LC domain is not an objective of the present study. Only climatic evaluations for the biggest source (M-2) will be demonstrated here (Table 1.). They are obtained by applying the mass-balance approach (1) to the calculated annual pollution roses, obtained in a cylindrical domain with a radius of 25km, around the source. It can be seen that the local pollution is caused only by a minor part of the emitted pollutants. The predominating part of the pollutants (about 98% for Sulphur and 99% for the Nitrogen compounds) leaves the local domain and thus becomes a subject of the larger transport scales.

Table 1 Annual pollution balance components for M-2 source for a cylindrical region with radius 25km

	Q	P	P_d	P_{out}	P_{in}	P_{wet}	F
NO[t(N)]	4637.0	0.167	0	38330	-34450	0	-755
NO_2[t(N)]	159.2	0.828	2.605	34890	-38380	0	-3652
HNO_3[t(N)]	0.	0.032	1.750	5	-206	8	-192
PAN[t(N)]	0	0.005	0.058	53	-181	0	-128
NO_3[t(N)]	0	0.098	0.302	7	-67	16	-42
SO_2[t(S)]	184100	41.620	866.8	4025	0	2973	-176200
SO_4[t(S)]	6460	2.505	9.860	0	-4025	574	-9897
total S [t]	4797	1.131	4.715	73280	-73280	24	-4768
total N [t]	190600	44.120	876.6	4025	-4025	3547	-186100

20 basic synoptic situations, representative for advective or none-gradient meteorological conditions, typical for the country, are chosen, after extended synoptic and climatic expertise, for studying the interaction of the LR and SG effects. Further extension of the scope of the basic synoptic situations is, of course, desirable.

Because of the limited volume of the paper only part of the basic situations, represented by a typical date (when the situation dose not change fast) or by a series of dates (when the evolution is faster) will be discussed in details.

Two different assessments of the pollution characteristics f (concentrations, depositions, acid rain, etc.) are made for the territory of the country:

$$f = f_{LR}, \tag{2}$$

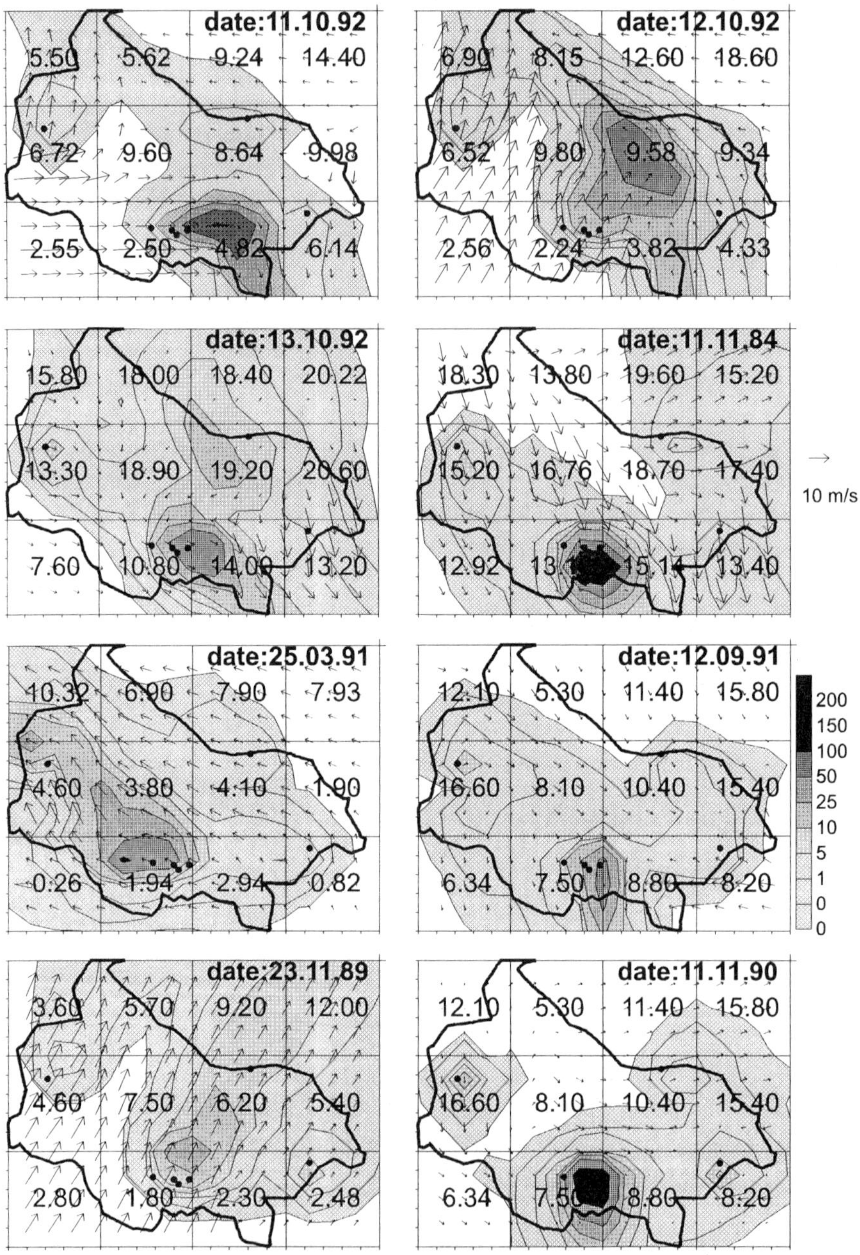

Figure 2. Isolines of the surface values of C_{SG} (for SO_2) and the values of C_{LR}^{red} (for SO_2) [$\mu g / m^3$] for the corresponding EMEP cells. The cell, which contains the four TPP from the "Maritza" complex is the (33,18) cell from the EMEP grid.

where f_{LR}, accounts for the long range transport only and is calculated by IMSM$_{LR}$, and the assessment specified with accounting for the SG effects:

$$f = f_{LR}^{red} + f_{SG}, \tag{3}$$

where f_{LR}^{red} is obtained like (2), with the only difference that emissions from the considered major TPP are subtracted from the EMEP emissions for the corresponding cells, and f_{SG} is obtained by the $IMSM_{SG}$, with a more detailed topography description, accounting only for the major (excluded from the f_{LR}^{red} simulations) Bulgarian TPP.

Table 2. Contribution of LR and SG transport scales to the formation of the mass-balance components [t] for the (33,18) cell of the EMEP grid, 11.11'90

		Q	P	P_d	P_g	P_{out}	P_{in}	F	P_o
NO	SG	54.1	0.2	0	1.4	1480	1430	-5	0.2
	LR	213.1	0.7	0	0	4380	4190	-31.7	0.7
	LR^{red}	159	0.5	0	0	3460	3330	-22.1	0.5
	$SG+LR_{red}$	213.1	0.7	0	1.4	4940	4760	-21.7	0.7
NO_2	SG	2.8	38.4	1	7.4	2230	2270	-26.2	27.7
	LR	11.2	109	5.9	0	6530	6720	-166	80.1
	LR^{red}	8.4	89.1	4.7	0	5170	5310	-117	62.4
	$SG+LR^{red}$	11.2	127.5	5.7	7.4	7400	7580	-143	89.1
HNO_3	SG	0	0	0.8	0.1	21.3	19.2	3	0
	LR	0	0	4.3	0	47.1	56	-4.6	0
	LR^{red}	0	0	3.4	0	37.9	44.6	-3.2	0
	$SG+LR^{red}$	0	0	4.2	0.1	59.2	63.8	-0.3	0
PAN	SG	0	12.3	0.7	4.3	6.5	32.4	-11.5	2.9
	LR	0	31.4	4.4	0	17.3	95.5	-49.3	6.9
	LR^{red}	0	28.6	3.7	0	14.6	75.7	-34.4	5.6
	$SG+LR^{red}$	0	40.9	4.4	4.3	21.1	108.1	-45.9	8.4
NO_3	SG	0	5.5	0.2	1.4	1.6	9.9	-3.9	2.7
	LR	0.	16.1	1.4	0	4.5	31.3	-17.2	7.9
	LR^{red}	0	14.1	1.2	0	3.8	24.9	-12.1	6.3
	$SG+LR^{red}$	0	19.6	1.4	1.4	5.4	34.8	-16.0	8.9
SO_2	SG	2170	1170	284	325	127	0	-985	728
	LR	2170	634	286	0	79	0	-1640	472
	LR^{red}	0	170	53.6	0	14.6	0	173	64.9
	$SG+LR^{red}$	2170	1340	337.6	325	141.6	0	-812	792.9
SO_4	SG	114	0	0.1	0	74.2	191	-231	0
	LR	114	0	0.1	0	47.9	118	-184	0
	LR^{red}	0	0	0	0	0.6	21.9	-21.4	0
	$SG+LR^{red}$	114	0	0.1	0	74.8	212.9	-252	0
SO_4 neutr.	SG	0	48.7	2.7	13.7	0	74.2	-39.7	30.6
	LR	0	17.6	2.5	0	0	47.9	-44.1	16.2
	LR^{red}	0	3.9	0.4	0	0	0.6	1.5	2.3
	$SG+LR^{red}$	0	52.6	3.2	13.7	0	74.8	38.2	32.9
NO_3 neutr.	SG	0	8.4	0.4	2.7	0.1	20.6	-10.3	1.3
	LR	0	21	2	0	0	44.4	-24.3	2.9
	LR^{red}	0	19.5	1.7	0	0	35.8	-16.9	2.4
	$SG+LR^{red}$	0	27.9	2.1	2.7	0.1	56.4	27.2	3.7

On the basis of inter-comparison of f_{LR}^{red} and f_{SG} ,correspondingly of f_{LR} and $f_{LR}^{red} + f_{SG}$, it can be detected the significance of the sub-grid correction to the long range estimations.

It is natural the analysis of the results to start with a discussion of the 24h averaged surface concentrations C (the SO_2 is chosen for demonstrations from the 10 simulated Sulphur and Nitrogen compounds), when (3) obtains the form $C = C_{LR}^{red} + C_{SG}$. The isolines of C_{SG}, together with the C_{LR}^{red} values for the corresponding EMEP cells are shone on Fig.2. The analysis of all the situations, including the demonstrated in Fig. 2., shows that the relation between C_{LR}^{red} and C_{SG} strongly depends on the synoptic conditions, when in most of the situations there are areas in which $C_{SG} >> C_{LR}^{red}$, i.e. the sub-grid corrections are very significant. These areas, usually with radius ~ 50-70 km, around ore shifted from the sources, in which the sub-grid correction is especially significant, can be.defined from the condition $C_{SG} \geq C_{LR}^{red}$. Generally speaking the maximal SG-effects can be observed when the wind direction coincides with the line formed by the 4 sources from the Maritza complex, at relatively low or medium wind speed, or blocking local circulation (alternated wind directions, sharp changes in wind speed or direction, etc.), caused by the complex topography influences. The SG effects are smaller at winds with bigger speed and direction perpendicular to the line around which the discussed sources are located, at ventilating canal topography effects, etc..

Table 3. Contribution of LR and SG transport scales to the formation of the SO_2 mass-balance components [t] for the (33,18) cell of the EMEP grid, 11.10.'92

		Q	P	P_d	P_g	P_{out}	P_{in}	F	P_o
SO_2	SG	2170	105	46.4	0.8	18	0	-2100	105
	LR	2170	324	186	0	48.7	0	-1900	295
	LR^{red}	0	62.5	20	0	6.6	0	53.1	36
	$SG+LR^{red}$	2170	167.5	66.4	0.8	24.6	0	-2047	141

Analogical effects can be followed for all the balance components (1). Each of them can be presented in the form (3). The balance components for a selected region-the air column above the (33,18) cell of the EMEP grid (chosen because the most powerful Sulphur sources are located there) are demonstrated at two different meteorological conditions. In the case shown at Table 2 the specified value of the mass-balance component $P_{LR}^{red} + P_{SG}$ is about 2 times bigger then P_{LR} - the corresponding assessment obtained with only the application of $IMSM_{LR}$. According to this the pollution outflows trough the side boundaries, are twice smaller when the SG effects are taken into account. As it can be seen at Fig. 2. The zone of maximal SG effects in this case is around the pollution sources. For the second case (for simplicity only the mass-balance components for SO_2 are shown in Table 3.) the tendency is opposite - $P_{LR}^{red} + P_{SG}$ is smaller then P_{LR}, while $F_{LR}^{red} + F_{SG}$ is bigger than F_{LR}. The explanation of this phenomena is that in the second case the zone of maximal SG effects is shifted from the pollution sources and out of the (33,18) EMEP cell (see Fig. 2.).

The SG acid rain component, which reflects mainly the SC effects of the pollution concentration field, but also the spotted structure of the precipitation are demonstrated at Fig, 3..

As demonstrated above the SG effects lead to significant corrections of the LR pollution estimations. That is why the problem of determination of the contribution of each of the considered TPP to the f_{SG} characteristics is of special interest. The problem is solved by the functions of influence branch of $IMSM_{SG}$ The functions of influence, obtained for two different situations - SE and NW invasions are shown at Fig. 4. The demonstrated functions of influence are calculated for the K_1 criteria for the "protected" region shown at the figure (estimation 1). The functions of influence, concerning the same criteria, but for the whole territory of Bulgaria taken as "protected" region (estimation 2), as well as concerning similar criteria, but each cell of the "protected" region weighted with

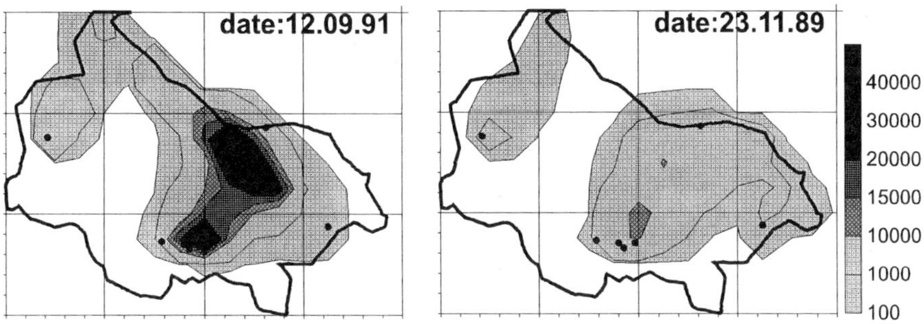

Figure 3. Isolines of the washed out SO_2 [$\mu g / m^2$] SG quantities.

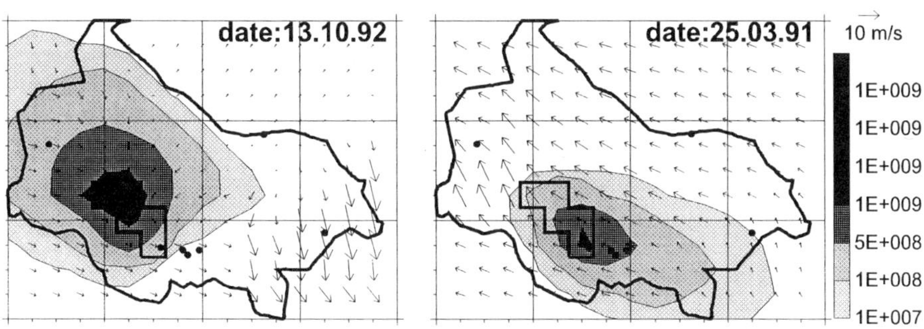

Figure 4. Function of influence at level $z = 200m$, calculated for K_1 criteria, concerning the "protected" region shown at the Figure (estimation 1).

Table 4. Contribution (in %) of the Bulgarian TPP to the air pollution, estimated with three different pollution criteria

date	esti-mation	BD	M-1	M-2	M-3	M-4	RS	VN	other
	2	0	1.3	58.3	15.1	23.5	0	0	1.8
25.03.	3	2.4	0.6	41.4	17.7	27.5	0.6	2.9	7.0
1991	1	1.1	0.5	42.6	16.7	20.6	0.7	3.9	8.4
	2	47.3	5.5	10.3	0.4	0.6	0	0	36.0
13.10	3	6.3	0.9	48.0	14.5	22.6	1.2	0.1	6.3
1992	1	6.9	0.7	40.0	14.4	22.4	1.1	0.8	13.7

the population density (estimation 3) are also calculated. The corresponding obtained TPP contributions are given in Table 4.

CONCLUSIONS

The following main conclusions can be made from the above considerations:

1. The importance of the accounting for the SG effects caused by powerful sources in regions with complex topography is convincingly demonstrated.

2. The proposed relatively simple procedure, which consists of the solution of the long-range transport problem with exclusion of the sources from the studied SG area and in addition a more precise treatment (in a nested grid and accounting for the topography effects) of the pollution from these sources, can be successfully applied for accounting for the SG effects.

3. For regions with proved significance of the SG effects it may be appropriate the environmental loads also to be treated in a more precise way, according to the approach followed in this work:

$$Load = Load_{LR}^{red} + Load_{SG} \tag{4}$$

4. It seems reasonable the approach followed in this work to be applied to more extensive studies (regular operational simulations), which can outline the statistically significant manifestation of the discussed SG effects for long time periods.

REFERENCES

1. E.Syrakov, K.Ganev, N.Godev, E.Georgieva and E.Bogdanov, A telescopic approach for coordinated study of the air pollution at long-range and regional transport over a complex terrain, in: *Special Environmental Report No.17 "Changing Composition of the Troposphere" - Extended Abstracts of the WMO Technical Conference. Sofia, Bulgaria, 23 - 27.10.1989* , WMO No.724, 189 - 192 (1989)
2. K.Ganev, E.Syrakov and E.Georgieva, An example for the interaction of local and synoptic transport scales - a case study, *Swedish-Bulgarian Workshop on Long-Range Air Pollution, Sozopol 19-23 October, 1997.* (1997)
3. E.Syrakov, K.Ganev and N.Godev, A trajectory model for studying long range pollution transport in PBL., in: *Proceedings of the WMO conference on air pollution modelling and its application., Leningrad, USSR, 19-24.05.1986*, WMO/TD No 187, v.I, 226-239.(1987)
4. E.Sirakov, K.Ganev and N.Godev, Long-range transport of sulfur and nitrogen, taking into account the influence of the synoptic conditions, orography and arranged vertical motions. (in Bulgarian), *Bulgarian Geoph. J.* v.XIV, 1, 11-27 (1988)
5. E.Sirakov, K.Ganev and N.Godev, Estimation of the mutual pollution of different regions at distant and regional transport. (in Bulgarian), *Bulgarian Geoph. J.* v.XIV, 4, 10-21 (1988)
6. E.Syrakov and K.Ganev, Modeling of sulphur and nitrogen compounds transport within the PBL under nocturnal and convective conditions. (in Russian), *Z. Meteorol.*, 39, 2, 81 - 88.(1989)
7. E.Syrakov and K.Ganev, *Study of the Environmental Impact of the Thermal Power Plants* (in Bulgarian), Annex No.12, Energoproect Archives, Sofia (1993)
8. E.Syrakov and K.Ganev, Transport, chemical transformations and characteristics of pollution in local scales. (in Bulgarian), *Bulgarian Geoph. J.*, v. XX, No 3, 5-25 (1994)
9. K.Ganev and E.Syrakov, Treatment of chemical transformations in atmospheric pollution modelling. (in Bulgarian), *Bulgarian Geoph. J.*, v. XX, No 4, 77-89 (1994)
10. K.Ganev and E.Syrakov, 1995: Application of the method of functions of influence on the problem of reduction of thermal power plants effects on the environment. *Bulgarian Geoph. J.*, v. XXI, No 1, 12-23.(1995)
11. J.-P.Tuovinen, K.Barret and H.Styve, *Transboundary Acidifying Pollution in Europe: Calculated Fields and Budgets 1985-93*, Meteorological synthesizing center-WEST, Oslo (1994)
12. E.Georgieva and N.Godev N., 1987, Interpolation of wind data taking into account the relief-some results and problems (in Bulgarian), *Bulgarian Geoph. J.* v.VIII, 4 (1987)

DISCUSSION

M. SOFIEV: While considering fine resolution model in comparison of large scale one - what have you found to be more important: redistribution of load inside the large cells, or change of pollution distribution between them?

K. GANEV: The higher horizontal resolution, of course, produces more detailed picture of the air pollution. As it can be seen from fig.2. in some synoptic situations zones with concentrations much higher than those produced by the large scale model, can be formed. But the more important sub-scale effects concern the air pollution characteristics averaged for the large scale (EMEP) grid cells. As shown at tables 2 and 3, the sub-scale corrections may lead, depending on the synoptic conditions, to loads (concentrations, depositions) almost twice bigger, or twice smaller then those obtained for the same 150x150km cell by the large scale model. The estimated outflows from the cell are correspondingly two times smaller or bigger, when the sub-scale effects are accounted for. So, both effects can be registered - redistribution of load inside the large cells, and change of pollution distribution between them. Our opinion is that the more interesting conclusion from the present work is that the sub-scale phenomena may significantly effect the pollution exchange between the large scale cells.

MESO-SCALE MODELLING OF BASE CATION DEPOSITION IN SWEDEN

Christer Persson[1], Gun Lövblad[2], and Elisabet Roos[1]

[1]Swedish Meteorological and Hydrological Institute (SMHI)
SE-601 76 Norrköping, Sweden
[2] Swedish Environmental Research Institute (IVL)
P.O. Box 47086
SE-402 58 Göteborg, Sweden

INTRODUCTION

The development of emission reduction policies for acidifying deposition, based on the critical load approach which is done in e.g. Sweden, requires detailed information about the deposition of both acidifying compounds and base cations (Ca^{2+}, Mg^{2+}, Na^{+} and K^{+}). Extensive studies have been performed for the deposition of acidifying compounds, while the atmospheric inputs of base cations to the ecosystem are still rather poorly known. However, base cations play a very important role in the chemical processes involved in acidification (e.g. Draaijers et al. 1997b). Deposition of base cations reduces, in indirect ways, the acidification effects for two reasons; it may lead to a net increase in base saturation of soils and it also represents an input of alkalinity with capacity to neutralise acidic deposition. Besides, the concentration of base cations in soil solution influences the effect of nitrogen on vegetation.

Results from monitoring (Hedin et al. 1994) have shown that in some areas of Sweden and also in other parts of Europe base cations deposition has decreased markedly over the past 20 years. This has to a large degree offset the simultaneous decline in sulphur deposition.

Base cation deposition maps for Europe have been presented by Draaijers et al. (1997a) based on the so-called inferential modelling technique, where precipitation chemistry measurements and a pure deposition model not including any description of emission or horizontal atmospheric transport processes, is used.

In the present study the Eulerian atmospheric transport modelling system MATCH is applied to a study of the base cation deposition over Sweden. The calculations of the Swedish contributions are based on a survey regarding anthropogenic Swedish emissions of base cations and a model estimate of soil erosion and its contribution to base cations in the atmosphere.

Air Pollution Modeling and Its Application XIII, edited by S.-E. Gryning and E. Batchvarova, Kluwer Academic/Plenum Publishers, 2000

MODELLING SYSTEM

The MATCH modelling system consists in this application of three parts: a regional atmospheric dispersion model including modules for emission and deposition, an objective analysis system for air- and precipitation chemistry data, and an objective analysis system for meteorological data.

Dispersion Model

The MATCH (Multi-scale Atmospheric Transport and CHemistry) model is presented in detail in (Robertson et al. 1998). In the present study, however, an earlier, somewhat simplified and less computer time demanding version of the MATCH model (Persson et al. 1996) has been applied, which is in operational use for annual air pollution assessment studies of sulphur and nitrogen over Sweden. MATCH is a three dimensional Eulerian atmospheric dispersion model. It is a so called "off-line" model, meaning that it requires meteorological data from an external archive at regular time intervals (three hours) in order to calculate transport and deposition. The horizontal resolution in this application over Sweden is 20 km. The base cations have been separated into five different particle-size classes according to Table 2 below. For the time being the hygroscopic particle growth is not considered in the model calculations.

Deposition

Wet scavenging of the different species is proportional to the precipitation rate and a scavenging coefficient. In this first tentative study we have used the same scavenging coefficient value for all base cations and all five particle-size classes, which is put equal to twice the value for sulphate.

For the larger particle-size classes sedimentation velocities were determined from Stoke's Law and included in the model calculations.

Dry deposition is proportional to the concentration and a dry deposition velocity at 1m height, which includes both turbulent exchange and sedimentation of coarse particles. Since the lowest model layer has a height of 75 m, the turbulent part of the dry deposition flux calculation is transformed to the middle of that layer using standard similarity theory for the atmospheric surface layer. Dry deposition velocities are specified for each particle-size class and as a function of the surface characteristics (fraction forest, field etc).

Long-range transport

The dispersion model described above, combined with national emission estimates and meteorological data, provides daily estimates of concentrations in air and precipitation as well as dry and wet deposition of the simulated base cations. These results refer to contributions from sources within Sweden. To derive distributions of the contribution from sources outside Sweden we have employed the following method. Model calculated daily contributions from Swedish sources are deducted from observed daily values of concentration in air and precipitation at background locations on a point by point basis. Four stations in Sweden for air concentrations and about 40 stations for concentration in precipitation in and around Sweden are used. The residual is termed long-range transport contribution. These residuals are analyzed using an optimum interpolation method to give distributions of long-range transport contributions of concentrations in air and precipitation over the whole modelling domain. The basic idea behind this method is that the long-range transport contributions to the concentrations can be expected to vary more smoothly in space than the total concentrations, which are affected to some extent by local sources, and should therefore be more suitable for interpolation. Long-range transport of wet deposition is then calculated by multiplying with the detailed observed precipitation field. Long-range transported dry deposition is calculated by running the long-range transport air concentrations through the dry deposition module of the dispersion model.

The same calculation procedure has earlier been applied with good results to assessment studies of sulphur and nitrogen over Sweden (Persson et al, 1996). For those compounds comprehensive measured data is available and considerable efforts have been spent on quality control. For base cations we still need more measured data for a reliable quality control.

EMISSION

Emissions of base cations derive mainly from three types of sources: a) industrial and combustion plants (anthropogenic), b) mechanical and wind erosion of soil particles from areas of uncovered soil and unpaved roads (natural and "semi-natural") and c) sea-salt. Wind erosion affecting the base cation deposition over Europe in general would be dominated by emissions from arid regions of southern Europe and the Saharan region while soil emissions from Sweden, with a humid climate, must be very much smaller. For a couple of regions in Sweden the local soil emissions can, however, play a significant role and has therefor been included in our model calculations together with the Swedish anthropogenic emissions.

Industrial and combustion emission

The anthropogenic base cation emission data for Sweden (1995/96), based on updating of the report (Kindbom et al, 1993), includes point source data (forest, steel and cement industry) and provincial data (burning of wood and other bio-fuel, oil and coal). Also estimates of the particle-size distribution of the emissions have been made. The provincial data were distributed into the 20 x 20 km model grid, with respect to the sources of the emission. Hence, the emission from burning of wood were distributed proportional to the population living in forested areas, the emission retrieved from burning of oil were distributed proportional to the population, and from burning of coal and bio-fuel, proportional to the population living in urban areas. The emissions used in the present study are given in Table 1 and estimated particle-size distributions are briefly indicated in Table 2.

Table 1. Anthropogenic base cation emissions (tonnes/year) from Sweden 1995/96

Source type	Calcium	Magnesium	Sodium	Potassium
Point sources				
Cement industry	230	5	5	5
Forest industry	480	30	820	30
Steel industry	80	30	20	25
Area sources				
Wood/bio-fuel	2200	210	65	540
Oil	180	45	265	30
Coal	120	25	25	10

The uncertainty for data in Table 1 is estimated to be less than a factor of two except for wood burning, for which it has been argued that the Swedish emissions could be as much as a factor of 5 larger than what is given in Table 1.

Table 2. Examples of estimated particle-size distribution (% per size-class) for some of the more important sources

Size	Burning of wood				Point sources, forest industry			
(μm)	Ca	Mg	Na	K	Ca	Mg	Na	K
0 – 2.5	60	60	60	60	75	65	70	65
2.5 – 6.0	10	10	10	10	4	4	5	4
6.0 – 10	10	10	10	10	3	5	3	5
10 – 15	10	10	10	10	3	3	3	3
> 15	10	10	10	10	15	23	19	23

The contribution from sea-salt has been estimated using measured sodium concentrations in precipitation and air assuming that sea-salt is the only source of Na^+ in air and precipitation. The sea-salt contribution to Mg^{2+}, Ca^{2+} and K^+ can then be expressed as $f_x[Na^+]$, where f_x equals 0.113, 0.0218 and 0.212 respectively, on a molar basis, and $[Na^+]$ is the sodium concentration. The effect of other sources than sea-salt for $[Na^+]$ over Sweden is discussed below.

Wind erosion emission

The present emission estimate includes only "natural" wind erosion from bare soils, while emissions caused by traffic on unpaved roads and agriculture activities, for the time being, are excluded. Several empirical studies show that the amount of uplifted dust from bare soils is depending on the wind speed or the friction velocity, with a threshold wind velocity at which the erosion starts. Global modelling of the dust flux have been presented by i.e. Tegen and Fung (1994). Gillette and Passi (1988) have discussed a model for estimates of the total dust production for the United States.

For the present meso-scale application of the MATCH model the methods given by Marticorena and Bergametti (1995) have been used. The following equations were introduced into the MATCH model:

Horizontal flux: $G = Cu_*^3(\rho_a/g)(1+u_{*t}/u_*)(1-u_{*t}^2/u_*^2)$ (1)

where C is a constant of proportionality with a value of 2.61, u_* = friction velocity, u_{*t} = threshold friction velocity, ρ_a = air density and g = acceleration of gravity.

Vertical flux: $F = \alpha G$ (2)

where $\alpha = 10^{-2} - 10^{-4}$ (in SI-units) for arid and semiarid areas.

The vertical flux, which is describing the emission of dust particles available for long-range transport, is thus put proportional to the horizontal flux. The horizontal flux calculations have been restricted to periods of dry weather, temperatures above zero and bare soil. For northern Europe very little information about the α-value is available. However, it is probably somewhat lower than for arid areas (Jönsson, 1994). In the present simulation α-values of $10^{-4} - 10^{-5}$ have been used.

The friction velocities available in the MATCH system have for the soil erosion calculations been re-calculated, using similarity theory, to represent wind-profiles over open fields with a roughness length of 0.5 cm.

Soil types in southern Sweden, for which erosion damage on agricultural crops have been reported (Nihlen, 1990), have been selected as possible sources for soil erosion emission in Sweden. The threshold friction velocities have been determined for these soil types in a rough way, based mainly on reported erosion and wind measurements in southern Sweden (Jönsson, 1994). The threshold friction velocities applied in the present model calculations are given in Table 3. Lack of quantitative measurements of the erosivity of the Swedish soils causes, however, substantial uncertainties.

Table 3. Soil types and threshold friction velocities applied in the calculations

Soil type	Threshold friction velocity (m/s)
Sandy soils	0.43 – 0.90
Fluvial sediments	0.50 – 2.00

The number distribution of dust particles with regard to particle radius, r, was assumed to be described by $dN/d\log r \sim r^{-2}$. No direct information for Swedish conditions is available. The corresponding mass distribution follows $dM/d\log r \sim r$. This expression then was used to distribute the soil emissions into the different particle-size classes in the MATCH model. The base cation content in the soil erosion was assumed to be 15mg/100g for Mg^{2+}, Na^+, K^+ and 220mg/100g for Ca^{2+}. These values are based on information on the cation exchange capacity for agricultural soils in southern Sweden.

Table 4. Preliminary model estimates of base cation emissions (tonnes/year) from Sweden 1996 caused by wind erosion

Calcium	Magnesium	Sodium	Potassium
20 – 200	1.5 - 15	1.5 - 15	1.5 – 15

The emission values in Table 4 are small compared to the total anthropogenic emissions from Sweden. However, the soil erosion emissions occur from very small areas and during only a few episodes per year. During 1996 only soil emission were obtained for the months April, May and to a small extent in September. Therefor, locally and during shorter periods the erosion can be of importance in Sweden.

RESULTS AND DISCUSSION

Examples of results for base cations over Sweden for 1996 are presented below. The results are, however, still tentative due to uncertainties in emission data, roughly estimated sink parameters and lack of atmospheric chemistry measurements. For example Swedish measurements of air concentrations for 1994 had to be used, since no later measurements are available.

Figure 1 shows a comparison between Swedish contribution, long-range transport and total wet+dry deposition of non-sea-salt Ca^{2+}. The Swedish contribution is calculated to be about 6-10% of the total deposition in most parts of southern Sweden and along the eastern coast. In areas with extensive wood burning values up to 15% are obtained. In northern Sweden, except along the coast, the Swedish contributions are less than a few percent. The highest total deposition values for non-sea-salt Ca^{2+} are found in south-eastern Sweden near the coasts.

Results for potassium are showed in Figure 2. The dry and the wet+dry depositions of non-sea-salt K^+ are compared. From these first tentative calculations the dry deposition should account for about 15 - 20% of the total deposition. This might, however, be an underestimate especially in forested areas, since we have assumed all long-range transport particles to be included in the two smallest particle-size classes. Semb et al. (1995) have showed air concentration measurements of base cations with a larger fraction in the coarse mode compared to our assumptions. Further studies of this will be performed. Comparisons of the non-sea-salt deposition with deposition including sea-salt, indicate that on the Swedish west-coast and in the westerly parts of the mountains in northern Sweden about 30% of the K^+ deposition is caused by sea-salt. In parts of eastern and central Sweden the value is less than 10%. The highest total deposition values for potassium, as well as for magnesium, are found in south-western and south-eastern Sweden near the coasts.

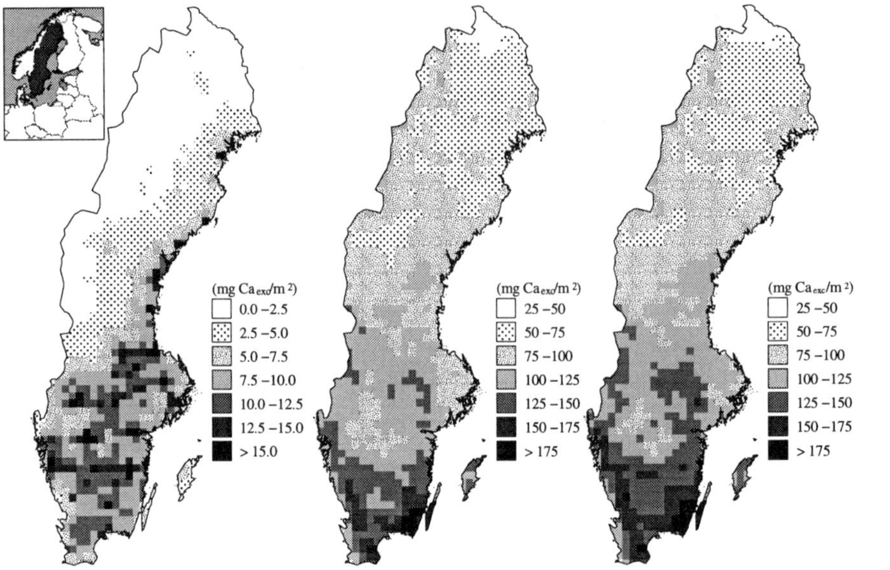

Figure 1. Calculated annual 1996 wet+dry deposition (mg/m^2) of non-sea-salt Ca^{2+} a) Swedish contribution, b) long-range transport and c) total.

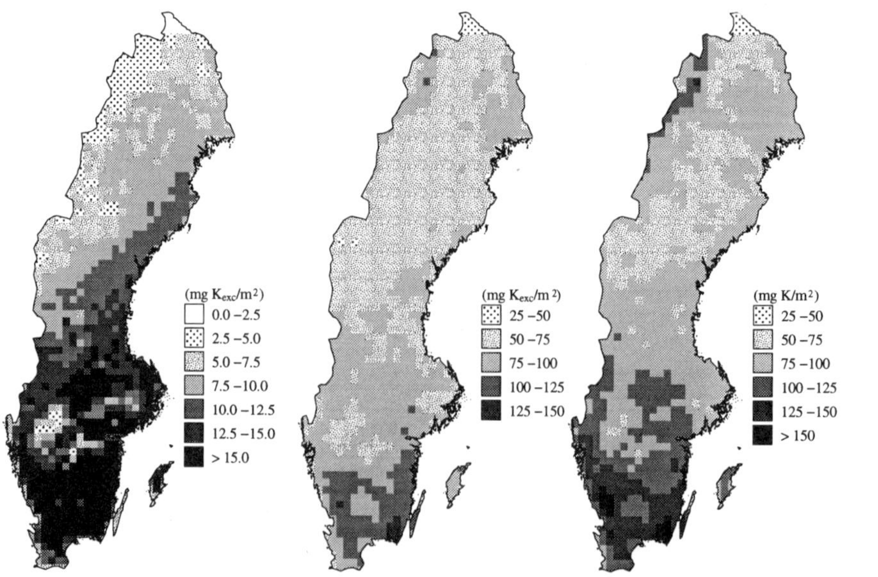

Figure 2. Calculated annual 1996 deposition (mg/m^2) of K^+: a) dry non-sea-salt, b) wet+dry non-sea-salt and c) wet+dry including sea-salt.

In Figure 3 the wet+dry depositions of non-sea-salt Mg^{2+}, of Na^+ and of the anion Cl^-, calculated in a similar way as for cations, are presented. The calculations indicate that Swedish anthropogenic sources account for less than 1 % of the Na^+-deposition on the Swedish western coast and in the westerly parts of the mountains in northern Sweden. On the eastern coast in northern Sweden the corresponding figures are 3-7 %. The non-sea-salt contribution to Na^+ is here mainly caused by the forest industry. This means that for the Swedish eastern coast at least 3-7 % of what is normally estimated to be sea-salt should not be regarded as such. The largest deposition values of sea-salt Na^+ and Cl^- occur in south-western Sweden and in the north-westerly mountains. However, the number of atmospheric chemistry stations is too small and the resolution of the model is too coarse to describe in detail the local gradients of sea-salt along the coast.

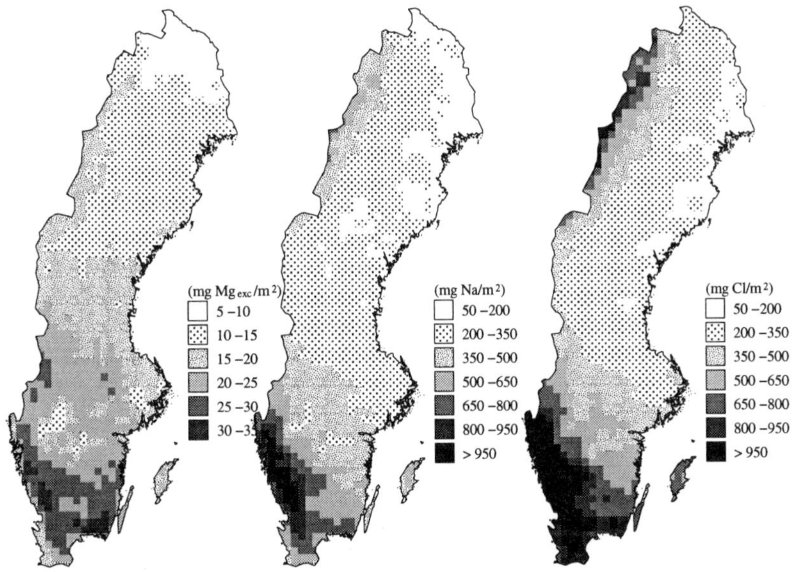

Figure 3. Calculated annual 1996 deposition (mg/m2): a) wet+dry non-sea-salt Mg^{2+}, b) wet+dry Na^+ and c) wet+dry Cl^-.

ACKNOWLEDGMENTS

This work has received financial support from the Swedish Environmental Protection Agency. Precipitation chemistry data from background stations in Norway and Finland were provided by Norwegian Institute for Air Research and Finnish Meteorological Institute, respectively. Karin Kindbom, IVL, has been much involved in the work concerning the Swedish base cation emissions

REFERENCES

Draaijers, G.P.J., van Leeuwen, E.P., De Jong, P.G.H. and Erisman, J.W. (1997a). Base cation deposition in Europe – Part I. Model description, results and uncertainties. *Atmospheric Environment Vol. 31, No. 24, 4139-4158.*

Draaijers, G.P.J., van Leeuwen, E.P., De Jong, P.G.H. and Erisman, J.W. (1997b). Base cation deposition in Europe – Part II. Acid neutralization capacity and contribution to forest nutrition. *Atmospheric Environment Vol. 31, No. 24, 4159-4168.*

Gillette, D.A. and Passi, R. (1988) Modeling dust emission caused by wind erosion. *J. Geophys. Res.*, **93,** 14,233-14,242.

Hedin, L.O., Granat, L., Likens, G.E., Buishand, T.A., Galloway, J.N., Butler, T.J. and Rodhe, H. (1994). Steep declines in atmospheric base cations in regions of Europe and North America. *Nature* **367**, 351-4.

Jönsson, P. (1994) Wind Cimate During the Instrumental Period and Recent Wind Erosion in Southern Scandinavia. *Meddelanden från Lunds universitets Geografiska Institutioner, avh. 121. Lund University Press.*

Kindbom, K., Sjöberg, K. and Lövblad, G. (1993). *Beräkning av ackumulerad syrabelastning från atmosfären. Delrapport 1: Emissioner av svavel, kväve och alkaliskt stoft i Sverige 1900 - 1990. IVL Rapport B 1109.*

Nihlen, T. (1990). Eolian Processes in Southern Scandinavia and the Mediterranean Area. *Meddelanden från Lunds universitets Geografiska Institutioner, avh. 110. Lund University Press.*

Persson, C., Langner, J. and Robertson, L. (1996) Air pollution assessment studies for Sweden based on the MATCH model and air pollution measurements. *Air Pollution Modeling and Its Application,* Vol **XI**. Eds Gryning S-E and Schiermeier F.A., Plenum Press, New York and London.

Robertson, L., Langner J. and Engardt M. (1998) *An Eulerian limited area atmospheric transport model. J. Applied Meteorology, accepted for publication.*

Semb, A., Hanssen, J.E., Francois, F., Maenhaut, W. and Pacyna, J.M. (1995) Long range transport and deposition of mineral matter as a source for base cations. *Water, Air and Soil Pollution **85**: 1933-1940.*

Tegen, I. and Fung, I. (1994). Modelling of mineral dust in the atmosphere sources, transport, and optical thickness. *Journal of Geophysical Research* **99**, 22.897-22.914.

DISCUSSION

G. S. RIVIN: You have two kinds of meteorological data. Why do you not use the aerological data? What is the method assimilation data you use?

C. PERSSON: In order to obtain the best possible quality of the input meteorological data, which is used to drive the atmospheric transport and dispersion model, we use two methods.

a) For near ground parameters an optimum interpolation technique is used, which makes use of all available ground-based routine meteorological observations. That method is used to derive a number of parameters including near ground horizontal wind and temperature, precipitation, friction velocity etc.

b) For 3-D wind and temperature fields we utilise meteorological analyses based on data assimilation or short (3h) forecasts from the operational numerical weather prediction model HIRLAM. In this data assimilation all available aerological observations within the model area are included.

The data assimilation within the HIRLAM system produces initial model states that not only makes optimal use of observational data, but also satisfies physical laws and is free of undesirable noise.

AREA OF INFLUENCE: IDENTIFYING REGIONS WHOSE SOURCES POTENTIALLY IMPACT DOWNWIND AIR QUALITY

James G. Wilkinson and Yueh-Jiun Yang

School of Civil and Environmental Engineering
Georgia Institute of Technology
Atlanta, GA 30332-0512 USA

INTRODUCTION

The process to develop emissions control strategies has always been hindered by the lack of knowledge about which sources impact air quality. "Expert" knowledge about which sources contribute to air quality problems has been used in past studies to guide decision makers in their efforts to develop effective emissions control strategies. In some cases, expert knowledge has been supplemented with information derived from brute force sensitivity analyses which indicate the emissions sources having the greatest impact on downwind air quality.

In this study, a method to identify the area of influence (AOI) for various subregions is developed and applied to the Los Angeles, California urban-scale airshed which is also known as the South Coast Air Basin (SoCAB). The AOI method is based on the decoupled direct method in three dimensions (DDM-3D) which allows for efficient computation of sensitivity coefficients to many air quality model parameters during a model run (Yang et al., 1998). In summary, sensitivity coefficients are computed at various points in the airshed due to small changes in the VOC and NO_X emissions, and the resulting sensitivity field is inverted which produces the area of influence. Based on this preliminary AOI, decision makers will be able to more effectively target emissions reductions programs in the emissions source regions which have the greatest impact on reducing peak ozone.

METHODS

The development of DDM-3D is briefly reviewed here. The evolution of chemical reacting species is described by the atmospheric diffusion equation (ADE), Equation 1, which is subject to initial conditions, Equation 2, and boundary conditions, Equations 3-6 (McRae et al., 1982a).

Air Pollution Modeling and Its Application XIII, edited by S.-E. Gryning
and E. Batchvarova, Kluwer Academic/Plenum Publishers, 2000

$$\frac{\partial c_i}{\partial t} = -\nabla(\mathbf{U}c_i) + \nabla(\mathbf{K}\nabla c_i) + R_i + S_i \qquad i = 1, 2, ..., N \tag{1}$$

$$c_i(t_o) = c_i^o \tag{2}$$

$$\mathbf{u}c_i - \mathrm{K}\nabla c_i = \mathbf{u}c_i^b \qquad \text{inflow boundary condition} \tag{3}$$

$$-\nabla c_i = 0 \qquad \text{outflow boundary condition} \tag{4}$$

$$v_g^i c_i - K_{zz}\frac{\partial c_i}{\partial z} = E_i \qquad \text{surface level } (z=0) \text{ boundary condition} \tag{5}$$

$$-\frac{\partial c_i}{\partial z} = 0 \qquad \text{domain top } (z=H) \text{ boundary condition} \tag{6}$$

$c_i(\boldsymbol{x},t)$ is the ensemble average concentration of species i at time t and location $\boldsymbol{x}$; $\mathbf{u}(\mathbf{x},t)$ is the wind field; $\boldsymbol{K}(\boldsymbol{x},t)$ is the second order turbulent diffusivity tensor; $R_i[c_1, c_2, ..., c_n, T, t]$ is the net rate of chemical production which is dependent on temperature (T); S_i is the elevated source rate; and N is the total number of species. c_i^o is the initial concentration of species i at time $t=0$; c_i^b is the concentration of species i at the lateral boundaries; v_g^i is the dry deposition velocity of species i; E_i is the surface level emissions rate of species i; and H is the top of the modeling domain.

The local sensitivity of a species to a model parameter is represented by Equation 7.

$$s_{ij}(t) = \frac{\partial c_i(t)}{\partial p_j} \tag{7}$$

$s_{ij}(t)$ is the local sensitivity of species i to model parameter (e.g. rate constant) or model input (e.g. on-road mobile source VOC emissions) p_j at time t. Because model parameters and model inputs can have large scale variations in space and time, a seminormalized sensitivity coefficient is computed, Equation 8.

$$s_{ij}^*(t) = P_j\frac{\partial c_i(t)}{\partial p_j} = P_j\frac{\partial c_i(t)}{\partial(\varepsilon_j P_j)} = \frac{\partial c_i(t)}{\partial \varepsilon_j} \tag{8}$$

P_j is the nominal value of the model parameter or model input, p_j, in time and space which is varied by a factor ε_j. By substituting Equation 8 into Equations 2-6, an auxiliary set of atmospheric sensitivity equations (ASEs), Equations 9-14, is obtained.

$$\frac{\partial s_{ij}^*}{\partial t} = -\nabla(\mathbf{u}s_{ij}^*) + \nabla(\mathbf{K}\nabla s_{ij}^*) + \mathbf{J}_{ik}s_{kj}^* + \frac{\partial R_i}{\partial \varepsilon_j} + \frac{\partial S_i}{\partial \varepsilon_j} - \nabla(\mathbf{u}c_i)\delta_{ij} + \nabla(\mathbf{K}\nabla c_i)\delta_{ij} \tag{9}$$

$$s_{ij}^*(t_o) = c_j^o\delta_{ij} \tag{10}$$

$$\mathbf{u}s_{ij}^* - \mathbf{K}\nabla s_{ij}^* = \mathbf{u}c_i^b - \mathbf{u}c_i\delta_{ij} + \mathbf{K}\nabla c_i\delta_{ij} \qquad \text{inflow boundary condition} \tag{11}$$

$$-\nabla s_{ij}^* = 0 \qquad \text{outflow boundary condition} \tag{12}$$

$$v_g^i s_{ij}^* - K_{zz}\frac{\partial s_{ij}^*}{\partial z} = v_g^i c_i \delta_{ij} + K_{zz}\frac{\partial c_i}{\partial z} + E_i \delta_{ij} \quad \text{surface level } (z=0) \text{ boundary condition} \qquad (13)$$

$$-\frac{\partial s_{ij}^*}{\partial z} = 0 \qquad \text{domain top } (z=H) \text{ boundary condition} \qquad (14)$$

J is the Jacobian matrix which is defined by $\mathbf{J}_{ij} = \frac{\partial R_i}{\partial c_k}$; δ_{ij} is a binary variable having a value of one if the parameter, p_j, is in the term and zero otherwise. By Equation 9, the sensitivity coefficients are coupled and are linearly dependent on the concentrations. However, because the concentrations do not depend on the sensitivity coefficients, the sensitivity equations can be solved separate from the ADE over the same time step. The numerical implementation of the ASE is discussed in Yang et al. (1998).

The AOI method uses the sensitivity coefficients that are computed via DDM-3D. In the current implementation of DDM-3D in the Caltech/Carnegie Institute of Technology (CIT) airshed model, the CIT computes temporally and spatially resolved sensitivity coefficients of all species concentrations to surface level VOC and NO_X emissions at user-defined points in the modeling domain (McRae et al. 1982b, Russell et al. 1988, McRae et al. 1992, Yang et al. 1998). To estimate the AOI, for example at the location of the predicted peak ozone, the sensitivity field is first rotated about this point via Equations 15 and 16.

$$x_o' = x_o + x_i - x_i' \qquad (15)$$

$$y_o' = y_o + y_i - y_i' \qquad (16)$$

(x_o', y_o') is the location of the rotated point; (x_o, y_o) is the location of the predicted peak ozone; (x_i, y_i) is the location where the emissions are perturbed; and (x_i', y_i') is the location where the sensitivity is estimated. Figure 1 graphically depicts this rotation.

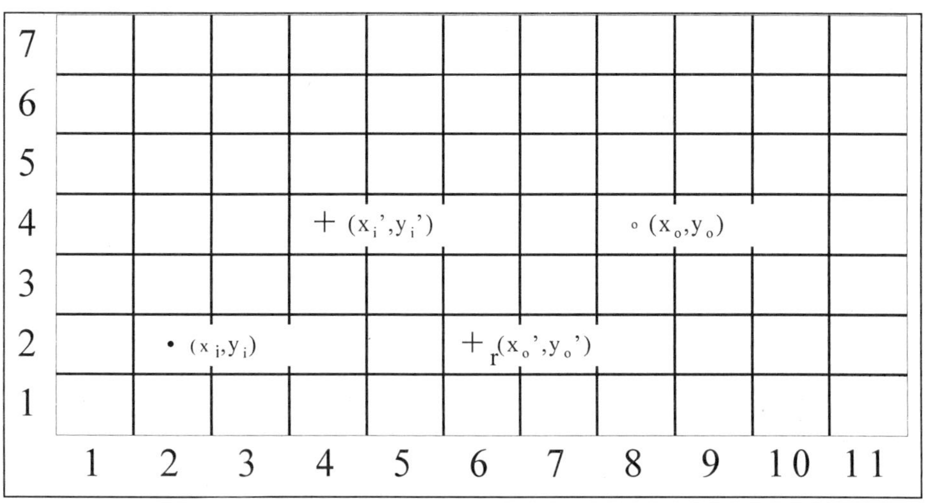

Figure 1. Rotation of the sensitivity field. Known locations are as follows: • (x_i,y_i), the point where emissions are perturbed; $+(x_i',y_i')$, the point where the sensitivity coefficient is estimated due to emissions at (x_i,y_i); and ⑤ (x_o,y_o), the location, for example, of predicted peak ozone. The rotated point is $+_r$ (x_o',y_o').

Once the field is rotated, the sensitivities at the rotated points (i.e. $+_r$ at $[x_o', y_o']$) are estimated via Equations 17-19.

$$s_r(x_o', y_o') = \sum_{i=1}^{n} s_i(x_i', y_i') \bullet w_i \tag{17}$$

$$w_i = \frac{\frac{1}{d_i^2}}{\sum_{i=1}^{n} \frac{1}{d_i^2}} \tag{18}$$

$$d_i^2 = (x_o - x_i)^2 + (y_o - y_i)^2 \tag{19}$$

s_r is the rotated sensitivity value at (x_o', y_o'); n is the number of sensitivities that have been computed at user-defined locations in the domain to surface level emissions; s_i is the sensitivity at (x_i', y_i'); w_i is the sensitivity weighting factor; d is the distance from location (x_i, y_i), the point where the emissions are perturbed, to location (x_o, y_o), the location, for example, of predicted peak ozone.

RESULTS AND DISCUSSION

The South Coast Air Basin (SoCAB) of Los Angeles, California is the only air basin in the United States to be classified in *extreme* nonattainment of the NAAQS for ozone which is 0.12 ppm on a one hour average basis (CFR, 1995). An air basin is in *extreme* nonattainment of the ozone NAAQS when the fourth highest, one hour average ozone value exceeds 0.28 ppm over three contiguous years. Although the number of days where ozone values above 0.12 ppm continues to decline in the SoCAB (e.g. 179 days in 1992, 147 days in 1993, 99 days in 1995 (SoCAB, 1998, 1994a)), the SoCAB continues to record the greatest number of days exceeding the ozone NAAQS of any area in the United States (SoCAB, 1994b). This is despite significant public and private sector efforts to improve air quality in the SoCAB over the last thirty plus years.

Because significant effort has been expended only to see improving air quality and not attainment of air quality standards, it seems prudent to investigate other, promising methods that may be applied to better understand the air pollution problem in the SoCAB and elsewhere. Here, the AOI method is applied to the 27-29 August 1987 ozone episode for the SoCAB with the goal of determining where the emissions sources are located which have the greatest impact on any arbitrary point in the modeling domain. As indicated earlier, the CIT was the photochemical air quality model that was used in this analysis. The inputs to the CIT for the 27-29 August 1987 episode are described in detail by Harley et al. (1992).

Figure 2 shows the predicted surface ozone field on 29 August 1987 at 1500 hours local time. The predicted peak ozone, 175 ppb, occurred at grid cell (59,27). Nine sensitivities were computed to VOC emissions. Figure 3 shows the location of the points where the VOC surface level emissions were perturbed. The nine ozone sensitivity fields to the VOC surface level emissions were rotated about grid cell (56,22), and the rotated sensitivities were estimated per the method that was previously described. The rotated sensitivities were plotted, and the resulting AOI on the predicted peak ozone was computed, Figure 3.

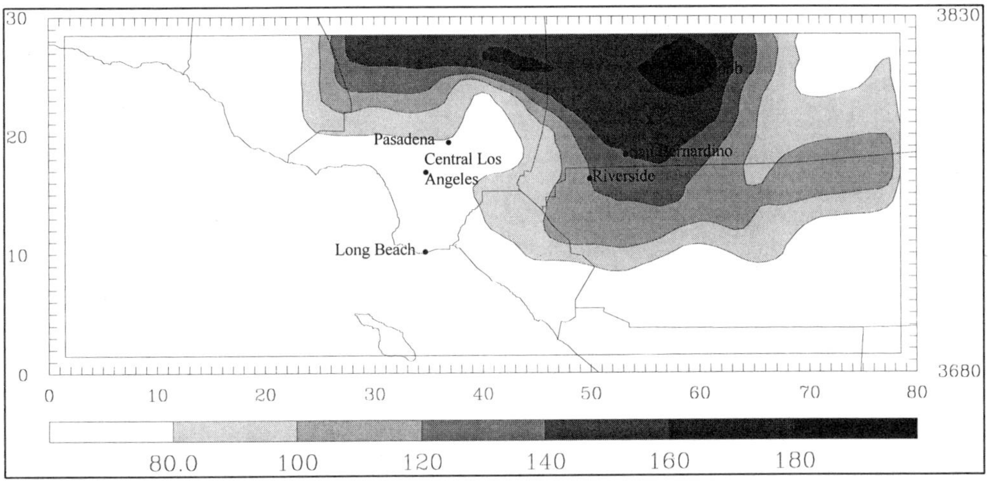

Figure 2. Predicted ozone field (ppb) at 1500 hours on 29 August 1987. Peak ozone at this hour was 175 ppb located at (59,27). x indicates location where AOI is computed (56,22).

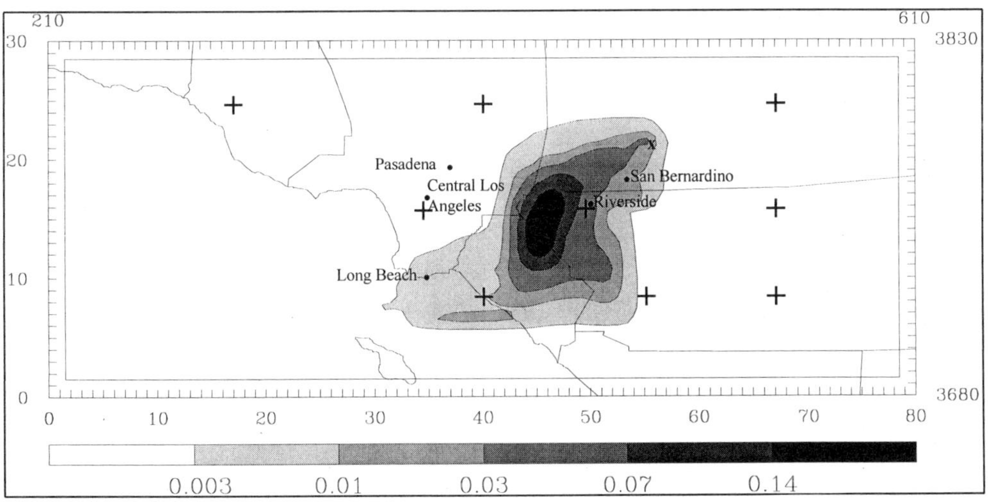

Figure 3. AOI on location (56,22), the predicted peak ozone, to ground level VOC emissions. + indicates a point where VOC emissions were perturbed to compute ozone sensitivity field. Results are presented in terms of ppb increase in ozone at grid cell (56,22) per one percent change in VOC emissions.

As formulated, ground level VOC emissions just east and south of Riverside and north and east of Long Beach have the greatest impact on the ozone concentration at (56,22). Ozone at (56,22) will increase on the order of 0.14 to 0.21 ppb per one percent increase in surface level VOC emissions emitted from all ground level sources in the area. Another way to interpret the results is that peak ozone will decrease 0.14 to 0.21 ppb per one percent decrease in surface level VOC emissions emitted from all ground level sources in the area.

LIMITATIONS

Also, the technique shares similar limitations to the "parent" model, CIT. While the accuracy of the sensitivity fields are dependent upon the accuracy of the inputs to the CIT, similar to the resulting air quality fields, initial tests suggest that the sensitivity fields do not vary as much due to changes in the emission fields. This is not surprising given the sensitivity work done on simpler systems (e.g. Bergin et al., 1995; Cho et al., 1987; Carmichael et al., 1997; Vuilleumier et al., 1997). This has important implications for developing AOIs. The sensitivity fields developed, particularly on a relative scale, are insensitive to the details of the basecase emissions. (Relative, in this case, refers to finding how one ton of emissions of a specific source type impacted air quality across the domain relative to a similar ton of emissions at another location.) Thus, the resulting AOI is more robust than might be found using other techniques.

The method is a local sensitivity technique, and is theoretically limited to assessing how small perturbations in the emissions impact air quality. However, in practice large perturbations in the emissions can be assessed since the sensitivity fields, as discussed above, do not vary that rapidly as the net base emissions fields are varied. This is in line with the results of Harley et al. (1997) and Chang et al. (1989). The size of the perturbation that can be treated depends upon the particular source in question. Results from Chang et al. (1989), suggest that if the source does not contribute more than about 30% to the total local emissions, the technique is very quantitative. Also, if the question being addressed is how specific source controls impact air quality (e.g., a 50% reduction in a specific source), the method should be very quantitative.

However, an exception may occur if the source is a very large NO_X point source, and the question is "how much is air quality impacted if that source is removed (i.e. shutdown)?" -- this issue is currently being explored. The study to date is actually clouded by the inaccuracies in how to quantify the impact of an individual point source. First, the traditional calculation of how an individual point source impacts air quality suffers from the same problems as the brute-force sensitivity technique: accurately finding the difference between large numbers in a calculation that has some numerical noise. Typically, air quality models show very little difference if one, large point source is added or subtracted from the emissions inventory. Thus, the absolute magnitude of the difference can have significant error, even when using a Plume-in-Grid (PIG) method. Second, current advection algorithms, particularly Smolarkiewicz (Smolarkiewicz, 1983) and Accurate Space Derivative (Chock, 1991; Dabdub and Seinfeld, 1994), though other techniques as well, can have significant problems in accurately treating a large, isolated source (Odman et al., 1996). It is not a *priori* obvious which approach, brute force removing a single point source or the direct sensitivity approach gives a more quantitative estimate of how removing that source would impact air quality, though the tests to date suggest that the sensitivity approach is more accurate, particularly for small sources.

CONCLUSION

A technique to identify an area of influence (AOI) about an arbitrary point has been developed which is based on the decoupled direct method in three dimensions (DDM-3D).

The AOI method has been applied to the 27-29 August 1987 ozone episode in Los Angeles, California to elicit an AOI to surface level VOC emissions. Further, limitations in the method have been discussed.

Future work will include further analysis of AOI sensitivities to large scale perturbations in the emissions inventory. Also, the AOI technique will be extended to regional-scale domains. Finally, the use of the AOI technique will be extended to apply to other air quality metrics such as eight hour average ozone.

REFERENCES

Bergin, M.S., Russell, A.G., Yang, Y.J., Milford, J.B., Kirchner, F. and Stockwell, W.R., 1995, *Effects of Uncertainty in SAPRC90 Rate Constants and Product Yields on Reactivity Adjustment Factors for Alternatively Fueled Vehicle Emissions, An Integrated Approach Using Formal Sensitivity and Uncertainty Analysis and Three-Dimensional Air Quality Modeling*, report prepared for the National Renewable Energy Laboratory (NREL) of the Department of Energy, Golden, CO, June.

Carmichael, G.R., Sandu, A. and Potra, F.A., 1997, Sensitivity Analysis for Atmospheric Models via Automatic Differentiation, *Atmos. Environ.*, **31**, 475-489.

CFR, 1995, *Code of Federal Regulations* (40 CFR Part 91), Office of the Federal Register, National Archives and Records Administration, U.S. Government Printing Office, Washington, DC.

Chang, T.Y., Rudy, S.J., Kuntasal, G. and Gorse, R.A., Jr., 1989, Impact of Methanol Vehicles on Ozone Air Quality, *Atmos. Environ.*, **23**, 1629-1644.

Cho, S-Y., Carmichael, G.R. and Rabitz, H., 1987, Sensitivity Analysis of the Atmospheric Reaction-Diffusion Equation, *Atmos. Environ.* **12**, 2589.

Chock, D. P., 1991, A comparison of numerical methods for solving the advection diffusion equation-III, *Atmospheric Environment*. 25A(5/6):853-871.

Dabdub, D. and Seinfeld, J. H., 1994, Numerical advective schemes used in air quality models - sequential and parallel implementation, Atmospheric Environment. 28(20):3369-3385.

Harley, R. A., Russell, A. G., McRae, G. J., McNair, L. A., Winner, D. A., Odman, M. T., Dabdub, D., Cass, G. R., and Seinfeld, J. H., 1992, Continued Development of a Photochemical Model and Application to the Southern California Air Quality Study (SCAQS) Intensive Monitoring Periods: Phase I, Final Report to the Coordinating Research Council Project: SCAQS - 8.

McRae, G. J., Goodin, W. R., and Seinfeld, J. H., 1982a, *Atmospheric Environment*, 31, 475.

McRae, G. J., Goodin, W. R., and Seinfeld, J. H., 1982b, Development of a Second Generation Mathematical Model for Urban Air Pollution, *Atmospheric Environment*, 16, 679.

McRae, G. J., Russell, A. G., and Harley, R. A., 1992, CIT Photochemical Airshed Model - Data Preparation Manual, February 1992.

Russell, A. G., McCue, K. F., and Cass, G. R., 1988, Mathematical Modeling of the Formation of Nitrate Containing Pollutants - I. Evaluation of an Eulerian Model, *ES&T*. 22, 263.

Odman, M.T., Wilkinson, J.G., McNair, L.A., Russell, A.G., Ingram, C.L. and Houyoux, M.R., 1996, *Horizontal Advection Solver Uncertainty in the Urban Airshed Model*, report prepared for California Air Resources Board, Sacramento, CA.

Smolarkiewicz, P. K., 1983, A simple positive definite advection transport scheme with small implicit diffusion, *Mon. Wea. Rev.* 111:479-486.

SoCAB, 1998, Air Quality Management Plan. http://www.aqmd.gov/aqmp/97aqmp/m-exec.html (5 July).

SoCAB, 1994a, Air Quality Trends, 1976-1993. Appendix II-B, September 1994 Air Quality Management Plan, South Coast Air Quality Management District, Diamond Bar, CA.

SoCAB, 1994b, Summary of Air Quality, 1990-1993. Appendix II-A, April 1994 Air Quality Management Plan, South Coast Air Quality Management District, Diamond Bar, CA.

Vuilleumier, L, Harley, R. and Brown, N., 1997, First- and Second- Order Sensitivity Analysis of a Photochemically Reactive System (a Greens Function Approach), *ES&T.* , **31**, 1206-1217.

Yang, YJ, Wilkinson, J. G., and Russell, A. G., 1998, Fast, Direct Sensitivity Analysis of Multidimensional Photochemical Models, *Environmental Science & Technology*. 31:10.

DISCUSSION

Z. ZLATEV: Could you extend your technique for finding optimal solutions? The optimal solution is here defined as finding which sources are to be reduced and by how much they are to be reduced in order to achieve the required reduction of the ozone concentrations in the cheapest way.

J. G. WILKINSON: Yes. Since the technique can quantify the extent to which sources contribute to downwind air quality problems, it would be feasible to use this information to develop an optimal emissions control strategy.

E. GENIKHOVICH: Why did you not use the adjoint operator technique which was applied to the solution of similar problems by Marchuk in the 1970s?

J. G. WILKINSON: At the time that this work was being conducted, we were not familiar with the work of Marchuk. We are currently evaluating Marchuk's techniques to see what value that they may lend to analyses such as this; however, it is likely that the application of any adjoint technique to a fully prescribed three dimensional, photochemically active air quality model would be difficult.

MULTIBLOCK DISPERSION MODELLING FOR URBAN AND REGIONAL SCALE INTERACTIONS

Clemens Mensink and Koen De Ridder

VITO, Centre for Remote Sensing and Atmospheric Processes (TAP)
Boeretang 200
B-2400 Mol
Belgium

1. INTRODUCTION

One of the challenges in regional scale air quality modelling is to represent the distribution of pollutant concentrations more accurately at locations where this is required, e.g. in urban areas, near concentrated emission sources, near strong gradients in the concentration field or in an area of particular interest, as defined by the user. A more detailed computation of concentration distributions is also needed when model results are compared with measurements from monitoring stations at locations corresponding with fixed points in the computational domain.

A typical regional scale air quality model provides a piece-wise constant concentration distribution over grid cells with dimensions of typically 60 x 60 km² (Mensink and Dumont, 1996). When these results are to be compared with measurements, the question arises which monitoring station(s) located within the 60 x 60 km² grid cell should be selected in order to be representative for that particular grid cell. This situation is illustrated in Figure 1, showing SO_2 monitoring stations in the Flanders region, with 22 stations in and around Antwerp located within an area bounded by the dimensions of a typical grid cell. The yearly averaged SO_2 concentration measured in this region varied in 1995 from 14 to 42 μg/m³ depending on the monitoring station selected, which demonstrates the large concentration gradient existing in the area.

It is obvious that grid refinement can help to reduce the gradient over an individual grid cell, so that a comparison of the measured concentrations with the modelled values, being piece-wise constant over the grid cell, can be carried out in a more accurate way. From a numerical point of view, grid refinement helps to reduce the amount of unnatural diffusion introduced by the advection scheme and by smearing out the emissions over the entire grid cell area. This implicitly introduced numerical diffusion contributes to the total diffusion in the model which, in contrast, is often parameterised explicitly. The numerical diffusion error introduced by the advection scheme is a function of Courant number and wavelength and can be reduced by means of grid refinement (Talat Odman, 1997).

Air Pollution Modeling and Its Application XIII, edited by S.-E. Gryning and E. Batchvarova, Kluwer Academic/Plenum Publishers, 2000

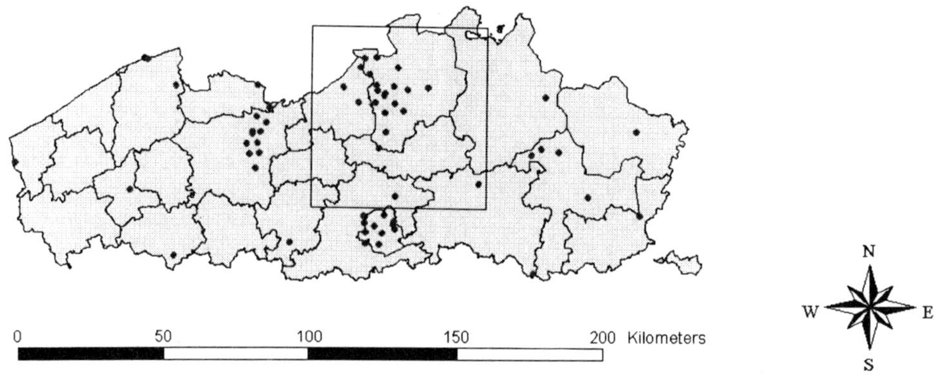

Figure 1. SO_2 monitoring stations inside a 60 x 60 km² grid cell.

Uniform refinement of the spatial resolution would result in too many grid points leading to unacceptable high computational costs. Therefore local grid refinement techniques are proposed: a higher grid resolution is only applied in certain areas of the model domain where it is needed or desired.

Several techniques exist to refine the grid locally. Grid *nesting* (Kunz and Moussiopoulos, 1997; Jacobs et al., 1995) combines a coarse grid domain with a finer grid domain using interpolation and integration techniques at the domain interfaces in order to pass information from one grid level to the other. When a strategy is used to follow the temporal behaviour or dynamics of the concentration field, an *adaptive* or *dynamic* grid refinement technique is used, which allows a more detailed resolution of temporal emission patterns and may follow more closely the dynamics and chemistry of smog episodes or accidental releases. Adaptive grid refinement techniques require a strong support from a dynamic data structure, which may lead to considerable computational overhead. Both structured adaptive grid refinement (Van Loon, 1996) and unstructured triangular mesh refinements (Tomlin et al., 1997) have been developed. More details on the characteristics of these grid refinement methods are given in section 2.

The *multiblock* method (Mensink, 1996) solves the governing equations on a set of non overlapping structured grid blocks that are connected in a conservative way. Grid line continuity over the block boundaries is not required, which easily allows local refinement. The multiblock approach aims at an improvement of the quality of the physical domain discretisation at low computational cost. The method is further discussed in section 3.

One of the important requirements of local grid refinement used in regional scale air quality modelling is the conservation of mass on the various grid levels. In the multiblock approach, this is obtained in a natural way by using a finite volume approach. Section 4 shows how the multiblock method is implemented and how mass conservation is obtained.

The potential of the multiblock method is to establish in a natural way the interactions between regional and urban scale modelling. This is demonstrated in section 5.

2. GRID REFINEMENT METHODS

Grid nesting approach

The grid nesting approach consists of either one way or two way nesting. One way grid nesting (Kunz and Moussiopoulos, 1997) considers only the influences from the larger scale (coarse grid) to the smaller scale (fine grid). Two way nesting (Jacobs et al., 1995) also considers the feed back from the smaller scale to the larger scale. At the grid interfaces the results obtained on the coarse grid may be used to define the lateral boundary conditions for the fine grid.

Besides an appropriate definition of the boundary conditions, one needs to exchange information from one grid to the other through *spatial* and *temporal* interpolations. This makes mass conservation a point of concern, especially if different time steps are used in the different (nested) domains. The interpolation will introduce numerical errors, especially at the interface, since both grids numerically evaluate the same quantities. Therefore diffusive filtering methods or relaxation techniques are needed to dampen these numerical errors, which will influence the solution or may introduce unwanted oscillations. Notice that in contrast to the unstructured or semi-unstructured (multiblock) approaches multiple grid levels are used which obviously require interpolation techniques.

Structured adaptive grid refinement

Local adaptive grid refinement is a dynamic approach: the domain may change after every (advection) time step, which can result in various multiple grid levels at different locations in the computational domain, as shown by van Loon (1996). Interpolation and integration are needed to go from one grid level to the other.

Like in the nesting approach, mass conservation is not automatically guaranteed, in particular when different time steps are used in different domains. The coarser grid provides initial and boundary conditions for the finer grid. Since the process of grid refinement is organised in a dynamic way, it requires strong support from a dynamic data structure as for example developed by Blom and Verwer (1993). As a strategy for the dynamic grid adaptation, refinement around the gradient or the local maximum in the concentration field is suggested. Alternatively, user defined criteria (flags) can be used as well.

Unstructured adaptive grid refinement

Unstructured grid refinement (Tomlin et al., 1997) has the potential to resolve in an accurate way steep spatial gradients by clustering of polygonal grid cells near that gradient. Triangular meshes are very suited for this purpose, especially when two dimensional flow patterns are studied. A grid extension to capture three dimensional flow patterns is much more complex than in a structured grid approach. As structured grids contain a regular pattern of co-ordinates and connectivities in which the relative position of the grid points is defined logically in function of the structure of the grid, unstructured grids do not have this logical connectivity and need additional information (complex data structure) to define the relative position of grid points.

Unstructured grids easily allow addition of new and removal of old grid points. This makes unstructured grids very suitable for complex domains or complex interactions of chemistry and flow. On the other hand, a constant reference to the connectivity matrix makes a solution on an unstructured mesh expensive and memory consuming.

3. MULTIBLOCK GRID REFINEMENT

In the multiblock grid refinement approach (Mensink, 1996) the main drawbacks of the structured grid approach are relieved and its strong properties are saved. Furthermore, the multiblock method is well suited for parallel computations on multiprocessor machines, as has been shown by Mensink and Deconinck (1992).

The discretisation technique used in the multiblock approach consists of a number of fixed domains or blocks (Figure 2). The mesh resolution in each block is independent of the resolution in other blocks, although large transitions should be avoided. There is only one (semi-unstructured) grid level that contains various non overlapping blocks with different grid resolutions. Neither interpolation, nor internal boundary conditions are needed to pass computational information from one block to the other, which enhances mass conservation. A time accurate solution can naturally be obtained if one single time step (smallest allowed) is applied throughout the different block domains. As a faster, but less accurate alternative, temporal interpolation techniques as used in the grid nesting technique can be applied.

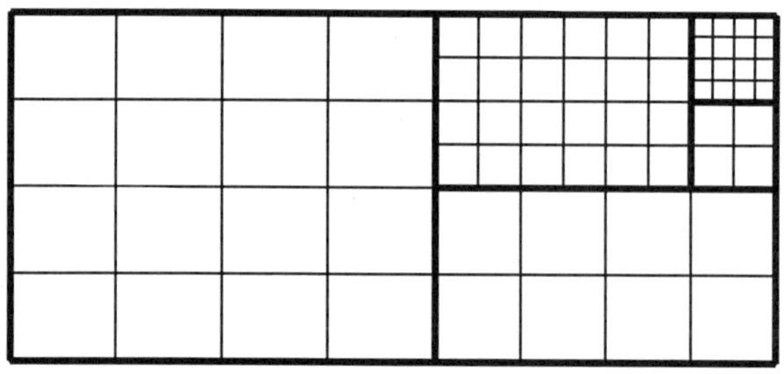

Figure 2. Example of a multiblock configuration containing 5 blocks and 3 grid levels.

The exchange between the block structured domains is carried out through (mass conservative) flux balances, which requires information in the form of a data structure describing the block configuration and block connectivity. However, the data structure needed to describe this block connection in only a fraction of the data structure needed for fully unstructured grids, since only the block boundary connections are to be described. The block connectivity data structure contains the following information (see also Figure 3):

$\forall$ blocks (m):

$\forall$ piecewise linear cell wall interfaces (k)
the following items are stored:

- number of the current block (m)
- cell indices & neighbouring indices in the current block (m) adjacent to cell wall interface k (i_f, j_f, i_s, j_s)
- number of the connected block (n)
- cell indices & neighbouring indices in the connected block (n) adjacent to cell wall interface k (p_f, q_f, p_s, q_s)
- boundary type number (continuous or discontinuous block boundary)
- cell wall interface lengths (Δx, Δy) from which the cell wall normal is computed

The connectivity array is created at the end of the grid generation process and is obtained by going along the block boundaries for every block. The information is necessary and sufficient to allow the multiblock code to handle both continuous and discontinuous block boundaries in a conservative way, as will be discussed in a next section.

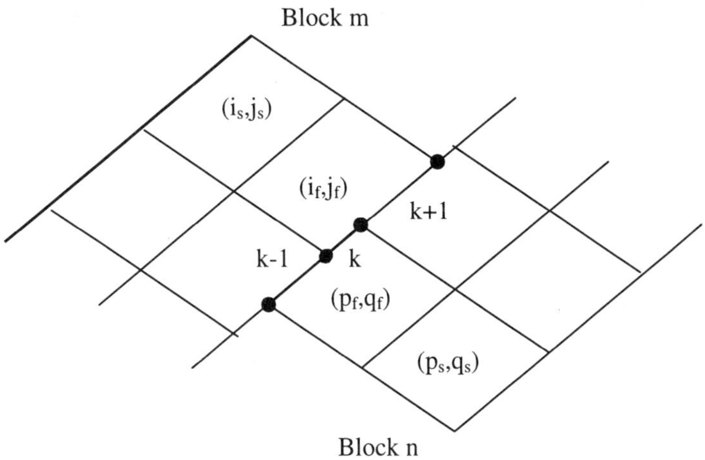

Figure 3. Construction of the block connectivity array.

4. IMPLEMENTATION AND MASS CONSERVATION

Mass conservation

In the grid refinement approaches it is required that the numerical discretisation schemes are *mass conservative,* i.e. mass remains conserved for an arbitrary control volume on an arbitrary discretisation (grid) level. Since the reduced cell sizes in the refined grid require a smaller time step in order to satisfy the Courant condition, the applied time step on the coarse grid might differ from the one on the refined grid. Due to the non-linearity of (some of) the operators applied in the governing equations, it can not be guaranteed that on the refined grid mass can be conserved in the same way as on the coarse grid. This is in particular the case for the nesting approach and the structured adaptive grid refinement approach. In the multiblock approach, this problem is avoided by applying one single time step (smallest allowed) on the different blocks.

Finite volume schemes are known to be mass conservative in a very natural way, because of the direct discretisation of the integral form of the conservation laws. On a volume Ω, bounded by an area A, the finite volume discretisation is given by :

$$\frac{\partial}{\partial t}\int_{\Omega}\mathbf{q}\,d\Omega = \oint_{A}(-\vec{\mathbf{v}}\cdot\vec{\mathbf{n}})\mathbf{q}\,dA + \int_{\Omega}s\,d\Omega \qquad \textbf{(1)}$$

where **q** is the solution vector, **v** the velocity vector, **n** the outward normal and s a source term. When applied to two dimensional advection, equation (1) can be rewritten as:

$$\Omega\frac{\partial q}{\partial t} + \sum_{i=1}^{n}\left(\mathbf{F}\cdot\mathbf{n}_x + \mathbf{G}\cdot\mathbf{n}_y\right)\Delta\Gamma_i = 0 \qquad \textbf{(2)}$$

where **F** and **G** are the horizontal and vertical (advective) flux vectors, as shown in Figure 4. On the left hand side of Figure 4, it is shown how the solution **q** can be updated at a new time step by applying a flux balance over the four cell wall interfaces Γ_1 to Γ_4. On the right hand side, the same procedure is applied to a multiblock grid cell. In this case the grid cell has five cell wall interfaces, because of the refined neighbouring block. However, mass conservation can still be obtained by simply applying equation (2), with n=5 instead of n=4. In this way boundary conditions or interpolation techniques are not needed in order to extend the advection scheme to a refined domain used in the multiblock context.

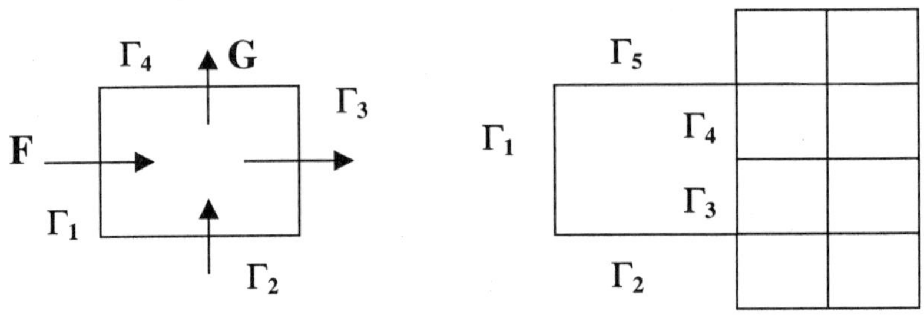

Figure 4. Extension of the finite volume approach towards a multiblock context.

Upwind finite volume schemes

Except for mass conservation, the advection scheme should also be positive definite, in order to prevent under- and overshoots. A scheme is called positive if it satisfies the following condition:

$$\frac{d}{dt}C_{i,j,k} \geq 0 \text{ if } C_{i,j,k} = 0 \tag{3}$$

A class of finite volume schemes that may have the properties of positive schemes, satisfying condition (3), is the class of upwind schemes. The restriction that has to be imposed on these schemes in order to make them positive, is that they have to be combined with an accurate time integration (e.g. Runge-Kutta). First order upwind schemes, however, are known to be very diffusive and not accurate enough. Higher order upwind schemes have to be combined with some limiting procedure in order to become positive. Examples are the TVD scheme and the third order kappa scheme (κ=1/3) with flux limiters. Other examples of schemes that may satisfy the condition of positivity are the Flux Corrected Transport (FCT) scheme, the Bott scheme (Bott,1989) and the McCormack scheme.

From numerical experiments carried out by van Loon (1996) and Syrakov et al. (1995), it was found that FCT schemes are relatively computational expensive and that the McCormack scheme gives less satisfying results. The κ=1/3 scheme was found to be accurate, mass conserving (on a regular grid) and fast. Also the Bott scheme (Bott, 1989) turned out to be efficient. Another interesting property of the Bott scheme is that it acts independently of the grid size, which makes it interesting for grid refinement applications.

The computational part of the multiblock method is based on an upwind finite volume discretisation (κ=1/3) using the flux difference splitting technique developed by Roe (1981) for the flux computation and the variable extrapolation technique proposed by Van Leer (1973) for second order space accuracy. An explicit 4th order accurate Runge-Kutta time integration is used in order to obtain a time accurate solution.

5. APPLICATIONS

The potential of the multiblock method to establish in a natural way the interactions between regional and urban scale modelling is demonstrated in Figure 5. The figure shows

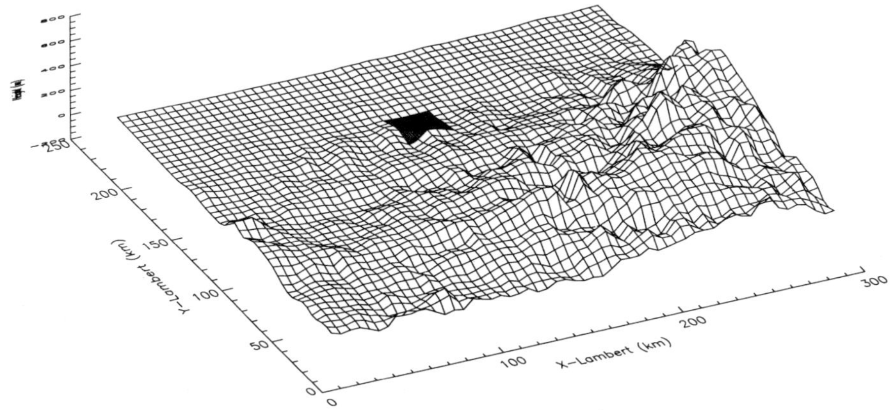

Figure 5. Multiblock grid (5 blocks) for Belgium with a refinement for the Brussels area.

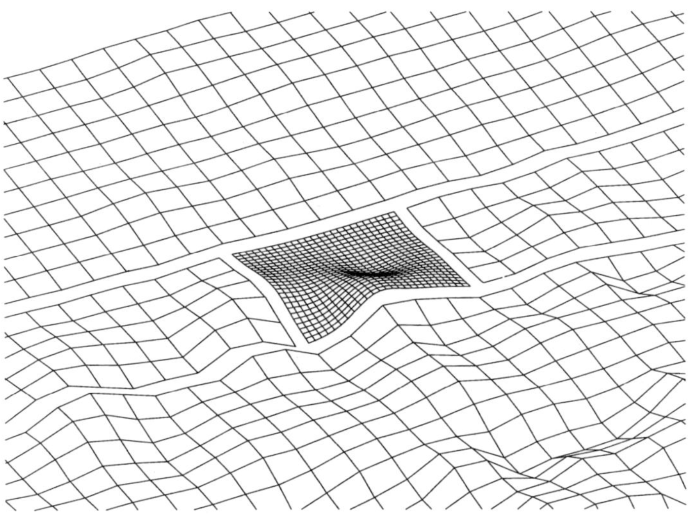

Figure 6. Detail of the refined block for the Brussels area (exploded view).

a digital elevation map of Belgium, covered by a three dimensional multiblock grid (bottom layer shown only). The grid covers a domain of approximately 300 km x 250 km (Lambert projection) and contained originally 55 grid cells in X-direction and 45 grid cells in Y-direction with a grid spacing of ± 5 km in both directions. During the grid refinement procedure, the mesh was refined in the Brussels area (25 x 25 km^2) with a grid spacing of 1 km in X- and Y-direction. After refinement, the domain is decomposed into 5 blocks, with the refined block containing 26 x 26 grid cells. Figure 6 shows a detail (exploded view) of the refined block and the four decomposed blocks.

6. CONCLUSIONS

Local grid refinement can be used to represent pollutant concentration distributions more accurately at locations where this is desired and to compute urban and regional scale interactions. Three possible approaches for local grid refinement have been investigated, namely the grid nesting approach, the local adaptive grid refinement approach and the multiblock approach. In the multiblock approach mass conservation can be obtained in a natural way, due to its finite volume implementation and provided that one single time step (smallest allowed) is applied on the different blocks.

Higher order upwind finite volume advection schemes are a means to obtain both mass conservation and positivity on the discretised domain. The $\kappa=1/3$ third order upwind scheme with limiter and the Bott scheme seem to be most attractive as advection schemes to be used in combination with grid refinement. Further development is needed to combine these advection schemes with subgrid parameterisations for the diffusion and dispersion of pollutants at smaller scales.

REFERENCES

Blom, J.G. and Verwer, J.G., 1993, VLUGR2*: A vectorized local uniform grid refinement code for PDE's in 2D*, CWI report NM-R9306, CWI, Amsterdam.

Bott, A., 1989, A positive definite advection scheme obtained by non-linear renormalization of advective fluxes, *Monthly Weather Review*, 117:1006-1015.

Jakobs, H. J., Feldmann, H., Hass H. and Memmesheimer, M., 1995, The use of nested models for air pollution studies: an application of the EURAD modell to a SANA episode, *Journal of Applied Meteorology,* 34:1301-1319.

Kunz. R. and Moussiopoulos, N., 1997, Implementation and assessment of a one-way nesting technique for high resolution wind flow simulations, *Atmospheric Environment*, 31:3167-3176.

Loon, M. van, 1996, *Numerical methods in smog prediction*, PhD thesis, University of Amsterdam.

Mensink, C. and Dumont, G., 1996, A critical evaluation and comparison of 6 European models for regional scale appliations, in: *Proc. EUROTRAC '96*, Vol I:559-564, P.M. Borell, P. Borell, T. Cvitas, K. Kelly, and W. Seiler, W. eds., Computer Mechanics Publications, Southampton.

Mensink, C., 1996, Numerical prediction of periodic vortex shedding in subsonic and transonic turbine cascade flows, *International Journal for Numerical Methods in Fluids*, 22:881-897.

Mensink, C. and Deconinck, H., 1992, A 2D parallel multiblock Navier-Stokes solver with applications on shared and distributed memory machines, in: *Computational Fluid Dynamics '92,* Vol 2:913- 920, Ch. Hirsch et al. eds., Elsevier Science, Amsterdam.

Roe, P. L., 1981, Approximate Riemann solvers, parameter vectors and difference schemes, *Journal of Computational Physics*, 43:357-372.

Syrakov, D., Andreev, V. and Valkov, N., eds., 1995, *EMEP MSC-E, Bulgarian contribution*, Annual report 1994, EMEP MSC-E, Sofia-Moscow.

Tomlin, A., Berzins, M., Ware, J., Smith, J. and Pilling, M.J., 1997, On the use of adaptive gridding methods for modelling chemical transport from multi-scale sources, *Atmospheric Environment*, 31:2945-2959.

Talat Odman, M., 1997, A quantitative analysis of numerical diffusion introduced by advection algorithms in air quality models, *Atmospheric Environment*, 31:1933-1940.

Van Leer, B., 1973, Towards the ultimate conservative difference scheme I. The quest of monotonicity. *Lecture notes in Physics*, 18:163-168.

DISCUSSION

T. IVERSEN: Normally global mass conservation is considered seriously when transport schemes are constructed. More rarely, local mass consistency implying individual conservation of mass mixing ratio, is considered. Have you thought about this problem?

C. MENSINK: By using a finite volume discretisation, mass can be conserved at any discretisation level, i.e. both locally and at a global level. This is due to the integral formulation. When a mass conservative scheme is applied to an arbitrary number of sub-volumes, the integrated volume containing all these sub-volumes is automatically mass conservative as well, provided that the total volume is equal to the sum of the sub-volumes.

G. S. RIVIN: You use a 3rd order scheme. But the wind from the most regional models has the second order. What is the reason of the using the 3rd order if in this case the full order of scheme will be the second order?

C. MENSINK: The 3rd order upwind scheme or kappa=1/3 scheme is used as a finite volume discretisation technique in combination with limiters and a Runge-Kutta method for the time integration. This combination leads to a positive scheme with good shape preserving properties, and low dispersion errors. It is also computational efficient. In theory, the kappa=1/3 scheme is 3rd order accurate. However, when it is applied to wind fields with a varying velocity, the accuracy of the scheme is reduced to second order. In situations where the limiter is applied, e.g. near strong gradients and in the vicinity of the boundaries, the accuracy of the scheme may be reduced to first order.

INTRODUCTION OF A PUFF-PARTICLE APPROACH FOR NEAR-SOURCE DISPERSION INTO THE CALPUFF MODEL

Peter de Haan,[1] Joseph S. Scire,[2] David G. Strimaitis,[2] and Mathias W. Rotach[1]

[1]Swiss Federal Institute of Technology, GIETH, Winterthurerstrasse 190, 8057 Zurich, Switzerland; dehaan@geo.umnw.ethz.ch
[2]Earth Tech, Inc., 196 Baker Ave., Concord, MA 01742, USA

Abstract—CALPUFF is a Lagrangian puff dispersion model for pollutant transport simulations for inhomogeneous and non-stationary conditions using a one-hour time step. Using so-called „absolute dispersion" parameterizations, the model predicts one-hour-averaged concentrations. In order to be able to predict not only „average" but also „possible" concentrations, i. e. the higher moments of the concentration probability density function, the „absolute dispersion" has to be separated in its puff-growth part and the additional contribution from the meandering of the plume during the averaging time. For this, the concept of the Puff-Particle Model (PPM) is introduced as a new near-source dispersion model within dispersion models such as CALPUFF. The PPM combines the advantages of both, puff and particle dispersion models by moving the center of mass of each puff along a trajectory which mimics the quickly changing turbulent flow field (artificial meandering). This trajectory is derived from the low-frequency part of trajectories as simulated by a Lagrangian stochastic particle model. It allows for a new puff-plume meandering scheme. The combination of CALPUFF and PPM improves the prediction of the highest possibly occurring near-source concentration, as well as the distance down-wind from the source where it occurs, while retaining the advantages of the CALPUFF model.

Key words: concentration pdf, puff models, plume meandering, relative dispersion.

INTRODUCTION

Within the field of dispersion modeling, puff models show a variety of advantages compared to Gaussian plume models. They can take into account the spatial variability of meteorological and dispersion conditions, causality effects, wet and dry deposition, low wind speed dispersion, etc. Lagrangian stochastic particle models, on the other hand, are the state-of-science of dispersion modeling, especially for the simulation of inhomogeneous (convective) turbulence. However, their demand of computing time generally limits their application to the simulation of short episodes rather than a full year. Furthermore, physical processes like dry and wet deposition, buoyant plume rise, and chemical transformations within a cluster, are much easier to implement within the framework of a puff model.

There are two types of puff models. The most common type can be called „plume segment“ puff models. Within the framework of these models, a puff (consisting of a center of mass, the actual velocity of the puff, and a 3D distribution of the total mass around the center) stands for the ensemble average of the concentration distribution belonging to a „piece“ (in time) of the pollutant release. These puffs do not correspond to a single cluster in nature, but are an ensemble average over many individual clusters. Dispersion parameterizations predicting such ensemble mean statistics are called „absolute dispersion“.

The second type is the group of „cluster dispersion“ puff models. These models identify the puff with an individual, physically realistic cluster of particles. They use the concept of relative dispersion (i. e. 2-particle statistics; Borgas and Sawford 1994): turbulent eddies smaller than the actual puff size will contribute to its growth (increasing the ensemble 2-particle separation), while larger eddies move the puff as a whole (without changing the separation of the particles). Hence, absolute dispersion is the combined effect of relative diffusion and of the meandering of the puff respective to a fixed point, caused by turbulent eddies larger than the puff.

The missing link between relative and absolute dispersion is the dispersing effect of meandering over time-scales shorter than the averaging time for which the absolute dispersion parameterization used applies (normally equaling the spectral gap, i. e. roughly one hour). For small (newly released) puffs, absolute dispersion is dominated by meandering rather than by puff-growth. Therefore, in contrast to „cluster dispersion“ models, „plume segment“ puff models are not suited to give correct concentration predictions for a single (instantaneous) release, especially in the near field.

In practice it is difficult to provide the flow field at a sufficiently high temporal rate. This means that the effect of meandering has to be simulated. For this, in the Puff-Particle Model (PPM) (de Haan and Rotach 1999), a „cluster dispersion“ puff model using relative diffusion, the meandering of the puffs' centers of mass is generated artificially. These meandering trajectories simulate the effect of all those eddies not resolved by the flow field but still larger than the puff. Models like the PPM are especially suited for the simulation of accidental hazardous releases. For preliminary risk assessments, the actual flow field is not known. But the artificial meandering scheme of the PPM allows the simulation of an ensemble of possible realizations, identifying worst-case scenarios. Thus, the PPM can determine the probability of a certain concentration level: a concentration probability density function (C-pdf).

The disadvantage of „cluster dispersion“ models is their need to simulate a large ensemble of individual releases to obtain the ensemble average. The additional information they generate is not needed all the time. In the present paper, the „cluster dispersion“ Puff-Particle Model (PPM) is introduced as a sub-model into the „plume segment“ CALPUFF model. In the next three sections, the basic philosophy and major features of CALPUFF and PPM are summarized. Then, we present the design of the interface between these two models, and describe how the concentration pdf are obtained. Finally. the form of such concentration pdf's for different down-wind distances and averaging times is discussed.

THE CALPUFF DISPERSION MODEL

CALPUFF is a Lagrangian puff dispersion model (Scire *et al.* 1995, 1997). Among its main fields of application are pollutant transport simulations for inhomogeneous and non-stationary conditions. The model is a member of the „plume segment“ puff model family, using "absolute dispersion", i. e. the time-averaged dispersion of an ensemble of pollutant releases. Together with the flow fields of its meteorological model, CALMET, CALPUFF is applicable to complex terrain and coastal situations.

The CALPUFF model is a non-steady-state puff dispersion model. The user-defined grid size allows for high-resolution simulation of episodes as well as for runs for one year or more with a one-hour time step for environmental impact assessments, and studies of air quality and pollutant transport on regional scales. By its puff-based formulation, it can account for a variety of effects such as spatial variability of meteorological and dispersion conditions, causality effects, dry deposition, plume fumigation, low wind speed dispersion, pollutant transformation, wet removal, and complex terrain effects. It has various algorithms for parameterizing dispersion processes, including the use of turbulence-based dispersion coefficients derived from similarity theory or observations. Evaluation studies of CALPUFF have been done for long range transport distances (U.S. EPA 1995), intermediate distances (Irwin 1998) and short to intermediate distances (Strimaitis *et al.* 1998).

THE PUFF-PARTICLE MODEL

The fact that instantaneous releases require puff models using relative dispersion, but that at the same time, the update frequency of the flow field information in almost all applications is too low to resolve all those turbulent eddies not covered anymore by the relative dispersion concept, gave rise to the development of the Puff-Particle Model (PPM). It represents the group of so-called „cluster dispersion“ puff models. The PPM in its current version is a research model for tracer pollutants, focusing on near-source dispersion, and neglecting deposition and chemical processes. It features a full stochastic Lagrangian particle dispersion model, which fulfills the well-mixed criterion (Thomson 1987). For convective conditions, the vertical component of the pdf is the same as in Luhar and Britter (1989). To provide a perfectly smooth transition between stable/neutral Gaussian turbulence to convective skewed turbulence, the transition function of Rotach *et al.* (1996) has been adopted. Further details on the PPM can be found in de Haan and Rotach (1999).

Every puff within the PPM follows a turbulent puff trajectory derived from a stochastic particle trajectory. The kinematic turbulent energy belonging to those eddies which are smaller than the puff's size-already covered within the concept of relative dispersion-is removed from the particle trajectories. For this, a Kalman low-frequency filter is used, where the cut-off frequency depends on the size of the puff (de Haan and Rotach 1998). Hence, every puff carries along its own position as well as the position and turbulent velocity components of the stochastic particle it 'belongs to'. The effect of meandering (caused by turbulent eddies larger than the puff but not resolved by the flow field) is simulated by the puff center trajectories, yielding a complete description of dispersion. It has been shown on the basis of tracer data, that the correct treatment and interpretation of the two contributions to the dispersion process is crucial for reproducing experimental results to a good correspondence (de Haan and Rotach 1998).

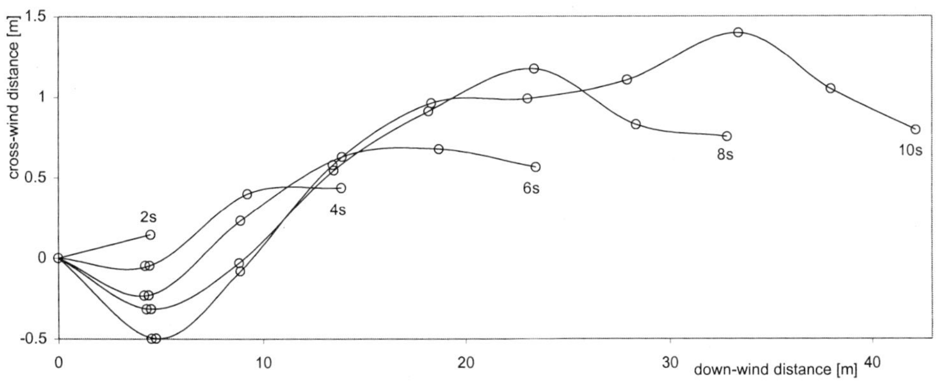

Figure 1. Illustration of the meandering puff-plume scheme. The released puff-plume is depicted 2, 4, 6, 8 and 10 seconds after its release. Puffs belonging to the same puff-plume are connected by a thin line. Circles indicate the position of the center of each puff (not the puff size). View from above.

THE PUFF-PLUME MEANDERING SCHEME

Within the PPM, the puffs use stochastic paths to artificially produce the correct meandering behavior. To obtain correct C-pdf's for any user-specified averaging time, however, that is not sufficient. „Neighbor" puffs should show „similar" meandering: the spatial and temporal correlation of turbulence has to be taken into account. If this is neglected, the most extreme concentration events will be underestimated, and the C-pdf will not be correct.

The PPM employs a plume meandering scheme based on the puff meandering scheme. Plumes are described as threads of puffs which are correlated in their turbulent movements to their next neighbors in the puff-plume. The PPM puff-plume meandering scheme is illustrated in Figure 1 and consists of two steps. *Step 1:* At the beginning of the time-step, each puff has its initial 3D stochastic velocities. After moving the puff-plume with the mean flow and by the stochastic velocities, the newest puff is released. Its stochastic velocities are correlated with those of the second-newest puff (its „mother" puff) by copying the mother's turbulent velocities and computing a puff trajectory over a time which corresponds to the spatial separation from this second-last puff to the source. *Step 2:* The size of the puffs is enlarged using relative dispersion, and new stochastic velocity components are computed (again, by computing a puff trajectory starting with the turbulent velocities from the „mother" puff). The first (most distant from the source) puff of the puff-plume follows a normal (non-correlated) PPM puff-center trajectory.

This meandering scheme can be illustrated by thinking of a spectrum of turbulent eddies „rolling back" towards the source along the puff-plume with the average mean wind speed (introducing the spatial correlation). Such a spectrum of eddies is „released towards the origin" from the front of the plume every time step, based on the puff path of the front puff (temporal correlation) (see Figure 2).

The Lagrangian particle model within the PPM yields a Lagrangian spectrum, when the turbulent velocity of a particle is followed over time. It has the correct „–1" slope of the inertial subrange, in contrast with the „–2/3" slope required for Eulerian spectra. The current approach will thus give a „–1" instead of a „–2/3"-inertial subrange slope when considering the turbulence at a fixed point in space. However, since these spectra are filtered and only their low-frequency part is used, this shortcoming has only a minor influence. The main difference between Eulerian and Lagrangian spectra is the location of the maximum of the spectrum, $f_{max,E}$ and $f_{max,L}$, respectively. This is corrected by shifting the Lagrangian spectrum towards higher frequencies by a factor $\beta = f_{max,E} / f_{max,L}$. So the turbulent velocities of a puff with a separation x to its „mother" are computed as a puff trajectory over a time $\beta x/U$ rather than x/U.

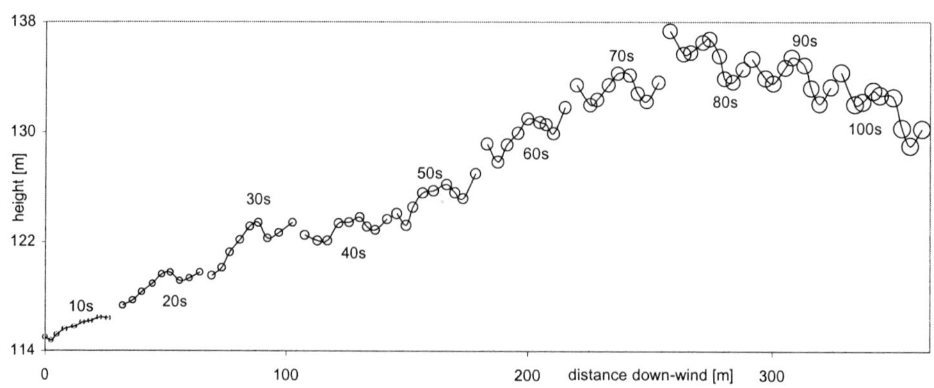

Figure 2. Possible evolution of a plume (consisting of 9 puffs released with a one second interval) over time, when using PPM's meandering puff-plume scheme. The position of the plume's puffs is shown every ten seconds. Example for convective conditions, release height 115 m. View from the side.

INTRODUCTION OF THE PUFF-PARTICLE MODULE INTO CALPUFF

The CALPUFF model has been developed for a broad range of applications, from micro-scale to meso-scale, from short-term to long-term, with treatments of wet and dry deposition, chemistry, coastal interactions, complex terrain, visibility impacts, etc. The PPM will be one additional option in this list, providing additional information on the near-source concentration moments. Due to the flexibility of the CALPUFF model on the side of the user, its code has reached a high grade of complexity. Therefore, one of the main objectives for the implementation of the Puff-Particle module into CALPUFF was to introduce as few interactions between the CALPUFF code and the PPM option as possible, in keeping with the module coding concept. Thus, the underlying thought of the concept adopted was that CALPUFF's puff treatment will not change; it is only within the Puff-Particle module that the simulation of artificial meandering of relatively dispersed puffs is computed.

Figure 3 illustrates the basic set-up of the PPM within CALPUFF:

- To every newly released puff, a so-called 'mirror ensemble' is attached. Such a mirror ensemble consists of a user-defined number, N, of puff-particles.
- In order to take into account the information from all grid cells through which the puff will pass in one hour, CALPUFF divides the one hour model time step into a varying number of sampling steps for each puff. For the duration of the sampling step of the parent puff within CALPUFF's main routine, the mirror-ensemble is advected with an internal user-defined PPM-timestep between, say, 1 and 10 seconds. For every internal PPM-timestep, new particle trajectories are computed, from which puff trajectories are derived. Additionally, the puff-particles are advected by the mean flow.
- At the end of the sampling step, from the mirror ensemble's first and second moments of the mass distribution, the position and size of the parent puff are calculated and handed back to CALPUFF's main routine. CALPUFF may then compute any physical process possibly changing the (parent) puff's mass or chemical composition, but not its size or location. The mirror ensemble remains in existence for use in the next sampling step.

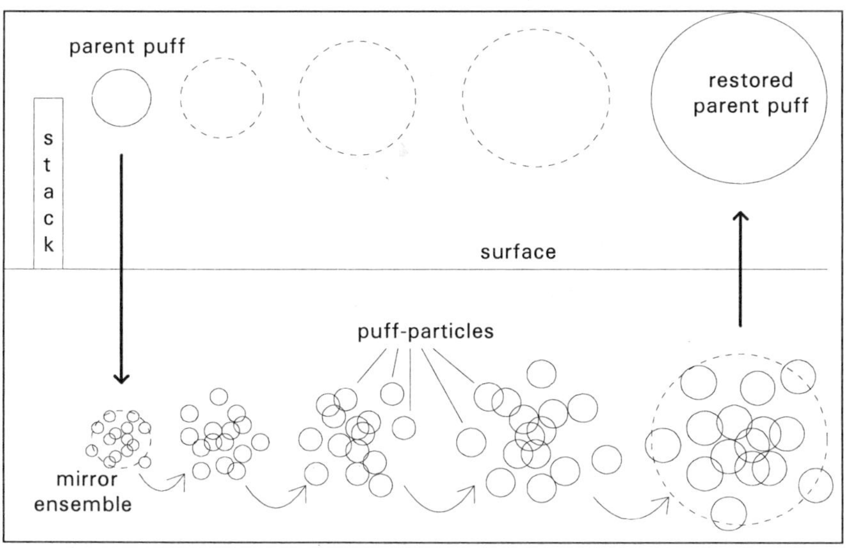

Figure 3. Illustration of the imbeddement of the PPM into the CALPUFF model. The procedure for one sampling time step is depicted. After „restoring“ the new location and size of the parent puff, the mirror-ensemble remains stored for use at the beginning of the next sampling step of the same puff, until the puff reaches a mature stage where the mirror-ensemble will be deleted.

- After a certain time, the size of the relatively dispersed particle-puffs in the mirror ensemble will be such that the largest part of the energy spectrum of turbulent eddies will be 'within' the puff-particle. This means that the relative and absolute dispersion for that ensemble become similar. The artificial meandering will show only little variation of the particle-puff's paths. Then, the 'parent puff' location and size is recomputed, the mirror ensemble deleted and the parent puff restored, which from then on is treated with the common absolute dispersion within the CALPUFF model.

The initial position of the particle-puffs at the time of creation of a mirror ensemble is taken randomly from a three-dimensional Gaussian distribution with mean at the center of the parent puff, and the standard deviations of the parent puff being the second moments of the Gaussian distribution. Analogously, the initial turbulent velocity components of the puff-particles of the mirror ensemble are taken from a three-dimensional Gaussian distribution with zero mean and with the ambient turbulence $\left(u_i'^2, i = 1,2,3\right)$ as the second moments.

DETERMINATION OF A CONCENTRATION PDF

The puff-particle approach gives a realistic picture of the transport (caused by the mean wind, provided by the flow field updates), the amount of meandering (covered by the stochastic puff center trajectories) and the diffusion of the release (caused by eddies smaller than the puff, and taken into account by relative diffusion). For each parent puff, an ensemble of mirror puff-particles is released. Since the meandering part of their motion is reigned by random particle motion, their trajectories will not be identical. Each of them is a possible realization of what might have happened to the parent puff in reality.

To obtain the mean (ensemble averaged) concentration at any receptor, the average over all mirror puff-particles can be taken, where each realization is assumed to represent $1/N$ of the total mass of the parent puff. In theory, the ensemble average concentration thus obtained should be close to the concentration estimation obtained by using the parent puff only, together with absolute dispersion. Of course, in practice, minor differences will appear. To let these differences vanish, it would be necessary to use the same set of parameterized spectra of turbulent energy as a foundation of the absolute as well as the relative dispersion parameterization and at the same time derive the particle pdf within PPM's particle model from this set of spectra.

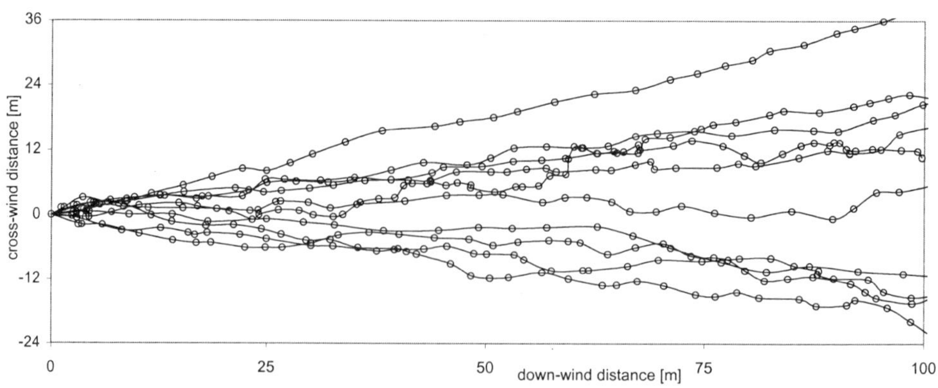

Figure 4. Method to construct a C-pdf. Many individual meandering puff-plumes are simulated. Each is treated as one possible realization that could actually have happened for the given meteorological data. The sorted concentration impacts (for a user-determined averaging time) at specified receptor locations from these many realizations constitute the C-pdf.

To obtain the pdf of concentration at such a receptor, every realization (i. e. mirror puff-particle) carries the whole mass of the parent puff, and the concentration due to this particular realization is computed (Figure 4). This leads to N possible concentrations, from which the concentration pdf easily can be derived. Of course, to make the concentrations thus computed comparable to the standard CALPUFF results, these N possible concentrations have to be computed using all puffs in the modeling domain impacting the receptor, whether they possess a mirror-ensemble or not. In the intermediate to far field, the will lead to a narrowing of the concentration pdf, reflecting the fact that well dispersed pollutant clusters will indeed lead to a higher degree of determination of the corresponding concentration. Therefore, the PPM option within CALPUFF will produce valuable results mainly in the near field, where the differences in puff position due to meandering potentially are of the same order of magnitude as the size of the puff.

CHARACTERISTICS OF CONCENTRATION PDF'S

When looking at the changes in the form of the C-pdf for increasing down-wind distances (Figure 5), two effects are occurring simultaneously. First, a shift of the mean (and the maximum) to the left, i. e. to lower ensemble average concentrations. Second, a narrowing of the C-pdf: the slopes to the left and to the right of the maximum are getting steeper. This means that the ensemble average becomes „better defined", due to the fact that the relevant time scales of fluctuation at these distances are much larger than the averaging time.

The form of the C-pdf also is a function of the averaging time T. Whereas a short averaging time of 60 s leads to a wide-spread C-pdf (in Figure 5, for 1000 m down-wind, the probability that the actual concentration is at least five times higher than the ensemble average is 5%), the C-pdf narrows for larger averaging times (without changing its mean, i. e. ensemble concentration). Additionally, C-pdf's for stable conditions are wider than for convective conditions, because more dispersion leads to a more well-defined ensemble mean, hence a narrowing of the C-pdf.

The form of the C-pdf often is assumed to be log-normal. For most models predicting concentration fluctuation probabilities, this is an input rather than a result. The results from the present model, however, suggest that the C-pdf is log-skewed towards lower concentrations.

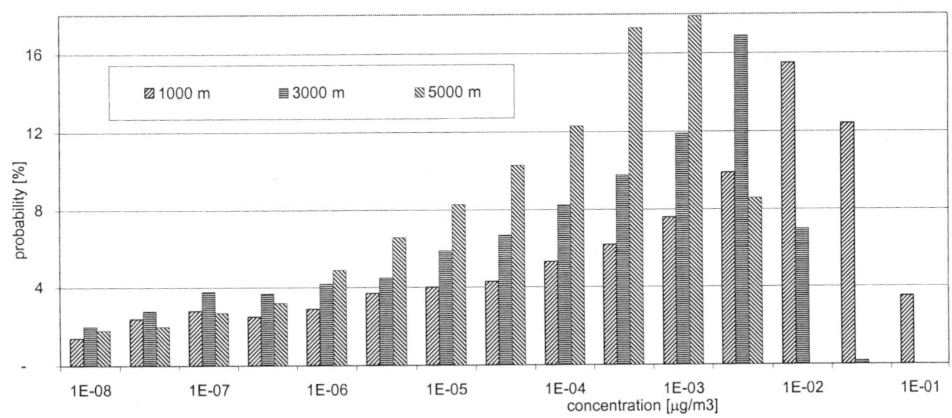

Figure 5. Example of C-pdf's for the plume centerline position at three distances down-wind from the source; averaging time 60 s. Example for the Copenhagen experiment (release height 115 m, conditions of forced convection).

SUMMARY AND CONCLUSIONS

The combination of two models, CALPUFF and PPM, using different approaches to describe the dispersion of pollutants, allows for the combination of the advantages of the CALPUFF model, with additional information on the higher moments of the concentration distribution at near-field receptors from the PPM module. This allows for an additional assessment of concentrations that actually could be observed at a given receptor, together with the corresponding probability. This additional feature is particular useful for the assessment of hazardous pollutants, where the question of the average exposure is less relevant than the highest exposure which could actually occur under the given meteorological conditions.

Acknowledgments—The present work was partly financed by the Swiss Agency of Education and Sciences (BBW) and the Swiss Agency of Environment, Forest and Landscape (BUWAL) through a project in the framework of COST 615 (citair), and by the Swiss National Science Foundation through Grant Nr. 21-46849.96.

REFERENCES

Borgas, M. S. and Sawford, B. L. (1994): A Family of Stochastic Models for Two-Particle Dispersion in Isotropic Homogeneous Stationary Turbulence. *J. Fluid. Mech.*, **279**, 69–99

de Haan, P., and Rotach, M. W. (1998): The Treatment of Relative Dispersion within a combined Puff-Particle Model (PPM). In: Air Pollution Modeling and Its Application XII, S.-E. Gryning and N. Chaumerliac (eds.), Plenum Press, New York, 389–396

de Haan, P., and Rotach, M. W. (1999): A novel approach to atmospheric dispersion modelling: the Puff-Particle Model (PPM). Accepted for publication in *Quart. J. Roy. Meteorol. Soc.*

de Haan, P., and Scire, J. S. (1999): Prediction of higher moments of near-source concentration by simulating the meandering of pollutant puffs. Preprints 13th Conf. Boundary Layers and Turbulence, Jan. 10–15, 1999, Dallas. American Meteorol. Soc., Boston, USA

Irwin, J. S. (1998): A comparison of CALPUFF modeling results with 1977 Inel field data results. In: Air Pollution Modeling and Its Application XII, S. E. Gryning and N. Chaumerliac (eds.), Plenum Press, New York, 143–151

Luhar, A. K. and Britter, R. E. (1989): A Random Walk Model for Dispersion in Inhomogeneous Turbulence in a Convective Boundary Layer. *Atmospheric Env.*, **23**, 1911–1924

Rotach, M. W., Gryning, S.-E. and Tassone, C. (1996): A Two-Dimensional Stochastic Lagrangian Dispersion Model for Daytime Conditions. *Quart. J. Roy. Meteorol. Soc.*, **122**, 367–389

Scire, J. S., Strimaitis, D. G., Yamartino, R. J., and Zhang, X. (1995): A User's Guide for the CALPUFF Dispersion Model. Prepared for the USDA Forest Service, Cadillac MI, by Earth Tech, Inc., Concord, MA 01742, USA.

Strimaitis, D. G., Scire, J. S., and Chang, J. C. (1998): Evaluation of the CALPUFF Dispersion Model with two Power Plant Data Sets. Preprints 10th Joint Conf. Appl. Air Poll. Meteorology, 11–16 Jan. 1998, Phoenix, Arizona.

Scire, J. S., Strimaitis, D. G., Yamartino, R. J., and Insley, E. M. (1997): A User's Guide for the CALPUFF Dispersion Model (Version 5.0). Earth Tech, Inc., Concord, MA 01742, USA.

Thomson, D. J. (1987): Criteria for the Selection of Stochastic Models of Particle Trajectories in Turbulent Flows. *J. Fluid Mech.*, **180**, 529–556

U.S. Environmental Protection Agency (1995): Testing of meteorological and dispersion models for use in regional air quality modeling. U.S. EPA, Research Triangle Park, NC.

DISCUSSION

D. WILSON: When using a discrete numerical model to simulate concentrations pdf's it is necessary to release a very large number of puffs to capture the extreme values of high concentration that occur at a fixed (Eulerian frame) receptor. Do you have a procedure for estimating the required number of puffs (N) needed to correctly simulate a concentration peak such as the C_{99} that is exceeded 1% of the time?

P. de HAAN: The current simulations were performed with $N = 1000$. This is believed to give reliable results for C_{95}. Hence $N = 10000$ would do for C_{99}. But it depends on the assumed form of the C-pdf. Statistically correct confidence limits, however, would also be a function of the averaging time T.

M. KAASIK: An experiment with measurements very near to the source is the Lillestrøm experiment. Will you use this data set?

P. de HAAN: The Lillestrøm data set consists of only four independent experiments. These took place under very stable conditions, but during sunrise. The concentration statistics are influenced by one very high observation. I consider the Prairie Grass data to be more adequate for near-source assessments.

S. RAFAILIDIS: A comment relating to the need of validation of your model. By the end of 1998 the University of Hamburg and the German "Umweltbundesamt" will release a comprehensive wind tunnel data set. It contains information on the spectral characteristics of the incoming flow, and concentration time series down-wind of the release (www.mi.uni-hamburg.de).

THE DESCRIPTION AND APPLICATION OF AN ODOUR DISPERSION MODEL

Tarja Savunen and Erkki Rantakrans

Finnish Meteorological Institute, Air Quality Research
Sahaajankatu 20 E, FIN-00810 Helsinki, Finland

INTRODUCTION

An odour dispersion model differs from other dispersion models by its capability to represent the short-term peak values of the concentrations of odorous compounds (Rantakrans and Savunen, 1995), which may vary with time scales less than one minute. The commonly-used hourly mean concentration is not adequate for describing the presence of odour (Hanna, 1986). The concentration of odorous compounds can exceed the odour threshold level, despite the fact that the hourly mean is clearly below the odour threshold (Fig. 1). We have developed an odour dispersion model, which can be used to estimate the short-term peak values of odorous compounds from one or multiple sources. The emissions may also vary with time. Both horizontal and vertical fluctuations of the plume are considered in the model. The model also includes computational methods for area- and volume sources. The model has been applied to provide dispersion estimates of odorous compounds for numerous domestic industrial facilities.

COMPUTATIONAL METHODS

The main goal in our model development has been to assess the concentration fluctuations caused by atmospheric turbulent diffusion. The odour dispersion model comprises two computational methods: the so-called meandering plume method for point sources and the internal fluctuation method (probability density function modelling) for area and volume sources (Hanna, 1986) (Fig. 2).

Air Pollution Modeling and Its Application XIII, edited by S.-E. Gryning and E. Batchvarova, Kluwer Academic/Plenum Publishers, 2000

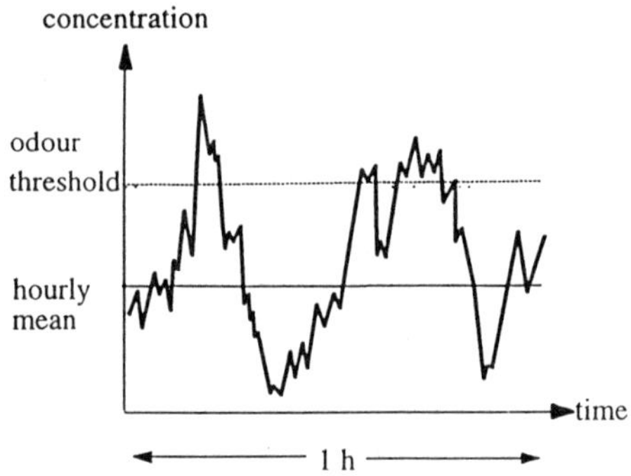

Figure 1. The odour concentration exceeds the odour threshold several times despite the hourly mean being below the threshold (Högström, 1972).

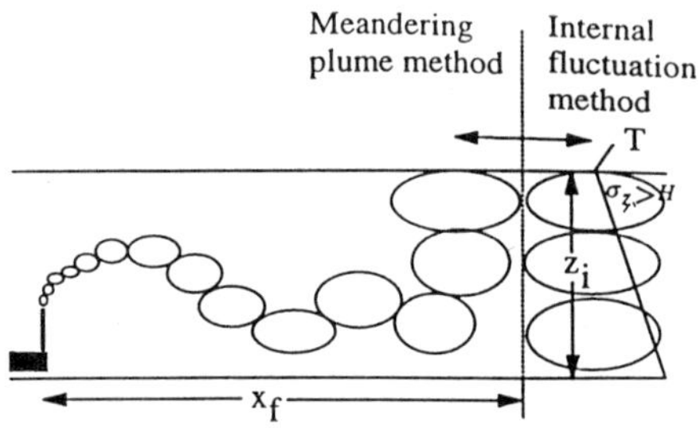

Figure 2. Plume dispersion so that in vertical direction it is reaching upper limit of mixing height and in distance x_f from a source.

The meandering method describes concentration fluctuations due only to the meandering of the plume:

$$(1) \qquad \sigma_{cT}^{2} = \sigma_{cI}^{2} + \sigma_{cM}^{2}$$

where σ_{cT}^{2} is the total standard deviation, σ_{cI}^{2} is the instantaneous standard deviation and σ_{cM}^{2} is the standard deviation of the meandering motions. In the internal fluctuation method it is assumed that the variance of the concentration is formed with respect to the mean flow both horizontally and vertically by the same kind of diffusion as the mean concentration field.

The model is based on the assumption that the concentration time series due to the meandering of the plume can be represented by half-minute periods in a one-hour sample. Each of the half-minute concentration values is assumed to correspond to the position of the axis of the plume with reference to the average airstream in both the horizontal and vertical direction at each of the points around the source (Fig 3). The instantaneous (30 s) concentration formula is:

$$(2) \qquad \overline{C}'(x,y,z) = \frac{Q}{2\pi \bar{u}_s \sigma_{yr}(x) \sigma_{zr}(x)} \exp\left(-\frac{y'^2}{2\sigma_{yr}^2(x)} \right) G(z;\sigma_{zr}(x),H',z_i)$$

where

Q	=	release rate (kg/s)
z	=	height from the ground (m)
$\bar{u}_s$	=	average wind speed (m/s) at height H_S
z_i	=	inversion or mixing height (m)
$\sigma_{yr}(x)$	=	instantaneous horizontal standard deviation (30 second average) (m)
$\sigma_{zr}(x)$	=	instantaneous vertical standard deviation (30 second average) (m)
y'	=	the rectangular y coordinate with respect to the instantaneous location of the plume axis (m)
H'	=	instantaneous height above ground of the plume axis (m)
G	=	vertical diffusion term
$\sigma_y(x)$	=	horizontal standard deviation (1 hour average) (m)
$\sigma_{yc}(x)$	=	horizontal standard deviation of plume centerline (m)

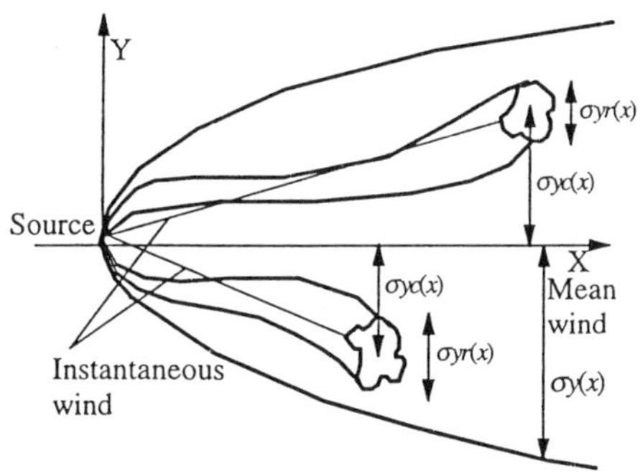

Figure 3. Plume meandering from a point source in horizontal plane (xy) (Beaman, 1988).

x = maximum = 12 %   • Emission source

Figure 4. Number of odour hours expressed as percentage caused by emissions of waste water handling plant. Stack height is 60 meters. Odour concentration of emission is 1000 odour units in a cubic meter of air.

The fluctuation of the plume axis is assumed to follow a normal distribution. To determine the odour, one must calculate the 30-second maximum concentration during one hour. This can be done either by the definite probability method or by the Monte-Carlo method. When the plume has spread and dispersed so widely that it reaches the upper limit of the mixing layer z_i (i.e. $\sigma_Z(x) > z_i$), the internal fluctuation method is used and equation (2) is replaced with the expression:

$$(3) \qquad \overline{C}(x,y,z) = \frac{Q}{2\pi z_i \bar{u}_s \sigma_y(x)} \exp\left(-\frac{y^2}{2\sigma_y^2(x)} \right)$$

The peak concentration is determined by a probability density function, which follows a log-normal distribution.

The odour dispersion model also contains calculation methods for area and volume sources for describing dispersed releases, which may typically occur from stacks or chimneys in roofs, from small pipes or holes in walls, etc. When these releases are observed, it is assumed that the release plume has already dispersed at the beginning so widely that the internal fluctuation method can be used. The peak concentration can be determined by a probability density function following a log-normal distribution.

Before calculating the instantaneous (30-second average) and the mean (1 hour average) concentrations needed in the model, the corresponding dispersion parameters and the dispersion of the plume axis have to be determined. Högström's method (Högström, 1972) is used to calculate the instantaneous (30 s average time) dispersion parameters and for the dispersion of the plume axis. Determination of the dispersion parameters required in the calculation of average concentrations is based on research on boundary layer meteorology (Businger et al. 1971; Cayghey et al. 1979; Hanna 1985; Holtslag 1984; Wratt 1987).

Peak concentrations caused by the effect of several sources together can be estimated in the odour dispersion model by the so-called background method or by the Monte-Carlo method. In the background method it is assumed that the hourly mean concentration caused by each source makes up the background concentration. At the same time each source causes its own peak concentration, to which is added the background concentration due to the other sources. In the Monte-Carlo method it is assumed that each source produces, during the course of a one hour sampling period, instantaneous concentrations (30 s average time), which are independent of the other sources, following, however, a certain probability distribution. From these random concentrations caused by each source the portion at each receptor point is calculated, and these are then combined; from this time series of concentrations the maximum concentration is then chosen.

In the actual odour dispersion model the program first reads in the initial meteorological situation and the corresponding release information. After this, the program calculates the algorithms of the equations governing the model, and the peak concentration can then be calculated. Following this the next hour is calculated using the corresponding release situation and so on, until the whole meteorological period (at least one year) has been employed. As a result, the model produces a time series of peak concentrations for each one-hour period. Ultimately, the time series is used to calculate the relative duration of odour as the number of times a specified threshold concentration has been exceeded. The odour dispersion model estimates odour frequencies as a percentage during one year (at least) and the arial distribution of these frequencies (Fig. 4).

ODOUR EMISSIONS - ODOUR CONCENTRATION - ODOUR THRESHOLD

The odour emission information required by the odour dispersion model is the amount of odour emissions per unit time. For this reason exhaust gas flow and the odour concentration are measured. The odour concentration is determined from an emission sample taken into a bag. The sample is passed to an olfactometer and sniffed by a group of people, called the odour panel. The olfactometer dilutes the original sample with clean air, forming a gas flow where the odour concentration is constant. The odour concentration of the emission is expressed as the number of odour units in a cubic meter of air. The number of odour units shows how many times a sample taken from the emission has to be diluted until it becomes odourless. An odour concentration of 1 odour unit in a cubic meter of air is called the odour threshold for the gas studied with the olfactometer. The odour dispersion model simulates the frequency of odour threshold exceedances in outdoor air, producing an odour distribution for the studied area.

The effect of odour emissions can be estimated even when the odour concentrations are not defined with an olfactometer. After a chemical analysis of the odorous emission, the model calculations can often be made with sufficient accuracy by studying e.g. the exceedances of the odour thresholds of the most easily smelling compounds in the area surrounding the emission source.

RESULTS AND INFLUENCE ESTIMATION OF ODOUR EMISSIONS

We have compared the odour frequencies computed with our model with experimental data. The comparison shows that the odour dispersion model is well compatible with the experimental results. The odour dispersion model has been applied to estimate the odour influence of e.g. food production and chemical plants as well as pulp mills and waste water handling facility.

The odour dispersion model registers as an odour hour every hour during which one 30 seconds odour concentration has exceeded the odour threshold i.e. odour is observed. Thus, the odour needs not to be present for the whole hour. Based on recent research results in Finland (Arnold, 1995) it is suggested, that odour frequencies of 3 - 9 % could be used as the odour guideline. The lower limit would be applied to very unpleasant odours with high inconvenience potential and the higher limit to odours with a more variable degree of pleasantness. Recommendations for odour frequency guidelines have also been given in some other countries for instance in Germany, Denmark and the Netherlands. The odour frequency values calculated by the odour dispersion model can be compared with the above mentioned suggested guidelines to estimate the air quality situation in areas with odour emissions.

REFERENCES

Arnold, M., 1995. *Basis for Odour Guidelines*. Technical Research Centre of Finland, Research Notes 711, Espoo, 83 p. + app. 14 p. (Text in Finnish,English Summary).

Barry, P.J., 1977. *Stochastic Properties of Atmospheric Diffusivity*. Sulphur and It's Inorganic Derivates in the Canadian Environment. National Research Council of Canada, p.313-358.

Beaman, A. L., 1988. A Novel Approach to estimate the Odour Concentration Distribution in the Community. *Atmos. Environ.* 22, p. 561 - 567.

Briggs, G. A., 1973. *Diffusion Estimation for Small Emissions*. Atmospheric Turbulence and Diffusion Laboratory Report No. 79, NOAA, Oak Ridge, Tenn.

Briggs, G. A., 1984. *Plume Rice and Boyancy Effects*. Ch. 8 in Atmospheric Science and Power Production. Ed. D. Sanderson, U. S. Dept. of Energy DOE/TIC - 27601, p. 327 - 366.

Businger, J.A., Wyngaard, J.C., Izumi, Y. and Bradley, E.F., 1971. Fluxprofile Relationships in the Atmospheric Surface Layer. *J. Atmos. Sci.* 28, p. 181 - 189.

Caughey, S.J., Wyngaard, J.C. and Kaimal, J.C., 1979. Turbulence in the Evolving Stable Boundary Layer. *J. Atmos. Sci.* 36, p. 1041 - 1052.

Hanna, S.R., 1984. Concentration Fluctuations in a Smoke Plume. *Atmos. Environ.* 18, p. 1091-1106.

Hanna, S.R., 1985. *Air Quality Modelling over Short Distances*. Handbook of Applied Meteorology, Chapter 25. Edited by Houghton, D.D., University of Wisconsin.

Hanna, S.R., 1986. Spectra of Concentration Fluctuations: The Two Time Scales of a Meandering Plume. *Atmos. Environ.* 20, p.1131 - 1137

Holtslag, A.A.M., 1984. Estimates of Diabatic Wind Speed Profiles from Near Surface Weather Observations. *Boundary-Layer Meteorol.* 29, p. 225 - 250.

Högström, U., 1972. A Method for Predicting Odour Frequencies from a Point Source. *Atmos. Environ.* 6, p. 103 - 121.

Rantakrans, E. and Savunen, T., 1995. *Modelling of the Dispersion of Odours*. Finnish Meteorological Institute, ISSN 0782-6095 Publications on Air Quality, 70 p. (Text in Finnish, English Summary)

Wilson, D.J., 1976. *Contamination of Building Air Intakes from Nearby Vents*. Dept. Mech. Eng. Report No. 1, University of Alberta, Edmonton, Canada.

Wratt, D.S., 1987. An Experimental Investigation of Some Methods of Estimating Turbulence Parameters for use in Dispersion Models. *Atmos. Environ.* 21, p. 2599 - 2608.

DISCUSSION

D. WILSON: What range of exposure times (e.g. 1 second to 3600 seconds) were you trying to simulate with the meandering plume model?

T. SAVUNEN: The odour dispersion model registers as an odour hour every hour which one 30 seconds odour concentration has exceeded the odour threshold i. e. odour is observed. Thus, the odour needs not to be present for the whole hour.

DEVELOPMENT AND INITIAL APPLICATION OF THE MICRO-CALGRID PHOTOCHEMICAL MODEL FOR HIGH-RESOLUTION STUDIES OF URBAN ENVIRONMENTS

R. Stern[1] and R. J. Yamartino[2]

[1]IVU, Gesellschaft für Informatik, Verkehrs- und Umweltplanung mbH
Burgweg 10 , D-79350 Sexau
[2]Atmospheric Studies Group, EARTH TECH, Concord MA 01742 USA

INTRODUCTION

In most major cities, air pollution is still a severe problem, and there are frequent exceedances of air quality standards for many primary and secondary pollutants. In particular, the high emissions from vehicles can lead to high concentrations within the urban canopy, where the atmospheric flow and turbulence is disturbed by buildings or other obstacles. The high emissions of vehicle NO_x from urban centers serves to lower O_3 levels within major cities. Nevertheless, NO_2 levels may be extremely high in such urban environments, and questions arise about the role of urban street canyons as "pre-reaction chambers" that may further enhance NO_2 formation. To understand the ramifications of proposed emission reduction strategies on the street canyon level, photochemical models for high resolution simulation of urban core air quality are needed. Previous modeling on the urban street canyon scale has either focussed on non-reactive pollutants or has invoked simple approximations for dealing with simplified NO-NO_2-O_3 chemistry (Yamartino and Wiegand, 1986; Berkowicz et al., 1997). While such simplifications may be appropriate within a single canyon for receptors nearby the emission sources, they are not reasonable for addressing questions associated with the series of canyons or quarters as are found in the urban core. This paper summarizes the development of a full photochemical model for high resolution simulation of urban core air and reports the results of the initial applications of this model, MICRO-CALGRID, to multiple and single street canyons. The model development was funded by the German Federal Environmental Agency (Umweltbundesamt).

MODEL OVERVIEW

Instead of building a completely new model, the urban-scale photochemical model CALGRID was used as starting point for the new micro-scale model. The CALGRID photochemical model (Yamartino et al., 1989; Yamartino, 1996) represents a fully second-

Air Pollution Modeling and Its Application XIII, edited by S.-E. Gryning
and E. Batchvarova, Kluwer Academic/Plenum Publishers, 2000

generation photochemical model for regulatory application. CALGRID's state of the science model features include:

- A vertical transport and diffusion scheme that incorporates the latest boundary layer formulations, correctly accounts for atmospheric density variation, permits several vertical level spacing approaches and accounts for all vertical flux components when employing either dynamic or fixed levels;
- A full resistance-based model for the computation of dry deposition rates as a function of geophysical parameters, meteorological conditions, and pollutant species;
- A chemical integration solver based on an adaptive time-step implementation of the QSSA-method. This solver can efficiently and accurately handle stiff modern schemes;
- Two modern photochemical reaction schemes, the SAPRC-90 and the CBM-IV chemical mechanism; and,
- A horizontal advection scheme based on spectrally-constrained cubics (Yamartino, 1993) that conserves mass exactly, prohibits negative concentrations, and exhibits low numerical diffusion.

For MICRO-CALGRID, a number of modifications to CALGRID were needed, including:

- Modification of CALGRID's framework to allow for simulation of 2-d domains (i.e., downwind and vertical planes assuming infinite crosswind domain extent), as well as the more traditional 3-d domains, thereby enabling modeling of large spatial domains with spatial resolutions down to a few meters;
- Conversion of the vertical grid system from terrain following coordinates to absolute vertical coordinates to accommodate the vertically walled buildings of urban cores, and allow for part of the modeling domain to be filled with impenetrable surfaces;
- Correction of CALGRID's operator-split advection methodology to account for the high wind gradients seen on very fine scales;
- Addition of vehicle-generated turbulence;
- Addition of a simple NO-NO_2-O_3-chemistry scheme that reflects the short traveling times of pollutants within modeling domains extending over only one or a just a few canyons;
- Linkage of MICRO-CALGRID to the flow model MISCAM that can consider geometries of arbitrary complexity and provides MICRO-CALGRID with the atmospheric flow and turbulence fields; and,
- Linkage of MICRO-CALGRID to CALGRID, thus allowing integrated (one-way nested) applications in the urban and the microscale.

Treatment of Vehicle Turbulence

Most of the treatments of vehicle-induced turbulence have been developed for simple, single lane, flat terrain roadway situations, where the complexities of sidewall interactions, curved and recirculating flow, and lanes of opposing flow traffic are absent. Such flat terrain treatments span a wide range of complexity. For example, the CALINE-4 model defines a "turbulent mixing zone" and simply assigns a turbulent vertical velocity, σ_w, for pollutants in this zone (Benson, 1984). At the other extreme, the vehicle wake theory of Eskridge and Hunt (1979) considers the velocity deficit 'plume' produced by a single vehicle moving at speed, V, relative to an arbitrary vector wind, **U**, and describes both the advective and diffusive portions of this plume. This theory was extended to a steady flow of vehicles and was incorporated into the ROADWAY model (Eskridge and Catalano, 1987). Other models, such as the CPB model of Yamartino and Wiegand (1986) have utilized the ROADWAY model predictions to develop a simple model of the size of the initial, vehicle induced mixing zone, $\sigma_z(0)$, based on vehicle size and speed. A similiar approach was used in a street canyon model developed by Berkowicz et al. (1997); however, such approaches cannot easily be transferred over to a numerical grid model, such as MICRO-CALGRID, as what is needed is the vehicle-induced enhancement to the diffusivity fields, K_x, K_y, K_z.

The MICRO-CALGRID approach considers the energy dissipated by a vehicle as it pushes its way through the ambient air. For a vehicle of frontal surface area, A, moving at velocity, V, relative to the ambient air, the force exerted on the vehicle is given as:

$$F = C_d A \ \rho V^2 / 2 \tag{1}$$

where ρ is the density of air and C_d is the drag coefficient of the vehicle relative to a flat plate perpendicular to the flow. F is also the energy per unit length (of along-roadway travel) that the vehicle expends in moving against the air resistance and this energy appears in the form of advective and turbulent air motions. Noting that only a fraction, f, of this energy will be turbulent TKE, that there is a steady flow of N vehicles per second, and that the x, z cell containing the vehicles has an area, $\Delta x \ \Delta z$, that fully encompasses the car frontal area A, one arrives at the turbulent energy production rate, P, per unit volume as:

$$P = f F N / (\Delta x \ \Delta z) \tag{2}$$

and a steady-state TKE within this production cell of:

$$TKE = P \tau \tag{3}$$

where τ is the lesser of the Eulerian time scale and the advective time scale $\Delta x/u$, and u is the x component of wind at the level of the vehicles and perpendicular to the traffic flow direction, y. Adding the definition of TKE as $TKE=(\sigma_u^2+\sigma_v^2+\sigma_w^2)/2$ and assuming TKE isotropy so that $\sigma_u=\sigma_v=\sigma_w$, one obtains the velocity variance in the initial cell as:

$$\sigma_w^2(0) = (f C_d /3) [A/(\Delta x \ \Delta z)] V^2 N \ \tau \tag{4}$$

Though a complete solution of the transport, dispersion, and dissipation of the TKE would require a Navier-Stokes solver (see for example Kuhlmann and Eichhorn, 1998), a first approximation of the vehicle-induced turbulence field for cells distant from the production could be:

$$\sigma_w^2(x,z) = (f C_d /3) [A \exp(-t/\tau_L)/d^2] V^2 (N \tau) \tag{5}$$

where $d^2 = x^2 + z^2 + \Delta x \ \Delta z$, t represents a 'transport' time from the production cell to the relative point (x,z), and τ_L is the Lagrangian time scale for the turbulent wake. Determining this 'transport' time in a complex flow environment would itself involve a model, but in the low wind speed environment of an urban street canyon, we shall approximate it as a diffusion time scale, given as: $t=d/\sigma_w(0)$. The Lagrangian time scale in the street canyon environment is estimated using the relation $\tau_L=(3/2) C_d (k/\varepsilon)$, where k and ε are the ambient TKE and TKE dissipation rates, respectively, as estimated from MISCAM runs. These runs yield a typical τ_L of 3-5 seconds. Diffusivities are then computed using the relation:

$$K_z = \sigma_w^2(x,z) \tau_L \tag{6}$$

Finally, we note that if the time scale τ introduced in Eq. (3) is first approximated as the Eulerian time scale, τ_E, it may then be related to the Lagrangian time scale for a single wavelength, λ, disturbance via $\tau_E = \tau_L / [1 + \tau_L u /(\alpha\lambda)]$, where u is the local wind speed and α represents some fraction, of order 1/4, of a wavelength. Knowing that two roll eddies are produced in a vehicle's wake and assuming the width of this wake region is Δx, one obtains $\alpha\lambda=\Delta x/4$, so that the Eulerian time scale can be calculated as

$$\tau_E = \tau_L / [1 + 4 \tau_L u / \Delta x] \tag{7}$$

which is now seen to embody both of the constraints mentioned in conjunction with the τ of Eq. (3).

The MISCAM Model

MISCAM (Micro-Scale Air Pollution Model), is a three-dimensional flow and dispersion model for inert species (Eichhorn, 1996). The flow model portion is used as the meteorological driver for MICRO-CALGRID and is based on the non-hydrostatic Eulerian equations of motion employing the anelastic Boussinesq approximation. Buildings and obstacles are treated explicitly as impenetrable surfaces. Turbulence closure is based on a k-ε model. Starting from a horizontally uniform wind field, the prognostic equations are solved on an arbitrary Cartesian grid until steady state conditions are reached. Inputs to the model include the three dimensional building geometry of the area considered, a wind speed and wind direction, and a vertical stability measure. Outputs of the model are the three dimensional flow and turbulence fields for the specified meteorological condition. These flow model output data are then used as input for MICRO-CALGRID.

MICRO-CALGRID MODEL APPLICATION TO A SIMPLE 2-D GEOMETRY

The simple 2-d geometry that is first considered is shown in Figure 1 and consists of three canyons 21m wide and bounded by buildings 21m high. The domain is modeled with a horizontal resolution of 3m and a vertical resolution of 3m near the surface, but increasing with height so as to cover the 100m vertical domain in 19 vertical layers.

MISCAM was used to generate the flow and turbulence fields based on a neutrally-stratified, perpendicular inflow of 5 m/s (i.e., defined at z=40m) assuming a roughness length of 1.0 m. Figure 1 also shows the resulting stationary windfield in the 2-d cross-section, with vortices in each canyon clearly visible.

Vehicle emissions are generated by four lanes of traffic within the first (i.e., leftmost) canyon only, and emission rates corresponding to traffic levels of approximately 40,000 vehicles per day were chosen to simulate a realistic situation. The vehicles were assumed to emit NO (90% of NO_x), NO_2 (10% of NO_x), CO and VOC speciated according to the SAPRC-90 chemical mechanism. The MICRO-CALGRID model run was started at 12 h LST assuming initial and boundary concentrations of: O_3 as 50 ppb, NO as 1 ppb, NO_2 as 1 ppb, and VOC and all other species as 0 ppb. In addition, clear sky insolation conditions and time dependent photolysis rates were considered. Figure 2 shows the calculated O_3, NO_2 and NO cross-sectional distributions after 600 s of simulation time, when quasi-stationarity is reached (n.b., small changes in the fields continue due to changes in the photolysis rates). After 600 s of simulation time, mass out of the first canyon is transported downwind through the horizontal grid. O_3 is consumed by NO in all three canyons to varying degrees. Compared to the background ozone concentration level of 50 ppb, one observes an average decrease of approximately 50% in the first canyon, where the precursors are emitted, and decreases of approximately 20% and 10% in the source-free second and third canyons, respectively. On the other hand there is a formation of NO_2 due to the reactions between O_3 and NO. The decrease in the maxima of the quasi-inert (i.e., on the time scale considered here) species CO from canyon 1 to canyon 3 is smaller (CO maxima in canyon 2 is 6%, in canyon 3 is only 3% of that in canyon 1) than the decrease of the NO maxima (NO maxima in canyon 2 is 2%, in canyon 3 is only 1% of that in canyon 1), because NO is oxidized to NO_2 by the reaction with O_3. Because of the photochemical formation of NO_2, NO_2 decreases from canyon 1 to canyon 3 are an order-of-magnitude smaller than corresponding NO decreases (NO_2 maxima in canyon 2 is 33%, in canyon 3 is 23% of that in canyon 1).

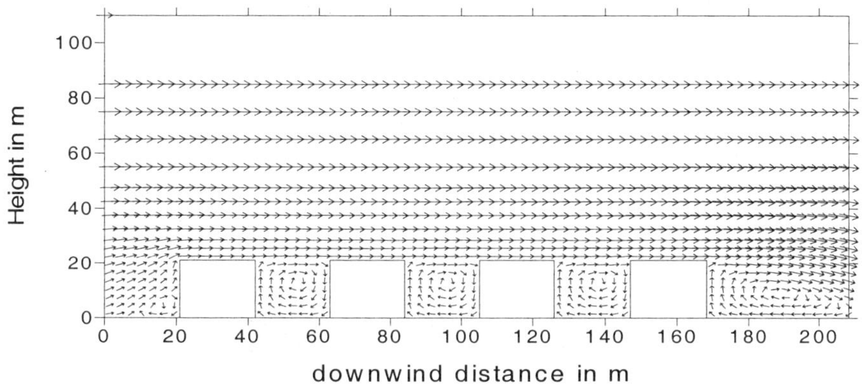

Figure 1. Vertical cross section of the MISCAM wind field used for the MICRO-CALGRID application to the 3 canyon case. Upwind wind speed at height of 40 m is 5 m/s.

LINKED MICROCALGRID-CALGRID APPLICATION

The urban scale photochemical model CALGRID with the SAPRC-90 mechanism has been applied in Germany to the greater Berlin domain using a horizontal resolution of 2 km (Stern et al., 1997; Umweltbundesamt, 1997). The calculated concentrations and the meteorological conditions within one grid column in the urban core of Berlin were extracted for one day of the modeled episode, and have been utilized as background conditions for a MICRO-CALGRID application to a series of street canyons embedded within the selected CALGRID urban grid cell. The meteorological input for MICRO-CALGRID was calculated with MISCAM and based on the grid-cell-specific meteorological information from CALGRID. The aim of the application was to get some insight into how CALGRID's urban scale concentrations would be modified within an embedded street canyon having high traffic emissions. Figure 3 shows the time series of NO, NO_2 and O_3 for a 24 hour, MICRO-CALGRID calculation using the aforementioned CALGRID results as initial and boundary conditions. The time series of the CALGRID calculations are also shown in Figure 3. The results are presented for two receptors, one on each side of the road. It is obvious that the urban scale concentrations of CALGRID are strongly affected by the sub-grid-scale emissions of the street canyon. Ozone concentrations in the street canyon are generally much smaller, and NO and NO_2 concentrations are generally much higher, than the respective urban scale concentrations. Which of the two receptors in the street canyon receives the higher NO_x concentrations and the lower O_3 concentrations depends on the prevailing above rooftop wind direction. In the case considered here, the urban scale wind direction changes in the early afternoon in such a way that the direction of the vortex in the street canyon reverses. Hence, the concentration maxima switch from the former downwind receptor 1 to the new downwind receptor 2 on the opposite side of the street.

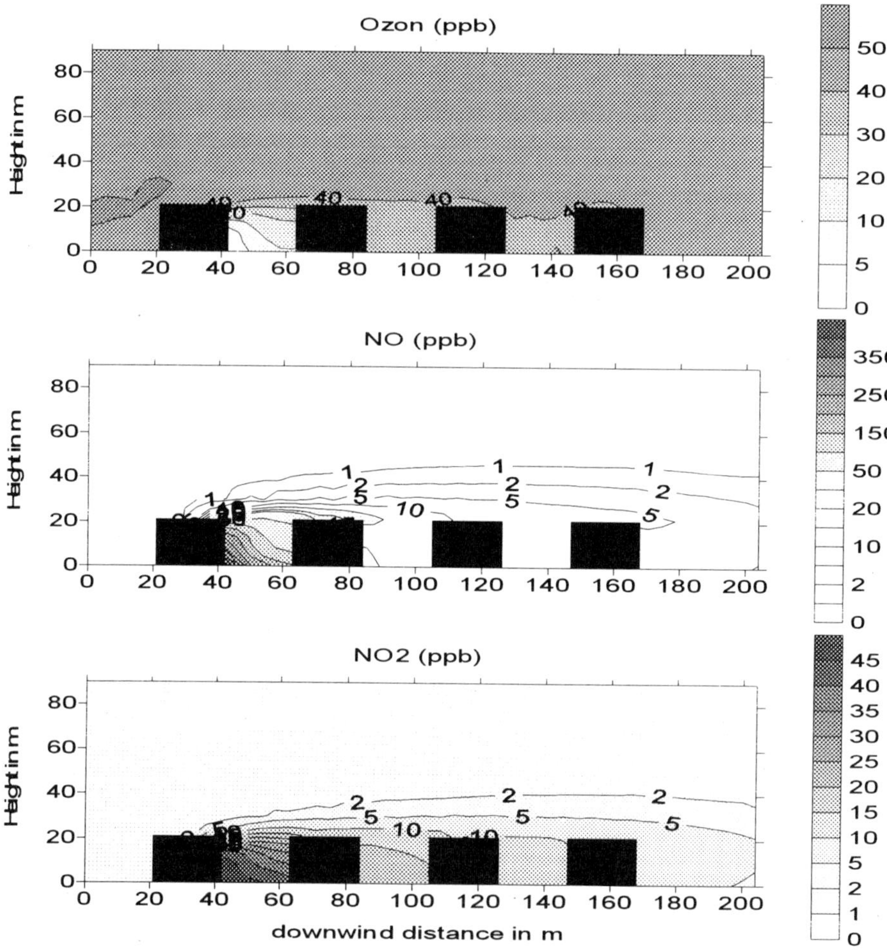

Figure 2. Vertical cross sections of O_3, NO, and NO_2 concentrations after 600 s of simulation time using MICRO-CALGRID with SAPRC-90. Inflow is from the left. Emissions occur only in the leftmost canyon. The underlying wind field is shown in Fig. 1.

INFLUENCE OF TRAFFIC INDUCED TURBULENCE

The effects of MICRO-CALGRID's parameterization for traffic induced turbulence on the concentrations in a street canyon were investigated for three different traffic scenarios that were developed from the traffic data of a busy four lane street canyon in Berlin. The three scenarios are: High traffic (HT) density with 750 vehicles per hour per lane at an average speed of 25 km/h; medium traffic (MT) density with 470 vehicles per hour per lane at a speed of 35 km/h; and low traffic (LT) density with 200 vehicles per hour per lane at a speed of 55 km/h. Results for the downwind roadside concentration ratio c(with vehicle turbulence)/c_0(without vehicle turbulence) of an inert species are shown in Figure 4 for a

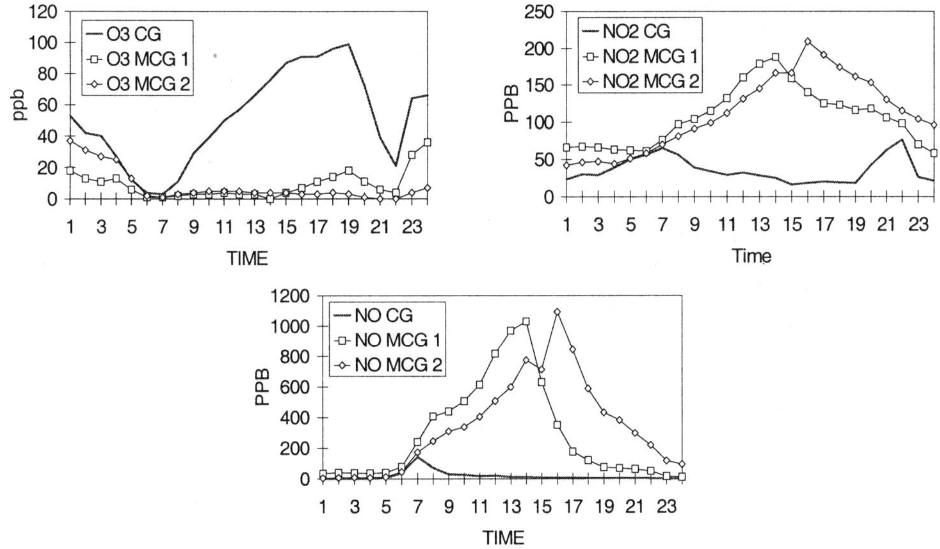

Figure 3. MICRO-CALGRID NO, NO_2 and O_3 time series for two roadside receptors (MCG 1 and MCG 2) in a street canyon embedded in a 2x2 km^2 grid cell of CALGRID. Also shown are the CALGRID time series (CG) for that cell.

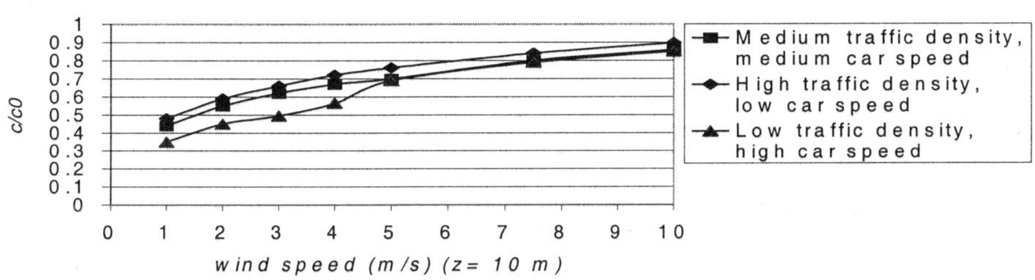

Figure 4. The concentration ratio, c(with vehicle turbulence)/c_0(without vehicle turbulence), at the downwind roadside as function of the inflow wind speed for three different traffic situations (see text).

number of different wind speeds u. The calculations have been carried out assuming that 75% of the energy expended by vehicles in moving against the air resistance converts to TKE and a Lagrangian time scale of 4 s. At low wind speeds, the roadside concentrations are reduced by a factor of two to three due to the traffic induced turbulence. The lowest ratio of c/c_0 is calculated for the the case of high car speed and low traffic density, because the vehicle induced turbulence, $\sigma_{w}^{2}{}_{(0)} \approx V^2*N$, is quadratic in the vehicle speed, V, but only linear in the vehicle density, N. At higher wind speeds, the significance of the vehicle turbulence diminishes relative to ambient turbulence levels. A first intercomparison with data obtained from a wind tunnel study (Kastner-Klein et al., 1998) is shown in Figure 5. In this figure, the ratio c/c_0 is plotted against the the velocity ratio, V/u, scaled by the dimensionless factor, $a^{1/3} = [(N/V)/0.067]^{1/3}$, which accounts for the traffic density and the wind tunnel to full-scale scaling ratio (see Kastner-Klein et al., 1998). Also shown in Figure 5 is the regression line, $c/c_0=1–0.18X$ (where $X=a^{1/3}V/u$), proposed by Kastner-

Klein et al. (1998) and labeled KK, and a revised formulation, $c/c_0 = 1/(1+.24X)$, labeled SY, which can be shown to be closely related to the physical picture of ambient turbulence acting on a vehicle-turbulence generated, well-mixed zone (or initial σ_z in the Gaussian plume sense) and which avoids the parametrization related problem of c/c_0 vanishing or becoming negative. This revised parameterization is seen to be appropriate out to values of X of order 10.

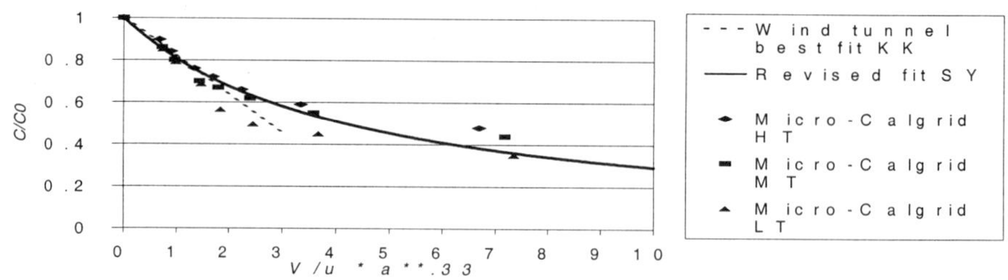

Figure 5. Influence of traffic induced turbulence. Comparison of MICRO-CALGRID simulations with wind tunnel data. See text for further explanations.

REFERENCES

Benson, P.E., 1984, CALINE 4 – A dispersion model for predicting air pollutant concentrations near roadways. FHWA User Guide, Report No FHWA/CA/TL-84/15, Trinity Consultants Inc., USA

Berkowicz, R., Hertel, O., Larsen, S., Sorensen, N.N., and Nielsen, M., 1997, Modelling traffic pollution in streets. National Environmental Research Institute, Roskilde, Denmark.

Eichhorn, J., 1996, Validation of a microscale pollution dispersal model. Air Pollution Modeling and ist Application XI. Plenum Press

Eskridge, R.E. and Catalano, J.A., 1987, Roadway- A numerical model for predicting air pollutants near highways. U.S. EPA Report EPA/600/8-87/010.

Eskridge, R.E. and Hunt, J.C.R., 1979, Highway Modeling. Part I: Prediction of velocity and turbulence fields in the wakes of vehicles. J. Appl. Meteor. 18, 387.

Kastner-Klein, P., Berkowicz, R., Rastetter, A. and Plate, E.J., 1998, Modelling of vehicle induced turbulence in air pollution studies for streets. 5th International conference on Harmonisation within Atmospheric Dispersion Modelling for Regulatory Purposes. 18-21 May 1998, Rhodes, Greece

Kuhlmann, R. and Eichorn, J.,1998, Parameterization of traffic induced turbulence in the three-dimensional air pollution model MISCAM. 5th International conference on Harmonisation within Atmospheric Dispersion Modelling for Regulatory Purposes. 18-21 May 1998, Rhodes, Greece

Stern, R., Scherer, B., Fath, J. Reimer, E., 1997, Ermittlung und Bewertung der Wirkung von Ozonminderungsmaßnahmen im Großraum Berlin-Brandenburg mit Hilfe des photochemischen Ausbreitungsmodells CALGRID. Freie Universität Berlin, Institut für Meteorologie. FE-Vorhabens 10402812/01 "Aktionsprogramm und Maßnahmenplan Ozon" des Umweltbundesamts.

Umweltbundesamt, 1998, Programme of control concepts and measures for ozone. www.umweltbundesamt.de/ozon

Yamartino, R. J., 1996, U.S. and German applications of the CALGRID model on the regional and urban scales. 5th US-German workshop on photochemical ozone problem and its control. Berlin Sept. 24-27, 1996

Yamartino, R. J., 1993, Nonnegative, conserved scalar transport using grid-cell-centered, spectrally constrained Blackman cubics for applications on a variable-thickness mesh. Mon. Wea. Rev., 121, 753-763.

Yamartino, R. J., Scire, J.S., Carmichael, G.R. Chang Y.S., 1989, The CALGRID mesoscale photochemical grid model-I. Model formulation. Atmos. Environ., 26A, 1493-1512.

Yamartino, R. J. and Wiegand, G., 1986, Development and evaluation of simple models for the flow, turbulence and pollutant concentration fields within an urban street canyon. Atmos. Environ., 20, 2137-2156.

DISCUSSION

P. SAHM: Car-induced turbulence is calculated for a single car and then multiplied by the number of cars. Do you account for the effect that with increasing traffic flow N (vehicles/s), the car-induced turbulence does not remain a linear function of N? That is, for $N \rightarrow \infty$ do you get $TKE \rightarrow \infty$?

R. YAMARTINO: In the current model, the TKE is simply proportional to N, but you are quite right. One could imagine that if N rises to the point that the cars are bumper-to-bumper - more like a train than a stream of separate vehicles - that the wake TKE might, in fact, be quite small. However, we think that for the typical traffic densities of interest, the approximation of independent wakes is not too bad. However, to really evaluate the effect you mention, the multiple vehicle wakes would have to be simulated in a full CFD model. Additionally, I should note that most traffic models would yield lower vehicle speeds, V, at higher traffic flows, N. As the TKE goes as N*V**2, the turbulence will be not be a simple linear function of N in any case.

X. CAI: In order to simulate the correct turbulence within the canyon, the boundary conditions for wind and turbulent kinetic energy may be very important. How does this model specify the boundary conditions and do you plan to validate the simulation of turbulence?

R. YAMARTINO: The ambient flow and turbulence fields used in Micro-CALGRID are taken directly from the output of the MISCAM Navier-Stokes flow generator employing k-ε closure. I agree that the flow/turbulence is sensitive to assumed boundary conditions, but the flow simulations I have seen from MISCAM look quite good in comparison with wind tunnel studies. In addition, the passive pollutant studies using MISCAM compare quite well with the more empirical street canyon model, CPB, that provides an excellent match to full-scale street canyon studies. We have no plans to validate MISCAM independently, but hope to do additional work on the vehicle turbulence module within Micro-CALGRID.

S. RAFAILIDIS: When you parameterize the vehicle-induced TKE, you must also include the thermal effects from the vehicles, as these are of order 500 times those due to solar insolation.

R. YAMARTINO: The MISCAM N-S flow and turbulence driver is primarily designed for neutral conditions and the current vehicle-induced turbulence algorithm in Micro-CALGRID ignores vehicle heat output, except that there is an overall multiplicative constant which scales the TKE. If one believes the TKE production rate is some significant fraction of the engine's thermal/mechanical power, then the thermal term could be incorporated into such a constant.

ON THE SUMMER EPISODES OF TOTAL SO_2 CONTENT IN THE AIR COLUMN OVER THE CITY OF THESSALONIKI, GREECE

C. Zerefos[1], K. Ganev[2], A. Vasaras[1], D. Syrakov[3], K. Kourtidis[1], M. Tzortziou[1], M. Prodanova[3], E. Georgieva[2]

[1]Laboratory of Atmospheric Physics, Aristotle University of Thessaloniki Thessaloniki 54006, Greece
[2]Institute of Geophysics, Bulgarian Academy of Sciences Akad. G. Bonchev str., block 3, Sofia 1113, Bulgaria
[3]National Institute of Meteorology and Hydrology, Bulgarian Academy of Sciences
Tzarigradsko Chausee 66, BG-1784, Sofia

ABSTRACT

The purpose of this study is to clarify the origin of columnar SO_2 episodes seen at Thessaloniki (40.5°N, 22.9°E) during the period 1994, to compare the columnar SO_2 values with surface measurements of SO_2 and explain the second maximum peak which appears in August in the columnar SO_2 time-series, recorded by a Brewer spectrophotometer since 1982. This continuous record provides an average columnar SO_2 of 2 m-atm-cm during the 15 years of operation. Although the maximum concentrations of columnar SO_2 originate form local sources in the winter, significant columnar SO_2 amounts are also seen in the summer. On a few occasions however, as high as 5-7 m-atm-cm of columnar SO_2 have been measured both in winter and summer period. Although the former originate from local sources in the city of Thessaloniki, it appears that the sources were not local during the summer. This study aims at identifying the origin of these high total columnar SO_2 sources which are formed in regions located at distances from 50 to 200 km, or more, to the north-northeast of Thessaloniki.

INTRODUCTION

Total columnar SO_2 over Thessaloniki is continuously monitored since March 1982 with the Brewer spectrophotometer 005 (Brewer, 1973; Zerefos et al., 1986; Bais et al., 1993) at the Laboratory of Atmospheric Physics, LAP, University of Thessaloniki, Greece, (40.5° N, 22.9° E). The daily time-series of columnar sulphur dioxide over the city of Thessaloniki for 1994 is plotted in Fig. 1, from which it appears that the maximum

Air Pollution Modeling and Its Application XIII, edited by S.-E. Gryning and E. Batchvarova, Kluwer Academic/Plenum Publishers, 2000

concentrations of SO_2 are found in February, averaging to 1.84 m-atm-cm (DU) and during the warmer part of the year.

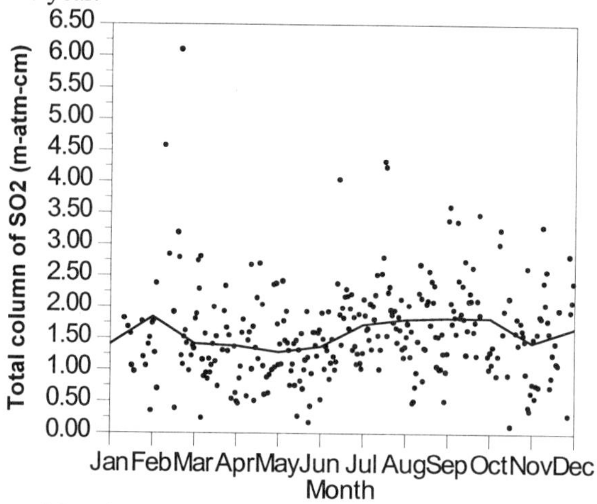

Figure 1. Daily columnar SO_2 values (m-atm-cm) from Brewer-LAP during 1994 and its seasonal variation (heavy line).

For the period between January and August of 1994, surface measurements of SO_2 have been recorded from a DOAS system (OPSIS, model AR 500) and are compared with the columnar SO_2 measurements for the same period. The DOAS (Differential Optical Absorption Spectroscopy) technique is described in detail elsewhere (e.g. Edner et al., 1993). The length of the monitoring path was 165 m and the mean height above ground was about 50 m, while the DOAS receiver was collocated with the Brewer at LAP.

For August 1994, ground-based measurements of SO_2 have been recorded from the UVF (Ultraviolet Fluorescence) method in two monitoring stations (Eptapirgio and Martiou) of the Municipality of Thessaloniki and are also compared with the columnar SO_2 measurements for the same month.

STATISTICAL ANALYSIS OF EXPERIMENTAL DATA

The first maximum peak in the time-series of columnar SO_2 (in February), Fig. 1, can be explained from the local emissions of the city which are increased during the winter time. However, the peaks during the warmer part of the year and particularly in August do not originate from local sources, as evidenced in Fig. 2. This figure shows the comparison

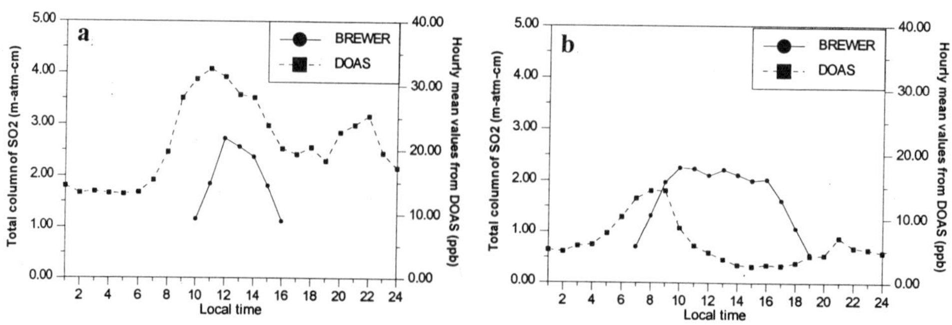

Figure 2. Comparison of diurnal variation of columnar and ground level SO_2 for: a) February 1994 and b) July-August 1994.

of diurnal variation of columnar and ground level SO_2, for February 1994 and for July-August of the same year. Although there is evidence for lag correlation between the diurnal variation of columnar and ground level SO_2 in the winter, there is evidence for high total columnar SO_2 values during the day which can not be attributed to local sources.

Looking at the wind directions at 850 hPa (12:00 UTC) for the winter and the summer months of 1994, and the corresponding daily mean concentrations of the total columnar SO_2, over Thessaloniki, we have constructed the rose diagrams seen in Fig. 3.

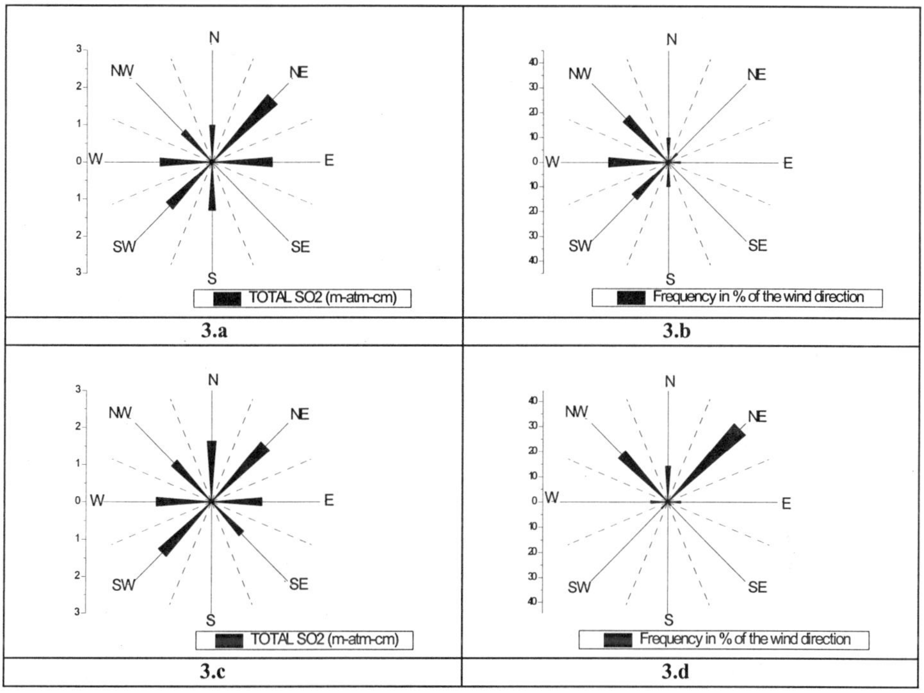

Figure 3. Rose diagrams of the mean total columnar SO_2 concentration for: a) January-February 1994 and c) July-August 1994; Rose diagrams of the wind direction for: b) January-February 1994 and d) July-August 1994 (Thessaloniki-LAP).

It is obvious that both in winter and summer the high columnar SO_2 concentrations, are observed in days with northeast winds, although the predominant winds are from NW for both winter and summer. So, it seems that the high total columnar SO_2 values recorded over Thessaloniki, especially during the summer (in July and August), are due to SO_2 plumes coming from north-northeast directions. In the winter time, also northeast winds contribute to the higher SO_2 concentrations, but local sources are the main contribution to the maximum values which form the winter peak.

A typical example of the contribution evidence of remote sources to the columnar SO_2 over Thessaloniki is shown in Fig. 4, for August 1994. Figure 4 compares ground level concentrations of SO_2 at three sites (LAP, Eptapirgio, Martiou Avenue) with columnar SO_2 measurements at LAP. The comparison shows that there is at least one period (early August) of high columnar SO_2 burden which can not be explained by local sources. This period is studied together with a period (end of August) of lower SO_2 column in the next paragraph.

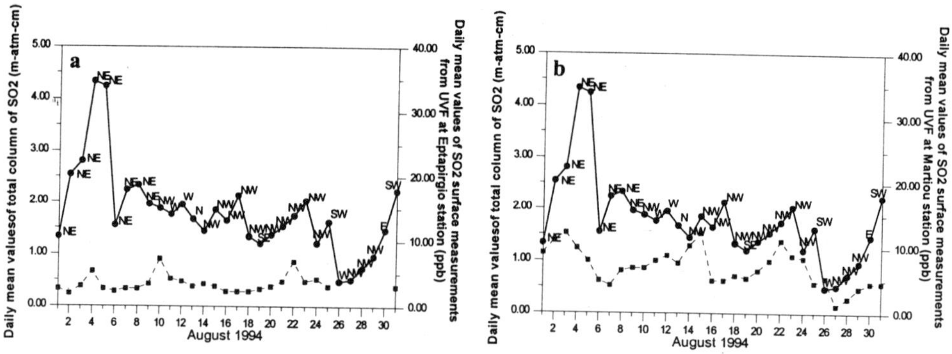

Figure 4. Comparison of the daily mean columnar SO_2 values from Brewer-LAP with ground level SO_2 values, recorded at: a) Eptapirgio, b) Martiou and the wind direction at 850 hPa (12:00 UTC) for August 1994.

NUMERICAL SIMULATIONS

The above stated messages from the experimental and statistical studies are clear enough. Some numerical studies of particular episodes are also carried out, in order the dynamic processes which form the columnar SO_2 in the city of Thessaloniki to be specified and it's origin to be determined. Two episodes are chosen for the purpose - a typical episode with high SO_2 contents at 4 and 5 August 1994 (episode 1) and a typical episode with low SO_2 contents at 26, 27 and 28 August 1994 (episode 2).

Models and Data

A multi-layer Eulerian model (Syrakov et al. 1988, 1989; Syrakov and Ganev 1989) is used for simulation of the pollution transport over the Balkan peninsula. The integration domain is a sub-region of the standard EMEP domain. The horizontal grid with steps 25km is nested in the EMEP's one. The meteorological information used for the simulations (wind field at 850 hPa, potential temperature difference between ground and 850hPa and rain intensity) is taken from the diagnosis of the Europa-Model of the RSMC-Offenbach, Germany and interpolated in the grid points of the domain of integration by a procedure described by Syrakov and Andreev (1995). The mentioned synoptic information is used for constructing the wind and turbulent coefficient profiles. Two ways for reconstructing the meteorological fields are applied - a simple one-dimensional PBL model (Yordanov et al., 1983), which further in the text will be referred to as model 1 and the diagnostic mass-consistent model DIAMO (Georgieva and Godev, 1987; Georgieva, 1990) which partially reflects the topography induced deformations by using the continuity equation as a constrain in the procedure of interpolation of the wind field (model 2).

For the Bulgarian Sulfur sources a detailed emission inventory is available (Syrakov and Andreev, 1995), while for the other countries in the domain the official EMEP inventory (Tuovinen et al., 1994) is used.

Discussion of the Results

The evolution of the simulated columnar SO_2 for the city of Thessaloniki, averaged for the period 10.00 - 15.00 h local time (8.00 - 13.00 UTC) is shown at Fig.5 together with the corresponding measured values. As it can be seen, there is a good qualitative and reasonable quantitative agreement between the simulated and measured values. The

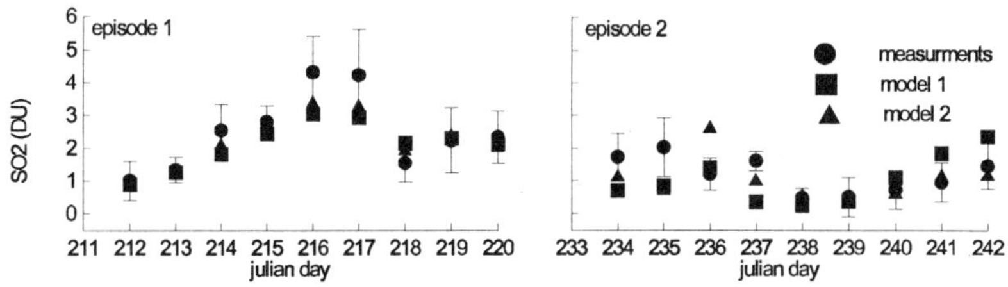

Figure.5. Evolution of the measured and simulated columnar SO_2 [DU] over the city of Thessaloniki for both the studied episodes. The error bars show the standard deviation of the measured columnar SO_2.

numerical experiments show that the application of the DIAMO model for reconstructing of the wind fields slightly improves the simulations, which is a demonstration of the significance of the topography effects on the air pollution transport in local scales. Unlike

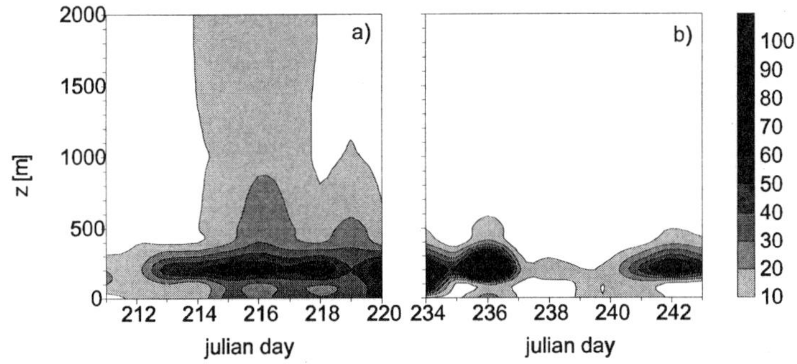

Figure 6. Evolution of the vertical distribution of the SO_2 concentration [$\mu g/m^3$] for episode 1 (a) and episode 2 (b).

some other case studies, however (Syrakov et al., 1989; Ganev et al., 1997), the improvement is not so significant. The possible reason is that for the present simulations data from ground based stations were not available, and it is well known that the diagnostic models of DIAMO type are as much effective as more data from ground observation is used in addition to the synoptic information. Anyway, the agreement with the measurements is good enough, so that the model can be evaluated as a sufficient one for studying the origin of the SO_2 pollution of Thessaloniki for the chosen episodes.

The evolution of the vertical profiles of SO_2 for the city of Thessaloniki for both episodes are shown at Fig. 6. The vertical SO_2 distribution changes in time in a rather complex way and there is no strong co-relation between the concentrations at different levels. This is in agreement with the experimental (see Fig.2.) facts and is an evidence that the SO_2 pollution of Thessaloniki is formed as a result of a complex interaction of different transport scales and dynamic processes.

Functions of Influence

The functions of influence, concerning the air column above Thessaloniki are simulated in addition to the SO_2 concentration field. The functions of influence are

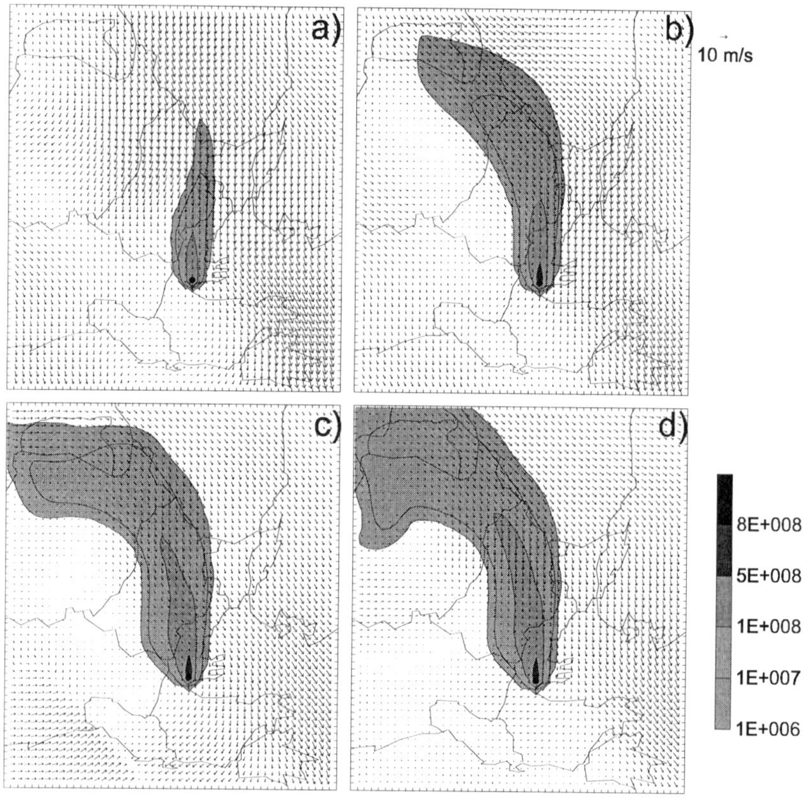

Figure 7. Evolution of the time integrated functions of influence at 00 UTC for 5 (a), 4 (b), 3 (c) and 2 (d) August, concerning the columnar SO_2 over Thessaloniki, averaged for the period 4-5 August 1994.

solutions of equation conjugated to the pollution transport equation (Marchuk, 1982; Ganev, 1991) . They are the Eulerian analog of the back-wind trajectories and give the possibility to evaluate the contribution of each source to the pollution in a region of interest (in this particular case - Thessaloniki).

The evolution of the time integrated functions of influence for the studied episodes are shown at Figs. 7,8. They, together with the corresponding wind fields at 850 hPa, demonstrate very well the origin of the columnar SO_2 in Thessaloniki for both the episodes. The evaluated contribution of the countries in the region to the SO_2 pollution of the city is given in Table 1. As it can be seen for episode 1 more than 80% of the SO_2 originates from foreign sources located northeast from Thessaloniki. For episode 2 the SO_2 originates from west-northwest from Thessaloniki and the contribution of the Greek sources is the most significant one.

CONCLUSIONS

From the above discussion the following conclusions can be drawn:

1) The total SO_2 content over the city of Thessaloniki adequately describes the SO_2 load either from and/or local and remote sources.

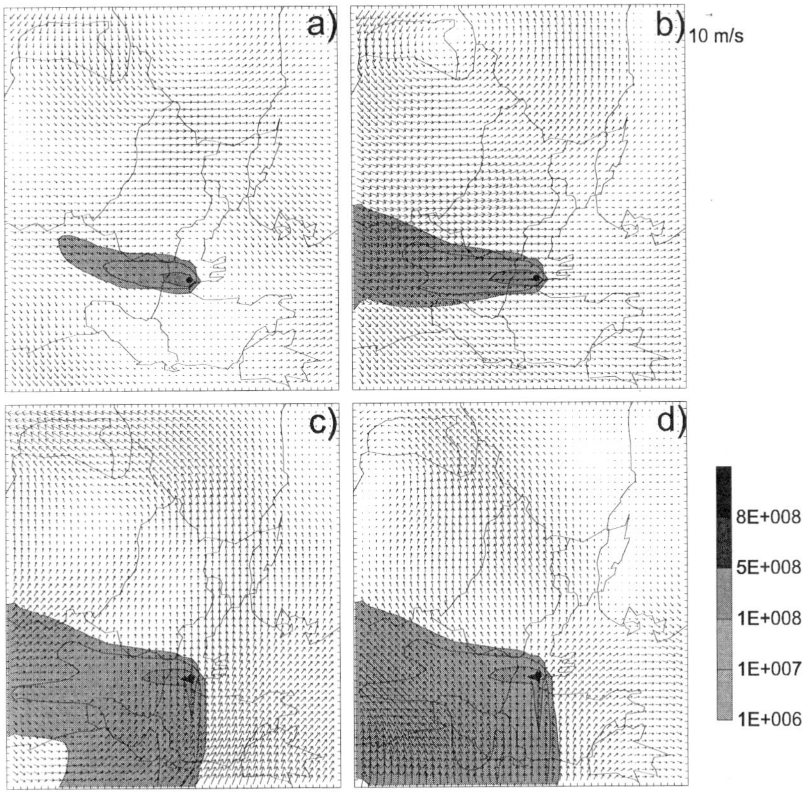

Figure 8. Evolution of the time integrated functions of influence at 00 UTC for 28 (a), 27 (b), 26 (c) and 25 (d) August, concerning the columnar SO_2 over Thessaloniki, averaged for the period 26-28 August 1994.

Table 1. Contribution [%] of the different countries to the columnar SO_2 in Thessaloniki, estimated for different periods in August 1994

period	BG	GR	TUR	ALB	YUG	ROM	MOL	UKR
4-5 August	44.46	18.53	0.13	0	0.04	26.56	5.98	4.36
26-28 August	0.12	77.9	0	1.55	19.13	1.29	0	0
4 August	51.17	15.48	0	0	0.07	18.44	7.18	5.66
5 August	35.46	21.23	0.27	0	0	35.15	4.94	2.95
26 August	0	98.15	0	1.55	0.30	0	0	0
27 August	0.01	68.96	0	8.48	22.55	0	0	0
28 August	0.32	60.8	0	0	35.56	3.34	0	0

2) During 1994, the average SO_2 columnar amount is about 1.6 m-atm-cm of SO_2, or equivalent of 6.85E-5 Kg/m^2 of SO_2. The standard deviation of the daily values is high, about 0.75 m-atm-cm, or equivalent of 3.21E-5 Kg/m^2 , which is a result of day-to-day variation of local sources and of contribution of remote areas.

3) The numerical estimations give the % contribution of different countries to the SO_2 column over Thessaloniki. Studying two episodes we arrived at the conclusions that about 80% of the SO_2 column originated during early August from northeast of Thessaloniki, while the origin of the sources of SO_2 total burden during the second episode is from NNW of the city.

ACKNOWLEDGEMENTS

The joint work on the present paper was possible because of a fellowship granted to one of the authors (K. Ganev) by the Greek Ministry of Education and Religion.

REFERENCES

Bais A., Zerefos C., Meleti C., Ziomas I. And Tourpali K., Spectral Measurements of Solar UVB Radiation and its Relations to Total Ozone, SO_2 and Clouds, *J. of Geoph. Res., Vol. 98, D3, 5199-5204, 1993.*

Brewer A., A replacement for the Dobson spectrophotometer? *PAGEOPH, 106-108:919-927,1973.*

Edner H., Ragnarson P., Spaennare S., and Svanberg S., Differential optical absorption spectroscopy (DOAS) system for urban atmospheric pollution monitoring, *Appl. Optics, 32, 327-333, 1993.*

Ganev K., 1991, Formulation of the influence function problem for a complex of interacting admixtures. *EUROTRAC workshop on regional and global modeling of minor atmospheric constituents. Wiesbaden, 23 - 24 April 1991 (Also published in Compt. Rend. Acad. bulg. Sci., 45, No3, 29-32).*

Ganev K., Syrakov E. and Georgieva E., 1997, An example for the interaction of local and synoptic transport scales - a case study, *Swedish-Bulgarian Workshop on Long-Range Air Pollution, Sozopol 19-23 October, 1997.*

Georgieva E. and Godev N., 1987, Interpolation of wind data taking into account the relief-some results and problems (in Bulgarian), *Bulgarian Geoph. J.* v.VIII, 4

Georgieva E., 1990, On the choice of the weight coefficients in the mass-consistent wind models (in Bulgarian), *Bulgarian Geoph. J.* v.XV, 2

Marchuk G.I., 1982, *Mathematical Modeling in Environmental Problems* (in Russian), Science, Moscow

Sirakov E., Ganev K. and Godev N., 1988, Long-range transport of sulfur and nitrogen, taking into account the influence of the synoptic conditions, orography and arranged vertical motions. (in Bulgarian), *Bulgarian Geoph. J.* v.XIV, 1, 11-27.

Sirakov E., Ganev K. and Godev N., 1988: Estimation of the mutual pollution of different regions at distant and regional transport. (in Bulgarian), *Bulgarian Geoph. J.* v.XIV, 4, 10-21.

Syrakov E. and Ganev K., 1989, Modeling of sulphur and nitrogen compounds transport within the PBL under nocturnal and convective conditions. (in Russian), *Z. Meteorol.*, 39, 2, 81 - 88.

Syrakov E., Ganev K., Godev N., Georgieva E. and Bogdanov E., 1989, A telescopic approach for coordinated study of the air pollution at long-range and regional transport over a complex terrain, in: *Special Environmental Report No.17 "Changing Composition of the Troposphere" - Extended Abstracts of the WMO Technical Conference. Sofia, Bulgaria, 23 - 27.10.1989* , WMO No.724, 189 - 192.

Syrakov D. and Andreev V., 1995, *Bulgarian Contribution - Annual Report 1994,* Meteorological synthesizing center-EAST, Sofia-Moscow

Tuovinen J.-P., Barret K. and Styve H., 1994, *Transboundary Acidifying Pollution in Europe: Calculated Fields and Budgets 1985-93*, Meteorological synthesizing center-WEST, Oslo

Yordanov D., Syrakov D. and Djolov G., 1983, A barotropic planetary boundary layer, *Boundary Layer Met.*, 25, 363-373

Zerefos C., Mantis H., Bais A., Ziomas I., and Zoumakis N., Solar Ultraviolet Absorption by Sulphur Dioxide in Thessaloniki, Greece, *Atmos. Ocean, 24(3), 292-300, 1986.*

Ozone Sensitivity and Uncertainty Analysis Using DDM-3D in A Photochemical Air Quality Model

Yueh-Jiun Yang, James G. Wilkinson,
M. Talat Odman, Armistead G. Russell

School of Civil and Environmental Engineering
Georgia Institute of Technology
Atlanta, GA 30332-0512, USA

ABSTRACT

Sensitivity analysis plays an important role in understanding the response of an environmental system to the variation of model inputs or parameters. This information can be further utilized to explore the model uncertainties introduced from these model inputs and parameters. However, sensitivity analysis has not been used as widely as desired in multidimensional models because of its complexity. A fast and formal sensitivity technique (DDM-3D) has been developed and implemented in the CIT (California/Carnegie Institute of Technology) airshed model to evaluate the sensitivity of predicted pollutant levels to the reaction rate constants of gas-phase photochemical mechanism. The study focuses on the chemical rate constants, which have been identified to be influential to the predicted ozone uncertainty in previous studies. The ozone sensitivities to rate parameters are then computed spatially in a multiday ozone episode, August 27-29, 1987, applied to the Los Angeles area, southern California. It was found that only a limited number of rate constants have a significant influence on ozone predictions. Combined with the sensitivity analysis, a first-order uncertainty was conducted and results indicate that uncertainty of reaction rate constants have significant impacts on the ozone levels. The uncertainty ($\pm 1\sigma$) in ozone ranged from 10-35% of predicted levels downwind of Los Angeles and 35-50% for the urban core during afternoon high-ozone hours. In addition to the first-order uncertainty analysis, a Monte Carlo simulation incorporated with Latin Hypercube Sampling (LHS) technique was conducted to assess the importance of non-linearities. It was found that the two approaches gave very similar results. The results also suggest that the overall uncertainty of predicted ozone is highly dominated by the uncertainty of rate constant of HNO_3 formation.

INTRODUCTION

Photochemical air quality models increasingly are being used to understand the atmospheric dynamics of air pollutants and as basis of emission control strategies. Of air quality models, an essential component is a chemical mechanism that describes the non-linear reactions responsible for formation of ozone and other secondary pollutants (such as ozone) from emitted precursors. Chemical mechanisms are assembled using the available literature on kinetics, mechanistic, and product data for the atmospherically important inorganic and organic reactions. In chemical mechanisms, numerous reactions are included by analogy with known reactions, such as rate constants of peroxy radical reactions and oxidation of some organic compounds. Although rapid progress has been made in understanding the reactions involved in polluted atmosphere, knowledge of more reactions still need be explored to improve the representative of the photochemical transformation in the atmosphere. One question raised as to use of chemical mechanisms in air quality models is how the uncertainty of photochemical kinetic parameters may affect the model predictions; e.g. predicted ozone levels. Quantification of uncertainty of reaction rate constants introduced to modeling results is thus important for improving air quality predictions and identifying effective control strategies.

Previous studies indicated that the uncertainty of reaction rate constants in chemical mechanisms can have an effect of 30% uncertainty on ozone predictions in urban area using a box model (Yang et al., 1995; Gao et al., 1995). However, the use of a box model does not account for the detailed physical transport of atmospheric pollutants. Two steps were used to quantify the uncertainties of rate constants of gas-phase chemical mechanism using CIT (California/Carnegie Institute of Technology) air quality model applying to Los Angeles area. First, a formal sensitivity analysis, DDM-3D (decoupled direct method in three dimension) (Yang et al., 1997) was conducted to identify the most influential rate constants to ozone predictions. The sensitivity results then were combined with rate constant uncertainty estimates to calculate the uncertainty contributions of influential rate constants to the ozone predictions using first order uncertainty analysis. Second, a formal uncertainty analysis was used to propagate the uncertainties of the influential rate constants through a set of Monte Carlo simulations. The model was applied to the August 27-29, 1987 Southern California Air Quality Study (SCAQS) episode, and incorporates SAPRC90 (Statewide Air Pollution Research Center 1990) chemical mechanism (Carter, 1994). The uncertainty estimates of rate constants in the SAPRC90 mechanism was primarily based Stockwell et al. study (1994), and updated from recent NASA panel data evaluation (DeMore et al., 1997) and work of Donahue et al. (1997).

METHODOLOGY

Multidimensional Sensitivity Analysis (DDM-3D)

Similar to other multidimensional photochemical air quality models, the formation and transport of chemically reacting atmospheric contaminants is described in the CIT model by the atmospheric diffusion equation:

$$\frac{\partial c_i}{\partial t} = -\nabla(\mathbf{u}c_i) + \nabla(\mathbf{K}\nabla c_i) + R_i + S_i \qquad i = 1,\ldots, N \qquad (1)$$

where $c_i(\mathbf{x}, t)$ is the ensemble average concentration of species i, $\mathbf{u}$ is the wind field, $\mathbf{K}$ is a second order turbulent diffusivity tensor, R_i is the net rate of chemical production, S_i is the elevated source rate of emissions of specie i, and N is the number of total species. This equation is solved numerically subject to initial conditions (IC) and boundary conditions (BC):

IC: $c_i(t_o) = c_i^o$

BC's: (1) $\mathbf{u}c_i - \mathbf{K}c_i = \mathbf{u}c_i^b$ inflow

(2) $-\nabla c_i = 0$ outflow

$$(3)\ v_g^i c_i - K_{zz}\frac{\partial c_i}{\partial z} = E_i \qquad \text{bottom of domain}$$

$$(4)\ -\frac{\partial c_i}{\partial z} = 0 \qquad \text{top of domain}$$

where c_i^b is the concentration of compound i at the boundary, v_g^i is the dry deposition velocity, E_i is the surface emission rate. The local semi-normalized sensitivity of pollutant c_i to a reaction rate constant k_j can be calculated as:

$$s_{ij}(\mathbf{x},t) = k_j \frac{\partial c_i}{\partial k_j} = \frac{\partial c_i}{\partial \varepsilon_j} \tag{2}$$

By substituting Eq. (2) into Eq. (1), the general equations of sensitivity coefficients and initial and boundary conditions are obtained as follows:

$$\frac{\partial s_{ij}}{\partial t} = -\nabla(\mathbf{u}s_{ij}) + \nabla(\mathbf{K}\nabla s_{ij}) + J_{ik}s_{ij} + \frac{\partial R_i}{\partial \varepsilon_j} + \frac{\partial S_i}{\partial \varepsilon_j} \tag{3}$$

where $J_{ik} = \partial R_i/\partial c_k$ is Jacobian. Similarly, the initial and boundary conditions of sensitivity coefficients, $s_{ij}(\mathbf{x}, 0) = s_{ij}^o(\mathbf{x})$ and $s_{ij}(\mathbf{x}, t) = s_{ij}(\mathbf{x}\in\Omega, t)$, can be derived by taking the derivatives of concentration initial and boundary conditions with respect to the individual semi-normalized parameters. The details of the implementation of DDM-3D is described elsewhere (Yang et al., 1997).

Uncertainty Estimates of Reaction Rate Constants of SAPRC90 Mechanism

This study focuses on uncertainties in the rate constants of gas-phase chemical mechanism. Table 1 shows the key reaction rate constants in SAPRC90 mechanism (Carter, 1994) that are considered important on the uncertainty of ozone predictions based on previous studies (Yang et al., 1995; Gao et al., 1995, Jiang et al., 1997; Hanna et al., 1997). The

Table 1. Key reactions of uncertainty contribution of rate constant to ozone predictions*

Reactions	Uncert. Factor**	Reactions	Uncert. Factor**
$NO_3 + HO \rightarrow$	1.3	MGLY + HO $\rightarrow$	1.6
$O_3 + NO \rightarrow$	1.2	$CCOO_2 + NO \rightarrow$	2.0
$NO_3 + h\nu \rightarrow$	2.0	$CCOO_2 + NO_2 \rightarrow$	2.0
$O_3 + h\nu \rightarrow$	1.4	PAN $\rightarrow$	2.0
$OSD + H_2O \rightarrow$	1.3	$C_2COO_2 + NO \rightarrow$	2.0
OSD + M $\rightarrow$	1.3	$C_2COO_2 + NO_2 \rightarrow$	2.0
$HO + NO_2 \rightarrow$	1.5	PPN $\rightarrow$	1.9
HO + CO $\rightarrow$	1.3	Ethene + HO $\rightarrow$	1.15
$HO_2 + NO \rightarrow$	1.3	HCHO + $h\nu \rightarrow 2HO_2$+CO	1.4
$HO_2 + HO_2 \rightarrow$	1.4	HCHO + $h\nu \rightarrow H_2$ + CO	1.4
$RO_2R + NO \rightarrow$	2.0	HCHO + HO $\rightarrow$	1.26
$R_2O_2 + NO \rightarrow$	2.0	Isoprene + HO $\rightarrow$	1.2
Cresol + HO $\rightarrow$	1.3	AAR1‡ + HO $\rightarrow$	1.2
AFG2 + HO $\rightarrow$	2.0	AAR2‡ + HO $\rightarrow$	1.2
AFG2 + $h\nu \rightarrow$	3.0	AAR3‡ + HO $\rightarrow$	1.2
CCHO + $h\nu \rightarrow$	1.4	OLE1† + HO $\rightarrow$	1.2
CCHO + HO $\rightarrow$	1.4	OLE2† + HO $\rightarrow$	1.2
RCHO + HO $\rightarrow$	1.4	OLE1† + $O_3 \rightarrow$	1.5
MEK + HO $\rightarrow$	1.3	OLE2† + $O_3 \rightarrow$	1.5
MEK + $h\nu \rightarrow$	1.5	Methane + HO $\rightarrow$	1.2

* Yang et al.,1995; Gao et al., 1995; Jiang et al., 1997; Hanna et al., 1997.
** estimated at 298°K, based on NASA definition (DeMore et al., 1997).
‡ Lumped species of alkanes and aromatics.
† Lumped species of alkenes.

uncertainty factors are listed adopting the NASA definition (DeMore et al., 1997). All of these rate constants were treated as random variables in the DDM-3D sensitivity analysis to quantify the most influential rate constants based on uncertainty contribution.

First Order Uncertainty Analysis and Monte Carlo Analysis

The first order Taylor series expansion for the an output variable of air quality models, $c_i(\mathbf{x}, t; \varepsilon_j)$, is :

$$c_i(\mathbf{x}, t; \varepsilon_j^{'}) = c_i(\mathbf{x}, t; \varepsilon_j) + \frac{\partial c_i}{\partial \varepsilon_j} \Delta\varepsilon_j \quad (4)$$

where $\Delta\varepsilon_j$ represents a perturbation of rate constant. Based on this approximation, the variance of $c_i(\mathbf{x}, t; \varepsilon_j)$ can be calculated as:

$$\sigma_{c_i}^2 = \sum_{j=1}^{m} \sigma_{\varepsilon_j}^2 \left(\frac{\partial c_i}{\partial \varepsilon_j} \right)^2 \quad (5)$$

Equation (5) gives an estimate of the uncertainty contribution of rate constants to the overall uncertainty of c_i. Here, covariance among rate constants is assumed insignificant for the overall uncertainty estimate. Yang et al. (1995) and Gao et al. (1995) found that the dependence across reaction rate constants in chemical mechanism is small and negligible for ozone predictions. Equation (5) also provides a basis for combining the sensitivities obtained from DDM-3D with preliminary estimates of the uncertainty or variance in rate constants, to evaluate the contribution of uncertainly in a given rate constant to uncertainty of an output variable. The uncertainty contribution of a reaction rate constant to ozone concentration then can be calculated as:

$$UC_i = \frac{\sigma_{\varepsilon_i}^2 \left(\frac{\partial O_3}{\partial \varepsilon_i} \right)^2}{\sum_{j=1}^{m} \sigma_{\varepsilon_j}^2 \left(\frac{\partial O_3}{\partial \varepsilon_j} \right)^2} \quad (6)$$

While the first order uncertainty analysis provides a quick view of how model parameter uncertainties have impacts on model predictions, a formal uncertainty analysis such as Monte Carlo analysis can properly propagate uncertainty sources through a set of simulation runs. However, a limitation to the use Monte Carlo analysis for complex air quality models is the fact that a large number of model runs may be needed to ensure that a representative sample is obtained for each parameter. In this study, Latin Hypercube sampling (LHS) technique was conducted for sampling random rate constants. The influential rate constants identified from the first order uncertainty analysis will be treated as random variables and assumed to be lognormally distributed.

RESULTS AND DISCUSSION

Figure 1 shows the surface ozone field at the time of peak predicted ozone (1500 h) on August 29, 1987 SCAQS multiday episode. As seen in the figure, the peak ozone occurs outside the urban core of Los Angeles, with a value of 176 ppb. Table 2 shows the 14 rate constants that contribute most to the uncertainty of ozone prediction where the ozone concentration is over 40 ppb using the first order uncertainty analysis. The two highest ozone uncertainty contributions are the rate constants of nitric acid formation and nitrogen dioxide photolysis, with 37% and 23% of total variance respectively. It was found that these 14 reactions account for 98% of total variance of ozone predictions due to rate constant uncertainties. The sequence of uncertainty apportionment shown in Table 2 is similar to the one found in a box model using an average episode of 39 cities across the United States (Yang et al., 1995).

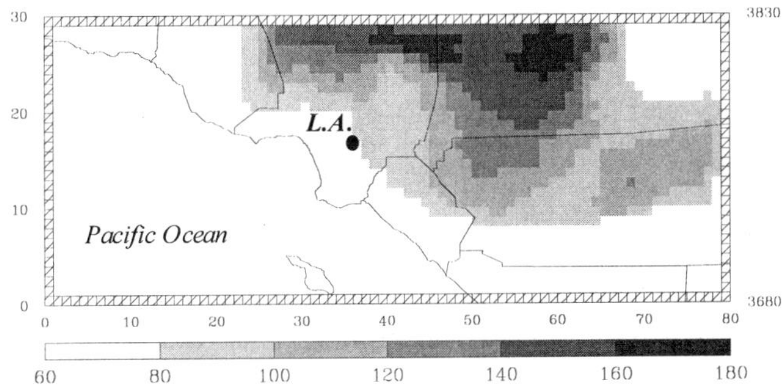

Figure 1. Spatial ozone distribution (ppb) at 1500-1600 h, August 9, 1987 SCAQS episode.

Table 2. Uncertainty apportionment of reaction rate constants of SAPRC90 mechanism using first order uncertainty analysis

Reactions	Uncertainty Contribution [*]
$NO_3 + HO \rightarrow$	0.37
$NO_2 + h\nu \rightarrow$	0.23
$O_3 + NO \rightarrow$	0.087
$CCOO_2 + NO \rightarrow$	0.077
$PAN \rightarrow$	0.048
$CCOO_2 + NO_2 \rightarrow$	0.046
$HCHO + h\nu \rightarrow$	0.023
$HO_2 + NO \rightarrow$	0.022
$RO2R + NO \rightarrow$	0.019
$AAR2 + HO \rightarrow$	0.018
$O_3 + h\nu \rightarrow$	0.013
$OSD + H_2O \rightarrow$	0.007
$OSD + M \rightarrow$	0.007
$AFG2 + h\nu \rightarrow$	0.005

[*] O_3 cutoff = 40 ppb

Figure 2 and 3 show the spatial distributions of ozone semi-normalized sensitivities due to the rate constants of HNO_3 formation and NO_2 photolysis, respectively. The uncertainties in these two rate constants account for about 60% of total uncertainty of predicted ozone based on the first order uncertainty analysis (Table 2). As shown in Figure 2, increasing the rate constant of NO_2+HO decreases ozone levels downwind because more HO radicals and NO_2 in system are removed through the reaction. The high sensitivities of ozone to the reaction rate constant of nitric acid formation also was found in previous studies (Yang et al., 1995; Gao et al., 1995) in a box model. On the contrary, increasing NO_2 photolysis rate increases ozone concentrations over domain since more oxygen atoms are available for ozone formation, illustrated in Figure 3.

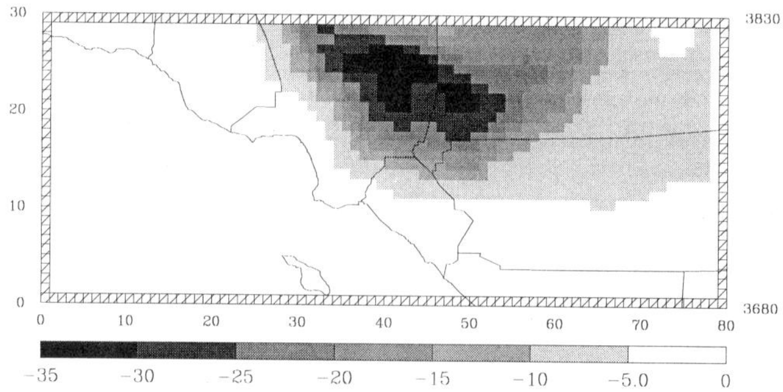

Figure 2. Ozone sensitivity (ppb of O_3) at 1500-1600 h, August 29, 1987, to $NO_2 + HO \rightarrow HNO_3$ rate constant.

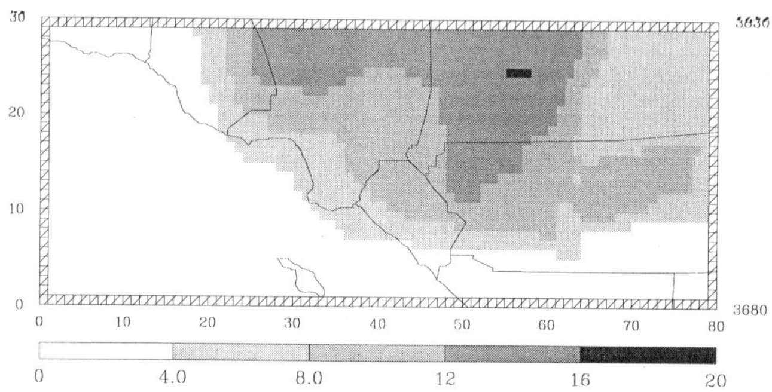

Figure 3. Ozone sensitivity (ppb of O_3) at 1500-1600 h, August 29, 1987, to NO_2 photolysis rate constant.

The results of first order uncertainty analysis that incorporates uncertainty estimates of 40 rate constants from Table 1 are presented in Figure 4 and 5. Figure 4 shows the distributed concentration of nominal ozone prediction plus an 1σ uncertainty calculated from Eq.(5). The maximum of (O_3+1σ) value is 209 ppb that is found in downwind area, but the location is not the same as the peak ozone location predicted in Figure 1. Figure 5 indicates the relative uncertainty of predicted ozone levels, i.e. $(1\sigma/O_3)$x100% , to the nominal ozone field shown in Figure 1. The ozone relative uncertainty ranges from 10 to 50% with a spatial average uncertainty of 20%. The area having higher ozone uncertainties is shown in the area immediately downwind of the urban core, up to 50% uncertainty to the nominal ozone levels.

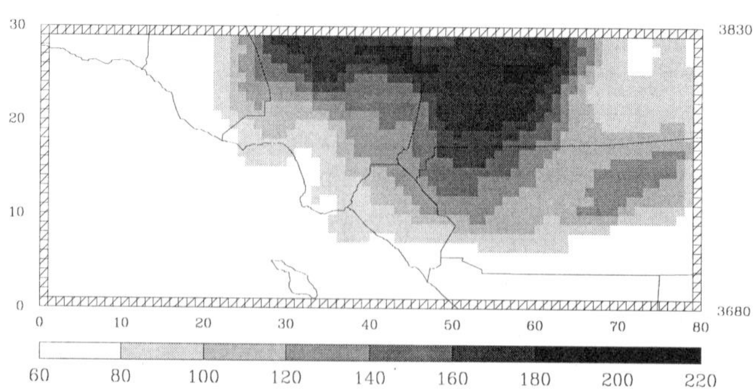

Figure 4. Spatial distribution of O_3+1σ (ppb), 1500-1600 h, August 29, 1987 SCAQS episode, using first order uncertainty analysis.

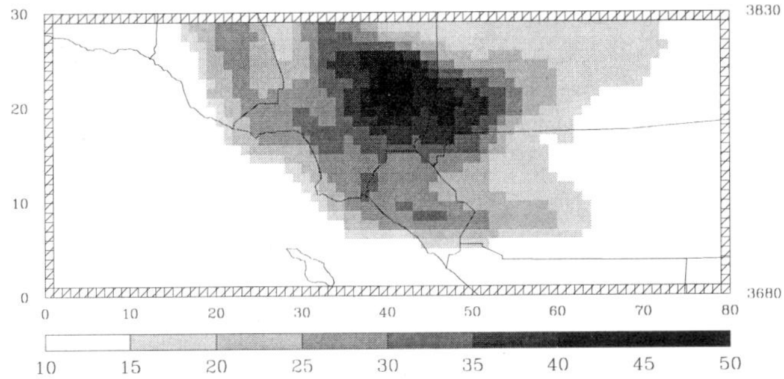

Figure 5. Spatial distribution of ozone relative uncertainty (%), 1500-1600 h, August 29, 1987 SCAQS episode, using first order uncertainty analysis.

Two sets of simulations were conducted for the Monte Carlo analysis with 50 and 100 simulation runs respectively. However, the differences (in terms of ozone concentrations) between two sets of analysis are insignificant. Figure 6 presents the 50-percentile ozone distribution from 100 simulation runs. The peak ozone is 172 ppb and 4 ppb lower than the nominal peak shown in Figure 1. Figure 7 and 8 present ozone distributed concentrations plus an 1σ deviation uncertainty as well as the relative uncertainty from 100 simulation runs. The maximum $O_3+1\sigma$ shown in Figure 7 is 208 ppb, and maximum relative uncertainty in Figure 8 is 49% of nominal ozone concentrations. The uncertainty distributions of ozone from the Monte Carlo analysis are pretty much comparable to the results from the first order approach in Figure 4 and 5. The first order approach has slightly higher spatial uncertainties in the range of 180 to 200 ppb as well as in the range of 25 to 35% compared to Figure 6 and 7.

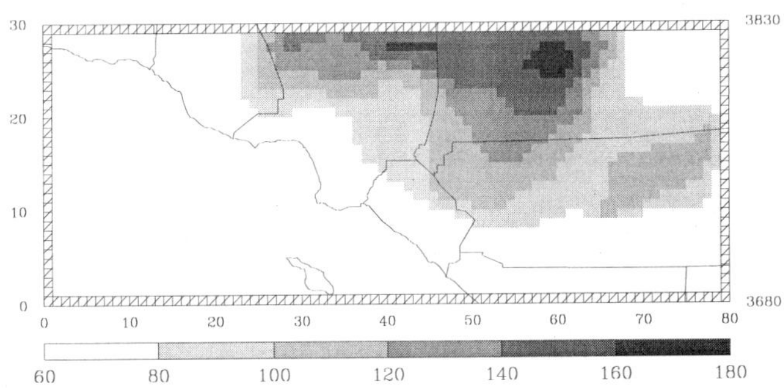

Figure 6. Spatial distribution of 50 percentile ozone (ppb), 1500-1600 h, August 29, 1987, SCAQS episode, from the Monte Carlo analysis.

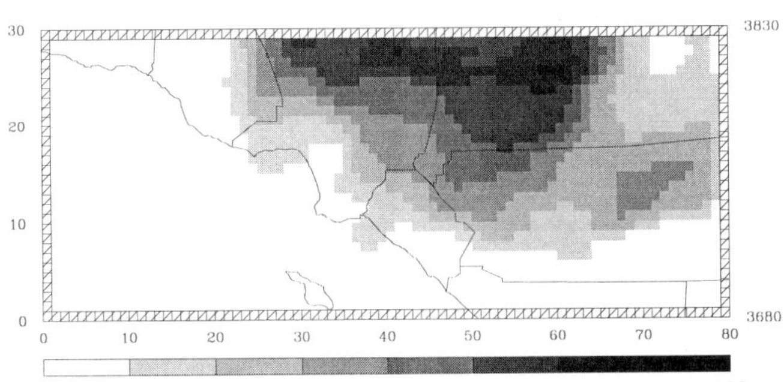

Figure 7. Spatial distribution of (mean O_3)+1σ (in ppb), 1500-1600 h, August 29, 1987 SCAQS episode, from the Monte Carlo analysis.

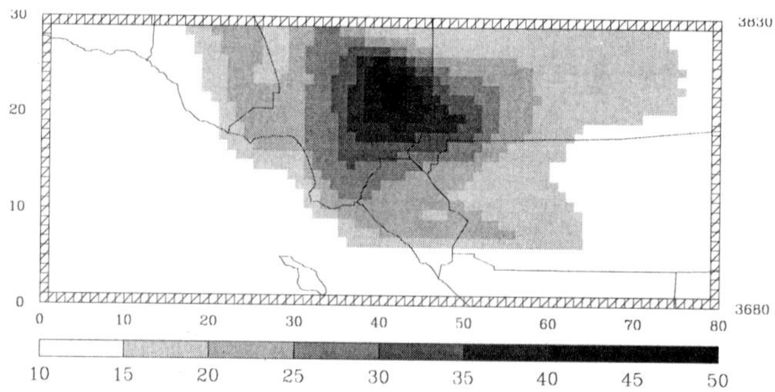

Figure 8. Spatial distribution of ozone relative uncertainty (%), 1500-1600 h, August 29, 1987 SCAQS episode, from the Monte Carlo analysis.

In Monte Carlo simulations, the random rate constants are sampled through the LHS; therefore, the perturbations of rate constants have an impact on the locations of where predicted peak ozone. Figure 9 shows the spatial variation in peak ozone on the third day (August 29) simulation. The cumulative density function of peak ozone is presented in Figure 10. The mean peak ozone is 176 ppb, same as nominal peak (shown in Figure 1), with a standard deviation of 33 ppb. The 1σ standard deviation (18% in relative scale) in peak ozone is much less than the relative uncertainty, i.e. 30%, found in the previous study using a box model (Yang et al., 1995).

CONCLUSIONS

Sensitivity and uncertainty analyses are able to not only identify the important model parameters but estimate the magnitude of uncertainty that have been introduced to the model predictions. The uncertainty of SAPRC90 chemical mechanism was analyzed in two different approaches, first order uncertainty analysis and Monte Carlo simulations. The only source of

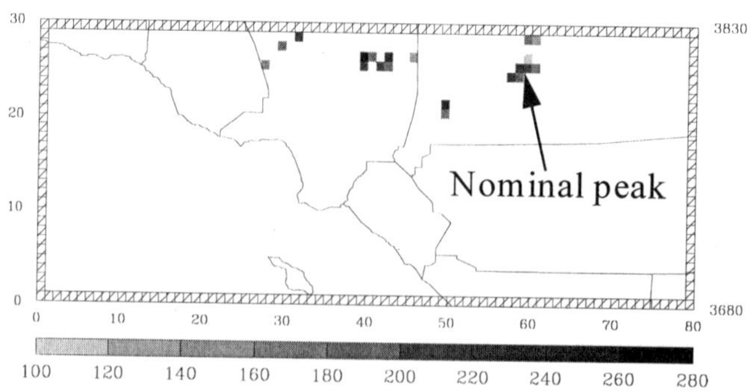

Figure 9. Spatial variation in peak ozone on August 29, 1987, from the Monte Carlo analysis.

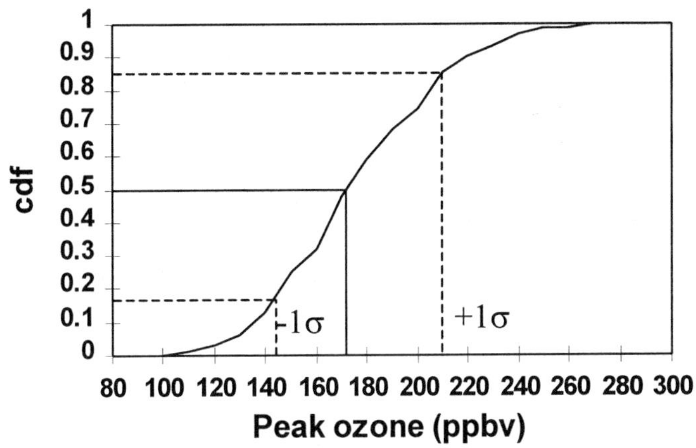

Figure 10. Cumulative density function of predicted peak ozone on August 29, 1987, from Monte Carlo analysis, shown with $\pm 1\sigma$ standard deviation.

uncertainty considered was the rate constants that have been identified to be important from previous studies. Other sources of uncertainty of chemical mechanism are not included in this study. The results from two approaches are much comparable in overall ozone uncertainty, but the first uncertainty analysis has a slightly higher ozone spatial uncertainty. The uncertainty ($\pm 1\sigma$) in ozone ranged from 10-35% of predicted levels downwind of Los Angeles and 35-50% for the urban core during afternoon high-ozone hours with a spatial average uncertainty of 20%. The highest ozone uncertainty is located in the area immediately downwind of urban core of Los Angeles with about up to 50% relative uncertainty. Only a limited number of rate constants are found to be influential for the uncertainty of ozone concentrations. The two highest ozone uncertainty contributions are from the rate constants of HNO_3 formation and NO_2 photolysis rate, accounting for 37% and 23% of overall ozone uncertainty respectively. The comparison between the first order and Monte Carlo approaches suggests that the overall uncertainty of predicted ozone is highly dominated by the uncertainty of rate constant of HNO_3 formation, and that the overall ozone uncertainty has nearly linear response to one standard deviation variability of the rate constants.

References

Carter, W.P.L., 1990, A Detailed Mechanism for the Gas-Phase Atmospheric Reactions of Organic Compounds, *Atmos. Environ.* 24A:481.

DeMore, W.B., Sander, S.P., Golden, D.M., Hampson, R.F., Kurylo, M.J., Howard, C.J., Ravishankara, A.R., Kolb, C.E., and Molina, M.J, *Chemical Kinetics and Photochemical Data for Use in Stratospheric Modeling*, Evaluation Number 12, JPL, NASA, 1997.

Donahue, N.M., Dubey, M.K., Mohrschladt, R., Demerjian, K. L., and Anderson, J.G., 1997, High-Pressure Flow Study of the Reactions OH + NO_x $\rightarrow$ $HONO_x$: Errors in the Falloff Region, J. *Geophys. Res.*, 102:6159.

Gao, D., Stockwell, W.R., and Milford, J.B., 1995, First Order Sensitivity and Uncertainty Analysis for a Regional-Scale Gas-Phase Chemical Mechanism, *J. Geophys. Res.*, 100:23153.

Hanna, S.R., Chnag, J.C., Fernau, M.E., and Hansen, D.A., 1998, Estimates of Sensitivity of Photochemical Grid Models to Uncertainties in Input Parameters as Applied to URM-IV on the New York Domain, printed in *Air Pollution Modeling and It's Application XII*, edited. by Gryning, S.E., and Chaumerliac, N., Plenum Press, New York.

Jiang, W., Singleton, D.L., McLaren, R., and Hedley, M., 1997, Sensitivity of Ozone Concentrations to Rate Constants in a Modified SAPRC90 Chemical Mechanism Used for Canadian Lower Fraser Valley Ozone Studies, *Atmos. Environ.*, 31:1195.

Stockwell, W.R., Yang, Y.J., and Milford, J.B., *A Compilation of Estimated Uncertainty Factors for Rate Constants in W.P.L. Carter's Detailed Mechanism*, Final Report for Auto/Oil Air Quality Improvement Research Program, Atlanta, GA, 1994.

Yang, Y.J., Wilkinson, J.G., and Russell, A.G., 1997, Fast, Direct Sensitivity Analysis of Multidimensional Photochemical Models, *Environ. Sci. Technol.*, 31:2859.

Yang, Y.J., Stockwell, W.R., and Milford, J.B., 1995, Uncertainty Analysis in Incremental Reactivities of Volatile Organic Compounds, *Environ. Sci. Technol.*, 29:1336.

DISCUSSION

E. GENIKHOVICH: Should your uncertainty analysis depend on the substance you are talking about (for example, ozone, PAN and so on)?

J. G. WILKINSON: This study focuses on the ozone uncertainty that is contributed from chemical reaction rate constants. The uncertainty range found for ozone predictions is not applicable to other species, such as PAN. However, a similar uncertainty analysis procedure can be conducted to any other compound in the chemical mechanism. As for the uncertainty analysis other than ozone, a good reference is the study by Gao et al. (JGR, 1995).

E. GENIKHOVICH: At such a high level of uncertainty which is in the model, do you think that the first-order analysis, i.e., linearization, is sufficient?

J. G. WILKINSON: The first-order uncertainty analysis should be considered as an approximate approach under any circumstance. Whether the approximate approach is sufficient (or good enough) or not, it can only be judged by the comparison against the results from a formal approach, such as Monte Carlo analysis. For instance, the first-order analysis in this study may be a fairly good alternative way to estimate ozone uncertainty due to chemical reaction rate constants within one standard deviation variability. But it will not be the case if other uncertainty sources other than chemical reaction rate constants are incorporated in the analysis.

B. FISHER: If one accepts that reaction rates are known to a better accuracy that photolysis rates (since radiation fluxes are rather uncertain), it would be of interest to restrict the sensitivity analysis just to those reactions involving photolysis?

J. G. WILKINSON: The sensitivity analysis is a technique to test how model outputs respond the model input or parameter small perturbations. However, the sensitivity coefficients are not necessary to be small for those parameters with small uncertainty estimates or to be large for those parameters with large uncertainty estimates. Several studies already show that some non-photolytic reaction rate constants can turn out to be very influential in terms of the uncertainty of model outputs. I would not think that applying the sensitivity analysis only to photolytic reactions would be informative in terms of uncertainty evaluation.

M. SOFIEV: It seems to me that presented analysis gives a very good lower estimate of uncertainty, because always there is a risk that we lost some reactions (and consequently, did not varied their rates). Accounting for these additional dimensions will result in increase of uncertainty estimate. Could you please comment?

J. G. WILKINSON: The uncertainty analysis performed in this study is based on the current knowledge available on the atmospheric chemistry, and the SAPRC mechanism used is one of the most comprehensive mechanisms in air quality models. Those uncorporated reactions in chemical mechanisms could have some uncertainty contribution to model outputs, but the uncertainty increase would be very limited. Note that uncertainty analysis should be updated if any better understanding of atmospheric chemistry is available.

PRELIMINARY SIMULATIONS OF THE BASYS 5 LAGRANGE PERIODS WITH THE HILATAR MODEL

Marke Hongisto, Kirsti Jylhä,and Virpi Flyktman

Finnish Meteorological Institute (FMI)
Air Quality Research
Vuorikatu 19
00101 Helsinki 10

INTRODUCTION

The FMI is participating in the EU MAST research program for regional seas: BASYS (Baltic Sea System Study) subproject 5: air pollution load. One primary objective of this subproject is to improve knowledge of the atmospheric input to the Baltic Sea by calculating and simultaneously measuring the fluxes of the relevant airborne pollutants, and several meteorological factors affecting to them, during specific field campaigns. The results are then used in improving the accuracy of some of the parameterization schemes used in the models for the physical and chemical processes. Long-period simulations will then be carried out with the improved model systems. Other goals are to study the marine atmospheric boundary layer in order to estimate the most important factors affecting atmospheric deposition over sea areas, and an examination of the representativeness of the routine coastal measurements of air pollutants (as to location and sampling period) for the total marine environment.

For this task, the FMI grid model HILATAR, constructed for simulating the transport, dispersion, chemical transformation and deposition of nitrogen and sulphur pollutants, has been extended to cover the most important European and West-Russian pollution areas between the latitudes of Spain and the Arctic Ocean with an approximately 56 km resolution. The EMEP 1995 50-km grid emission inventory for oxidized and reduced nitrogen and sulphur has been used in simulating the July 1997 and March 1998 two-week Lagrange experiment campaigns. Preliminary model results for the periods are described, including meteorological analyses. The present model structure, with verification and simulation results for the year 1993, is described in Hongisto, 1998.

HILATAR is a nested model set covering three geographical scales with a variable horizontal (56-28-11 km, 0.5°-0.25°-0.1°) grid resolution: the European area, the Baltic Sea with its surrounding countries, and Finland. The long-range transported background for the sub-area models is estimated by adding either the concentrations from the lower-resolution model, or those calculated with the EMEP model (Berge, 1997) at the EMEP MSC-W centre, to the air flowing into the simulation area through the boundaries.

The meteorological input parameters are taken from the operational 6-hour predictions of the HIRLAM (HIgh Resolution Limited Area Model) numerical weather model of the FMI. The turbulence parameters are recalculated from the temperature, wind and specific humidity profiles using two boundary-layer models. The meteorological data base with a 1-hour time resolution and a boundary-layer model are also used in the simulations carried out at the German Institute GKSS with the ADOM model. (Kruger and Petersen, 1998)

THE STRUCTURE OF THE HILATAR MODEL

HILATAR is an Eulerian grid-point model in which the concentrations $c(\vec{x},t)$ at time t and place $\vec{x} = x, y, z$ are calculated by numerically solving the transport equation (1)

$$\frac{\partial}{\partial t}c(\vec{x},t) + [\vec{V}(\vec{x},t)\cdot\nabla c(\vec{x},t)] = \nabla\cdot\left[\vec{K}(\vec{x},t)\cdot\nabla c(\vec{x},t)\right] + S(\vec{x},t) \qquad (1)$$

where the second term on the left-hand side represents advection, $V(\vec{x},t)$ is the mean wind velocity and $S(\vec{x},t)$ includes emissions, and the source and sink terms of chemical transformation and deposition. The turbulent diffusion term is written using gradient transport theory. Equation (1) is solved by the method of fractional steps with the following boundary conditions:

a) the initial and advected concentrations are known along all grid borders,
b) the dry deposition flux defines the lowest vertical layer boundary condition, and
c) the mass of the air, used for the determination of the vertical velocity, is conserved.

The model uses the HIRLAM rotated spherical grid coordinates as its frame of reference in the horizontal, and hybrid η terrain-following coordinates vertically (HIRLAM, 1996). 10-13 levels below 3-10 km are used.

DESCRIPTION OF THE BOUNDARY LAYER

The turbulence parameters are calculated from the HIRLAM temperature, wind and specific humidity profiles. Over land, the FMI meteorological preprocessor (Karppinen et al., 1997), with diabatic correction functions from van Ulden & Holtslag (1985) is used, over seas, the preprocessor described in Lindfors et al. (1991) is employed.

The mixing layer height hmix is calculated as in the FMI meteorological preprocessor for local dispersion model studies. During unstable situations the first inversion level is searched for. In stable and near-neutral situations hmix is a function of the gradient $\delta\theta_v/\delta z$ =G1 between the lowest model level and 2m. During such situations, hmix can vary from 900 m to a minimum height of 150 m, depending on the inversion strength.

The vertical velocity is calculated from the HIRLAM horizontal wind fields, starting from the surface, and forcing the mass flux sum across all sides of each model box boundary to be conserved. The eddy diffusion coefficient parametrization is based on the scaling regimes of

Holtslag & Nieuwstadt (1986) as in Hass (1991). Weak mixing across the boundary layer was allowed.

In the 0.5° and 0.25° resolution models, the surface classification and the roughness length z_o are directly extended from those of the forecast model with the respective resolution. In the 0.1° model, the the GRID Arendal on-line GIS Baltic Sea Drainage Basin land cover database, from http:://www.grida.no/baltic/, is used. z_o is fixed for each land use category and season, and over sea and lake areas a wind-dependent formula is used.

CHEMISTRY, DRY AND WET DEPOSITION AND NUMERICAL METHODS

The chemistry module is the EMEP-MSC-W code (Iversen et al., 1989) with updated rate coefficients and the following modifications:

1. Sulphur dioxide is assumed to be emitted in non-scavengable form for correction of the local wet deposition, and transformed to a scavengable SO_2 with a half-life of 1.5 hours;
2. No local deposition correction coefficients are used;
3. Inside clouds, the sulphate production rate differs from the clear sky conversion rate;
4. Nitrate and sulphur particles are split into two size classes,
5. Dry deposition is used as a boundary coefficient in the vertical diffusion equation; only the gravitational settling of the particles is described in the chemistry submodel; and
6. The formulas for the scavenging and dry deposition rates have been rewritten.

The rate constants have been updated following the recommendations of Wirtz et al., (1994). The night-time chemistry of nitrate, as well as the ammonium sulphate and nitrate formation rates, are left unchanged from the EMEP model. (Iversen et al., 1989). The conversion rate of SO_2 to sulphate in cloud is a function of the estimated liquid water content, temperature, season, latitude, etc. The cloud cover is parameterized, as in HIRLAM (1990), in terms of relative humidity. The liquid water content of the clouds is assumed to be 0.1–1. g m^{-3} depending on the cloud type, precipitation type and temperature.

The resistance analogy, $v_d = (r_a + r_b + r_c)^{-1}$ is used in estimating dry deposition velocities v_d. In the parametrization, the articles of Voldner et al. (1986), Wesely (1989) and Erisman (1994), and over sea areas the Lindfors et al. (1991) method are mainly used. The surface resistance of canopies depends on the solar radiation, surface temperature, relative humidity, amount of rain, dew or fog and pollutant exposure time. A specific resistance for snow and wet leaves is included. In order to avoid too high dry deposition fluxes a minimum surface resistance of 0.1 s cm^{-1} for all gases and surfaces is assumed.

In Lindfors et al. (1991) resistance method for gases, the diabatic correction function ψ_e and roughness z_{oe} for moisture, are used, and r_c was set to zero for soluble gases like HNO_3, NH_3 and SO_2 and also over a frozen sea. The dry flux of particles is estimated using Slinn et al., (1978); Slinn & Slinn, (1980) and Williams, (1982). The terminal settling velocity in a humid region is obtained keeping in mind the fact that, when the relative humidity exceeds 80 %, condensation increases the dry particle diameter.

Precipitation is assumed to be always to be rain when the temperature T(z) exceeds 270 K, and snow when T(z) is below 260 K, varying linearly between these limits. Its height dependency varies with time according to Chang (1986).

The scavenging rates are from Chang (1984,1986) and Asman & Janssen (1987). The SO_2 removal rates in and below-cloud are assumed to vary with the respective HNO_3 scavenging rate, according to an expression which gives the same summer/winter sink ratios as that of the EMEP model of Jonsen & Berge (1995). The emitted SO_2 is described by a non-scavengeable variable which is, downwind of the emission sources, linearly converted into a scavengeable form with a half-time of one-and-a-half hours.

The area-preserving flux-form advection algorithm of Bott (1989) has been used for the horizontal advection. Vertical transport is described by a simple mass-transfer balance discretization method. The QSSA (quasi steady-state approximation) method, employed in the EMEP model, is used in solving the chemistry equations. The vertical diffusion is solved by the Crank-Nicholson differentiation algorithm with a staggered grid (Tuovinen 1992).

SIMULATION RESULTS

Within Basys 5 during the summer and winter Lagrangian study campaigns intensive measurements were carried out to study the two-week periods of 2-15 July 1997 and 2-15 March, 1998. During these periods, two ships and four coastal stations (Hel, Hoburg, Kap Arkona and Preila), using intercalibrated instruments, made simultaneous measurements of concentration and deposition patterns. The two research vessels from IOW Warnemunde (RV. Alexander von Humbolt and RV. Prof Alfred Peck), and occasionally one coastal station, were attempted to be situated along the same trajectory calculated from the HIRLAM forecasts at the FMI; the analyzed forecasts were also transmitted to the ships and in March to the stations and to the GKSS for further analysis of the ship locations.

Both periods were first simulated with the 0.25° HILATAR model without long-range transported concentrations (in March 98 we tried to predict concentrations with an old emission inventory; however, the data connections to the ships were very bad, and the sending of mereorological data was considered more important). Accumulated depositions during the periods, simulated with the the 0.5° model after the experiment, are presented in Fig 1. Re-calculations with the finer-scale model including slight modifications are in progress.

3-15. JULY 1997

At the beginning of July 1997 the large-scale weather pattern over Europe was characterized by a low over the British Isles and a ridge extending from the Mediterranean to northeastern Europe. Although Scandinavia was affected by several cyclones, the precipitation amounts above the sea remained rather low. Hydrostatically stable conditions prevailed during the first week, which is clearly seen in the time series of the friction velocity and mixing height. Apart from the first 3 days, the prevailing wind direction was northerly or north-easterly; especially after the passage of a cold front over Scandinavia on the 8th of July, when a large part of Northern Europe came under the influence of an extended high-pressure area without rain.

During the second week the modelled concentrations over the Southern Baltic Sea were low. At the end of the period the ships sailed northwards in order to intercept the Stockholm plume. The best tracer of pollution transport modelled in the 26 km grid has been found to be sulphur dioxide. During the 11th and 12th of July, sulphur plumes from the Kola peninsula can be shown to have been transported relatively far south over the Baltic sea. According to the simulations, the Stockholm plume seems to have reached the Alexandre von Humbolt on the 11th of July; although the horizontal resolution of the emission inventory was rather poor, to clearly distinguish a plume of the town.

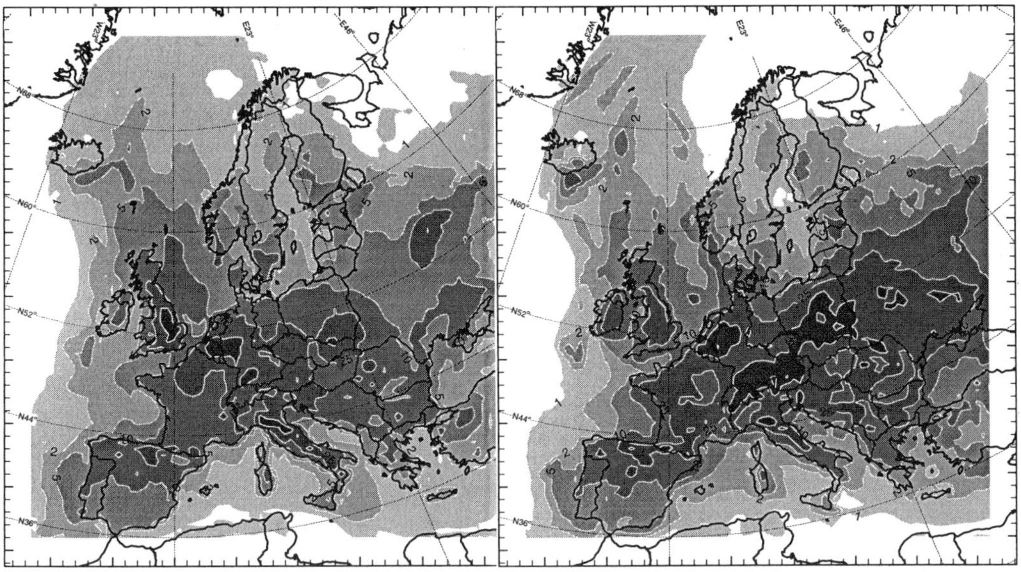

Oxidized Nitrogen, July 1-15, 1997. Reduced Nitrogen, July 1-15, 1997.

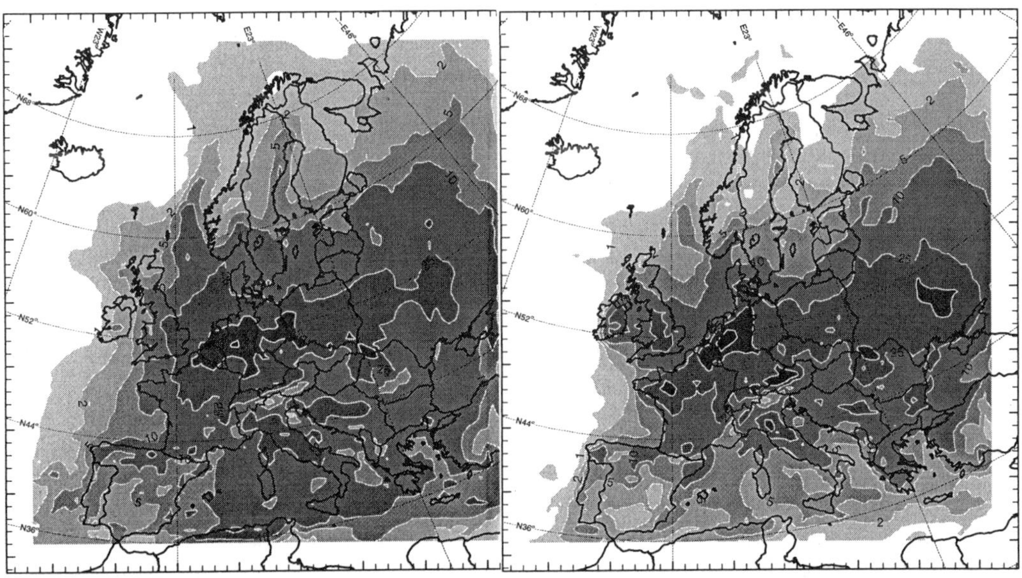

Oxidized Nitrogen, March 1-15, 1998. Reduced Nitrogen, March 1-15, 1998.

Fig.1. Cumulative deposition of oxidized and reduced nitrogen during the Basys 5 Lagrange periods.

The time series of the deposition distribution at 6 h intervals shows that the grid average precipitation might be overestimated at the beginning of the month (no rain was detected on board the ships - the modelled grid average wet deposition amounts were, however, very low). The period was rather dry, wet deposition, received mainly during the first days when the ships were not at yet sailing, presenting only less than one third of the total deposition of both NO_x and NH_3. Dry deposition also had a clear gradient at the coastal line, the explanation being the summertime dry deposition velocity differences over water and land surfaces, because episodic concentrations were transported from the land out over the sea. The time series shows that the wet deposition was episodic and did depend on the weather systems; during the study period the sea seemed neither to weaken (as in spring, except over some places with convective rain) nor strengthen the rain amounts (as in autumn). Over the Southern Baltic the reduced nitrogen deposition clearly exceeded that of oxidized nitrogen because of its stronger dry deposition flux to water.

2-14. MARCH 1998

During the second period, there was an extended low-pressure area over northern Europe with a rather strong westerly flow prevaileing over the southern Baltic. Cyclones related to the low-pressure area moved northeastwards, frequently changing the wind direction and origin of the air masses at the measurement places by 180° after the 6th of March. On 10th of March, a major high pressure area formed, extending from the western Mediterranean to Scandinavia. This feature moved slowly eastwards and on 14th March was followed by a cold trough and an associated low-pressure area with rather strong winds over the Baltic Sea.

In March, about two-thirds of the total nitrogen deposition was oxidized nitrogen, consisting mainly of the wet component, reflecting the rainy weather conditions. The dry deposition of reduced nitrogen was about half that of the wet. The dry deposition distribution of reduced nitrogen does not seem to decrease when going from the coast towards the open sea; the increase in the dry deposition velocities over open sea areas compensates the dilution of the concentrations. Total deposition did not have a clear or strong gradient towards the open sea for neither nitrogen components, though generally the deposition gradient was directed from south to north, also depending however on the emission areas.

Because of much higher rain amounts and greater wintertime emissions, the total NO_x deposition was up to five times higher than that during the respective summer period. The winter deposition of reduced nitrogen was 2-3 times higher than in summer over the Baltic Sea.

DISCUSSION

The simulated 1-hour time series at the stations and ships will be added to the Basys-5 data base maintained at the University of Hamburg, for comparison with the measurements. Comparison results with the 11-km resolution model for several EMEP stations, partly presented in Hongisto (1998) show, however, that the 56-km grid resolution might be too low to reproduce the measured deposition or comparison patterns. The most useful information for the model results might be the analysis of the time series and the animation, which explains the development of the weather systems, the three-dimensional mass transfer and transformation, origin and dynamical development of the pollution episodes.

The aim of the Lagrangian study was to estimate, from measurements made on the ships and at coastal stations, the chemical transformation during the transport, as well as the air-sea exchange, mass dispersion and gradients along the trajectories. This kind of analysis will be carried out when the emission data base for the models with higher spatial resolution has been updated. There will also be a need for a more detailed time variation of the EMEP emissions, and for some improvements to the model structure, which may result from the model measurement comparison. For example, interactions with sea salt may cause the particle-size distribution associated with the sulphur and nitrogen compounds to be shifted towards larger sizes, thereby increasing the dry deposition. Chemical interactions also create other marine-specific features, which are not included in the model parameterization as yet.

For the dry flux of ammonia the model can give only an upper limit estimate, because the atmospheric $NH_3(g)$ is in Henrys law equilibrium with the dissolved $NH_3(aq)$, NH_4^+ and H^+ in sea water. It has been experimentally proved (ASMAN et al., 1994), that near the coast and esturies, if there are significant nitrogen discharges through the rivers and low gas-phase concentrations, the dry flux of ammonia is directed upwards (which is common over remote and temperate marine areas), and the dry deposition of reduced nitrogen is zero.

The surface concentrations are very sensitive to the value of hmix, which, when estimated from the HIRLAM profiles, seems to be lower than that estimated using on-board soundings (Savolainen, 1997). Comparison of the HIRLAM parameters over sea areas with soundings has been done at the Institute of Marine Research (Uotila et al., 1997) and this work continues at the FMI.

REFERENCES

Asman W. A. H. & Janssen A. J., 1987. A long range transport model for ammonia and ammonium for Europe. *Atmos. Environ. Vol. 21, No. 10* p. 2099-2119

Berge E., (ed.) (1997). Transboundary air pollution in Europe. Part 1. Emissions, dispersion and trends of acidifying and eutrophying agents. EMEP/MSC-W Report 1/97. 108 p.

Bott A., (1989). A positive definite advection scheme obtained by nonlinear renormalization of the advective fluxes. *Mon. Wea. Rev. Vol. 117, May 1989*, p. 1006-1015.

Chang T. Y., 1984. Rain and snow scavenging of HNO3 vapour in the atmosphere. *Atm. Env. Vol. 18, No. 1* p. 191-197.

Chang T.,Y., 1986. Estimates of nitrate formation in rain and snow systems. *J. Geophys. Res*, Vol. 91, No. D2, p. 2805-2818.

Erisman, J.,W, 1994. Evaluation of a surface resistance parametrization of sulphur dioxide. *Atm. Environ. Vol. 28, No. 16*, p. 2583-2594.

Hass H., 1991. Description of the EURAD Chemistry-Transport-Model Version 2 (CTM2). *Mittailungen aus dem Institut fur Geophysik und Meteorologie der Universität zu Köln. Herausgeber A. Ebel, F.M. Neubauer, P. Speth. Heft 83*, 100 p.

HIRLAM, 1990. Hirlam Forecast model level 1. On-line documentation manual. Per Kållberg, (ed.), *SMHI S - 60176*, Norrköping, Sweden, June 1990.

HIRLAM 1996. HIRLAM documentation manual, System 2.5. Källen E.,(ed.), June 1996.

Holtslag A. A. M. & Nieuwstadt F. T. M., 1986. Scaling the atmospheric boundary layer. *Boundary-Layer Meteorology 36*, 201-209.

Hongisto M., 1998 HILATAR, a regional scale grid model for the transport of sulphur and nitrogen compounds. Description of the model and simulation results for the year 1993. *Finnish Meteorol. Inst. Contributions No. 21*, Yliopistopaino, Helsinki, 152 p.

Iversen T., Saltbones J., Sandnes H., Eliassen A. and Hov Ø., 1989. Airborne transboundary transport of sulphur and nitrogen over Europe - Model descriptions and calculations. *EMEP MSC-W Report 2/89. DNMI Technical Report No. 80.* DNMI, Oslo, Norway, 92 p.

Jonsen J.,E. & Berge E., 1995. Some preliminary results on transport and deposition of nitrogen compounds by use of the Multilayer Eulerian Model. *EMEP/MSC-W, Note 4/95*, 25 p.

Karppinen A., Joffre S. and Vaajama P., 1997. Boundary layer parametrization for Finnish Regulatory dispersion models. *International Journal of Environment and Pollution* (in print).

Krueger and Petersen , 1998. The ADOM model system for mercury and persistent organic pollutants. In Borrell B. (ed.) EUROTRAC-2 Symposium 1998, Transport and chemical transformation in the troposphere. EUROTRAC International Scientific Secretariat, Munich 1998.

Lindfors V., Joffre S.M. and Damski J., 1991. Determination of the wet and dry deposition of sulphur and nitrogen compounds over the Baltic Sea using actual meteorological data. *Finnish Meteorological Institute, Contributions No. 4*, 111p.

Savolainen V., 1997. Sounding data from Humbolt. Working paper for the Basys Sp. 5, Atmospheric load. 9 p.

Slinn W.G.N., Hasse L., Hick B.B., Hogan A.W., Lal D., Liss P.S., Munnich K.O., Sehmel G.A. and Vittori O., 1978. Some aspects of the transfer of atmospheric trace constituents past the air-sea interface. *Atmos. Environ. 12*, p. 2055-2087.

Slinn & Slinn, (1980)

Tuovinen J-P., 1992. A dispersion model of air pollutants based on the K theory of turbulentDiffusion. Teknillinen korkeakoulu, Tietotekniikan osasto, Teknillisen fysiikan koulutusohjelma. Diplomityö. 109 p. (in Finnish)

van Ulden A.P. & Holtslag A.A.M., 1985. Estimation of atmospheric boundary layer parameters for diffusion applications. *J. Climate Appl. Meteor. Vol. 24,* p.1196-1207.

Uotila J., Vihma T & Launiainen J., 1997. Marine meteorological radiosoundings in the Northern Baltic Sea from R/V Aranda in 1994-1995. Meri, Report Series of the Institute of Marine Research No. 30, 57 p.

Wesely M.L. 1989. Parametrization of surface resistances to gaseous dry deposition in regional-scale numerical models. *Atm. Environ., Vol 23, No. 6,* p. 1293-1304.

Williams R.M., 1982. A model for the dry deposition of particles to natural water surfaces. *Atmos. Environ. 16:8*, p. 1933-1938.

Wirtz K., Roehl C., Hayman G.D., together with Jenking M. E. and the LACTOZ Steering Group, 1994. LACTOZ re-evaluation of the EMEP MSC-W photo-oxidant model. EUROTRAC International Scientific Secretariat, Garmish-Partenkirchen, 45 p.

Voldner, E.C., Barrie, L.A. & Sirois, A., 1986. A literature review of dry deposition of oxides of sulphur and nitrogen with emphasis on longrange transport modelling in North America. *Atmos. Environ. Vol. 20, No. 11,* p.2101-2123.

DISCUSSION

Z. ZLATEV: The uncertainty of the contribution from the air to the pollution in the Baltic Sea is very big. Do you expect to reduce this uncertainty during the work of this big international project?

M. HONGISTO: This large variation, both in internal cycling or in external load through rivers and air in the annual level was presented last week in the Basys annual conference in Stockholm. The load depends on the meteorological conditions during the year, on the emission data base as well as on the estimation method used. Open sea conditions are not suitable for continuous monitoring and instead of direct measurements, several estimation methods are used (extrapolation from coastal measurements, 1-3D models with different structure and resolution etc.). Although the models might give rather similar annual deposition distribution, interannual variation is big, and dry- and wet deposition differences in the results increase the uncertainty. And the declining trend of the airborne nutrient load, calculated by the EMEP MSC-W center, is not supported by the coastal measurements of the concentrations of all precursors (NO_2) in the air. The aim of the Basys subproject 5 Atmospheric load is to reduce this uncertainty in the calculation methods by iterative process studies and model-measurement intercomparison, with open sea, coastal and profile measurements.

J. FEICHTER: Comment on NO_x emission from ships. The insurance company Lloyds has data available about ship traffic activity worldwide. IGAC/GEIA tries to obtain this data.

M. HONGISTO: Thanks for the hint. The fact, that in international inventories there exists wide areas totally without (Mediterranean Sea) or underestimated (Baltic Sea) emissions, against common knowledge of the traffic activity, makes all model estimates, especially those of coastal ozone or nutrient loads to the seas, rather unreliable. Because over the Baltic Sea the missing emissions are large especially during the summer, they might have important contribution to the eutrophication and unwanted growth of phytoplancton species.

GLOBAL AND LONG-RANGE TRANSPORT

chairmen:	B. Fisher M. Rotach
rapporteurs:	M. Kolarova J. H. Sørensen

Calculated Global Distribution, Radiative Forcing and Climate Response of Sulfate and Carbonaceous Aerosol Particles from Fossil Fuel Combustion

Johann Feichter[1] and William F. Cooke[2]

[1]Max-Planck-Institute for Meteorology
Hamburg, Germany
[2]Centre des Faibles Radioactivites, Laboratoire Mixte'/CNRS-CEA
Gif sur Yvette, France

1. Introduction

Combustion of fossil fuels contributes significantly to the atmospheric aerosol load due to emissions of carbonaceous particles and of sulfur dioxide which forms sulfate particles. The radiative forcing due to the direct effect of anthropogenic sulfate aerosols has been studied extensively over the past few years. Studies by Charlson et al [1991], Penner et al. [1992], Kiehl and Briegleb [1993] and Feichter et al. [1996] show a range of -0.3 to -0.9 W m^{-2}. The effect of an aerosol mixture, consisting of carbonaceous particles which may exert a warming and sulfate particles which exert a cooling, has been investigated in several studies [Haywood and Shine, 1995; Haywood et al., 1997; Schult et al., 1997; Cooke et al., 1998] and has been estimated to be in the range of -0.13 to -0.20 W m^{-2} ; but there is still great uncertainty over the magnitude of the climate forcing.

Hansen et al. [1997] published a study of the response of the climate system due to perturbations by absorbing aerosols using the relatively simple "Wonderland" model. Hansen et al. examined the climate response of a wide range of different kinds and different distributions of prescribed forcings. They pointed out that the forcing-response relationship is nearly linear except for absorbing aerosols and for some changes in the ozone distribution. Hansen point out that absorbing aerosols warm the aerosol layer and cool the surface. This results in a more stable layering and decreases the cloud cover which in turn enhances the warming.

All of the studies mentioned above have not allowed any interaction, or feedback, between the aerosols and the climate system. In this study we utilise a fully-coupled atmosphere/chemistry/mixed-layer-ocean/sea-ice model with quite advanced cloud physics to study the radiative forcing and the climate response due to the direct effect of an external mixture of sulfate, organic (OC) and black carbon (BC) aerosols. The implementation of emission, transport and deposition of aerosols in a global circulation model and a comparison with measurements have previously been reported by Feichter et al. [1996; 1997] and Cooke et al. [1998] for sulfate and carbonaceous aerosols, respectively. The advance made in this work is that the aerosol and the meteorology interact. We consider only fossil fuel sources of sulfate, BC and OC and realise that this is not a comprehensive study of all anthropogenic climate forcing due to aerosols. The focus of this study

Air Pollution Modeling and Its Application XIII, edited by S.-E. Gryning
and E. Batchvarova, Kluwer Academic/Plenum Publishers, 2000

is to examine the relationship between radiative forcing and climate response due to externally mixed scattering and absorbing aerosols.

We performed a 40 year integration and present results of the last 20 years and the differences compared to a control experiment without these aerosol forcings as well. Firstly, we will describe the model compounds and the experimental set-up in section 2. The spatial distributions of the aerosol compounds will be discussed in section 3 and the radiative forcings of the external mixture and the single aerosol compounds and the calculated climate response in section 4. The major findings will be summarized in section 5.

2. Model Description and Experiments

A global coupled atmospheric circulation and mixed layer ocean model has been developed for climate sensitivity studies. The model also considers emission, transport, chemical transformations and removal of sulfur and carbonaceous species as well as the effect from the individual greenhouse gases and from aerosol particles on the radiative fluxes. The mass specific optical properties have been prescribed for each aerosol compound.

The atmospheric general circulation model has been adopted from the European Centre for Medium-Range Weather Forecasts (ECMWF) model [Roeckner et al., 1996]. Prognostic variables are vorticity, divergence, temperature, (logarithm of) surface pressure, and the mass mixing ratios of water vapor, and total cloud water (liquid and ice together). The model equations are solved on 19 vertical levels in a hybrid p-s-system by using the spectral transform method with triangular truncation at wavenumber 30 (T30). Nonlinear terms and physical processes are evaluated at grid points of a "Gaussian grid" providing a nominal resolution of 3.75^{o} x 3.75^{o}. A semi-implicit leap frog time integration scheme with a timestep of 30 minutes is used for the simulation with T30 resolution. Cumulus clouds are represented by a bulk model including the effects of entrainment and detrainment on the updraft and downdraft convective mass fluxes [Tiedtke, 1989]. An adjustment closure based on the convective available potential energy (CAPE) is used [Nordeng, 1995]. Organized entrainment is assumed to depend on buoyancy, and the parameterization of organized detrainment is based upon a cloud population hypothesis. The turbulent transfer of momentum, heat, water vapor, and total cloud water is calculated on the basis of a higher-order closure scheme [Brinkop and Roeckner, 1995]. The radiation code has been adopted from the ECMWF model [Morcrette, 1991] with a few modifications such as the consideration of additional greenhouse gases (methane, nitrous oxide and 16 CFCs) and various types of aerosols [Koepke et al., 1997]. The water vapor continuum has been revised to include temperature weighted band averages of e-type absorption and also a band dependent ratio of p-e-type to e-type continuum absorption [Giorgetta and Wild, 1995]. The single scattering properties of cloud droplets and approximately "equivalent" ice crystals are derived from Mie theory with suitable adaptation to the broad-band model [Rockel et al., 1991]. The effective radius of droplets and ice crystals is parameterized in terms of the liquid and ice water content, respectively.

The aerosol mass mixing ratio for each species is transformed into a particle number concentration assuming a mono-modal log-normal distribution with a prescribed mean particle radius. The optical properties -specific extinction, single scattering albedo and asymmetry factor- for different aerosol compounds are prescribed for every compound, for different relative humidity classes and for the model's spectral intervals. The optical and physical properties of the aerosol components and their dependency on the ambient relative humidity have been taken from Koepke et al. (1997) and are shown in Table 1. Growth of the particles due to water uptake has been assumed for sulfate and the hydrophilic fraction of the organic carbon aerosols only but not for BC and the hydrophobic OC.

To calculate the radiative forcing of sulfate we use the optical and physical properties in Table 1 as given for sulfuric acid. We assume that sulfate is mainly present in the form of ammoniumsulfate and multiply our calculated sulfate mass mixing-ratio by the ratio of the molar weight

Table 1. Physical and optical aerosol properties for dry aerosol according Koepke et al. (1997) (OP denotes the density used to calculate the optical properties, PN to calculate the particle number)

	Relative humidity = 0%			Wavelength = 0.5 μm		
Aerosol compound	Mode radius [μm]	Standard deviation	Specific density [g cm^{-3}]	Mass extinction efficiency[m^2 g^{-1}]	Single scattering albedo	Asymmetry factor
Sulfuric acid	0.0695	2.03	1.7	3.8	1.00	0.77
Water-soluble	0.0212	2.24	1.8	3.6	0.98	0.68
Soot	0.0118	2.00	1.0 OP 2.3 PN	11.0	0.23	0.35

of $(NH_4)_2SO_4/SO_4$. To consider the effects of OC we use the aerosol properties as given for water-soluble particles and for BC the properties of soot. Although BC forms clusters with radii much larger than given in Table 1, the optical properties are assumed to be equal to that of an individual particle, of which the cluster is composed (Chylek et al., 1995). The optical properties have been adapted to the model's spectral resolution by Dorland and Schult (pers.comm.).

We assume an external mixture of the three aerosol compounds when calculating the climate response and the radiative forcing. In fact aerosols in the real atmosphere are externally and internally mixed. However, the spatial and temporal variability of the mixture of the aerosol and the physical and optical properties of the internal mixture are unknown. Moreover, internal mixtures may contain additional compounds not considered in this study. Regarding these uncertainties we think that an external mixture is the best assumption which we can make.

The mixed layer ocean model (Roeckner et al., 1995) is a simple slab model with a depth of 50 m and accounts for the heat exchange between the atmosphere and the ocean. It is coupled to a thermodynamic sea-ice model to reproduce the seasonal variation of the sea-ice extent. Any effects that might result from a change in the ocean circulation in a changing climate cannot be considered.

The parametrization of the sulfur chemistry is described in detail by Feichter et al. [1996]. Transport, dry and wet deposition and chemical transformations of the constituents are calculated on-line with the general circulation model. Prognostic variables are mass mixing-ratios of sulfur dioxide (SO_2) as gas and sulfate as an aerosol. The transport of these species due to advection, vertical diffusion and convection is treated in the same way as the transport of water vapor. Emissions from combustion of fossil fuel and from smelting occur as SO_2 [Benkovitz et al., 1996]. These data are available as seasonal averages. As shown in Table 2, anthropogenic sulfur emissions for the year 1985 amounts globally to 66.7 Tg sulfur per year. Dry deposition of SO_2 and of SO_4^{2-} at ground is assumed to be proportional to the mixing-ratio in the lowest model level and to a prescribed deposition velocity. Dry deposition velocities used in this study are based on Ganzeveld et al. (1998) and Hauglustaine (pers. comm.) and are given in Table 3. Removal of SO_2 and sulfate by precipitation is calculated explicitly in terms of the model's precipitation formation rate (Feichter et al., 1996). SO_2 in the gaseous phase is oxidized by reaction with hydroxyl (OH)

Table 2. Global emissions from fossil fuel use and smeltering per year and calculated annual and global mean budgets

Species	Reference	Source Strength	Calculated Atmospheric Burden	Atmospheric Residence Time
SO_2	Benkovitz et al., 1996	66.7 Tg S	0.22 Tg S	1.8 d
SO_4^{2-}	Oxidation of SO_2	40.0 Tg S	0.33 Tg S	3.0 d
BC	Cooke et al., 1998	5.1 Tg C	0.076 Tg C	5.4 d
Primary OC	Cooke et al., 1998	7.0 Tg C	0.085 Tg C	4.4 d

during the day. Dissolution of SO_2 within cloud water is calculated according to Henry's law. In the aqueous phase we consider oxidation of SO_2 by hydrogen peroxide (H_2O_2) and ozone (O_3). Three-dimensional monthly mean oxidant concentrations are prescribed based on calculations with ECHAM and a more comprehensive chemical model (Roelofs and Lelieveld, 1995). The calculation of the reaction rates, and of the dissolution and dissociation rates of SO_2, requires assumptions about the pH of cloud water. Hence, we diagnostically estimate the cloud water pH as a function of the sulfate concentrations. The end product of the gaseous and the aqueous oxidation of SO_2 is sulfate (SO_4^{2-}).

Table 3. Dry deposition velocity in cm s^{-1} for gases and aerosols

Surface	SO_2	Hydrophobic aerosol	hydrophilic aerosol
Ice and snow	0.1	0.025	0.025
Water or wetted land surfaces; forests	0.8	0.025	0.200
Dry or frozen land surfaces	0.2	0.025	0.025

Emissions of primary carbonaceous aerosols from fossil fuel combustion are prescribed using an inventory of black carbon and organic carbon aerosol from Cooke et al. (1998). In that work a source strength of both the bulk and sub-micron emission was derived. As only sub-micron particles undergo long-range transport, the source strength of sub-micron particles is used in this study (see Table 1). For the northern hemisphere we assumed a seasonality of the emissions with maximum emissions in January and minimum in August (Rotty, 1987). No seasonality was prescribed in the southern hemisphere. A certain fraction of the freshly emitted carbonaceous aerosol is hydrophobic. These aerosols become hydrophilic after soluble compounds become attached to the aerosol surface. We assumed that 80% of BC and 50% of OC is emitted in the hydrophobic form. The typical exponential aging time, for a hydrophobic aerosol to become hydrophilic, has been assumed as 40 hours (Cooke et al. 1998). Sinks for fresh carbonaceous aerosol are only dry deposition applying a dry deposition of 0.025 cm s^{-1}; aged carbonaceous aerosol is treated like sulfate. We should keep in mind that primary organic aerosols represent only a fraction of the organic aerosols. Secondary production accounts for approximately 50% of the organic aerosol close to the anthropogenic source regions and even more at remote sites (Cooke et al.,1998).

3. Distribution of the Aerosol Compounds

The calculated surface distributions have been compared to observations by Feichter et al. (1996) and Cooke et al. (1998). The model tends to underpredict the aerosol concentrations in the Arctic during the Arctic haze season and sulfate in winter over Europe. Black carbon is underestimated at remote and urban sites by 30% and 40%, respectively, and overestimated in rural areas by about 20%. BC in precipitation is underestimated in all regions where measurements are available indicating that the emission strength may be too low. Organic aerosol concentrations are underestimated by at least a factor of two because only primary production has been considered.

A comparison of the calculated BC mixing-ratios in the free troposphere with aircraft measurements (Cooke et al, 1998) has shown that the model performs well up to six km altitude but overestimates the observed mass mixing-ratio by about one order of magnitude higher up. This discrepancy may be partly explained by the fact that the model results represent monthly means whereas the observations are samples over very short time periods. However, we cannot exclude that the assumed aging time until carbonaceous particles become hydrophilic is too long. Such an overestimate of this aging time would result in too high mixing ratios at higher altitudes and in remote regions.

Because the net effect on the radiation balance depends on the relative amount of scattering versus absorbing aerosols, in the following section we will discuss the ratio between the scattering and the absorbing aerosol mass ($M_{C/SO4}$). (Please note, the ratio is given in mass carbon / mass sulfate; in other paragraphs we used mass sulfur). Haywood and Shine (1995) (HS95) assumed for their forcing calculations a constant $M_{C/SO4}$ based on surface measurements. HS95 discuss two data compilation, one of Sisler and Malm (1994), who gives based on three years of surface observations over the US a $M_{C/SO4}$ of 0.16 in air, and one of Ogren and Charlson (1984), who found a mass ratio of 0.07 in precipitation in rural Sweden. HS95 applied the ratio from Ogren and Charlson assuming that the ratio in precipitation is representative for the vertical column and used globally a $M_{C/SO4}$ from fossil fuel combustion of 0.075. Schult et al. (1997) derived $M_{C/SO4}$ from two model simulations calculating sulfate (Langner and Rodhe, 1991) and BC (Cooke and Wilson, 1996) both performed with the transport model MOGUNTIA (Zimmermann, 1987). They found a $M_{C/SO4}$ of 0.12 globally averaged from the surface up to 100 hPa. At the surface over the US their $M_{C/SO4}$ ranges between 0.10 - 0.25 in January and between 0.05 and 0.10 in July. This

Table 4. Calculated annual mean mass ratio of BC/Sulfate in [g C/ g SO_4^{2-}]

Region	US	Southern Scandinavia	North of 30°N	Global
Vertical Column	0.09	0.10	0.09	0.08
Surface	0.25	0.40	0.24	0.13

study calculates an annually and globally averaged ratio of BC and SO_4^{2-} of 0.08. Due to the longer lifetime of BC this ratio is higher than the ratio of the emission sources, which is 0.0425. The global mean ratio near the surface has been calculated as 0.13. North of 30° N, where biomass burning emissions -which have been neglected in this study- may not be very important, we found a surface ratio of 0.24. Figure1 shows the annual mean ratio between black carbon and sulfate of the surface mixing-ratio and the vertical column burden. Close to the source regions the ratio is relatively high (> 0.2) due to the fact that BC is emitted as a primary aerosol but sulfate as a secondary aerosol which is formed after the oxidation of SO_2. Over the oceans the ratio is in the range between 0.01 and 0.1. Although the emission of both species occurs mainly in the NH, the

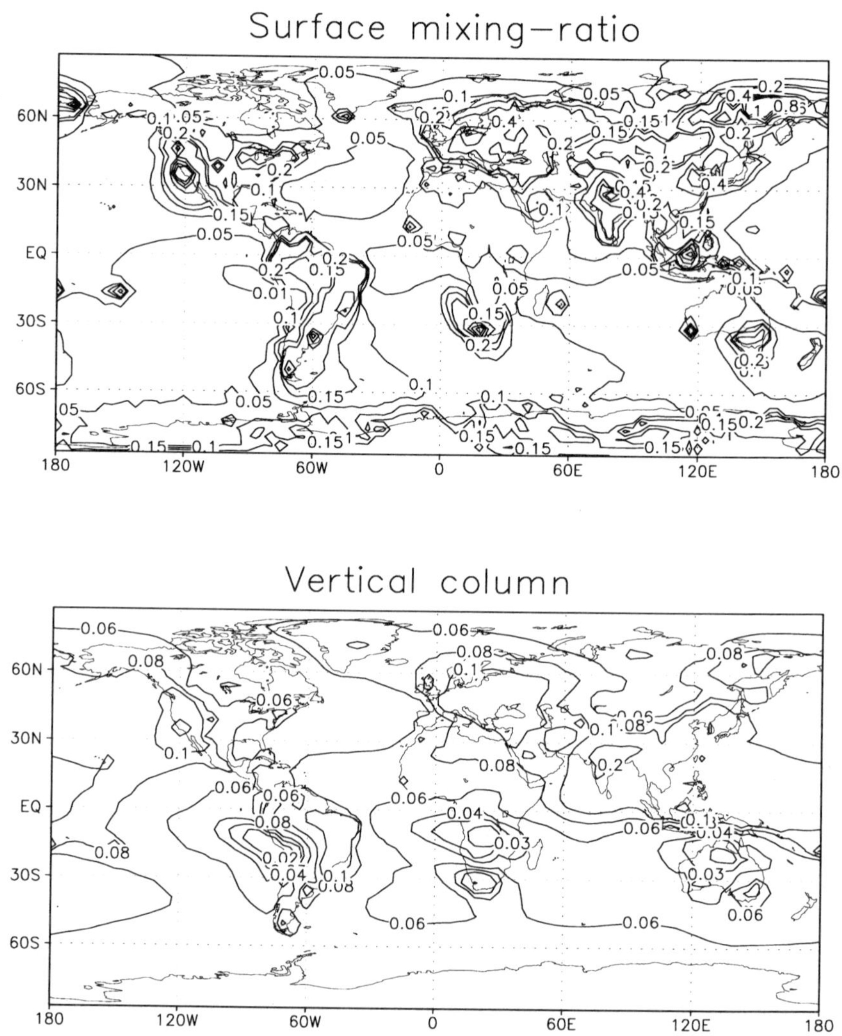

Figure 1. *Annual mean ratio between BC and sulfate of the surface mixing ratio (upper graph) and of the vertical burden (lower graph)*

differences in the ratio between the NH and the SH are surprisingly small. The ratio of the vertical column burden shows maxima of about 0.1 downwind of the source regions and ratios between 0.02 and 0.04 over the oceans.

Table 4 shows the calculated ratio $M_{C/SO4}$ in some specific regions. At the surface over the US the model calculated ratio of 0.25 is higher than that of 0.16 reported by Sisler and Malm (1994). Chin et al. (1996) report annual mean sulfate mixing ratios in industrialised regions in the range between 3000 and 8000 ng m^{-3} and in the Arctic and Subarctic between 600 and 1900 ng m^{-3}. If we compare these values to observed BC mixing ratios given by Cooke et al. (1998) (600-4600 ng m^{-3} in industrialised regions and 3-300ng m^{-3} in the Arctic) we obtain ratios in the order of 0.2 to 0.6 in industrialised regions and 0.003 to 0.16 in the Arctic and Subarctic. These ratios are much closer to the model calculation (see Figure 1a). The ratio observed in precipitation in Scandinavia of 0.07 agrees quite well with the model calculated vertical column burden of 0.10 indicating that the ratio in precipitation reflects the ratio in the atmosphere.

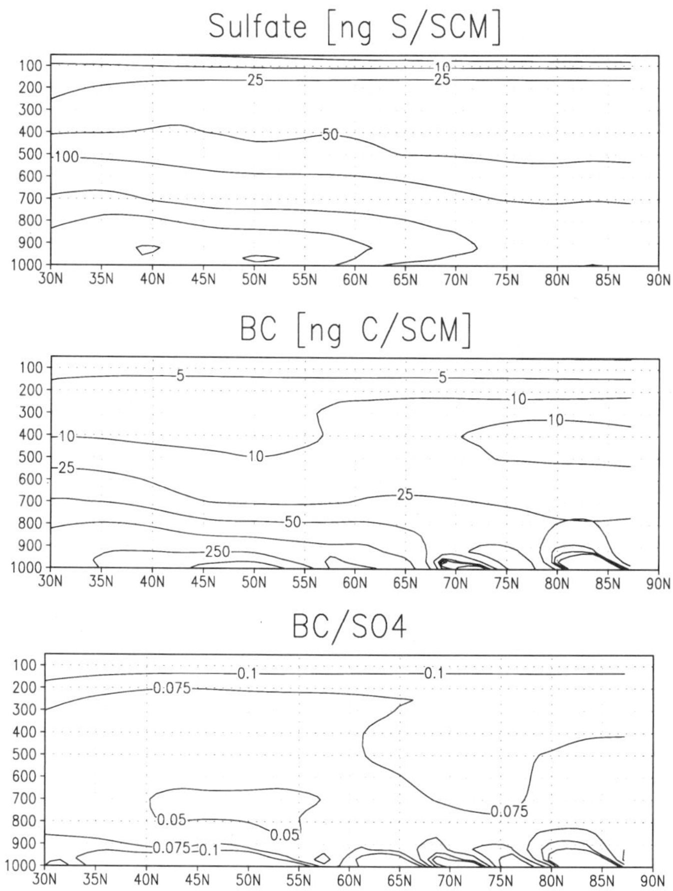

Figure 2. *Vertical cross sections of the annual and zonal mean sulfate mixing ratio (upper graph), the BC mixing ratio (middle graph), and the ratio between BC and sulfate (lower graph).*

The vertical distribution of the ratio $M_{C/SO4}$ averaged north of 30°N peaks at the surface, decreases higher up with a minimum in 700 hPa of 0.08 and increases again up to 400 hPa. Figure 2 shows the annual and zonal mean sulfate and BC mixing-ratios and the ratio $M_{C/SO4}$ depending on latitude and height. Sulfate mixing ratios are highest between 975 and 925 hPa whereas BC exhibits the maximum at the surface. The ratio between BC and sulfate is highest at the surface and in the uppermost model levels and has a minimum between 800 and 700 hPa (ca. 2-3 km). As discussed above BC mixing ratios are, compared to that of sulfate, relatively higher close to the sources and in remote areas. This may be due to the longer residence time of BC and due to the fact that BC is released as a primary aerosol.

4. Radiative Forcing and Climate Response

We define the radiative forcing as a change in the net radiative fluxes at some level in the atmosphere due to scattering and absorbing of aerosols. The calculations of the forcing of single compounds has been done in the following way: First we calculated the forcing of the external mixture of all the three compounds together (EXP1); then we did an experiment where only the effect of two compounds have been considered (EXP2). The difference between these two experiments represents the forcing of the aerosol compound neglected in EXP2. Each of these experiments have been integrated over one year and the forcing has been calculated every two hours.

In Table 5 we present the radiative forcings of the external mixture at the surface and at the top of the troposphere (at 200 hPa). Long- and short-wave forcing at the top of the troposphere amounts to -0.13 Wm^{-2} which is within the range of recent estimates (Haywood and Shine, 1995; Haywood et al., 1997; Schult et al., 1997). BC aerosols exert a cooling in the lower troposphere and a warming above 700 hPa (3 km). The long-wave radiative forcing is positive in all vertical levels and amounts to about 20% of the short-wave forcing at the top of the troposphere and at the surface as well. This absorption in the long-wave range is mainly due to the sulfate aerosols. The effect of OC on the long-wave radiation is negligible; the effect on the short-wave radiation is 7% of the total forcing at the top of the atmosphere and 4% at the surface. The geographical distribution of the annual mean radiative forcing at the top of the troposphere is shown in Figure 3b. The model calculates negative forcing in most of the northern hemisphere with maxima over the industrialized regions ranging between -0.5 and -2 W m^{-2}. Small positive forcing is found over the Arctic and over deserts where the surface albedo is high. Comparing Figure3b to the forcing of a clear sky atmosphere (Figure 3a) shows a distinct higher negative forcing when the effect of clouds is neglected. The annually and globally averaged clear sky forcing at the top of the troposphere is, compared to the forcing with clouds present, by a factor of 3.2 more negative for the solar radiation and by a factor of 1.6 more positive for the thermal radiation. The presence of clouds affects the solar aerosol forcing in two ways: Firstly, mid- and high level clouds shield the underlying aerosol layer and reduce the cooling effect of scattering particles. Secondly, due to their high albedo, low level clouds increase the absorbing effect of an overlying BC aerosol layer.

Table 5. Calculated forcing [W m^{-2}]

Forcing	Surface				Top of the troposphere			
	Global annual mean	NH annual mean	Global winter	Global summer	Global annual mean	NH annual mean	Global winter	Global summer
Long-wave	0.10	0.19	0.08	0.11	0.04	0.08	0.03	0.06
Short-wave	-0.52	-0.94	-0.36	-0.70	-0.17	-0.28	-0.12	-0.23

Absorption is in particular increased in areas with low surface albedo as over the ocean and over vegetated continents. However, the ratio between the clear sky and the cloudy sky forcing of the single aerosol compounds indicates that most of the cloud effect can be attributed to sulfate (ratio=2.5) rather than to BC (ratio=0.9).

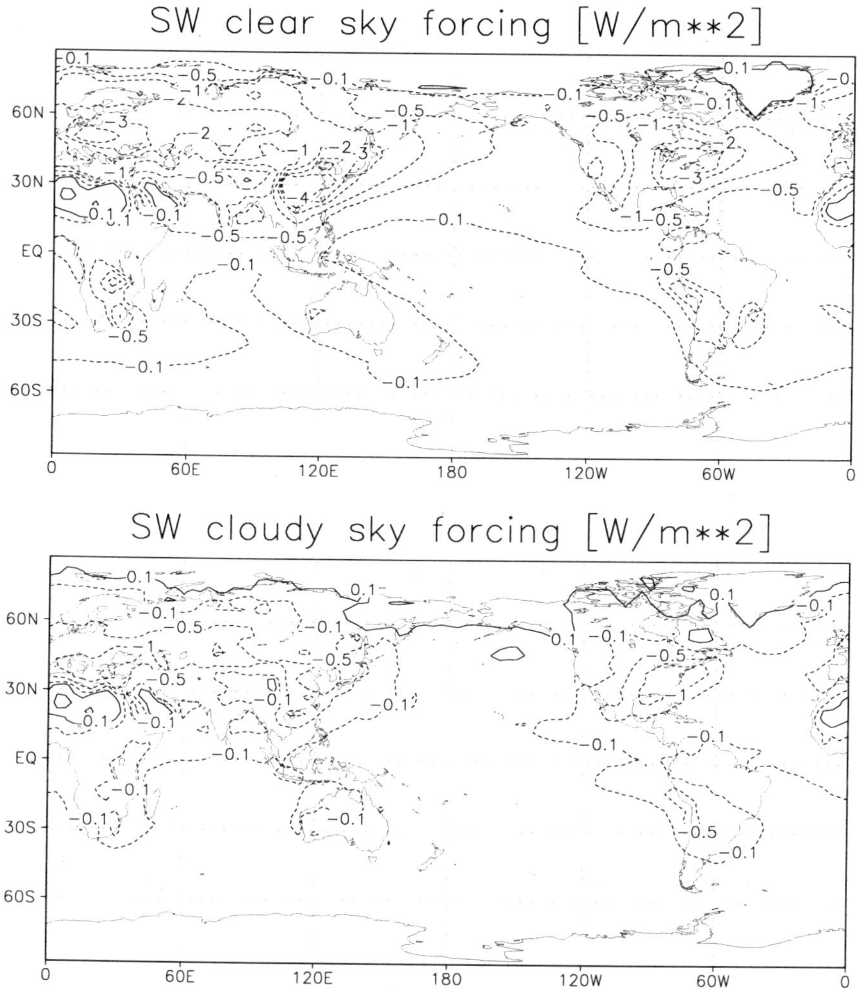

Figure 3. *Annual mean short-wave forcing in W m*$^{-2}$.

The response of the climate model to the forcing of the externally mixed aerosols is discussed next. Figure 4 shows the calculated anomaly (defined as aerosol experiment minus control experiment) of the annual mean 2-m temperature. For most regions we find a cooling with maxima of about -1° K at high latitudes over the northern hemisphere (NH) continents and a few smaller areas with a warming of about 0.1 to 0.2^{o}K. In general, the cooling over the NH continents is stronger than -0.2^{o}K. The cooling amounts globally to -0.15^{o} K and in the NH to -0.17^{o}K. Latitudinally the cooling is strongest between 65 and 75°N (-0.5^{o}K) and close to zero between 60 and 65°S. However, the temperature anomalies are in the range of the internal variability of the model and cannot be considered as significant. The forcing in Figure 3b does not bear much resemblance with the temperature change (Figure 4). The forcing is close to zero in the regions of strongest cooling. Associated with the surface cooling of 1% in the NH, where most of the aerosols from fossil fuel use are released, we also expect some changes in the hydrometeorology. In the NH the evaporation is reduced by 0.7% and the surface latent heat fluxes by 0.6%. The total precipitation was reduced by 0.2% but snow fall increased, due to the lower temperatures, by 1.5% and the sea ice cover extent by 2.9%. The lower temperatures also resulted in lower water vapor content (-0.8%) and a lower cloud water amount (-0.6%) whereas changes in the cloud cover have been negligible.

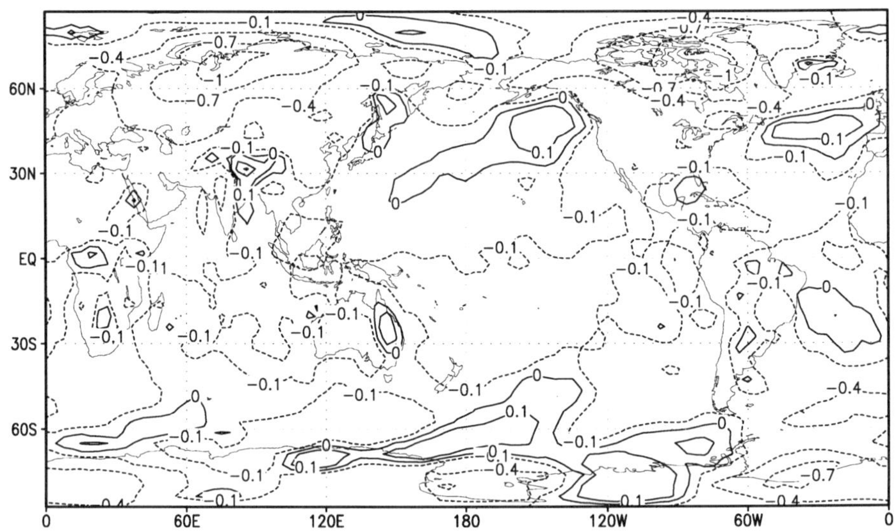

Figure 4. *Calculated annual mean temperature anomalies in deg K.*

Changes in the free troposphere have the same sign as at the surface. Figure 5 presents annual and zonal means of the temperature, the specific humidity and the cloud cover anomalies from the surface up to 10 hPa. Temperature anomalies are negative in the troposphere and positive in the lower stratosphere. However, the changes are relatively small and are not restricted to the NH where the forcing is strongest. This cooling is associated with a decrease of the specific humidity and negligible changes in cloud cover. Since the anomalies of temperature and specific humidity have the same sign in the entire troposphere, we consider it as significant even if the magnitude is quite small.

5. Discussion and Conclusion

This model study investigated the effect of anthropogenic emissions of fossil fuel sulfur, organic and black carbon aerosols on the climate system. So far climate response simulations studying the aerosol effect have only considered sulfate aerosols or prescribed aerosol distributions. We introduced sulfur chemistry and aerosol physics in a general circulation model which has been coupled to a mixed-layer ocean and a sea-ice model. This model allows the representation of the spatial and temporal heterogeneity and variability of the aerosol distribution in a way which is consistent with the meteorology.

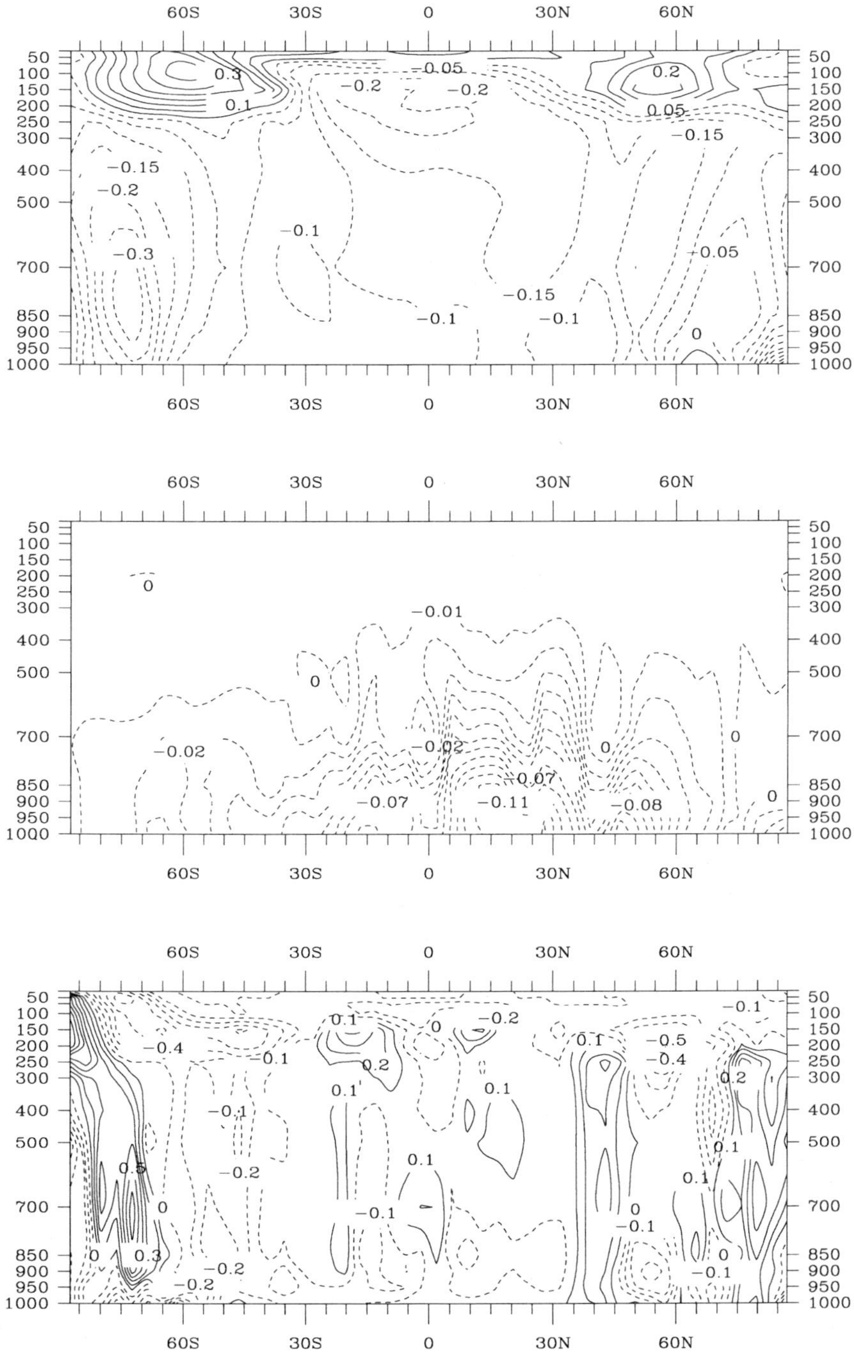

Figure 5. *Vertical cross sections of the calculated annual and zonal mean anomalies of the temperature (upper graph) in K, of the specific humidity (middle graph) in g/kg and of the fractional cloud cover (lower graph) in %.*

Since we only consider carbonaceous and sulfate aerosols from fossil fuel sources, we cannot expect to obtain a realistic estimate of the anthropogenic impact on climate. However, we attempt to study the effect of a mixture of both absorbing and scattering aerosols on circulation and hydrology.

The calculated mass mixing-ratios of sulfate and black carbon (BC) agree quite well with observations, at least in the northern hemisphere where emissions from fossil fuel sources dominate. Organic carbon (OC) mass mixing-ratios are underestimated by a factor of two because we only considered OC aerosols directly produced by combustion sources but ignored secondary gas-to-particle conversion of hydrocarbons. The ratio of BC to sulfate integrated over the whole model domain is 0.08 in agreement with the ratio observed in precipitation (Ogren and Charlson, 1984) and is 0.13 at the lowest model level (30m). Over the US the ratio amounts to 0.25 which is higher than observed by Sisler and Malm (1994). Generally, the ratio is high close to the sources and is lower at remote areas because BC is emitted as a primary particle whereas sulfate is formed in the atmosphere by oxidation of sulfur dioxide. The global and annual average atmospheric burden of sulfate amounts to 1 Tg (SO_4^{2-}), that of BC to 0.076 Tg (C) and that of OC to 0.085 Tg (C).

The calculated forcing of the externally mixed aerosols is negative except over surfaces with high albedo, such as deserts or glaciers. The forcing at the top of the troposphere amounts to -0.13 W m^{-2} (solar forcing -0.17, terrestrial +0.04 W m^{-2}) as a global and annual mean which is in the range of recent estimates. Haywood and Shine (1995) calculated for an external mixture of BC and sulfate a solar forcing of -0.29 W m^{-2}, Schult et al. (1997) for clear sky conditions -0.21 W m^{-2} (average of January and July forcing) compared to -0.53 W m^{-2} in the current study and Haywood et al.(1997) report a solar forcing of -0.18 W m^{-2}. The forcing in the thermal range is generally positive and mainly due to sulfate aerosols. The solar forcing of BC is negative in the boundary layer and lower troposphere and is positive higher up. The cooling of sulfate aerosols at the top of the troposphere is reduced by the presence of BC by 26% or 0.06 W m^{-2}, corresponding to a reduction of about 3.4% per 1% BC to sulfate ratio. Chylek et al. (1995) found a reduction of the sulfate cooling by BC of 2.6% per 1% BC to sulfate mixing ratio. This calculation is based on a much higher sulfate forcing of -1.33 W m^{-2} and a smaller BC amount of 0.5% of the sulfate mass. The effect of OC is relatively small and amounts to 7% of the forcing at the top of the troposphere with negative sign, whereas BC exerts a quite strong effect taking into account the small mass fraction.

The externally mixed sulfate and carbonaceous aerosols result in a small cooling in most regions of the globe with an annual and global mean 2-m temperature reduction of -0.15°. We generally found a cooling in the troposphere and a warming in the stratosphere. The climate sensitivity parameter, defined as the ratio between the global change of the 2-m temperature and the radiative forcing, is 0.89, within the range of other forcings (Roeckner, pers. comm.). Because the surface forcing of -0.43 W m^{-2} (solar and terrestrial) is much higher than the forcing at the top of the troposphere, aerosols reduce the evaporation, precipitation and latent heat fluxes. Also the specific humidity is reduced in the whole troposphere whereas the cloud cover remains unchanged. This damping of the hydrological cycle has also been found in experiments only considering the effect of sulfate aerosols (Roeckner et al., 1998). This damping has also been discussed in Mitchell (1971) and Hansen et al. (1997) but these authors expected additionally a decrease in cloud cover which has not been found in the current study. However, one should note that, although aerosol physics and meteorology are calculated in a consistent way, the uncertainties are still very large.

References

Benkovitz, C. M., M. T. Scholtz, J. Pacyna,L. Tarrason, J. Dignon, E. C. Voldner, P. A. Spiro, P J. A. Logan, T. E. Graedel, Global gridded inventories of anthropogenic emissions of sulfur and nitrogen, *J. Geophys. Res., 101*, 29,239-29,254, 1996.

Brinkop, S., and E. Roeckner, Sensitivity of a general circulation model to parameterizations of cloud-turbulence interactions in the atmsopheric boundary layer, *Tellus*, *47A*, 197-220, 1995.
Charlson, R.J., J. Langner, H. Rodhe, C.B. Leovy, and S.G. Warren, Perturbation of the Northern Hemisphere radiative balance by backscattering from anthropogenic aerosols, *Tellus*, *43AB*, 152-163, 1991.
Chin, M., D. J. Jacob, G. M. Gardner, M. S. Foreman-Fowler, P. A. Spiro, and D. L. Savoie, A global three-dimensional model of tropospheric sulfate, *J. Geophys. Res., 101*, 18,667-18,690, 1996.
Chylek, P. , G. Videen, D. Ngo, R. G. Pinnick, J. D. Klett, Effect of black carbon on the optical properties and climate forcing of sulfate aerosols, *J. Geophys. Res., 100,* 16,325-16,332, 1995.
Cooke, W. F., and J. J. N. Wilson, A global black carbon aerosol model, J. *Geophys. Res., 101,* 19,295-19,409, 1996.
Cooke, W. F., C. Liousse, H. Cachier and J. Feichter, Construction of a 1^ox1^o fossil fuel emission dataset for carbonaceous aerosols and implementation into the ECHAM4 model, subm. to J. Geophys. Res., 1998.
Feichter, J., E. Kjellström, H. Rodhe, F. Dentener, J. Lelieveld, and G.J. Roelofs, Simulation of the tropospheric sulfur cycle in a global climate model, *Atmos. Env.,* 30, 1693-1707, 1996.
Feichter, J., U. Lohmann, and I. Schult, The atmospheric sulfur cycle and its impact on the shortwave radiation, *Clim. Dyn.,* 13, 235-246, 1997.
Ganzeveld,L., J. Lelieveld and G-J. Roelofs, A dry deposition parameterization for sulfur oxides in a chemistry and general circulation model, *J. Geophys.Res., 103,* 5679-5694, 1998.
Giorgetta, M., and M. Wild, The water vapour continuum and its representation in ECHAM4, Max-Planck-Institute for Meteorology, Hamburg, Germany, 1995.
Graf, H., J. Feichter and B. Langmann, Volcanic sulfur emissions: Estimates of source strength and its contribution to the global sulfate distribution, *J. Geophy. Res.*, 102, 10,727-738, 1997.
Hansen, J., M. Sato and R. Ruedy, Radiative forcing and climate response, *J. Geophys. Res., 102,* 6831-6864, 1997.
Haywood, J. M., and K. P. Shine, The effect of anthropogenic sulfate and soot aerosols on the clear sky planetary radiation budget, *Geophys. Res. Letters, 22*, 603-606, 1995.
Haywood, J. M., D. L. Roberts, A. Slingo, J. M. Edwards and K. P. Shine, General circulation model calculations of the direct radiative forcing by anthropogenic sulfate and fossil-fuel soot aerosol, *J. Clim., 10*, 1562-1576, 1997.
Kiehl, J.T., and B.P. Briegleb, The relative roles of sulfate aerosols and greenhouse gases in climate forcing, *Science*, *260*, 311-314, 1993.
Koepke, P., M. Hess, I. Schult and E. Shettle, Global aerosol data set, Report No. 243, Max-Planck-Institute for Meteorology, Hamburg, Germany, 1997.
Langner, J., and H. Rodhe, A global three-dimensional model of the global sulfur cycle, *J. Atmos. Chem.*, *13*, 225-263, 1991.
Mitchell Jr., J. M., The effect of atmospheric aerosolson climate with special reference to temperature near the Earth's surface, *J. Appl. Meteorol., 10,* 703-714, 1971.
Morcrette, J.-J., Radiation and Cloud Radiative Properties in the European Centre Medium Range Weather Forecasts Forecasting System, *J. Geophys. Res.*, *96* (D5), 9121, 1991.
Nordeng, T.E., Extended versions of the convective parameterization scheme at ECMWF and their impact on the mean and transient activity of the model in the tropics, *J. R. Meteorol. Soc., (submitted)*, 1995.
Ogren, J. A. and R. J. Charlson, Wet deposition of elemental carbon and sulfate in Sweden, *Tellus, 36B,* 262-271, 1984.
Penner, J. E., R. Dickinson and C. O'Neill, Effects of aerosol from biomass burning on the global radiation budget, *Science, 256*, 1432-1434, 1992.

Rockel, B., E. Raschke, and B. Weyres, A parameterization of broad band radiative transfer properties of water, ice and mixed clouds, *Beitr. Phys. Atmosph., 64,* 1-12, 1991.

Roeckner E., T. Siebert and J. Feichter, Climatic response to anthropogenic sulfate forcing simulated with a general circulation model, In: *Aerosol Forcing of Climate,* ed.: R. J. Charlson and J. Heintzenberg, John Wiley&Sons Ltd., 1995.

Roeckner, E., K. Arpe, L. Bengtsson, M. Christoph, M. Claussen, L. Dümenil, M. Esch, M. Giorgetta, U. Schlese, and U. Schulzweida, The atmospheric general circulation model ECHAM-4: Model description and simulation of present-day climate, Max-Planck-Institute for Meteorology, Hamburg, Germany, 1996.

Roeckner, E., L. Bengtsson, J. Feichter, J. Lelieveld and H. Rodhe, Transient climate change simulations with a coupled atmosphere-ocean GCM including the tropospheric sulfur cycle, Report No. 243, Max-Planck-Institute for Meteorology, Hamburg, Germany, 1998.

Roelofs, G.-J., and J. Lelieveld, Distribution and budget of O_3 in the troposphere calculated with a chemistry general circulation model, *J. Geophys. Res., 100* (D10), 20,983-20,998, 1995.

Rotty, R. M., Estimates of seasonal variation in fossil fuel CO_2 emissions, *Tellus, 39B,* 184-202, 1987.

Schult I., J. Feichter and W. F. Cooke, The effect of black carbon and sulfate aerosols on the global radiation budget, *J. Geophys. Res., 102,* 30,107-30,117, 1997.

Sisler, J. F. and W. C. Malm, The relative importance of soluble aerosols to spatial and seasonal trends of impaired visibility in the United States, *Atmos. Environ., 28,* 850-862, 1994.

Tiedtke, M., A comprehensive mass flux scheme for cumulus parameterization in large-scale models, *Mon. Wea. Rev., 117,* 1779-1800, 1989.

Zimmermann, P. H., Moguntia: A handy global tracer model, Proceedings of the 16th NATO/CCMS, *International Meeting on Air Pollution Modeling and its Application*, Lindau, April 6-10, D. Reidel, Dordrecht, 1987.

DISCUSSION

T. IVERSEN: Do you not find it inconsistent that you assume a turnover from hydrophobic to hydrophilic BC at the same time as assuming BC and SO_4 externally mixed.

J. FEICHTER: Yes. I agree there is an inconsistency. In order to calculate a reliable BC burden, you have to remove BC by wet deposition. We assumed that BC becomes internally mixed and with that hydrophilic about one day after emission. On the other hand, the aerosol effect on the radiation fluxes was calculated assuming an externally mixed aerosol. We think, that regarding the temporal and spatial variability of the mixture of aerosols and the uncertainty in the optical properties the assumption of an externally mixed aerosol is the most acceptable one.

RUNNING LARGE-SCALE AIR POLLUTION MODELS ON PARALLEL COMPUTERS

K. Georgiev[1] and Z. Zlatev[2]

[1] Bulgarian Academy of Sciences
Acad. G. Bonchev str., Bl.25A
1113 Sofia, Bulgaria
[2] National Environmental Research Institute
Frederiksborgvej 399
DK-4000 Roskilde, Denmark

INTRODUCTION

The protection of our environment is one of the most important problems in the modern society (and its importance is steadily increasing). Reliable and robust control strategies for keeping the pollution caused by harmful chemical compounds under certain safe levels have to be developed and used in a routine way. Large mathematical models, in which all physical and chemical processes are adequately described, can successfully be used in the solution of this task. During the transport of the species the full spectrum of chemistry and physics is involved. The gas molecule may react chemically, may be absorbed into a particle or droplet, may be transported into the stratosphere or may be removed by interaction with the earth's surface. However, the adequate description of all physical and chemical processes leads to huge computational tasks, which cause great difficulties even when modern high-speed computers are available. Several hundred runs (and sometimes even several thousand runs) have to be performed in a typical simulation. One has as a rule to carry out several thousand time-steps in each of these runs. Several systems of ordinary differential equations containing up to several millions equations must be handled numerically at each time-step. Therefore, one has either to make some non-physical assumptions in order to simplify the model and, thus, to make it tractable on the computers available or to develop codes that exploit fully the great potential power of the new modern supercomputers. An attempt to solve the computational problems by using the second approach, which is the difficult one, will be discussed in this paper.

A parallel algorithm has been developed and applied to treat numerically the two dimensional Danish Eulerian Model (a large-scale model for studying air pollution levels in Europe developed at the National Environmental Research Institute, Roskilde, Denmark). With this model *sulfur-containing, nitrogen containing and carbon-containing air pollutants*, total of 56

Air Pollution Modeling and Its Application XIII, edited by S.-E. Gryning and E. Batchvarova, Kluwer Academic/Plenum Publishers, 2000

can be studied. Among the components are: $NO, NO_2, HNO_3, NO_3^-, NH_3, NH_4^+, PAN, O_3, SO_2, SO_4^=, OH, H_2O_2$ and many *hydrocarbons.* The chemical transformation are performed by using more than 70 chemical reactions. Most of them are **nonlinear** and some of them are **photochemical**.

The algorithm has actually been used on an IBM SP computer. However, the implementation is based on MPI (Message Passing Interface), which is a standard tool for developing parallel codes. Therefore, most of the main ideas can also be used in connection with other message passing computers, and also in connection with other air pollution models. The algorithm is based on partitioning of the computational domain in several subdomains. The number of the subdomains is equal to the number of the processors which are available. The new algorithm is highly parallelizable.

Numerical results, obtained when up to 32 processors on the IBM SP computer at the Danish Computing Centre for Research and Education (UNI-C) are used, will be presented and discussed. Some ways for a further improvement of the results will briefly be outlined.

THE MATHEMATICAL MODEL

The following **physical processes**: *advection* (the transport due to the wind), *diffusion, deposition, emission* and the **chemical reactions** have to be considered. There are two different ways to describe mathematically the air pollution phenomena. The first is the Eulerian approach in which the behavior of the species is described relative to a fixed coordinate system. This description is a common way of treating heat and mass transfer. The second approach is the Lagrangian in which the changes of the concentrations are described relative to the moving fluid. Hereafter, the Eulerian approach will be used. It is very useful because the Eulerian statistics are readily measurable and the mathematical expressions are directly applicable to situations in which the chemical reactions take place.

Let us denote with q the number of the pollutants under consideration. The numerical experiments discussed in this paper have been carried out by using a chemical scheme with 56 compounds, i.e. $q = 56$. Let Ω be a bounded domain in the plane Oxy with a boundary $\Gamma \equiv \partial\Omega$. The Eulerian model for an adequately description of the physical and chemical processes involved in the long-range transport of air pollutants in two dimensions can be represented by the following system of partial differential equations (PDE's) [16, 17]:

$$\begin{aligned} \frac{\partial c_s}{\partial t} &= -\frac{\partial(uc_s)}{\partial x} - \frac{\partial(vc_s)}{\partial y} + \frac{\partial}{\partial x}\left(K_x\frac{\partial c_s}{\partial x}\right) + \frac{\partial}{\partial y}\left(K_y\frac{\partial c_s}{\partial y}\right) + \\ &+ E_s + Q_s(c_1, c_2, \dots, c_q) - (k_{1s} + k_{2s})\, c_s, \qquad s = 1, 2, \dots, q, \end{aligned} \tag{1}$$

where c_s are the concentrations; u and v are the wind velocities; K_x and K_y are the diffusion coefficients; the sources are described by the functions E_s; k_{1s} and k_{2s} are the deposition coefficients (wet and dry deposition); the chemical reactions are described by the functions Q_s. Very often advection part of the equations can be simplified by assuming that a conservation law is satisfied for the wind velocities in the lower parts of the atmosphere. The **advection** is described by the first two terms while the **diffusion** - by the third and fourth terms in the first line in (1). The terms E_s describe the **emission** sources in the space domain under consideration. The **chemical processes**, presented by the terms Q_s, play a special role in the model. The equations in the system (1) are coupled only through the chemical reactions. From the other hand the chemistry introduces nonlinearity in the model. As a rule, the chemical reactions are represented by the following formulae:

$$Q_s(c_1, c_2, \dots, c_q) = -\sum_{i=1}^{q} \alpha_i c_i + \sum_{i=1}^{q}\sum_{j=1}^{q} \beta_{ij} c_i c_j \tag{2}$$

The nonlinear chemical reactions in large long-range study of air pollution are discussed in [13, 14] and the photochemical mechanism is described in [4].

Initial and boundary conditions are added to the system of PDE's (1).

It is difficult to treat numerically the PDE's system (1) directly due to the difficulties in finding a common, effective and fast, numerical method. The main reason is that the different terms of the right-hand side of the system come from very different physical processes, and they have different properties from the mathematical point of view. Therefore, it is very difficult to satisfy all requirements of the different terms simultaneously. The way to solve this problem is to apply some splitting procedure. Following ideas in Marchuk [8] and McRae et al. [10] the model is splited in four submodels according to the different physical processes that are involved in: *advection, diffusion, chemistry* (together with *source terms*) and *deposition.*

$$\frac{\partial c_s^{(1)}}{\partial t} = -\frac{\partial (uc_s)^{(1)}}{\partial x} - \frac{\partial (vc_s)^{(1)}}{\partial y} \qquad s=1,2,\ldots,q, \tag{3}$$

$$\frac{\partial c_s^{(2)}}{\partial t} = \frac{\partial}{\partial x}\left(K_x\frac{\partial c_s^{(2)}}{\partial x}\right) + \frac{\partial}{\partial y}\left(K_y\frac{\partial c_s^{(2)}}{\partial y}\right) \qquad s=1,2,\ldots,q, \tag{4}$$

$$\frac{\partial c_s^{(3)}}{\partial t} = E_s + Q_s\left(c_1^{(3)}, c_2^{(3)}, \ldots, c_q^{(3)}\right) \qquad s=1,2,\ldots,q, \tag{5}$$

$$\frac{\partial c_s^{(4)}}{\partial t} = -\left(k_{1s}+k_{2s}\right)c_s^{(4)}, \qquad s=1,2,\ldots,q. \tag{6}$$

The initial values of the concentrations are used as an initial condition for the first system of PDE's (3) at the first time step. The process of coupling, which has to be carried out at every time-step, can be performed as follows: Assume that we have somehow obtained the values $c_s(t)$ of the concentrations of all pollutants in all the grid points at the time t. Now we have to find the values of the concentrations at time $(t+\Delta t)$. The values of $c_s(t)$ are used as an initial condition for the first system. As a result we will obtain the values of $c_s^{(1)}(t+\Delta t)$ and these values are used as an initial condition for the second system to get approximations to $c_s^{(2)}(t+\Delta t)$. This process can be continued for the 3-d and 4-th system of PDE's. Finally, solving the fourth system (dry and wet deposition) we will obtain the values of the concentrations at the time $(t+\Delta t)$ and we are ready to repeat this procedure for the next time-step.

The submodels (3)-(4) can be discretized in the space by using different discretization algorithms and treated successively at each time-step (see for example [15, 16]). In the algorithm considered in this paper, the spatial derivatives in (3) and (4) are discretized by using one-dimensional linear finite elements. After this space discretization the following four systems of ordinary differential equations (ODE) have to be solved:

$$\frac{dg^{(i)}}{dt} = f^{(i)}\left(t, g^{(i)}\right), \quad g^{(i)} \in R^{N_x\times N_y\times N_s}, \quad f^{(i)} \in R^{N_x\times N_y\times N_s}, \quad i=1,\ldots,4, \tag{7}$$

where N_x and N_y are the numbers of the grid points along the coordinate axes and $N_s = q$ is the number of the chemical species. The functions $f^{(i)}$, $i = 1,2,3,4$ depend on the particular descritazion methods used (for more details see [16]). Predictor-corrector methods with several different correctors are used in the solution of the first and second subsystems. The correctors are chosen so that the stability of the method is enhanced [12]. The simple and relatively stable *quasi-steady-state algorithm* (QSSA) is used in the experiments described here with a small stepsize in order to improve its accuracy. The ODE's in the last subsystem are linear and therefore they are solved exactly. Since we are now using version with 56 chemical species and 96×96 grid, each of these four systems of ODE's contains 516096 equations. The time-period for the model is often *one month + five days* (to start-up the model, see [16]) and then 3456 time-steps are needed, assuming that the time-step for the advection submodel is $900 sec$. Moreover, the systems of ODE's arising into the chemistry sub-model are stiff and therefore they have to be treated with smaller time-step (150*sec.* is actually used). It is clear that the computational task has to be solved is enormous and it causes great difficulties even when big and fast modern supercomputers are used. Therefore it is essential to select both sufficiently accurate and fast numerical algorithms and optimize the code realizing these algorithms for runs on parallel computers (with shared and distributed memory) and vector computers. Hereafter we will present an algorithm based on a partitioning of the space domain in several subdomains for the 2D version of the Danish Eulerian Model which is for use on parallel computers.

THE PARALLEL ALGORITHM

Parallel Computations by Using Partitioning of the Computational Domain

Let us consider a parallel computer with p processors. Then we divide the computational domain Ω, which is a square now, in p subdomains in Y-direction. Similar algorithm one can get if the dividing is in X-direction. The combination of these two approaches will lead to a new algorithm. As was mentioned above, different numerical algorithms are used in the different submodels. Therefore, two types of subdomains are appeared. *Nonoverlapping* subdomains - Ω_i^{ch}, for the chemistry and deposition submodels and *overlapping* - Ω_i^{adv}, subdomains for advection and diffusion submodels (see Fig.1).

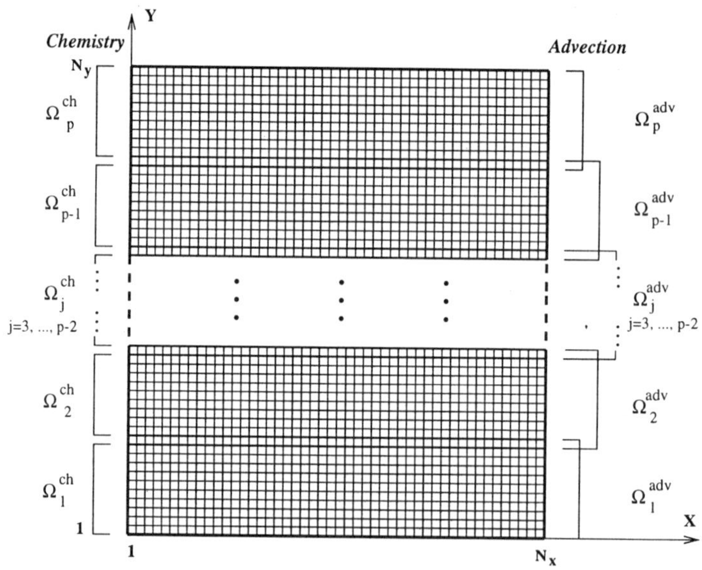

Figure 1. Partitioning of the computational domain.

During the advection and diffusion processes the values of the concentrations inside Ω_i^{adv} are updated but we need the values of the concentrations on the neighbor rows which are used as inner boundary conditions. In the chemistry and deposition submodels we do not need any boundary conditions and therefore there is no overlapping of the subdomains in these two submodels. Hence some communications are needed after the chemical and deposition part of the model. The number of communications is very small and communication time does not depend on the number of the subdomains when $p > 2$. The number of real numbers have to be communicated among the neighbor processors is two times the number of grid points belong to one grid line. If the processor i is not the first or the last one then during the communication step it has to send to its neighbors in "north" and "south" directions the values of the concentrations in the grid-points belong to the second and last but one grid-line of the subdomain it is responsible. At the same time, it has to receive from the same neighbors the corresponding data and put these data on the grid-points which belong to the first and last grid-line of its own subdomain. These updated values of the concentrations will be used as boundary conditions for the next advection time-step. It is clear that the first processor has not a neighbor from bellow and the last processor has not a neighbor from above. Therefore, they will send and receive information only in "north" or "south" direction, correspondingly. Some additional communications are needed in the end of the algorithm when the values of the concentrations of all pollutants computed in the different processors have to be sent to the root processor which will prepare needed output data in the whole computational domain.

For the communication part of the algorithm the Message Passing Interface (MPI) standard (see i.e. [5, 11]) is used on both shared and distributed memory parallel computer platforms. The use of MPI is a key component in the development of concurrent computing environment in which the applications and tools can be transparently ported between different computers.

Input Data and Output Data

A big amount of input data is used in the two dimensional Danish Eulerian Model. Mainly, these input data may be divited as *emission* and *meteorological* data. The emission data are organized in different files and consist: *ammonia, sulphur, nitrogen* and *antropogenic VOC emissions.* The input information about the *forests* is used for producing the *natural VOC emissions* by the algorithm of Lübkert and Schöp. In addition there is a file with the *latitudes* and *longitudes* of the grid points. The meteorological data are organized in eight different files: *cloud covers* (in percent), *humidity* (in percent), *mixing hight, precipitation, temperature, surface temperature, vertical wind, horizontal components of the wind field.* The total amount of the input data for one month period is more than 54 MBytes. These data which is for the grid points into the whole computational domain, is obtained from the Norwegian Meteorological Institute. For an efficient use in parallel computations these data have to be reorganized and prepared for separate use from each processor. This is done into the so-called preprocessing procedure where the data for each subdomain are prepared and sent to the local memories of the different processors.

The output data from the Danish Eulerian Model are even more than the input data. Experiments with different emission scenarious resulted in collecting very large output data sets. The total amount of these data is greater than 100 MBytes (see e.g. [3]). These data include the meanvalues of the: *concentrations of all pollutants per month, per day* and *per hour, concentrations in precipitations of all pollutants per month and day* and *wet depositions of all pollutants per month.* The gathering of the data from the slave processors to the root processor is done into the so-called postprocessing procedure.

The new parallel algorithm is as follows:

- **Preprocessing:**
- **Initial stage:**
  - Read input data: *emissions, meteorological data, other data (forests, longitude, latitudes)*;
  - Calculation the addresses of the local arrays;
  - Perform the splitting procedure.
- **Time cycle:**
  - Read new data for the *wind velocities*;
  - Perform the *advection* submodel;
  - Read new *meteorological data*;

- Perform the *chemistry* and *deposition* submodels;
- Perform the *communications* and *update* the values on the inner boundaries.

- **Postprocessing**

NUMERICAL EXPERIMENTS

The most time consuming parts of the Danish Eulerian Model are *advection* and *chemistry* submodels. A module consisting of these two submodels was constructed and tested on different parallel computers (see [7, 16]). Such an advection-chemistry module is a good tool for testing the new algorithm according to the performance and accuracy in each of the processes and both together as well as some effects of the splitting procedure used in the last case. The same module as in [7, 16] will be used in this paper as a test-example.

The numerical experiments for the transport-reaction scheme were performed on the parallel computer with distributed memory IBM SP Power 2 (up to 32 × 120 MHz RS/6000 processors). The top performance of each processor is 0.48 GFlops (billions of floating point computations per second).

The results from the numerical experiments when the *advection-chemistry* module was run on different number of processors can be found in the following table. One is able to see: the *computing times* (T_p) measured in seconds; the *speedups* defined as the computing time on p processors divided by the computing time on two processors ($S_p = T_p/T_2$); the *parallel efficiency* ($E_p = S_p/p$) the *computational speeds* (CS_p) measured in GFlops and the *computational efficiency* (in percent) (CE_p) defined as $CE_p = 100 * \text{GFlops}/p * \text{top-performance}$.

Table 1. Results from the numerical experiments for an *advection-chemistry* module obtained on IBM SP

Number proc.	Comput. time (T_p)	Speedup (S_p)	Parallel efficiency (E_p)	Comput. speed (CS_p)	Comput. efficiency (CE_p)
2	2683	-	-	0.18	18.0
4	1321	2.03	1.02	0.37	18.3
8	689	3.89	0.97	0.70	17.6
16	389	6.89	0.85	1.24	15.5
32	218	12.30	0.77	2.22	13.9

As it is well seen from this table the new algorithm leads to a considerable reduction of the computational times. However, the efficiency which is measured by the ratio of the GFlops actually achieved and the maximal number of GFlops that can be achieved on the configuration used, is still low. One of the main reasons for these results is that the realization of this algorithm does not lead to an optimal use of the cache memory of the processors.

Table 2. Results obtained on CRAY C92A computer

Parameters measured	Results
Computing time (in seconds)	1050
Computational speed (in GFlops)	0.46
Computational efficiency (in percent)	51

The code realizing the same algorithm was run on the vector computer CRAY C92A with the top performance of the processor of 0.9 GFlops. The results from this run are given in Table 2. It is seen that the best computing time achieved on the IBM SP computer platform is about five times less than the the computing time obtained on the IBM SP computer. However, the efficiency of the vector variant of the algorithm, achieved on the CRAY C92A computer is much better than the efficiency of the parallel variant on the IBM SP computer.

CONCLUSIONS AND OUTLOOK

In this paper we discuss a new parallel algorithm for an implementation of the two-dimensional version of the Danish Eulerian model for long-range transport of air pollutants in Europe. The same idea is applicable to the big three-dimensional version of the model. The numerical results reported here are for the most time consuming parts of the model: advection and chemistry submodels, but approximately the same results (for the reduction of the computing time in percent, speedups and parallel efficiency) can be expected for the whole model. These results show that the partitioning of the computational domain in several sub-domains (as a rule the number of subdomains is equal to the number of the processors used) allows to solve efficiently large-scale problems in air pollution modelling. The algorithms based on the domain partitioning are a good tool because they lead to a considerable reduction of the computational time as well as to a possibility to store different parts of the input and output data, and different parts of the coefficient matrices after the discretization procedures (space and time discretizations) into the memories of the different processors, where distributed memory parallel computers are used and, hence, to solve efficiently huge real-life problems, when the problem does not fit in the memory of one processor. The considerable reduction of the computational times (proportional to the number of the processors used) is very important in order to use the computer implementation of the Danish Eulerian Model for operational purposes even in its three dimensional version. The use of the standard message passing interface (MPI) is a key component in the development of this concurrent computing environment in which applications and tools can be transparently ported between different computers. The developed code is a portable code for parallel computers with distributed and shared memories, symmetric multiprocessor computers and clusters of workstations with MPI installation.

ACKNOWLEDGMENTS

The research reported in this paper is partly supported by NATO Scientific Affairs Division and the Bulgarian Ministry of Education and Science under Grant I-505/95. Furthermore, a grant from the Danish Natural Science Research Council gave us access to all Danish suppercomputers.

REFERENCES

[1] V. Alexandrov, A. Sameh, Y. Siddique, Z. Zlatev, *Numerical integration of chemical ODE problems arising in air pollution models*, Environmental Modelling and Assessment, **2** (1997), 365-377

[2] W. Crowley, *Numerical advection experiments*, Mon. Weath. Rev., **96** (1968) 1-11

[3] I. Dimov, K. Georgiev, Z. Zlatev, *Long-range transport of air pollutants and source-receptor relations*, Notes on Numerical Fluid Mechanics, **62** (1998), 155-166

[4] M. Gery, G. Whitten, J. Killus, M. Dodge, *A photochemical kinetics mechanism for urban and regional computer modelling*, J. Geophys. Res., **94** (1989), 12925-12956

[5] W. Gropp, E. Lusk, A. Skjellum, *Using MPI: Portable programming with message passing interface*, MIT Press, Cambridge, MA (1994)

[6] O. Hertel, R. Berkowicz, J. Christensen, Ø. Hov, *Test of two numerical schemes for use in atmospheric transport-chemistry models*, Atmospheric Environment, **27A** (1993), 2591-2611

[7] Ø. Hov, Z. Zlatev, R. Berkowicz, A. Eliassen, L. Prahm, *Comparison of numerical techniques for use in air pollution models with nonlinear chemical reactions*, Int. J. Chemical Kinetics, **10** (1978), 971-994

[8] G. Marchuk, *Mathematical modelling for the problem of the environment*, Studies in Mathematics and Applications, **16** North-Holland, Amsterdam (1985)

[9] C. Molenkampf, *Accuracy of finite difference methods applied to the advection equation*, J. Appl. Meteor., **7** (1968), 160-167

[10] G. McRae, W. Goodin, J. Seinfeld, *Numerical solution of the atmospheric diffusion equations for chemically reacting flows*, J. Comp. Physics, **45** (1984), 1-42

[11] M. Snir, St. Otto, St. Huss-Lederman, D. Walker, J. Dongara, *MPI: The Complete Reference*, MIT Press, Cambridge, MA (1996)

[12] Z. Zlatev, *Application of predictor-corrector schemes with several correctors in solving air pollution problems*, BIT, **24** (1984), 700-715

[13] Z. Zlatev, *Advances in the theory of variable stepsize variable formula methods for ordinary differential equations*, Appl. Math. and Comp., **31** (1989), 209-249

[14] Z. Zlatev, J. Christensen, Ø. Hov, *An Eulerian air pollution model for Europe with nonlinear chemistry*, J. Atmos. Chemistry, **15** (1992), 1-37

[15] Z. Zlatev, I. Dimov, K. Georgiev, *Modeling the long-range transport of air pollutants*, IEEE Computational Sciences & Engineering, **3**, (1994), 45-52

[16] Z. Zlatev, *Computer treatment of large air pollution models*, Kluwer Academic Publishers, Dordrecht-Boston-London (1995)

[17] Z. Zlatev, I. Dimov, K. Georgiev, *Three-dimensional Version of the Danish Eulerian Model*, ZAMM · **Z**eitung für **A**ngewandte **M**athematik und **M**echanik, **76** (1996), 473-476

DISCUSSION

T. ODMAN: Looking at your efficiency numbers, it seems the computing load was equally distributed between the subdomains and the losses were only due to communications. In actual applications, the chemistry for one subdomain may be more stiff than the chemistry in another subdomain. So one subdomain may require a smaller time step and more computations. The other subdomain would finish its computations earlier and stay idle for a while. How did you balance the computational load such that each subdomain receives an equal amount of load?

K. GEORGIEV: The load balance depends crucially on the ability of the numerical method to be run with a constant stepsize. If the Quasi-Steady-State-Approximation algorithm is used, then there are no problems. This method is normally run with a constant stepsize in all large air pollution models. If a classical numerical method (because of the stiffness such a method has to be implicit) is used, then the situation becomes more complicated. However, even in this case the model can be run with a constant stepsize during most of the time period and, therefore, the loading balance is rather good also in this case. Of course, this is under an assumption that the stepsize used is sufficiently small. If this is not the case, one probably should take care for a possible load imbalance and then more advanced techniques for keeping the load balance will be needed. The numerical results presented in this talk are for a generalization of the well known rotation test proposed by Molenkampf, where the chemical reactions which take place are added. Our first results with parallel runs of the whole 2D-version of the Danish Eulerian Model show the same good load balance.

S. RAFAILIDIS: Have you looked into splitting the marching of the model with different time steps? Can you, for example, run the chemistry parts with smaller time steps than the advection, without the whole algorithm collapsing?

K. GEORGIEV: Indeed, the chemical submodel which consists of stiff ordinary differential equations (mainly due to the photo-chemical reactions) is run with a stepsize, which is six times smaller than the stepsize used in the other submodels. The perfect situation will be to determine the stepsize for the advection part by using some stability conditions based on some norm of the wind velocity vector. Then the chemical part can be adjusted according to the degree of stiffness. This will allow us to run the model with nearly optimal stepsize. However, such a strategy demands some administration rules, which may deteriorate the speed of computations. Also, it will be needed to apply more complicated devices in the parallel algorithms. This has not been done yet, but in principle, it is possible to develop such an algorithm.

A LONG-TERM MODELLING OF DISTRIBUTION AND ACCUMULATION OF THE MERCURY SPECIES IN THE NORTHERN HEMISPHERE

Mikhail Sofiev

Institute of Program Systems, Russian Academy of Sciences
Cont.: Stroiteley str. 4-1-18, Moscow 117311, Russia

INTRODUCTION

Mercury is one of toxic pollutants with a long atmospheric lifetime, which is resulted from both rather slow deposition and high re-evaporation capabilities. As a result, mercury pollution is characterized by rather homogenous distribution of the substances in the atmosphere (in comparison with e.g. acidic species or other heavy metals like Pb and Cd). It is known that total mercury concentrations in air in the Northern Hemisphere are mostly within the range of 1-1.5 ng per cubic meter with scarce peaks around large emission sources. The measurements of vertical concentration profiles showed that up to several kilometers the concentrations are again rather homogenous.

Antropogenic emission of mercury in the Northern Hemisphere is estimated within the limit of 1.5 kt per year (Pacyna, *et al.,* 1993). Several model simulations have confirmed that this amount of annual emission in combination with known natural emission (e.g. volcanic) can not provide above mentioned background level. In order to meet the measured concentrations it is necessary either to introduce artificial background level (Petersen *et al.,* 1995) or consider some "natural" emission sources with unknown origin (Galperin *et al.,* 1994).

To the best knowledge of the author this uncertainty has never been addressed before. The computer experiment presented below is aimed at checking of one of possible explanations of the mercury background concentrations.

THE HYPOTHESIS OF MERCURY ACCUMULATION

One of important features of Hg is high evaporation capability of mercury-containing compounds. According to available information only about 1-2% of the depositions onto the oceans are fixed in the bottom layers (Mason *et al.,* 1995). For terrestrial systems this estimate varies from 20% up to 80% (e.g. (Johnson & Lindberg, 1995), (Mason *et al.,* 1995), etc.). Remained pollution can be re-emitted to the atmosphere in case of appropriate conditions. This results in appearing of the pollution cycle: emission - atmospheric transport - deposition - partial fixation / re-emission back to the atmosphere.

This cycle can not be recognized by measurements because upward flux from soil of oceanic surfaces is normally treated as a natural emission. Separation of true natural component from the secondary emission of previously deposited substances is impossible.

The intensity of secondary flux is definitely highly dependent upon the meteorological conditions (first of all, temperature) because one of important mechanisms is direct evaporation / de-sorption of mercury from the ground. In addition, there is a parallel process of diffusion of mercury to the deep layers of water, sediments or soils, where it can be stored for a considerable period if time until decreasing of surface concentrations causes upward diffusion fluxes with further emission to the air.

This mechanism of accumulation of pollution from air, which results in general enrichment of the environment by mercury and, consequently, creating of rather homogeneous background level of atmospheric concentrations.

CONTENT OF THE COMPUTER EXPERIMENT

As it was mentioned above, detection of secondary emission and it separation from natural sources is hardly possible without numerical modelling of the processes.

The idea of the computer experiment was to simulate the main processes taken place in atmosphere, soil and Ocean and than check if long-term calculations confirms the enrichment of the environment by mercury and what would be calculated background concentration. In case of successful reproduction of the main characteristics of the background level, the model will be validated against available measurements also in industrial regions (where the observed values are up to 10 ng per cubic meter).

The first part of the experiment was based on hemispheric approach. The 3D Eulerian model was applied to actual meteorological data extracted from long-term meteorological archive. Only antropogenic emission in the Northern Hemisphere was taken into account. Period of calculations covered 9 years starting from 1967 up to 1975 inclusive.

The simplified atmospheric chemistry scheme included only processes of sequential oxidation of mercury, dry and wet deposition. Mechanisms of re-emission and fixation in soil and oceans were also taken into account.

The output of the model contains monthly maps of mercury air concentrations both for metallic and oxidized forms, dry and wet deposition amount, fixed and re-emitted mercury. These maps enable to calculate all integral characteristics of the model, in particular, mass balance for every layer and for the whole 3-D calculation domain. Amount of pollution transported outside the calculation area was also stored and included into the total mass balance.

MODEL DESCRIPTION

The model designed for the experiment is based on Eulerian 3-D approach with horizontal resolution 150 km linear size. It is based on the large-scale model of heavy metal pollution described in (Galperin *et al.,* 1995) and on mercury model (Galperin, *et al.,* 1996).

Computational grid covers Northern Hemisphere to the north of 10^0 N with horizontal resolution approximately 150 km $\times$ 150 km in polar stereographic projection. The grid contains 99 $\times$ 99 cells with northern pole in its centre. Model input contains geographical data taken form (Global data set), antropogenic emission (Pacyna, *et al.* 1993) and meteorological data (Grid point data set) - Table 1.

In comparison with these models current version has multi-layer structure (the current realization enables to use up to 10 layers). The presented calculations used 6 layers, which structure is shown in Figure 1.

The scheme of atmospheric transformations is presented in Figure 3. This scheme was made on the basis of (Mason *et al.,* 1995) and (Fitzgerald 1995). Maximum possible simplifications were introduced in order to speed-up the model.

Table 1. Meteorological parameters used in calculations

Parameter	*Level, mbar*	*Source of information*
Wind speed	1000, 850, 700	NMC/NCAR+reproducing procedure
Temperature	1000, 850, 700	NMC/NCAR+reproducing procedure
Precipitation	Surface	DB "Prognoz" +reproducing procedure
Geopotential	850, 700	NMC/NCAR

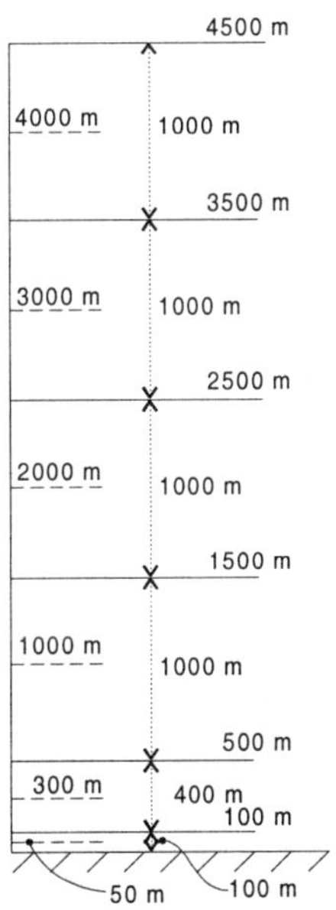

Figure 1. Vertical structure of the model.

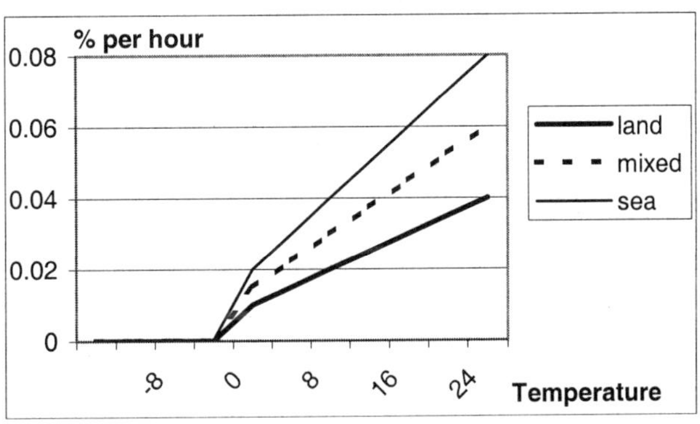

Figure 2. Re-emission rates for different surface types.

Mercury is emitted primarily in metallic form (80%) and about 20% in oxidized form. Metallic mercury Hg^0 is considered to be almost inert substance. Its dry deposition velocity is not more than 0.01 cm s^{-1} , and scavenging with precipitation is negligible. After deposition it rapidly re-emits to the atmosphere. The heterogeneous oxidation reaction

$$Hg^0 \rightarrow Hg(II)$$

is considered with rate of 0.01% per hour.

Oxidized mercury Hg(II) is subject to dry and wet deposition with the characteristics similar to those of fine aerosols (smaller than 1 μm in diameter). Fixation rates for soil and water are sown on Figure 3 and considered to be constant. The deposited substances can be reduced to metallic form and re-emitted with rates dependent upon the type and state of the surface and atmospheric conditions (Figure 2).

THE RESULTS OF MODEL SIMULATIONS

Some examples of the first step of the experiment are shown in Figure 4 - Figure 9 . Figure 4 - Figure 6 present maps of monthly mean concentrations of total (metallic and oxidized) mercury for the 1-st, 25-th and 108-th months correspondingly. It is well seen that at the beginning the pattern of pollution is quite similar to that of many "normal" substances like oxidized sulphur or nitrogen. The picture changes after several years of continuous calculations. Values of peaks remain mostly the same while the concentrations in remote regions become higher and more homogeneously distributed.

Figure 7 - Figure 9 show the integrated characteristics of the model. The first two figures present the mean over grid concentrations of total and oxidized mercury correspondingly. These charts are created for every of 6 model layers and provide an illustration of the vertical structure of the concentrations. Low layers 1, 2 and 3 are mostly located inside mixing layer and consequently their concentrations are practically the same. Some difference exists only for oxidized mercury, which is caused by temperature gradient along the height. Seasonal variations are also well pronounced for both types of substances.

Figure 9 presents the mass balance of the model. It shows the emission intensity, total deposition inside the domain, re-emission, and transport outside the grid. Sum of these values gives a net atmospheric mercury balance, which is also shown.

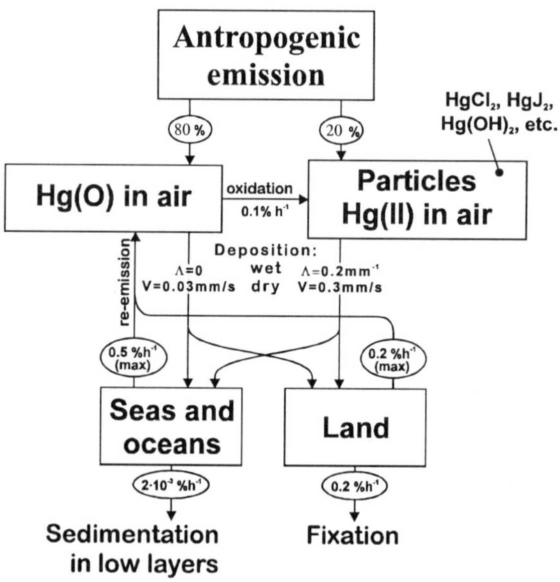

Figure 3. Chemical reactions and deposition processes considered in the model.

DISCUSSION

The presented results show that current model setup leads to rather fast establishment of equilibrium between antropogenic emission and re-emission from one side and deposition, fixation and transport outside from the other side. The dynamic equilibrium is observed after 3-4 years of simulations. After that only deviations mainly connected with seasonal motions take place. That moment is also characterized by establishing of rather homogeneous background level of concentrations (with except of areas close to the grid boundaries). So, qualitatively the model behavior is quite similar to what was expected.

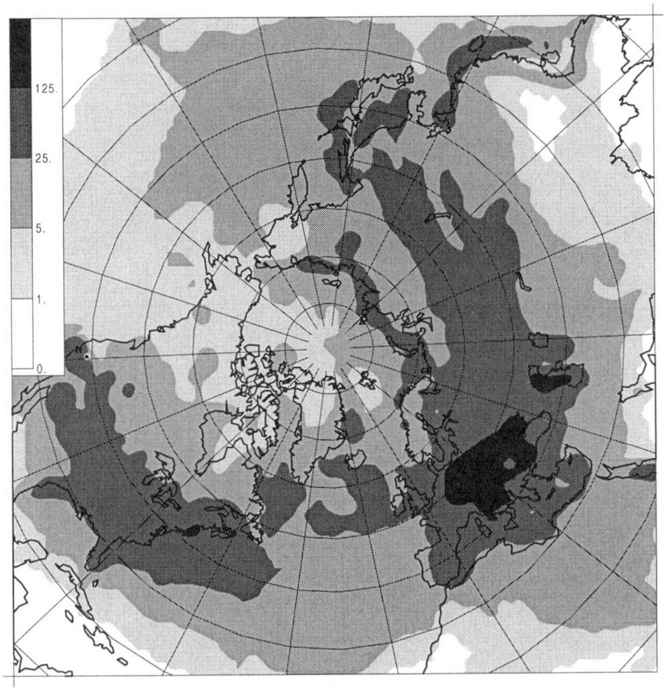

Figure 4. Hg total concentrations for the 1st month of simulations. *Unit = 10 pg Hg m*$^{-3}$.

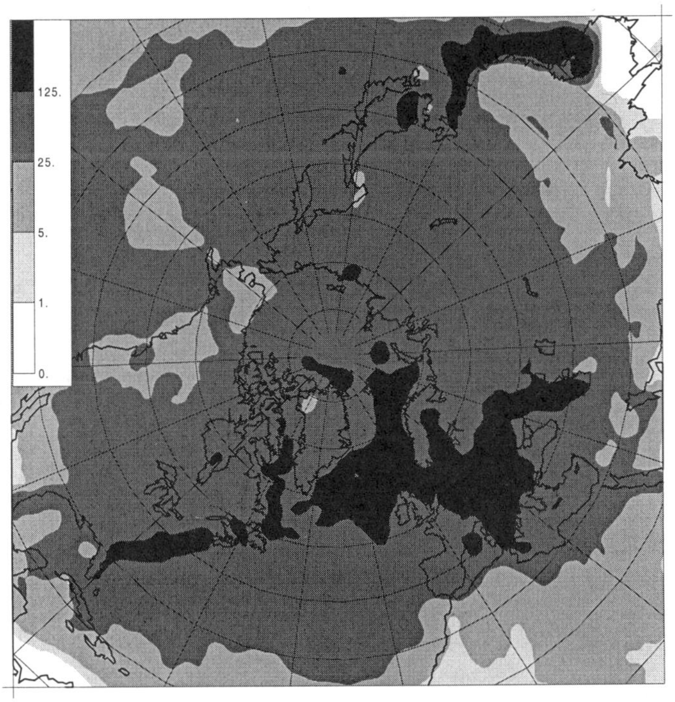

Figure 5. Hg total concentrations for the 25th month of simulations. *Unit = 10 pg Hg m*$^{-3}$.

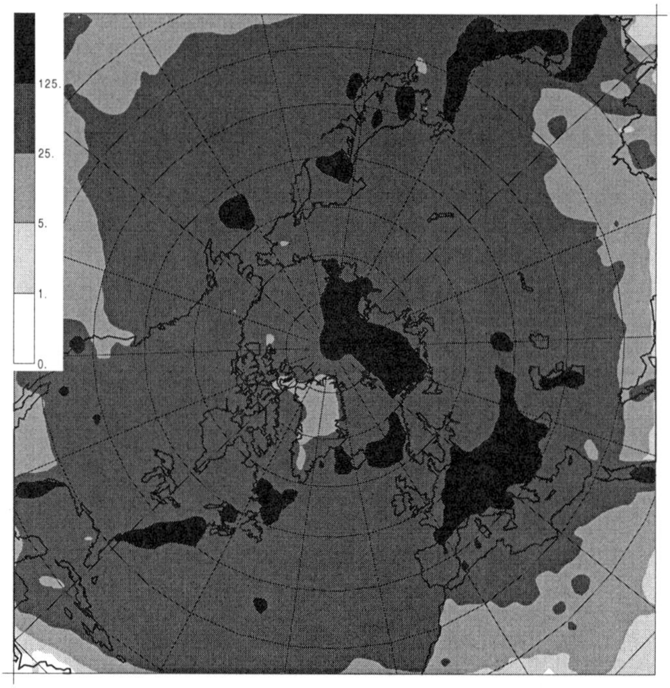

Figure 6. Hg total concentrations for the 108th month of simulations. *Unit = 10 pg Hg m^{-3}.*

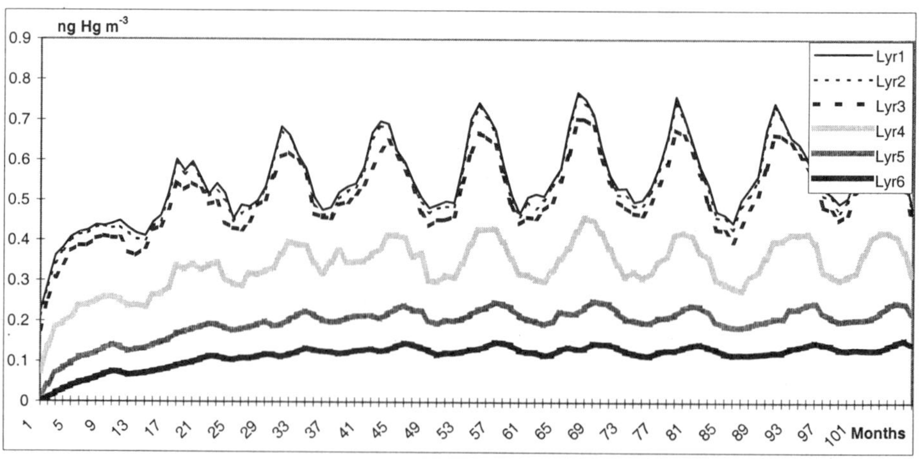

Figure 7. Mean over grid of total Hg concentrations. *Unit = 1 ng Hg m^{-3}.*

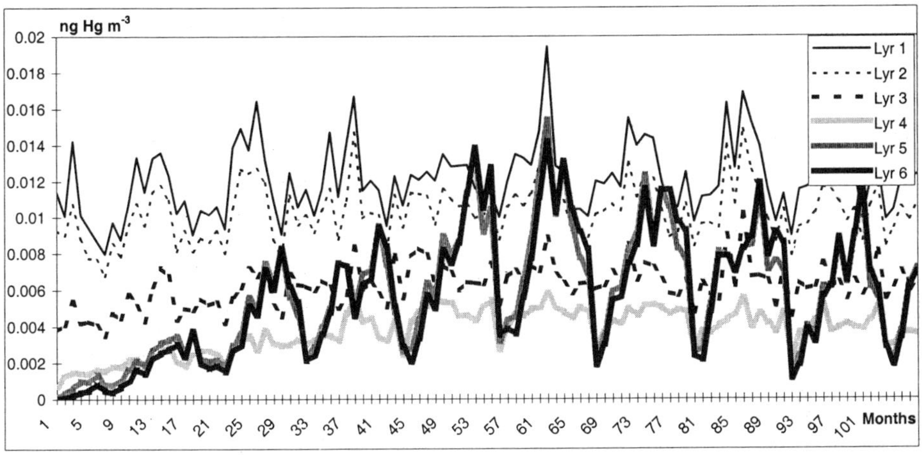

Figure 8. Mean over grid of particulate Hg concentrations. *Unit = 1 ng Hg m⁻³.*

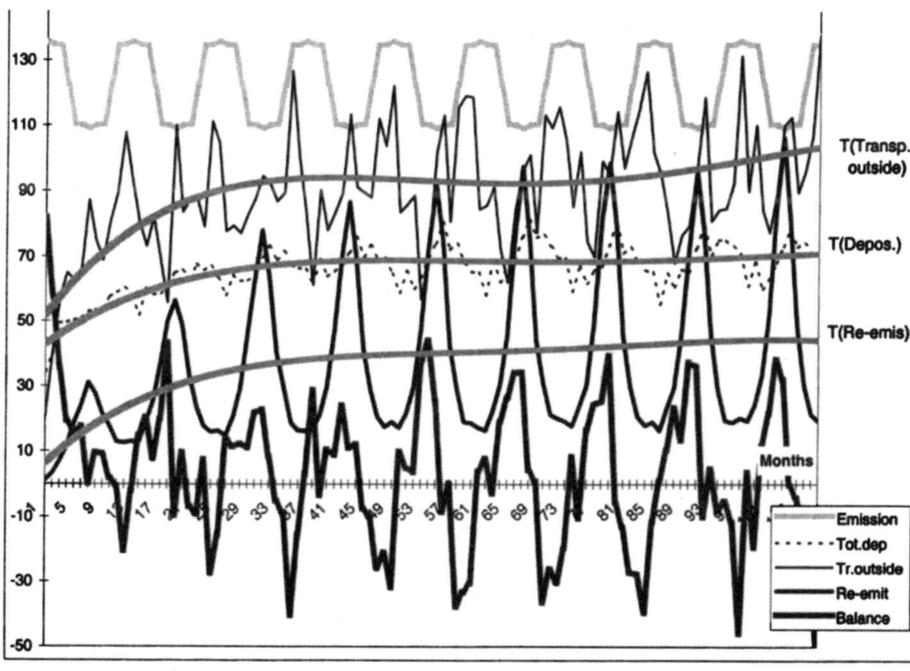

Figure 9. Monthly mean of integral mass balance. *Unit = 1 t Hg month⁻¹.*

Unfortunately quantitative characteristics are far from observed values. In particular, average value of background concentration is about 0.5 – 0.7 ng m^{-3} , which is twice lower than observations. Even more considerable underestimation of concentrations takes place in upper layers of the model. Observations show that mercury is distributed almost uniformly up to height of 3-4 km, while the model demonstrates only 1.5 km with considerable decrease of concentrations above this level.

Two reasons of such behavior can be noted. The first one is large transport outside the calculation domain (which is much larger than the deposition inside). The fastest transport through the boundaries occurs in the upper levels, which leads to quick removing of mercury from these layers. The second reason is oversimplified consideration of soil and oceanic processes. More accurate simulation should include diffusion in deeper water layers and some basic chemical and physical processes in the soils. These considerations would justify the fixation and long-term accumulation rates, which are very rough so far.

One more potential future problem is too low upper boundary of the model. Currently the highest level ends at 4.5 km height, which well corresponds to the expected height of homogeneous mercury distribution along the vertical, but will be insufficient for reproduction of the whole profile.

CONCLUSIONS AND FUTURE EFFORTS

1. At a qualitative level the hypothesis of possible mercury accumulation in the environment is not refused by the first results. Simple model appears to be capable to reproduce many of characteristics derived from the hypothesis.
2. Quantitatively the model is far below the observed values. The underestimation is between factor of 2 and 4 for different characteristics. The evident reasons are: insufficient calculation domain, which causes massive transport of pollution outside, and oversimplified consideration of the processes under the surface. It is also expected that current upper boundary of the model (4.5 km) is insufficient for further calculations.

REFERENCES

W.F. Fitzgerald Is mercury increasing in the atmosphere? The need for an atmospheric network *W.A.S.P.* **80**, pp. 245-254.

M.Galperin, M.Sofiev, O.Afinogenova (1995) Long-term modelling of airborne pollution within the Northern Hemisphere, *W. A. S. P.* **85**, pp.2051-2056.

M.Galperin, M.Sofiev, E.Mantseva (1996) A model of the chemical transformation of mercury and its long-range atmospheric transport. In: Global and Regional Mercury cycles: Sources, Fluxes and Mass Balances, eds. Bayens W., R.Ebinghaus, O.Vasiliev, NATO ASI Series 2: Environment- **21**, Kluwer Academic Publishers, the Netherlands, pp.219-227.

Global Data Set For Land-Atmosphere Models, USA, NASA, GDAAC ISLSCP (1987-88) Initiative 1 **1-5**.

Grid point data set, v.II. Gridded MNC analysis for the Northern Hemisphere. National Meteorological Centre , Atmospheric Sci.Univ.of Washington, data support section NCAR.

D.W.Johnson, S.E.Lindberg (1995) The biochemical cycling of Hg in forests: alternative methods for quantifying total deposition and soil emission W.A.S.P. v.80, pp.1069-1077.

R.P.Mason, K.R.Rolthus and W.F.Fitzgerald (1995) Methylated and elemental mercury cycling in surface and deep ocean waters of the North Atlantic, W.A.S.P., v.80, pp.665-677.

J.Munthe (1995) Mercury in the atmosphere - emissions, transformations, depositions and effects, Proc. of the first Workshop on Emissions and Modelling of Atmospheric Transport of Persistent Organic Pollutants and Heavy Metals, Durham, NC, USA, EMEP / CCC report 7/93, pp.225-254.

J.M.Pacyna, E.Voldner, T.Bidleman, G.Evans, G.J.Keeler, 1993 Emissions, atmospheric transport and deposition of heavy metals and persistent organic pollutants, Proc. of the first Workshop on Emissions and Modelling of Atmospheric Transport of Persistent Organic Pollutants and Heavy Metals, Durham, NC, USA, EMEP / CCC report 7/93, pp. 45-138.

DISCUSSION

M. HONGISTO: GKSS in Germany has estimated, using local scale model and short time simulations, that convective transport to upper layers and chemical transformation in clouds is important for Hg distribution. Can you benefit from this simulations in your long-term simulations?

M. A. SOFIEV: Yes, certainly. Actually the chemical scheme of the presented model was discussed with Gerhard Petersen. It contains less reactions than scheme applied in GKSS model, and roughly can be considered as its aggregated version with some specific required for long-term large-scale modelling (first of all, re-emission processes are of primary importance here).

B.FISHER: Can you explain how the re-emission process is parametrised in the model and what improvements are planned?

M. A. SOFIEV: Now there are two main parameters. The model takes into account 3 types of underlying surface (sea, land and mixed). The second key parameter is near-ground temperature - for example, if it is below zero then no re-emission process is assumed. The main improvement could be consideration of mercury diffusion in soil and water, which requires several underground layers to be entered into the model. Such a description enables to define the concentration of mercury close to surface both below and above the ground and then simulate the exchange between ground and air in accordance with concentration values.

INVERSE TRACES FOR THE PASSIVE POLLUTANTS

Ekaterina G.Klimova and Gdaly S.Rivin

Laboratory of the Modelling of Atmospheric Processes
Institute of Computational Technologies SB RAS,
6, Ac. Lavrentjev Ave., Novosibirsk, 630090 Russia
klimova@ net.ict.nsc.ru
rivin@adm.ict.nsc.ru

INTRODUCTION

One of main ecological problems is the calculation of inverse traces for the passive pollutants for some points. It is important to identify the region from which the passive pollution is transported for the concrete data and also to calculate the seasonal mean traces. The parer is devoted to research of algorithms of an evaluation of influence of actual and climatic meteorological conditions on propagation of aerosols in Siberian region. Such evaluation is made because of constructions of inverse traces under the actual and climatic meteorologic data. The data of the International project "Reanalyses" (Kalnay et al.,1996), mean for each month for period 1982-1994 are used as the climatic meteorological information. The methodology of calculating of the inverse traces based on the Lagrangian method. In our researches we attract various components the system MAP for Siberian region, being developed at ICT SB RAS (Rivin, 1996; Klimova and Rivin, 1996; Rivin et al., 1997).

In the article are described the system MAP for Siberian region, used in experiments method finding of inverse traces, possible picture of distribution of aerosols in Siberian region, in the Conclusion the brief conclusions obtained in the article and the plan of further activity are adduced.

SYSTEM MAP

The system MAP for modeling of the atmosphere processes and pollutant transport in Siberian region includes the following main units: scheme of numerical analysis of the observed meteorological data by the three-dimensional box-variant of optimal interpolation; the nonlinear normal-mode initialization scheme; model of the atmosphere based on the splitting-up method with 15 levels (1000 - 10 hPa) and the horizontal model grid is D-grid in the notations of A.Arakawa); model of the advection pollutions with 19 levels and 30 km grid step; modules of visualization. This system makes possible to

Air Pollution Modeling and Its Application XIII, edited by S.-E. Gryning
and E. Batchvarova, Kluwer Academic/Plenum Publishers, 2000

investigate methods of the mathematical modelling, observation data analysis, data assimilation under quasioperative conditions. The numerical experiments carried do under the quasioperative conditions demonstrate its effectiveness and high quality. Now the work is being conducted on the further development of the components of the system described above and their combination.

METHODS OF FINDING THE INVERSE TRACES

For an evaluation of areas of influence as the climatic meteorologic information the data from the indicated above archive "Reanalyses", mean for each month for period 1982-1994 are used. Thus, the traces, constructed on such data, allow to evaluate typical of each month climatic of area of influence on the aerosol picture in some region. Let's mark, that the full data archive "Reanalyses" contains the meteorologic information for 1956 - 1996 years.

The construction of inverse traces is possible as with the help of analysis of synoptic maps and with the help of of methods of mathematical modelling under the data of the objective analysis. The first method based on the analysis of synoptic maps, the apart from of positive properties, has certain defects connected that the inverse traces of a flow of air weights are restored, instead of partiáles, the speed of which carry (especially on a vertical) differs from a flow velocity of an air. For this reason in the given activity the second method is used, namely, the method of the Lagrange is applied for construction of inverse traces of an atmospheric aerosol.

Let's describe briefly algorithm used for calculation of inverse traces. Let (λ, θ, p) - three-dimensional coordinates of a point in spherical coordinate system, p - pressure. Then the equation for change of coordinates on time has a kind:

$$\frac{d\lambda}{dt} = \frac{U(\lambda, \theta, p, t)}{r\cos\theta}, \tag{1}$$

$$\frac{d\theta}{dt} = \frac{V(\lambda, \theta, p, t)}{r}, \tag{2}$$

$$\frac{dp}{dt} = W(\lambda, \theta, p, t). \tag{3}$$

Let's present equations (1) - (3) in a finite difference kind:

$$\lambda^{n+1} - \lambda^{n} = \tau U(\lambda^{n+1/2}, \theta^{n+1/2} p^{n+1/2}, t^{n+1/2}) / (r\cos\theta^{n+1/2}), \tag{4}$$

$$\theta^{n+1} - \theta^{n} = \tau V(\lambda^{n+1/2}, \theta^{n+1/2} p^{n+1/2}, t^{n+1/2}) / r, \tag{5}$$

$$p^{n+1} - p^{n} = \tau W(\lambda^{n+1/2}, \theta^{n+1/2} p^{n+1/2}, t^{n+1/2}), \tag{6}$$

where τ - step on time.

From (4) - (6) we obtain:

$$\lambda^n = \lambda^{n+1} - \tau U\left(\frac{\lambda^n + \lambda^{n+1}}{2}, \frac{\theta^n + \theta^{n+1}}{2}, \frac{p^n + p^{n+1}}{2}, t^n + \frac{\tau}{2}\right) / \left(r\cos\frac{\theta^n + \theta^{n+1}}{2}\right), \tag{7}$$

$$\theta^n = \theta^{n+1} - \tau V\left(\frac{\lambda^n + \lambda^{n+1}}{2}, \frac{\theta^n + \theta^{n+1}}{2}, \frac{p^n + p^{n+1}}{2}, t^n + \frac{\tau}{2}\right) / r, \tag{8}$$

$$p^n = p^{n+1} - \tau W\left(\frac{\lambda^n + \lambda^{n+1}}{2}, \frac{\theta^n + \theta^{n+1}}{2}, \frac{p^n + p^{n+1}}{2}, t^n + \frac{\tau}{2}\right). \tag{9}$$

The values U,V,W were obtained from the climatic data of archive "Reanalyses". The implicit equations (7) - (9) were solved by a method of simple iteration. The values U,V,W in a point $(\lambda^{n+1/2},\theta^{n+1/2},p^{n+1/2})$ were calculated by a method interpolation with cubic spline. On vertical coordinate the linear interpolation on $\ln(p)$ was made.

NUMERICAL EXPERIMENTS

Because of algorithm, circumscribed in the previous item, the calculations of inverse traces under the climatic data were made. Climatic data are understood as the data for a calendar month, mean for period 1982-1994. The traces, obtained on such data, are close to "mean" on the same period to traces. They allow to evaluate characteristic of each month of area of influence on distribution of an aerosol in city.

The results of numerical experiments with Reanalysis data for the different seasons for the Novosibirsk region are published (Rivin and Klimova, 1997). The inverse climatic traces for north region of Siberia were calculated too to obtain the possible regions of influence on the aerosol and radiation picture in this region.

The calculation of inverse traces have produced for 10 days back with $\tau = 1$ hour. The results are presented in Figures 1-3 for periods "January - April", "May - August", "September - December". On these figures the numbers designate numbers of months. As it is visible from Figures 1-3, the motion of air mass from a west on east is characteristic of the first and third periods, north-west direction of motion of air weights is characteristic of the second period. The conducted experiments show, that probable sources of aerosols and radionuclides, detected in northern part of Siberian region, can be regions of islands New Lands (in Russian: Novaia Zemlia) and Scandinavia.

CONCLUSIONS

The tentative estimation of probable zones of influence on allocation of aerosols in Siberian region was conducted. More detail learning of a technique of obtaining of inverse traces with allowance for of speeds of a settling of aerosol particles of the different sizes and mechanism of turbulent intermixing hereinafter is supposed. Except calculations with climatic data calculations on actual meteorological data also will be conducted.

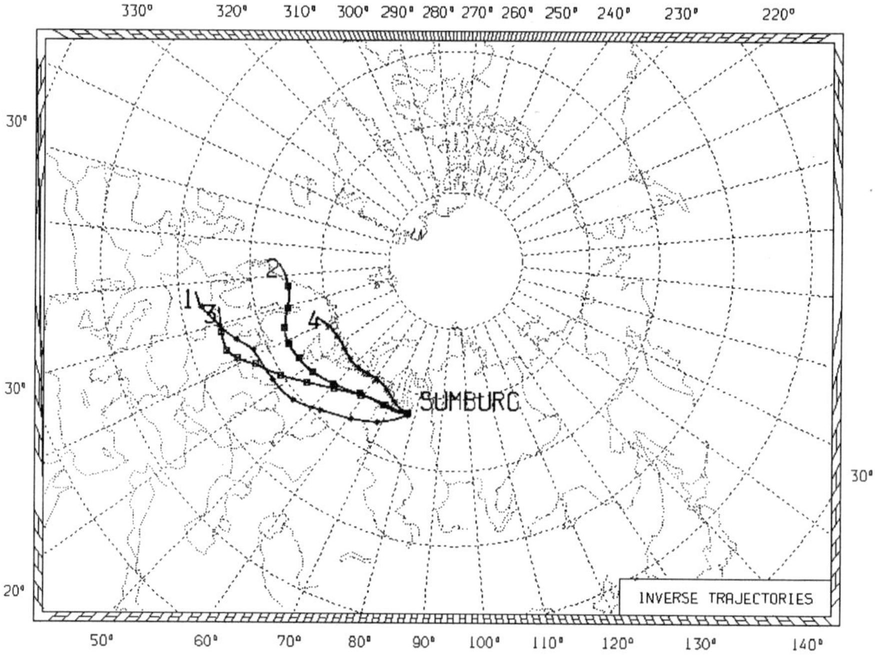

Figure 1. The inverse climatic traces for north region of Siberia for period "January - April".

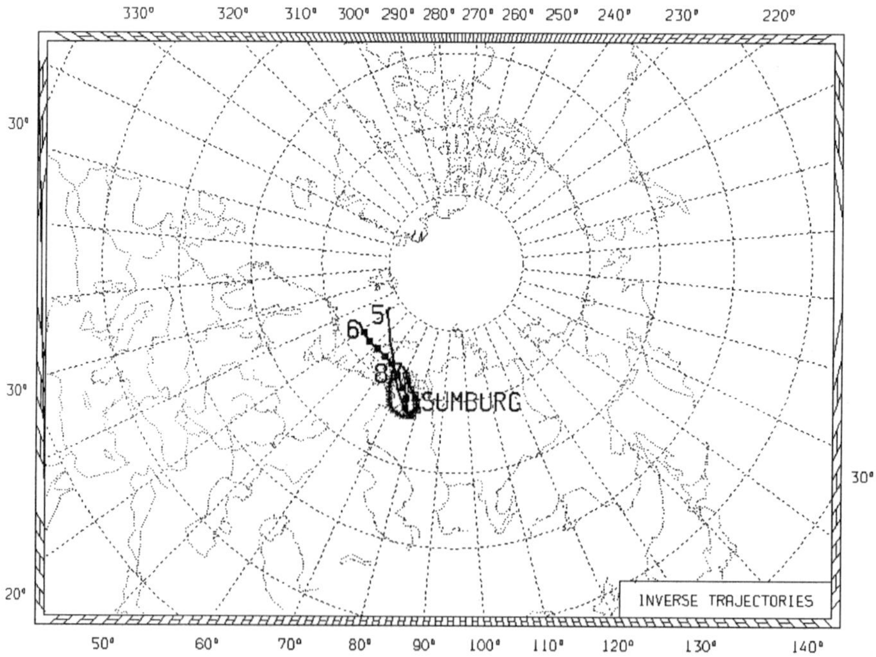

Figure 2. As in Figure 1, but for period "May - August".

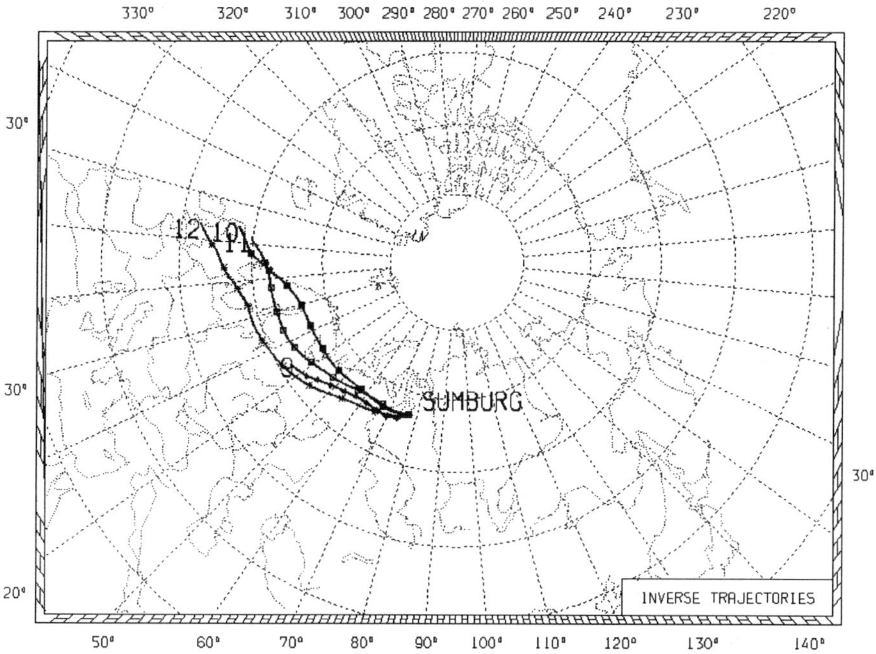

Figure 3. As in Figure 1, but for period "September - December".

ACKNOWLEGEMENTS

We consider with the pleasant debt to express thanks all researches, who done project "Reanalyses", Dr. E. Kalnay for the publications, submitted to us on Reanalysis, and Prof. K.P.Koutsenogii for useful arguing of aspects of operation connected to transfer of an aerosol in Siberian region.

This work was partly funded by the Foundation of Fundamental Research of the Russian Federation under Grant 98-05-65302.

REFERENCES

Kalnay E., Kanamitsu M., Kistler R., Collins W., Deaven D., Gandin L., Iredell M., Saha S., White G., Woollen J., Zhu Y.,. Chellian M., Ebisuzaki W., Higgins W., Janowiak J., Mo K.C., Ropelewski C., Wang J., Leetmaa A., Reynolds R., Jenne Roy , Joseph. Dennis, 1996, The NCEP/NCAR 40-Year Reanalysis Project, *Bulletin of the American Meteorological Society*, 77:437.

Rivin G.S., 1996, Numerical modelling of background atmospheric processes and the problem of aerosol transport in Siberian region, *Atmospheric and Oceanic Optics*, vol. 9, n. 6, p. 493 - 496.

Klimova E.G., Rivin G.S., 1996, The system of meteorological data assimilation for Siberia: numerical experiments with real data, *Meteorology and hydrology*, 12:19 (In Russian).

Rivin G.S., Koutsenogii K.P., Klimova E.G., Voronina P.V., Smirnova A.I., 1997, Models to describe meteofields and fields of concentration of gaseous and aerosol impuriries in Siberian region, *Atmospheric and Oceanic Optics*, vol. 10, n. 6, p. 378 - 381.

Rivin G.S., Klimova E.G., 1997, Assimilation system, atmospherical modelling and ecological problems for the Siberian region using NCEP reanalysis data, in: *Abstract Digest of the First WCRP International Conference on Reanalysis, Washington, DC, USA, 25 - 31 October, 1997.*

DISCUSSION

M. ROTACH: What is the size of aerosols, the model you presented is able to deal with?

G. RIVIN: 0.5 μm or smaller.

D. SYRAKOV: How do you calculate the single trajectories, presented in graphs - averaged individual trajectories or using the mean flow?

G. RIVIN: The mean (for 1982-1994) flow for corresponding month.

IMPACT OF THE GREAT CHINA FIRE OF 1987 ON THE TROPOSPHERIC CHEMISTRY OF EAST ASIA

Mahesh J. Phadnis and Gregory R. Carmichael

Center for Global and Regional Environmental Research &
Department of Chemical and Biochemical Engineering
The University of Iowa
Iowa City, IA 52240

INTRODUCTION

There is a growing concern that biomass burning as a consequence of anthropogenic activities has significant impact on the atmospheric chemistry, climate and on the global biogeochemical cycles. Since the late seventies, when Crutzen et al. (1979) first proposed that the emissions of trace gases from biomass burning can make an important contribution to their budgets in the atmosphere, there has been an increase in the number of research activities in parts of the world with extensive biomass burning (Andreae, 1991). It is observed that biomass burning occurs mostly in the continental tropics coinciding with the local dry season (Crutzen et al., 1985). Trace gases such as carbon monoxide (CO), methane (CH_4), nitrogen oxides (NO_x = NO + NO_2) and non-methane hydrocarbons (NMHCs) which are emitted from the burning fires play important roles in the production of ozone, thereby impacting the tropospheric photochemical oxidant cycle. Tropospheric ozone derived from satellite data is shown to be greater than 40 Dobson Units (DU) over southern Africa and tropical south Atlantic (Fishman et al., 1990; Fishman et al., 1991). Studies on the extent of the perturbation of this biomass signal on the composition of the atmosphere have been done in the past (Fishman et al., 1993). More recently, measurements were taken during the dry season of September-October 1992 in the Transport and Atmospheric Chemistry near the Equator (TRACE) Experiment (Fishman et al. 1996). The conclusions from this activity were that the widespread biomass burning in both South America and southern Africa is the dominant source of the precursor gases responsible for the huge amounts of ozone over the South Atlantic Ocean. Efforts in understanding the spatial and vertical distribution of the trace gases emitted from biomass burning have resulted in numerous insitu field campaigns and observational data. Measurements techniques such as satellite remote sensing, radiosondes and aircraft measurements are commonly being used to probe the atmosphere. On the other hand, synoptic-scale modeling of the influence of the fires on the atmosphere has been minimal, with the majority of work done in modeling the convective transport and redistribution of biomass burning emissions (Crutzen and Carmichael, 1993; Pickering et. al, 1996; Chatfield et. al, 1996).

In contrast to the tropical regions, large areas of the world's boreal forest zone, also burn annually. The use of these forests for industrial, recreational and other purposes has increased dramatically over the past couple of decades. The fire that developed in northeastern China and southern Siberia in the spring of 1987 is one example. In this paper, we explore the impact of the emissions of trace gases from forest fire on the photochemical oxidant cycle in east Asia. Specifically we will investigate the production and transport of ozone on a regional scale.

METEOROLOGICAL CONDITIONS AND THE FOREST FIRES OF SPRING 1987

On May 6, 1987, three fires ignited in the boreal forests of the Helongjing province of the People's Republic of China. The Helongjing Province borders the former Soviet Union along the Amur river. These fires combined and continued to burn for about three weeks burning in the excess of 1.3 million hectares of prime forest and resulting in the loss of over 200 lives and 50000 homes (Cahoon et. al. 1991). The widespread fire activity was evidence that the meteorological conditions responsible for the rapid growth of the fire were not just a local phenomena, but extended over a large area. The climate in the extreme northeastern region of China and in southeastern Siberia is continental cold-temperate, with long, cold, dry winters and short, warm, humid summers. The annual climatic cycle in this region includes the formation of a dominant high pressure system in winter, which breaks down in the spring as the temperatures increase. This breakdown is followed by intense storms sweeping across the east Asian region from the west to the east. The summers are mostly humid with precipitation occurring most of the time, while the fall is like the spring season, but lacking the intense storm activity. More details about the climatic features can be found elsewhere (Cahoon et al., 1994).

The conditions prevailing in northern China and southeastern Siberia before the spring of 1987 were just right for the ignition of a forest fire. The snowfall during the previous winter was light, which disappeared quickly as spring approached. Throughout April, the region experienced low temperatures and relative humidity, with little precipitation. The cumulative result was an extremely dry forest fuel situation in this region by the beginning of May 1987, which combined with increasing temperatures and strong winds to produce uncontrollable forest fires.

MODEL DESCRIPTION

The production and transport of ozone in east Asia was investigated by use of the STEM-II regional-scale model. The STEM-II model is a three-dimensional, Eulerian numerical model which accounts for the transport, chemical transformation and deposition of atmospheric pollutants in gas, cloud, rain, snow and aerosol. The current 3-D version of the model is based on the chemical mechanism of Lurmann et al. (1986) and Atkinson et al. (1989), and modified to include low NO_x conditions and explicit treatment of isoprene phases (Carmichael et al., 1991). The numerical integration of the chemistry is done by a fully implicit integrator based on the Runge-Kutta-Rosenbrock type of solvers (Sandu et al., 1997). The complete mechanisms with lists of species and reactions in the model are described in detail in Carmichael et al. (1986; 1991; 1998). The study domain, with a spatial resolution of $1^o x 1^o$ includes all of Japan, North and South Korea, Taiwan, most of

China, and parts of Russia and Mongolia. The vertical domain covers from the surface to 10 km with an unequal vertical resolution consisting of more layers in the boundary layer and less in the middle troposphere. More details can be found in Carmichael et al. (1998).

The emissions used in the model are summarized in Table 1. NO_x and SO_2 emissions for the region are available from several sources including Fujita et al., (1992) and Akimoto and Narita (1994). Anthropogenic hydrocarbon emissions are interpolated from Piccot et al.'s (1992) 10^ox10^o global inventory of volatile organic compound emissions. In the case of natural emissions of NMHCs for species like isoprene, emissions are taken from Guenther et al. (1995). The total emissions of the non-methane volatile hydrocarbons are presented in Table 1. Carbon monoxide from both anthropogenic sources and biomass burning and methane emissions were obtained from Saylor et al. (1991).

The forest fire emissions of the trace gases such as CO, CH_4 and NMHCs, were estimated from the work of Cahoon et al. (1994). The NO_x, SO_2 and particulate emissions were calculated according to the emission factors given in Crutzen and Andreae (1990). The emissions of the individual NMHCs were estimated using emission factors of Lobert et al. (1991) and preliminary calculations of Carmichael (1997). The estimates of these emissions are shown in Table 1. It can be seen in Table 1 that the amounts of primary gases like CO, NO_x and SO_2 emitted by the forest fire are much greater than the total anthropogenic fossil fuel emissions in east Asia. These emissions thus have a significant potential to alter the regional photochemical oxidant cycle in east Asia.

The three-dimensional meteorological fields needed by the model were obtained from ECMWF analyzed winds. The vertical wind fields were recalculated from the given horizontal data using the divergence. Other meteorological data used in the simulation were temperature, pressure, relative humidity and geopotential fields. Vertical eddy diffusivities used varied spatially and temporally, and were determined by a diagnostic analysis of the planetary boundary layer. The horizontal eddy diffusivity was held constant. Dry deposition velocities varied diurnally and differences in continental and marine deposition velocities were taken into account. Further details are presented in Zhang (1994) and Kotamarthi and Carmichael (1990).

The simulated period was May 1-30, 1987. The starting time was 0 Z on the May 1, i.e. 9 A.M. JST (Japan Standard Time), and the transport time step was 15 minutes and the gas-phase chemical reactions were calculated every 3600 seconds. This large timestep for chemical integration was possible because of the use of an implicit solver with a dynamically variable step size in the steady state calculations.

Table 1. Comparison between Anthropogenic and Forest Fire Emissions in east Asia

	China	Japan	Korea	Taiwan	Other	Total (Anthro)	Forest Fire
NO_X (Tg N/yr)	2.8	0.60	0.41	0.16		3.97	0.35
SO_2 (Tg S/yr)	9.92	0.63	0.94	0.24		11.72	0.021
CO (Tg C/yr)	12.37	3.07	2.00	0.52		17.95	15.34
NMHC (Tg/yr)	7.28	3.45	1.19	1.23		13.14	1.46
CH_4 (Tg C/yr)	11.68	0.72	0.44	0.11	0.33	13.28	2.03

RESULTS AND DISCUSSION

During the spring of 1987, surface based observations for a variety of gaseous species were carried out over most of Japan. A time series profile of the model predicted ozone

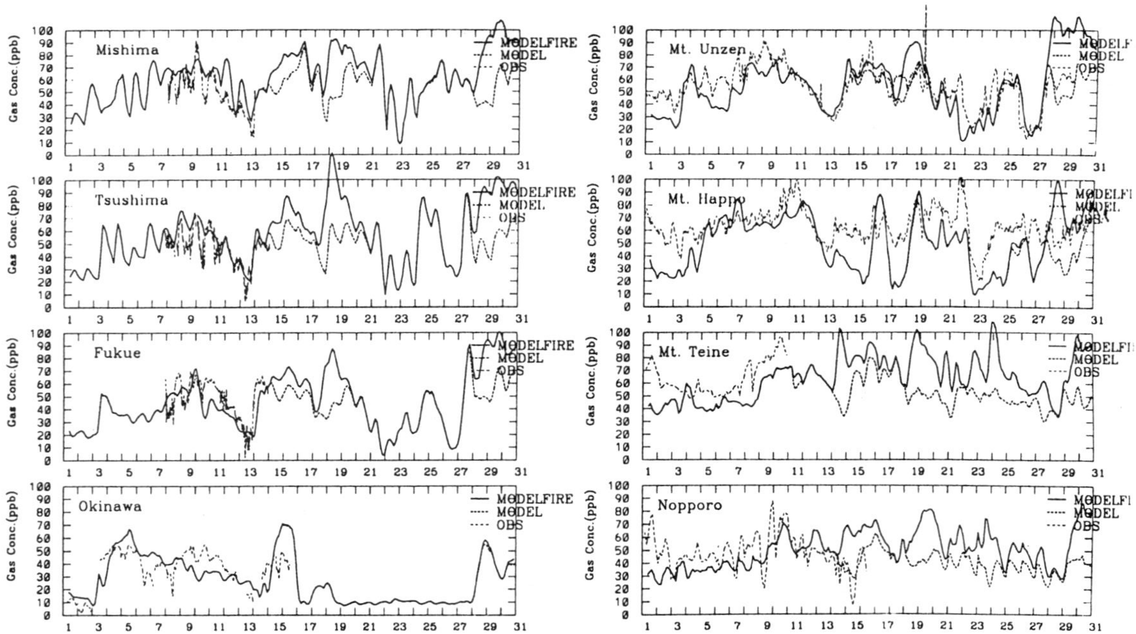

Figure 1. Time Series Profile of Model Predicted and Observed Ozone over Japan. MODELFIRE and MODEL represent the model predicted concentrations with and without the forest fire respectively. OBS is the observed ozone concentration.

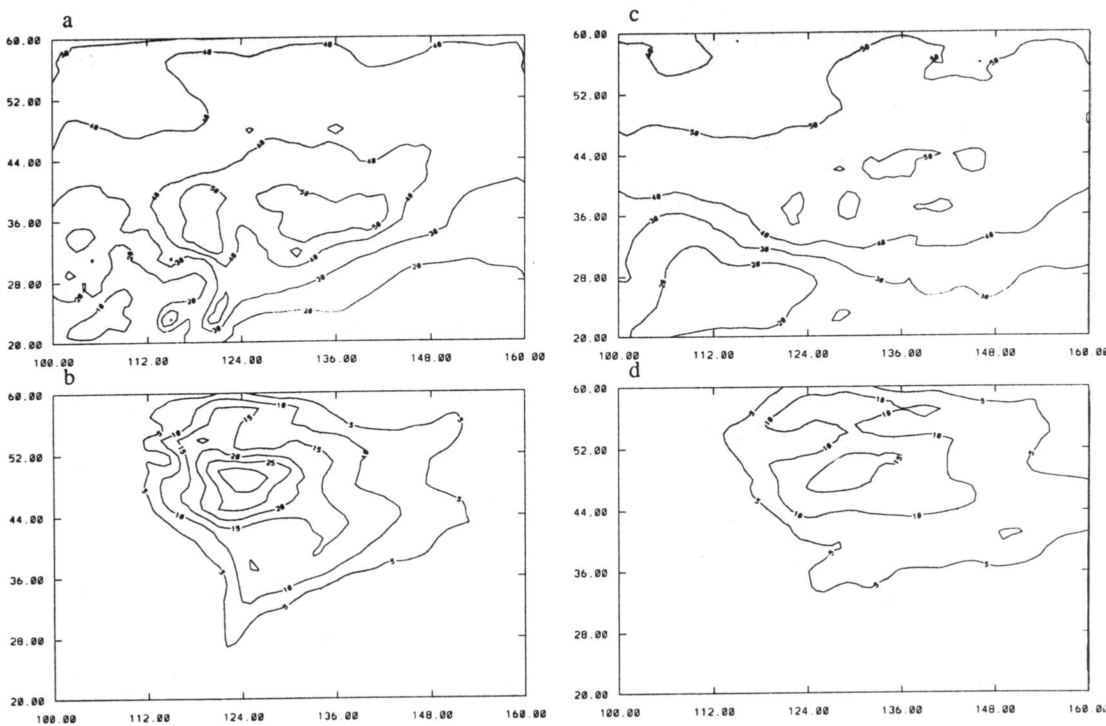

Figure 2. Spatial Distribution of the monthly average ozone during May 1987. a) Ozone levels in the absence of forest fire at 400 m. b) Difference in the Ozone levels due to the fire. c) Ozone at 2000 m in absence of the fire. d) Ozone differences at 2000 m due to the fire. All contours are in ppb.

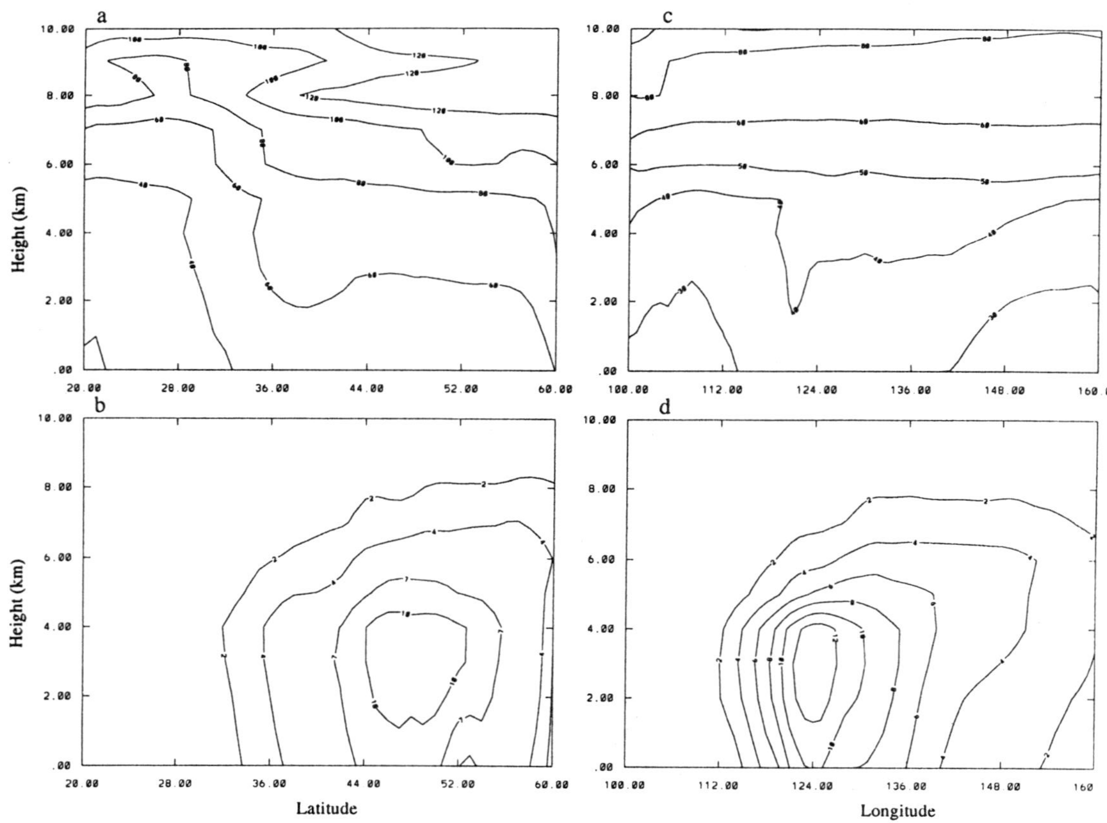

Figure 3. Vertical Distribution of Ozone during May 1987. a) and c) Ozone levels without the presence of the forest fire. b) and d) Difference in Ozone levels due to the forest fire. All contours in ppb.

concentration along with the observed ozone at the eight observation stations is presented in Figure 1. There are two simulated cases shown in the figure. First, a base case, in which there is photochemical production of ozone due to the anthropogenic emissions and normal biomass burning activities. The other one, in which the influence of the forest fire is also considered. The model predicted ozone is in reasonable agreement with the observed values. The temporal nature of the trends in the ozone levels is captured nicely by the model. For example, at Tsushima, the periodic cycle of ozone as indicated by the observations is also seen in the model predictions. Although in this case the model tends to systematically overpredict the observed values by about 1-2 ppb. On the other hand, at places like Mt. Unzen and Napporo, the model tends to underpredict the observed values. These disparities might be due to twofold factors - uncertainty in the emissions of NMHCs and transport from the upper troposphere/lower stratosphere to the lower troposphere. Unfortunately, we do not have sufficient observation data for ozone at these locations throughout the month of May 1987 for a proper comparison with the model predictions. Only two stations, viz. Mt. Unzen and Mt. Happo can be used for a month-long comparison.

There are a few characteristic features of the influence of the forest fire on the model predicted ozone concentrations at the given locations in Japan. For the southern stations of Mishima, Tsushima, Fukue and Okinawa, the emissions from the forest fire do not have a significant effect on the model predicted levels of ozone. Episodes of elevated levels of ozone are experienced between May 13 and May 21. This corresponds to the arrival of a front from the north. As seen in the figure, the model predicts an average increase of 15-20 ppb (and at some instances, more than 50 ppb) at these sites. Okinawa, located at the extreme south, is not influenced at all by the fire.

While, for the elevated locations of Mt. Happo and Mt. Unzen, in the central and south central part of Japan respectively, one cannot see the effect of the forest fire during most of the time. Mt. Happo and Mt. Unzen are located at elevations of between 1000m and 2000m. For these places to experience any effect of the fire, which is located much to the north, the ozone and its precursors have to be convected significantly above the boundary layer, and then to be advected southward to eventually reach Happo and Unzen. In east Asia, during spring, this is not the case most of the time. Hence the trends are as shown in the figure. Lastly, for Mt. Teine and Napporo, which are situated on the northern island of Hokkaido, the frequency of arrival of the air trajectories from the location of the forest fire is more than that at the other stations. Hence, one can see a systematic increase in ozone (10-20 ppb) during most of May 1987 due to the forest fire.

The spatial distribution of ozone for the whole region can give us an idea of the location of the anthropogenic precursor emissions of ozone. In Figure 2, the monthly average ozone concentration predicted by the model is shown. In the upper part of the figure, the contours represent the ozone levels in the boundary layer (400 m) and the lower troposphere (2000 m). The values shown are for the base case in the absence of the forest fire. As seen in the figure, in the boundary layer, the ozone levels are enhanced close to the regions with large anthropogenic emissions of ozone precursors such as NO_x and NMHCs. For example, one can see the concentration of ozone in the excess of 50 ppb over the eastern parts of China, over the Korean peninsula, and extending further east towards Japan. A similar picture can be seen for the ozone in the lower troposphere, though the enhancements are not so prominent.

In the lower part of Figure 2, the monthly average increase in ozone due to the presence of the forest fire is shown. In the boundary layer, the difference in the model predicted ozone is nearly 30 ppb over northern parts of China (120°E-130°E, 44°N-52°N), while in the lower troposphere, the increase is 15 ppb. The figure also shows the change follows the synoptic weather pattern during this season in east Asia. In spring, the wind is mostly westerly and hence the pollutants tend to be transported eastward from the location of the forest fire towards northern Japan. Consequently, over most parts of continental east Asia, there is no substantial increase in ozone due to the fire. Over Japan, the increase due to the fire is only about 5 ppb (10%). At higher elevations, there is more spread in the concentration contours, with a prominent eastward movement. Once the ozone from the boundary layer is lifted into the middle troposphere, it is rapidly transported eastward by the fast moving westerlies. One can see an increase of ~5% as far as the Kamchatka peninsula in Russia.

The monthly average meridional and zonal profiles of ozone are presented in the upper part of Figure 3. Some general features can be observed in this vertical distribution. In the zonal average, high ozone values are found at higher latitudes, corresponding with the lower tropopause and influence of the stratospheric ozone at higher elevations. In the lower troposphere, increased ozone is found between 35°N and 45°N. A similar feature is observed in the meridional average also. Higher ozone values between 120°E and 140°E. The box bounded by these four segments represents an area of very high emissions of ozone precursors. The influence of the forest fire on the vertical profile of ozone is depicted in the lower part of Figure 3. There is an increase in the ozone concentration between 2 and 4 km around the source of the fire (~48°N). The ozone increases by more than 10 ppb in this region. This feature reflects the uplifting of boundary layer ozone and advection in the lower troposphere. This phenomenon is explained further in Figure 3d. The zonal profile of the ozone increase again shows a high between 2 and 4 km. Also, as discussed before, there is a predominant eastward movement of ozone in the lower and middle troposphere. Increase in the ozone levels upto 2 ppb are observed even at higher elevations of about 8 km.

The model predictions discussed so far were only some of monthly snapshots of the extent of the perturbation of the forest fire on the photochemical oxidant cycle of the east Asian region. There might have been local episodes wherein the ozone levels would have increased significantly and got transported large distances. Such types of episodes must be studied using instantaneous values and other techniques such as trajectory analysis. These investigations would be a part of the future work.

There is limited data available to compare the model predictions with observations. Satellite data analysis like the tropospheric residual ozone from TOMS and SAGE measurements, as suggested by Fishman et al. (1990) cannot be used for this purpose. The reason being, at mid-latitudes, there is variability in the stratospheric column derived from SAGE measurements. Hence, the variability in the TOMS signal normally reflects the stratospheric influences rather than something in the lower troposphere. Ozonesonde measurements have been carried out for a long period of time in Japan. But the ozonesonde stations are more towards the south to be influenced by the forest fire. The only promising aspect currently available is the detection of smoke from the fire by satellite methods. This technique from the TOMS instrument for absorbing aerosols is explained in Herman et al. (1997). This analysis and the model predictions of absorbing aerosols (smoke) during this forest fire episode will be a subject of future study.

CONCLUSIONS

In this paper, we investigated the impact of the trace gas emissions from a forest fire on the photochemical oxidant cycle in east Asia. The CO, NO_x and NMHC emissions from the forest fire were influential in the production of ozone in the boundary layer near and downwind of the source of the fire. The model predicted ozone at locations in southern parts of Japan showed little increase, while in the north, more than 50% increase was predicted. Based on monthly average, the increase in the ozone in the boundary was predicted to be more than 30 ppb (50 %) due to the fire. While in the lower troposphere, the increase was just 10 %. Vertical distribution of ozone showed enhancements of the order of 10-20 ppb (25-30 %) between 2 and 4 km. More work needs to be done in understanding the localized and transient impacts of the fire emissions. The effects of the emissions of the particulate matter need to be investigated in context of their radiative impacts. Also, newer tools of analysis, such as satellite observations are required to obtain a clear picture of the extent of the damage due to the forest fires.

REFERENCES

Akimoto, H. and Narita, H., 1994, Distribution of SO_2, NO_x, and CO_2 Emissions from Fuel Combustion and Industrial Activities in Asia with $1^o x 1^o$ Resolution., *Atmos. Environ.* 28 (2), 213-225.

Andreae, M. O., 1991, Biomass Burning: Its History, Use, and Distribution and Its Impact on Environmental Quality and Global Climate, in: *Global Biomass Burning* 3-21, J. S. Levine, ed., MIT Press, Cambridge, Mass.

Atkinson, R., Baulch, D. L., Cox, R. A., Hampson, R. F., Kerr, J. A. and Troe, J., 1989, Evaluated Kinetic and Photochemical Data for Atmospheric Chemistry: Supplement III, *Int. J. Chem. Kin.* 21:115-150

Cahoon, D. R., Stocks, B. J., Levine, J. S., Cofer, W. R. and Pierson, J. M., 1994, Satellite analysis of the severe 1987 forest fires in northern China and southeastern Siberia, *J. Geophys. Res.* 99(D9):18627-18638.

Cahoon, D. R. et al., 1991, The Great Chinese Fire of 1987: A View from Space, in: *Global Biomass Burning* 61-66, J. S. Levine, ed., MIT Press, Cambridge, Mass.

Carmichael, G. R., 1997, personal communications.

Carmichael, G. R., Peters, L. K. and Kitada, T., 1986, A second generation model for regional-scale transport/chemistry/deposition, *Atmos. Environ.* 20(1):173-188.

Carmichael, G. R., Peters, L. K. and Saylor, R. D., 1991, The STEM-II regional scale acid deposition and photochemical oxidation model - I, An overview of model development and applications, *Atmos. Environ.* 25A(10):2077-2090.

Carmichael, G. R., Uno, I., Phadnis, M. J., Zhang, Y. and Sunwoo, Y., 1998, Tropospheric ozone production and transport in the springtime in east Asia., *J. Geophys. Res.* 103(D9):10649-10672.

Chatfield, R. B., Vastano, J. A., Singh, H. B. and Sachse, G., 1996, A general model of how fire emissions and chemistry produce African oceanic plumes (O_3, CO, PAN, smoke) in TRACE A, *J. Geophys. Res.* 101(D19):24279-24306.

Crutzen, P. J. et al., 1979, Biomass burning as a source of atmospheric gases CO, H_2, N_2O, NO, CH_3Cl and COS, *Nature* 282:253-256.

Crutzen, P. J. and Andreae, M. O., Biomass Burning in the Tropics: Impact on Atmospheric Chemistry and Biogeochemical Cycles, *Science* 250:1669-1678.

Crutzen, P. J. and Carmichael, G. R., 1993, Modeling the Influence of Fires on Atmospheric Chemistry, in: *Fire in the Environment: The Ecological, Atmospheric and Climatic Importance of Vegetation Fires* 89-105, P. J. Crutzen and J. G. Goldammer, eds., John Wiley & Sons Ltd., New York.

Crutzen, P. J. et al., 1985, Tropospheric Chemical Composition Measurements in Brazil During the Dry Season, *J. Atmos. Chem.* 2:233-256.

Fishman, J., Fakhruzzaman, K., Cros, B. and Nganga, D., 1991, Identification of Widespread Pollution in the Southern Hemisphere Deduced from Satellite Analysis, *Science* 252:1693-1696.

Fishman, J., Hoell, J. M., Bendura, R. D., McNeal, R. J. and Kirchoff, V. W. J. H., 1996, NASA GTE TRACE A Experiment (Septembet-October 1992): Overview, *J. Geophys. Res.* 101(D19):23865-23879.

Fishman, J., Watson, C. E., Larsen, J. C. and Logan, J. A., 1990, Distribution of Tropospheric Ozone Determined From Satellite Data, *J. Geophys. Res.* 95(D4):3599-3617.

Fishman, J. et al., 1993, Group Report: What is the Impact of Fires on Atmospheric Chemistry, Climate, and Biogeochemical Cycles?, in: *Fire in the Environment: The Ecological, Atmospheric and Climatic Importance of Vegetation Fires* 345-356, P. J. Crutzen and J. G. Goldammer, eds., John Wiley & Sons Ltd., New York.

Fujita, S., 1992, Acid Deposition in Japan, in: *Report of the Central Research Institute of Electric Power Industry*, Japan.

Guenther, A. et al., 1995, A global model of natural volatile organic compound emissions, *J. Geophys. Res.* 100(D5):8873-8892.

Herman, J., Bhartia, P. K., Torres, O., Hsu, C., Seftor, C. and Celarier, E., 1997, Global distribution of UV-absorbing aerosols from Nimbus 7/TOMS data, *J. Geophys. Res.* 102(D14):16911-16922.

Kotamarthi, V. R. and Carmichael, G. R., 1990, The Long Range Transport of Pollutants in the Pacific Rim Region, *Atmos. Environ.* 24A(1990):1521-1534.

Lobert, J. M. et al., 1991, Experimental evaluation of biomass burning emissions: Nitrogen and carbon containing compounds, in: *Global Biomass Burning* 289-404, J. S. Levine, ed., MIT Press, Cambridge, Mass.

Lurmann, F., Lloyd, A. and Atkinson, A., 1986, A chemical mechanism for use in long range transport/acid deposition computer modeling, *J. Geophys. Res.* 91(D10):10905-10936.

Piccot, S. D., Watson, J. J. and Jones, J. W., 1992, A global inventory of volatile organic compounds emissions from anthropogenic sources, *J. Geophys. Res.* 97:9897-9912.

Pickering, K. E. et al., 1996, Convective transport of biomass burning emissions over Brazil during TRACE A, *J. Geophys. Res.* 101(D19):23993-24012.

Sandu, A., Verwer, J. G., Blom, J. G., Spee, E. J., Carmichael, G. R. and Potra, F. A., 1997, Benchmarking Stiff ODE Solvers for Atmospheric Chemistry Problems II: Rosenbrock Solvers, *Atmos. Environ.* 31:3459-3472.

Zhang, Y., 1994, The chemical role of mineral aerosols in the troposphere in east Asia. Ph.D. Thesis, Dept. of Chem. & Biochem. Eng., Univ. of Iowa, Iowa City.

A HEMISPHERIC DUST STORM IN APRIL 1994 — OBSERVATIONS, MODELLING AND ANALYSES

Emin Özsoy[1], Slobodan Nickovic[2], Nilgün Kubilay[1], Cemal Saydam[3]

[1]Institute of Marine Sciences, Middle East Technical University
P.K. 28 Erdemli - İçel 33731 Turkey
[2]Department of Applied Physics, University of Athens,
Panepistimoupolis Bldg., Phys-V, Athens 15784, Greece
[3]Middle East Technical University and
Turkish Scientific and Technical Research Council, Ankara, Turkey

INTRODUCTION

Atmospheric inputs of aerosol dust from deserts are important in the climate system, primarily because of their effects to modify the global (Charlson and Heintzenberg, 1995; Hansen *et al.*, 1993; Tegen *et al.*, 1996) and regional radiative heat budgets (*e.g.* semi-enclosed seas, Gilman and Garrett, 1994, Gilman and Outerbridge, 1994), or to influence the heterogeneous chemistry of the tropospheric and greenhouse gases (Dentener *et al.*, 1996), as well as the biogeochemical cycles in the marine environment (Duce *et al.*, 1991).

Mediterranean dust events are typically associated with the so-called 'Sharav' cyclones (Alpert and Ziv, 1989) created south of the Atlas Mountains (Reiter, 1979; Brody and Nestor, 1980). On the other hand, Sahara dust is transported into the Atlantic Ocean (Prospero, 1981) with the Equatorial Trade Winds. Large scale atmospheric / climatic controls are suggested by the simultaneous occurence and the similar interannual dependence of the Mediterranean and Atlantic transports (Moulin *et al.*, 1997), demonstrated through annual correlations with the North Atlantic Oscillation (NAO, Hurrell, 1995). The Saharan dust arriving at Barbados Islands appears well correlated with African drought conditions, but some winter events have been associated with the El Niño / Southern Oscillation (ENSO, Prospero and Nees, 1986). During exceptional cases, such as in early April 1994 (Li *et al.*, 1996; Andreae, 1996) also studied in the present paper, significant alterations are found in the radiative energy fluxes on a hemispheric scale.

The spectacular dust event in early April 1994 creating simultaneous transport into the Tropical Atlantic (Li *et al.*, 1996; Andreae, 1996) and the Eastern Mediterranean, and producing the 30 years' maximum concentrations on Barbados Island (Li *et al.*, 1996), and a marked decrease of visibility in the eastern Mediterranean, is analyzed with the aid of ground and satellite measurements and numerical modeling, for a case study establishing the roles of synoptic meteorology and large scale controls.

THE DATA AND THE METHODS

Aerosol measurements at Erdemli, Turkey (36°N, 34°E, height: 20 *m*, Kubilay, 1996) and published data from the tropical Atlantic (Barbados, 13°N, 59°W, Li *et al.*, 1996; Sal Island, 16.7°N, 23.0°W, height: 125 *m*, Chiapello *et al.*, 1995, 1997; Tenerife, 28.3°N, 16.5°W, height: 2500 *m*, Arimoto *et al.*, 1995) are used to describe the transport and verify the modelling results. Rawinsonde profiles of temperature and humidity acquired from the National Climatic Data Center (NCDC) of the National Oceanic and Atmospheric Administration (NOAA), and visibility data obtained from the

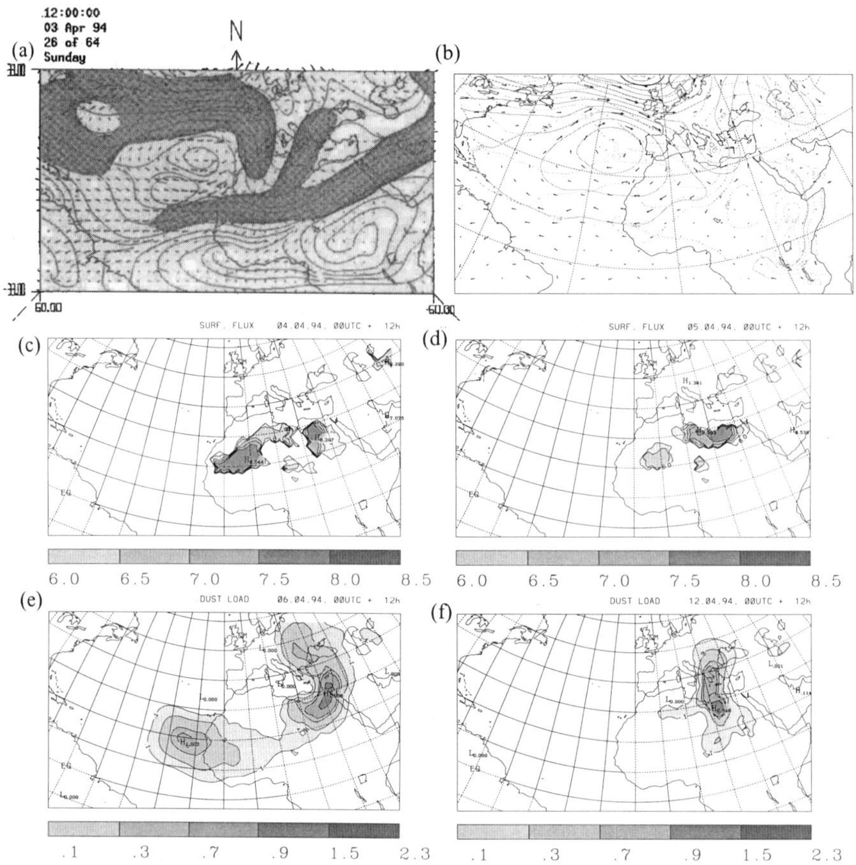

Figure 1. (a) The upper air streamfunction at 8.35 km and wind vectors at 0.5 km elevations, plotted together with the projected polar and subtropical jet streams (shaded areas where upper air speeds are in excess of 30 m/s), April 3, 1994, 12:00 UTC, and (b) geopotential height, temperature and wind vectors on the 850Hpa isosurface, (c) dust flux from the surface on April 4, 12:00 UTC, and (d) on April 5, 12:00 UTC, (e) dust load, on April 6, 1994, 12:00 UTC and (f) on April 12, 1994, 12:00 UTC. The dust flux contours in (c) and (d) are for $\log_{10}(S \times 10^6)$ where S is the surface flux in $\mu g m^{-2} s^{-1}$ (*e.g.* the contour value for for 1 $\mu g m^{-2} s^{-1}$ is 6).

African Center of Meteorological Applications for Development (ACMAD) have also been used in the analyses.

Three dimensional, 3-day back trajectories terminating at 900, 850, 700 and 500 hPa final levels in Erdemli obtained from the European Center for Medium-Range Weather Forecasts (ECMWF), and isentropic, 10-day back trajectories terminating at Barbados, Sal and Tenerife Islands obtained from the Climate Monitoring and Diagnostic Laboratory (CMDL) of NOAA, have been utilized to trace dust transport and identify possible source regions.

The 'ETA' regional atmospheric model, based on a generalized terrain following coordinate system, originally developed by the University of Belgrade and the Federal Hydrometeorological Institute of Yugoslavia, and further enhanced by the National Center for Environmental Prediction (NCEP), USA (Mesinger, *et al.*, 1988; Janjic, 1984, 1990, 1994; Janjic and Mesinger, 1989), is used for simulation. In addition to the sophisticated numerical techniques and physical parameterizations, the model also supports complex formulations for turbulent mixing, viscous sub-layer and convection (Mellor and Yamada, 1982; Betts and Miller, 1986; Janjic, 1994).

The model has been upgraded to include the suspension, transport and deposition of dust (Nickovic and Dobricic, 1996; Nickovic et al., 1997), such that the conservation equation for passive dust concentration is integrated on-line, using numerical schemes valid for other scalar variables, and the viscous sublayer model (Janjic, 1994) for production (suspension) of dust, exploiting the physical similarity between turbulent mixing over the sea and over deserts (Chamberlain, 1983; Segal, 1990). The

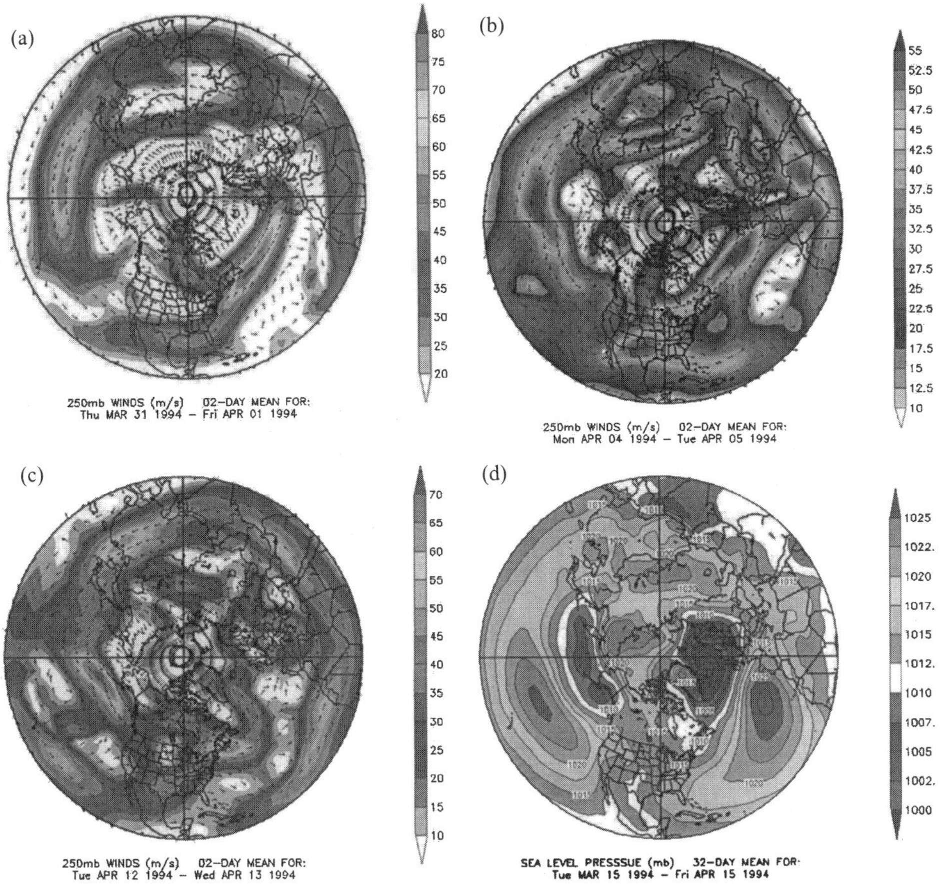

Figure 2. Northern hemisphere upper atmospheric circulation showing the evolution of the polar and subtropical jets. $250 hPa$ wind speed (shading) and direction (vectors) based on two day averages for (a) March 31 - April 1, (b) April 4 - 5, 1994, (c) April 12 - 13, 1994 and (d) the monthly average sea level pressure for the priod March 15 - April 15, 1994. The source for the data is NOAA Climate Diagnostic Center (plotting page web adress http://www.cdc.noaa.gov/HistData/)

effects of ground wetness, wet/dry deposition are also accounted for. The surface dust source distribution is based on the Wilson and Henderson-Sellers (1984) vegetation data set, specifying desert areas as the source for dust.

The model, with a coverage of the Sahara and Atlantic regions, with horizontal resolution of 0.75° and a vertical representation at 32 levels was run from March 28 until April 15, 1994, initialized every 24 hr with meteorological fields with the dust field updated from the previous day's results, and lateral boundary conditions updated every 6 hr, using forecast data obtained from the European Center for Medium-Range Forecasts (ECMWF).

JET INTERACTIONS AND STORM DEVELOPMENT

The prevailing synoptic situation in the beginning of April 1994 was characterized by the subtropical high pressure centered at 35 N, the subpolar (Icelandic) low pressure centered at 65 N in the Atlantic Ocean, and the equatorial low pressure extending across the Tropical Atlantic and the African continent, representing typical winter conditions (Tucker and Barry, 1984). Near the surface, the Trade Wind circulation dominated over the Atlantic Ocean. The Siberian high pressure dominated the Eurasian continent.

It appears that the active developments to follow were linked with a blocking circulation and the interactions of jet streams in the upper troposphere (Fig.s 1a and 2). An upper air modon structure

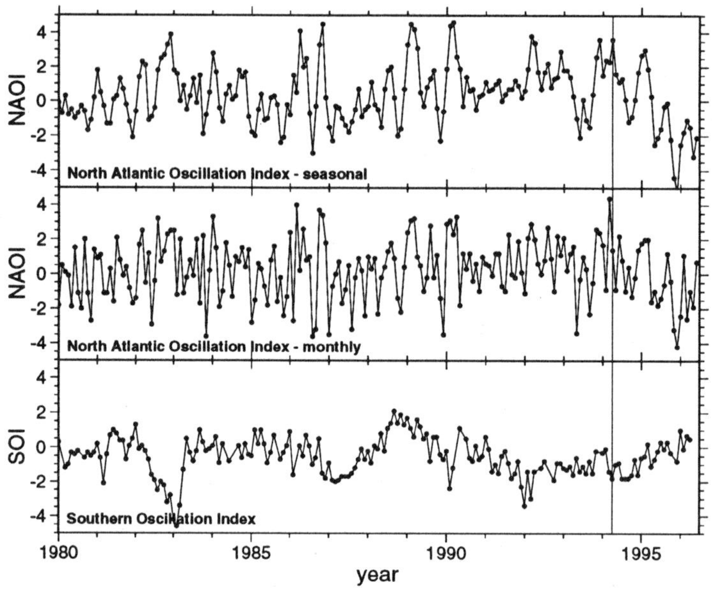

Figure 3. Time variation of (a) seasonal, (b) monthly NAO and (c) SO indices for the last two decades. The vertical line marks the April 1994 event.)

(a coherent dipole eddy with balanced weak nonlinearity and dispersion, often embedded in a zonal flow, *e.g.* McWilliams, 1980; Wallace and Blackmon, 1983) accompanied the surface anticyclone in the Atlantic Ocean, and its approach towards the Mediterranean first resulted in an intensification of Trade winds and a displacement of the polar jet stream to the south towards the subtropical jet stream. While the the two jet streams communicated through filaments earlier (March 29), the coming into contact of the two jet streams around 2-4 April (Fig. 1a) resulted in the development of strong meridional circulations over Europe and the Mediterranean area. An explosive cyclone development took place in the north central Sahara desert (Fig. 1b) in the lee of the Atlas Mountains, and was followed by its passage to the Balkans on April 6.

Initially, on March 31 - April 1, the when the two jet streams were separate, the upper air circulation pattern was defined mainly by the zonal jets (Fig. 2a). In the period following the Mediterranean activity and Sahara dust pulse initiated on around April 3-4, strong meridional circulation patterns developed over much of the northern hemisphere (Fig. 2b). The meandering hemispherical jet pattern continued for about 10 days, and on April 12-13, renewed cyclogenesis resulted from continuing jet interactions (Fig. 2c), consequently leading to a second instance of Saharan dust transport into the eastern Mediterranean, verified by observations and the model results.

The average sea level pressure during March 15-April 15 1994 (Fig. 2d) indicates a dipole of Icelandic low pressure and Azores high pressure typical of the North Atlantic Ocsillation (NAO). The NAO index had a high value of $\sim$4 (seasonal index values of the same magnitude were reached only 8 times during the 1980-1997 period) as shown in Fig. 3, where a comparison is also made with the Southern Oscillation (SO) index. It is significant that a high NAO index circulation created the hemispheric disturbances leading to the anomalous dust event.

The interaction of the two jets, shown in the present case, has been suggested (e.g. Reiter, 1975) as a principal mechanism of Mediterranean cyclogenesis, although only limited cases have been demonstrated earlier (Karein, 1979). The Mediterranean / North African region is considered as one of the three global areas of potential jet interactions (Reiter, 1975; Özsoy, 1981), the other regions being in the lee of Rocky and Himalaya mountain ranges.

An 'Omega High' type of anticyclonic circulation often characterizes northwesterly Mistral winds in the western Mediterranean, in some cases leading to meridional flow that results in 'Genoa cyclogenesis' (Reiter, 1975; Brody and Nestor, 1980). The present case appears to be a rare extension of the strong Mistral winds and the associated subsidence (with subsequent surface divergence) reaching into the West African desert.

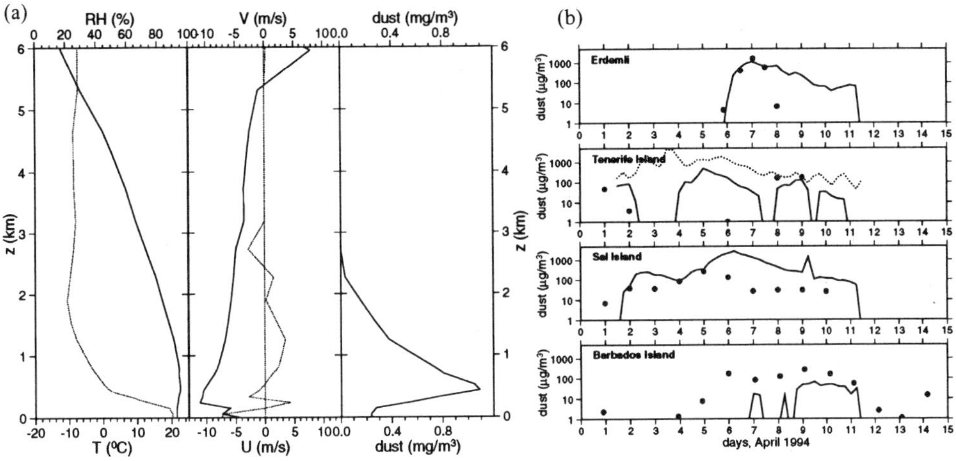

Figure 4. (a) The vertical structure of the trade wind system: Temperature, relative humidity (dotted line), east and north (dotted line) wind speed and dust concentration profiles at Sal Island on April 4 1994 12 GMT. (b) model predictions (solid line) and measurements (dots) of dust concentration at various locations in the Atlantic and the Mediterranean.

DUST SUSPENSION AND TRANSPORT

The synoptic situation in early April led to the generation and dispersal of desert dust in the atmosphere, confirmed by measurements and model simulations. Backward air mass trajectories (not shown) ending at Erdemli, Tenerife and Sal Island measurement stations on April 6 and Barbados Islands on April 9 indicated sources of air originating in the north Atlantic, passing over the Sahara desert and diverging towards the respective sites.

The model surface dust fluxes on April 4 and 5 are shown in Fig.s 1c and 1d, where the two main sources in the western and eastern Sahara can be identified. The dust source in the west was created on April 3 by the descending motions into western Sahara, and continued to be active during the next few days. Then on April 4 and 5, a second source was created in the eastern Sahara, as a result of the surface wind activity of the cyclone. An exceptional pattern of dust transport was then created, with one part of the cloud moving anticyclonically to the Atlantic, and the other part moving north towards the Eastern Mediterranean and Balkans with the cyclonic depression (Fig. 1e).

The column integrated dust load, a parameter directly correlated with the aerosol optical depth, in Fig. 1e shows the localization of dust into two separate clouds moving towards the Mediterranean and the Atlantic Ocean. A second dust pulse was created on 12-13 April (Fig. 1f), by a repeated cyclogenesis in the eastern Mediterranean by jet interactions shown in Figure 2b.

The subsidence south of the Atlas mountains on April 4 created a shallow suspension of dust flattened by divergence, transfering dust to the Atlantic Ocean with the intensified Trade Winds on later days, confined within the first few kilometers of the atmosphere. This structure of the Trade Winds, supplied by subsidence along the periphery of the North Atlantic anticyclone is well known (Tucker and Barry, 1984). On the other hand, the dust put into motion by the Sirte cyclone penetrated upwards along the warm front by convection and frontal uplifting. The model results also showed a large part of the domain affected by transport to be generally free of clouds, except in the east where frontal uplifting of the dust coincides with convective clouds of the cyclone.

In the subtropical Atlantic, the troposphere can generally be divided into a lower moist and an upper dry layer, the moist layer being capped by the Trade Wind inversion. The shallow Trade Inversion near the African coast (around 15° N) is strengthened by the effects of subsidence on the eastern side of the subtropical anticyclone, and cooling from below by the cold upwelled water along the African coast (Riehl, 1954). The inversion in the eastern subtropical Atlantic does not coincide with the top of the Trade Wind regime, but is usually much shallower, at about an elevation of 500 m near Africa. The thickness of the moist surface layer depends on the balance of convection and subsidence, and sharply rises to 2-3 km in the west, characterized with Trade Wind cumuli formed near the Caribbean (Byers, 1959). These features were also verified with model results (not shown). The dust maximum occurs immediately above the shallow Trade inversion.

(a) (b) (c)

Figure 5. NOAA AVHRR satellite visible data on (a) April 6, 07:16 GMT, (b) April 6, 14:28 GMT and (c) April 7, 07:43 GMT, showing dust coming from Libya and Egypt and accumulated along the Levant coast,

The typical vertical structure of the Trade Wind system, and the dust concentration profiles at Sal Island are shown in Fig. 4a. The dust maximum occurs immediately above the shallow Trade inversion, which is well below the Trade Wind boundary.

The model simulated instantaneous dust concentrations are compared with the available observations in Fig. 4b. At the Erdemli station, the observed peak concentration of 1.6 mg/m^3 on the evening of April 6 and its decrease on April 7 agrees well with the model calculated concentrations (Fig. 4b), and with the satellite observations of Fig. 5. At the sampling site on Tenerife Island, ground observations only existed on April 8 and 9 following a failure of the equipment for two days, reaching a level of 160 $\mu g/m^3$ April on 8 and 9 (R. Arimoto, personal communication). The model results at 2.5 km elevation (solid lines, Fig. 4b, second panel from top), corresponding to the location of the measurement station indicate an order of magnitude agreement with model results. It is interesting to note that the measurement station at this height would have missed the much higher values at the lowest model level (dotted line, Fig. 4b), as a result of the shallow, boundary layer structure of the dust plume. At Sal Island, the available measurements indicated dust concentrations reaching a peak of about 260 $\mu g/m^3$ on April 5, which appeared approximately at the same time, but much smaller than the model predicted maximum concentrations. Considering the shallow structure of dust close to the source region and the relatively poor resolution of the present model in the lower layers, this agreement is considered to be satisfactory. Throughout the monitoring study at Sal Island, concentrations in excess of 15 $\mu g/m^3$ were considered as dusty days. Peak dust concentrations of about 300 $\mu g/m^3$ usually occurred in the winter period (December and January) at an altitude of 1.5 to 3 km as a result of dust transport by Trade Winds. In 1992 and 1993 the peak dust concentrations in April were 45 $\mu g/m^3$, while exceptional levels were recorded in 1994 (Chiapello *et al.*, 1995; 1997). At Barbados Islands, the model predictions were considerably lower than the measurements, although the timing of the peak period was not dramatically inconsistent. The maximum measured concentrations of 280 $\mu g/m^3$ represented an all-time high in the observation period of 30 years (Li et al., 1997). The seasonal cycle of dust measured at Barbados Islands reveals a maximum in the summer (June - August), whereas the seasonal dust emission from the Saharan desert is estimated to have a maximum in the spring (March - May) in Africa (d'Almeida, 1986) indicating the importance of meteorological conditions in the long-range transport of dust in comparison to the availability of dust at the source region.

The high concentrations at the center of the dust plume moving towards the Atlantic decreased rapidly by settling in the marine region off the African coast. On the other hand, in the Eastern Mediterranean, the massive dust incursion created by the cyclone resulted in high concentrations in Egypt and along the Levant coast on April 6 (Fig.s 5a-c 3), which persisted on April 7 along the same coast, and also moved across Anatolia to the Black Sea.

Limited comparisons were made with visibility data (not shown) at selected stations from Nigeria, on the south side of the Sahara desert. Visibility lower than $10km$ is generally accepted as a sign of atmospheric dust (Legrand *et al.*, 1989), although other variables such as humidity are interfering factors. The observations suggest heterogeneous behavior over a relatively small region, and possible signs of the dust storms during April 1-15 (Figures 10a-d) and 23-27, 1994.

CONCLUSIONS

The maximum concentration of dust measured in April 1994 at Barbados Island has been found to represent an extreme event, possibly carrying the signatures of a large amplitude climatic fluctuation. The limitless possibilities leading to the creation of an interdecadal climatic variation has not been tested in the present paper, leaving it open for further query. However, we emphasize the role of large scale atmospheric control in creating the dust event, and show that the sequence of events leading from a strong case of atmospheric blocking, to undulations of the polar jet, its interactions and exchange of mass and energy with the subtropical jet, cyclogenesis driven by barotropic / baroclinic instabilities, subsidence and strengthening of Trade Winds, *etc.*, make up the ingredients of the studied dust transport event. We also find that this kind of hemispherical weather is associated with a prominent example for the NAO transient, and hence strengthen the earlier attempts to correlate dust transport with characteristic teleconnection patterns; however we do that on an event basis, rather than on annual to seasonal time scales considered earlier.

Available ground measurements and satellite data confirm the modeling results and analyses. In particular, it is observed that the incursion of the jet circulation into Africa results in surface suspensions of dust propagating simultaneously but in two different directions into the Atlantic and Mediterranean regions. The two pulses are dissimilar in their characteristics representing different meteorological conditions, and mesoscale characteristics. The order of magnitude agreement between model results and observations is generally satisfactory. Dust reaching the Barbados Islands is underestimated, as a result of specific planetary boundary layer processes of the Trade Wind and the Trade inversion regimes not fully resolved in the model.

ACKNOWLEDGEMENTS

Modeling was performed at the Institute of Marine Sciences of the Middle East Technical University (IMS-METU), with support from the NATO Science for Stability (SfS) Program, TU-REMOSENS Project, thanks to C. Saydam. AVHRR satellite data for dust events were collected at the IMS-METU, with support from the NATO SfS Project TU-BLACK SEA. Soil dust ground observations at Barbados and Tenerife obtained as a part of AEROCE program kindly have been supplied by J. Prospero (University of Miami) and R. Arimoto (New Mexico State University). Ground truth measurements at Sal Island were supplied by L. Gomes of the Université de Paris, France. Isentropic back trajectories were provided by J. Harris, Climate Monitoring and Diagnostic Laboratory of NOAA. Visibility measurements obtained at the meteorological offices of Niger were provided by M. Boulahya, Niamey. We also thank H. Virji, the Deputy Director of the International START Secretariat, Washington D.C., and J.-L. Fellous, the Director of Medias-France, for their kind support during this study.

REFERENCES

Alpert, P., and B. Ziv, The Sharav cyclone: observations and some theoretical considerations, *J. Geophys. Res.*, **94**, 18495-18515, 1989.

Andreae, M. O., Raising dust in the greenhouse, *Nature*, **380**, 389-390, 1996.

Arimoto, R., R. A. Duce, B. J. Ray, W. G. Ellis Jr., J. D. Cullen and J. T. Merrill, Trace Elements in the Atmosphere over the North Atlantic, *J. Goephys. Res.*, **100**, 1199-1214, 1995.

Betts, A. K. and M. J. Miller, A new convective adjustment scheme. Part II: Single column tests using GATE wave, BOMEX, ATEX and Arctic Air mass data sets. *Quart. J. R. Met. Soc.*, **112**, 693-709, 1986.

Brody, L. R. and M. J. R. Nestor. Regional Forecasting Aids for the Mediterranean Basin, Handbook for Forecasters in the Mediterranean, Part 2, Naval Environmental Prediction Research Facility, Monterey, California, Technical Report TR 80-10, 178pp, 1980.

Byers, H. R. *General Meteorology*, McGraw-Hill, New York, 1959.

Charlson, R., J. and J. Heintzenberg, *Aerosol Forcing of Climate*, John Wiley, New York, 416 pp., 1995.

Chamberlain, A. C. Roughness length of sea, sand and snow. *Boundary-Layer Meteor.*, **25**, 405-409, 1986.

Chiapello, I., G. Bergametti, L. Gomes, B. Chatenet, F. Dulac, J. Pimenta and E. Santos Suares, An additional low layer transport of Sahelian and Saharan dust over the north-eastern tropical Atlantic, *Geophys. Res. Lett.*, **22**, 3191-3194, 1995.

D'Almeida, G. A., On the variability of desert aerosol radiative characteristics, *J. Geophys. Res.*, **92**, 3017-3026, 1987.

Dentener, F. J., G. R. Carmichael, Y. Zhang, J. Lelieveld and P. J. Crutzen, Role of mineral aerosol as a reactive surface in the global troposphere, *J. Geophys. Res.*, **101**, 22,869-22,889, 1996.

Duce *et al.*, The atmospheric input of trace species to the world ocean, *Global Biogeochem. Cycles*, **5**, 193-256, 1991.

Gilman, C. and Garrett C., Heat flux parameterizations for the Mediterranean Sea: The role of atmospheric aerosols and constraints from the water budget, *J. Geophys. Res.*, **99**, 5119-5134, 1994.

Gilman, C. and A. M. Outerbridge, The Red Sea as a Climate Test Basin: Aereosols and the Air-Sea Budget, 1994.

Hansen, A. D. A., V. N. Kapustin, V. M. Kopeikin, D. A. Gillette and B. A. Bodhaine, Optical absorption by aerosol black carbon and dust in a desert region of central Asia, *Atmos. Env.*, **27A**, 2527-2531,1993.

Hurrell, J. W., Decadal trends in the North Atlantic Oscillation: Regional temperatures and precipitation, *Science*, **269**, 676-679, 1995.

Li, X., H. Maring, D. Savoie, K. Voss and J. M. Prospero, Dominance of mineral dust in aerosol light scattering in the North Atlantic trade winds, *Nature*, **380**, 416-419, 1996.

Janjic, Z. I., Non-linear advection schemes and energy cascade on semi-staggered grids, *Mon. Wea. Rev.*, **112**, 1234-1245, 1984.

Janjic, Z. I., The step-mountain coordinate: physical package, *Mon. Wea. Rev.*, **118**, 1429-1443, 1990.

Janjic, Z. I., The Step-mountain Eta Coordinate Model: Further developments of the convection, viscous sublayer and turbulence closure schemes, *Mon. Wea. Rev.*, **122**, 927-945, 1994.

Janjic, Z. I. and F. Mesinger, Response to small-scale forcing on two staggered grids used in finite-difference models of the atmosphere. *Quart. J. R. Met. Soc.*, **115**, 1167-1176, 1989.

Legrand, M., J. J. Bertrand, M. Desbois, L. Menenger and Y. Fouquart, The potential of infrared satellite data for the retrieval of Saharan-dust optical depth over Africa, *J. Appl. Meteor.*, **28**, 1989.

McWilliams, J. C., An application of equivalent modons to atmospheric blocking, *Dyn. Atm. Oceans*, **5**, 43-66, 1980.

Mellor, G. L. and T. Yamada, Development of a turbulence closure model for geophysical fluid problems, *Rev. Geophys. Space Phys.*, **20**, 851-875, 1982.

Mesinger, F., Z. I. Janjic, S. Nickovic, D. Gavrilov and D. G. Deaven, The step-mountain coordinate: model description and performance for cases of Alpine lee cyclogenesis and for a case of an Appalachian redevelopment, *Mon. Wea. Rev.*, **116**, 1493-1518, 1988.

Nickovic, S., and S. Dobricic, A model for long-range transport of desert dust, *Mon. Wea. Rev.*, **124**, 2537-2544, 1996.

Nickovic, S., D. Jovic, O. Kakaliagou, and G. Kallos, Production and long-range transport of desert dust in the Mediterranean region: Eta model simulations. 22nd NATO/CCMS International Technical Meeting on Air Pollution Modelling and Its Applications, 2-6 June 1997, Clermont-Ferrand, France, 1997.

Özsoy, E., On the Atmospheric Factors Affecting the Levantine Sea, *Tech. Rep. 25*, European Centre for Medium Range Weather Forecasts (ECMWF), Reading, England, 30 pp., 1981.

Prospero, J.M., Arid regions as sources of mineral aerosols in the marine atmosphere, *Geological Societ of America, Special Paper 186*, 71-86, 1981.

Prospero, J. M. and R. T. Nees, Impact of the Norh African drought and El Niño on mineral dust in the Barbados Trade Winds, *Nature*, **320**, 735-738, 1986.

Reiter, E. R. Handbook for Forecasters in the Mediterranean; Weather Phenomena of the Mediterranean Basin; Part 1: General Desciription of the Meteorological Processes, *Tech. Pap. 5-75*, 344 pp., Environmental Prediction Research Facility, Naval Postgraduate School, Monterey, California, 1979.

Riehl, H., *Tropical Meteorology*, McGraw-Hill, New York, 392 pp., 1954.

Segal, M., On the impact of thermal stability on some rough flow effects over mobile surfaces, *Boundary-Layer Meteor.*, **52**, 193-198, 1990.

Tegen, I., A. A. Lacis and I. Fung, The influence on climate forcing of mineral aerosols from disturbed soils, *Nature*, **380**, 419-422, 1996.

Tucker, G. B. and R. G. Barry, Climate of the North Atlantic Ocean, in *Climates of the Oceans, World Survey of Climatology*, edited by H. Van Loon, *15*, Elsevier, 193-257, 1984.

Wilson, M. F., and A. Henderson-Sellers, Land cover and soils data sets for use in general circulation climate models, *J. Climatol.*, **5**, 119-143, 1984.

Wallace, J. M. and M. L. Blackmon, Observations of low-frequency atmospheric variability, in *Large-Scale Dynamical Processes in the Atmosphere*, edited by B. J. Hoskins and R. P. Pearce, pp. 54-95, Academic Press, 1983.

DISCUSSION

C. PERSSON: How did you calculate the emission of dust in your model calculations?

E. ÖZSOY: The surface sources of dust are specified according to the Wilson and Henderson-Sellers (Land cover and soils data sets for use in general circulation climate models, J. Climatol., 5, 119-143, 1984) vegetation data set, with the source area coinciding with areas specified as desert. The dust production scheme is based on turbulent mixing, following a viscous sublayer formulation in the smooth regime, and with a transition to the rough regime under fully developed turbulence. Other details of the dust suspension, transport and deposition formulations (ground wetness effects, wet/dry deposition and advection processes) can be found in Nickovic and Dobricic (A model for long-range transport of desert dust, Mon. Wea. Rev., 124, 2537-2544, 1996).

ETEX SESSION

chairman: P. Builtjes

rapporteur: A. K. V. Falk

EVIDENCE FOR MESOSCALE INFLUENCE ON LONG-RANGE DISPERSION

Jens Havskov Sørensen,[1] Alix Rasmussen,[1]
Thomas Ellermann[2], and Erik Lyck[2]

[1]Meteorological Research Division
Danish Meteorological Institute (DMI), Denmark
[2]Department of Atmospheric Environment
National Environmental Research Institute (NERI), Denmark

INTRODUCTION

In autumn 1994, a large-scale atmospheric dispersion experiment, the European Tracer Experiment (ETEX) (Graziani *et al.,* 1998; Archer *et al.,* 1997), was carried out. From a site in Brittany, France, a controlled release of an inactive non-depositing tracer gas, a perfluorocarbon (PFC) compound, was performed for a period of 12 hours starting on October 23 at 16 UTC. The release rate was 7.9 g/s.

In the days following the release, the dispersing tracer gas was monitored over a range of 2000 kilometers by 168 samplers set up mainly at WMO synoptic stations in Europe. In addition, the tracer gas was monitored with hourly resolution (Sørensen *et al.,* 1998; Ellermann and Lyck, 1997) by the Danish National Environmental Research Institute (NERI) using two sampling techniques at Risø, Denmark, around 1,400 kilometers from the release point.

DANISH EMERGENCY RESPONSE MODEL OF THE ATMOSPHERE (DERMA)

The Danish Emergency Response Model of the Atmosphere (DERMA), which is a three-dimensional Lagrangian long-range transport model (Sørensen, 1997, 1998; Sørensen *et al.,* 1998), utilises a multi-level puff diffusion parametrisation. The concentration field is established as the sum of contributions from individual puffs, i.e., concentration fields surrounding each particle. In the horizontal, Gaussian distributions are assumed for the individual puffs. In the vertical, a well-mixed assumption is employed for puff centres inside the boundary layer, while Gaussian distributions are used for puff centres above this layer. In accordance with the assumption of complete mixing, a cold release from the surface is simulated by instantaneous vertical mixing, i.e., particles are emitted at equidistant heights between ground and the boundary layer top.

Consider a particle, p, positioned at point (x_p, y_p, z_p) above the boundary layer of height h at this position $(z_p > h)$. By using a standard Gaussian formula (Zannetti, 1990), assuming

Air Pollution Modeling and Its Application XIII, edited by S.-E. Gryning
and E. Batchvarova, Kluwer Academic/Plenum Publishers, 2000

total reflection at the ground, the contribution, C_p, from the puff to the total concentration field at point (x,y,z) is given by

$$C_p = \frac{Q_p}{(2\pi)^{3/2}\sigma_{\mathrm{H}}^2\sigma_z}\exp\left\{-\frac{1}{2}\left(\frac{x-x_p}{\sigma_{\mathrm{H}}}\right)^2-\frac{1}{2}\left(\frac{y-y_p}{\sigma_{\mathrm{H}}}\right)^2\right\}$$
$$\times\left(\exp\left\{-\frac{1}{2}\left(\frac{z-z_p}{\sigma_z}\right)^2\right\}+\exp\left\{-\frac{1}{2}\left(\frac{z+z_p}{\sigma_z}\right)^2\right\}\right). \quad (1)$$

The quantities σ_{H} and σ_z are the horizontal and vertical standard deviations, respectively, of the spatial distribution of concentration. The quantity Q_p is the amount of tracer gas associated with puff p depending on the (time varying) emission rate.

Now consider a puff centre inside the boundary layer ($z_p \le h$). Assuming complete mixing in this layer, i.e., a vertical cylindrical distribution of the puff concentration field, we arrive at the following relation,

$$C_p = \frac{Q_p}{2\pi\sigma_{\mathrm{H}}^2 h}\exp\left\{-\frac{1}{2}\left(\frac{x-x_p}{\sigma_{\mathrm{H}}}\right)^2-\frac{1}{2}\left(\frac{y-y_p}{\sigma_{\mathrm{H}}}\right)^2\right\}\delta(z,h), \quad (2)$$

where the step-function $\delta(z,h)$ is given by

$$\delta(z,h) = \begin{cases} 1 & \text{for } z \le h, \\ 0 & \text{for } z > h. \end{cases} \quad (3)$$

From Gifford's random-force theory, or Langevin model, of atmospheric diffusion (Gifford, 1982 and 1984), the following expression is obtained for the variance σ_{H}^2 of relative horizontal diffusion,

$$\sigma_{\mathrm{H}}^2 = 2K_{\mathrm{H}}t_{\mathrm{L}}\left\{\tau-\left(1-\mathrm{e}^{-\tau}\right)-\tfrac{1}{2}\left(1-\mathrm{e}^{-\tau}\right)^2\right\}. \quad (4)$$

The parameter σ_{H} enters in the contributions (1) and (2) from the individual puffs to the concentration field. The parameter τ is the travel time, t, in units of the Lagrangian time scale of atmospheric turbulence, t_{L}, ($\tau \equiv t/t_{\mathrm{L}}$).

Eq. (4) has the following asymptotic expressions,

$$\sigma_{\mathrm{H}}^2 = \begin{cases} \frac{2}{3}K_{\mathrm{H}}t_{\mathrm{L}}^{-2}t^3 & \text{for } t \ll t_{\mathrm{L}}\,, \\ 2K_{\mathrm{H}}t & \text{for } t \gg t_{\mathrm{L}}. \end{cases} \quad (5)$$

Carras and Williams (1988) analysed observations of Australian smelter plumes. Their findings support the $t^{1/2}$ behaviour of σ_{H} at plume travel times, t, greater than 10 hours, and also the $t^{3/2}$ behaviour at travel times less than a few hours. The asymptotic behaviour of σ_{H} at large travel times is the Fickian diffusion expression, and it also comes out as the limit of Taylor's (1921) statistical theory of diffusion. The short-time $t^{3/2}$ asymptotic expression of σ_{H} is derived also by Smith (1968) by using the conditioned particle motion theory assuming an exponential auto-correlation.

For the horizontal eddy diffusivity, a value of 6×10^3 m^2 s^{-1} is used. Sørensen (1998) found that values in the range 4–8 $\times 10^3$ m^2 s^{-1} provide the best performance of DERMA against the ETEX measurements. This range of values is in good agreement with findings by Barr and Gifford (1987). For the Lagrangian time scale, Sørensen (1998) found that a value of 2–4 h provide the best performance of DERMA. Again, this is in good agreement with the value suggested by Barr and Gifford (1987) of 10^4 s (2.8 h).

For puff centres above the boundary layer, a Gaussian distribution is assumed also for the vertical spatial distribution, cf. Eq. (1). In this case, an expression similar to (4) is used for the vertical standard deviation, σ_z,

$$\sigma_z^2 = 2K_z t_L \left\{ \tau - \left(1 - e^{-\tau}\right) - \tfrac{1}{2}\left(1 - e^{-\tau}\right)^2 \right\}. \tag{6}$$

A value of 1 $m^2 s^{-1}$ is used for the vertical eddy diffusivity, K_z. For the ETEX simulations, the calculated surface concentrations are not sensitive to the value of K_z.

Sørensen (1998) showed that the DERMA model has a much more pronounced sensitivity to boundary-layer diffusion parameters when using high-resolution meteorological input from the DMI-HIRLAM model than when using lower-resolution data from the global model of the European Centre for Medium-Range Weather Forecast (ECMWF). Besides, DERMA is more sensitive to the boundary-layer height parametrisation than to the Gaussian diffusion parametrisation.

For atmospheric dispersion modelling it is of great significance to estimate the height of the boundary layer (the mixing height) well. In DERMA, the mixing height is estimated by a bulk Richardson number approach (Sørensen *et al.,* 1996; Sørensen and Rasmussen, 1997; Vogelezang and Holtslag, 1996). The method, which is based on work by André and Mahrt (1982), Troen and Mahrt (1986) and Holtslag and Moeng (1991) is robust and fairly accurate, and it is suited for use in situations, where the vertical resolution of temperature and wind is limited as, e.g., output from numerical weather prediction (NWP) models models. The bulk Richardson number at a given geographical location for the layer between the surface and height z above ground is given by the following expression,

$$\mathrm{Ri_B} = \frac{gz(\theta_v - \theta_s)}{\theta_s (u^2 + v^2)}. \tag{7}$$

This expression is obtained from the gradient Richardson number by using finite differences between values of wind and virtual potential temperature at height z, and corresponding surface values. The quantities θ_s and θ_v are the virtual potential temperature at surface and at height z, respectively, u and v are the horizontal wind components at height z, and g is the gravitational acceleration. The height of the boundary layer is given by the smallest height z at which the bulk Richardson number takes on a prescribed critical value.

For operational use of long-range dispersion modelling, one is faced with using somewhat inaccurate NWP model vertical profiles of wind, temperature and humidity with rather low resolution. For such modelling, it is thus essential to employ a robust method for calculation of the mixing height. Contrary to local methods, e.g., gradient Richardson number approaches, which are sensitive to details of the profiles, the bulk Richardson number method is robust. Besides, use of this method ensures that the calculated boundary-layer heights are dynamically consistent with the vertical structure of the model atmosphere.

In a study involving comparisons with a large number of radiosoundings under different meteorological conditions (Sørensen *et al.,* 1996; Sørensen and Rasmussen, 1997), a critical value of 0.24 was found most appropriate (with a correlation of 68%) for the DMI-HIRLAM model. From this study it furthermore appears that critical values of the bulk Richardson number in the range 0.15–0.35 are adequate for DMI-HIRLAM data. Vogelezang and Holtslag (1996) found a value of 0.22 by similar methods. Sørensen (1998) found that a critical value of 0.25 for DMI-HIRLAM data, and a value of 0.45 for ECMWF data, gives the best performance of DERMA against the ETEX measurements.

In the real-time long-range dispersion model evaluation of the first ETEX release (Graziani *et al.,* 1998), DERMA belongs to the group of models with "excellent performance".

NUMERICAL WEATHER PREDICTION MODELS

At the Danish Meteorological Institute (DMI), a comprehensive three-dimensional regional atmospheric forecasting system is operational. The system is based on the HIRLAM model (Källén *et al.*, 1996), which is a primitive-equation NWP model using a grid-point representation with second-order difference approximations for the spatial derivatives. The Arakawa C grid (Arakawa, 1966) is used as horizontal grid in a rotated spherical coordinate system, while the vertical coordinate is a terrain-influenced hybrid coordinate (Simmons and Burridge, 1981). The present analysis scheme is based on the optimum interpolation technique extended to three-dimensional multi-variate interpolation of observed deviations from forecast first-guess fields.

The DMI-HIRLAM forecasting system which was operational in 1994 (Rasmussen and Sørensen, 1997; Sass, 1994) was run on two different limited areas. The boundary fields for the large-area version (G) are obtained from the global model run by the ECMWF. The G version covers Europe, the North Atlantic, Greenland and parts of USA, Canada and the Arctic Ocean. The E version covering Europe is nested in the G version, which provides the boundary values. The horizontal resolution is 0.42° (46 km) for G and 0.21° (23 km) for E, and the forecast lengths are 48 and 36 hours for G and E, respectively. Both models are run with the same vertical resolution (31 levels). As an example, the models have nine model levels available for resolving a boundary layer with a height of 1500 m. The model versions are run with their own six-hourly data-assimilation cycle. The HIRLAM forecasting system consists of pre-processing, analysis, initialisation, forecast, post-processing and verification.

As for the DMI-HIRLAM model, the ECMWF global spectral model (T213) uses a hybrid vertical coordinate with 31 terrain-influenced levels. The DMI-HIRLAM models have more levels in the boundary layer, though.

MEASUREMENTS

The passage of the tracer gas plume was monitored at Risø, (55°39′N, 12°07′E), Denmark, by using two different sampling techniques (Sørensen *et al.*, 1998; Ellermann and Lyck, 1997). Every hour, an air sample was taken in a Saran bag, and a six-minute sample was taken by means of a pump drawing air through an adsorption tube produced for ETEX by the Joint Research Centre (JRC) at Ispra, Italy. The flow rate (5 l/min) was constant throughout the sampling period. The air samples were collected at ground level (at a height of approximately 1 m) upwind of the tracer laboratory in order to avoid possible contamination from the laboratory. The techniques are described in detail by Ellermann and Lyck (1997).

The samples were later analysed by gas chromatography. The resulting concentrations of perfluoromethyl cyclohexane (PMCH), which was used as a tracer gas for ETEX, show a variation well above the background value of about 0.1 $\mathrm{ng\,m^{-3}}$ whereas the concentrations of other PFC compounds, perfluoromethyl cyclopentane (PMCP) and perfluorodimethyl cyclohexane (PDCH), remained constant through the passage of the tracer. The concentrations of the three tracers in ambient air were determined prior to the ETEX release. The uncertainty of the measured concentrations is estimated to $\pm 25\%$.

The measured concentration time series are shown in Fig. 1. The two sampling methods provided results, which agree well. Due to an error in handling the ETEX tubes, a few recordings are missing in the corresponding time series. The measurements show that the tracer plume arrived at Risø at 44 hours after the start of the release, and that the duration of the passage was 20 hours. The concentration time series has a double-peak structure with peak concentrations at 13 and 5 times the background concentration.

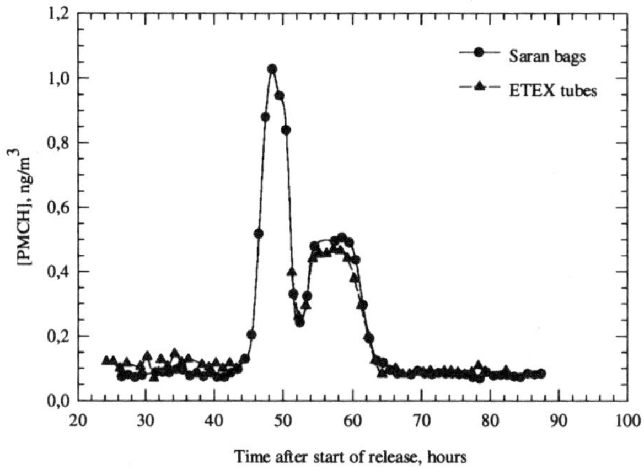

Figure 1. ETEX tracer gas concentrations measured at Risø, Denmark, shown as a function of time since the start of the release. Circles correspond to the Saran bag sampling method, triangles to the ETEX tube method.

MODEL RESULTS

The tracer gas plume of the first ETEX release is simulated by the DERMA atmospheric long-range dispersion model on a range of several thousand kilometers covering a period of four days. The DERMA model has been run using NWP data of 0.5° horizontal resolution from the global model of ECMWF, or data of 0.21° horizontal resolution from the Europe version of the DMI-HIRLAM model. In case of ECMWF data, use was made of six-hour analyses only, whereas for DMI-HIRLAM data, three-hour forecasts were used in between the analyses.

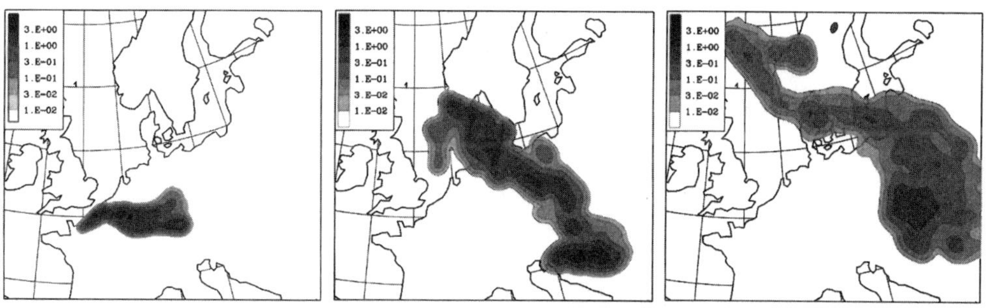

Figure 2. DERMA simulation of the first ETEX experiment using analysed DMI-HIRLAM data. One-hour average surface concentrations are shown in units of ng m^{-3} at 24, 48 and 72 hours after the start of the release.

Each five minutes a batch of forty puffs was released thereby making up the plume resulting from the 12-hour release by 5,800 puffs in total. In Fig. 2 calculations of one-hour average surface concentrations are shown in units of ng m^{-3} at 24, 48 and 72 hours after the start of the release. In this simulation, DERMA has been using analysed DMI-HIRLAM data.

In Fig. 3 model results are compared with the measurements at Risø for an 88-hour period. For the left sub-figure use has been made of ECMWF model data. As is seen from the figure, the duration and maximum concentration are in good agreement with the measurements, but the model plume arrives a few hours too late, and the model fails in reproducing the double peak. However, using higher-resolution DMI-HIRLAM data the observed double-peak structure is nicely reproduced.

At some of the neighbouring ETEX stations double-peak structures corresponding to the one seen at Risø are also observed (Sørensen *et al.,* 1998). Unfortunately, the measure-

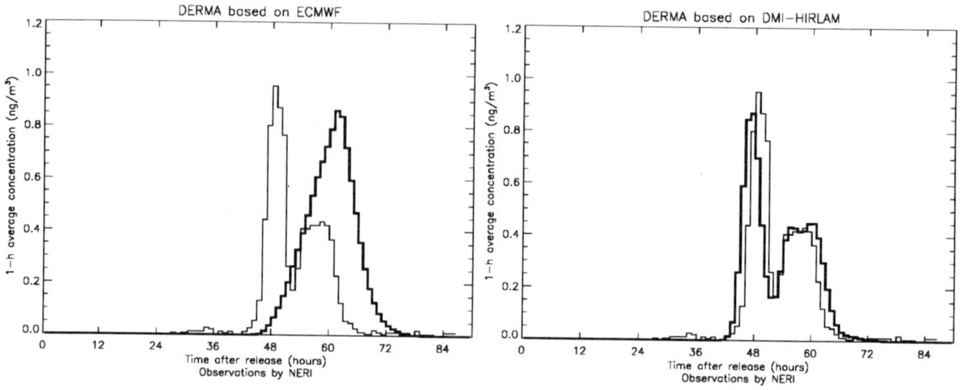

Figure 3. With thin curves the one-hour average concentrations measured at Risø are shown as a function of time after the start of the release. The background concentration has been subtracted. With thick curves corresponding DERMA results using analysed NWP data are shown. On the left figure, ECMWF model data are used, on the right figure, DMI-HIRLAM data are used.

ments from ETEX station DK6, which is the station closest to Risø, cf. Fig. 5, are erroneous (K. Nodop, private communication).

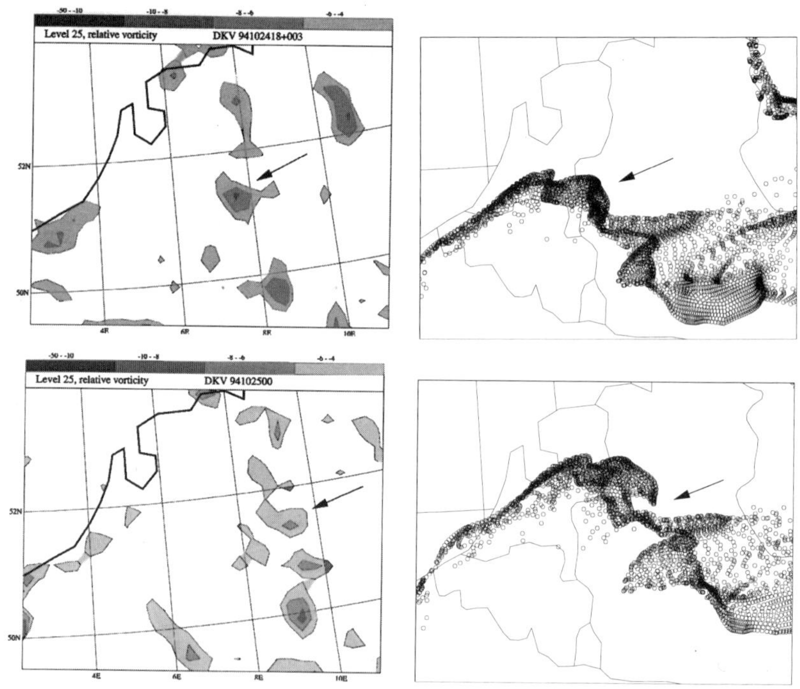

Figure 4. On the left-hand side, relative vorticity is shown in units of s^{-1} at about 900 meters above ground at 29 and 32 hours after the start of the release. On the right-hand side, particle positions are shown corresponding to the valid times of the left-hand side sub-figures. The arrows indicate the anti-cyclonic mesoscale eddy causing the formation of the toroidal perturbation of the dispersing plume as well as the toroid itself.

According to DERMA using DMI-HIRLAM data, the double-peak structure of Fig. 3 is due to a mesoscale horizontal eddy formed at an earlier stage of the dispersion process. The left panel of Fig. 4 shows the relative vorticity of DMI-HIRLAM about 900 meters above ground at 29 and 32 hours after the start of the release. The anti-cyclonic eddy is seen as a pronounced minimum occurring about 700 km from Risø. The eddy persists at least 18 hours in the DMI-HIRLAM model with a horizontal extent on the order of 100 km and more, corresponding to at least five grid lengths, and extending from about 200 to 2000 m above ground with a maximum at 1500 m. The model description of the eddy is thus above the noise level. As the model eddy does not reach the ground, it cannot be detected in the synoptic data.

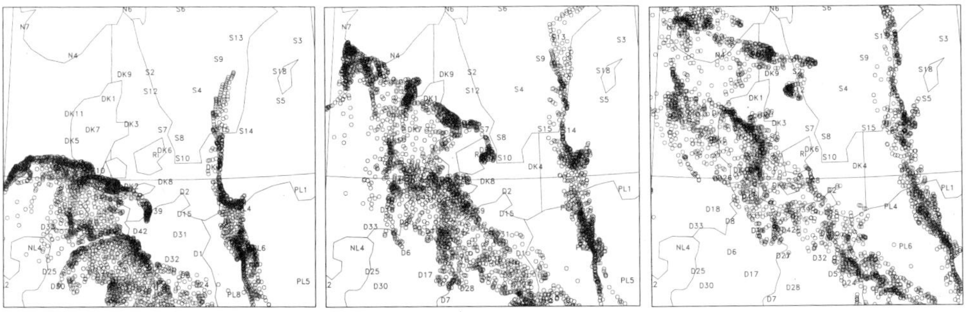

Figure 5. Particle positions (puff centres) at 42, 48 and 54 hours after the start of the release. The DERMA simulation is performed using analysed DMI-HIRLAM data. ETEX monitoring stations are indicated by numbers, and Risø, which is located in the centres of the sub-figures, is marked by Ri.

The anti-cyclonic eddy, which is superimposed on the general flow, influences the dispersing tracer gas for around 12 hours causing a toroidal perturbation. This is seen from the right panel of Fig. 4 which shows positions of released particles (puff centres) corresponding to the valid times of the left panel. Initially, the toroidal perturbation is quite small. In the ensuing days, it grows in size until it becomes so diffuse that it looses its identity. The period in which the plume is influenced by the eddy is smaller than the period of revolution. This implies that the toroid does not close as seen from Figs. 4 and 5.

The influence of the toroidal perturbation on the concentration time series observed at Risø is seen from Fig. 5, which shows positions of released particles as the plume passes by Risø. As it appears from the figure, the toroidal perturbation is the reason for the first peak observed at around 48 hours after the start of the release. The second broader peak is due to the "bulk" of the plume passing over Risø. The toroidal perturbation had similar effects at a number of official ETEX observation sites (Sørensen *et al.,* 1998).

CONCLUSION

The dispersing tracer gas from the first ETEX release was monitored at Risø, Denmark, by the Danish National Environmental Research Institute (NERI). The two methods, which were used to sample the tracer gas, are described, and the time series obtained presented. The two methods gave very similar results. The measurements show the passage of the plume at a clearly detectable level, and the time series has the shape of a double peak.

Predictions by the Danish Emergency Response Model of the Atmosphere (DERMA) are presented corresponding to the first ETEX release. Results are shown using high resolution NWP model data from the DMI-HIRLAM model as well as using coarser data of the global model of ECMWF. Model results using the coarser data represent fairly well the gross features of the measurements by NERI such as time of arrival, duration and maximum concentration, but fail to describe the detailed structure. However, using high-resolution DMI-HIRLAM data the double-peak structure of the measurements is nicely reproduced. The double-peak structure is explained by the influence of an anti-cyclonic mesoscale eddy on the dispersing tracer gas.

In conclusion it is shown that mesoscale features may influence atmospheric dispersion on a scale of up to thousand kilometers causing significantly distorted concentration patterns.

REFERENCES

André, J.C., and Mahrt, L., 1982, The nocturnal surface inversion and influence of clear-air radiative cooling. *J. Atmos. Sci.* **39,** 864–878

Arakawa, A., 1966, Computational design for long-term numerical integrations of the equations of atmospheric motion. *J. Computational Phys.* **1,** 119–143

Archer, G., Girardi, F., Graziani, G., Klug, W., Mosca, S., Nodop, K., and Stingele, A., 1997, The European long range tracer experiment ETEX, a data base for comparing model results and for harmonisation. "4th Workshop on Harmonisation within Atmospheric Dispersion Modelling for Regulatory Purposes", Oostende, Belgium, 6–9 May, 1996. In: *Int. J. Environment and Pollution,* Vol **8.**, Nos. 3–6, pp. 260–263

Barr, S., and Gifford, F.A., 1987, The random force theory applied to regional scale tropospheric diffusion. *Atmos. Environ.* **21,** 1737–1741

Carras, J.N., and Williams, D.J., 1988, Measurements of relative σ_y up to 1800 km from a single source. *Atmos. Environ.* **22,** 1061–1069

Ellermann, T., and Lyck, E., 1997, Tracer measurements during ETEX at Roskilde, Denmark. In: *Proceedings of the ETEX Symposium,* Vienna, Austria, 13–16 May, 1997, pp. 91–94

Gifford, F.A., 1982, Horizontal diffusion in the atmosphere: A Lagrangian-dynamical theory. *Atmos. Environ.* **16,** 505–512

Gifford, F.A., 1984, The random-force theory: Application to meso- and large-scale atmospheric diffusion. *Boundary-Layer Meteorol.* **30,** 159–175

Graziani, G., Klug, W., and Mosca, S., 1998, Real-Time Long-Range Dispersion Model Evaluation of the First ETEX Release. EUR 17754 *Available from the Office for Official Publications of the European Communities, Luxembourg,* ISBN 92-828-3657-6

Holtslag, A.A.M., and Moeng, C.H., 1991, Eddy diffusivity and countergradient transport in the convective atmospheric boundary layer. *J. Atmos. Sci.* **48,** 1690–1698

Källén, E., (Ed.), 1996, HIRLAM Documentation Manual, System 2.5. *Available from the Swedish Meteorological and Hydrological Institute (SMHI)*

Rasmussen, A., and Sørensen, J.H., 1997, Quality validation of analyzed and forecast vertical profiles of wind and temperature from the DMI-HIRLAM model in comparison with radiosoundings. In: *Proceedings of the Sixth Topical Meeting on Emergency Preparedness and Response.* San Francisco, California, April 22–25, 1997, pp. 31–34, ISBN 0-89448-623-3

Sass, B.H., 1994, The DMI Operational HIRLAM Forecasting System, Version 2.3. *DMI Tech. Rep.* **94–8**

Simmons, A.J., and Burridge, D.M., 1981, An energy and angular momentum conserving vertical finite-difference scheme and hybrid vertical coordinates. *Mon. Wea. Rev.* **109,** 758–766

Smith, F.B., 1968, Conditioned particle motion in a homogeneous turbulent field. *Atmos. Environ.* **2,** 491–508

Sørensen, J.H., 1997, Sensitivity of DERMA to boundary-layer parameters, and evidence for mesoscale influence on long-range transport. In: *Proceedings of the ETEX Symposium,* Vienna, Austria, 13–16 May, 1997, pp. 207–210

Sørensen, J.H., 1998, Sensitivity of the DERMA long-range Gaussian dispersion model to meteorological input and diffusion parameters. Accepted for publication in *Atmos. Environ.*

Sørensen, J.H., and Rasmussen, A., 1997, Method for calculation of atmospheric boundary-layer height used in ETEX dispersion modeling. In: *Proceedings of the Sixth Topical Meeting on Emergency Preparedness and Response,* San Francisco, California, pp. 503–506

Sørensen, J.H., Rasmussen, A., and Svensmark, H., 1996, Forecast of atmospheric boundary layer height utilised for ETEX real-time dispersion modelling. *Phys. Chem. Earth* **21,** 435–439

Sørensen, J.H., Rasmussen, A., Ellermann, T., and Lyck, E., 1998, Mesoscale influence on long-range transport; evidence from ETEX modelling and observations. Accepted for publication in *Atmos. Environ.*

Taylor, G.I., 1921, Diffusion by continuous movements. *Proc. London Math. Soc.* **20,** 196–211

Troen, I., and Mahrt, L., 1986, A simple model of the atmospheric boundary layer; sensitivity to surface evaporation. *Boundary-Layer Meteorol.* **37,** 129–148

Vogelezang, D.H.P., and Holtslag, A.A.M., 1996, Evaluation and model impacts of alternative boundary-layer height formulations. *Boundary-Layer Meteorol.* **81,** 245–269

Zannetti, P., 1990, *Air Pollution Modeling,* pp. 125–130 and 166–168. Computational Mechanics Publications, Southampton Boston, Van Norstrand Reinhold, New York

DISCUSSION

M. GARDNER: How generalizable are the results obtained in this tracer study to the pollutants of more importance in "Air Quality", e.g. ozone, particles and NO_x?

J. H. SØRENSEN: The background for ETEX is nuclear emergency preparedness. Therefore, the tracer experiment was set up to resemble an accidental release of a large amount of radioactive material from a nuclear power plant. And thus the point source of short duration (12 hours) does not resemble an emission inventory for conventional air pollutants. ETEX is, however, of great value also to regional air-pollution models since the experiment particularly addresses the effects of advection and diffusion, which are essential also for air-quality modelling.

D. SYRAKOV: Have you calculated 3-hour averages of the Risø 1-hour data series, and did this gap exist on the new data set?

J. H. SØRENSEN: Yes I have, and the gap in the concentration time series, or the double-peak structure, is still clearly evident

SIMULATION OF THE ETEX FIRST RELEASE BY BULGARIAN EMERGENCY RESPONSE MODELS

D. Syrakov and M. Prodanova

National Institute of Meteorology and Hydrology, Sofia 1784, Bulgaria

INTRODUCTION

The Bulgarian team takes part in the ATMES-II with two models. ATMES-II (Atmospheric Transport Model Evaluation Study) is an international verification of long-range dispersion models using the results of European Tracer Experiments (ETEX) first release and a common meteorological data set, prepared by ECMWF. The first Bulgarian model is the puff-type model LED currently used in the Bulgarian Emergency Response System (BERS) the creation and development of which was highly stimulated by ETEX. The second model is the newly created Eulerian multi-level model EMAP.

Both models possess two important features - they are PC-oriented and use quite limited amount of input meteorological information. The development of these models has been urged by the specific conditions in the country, namely the limited possibility to use high performance computers and the lack of detailed meteorological information. Relatively simple approaches are searched for describing air dispersion processes and special attention is paid to time and space optimisation of the calculation schemes. The lack of own numerical weather forecast and the limited possibility to get remote meteorological information make the Global Telecommunication System (GTS) of the World Meteorological Organisation the only source of meteorological data. This predetermines the second important feature of Bulgarian air pollution models - they use only synoptic input information. As the GTS distributed meteo-information has neither the time nor the space resolution needed by air pollution applications, a simple, but very effective, one-dimensional Atmospheric Boundary Layer (ABL) model (Yordanov et al., 1983) is built in both models. The wind and the potential temperature at 850 hPa level, as well as the potential temperature at the ground level are input for this ABL model which reconstructs the vertical profiles of the wind and the vertical diffusion coefficient at each grid point of the model area.

In this paper, the results of different runs, simulating ETEX first release are presented and commented.

Air Pollution Modeling and Its Application XIII, edited by S.-E. Gryning
and E. Batchvarova, Kluwer Academic/Plenum Publishers, 2000

MODEL AREA AND INPUT INFORMATION

A rectangular area of 2500×2450 km, covering Europe, is chosen on the stereographic map of the Northern hemisphere. Cartesian co-ordinate system with y-axes parallel to the 10°E meridian is introduced in it. This area is discretized by 101x99 points with 25 km horizontal step.

Three types of input information are needed for the models' performance:

A. Source information. Source strength, duration, release height, source location from the ETEX first release are used in this study (Graziani et al., 1998).

B. Meteorological data. 850 hPa level wind and potential temperature, and surface level potential temperature are the input data for the ABL model. Four sets of meteorological data are dealt with in this study:

1. Data set prepared for ATMES-II by ECMWF. This is analysed data with space discretization of 0.5°x0.5°, and time resolution $Dt_m = 6$ h. The input met-file produced by these data is called **ECMWF_a**.
2. Analysed data from the ARPEGE model system of METEOFRANCE, Toulouse. Space resolution 0.5°x0.5°; time resolution $Dt_m = 3$ h; input met-file **MetFr_a**.
3. Numerical production of the Global Model of the Met Office in Bracknell, England - analysed data. Space resolution 2.5°x2.5°; time resolution $Dt_m = 12$ h; input met-file **Brack_a**.
4. Global Model forecast. Space and time resolution as 3; input met-file **Brack_p**.

C. Land-use information. It includes roughness field and sea-land mask. Roughness is necessary for the ABL model and it is specified for each grid point. Two constants, sea- and land-specific dry deposition velocity, are introduced initially and then the sea-land mask is used to form the respective field.

MEASUREMENTS, STATISTICS AND ESTIMATING PROCEDURES

The data set of the ETEX first release, collected by 168 sampling stations and analysed at EC Joint Research Center in Ispra, Italy, is used in this study. Statistical analysis is performed on a set of concentration data with lower boundary of 0.01 ng m^{-3} (the detectable limit) including all measured and calculated values (measured zeros included) but for the lost, non-sampled or non-analysed measurements.

To estimate how well the measured and the model-calculated concentrations fit, the statistical methods developed for ATMES (Klug et al., 1992) and adopted in ATMES-II (Mosca et al., 1998) are used.

The following statistical indexes are chosen from the ATMES-II set:

- global analysis - Normalised Mean Square Error (NMSE), bias (BIAS), Pearson's correlation coefficient (CORR), fraction of over-estimation (FOEX), fraction of in-factor α fits (FAα);
- space analysis - Figure of Merit in Space (FMS) for 24, 36, 48 and 60 hours after the start of the release, boundaries of the tracer spots of 0.1 ng m^{-3};
- time analysis - time series and Figure of Merit in Time (FMT).

After the statistical indexes of a set of runs are determined, a ranking procedure similar to that in Mosca et al.(1998) is run. First, an index-by-index ranking is performed. Then, the total rank for each run is obtained as a sum of all ranks, whereas each statistical index is assigned the same weight. The run with the smallest total rank is considered best. Here, the used statistical indexes are: NMSE, BIAS, CORR, FA2, FA5, FOEX, FMS24, FMS36, FMS48, FMS60.

SIMULATION OF ETEX FIRST RELEASE BY THE LED MODEL

LED (Lagrangean-Eulerian Diffusion), a typical puff model, is described in Djolov et al., (1987). The action of each continuous pollutant source is presented as a release of successive clouds (puffs), transported and dispersed independently. The wind components and the vertical velocities are used in the simplest Euler-forward 3D scheme to calculate the position of the puff centre at each time step. A volume-source type analytical solution of the Fick equation is applied to parameterize horizontal diffusion. The vertical distribution of the pollutant is described by four analytical solutions of the vertical diffusion equation, obtained using different boundary conditions. They are applied depending on puff dimensions, centre position and mixing height.

For the ATMES-II exercise, a set of LED runs was performed to model the pollution source. The RISO meteorological group reported at the Prague'95 ETEX meeting that, during the release, the atmosphere was extremely unstable. This means fast elevation of the tracer, just at the release point. A number of presentations at that meeting also manifested better results in case the source is modelled as tracer release at significant heights rather than at the ground level. A set of runs is completed, the source being presented as a cluster of initial puffs with different number of members, different heights, dimensions and amounts.

The comparing procedure shows that the run in which the source is presented as a cluster of 3 puffs with center height of 50, 150 and 300 m fits the measurements best. The overlapping of the measured and the calculated tracer spots for the same run is shown in Figure 1. It can be seen that in this case LED provides a good description of the tracer distribution during the first 48 h after the start of the release, but later on a faster movement of the calculated spot in comparison with the measured one is observed. In addition, the discrete structure of the calculated pollution field, generated by an ensemble of puffs, is well manifested for long periods.

This run took part in the ATMES-II exercise (Mosca et al., 1998), referred to as model 104 and classified 14th by the Evaluation team among 34 models. In spite of the satisfactory results (they are rather good, bearing in mind the poor meteorological information), the defects of the LED performance have to be taken into consideration. Further on, much more attention is paid to the second model EMAP, which can be considered as LED alternative in BERS.

SIMULATION OF ETEX FIRST RELEASE BY THE EMAP MODEL

EMAP stands for Eulerian Model for Air Pollution. The full description of the model, as well as some tests of the sensitivity analysis can be found in Syrakov (1995). The model is realised using the time splitting approach. The model domain is ABL, approximated by a log-linear terrain-following staggered Z-coordinate system. A staggered Arakawa C-type grid is used in the horizontal plain. A new Bott-type advection scheme, called TRAP, is built in the model. It possesses the same properties as Bott's scheme (Bott, 1989) but is at least twice faster (Syrakov, 1997). The simplest 2-order time forward explicit scheme is used for describing horizontal diffusion. The vertical diffusion block uses the simplest implicit scheme with changeable grid steps. The dry deposition flux is set as a bottom boundary condition. The top boundary condition is of the "open" type and an additional upper layer is introduced, accounting parameterically for free atmosphere. A proper surface layer parameterization (Syrakov and Yordanov, 1997) allows to have the first computational level at the top of this layer.

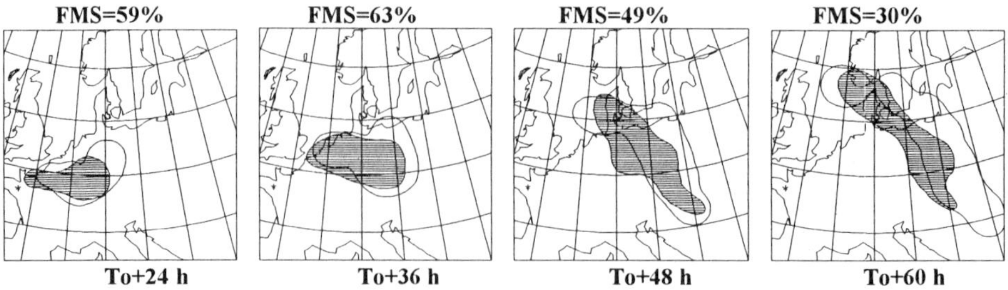

Figure 1. LED's best run spatial distribution of measured (shadowed) and calculated tracer spots.

For ATMES-II simulations, a four-layer version of EMAP is used. The boundaries of the layers are 20, 100, 375, 995 and 1930 m; the layer's thickness are 80, 275, 620 and 935 m, respectively. The time step is Dt = 15 min.

Five sets of runs are made with EMAP, testing different aspects of the model's performance and input information:

Set 1. The aim of this set of runs is to obtain a source presentation producing a description which fits the measurements best. The experiments are made with meteorological data provided by ECMWF. As shown in Table 1, the source is considered a column of grid boxes placed above the source point, having various vertical mass distribution. Runs are combinations from a one-layer box to a column spanning all layers. The released masses are determined in such a way that the source concentration is constant with height. Run 1.3 presented EMAP performance in ATMES-II (Syrakov and Prodanova, 1997a). It was coded as model 103 and ranked 9th among 34 models (Mosca et al., 1998). In runs 1.6 - 1.8, an attempt to take the meteorological conditions at the release site into consideration is made. The concept of mixing height and its evolution is used. The mixing height estimated from the radiosoundings on October 23 was 750 m at 12 UTC and about 950 m at 15 UTC (Gryning and Batchvarova, 1997). The modelling, made in the same paper, shows that, typically, daytime mixing height reaches 1000 m in the late afternoon and night-time mixing height varies between 100 m and 750 m. This is taken into account when constructing the source presentations in the last three runs.

Table 1. Description of EMAP runs (Set 1), varying the source presentation

Runs	Source layers	Q [g/s]	Cs [$ng/m^3.s$]	Runs	Time	Source layers
1.1	1	7.95	16.52			
1.2	2	6.212	3.614		from 16:00 till 20:00	as run 1.3
	1	1.739	3.614	1.6	from 20:00 till 24.00	as run 1.2
1.3	3	5.510	1.420		from 00:00 till 04:00	as run 1.1
	2	2.440	1.420			
	3	5.074	1.307			
1.4	2	2.247	1.307	1.7	from 16:00 till 20:00	as run 1.3
	1	0.629	1.307		from 20:00 till 04:00	as run 1.2
	4	3.898	0.666			
1.5	3	2.586	0.666	1.8	from 16:00 till 18:00	as run 1.3
	2	1.145	0.666		from 18:00 till 04:00	as run 1.2
	1	0.321	0.666			

Table 2. Statistical estimates and ranking of EMAP runs (Set 1). Met-file **ECMWF_a**, runs described in Table 1.

Run	NMSE	BIAS	CORR	FA2	FA5	FOEX	FMS24	FMS36	FMS48	FMS60	Total	Rank
1.1	445.0	0.435	0.611	21	43	10	52.7	56.7	52.7	48.8	46	7
1.2	52.2	0.258	0.574	23	45	8	50.8	54.0	51.3	45.4	52	8
1.3	11.7	0.044	0.559	27	50	-2	52.1	49.7	49.8	40.9	43	6
1.4	13.3	0.075	0.582	27	51	1	54.1	53.1	51.3	42.9	34	3
1.5	11.2	-0.003	0.582	30	54	-10	55.5	50.4	48.6	36.5	37	4
1.6	236.0	0.251	0.622	24	46	9	55.5	58.1	52.6	49.4	32	2
1.7	45.3	0.183	0.611	24	46	7	54.1	56.9	53.0	47.9	28	1
1.8	48.5	0.219	0.596	23	45	7	53.0	56.0	52.5	46.8	40	5

Table 3. Statistical estimates and ranking of EMAP runs with different meteorological inputs. Run 3.1 - **ECMWF_a**; run 3.2 - **MetFr_a**; run 3.3 - **Brack_a**; run 3.4 - **Brack_p**.

Run	NMSE	BIAS	CORR	FA2	FA5	FOEX	FMS24	FMS36	FMS48	FMS60	Total	Rank
3.1	45.3	.183	.611	24	46	7	54.1	56.9	53.0	47.9	21	2
3.2	43.8	.193	.632	24	47	10	58.5	60.8	53.4	51.9	15	1
3.3	72.3	.118	.505	19	37	-15	60.5	40.1	34.3	20.8	30	3
3.4	66.5	.151	.489	19	40	-11	56.3	42.9	27.6	25.6	30	3

Table 4. Statistical estimates and ranking of EMAP runs varying the interpolation frequency of meteorological input. Met-file **ECMWF_a**, $Dt_m = 6$ h.

Run	Dt_i[h]	NMSE	BIAS	CORR	FA2	FA5	FOEX	FMS24	FMS36	FMS48	FMS60	Total	Rank
4.1	6	45.3	.183	.611	24	46	7	54.1	56.9	53.0	47.9	27	5
4.2	3	45.5	.185	.663	24	47	7	62.0	66.8	57.0	44.4	26	4
4.3	1	45.4	.183	.669	24	47	8	63.0	68.3	57.1	42.7	22	3
4.4	0.5	45.3	.182	.670	24	46	8	63.4	68.8	57.1	41.8	19	2
4.5	0.25	45.3	.181	.671	24	46	8	63.5	68.9	57.1	41.3	16	1

Table 5. Statistical estimates and ranking of EMAP runs with different meteorological inputs and minimal interpolation time Dt_i= Dt. Run 5.1 - **ECMWF_a**; run 5.2 - **MetFr_a**; run 5.3 - **Brack_a**; run 5.4 - **Brack_p**.

Run	NMSE	BIAS	CORR	FA2	FA5	FOEX	FMS24	FMS36	FMS48	FMS60	Total	Rank
5.1	45.3	.181	.671	24	46	8	63.5	68.9	57.1	41.3	15	1
5.2	44.1	.189	.656	24	46	9	61.9	66.5	55.7	47.8	19	2
5.3	80.6	.110	.601	18	38	-18	59.4	57.8	34.5	32.2	30	4
5.4	73.2	.124	.558	20	38	-13	54.5	47.8	31.1	49.0	29	3

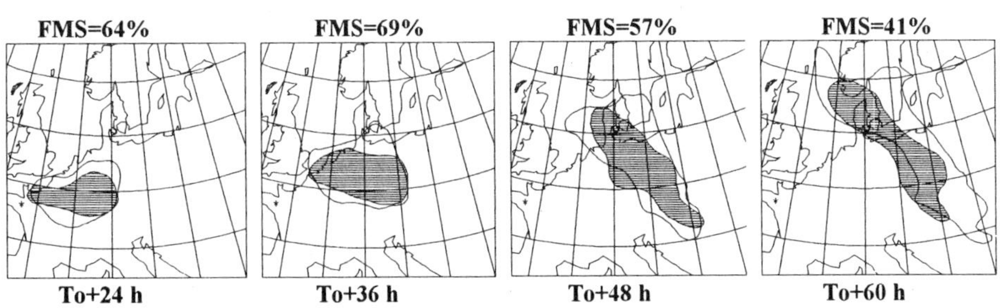

Figure 2. Spatial distribution of measured (shadowed) and EMAP calculated tracer spots (Run 5.1).

The statistical estimates, total ranks and ranking are displayed in Table 2. Run 1.7, where the released mass is distributed homogeneously up to 995 m from the beginning of the release till 20:00 UTC and up to 620 m during the night, occurs to be the best. In the frame of the model's vertical structure, this coincides with the behaviour of mixing height at the release site as shown in Figure 4 of Gryning and Batchvarova (1997). The source presentation from Run 1.7 is used in all other sets of runs.

Set 2. This set of runs tests the horizontal diffusion coefficient value K_h, which in EMAP is constant and is prescribed in the initialisation setting. The final ranking shows that values of K_h in the range 3.10^4 - 6.10^4 m^2 s^{-1} produce fields with equal simulation quality. This coincides with the results shown in Syrakov and Prodanova (1997b). The value of $K_h = 5.10^4$ m^2 s^{-1} is used in the next calculations.

Set 3. This set is to check the quality of the different meteorological data sets. As already mentioned, these data have different space and time resolution. In Table 3, all statistical estimates are displayed together with their total ranks and final ranking. The best run is the one with **MetFr_a** as input met-file. The second one is the **ECMWF_a** run. The both Bracknell meteorological sets have equal ranks with much higher totals.

As ECMWF and METEOFRANCE data sets have equal space resolution, it seems that the ranking is a result of the different time resolution, i.e. the higher time resolution, the better simulation. The coarse space and time resolutions of the Bracknell data make the corresponding statistical estimates much worse.

Set 4. Following Set 3 results, an attempt to increase the time resolution of the meteorological input is done. For the purpose, EMAP was reconstructed so as to provide for time interpolation of meteorological data. The synoptic input met-data is provided at time intervals Dt_m. It is subjected to interpolation for the intermediate moments of time with discretization Dt_i, specified by the user. The interpolated values are then introduced in the ABL model which produces 3D fields with time resolution Dt_i. In Table 4, the statistical estimates and ranking are displayed, using the met-file **ECMWF_a**. Runs are made by decreasing interpolation time from its maximal value ($Dt_i = Dt_m$) to the minimal one ($Dt_i = Dt$). The best run is the one with minimal Dt_i. This result shows that the bad resolution can be overcome by simple linear interpolation for the intermediate moments between the synoptic terms. This fact is very important for BERS which gets its meteorological input from GTS with time resolution of 12 hours.

Set 5. The final ranking, shown in Table 5, is to check the quality of the different meteorological data sets decreasing the interpolation time to the minimal one $Dt_i = Dt$. The best run is the one with **ECMWF_a** input met-file. The second one is the **MetFr_a** run. Bracknell meteorological sets rank almost equally.

The quality of EMAP simulation is demonstrated on the best run from Set 5, Run 5.1. Figure 2 presents the overlapping of the measured and the calculated clouds and the respective FMSs at different moments. It can be seen that EMAP reproduces the evolution of the observed tracer spot pretty well but for some outrun of the calculated cloud after a long period of time.

In Table 6 the estimates from the time-dependent analysis at 11 specific sites, placed approximately along two arcs at different distance from the source, are shown. The first arc is closer to the source, located approximately along the East border of Belgium. The second arc extends from Denmark to Hungary. One can see from Table 6, that the model behaviour is generally better far from the source (second arc) than close to the source (first arc, with the exception of NL05). The worst NMSE predictions occur only at two stations B05 (18.0) and D44 (20.30), and a very good value of NMSE has been recorded at station DK02 (0.31). The worst bias prediction occurs for station D44 (0.645) and a very good one is for DK02 (0.069). All the stations have good values of the correlation

Table 6. Time dependent analysis at 11 selected sampler locations (Run 5.1)

TIME ANALYSIS	First arc of stations				Second arc of stations						
	NL05	B05	NL01	D44	DK05	DK02	D42	D05	CR03	PL03	H02
NMSE	1.82	18.00	5.44	20.30	2.27	0.310	1.19	2.87	5.37	5.13	3.81
BIAS	0.051	0.293	0.312	0.645	-0.115	0.069	-0.016	0.275	0.128	0.150	0.077
CORR	0.593	0.677	0.682	0.949	0.921	0.852	0.782	0.862	0.952	0.767	0.957
FMT	52	15	31	15	51	72	48	40	35	31	44
peak conc. (ng/m^3)											
predicted	1.84	2.58	2.64	4.53	0.85	1.31	1.63	2.22	1.50	1.70	1.09
measured	1.99	0.70	1.74	0.68	2.01	1.03	2.02	1.02	0.57	0.72	0.52
time of peak (h)											
predicted	30	27	30	24	48	45	42	42	39	42	45
measured	30	27	33	24	45	48	45	42	39	48	42
time of arrival (h)											
predicted	12	12	12	12	33	27	24	27	27	33	36
measured	18	15	18	15	36	39	30	30	30	39	36
duration (h)											
predicted	30	30	33	30	39	36	33	30	30	30	39
measured	33	21	24	21	51	51	42	27	45	51	51

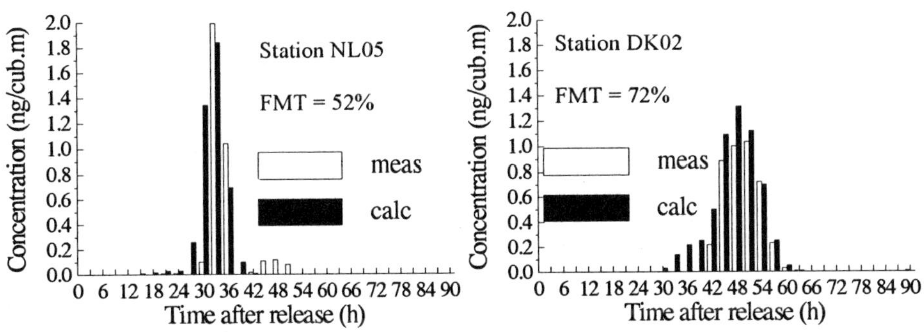

Figure 3. Time evolution of measured and EMAP calculated concentrations at stations NL05 and DK02. Run 4.1, met-file **ECMWF_a**, Dt_m = 6 h, Dt_i= 15 min.

coefficient, particularly good at station H02 (0.957).The minimum FMTs, equal to 15% are calculated at B05 and D44. For the stations of the second arc, FMT is from 31% to 72%, the latter achieved at DK02. Almost always the calculated peak concentration exceed the measured ones, the time of peak lying in the interval ±3 h. The arrival time differences are mainly in a 6 h interval, and the duration is generally underpredicted.

In Figure 3, the time evolution of the observed and simulated concentrations in two sampling stations representing both arcs are shown. It can be seen that the measured and calculated peak concentration are almost the same (pay attention that the y-axes is linear), the peak for NL05 being calculated at the right moment and for DK02 - 3 h earlier.

CONCLUSION

The main results from the experiments with the ETEX data base are:

- The LED puff model behaves quite well but some defects of its show that the model needs serious reconstruction or it should be replaced by some other model.

- The Eulerian model EMAP shows pretty good simulation ability but it can not describe adequately pollution distribution near the source. This defect is typical for all Eulerian models and measures for its improvement must be taken.
- The meteorological conditions at the release point are of great importance for the dispersion process.
- The quality of input meteorological data is of great importance for the model's performance. In case of bad resolution, a simple linear interpolation can greatly improve the model's performance.

ACKNOWLEDGEMENT

Bulgaria's inclusion in ETEX stimulated the development of the Bulgarian Emergency Response System. The EC COPERNICUS project CIPA-CT94-0110 "ETEX - second phase" allowed for EMAP's further development and validation. Special thanks are due to G.Graziani and S.Mosca from EC Joint Research Centre, Ispra, Italy, for the useful consultations.

REFERENCES

Bott A., 1989, A positive definite advection scheme obtained by nonlinear renormalization of the advective fluxes, *Mon.Wea.Rev.* **117**, 1006-1015.

Djolov G., Yordanov D., Syrakov D., 1987, Modelling the long range transport of air pollutants with atmospheric boundary layer chemistry, *Boundary Layer Meteorology* **41**, 407-416.

Graziani G., Gug W. and Mosca S., 1998, The first ETEX release.Experimental part, in: Real-time Long-Range Dispersion Model Evaluation of the ETEX First Release, 43-44, Office for Official Publications of the European Communities, Luxembourg, EUR 17754 EN

Gryning S.-E and Batchvarova E., 1997, Boundary layer parameters at the release site, in: *Proceedings of ETEX Symposium on Long-range Atmospheric Transport, Model. Verification and Emergency Response, 13-16 May 1967, Vienna (Austria)*, ed. K. Nodop, 71-74, Office for Official Publications of the European Communities, Luxembourg, ISBN 92-828-0669-3.

Klug W., Graziani G., Grippa G., Pierce D. and Tassone C., <Eds.> 1992, *Evaluation of Long Range Atmospheric Transport Models using Environmental Radioactivity Data from Chernobyl Accident, The ATMES Report*, 18-28, Elsevier Applied Science, London and N.Y.

Mosca S., Graziani G., Klug W., Bellasio R and Bianconi R., 1998, ATMES-II - Evaluation of Long-Range Dispersion Models Using 1st ETEX Release Data, EUR 17756/EN.

Syrakov D., 1995, On a PC-oriented Eulerian multi-level model for long-term calculations of the regional sulphur deposition, in: *Air Pollution Modelling and its Application XI,* eds. S.E. Gryning and F.A. Schiermeier, Volume 21, 645-646, Plenum Press, N.Y. and London.

Syrakov D., 1997, On the TRAP advection scheme - description, tests and applications, in: *Regional Modelling of Air Pollution in Europe. Proceedings of the first REMAPE Workshop, Copenhagen, Denmark, September 1996*, eds. G. Geernaert, A. Walloe-Hansen and Z. Zlatev, 141-152, National Environmental Research Institute, Denmark.

Syrakov D. and Prodanova M., 1997a, Bulgarian long-range transport models - simulation of ETEX first release, in: *Proceedings of ETEX Symposium on Long-range Atmospheric Transport, Model. Verification and Emergency Response, 13-16 May 1967, Vienna (Austria)*, ed. K. Nodop, 141-144, Office for Official Publications of the European Communities, Luxembourg, ISBN 92-828-0669-3.

Syrakov D. and Prodanova M., 1997b, 1997, Testing LED and EMAP dispersion models against ETEX release one data set, in: *Proceedings of the Swedish-Bulgarian Workshop on Long-Range Air Pollution: from Models to Policies*, 19-23 Oct., 1997, Sozopol

Syrakov D. and Yordanov D., 1997, Parameterization of SL diffusion processes accounting for surface source action, in: *Proceedings of 21st NATO/CCMS International Technical Meeting on Air Pollution Modelling and its Application, 2-6 June 1997, Clermont-Ferrand, France,*. 111-118.

Yordanov D., Syrakov D., Djolov G., 1983, A barotropic planetary boundary layer, *Boundary Layer Meteorology* **25**, 363-373.

DISCUSSION

B. FISHER: What were the limitations of the puff model LED which let to its replacement by the Eulerian EMAP model?

M. PRODANOVA: LED model presents a pretty good simulation of the ETEX release but the performance of the EMAP model is better. This is proved by the ATMES-II ranking of models. The shortcomings of LED simulations are seen better for long times - the puff structure of the simulated spot is well manifested.

T. ODMAN: The diffusion of the TRAP scheme may be small, but every Eulerian model will lead to excessive artificial diffusion near source. Could the poor performance near the source be due to numerical diffusion? What are some other possibilities for this behavior (poor performance near the source) of your Eulerian model (EMAP)? Why does the performance get better further downwind?

M. PRODANOVA: In principle, the Eulerian models have poor presentation of small scale features, especially a point source. This is due to the discretization, not to numerical diffusion of the advection scheme. Further, the numerical and modeled diffusion, simulate together the real diffusion and results can improve. It is difficult to say what is the simulation quality of all models near the source during ETEX because measurements are missing (the sampling network is rather coarse). The situation with poor presentation of point sources can be overcome by using the so called "plum in cell" approach.

EVALUATION OF A SIMPLE AIR MASS TRAJECTORIES MODEL AGAINST ETEX DATA BASE

François Bompay, Béatrice Audard, Anne-Claire Fontan, Sophie Voirin, and Alain Pinault

Meteo France
SCEM/SERV/ENV
42 avenue G. Coriolis
31057 TOULOUSE Cedex, France

INTRODUCTION

Meteo-France took part in the European Tracer experiment (ETEX[1]) in 1994 in order to evaluate its atmospheric transport and dispersion model using actual meteorological data facing a real event. The results obtained were very informative and showed that the model provided valuable information for decision makers. The simple air mass trajectories model, which is a component of the emergency response system, was not tested during the experiment. The purpose of this paper is to present an evaluation of that model against ETEX release one data base. The model is briefly described in a first part. The evaluation of forward trajectories is summarised in a second part. The last part is devoted to the evaluation of backward trajectories.

DESCRIPTION OF THE AIR MASS TRAJECTORIES MODEL

The code is a simple three dimensional trajectories computation system which solves the motion equation :

$$X = U\,T + X0$$

In this code the velocity is the wind vector directly provided by the operational numerical weather prediction model. The main problem is to estimate the wind components at specific location and time. In the code, simple linear interpolations are used to get the three components of the wind and then time evolution is computed using a leap-frog scheme. A complete description of the code can be found in Martin et al[2] (1987).

Three dimensional trajectories are computed using the analysed meteorological data from ECMWF (European Centre for Meteorological Weather Forecast). Distances from

Air Pollution Modeling and Its Application XIII, edited by S.-E. Gryning and E. Batchvarova, Kluwer Academic/Plenum Publishers, 2000

trajectories to ETEX sampling stations are calculated to estimate the quality of the trajectories model.

FORWARD TRAJECTORIES EVALUATION

The first release occurred in Monterfil, western part of France, from 23 October 1994 16 UTC to 24 October 1994 04 UTC. During the release period hourly trajectories starting from Monterfil at four different vertical levels (1000, 975, 950 and 925 hPa) are computed. Figure 1 shows the area covered by the trajectories 48 hours after the release in connection with the sampling stations where tracer is detected at the same period. The trajectories are close to sampling stations and the area covered by trajectories seems to match the observations.

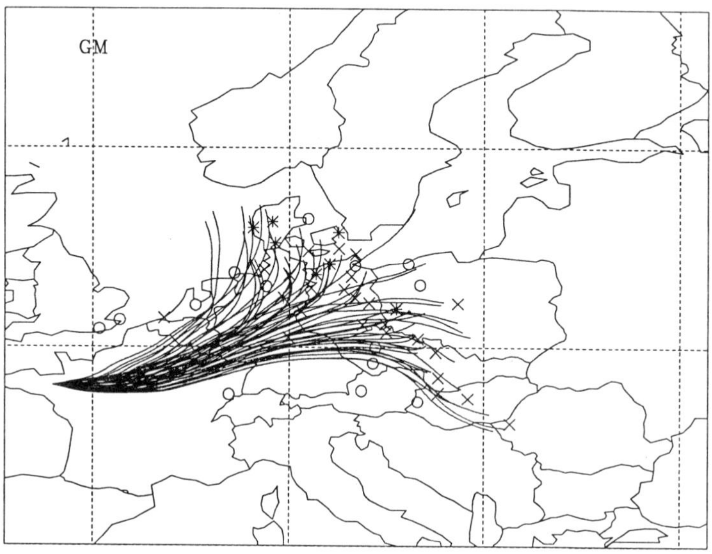

Figure 1. Forward trajectories at 48 hours and sampling stations detecting tracer 48 hours after the release.

Then distances from the 52 trajectories to sampling stations where the tracer is detected are calculated at different time step. In order to estimate the quality of the trajectories and to define when a sampling station is covered by a trajectory some parameters are computed. ADNS is the average distance between the 52 computed trajectories and the nearest sampling station where the tracer is detected. That parameter indicates an overall estimation of the quality of the trajectories model. ADMNS is the average minimum distance between the 13 hourly trajectories and the nearest sampling stations. Assuming that the model is good if one of the 4 vertical trajectories is within a required distance to the pollutant, that parameter indicates the absolute quality of the model. DSM is the distance between the nearest sampling station to trajectories and Monterfil. The ratio ADNS/DSM indicates the error in percentage between the trajectories and the detected tracer. A station is considered covered by a trajectory if ADMNS is less than 40 kilometres or if ADNS/DSM is less or equal to 5 percent. RSC is the ratio between the number of stations covered by a trajectory and the number of stations detecting the tracer.

At any time from 12 to 48 hours at least one of the 52 trajectories is within 30 kilometres of a sampling station where the pollutant is detected. The average distance is less than 60 kilometres. The average error is less or equal to 5 percent. The conclusion is that up to 48 hours the information provided by the trajectories is reliable.

From 54 to 72 hours the results deteriorate rapidly with an average distance equal to 291 kilometres at 72 hours. The model roughly describes the area where the pollutant is suspected to be but the information is not accurate enough. Looking further and comparing distances at time step plus or minus 6 hours the results are better. ADMNS is then 45 kilometres and ADNS 80 kilometres at 60 hours.

Table 1. Evolution of different parameters according to time step

Time step	ADNS (km)	ADMNS	DSM (km)	ADNS/DSM (%)	RSC
12	14	9	280	5	2/8
18	26	19	405	6	4/14
24	43	29	663	6	8/24
30	50	30	877	4	11/31
36	51	26	1058	4	15/44
42	59	25	1195	5	22/43
48	67	28	1311	5	19/45
54	94	41	1421	6	18/43
60	127	56	1538	7	9/43
66	194	77	1648	11	5/29
72	291	120	1754	16	2/24

The value of RSC indicates that trajectories model can only explain a part of the pollution. From 25 to 50 percent of the stations measuring the tracer up to 54 hours are considered covered by the trajectories. One explanation is that the model only computes the advection of pollutant, the diffusion process is not taken into account by this simple model. Another explanation is that trajectories model and measurements are not synchronous.

Regardless of timing 75 percent of the 112 stations detecting the tracer from the emission up to 72 hours are within 40 kilometres of one of the 52 trajectories at one moment of the period. Considering the stations which detected more than 0.5 ng/m^3 the rate reaches 90 percent. There is a shift in time between the trajectories and the measurements but the area covered by the trajectories is reliable. That result is very informative as regard the usefulness of trajectories computation in case of emergency response.

An element of explanation of this shift in time can be found when considering the vertical distribution of the trajectories. As shown in figure two the four initial levels are clearly identified. The trajectories rapidly covered a layer from ground level up to 800 hPa. At 48 hours the major part of the trajectories is in altitude. Measurements are made at ground level where the wind is generally lower than in altitude. That can explain why the quality of the results deteriorates from 48 to 72 hours, when considering location of the air parcels and measurement simultaneously. The model overestimates the displacement of the air parcel due to an overestimation of the wind velocity.

The vertical distribution of trajectories should be kept in mind during the evaluation of the backward trajectories, as well as on the use of trajectories for backtracking purposes, trying to find the release area connected to a pollution peak detected by a network of stations.

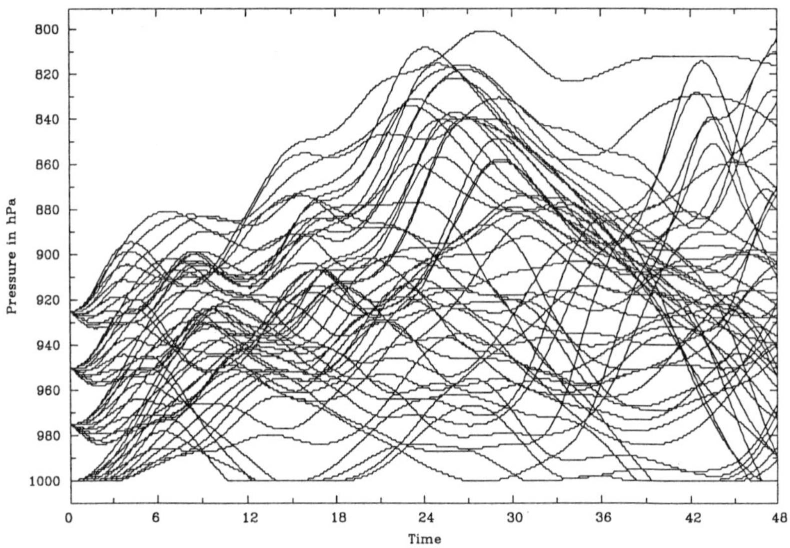

Figure 2. Vertical cross section of the trajectories at 48 hours.

BACKWARD TRAJECTORIES EVALUATION

One advantage of the trajectories model is that it can be operated backward, using the meteorological data from present time to the past. In order to evaluate the performance of backward trajectories, a selection is first made between all the stations which detected the tracer. Then backward trajectories at four vertical levels are computed starting at the moment when the tracer concentration is more than 0.5 ng/m^3.

Figure three shows backward trajectories starting 48 hours after the release from the 14 selected stations. Major part of the trajectories is located near Monterfil during the release period. The result is the same for all selected stations from 12 up to 48 hours. Generally the best trajectories start at 950 hPa if measurement is made during daytime and at 1000 hPa if measurement is done during the night. This can be explained by the diurnal variation of the boundary layer depth and suggests that the boundary layer depth is to be estimated at stations detecting a pollution before computing backward trajectories for backtracking purposes.

Distance from trajectories to the release site (DTM) are calculated and the main results are reported in table 2. Ratio of accordance (RA) is defined as the ratio between DTM and distance of sampling station to Monterfil. The trajectory is considered reliable if the value of RA is less or equal to 5 percent. That criterion of accordance (CA) is translated in kilometres in table 2.

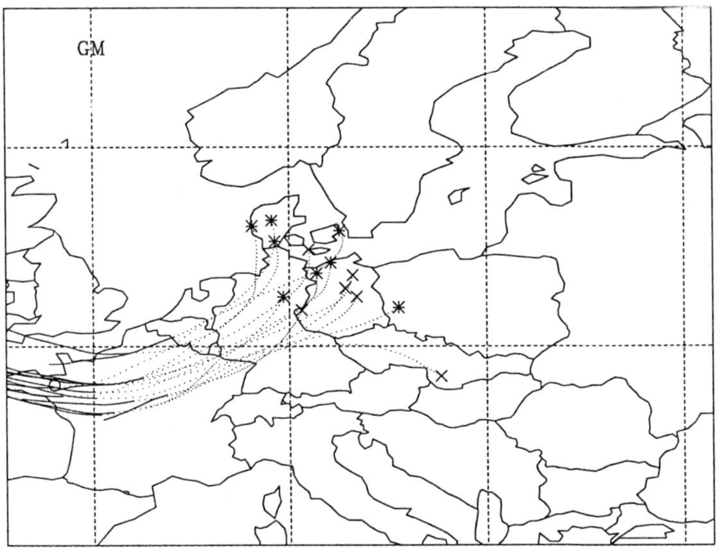

Figure 3. Backward trajectories staring from selected stations 48 hours after the release. The release period is in solid line.

Table 2. Evolution of different parameters according to time step

Time step	DTM (km) min	DTM (km) max	DTM (km) average	CA (km)	Percentage of accordance
12	0	53	19	20	75
24	4	96	33	30	66
36	5	383	94	45	66
48	12	230	75	60	57
60	160	462	289	75	0

Up to 36 hours most of backward trajectories are near the release site during the release period. At 48 hours 57 percent of backward trajectories fit the criterion of accordance. After 48 hours none of the trajectories match the criterion during the release period. Same comments as for forward trajectories can be outlined, there is a shift in time after 48 hours. For example at 60 hours backward trajectories are located within 75 kilometres from Monterfil but delayed by 12 to 18 hours.

Figure four shows the cumulated percentage of trajectories starting of selected stations at any time step from 12 to 72 hours which approach the release site by specific distance. It reveals that 60 percent of the trajectories are within 60 kilometres of the release site and that almost 30 percent are located at more than 100 kilometres.

So information provided by backward trajectories is reliable and the model can be useful for backtracking. It delivers information about area where the emission source is probably located at different time step and should be complemented by direct simulations with atmospheric dispersion model to confirm different assumed scenarii.

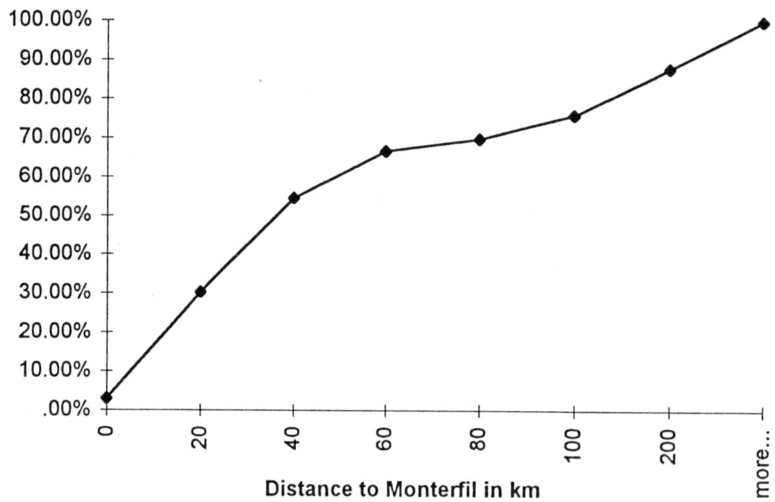

Figure 4. Cumulated percentage of stations measuring more than 0.5 ng/m3 according to distance to Monterfil.

CONCLUSION AND PERSPECTIVE

The evaluation of forward and backward trajectories against ETEX release one data base can not raise definitive conclusion. Nevertheless interesting information can be retrieved from that experiment. The simple air mass trajectories model performed well up to 48 hours. The area covered by forward trajectories is reliable and error in location is about 5 percent. Regardless of synchronization 75 percent of the sampling stations which detected the tracer are within 40 kilometres of the trajectories. As for backtracking purposes the model provides accurate information up to 48 hours. Most of backward trajectories starting from stations detecting more than 0.5 ng/m^3 are located near Monterfil during the release period.

There is a shift in time after 48 hours which is probably due to the elevation reached by the three dimensional trajectories during the computation. In the meteorological situation of ETEX release one, both forward and backward trajectories after 48 hours provide only a roughly estimation of the location of the tracer, even if the meteorological data used are analysis.

The vertical distribution of trajectories is important even for a release at ground level, indicating that an estimation of the boundary layer depth is to be made before computing trajectories. That should be kept in mind in case of emergency response.

The evaluation of the model against ETEX release two is not yet achieved but will be done soon. Some of the above statements could then be confirmed.

REFERENCES

1. K. Nodop. *ETEX Symposium On Long-Range Atmospheric Transport, Model Verification And Emergency Response - Proceedings,* Office for Official Publications of the European Communities, Luxembourg (1997).
2. D. Martin, C. Mithieux and B. Strauss, On the use of the synoptic vertical wind component in a transport trajectory model, *Atmospheric Environment* 21:45-52 (1987)

DISCUSSION

G. SCHAYES: The trajectory determination is very sensitive to the time step. What is the time-step used?

F. BOMPAY: We use a time step of 5 minutes in the computation of trajectories. For the application to ETEX we used analysed meteorological data from ECMWF every 6 hours and then linear interpolation.

ON THE ISSUE OF QUALITY CONTROL IN DATA ASSIMILATION

Lennart Robertson and Joakim Langner

Swedish Meteorological and Hydrological Institute (SMHI)
SE-601 76 Norrkoping
Sweden

ABSTRACT

Attempts to use ETEX-I data to reconstruct the source term of the experiment has been shown to be rather successful by means of variational data assimilation using the Eulerian transport model MATCH (Robertson and Langner, 1998). We have taken one step further and investigated the possibilities of quality control of data within the concept of variational data assimilation. With the adopted strategy it is found that the quality control mainly skip zero observed values, that does not agree with the model. To some extent is this justified by the representative error in the measurements in relation to the scale over which the model is valid. So far we have not gained anything but experience from this quality control approach. The quality control does not improve the model performance which never the less encourage us to further investigate the use of quality control.

INTRODUCTION

Quality control (QC) of observations is a crucial part of any data assimilation procedure. In earlier attempts by the authors to apply variational data assimilation in the Eulerian transport model MATCH using ETEX-I data (Nodop *et al.*, 1998) quality control procedures was not included (Robertson and Langner, 1998). However, we may learn from the numerical weather forecast community where large efforts have been devoted to quality control in prediction and analysis systems (see e.g. Lorenc, 1986). Variational data assimilation systems are running in a growing number of weather centers (see e.g. Courtier *et al.*, 1993), which has prompted for QC incorporated in the variational approach. Dharssi *et al.* (1992), Ingleby and Lorenc (1993) and Andersson *et al.* (1998) among others have described quality control schemes incorporated in the variational data assimilation approach.

We distinguish between two types of observational errors, measurement errors and gross errors. There are several measurement errors that have to take into account such as instrumental errors, analysis uncertainty, representativity errors (in relation to the scales the model could resolve). The gross errors represent more severe errors that would make the observation contain less useful information such as errors in registration of chemical analyses results, contamination of analysis samples, transfer errors in the telecommunication system. Our objective with QC is to identify and exclude observations with gross errors.

In Robertson and Langner (1998) the model adjustment to observations was made by adjusting the source term, and the model fit was measured by the square difference between model and measurements, i.e. a least square fit to the observations was adopted. Neither model or observation errors were considered. A quality control procedure obviously lead to that observation and model errors have to be included.

METHODOLOGY

In contrast to previous attempts by the authors on variational data assimilation (Robertson and Langner, 1998), where a more simple least square fit was adopted a more comprehensive statistical theory has to be implemented in order to include QC. Following the ideas of e.g. Dharssi *et al.* (1992), Ingleby and Lorenc (1993) and Andersson *et al.* (1998) the Bayesian probability theory is basis for QC in variational data assimilation. Let us assume Q to be the sought source term vector, and $Y=(y_1, y_2, \ldots, y_N)$ is a set of observations in a space and time volume. The Bayesian theory may briefly be phrased as the conditional probability that a model solution is correct, given a set of observations ($P(Q|Y)$), is proportional to the joint probability of the model result to be correct by itself ($P(Q)$) and the conditional probability that the observations are correct given the model simulation ($P(Y|Q)$),

$$P(Q \mid Y) \propto P(Q)P(Y \mid Q) = P(Q)\prod_i^N P(y_i \mid Q) \qquad (1)$$

where we have assumed the observation errors to be uncorrelated. The optimum source term vector is the one that maximize the above probability. The maximum probability may equally be found by minimizing the logarithm of the probability, where the latter in variational analysis is denoted as cost function (Lorenc, 1986),

$$J(Q) = -\ln P(Q \mid Y) + c \qquad (2)$$

where c is an arbitrary constant. We may assume the error in the source term estimate to be Gaussian distributed about some guess or background value Q_B,

$$P(Q) = p(Q)dQ \propto \exp\left(\frac{1}{2}(Q - Q_B)^T B^{-1}(Q - Q_B)\right)dQ \qquad (3)$$

where B is the background source error covariance matrix. The conditional probability that the observations are correct may take various forms dependent of the observation error structures. Let us first assume the observation errors to be Gaussian distributed,

$$P(Y \mid Q) = p(Y \mid Q)dY = \prod_i^N p(y_i \mid Q)dy_i \propto \prod_i^N \exp\left(-\tfrac{1}{2}z_i^2\right)dy_i \qquad (4a)$$

where

$$z_i = \frac{y_i - K_i(Q)}{\sigma_i} \tag{4b}$$

is the normalized observation increment, σ_i is the observation error standard deviation, and $K_i(Q)$ is the observation operator that include a model integration to the time-point of the observation. The cost function for normal distributed model and observation errors is then given by,

$$J_N(Q) = -\ln P(Q \mid Y) + c = \frac{1}{2}(Q - Q_B)^T B^{-1}(Q - Q_B) + \sum_i^N \frac{1}{2} z_i^2$$
$$J_N(Q) = J_N^B + J_N^O \tag{5}$$

If good background information is not at hand the background term, J^B, has to be assigned a large error and is thus small in comparison with the observation error term, J^O. For the ETEX experiment we are assuming that the source term is not a priori known and it may therefor be justified at this stage to omit the background term. The cost function in (5) is then very similar to the square difference term earlier used by the authors (Robertson and Langner, 1998). The background term would as well vanish if we assume the model to be perfect $P(Q)$=1 which is a contradiction to the argument above. Omitting the background term implies that the conditional probability that the source term is the correct one, given the observations, is just proportional to the conditional probability that the observations are correct, and the model uncertainty is then not sufficiently taken into account. It is a matter of further investigation to include the background term and evaluate the impact on the data assimilation aiming to reconstruct a source term.

Several algorithms may be utilized for minimizing the cost function. We apply the so called quasi-Newton conjugate gradient method with successive updates of the source term,

$$Q^{n-1} = Q^n - (J''(Q^n))^{-1} \nabla J(Q^n) \tag{6}$$

where the second derivative of the cost function is determined in an approximate fashion (Gilbert and Lemaréchal , 1989). The gradient of the cost function with respect to the source term ($\nabla J(Q)$) implies a more or less complex model transfer function from the observation to the source term both in time and space which is derived by means of so called adjoint model equations (see e.g. Robertson and Persson, 1993). We will restrict the presentation to just the gradient of the cost function with respect to the observation increment, keeping in mind that this is only the first step for updating the source term,

$$\nabla_z J_N^O = z_i \tag{7}$$

and thus the gradient is simply the normalized observation increment for a Gaussian distributed observation error probability function. Non-Gaussian description of the observation errors are presented by Dharssi *et al.* (1992), Ingleby and Lorenc (1993) and Andersson (1996) where the observation error probability function is composed of different functions for gross errors and measure errors, respectively, combined by the a priory probability for gross errors and no gross errors to occur,

$$p(y_i \mid Q) = p(y_i \cap G \mid Q)P(G) + p(y_i \cap \overline{G} \mid Q)P(\overline{G}) \tag{8}$$

where G is the event of gross error, $\overline{G}$ the event of no gross errors, and $P(G)$ is the a priori probability for gross errors to occur. Note that the complementary probability

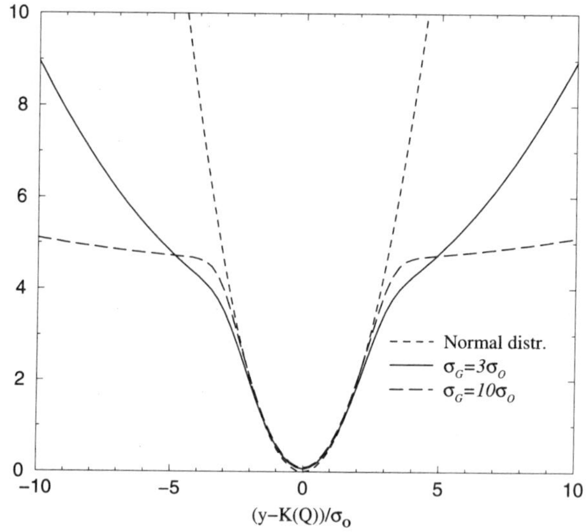

Figure 1. The contribution to the cost function from one single observation as a function of the normalized observation increment. Three different expressions of the cost function are shown, normal distributed observation errors (dashed line), expression including gross error distribution with two different gross error standard deviations, $\sigma_G=3\sigma_O$ (full line) and $\sigma_G=10\sigma_O$ (long lashed line), respectively. In both the latter cases the a priori probability of gross errors is set to 10%.

$P(\overline{G}) = 1 - P(G)$. Let us assume normal probability distributions of gross and measure errors,

$$p(y_i \mid Q) = \frac{P(G)}{\sqrt{2\Pi}\sigma_G}\exp\left(-\frac{1}{2}(z_i\beta)^2\right) + \frac{1-P(G)}{\sqrt{2\Pi}\sigma_i}\exp\left(-\frac{1}{2}(z_i)^2\right) \tag{9}$$

where $\beta=\sigma_i/\sigma_G$, and σ_G is the gross error standard deviation. With some manipulations the cost function is given by,

$$J^O(Q) = -\ln P(Y \mid Q) + c = \prod_i^N \frac{1}{2}z_i^2 - \ln\left[\frac{P(G)}{\sigma_G}\exp\left\{\frac{1}{2}z_i^2(1-\beta^2)\right\} - \frac{1-P(G)}{\sigma_i}\right] \tag{10}$$

and its gradient with respect to z_i is

$$\nabla_z J^O = z_i\left(1 - \frac{P(G)(1-\beta^2)}{P(G) + (1-P(G))\beta^{-1}\exp\left(-\frac{1}{2}z_i^2(1-\beta^2)\right)}\right) = z_i\Lambda_i \tag{11}$$

In Figure 1 the contribution to the cost function from a single observation is shown for normal distributed observation errors together with two different distributions accounting for gross errors, with σ_G set to 3 and 10 times the observation standard deviation, respectively. In the latter cases the a priori probability of gross errors is set to 10% ($P(G)$=0.1). It is clear from Figure 1 that the influence from observations far from the model solution is significantly decreased when including probability for gross errors in comparison with the normal distributed observation error, which is a desirable feature if we believe that such observations may be erroneous.

The term Λ_i in Eq, (11) which modifies the gradient in Eq. (7) can be shown to represent the *a-posterior* likelihood that the observation is correct if we assume the source term to bee correct (see Ingleby and Lorenc, 1993, *section 2g*). Λ_i mainly constitute the

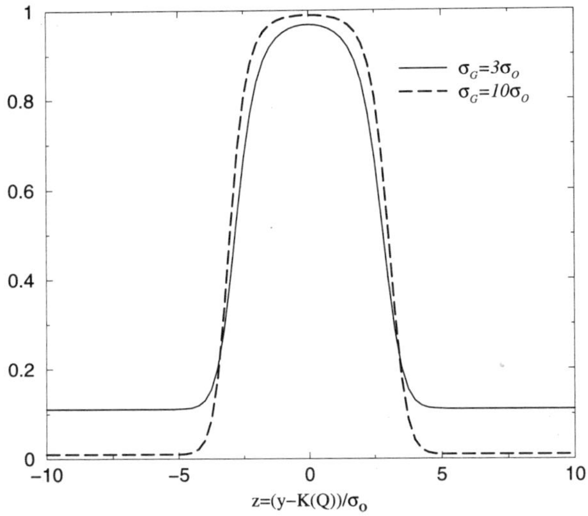

Figure 2. The probability of a correct observation (Λ_i in Eq. 11) as a function of the normalized observation increment, with two different gross error standard deviations, $\sigma_G=3\sigma_O$ (full line) and $\sigma_G=10\sigma_O$ (long lashed line), respectively. In both cases the a priori probability of gross errors is set to 10%.

quality control within variational data assimilation, see Figure 2. From the above description the quality control is governed by the departure from the model solution on an observation by observation basis. It is however important that the quality control as well should be influenced by other observations adjacent in space or time. This is achieved by the iterative nature of solving Eq. (6) for which $P(G)$ in Eq. (10) and (11) is set to zero during the first iterations. The observation increments (Eq. (4b)) thereafter will implicitly be the increments to the model solution given by other observations.

EXPERIMENTAL SETUP

The data assimilation procedure was in this exercise similar to Robertson and Langner (1998) in terms of the simulation period, source term vertical extension and timing, and data used from the ETEX-I tracer experiment. A 24 h simulation period was selected and the source term lasted 12 h within this period. Data beyond the assimilation period (up to 92 hours) were used as independent data in the statistical evaluation. The following procedure was adopted for quality control,

1. one first single iteration of Eq. (6) was performed setting the initial source term to zero and all the observation standard deviations equal to 1,
2. the gradient from the first iteration in item 1 was regarded as a first estimate of the source term, i.e. $Q^1=-\nabla J(0)$ (cf Eq. (6)),
3. one forward run with the source term in item 2 and checking for bias error,
4. the source term was corrected for bias error,
5. one forward model run with the bias corrected source term in item 4 was performed during which the quality control weights, Λ_i, were determined. The probability of gross

errors was set to 10% ($P(G)$=0.1), and the gross error standard deviation was set to 3 times the observation standard deviation (β=1/3), which in turn was set to twice the observed value ($\sigma_i = 2y_i$), see discussion below,

6. the quality control weights, Λ_i, were then kept constant during the remaining iterations (quasi-Newton conjugate gradient method (Gilbert and Lemaréchal , 1989)).

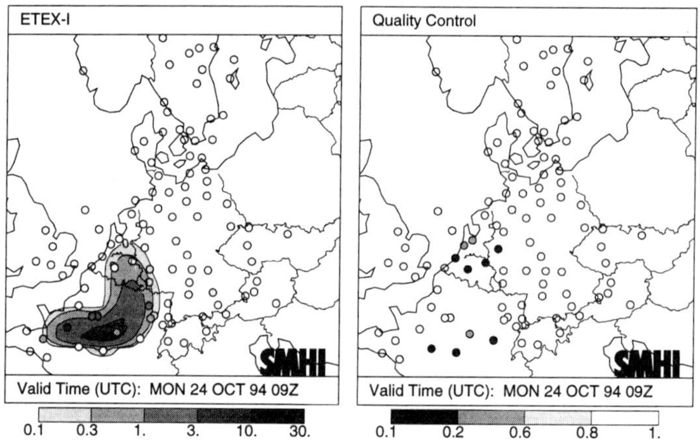

Figure 3. The MATCH model result, with the assimilated source term including QC, together with the measurements at one instant of the ETEX-I experiment (left, units ng m^{-3}), and quality flags determined by Λ_i in Eq. 11 (right, unit 0-1). In the left panel the observations are shaded with the same shading scale as the concentration field. Empty circles meaning zero observed values.

Several attempt were made to define the observation error standard deviation in a proper way. It should be stressed that the observations are ranging over several orders of magnitudes (0.01 – 20 ng/m^3) and a single global standard deviation did not appear to be a good choice. We came to the conclusion that the observation error should be a function of the observed value itself, i.e. the uncertainty in the observation increases with increased measured value. We rather arbitrary adopted that the observation error was twice the measured value with a lower limit at the detection level of 0.01 ng m^{-3}. This implies that however that small observed values will be imposed a tougher quality control than larger ones.

RESULTS

Figure 3 shows an example of the result of the quality control at one instant during the simulation period. The left panel shows the model simulation of the release with source term gained by quality control. The right panel shows the quality control weights given by Λ_i in Eq. 11 and demonstrated in Fig. 2. In general it is observed that zero measurements close to or in the middle of the simulated cloud were excluded from the variational analysis.

Figure 4 shows the temporal evolution of the true ETEX-I release, with and without QC, and the bias corrected source term (see item 1 to 3 in previous section). The initial source term is very much bell-shaped while the source term without QC has a tendency to a more sharp source term with a dip in the source intensity and a exaggerated second peak.

The source term with QC appears to be something in the middle, and subjectively looks more realistic than without QC.

Table 1 and 2 shows some statistics for dependent data during the assimilation period and for independent data beyond the assimilation period and up to the end of the ETEX-I tracer experiment. Observations that were rejected by the QC were still included in the statistics in Table 1, which may explain why the run without QC appears to have better statistical scores for the assimilation period. For independent data, Table 2, the variational

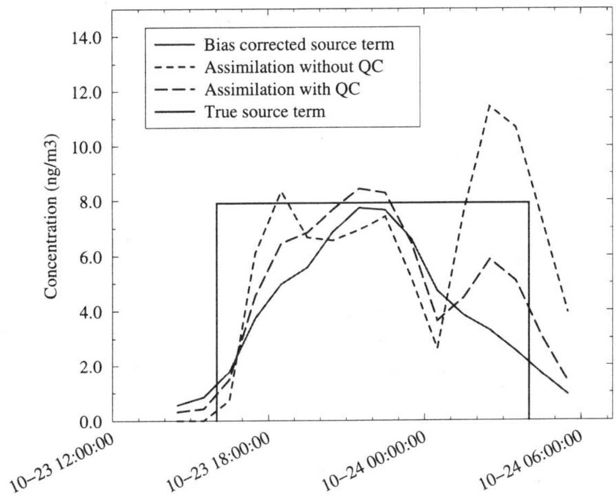

Figure 4. The temporal evolution of the ETEX-I source term for the bias correction of initial source guess (light solid line), variational assimilation without QC (dashed line) and with QC (long dashed line), together with the true temporal evolution of the tracer release (dark solid line).

assimilation with QC gives slightly smaller bias and RMS error, but the correlation is higher using the source term without QC. The full variational data assimilation, with or without QC, appears to improve the results over the initial only bias corrected source term, even though the signal is not that clear as the RMS error is smallest (with independent data) when only the bias corrected source term is used.

CONCLUSIONS

A procedure for quality control, incorporated in the variational data assimilation approach, has been tested. There is not a clear signal that the quality control really improves the model performance, but the temporal evolution of the source intensity appears subjectively to be more realistic than without quality control.

The main effect of the quality control is that zero measurements, inside or in the vicinity of the model tracer cloud, are excluded from the analysis. The exclusion of zero values may to some extent be justified by the representativity errors that such measurements represent in relation to the resolvable scale by the model (in this case ~50km). In this respect the model could not describe narrow plumes that are indicated by some of the zero measured values. The rejection of zero values is mainly a result of the observation error model adapted. As the observation error is set to twice the measured value (with a lower limit at the detection level), small measured values have a more narrow range within assumed correct (see Fig. 2).

In conclusion the model performance does not improve with the quality control scheme applied. The scheme is not complete in the sense that the so called background term (see Eq. (5)) is excluded, and thus the model errors are not properly taken into account. In further studies the background term should be included in the cost function, which in turn implies that spatial model error structures for the background covariance matrix have to be determined. Determining such error structures is by no means trivial for a regional transport and diffusion model.

Table 1. Statistical evaluation of the variational data assimilation, for the assimilation period (dependent data), from runs with and without quality control (QC) together with statistics for the bias corrected source term which was used as initial source term in the variational data assimilation. Bias and RMS are in units ng m^{-3}. The best scores are underlined.

	Bias corrected source term	Assimilation without QC	Assimilation with QC
Bias	-5.5e-2	0.125	<u>2.13e-2</u>
Correlation	0.805	<u>0.907</u>	0.864
RMS	0.969	<u>0.785</u>	0.826

Table 2. Statistical evaluation of the variational data assimilation, for data *beyond* the assimilation period (independent data), from runs with and without quality control (QC) together with statistics for the bias corrected source term which was used as initial source term in the variational data assimilation. Bias and RMS are in units ng m^{-3}. The best scores are underlined.

	Bias corrected source term	Assimilation without QC	Assimilation with QC
Bias	-5.5e-3	9.0e-2	<u>3.2e-2</u>
Correlation	0.337	<u>0.436</u>	0.371
RMS	<u>0.452</u>	0.495	0.463

ACKNOWLEDGMENT

This work was supported by the Swedish Radiation Protection Institute, and by the CEC Radiation Protection Research Programme under the contract No. F14P-CT95-0007.

REFERENCES

Andersson, E.. Implementation of variational quality control. Pp 633:643 in proceedings of ECMWF Workshop on 'Non-linear aspects of data assimilation', 9-11 September 1996 (Avaliable from ECMWF, Reading, UK).

Courtier, P., Derber, J., Errico, R. Louis, J-F. and T. Vukicevic. Important literature on the use of adjoint variational methods and the Kalman filter in meteorology. *Tellus* **45A**(5), 342:357 (1993).

Dharssi, I., Lorenc, A. C. and N. B. Ingleby. Treatment of gross errors using maximum probability theory. *Q. J. R.. Meteorol Soc.,* **118**, 1017:1036 (1992).

Ingleby, N,B. and A. C. Lorenc. Bayesian quality control using multivariate normal distributions. *Q. J. R. Meteorol. Soc.*, **119**, 1195:1225 (1993).

Lorenc, A. C.. Analysis methods for numerical weather prediction. *Q. J. R. Meteorol. Soc..* **112**, 1171:1194 (1996)

Nodop, K., Connolly, R. and F. Girardi. The field campain of the European Tracer Experiment (ETEX): Overview and results. *Atm. Environ.,* Vol. 32, No.24, 4095:4108 (1998).

Robertson, L. and C. Persson. Attempt to apply four dimensional data assimilation of radiological data using the adjoint technique. *Radiation Protection Dosimetri,* **50**(2-4), 333:337 (1993).

Robertson, L. and J. Langner. Source function estimate by means of variational data assimilation applied to ETEX-I tracer data. *Atm. Environ.,* Vol. 32, No.24, 4219:4225 (1998).

DISCUSSION

Z. ZLATEV: You mentioned already that the results of your procedure depend on the model that is used. However, they depend also on the set of measurements, which is applied. How does the uncertainty of the measurements influence the results?

L. ROBERTSON: There are different aspects around this question. The first one is the real observation error and its impact on the data assimilation, and the second one is how we model the observation uncertainty. For the first part we do not really have a good picture of the impact of observational errors, but it might be fruitful to utilize various fractions of the available ETEX data which then to some extent would respond to various errors in the data set. For the second part we have tried to use a global measure of the observational error but the QC behave very strange presumably because the measured values are with in a range of several orders of magnitude. The quality control in this case tended to reject observations that we had reasons to believe were correct. The model we came up with was to assign an observational error that was a function of each observed value. This on the other hand made zero measurements very sensitive for rejection.

G. RIVIN: Have you done attempts with other methods of QC?

L. ROBERTSON: No.

G. RIVIN: If you have not so good first guess you can receive local minima, not global. What do you do for a good estimation of the first guess?

L. ROBERTSON: The first guess was derived from the observations by making one iteration with a zero initial source term. We assume that this first guess would be close to the global minima. Still the temporal variation of the assimilated source, with or without QC, makes us suspicious that we might have hit a local minima.

D. ANFOSSI: Which is the influence of the model used for computing the dispersion on your results. Does a better model produce better results?

L. ROBERTSON: We have only applied the MATCH model so we do not have any experience of other model concepts. In general one would suspect that this could be case dependent. If a model do not perform well for a given case, the derived source term would be the one that would also correct for other model errors, like errors in the weather data just to take one example.

THE EFFECT OF UNRESOLVED MESOSCALE WIND VELOCITY FLUCTUATIONS ON DISPERSION MODEL RESULTS

Andreas Stohl

Lehrstuhl für Bioklimatologie und Immissionsforschung
Ludwig-Maximilians-Universität München
Am Hochanger 13, D-85354 Freising-Weihenstephan, Germany

INTRODUCTION

Large-scale dispersion models frequently use the output of numerical weather prediction models with rather coarse horizontal and temporal resolution (typically 100 km and 3 h). Processes acting on smaller or shorter scales cannot be resolved explicitly and must therefore be parameterized. Most important is the parameterization of turbulence in the boundary layer, which is usually based on observations of high-frequency wind fluctuations. It is often not explicitly stated (and hence unknown) for which scales the parameterizations available from the literature have been developed actually. However, most of them are valid only at time scales on the order of 10 minutes (and correspondingly short length scales). Wind fluctuations with periods longer than this but shorter than the time scale of the model (twice the time resolution of the wind fields) are contained neither in the turbulence parameterizations nor in the model-resolved data (Maryon, 1998).

It can be argued that these low-frequency wind fluctuations are of minor importance, because they fall into the energy gap in the turbulence spectrum (Stull, 1988) between the large-scale Rossby-waves and the diurnal cycle (resolved by the model), and the small-scale turbulence (parameterized in the model). Nevertheless, neglect of wind velocity variance on these intermediate (meso-) scales leads to a systematic underestimation of plume width in dispersion models. This was demonstrated by Gupta *et al.* (1997) who drove a particle dispersion model with output from a one-dimensional boundary layer model with and without nudging of radar profiler data. The nudging incorporated observed wind fluctuations with periods of 2 to 10 h into the boundary layer model that were otherwise missing, and this enhanced the widths of the plumes in the particle model substantially. In this paper it is studied how a large-scale particle dispersion model estimates plume growth in comparison to experimental data from several large scale tracer experiments. The effect of two parameterizations for low-frequency wind fluctuations on plume growth is investigated.

Air Pollution Modeling and Its Application XIII, edited by S.-E. Gryning
and E. Batchvarova, Kluwer Academic/Plenum Publishers, 2000

THE DISPERSION MODEL

The FLEXPART model (Stohl, 1996; Stohl *et al.,* 1998; Stohl and Thomson, 1998), a Langevin-equation (Thomson, 1987) type particle model based on data from the European Centre for Medium-Range Weather Forecasts (ECMWF), is employed in this study. The need to parameterize low-frequency wind fluctuations in order not to underestimate plume width was recognized in the FLEXPART model development from the beginning, but due to the general lack of studies on this subject, the method used hitherto is rather ad hoc and can be justified only *a posteriori* from comparisons with measurement data. It is assumed that the wind used to advect a particle is composed of three independent components: $\mathbf{v} = \overline{\mathbf{v}} + \mathbf{v}'_t + \mathbf{v}'_m$, where $\overline{\mathbf{v}}$ is the grid scale wind, $\mathbf{v}'_t$ the turbulent wind captured by a modified Hanna (1982) turbulence parameterization and $\mathbf{v}'_m$ are the low-frequency wind variations under study in this paper. To calculate $\mathbf{v}'_t$ and $\mathbf{v}'_m$, two independent Langevin equations are solved. This separation is valid if the processes acting on the different scales do not strongly interact with each other, which seems reasonable. For obtaining $\mathbf{v}'_m$, it is assumed that the variance observed at the grid scale provides some information on its subgrid variance; the wind velocity standard deviation used in the Langevin equation is set to half the standard deviation of the wind at the 16 grid points surrounding the particles' position in space and time. The corresponding Lagrangian decorrelation time scale used is half the interval of the meteorological input data.

Maryon (1998), furtheron referred to as M98, recently added low-frequency wind variations (meandering in his terms) in a similar way (i.e. using a second Langevin equation) to the U.K. NAME model. He developed a more physical parameterization of cross-wind velocity variance, based on a spectral analysis of 10 minute average wind registrations at a site in England, and derived the formulation

$$\sigma_{v,l}^2 = 2c_l u_{10} \Delta T_F \ , \tag{1}$$

where $\sigma_{v,l}$ is the cross-wind velocity variance, $c_l = 0.03$ ms^{-1}h^{-1} is a coefficient determined from the measurement data, u_{10} is the mean wind at 10 m taken from the weather prediction model and ΔT_F is the time resolution of the model wind fields in hours. Since the autocorrelation of the measurement data did not show an exponential fall-off to zero, the derivation of the corresponding time scale was not straightforward and M98 suggested several different ones which varied by almost an order of magnitude. Of these, the longest time scale (171 and 267 minutes for ΔT_F being 3 h and 6 h, respectively) seemed to have the most beneficial impact on dispersion model results (Maryon, 1998, personal communication). In M98, low-frequency variance of the along-wind and of the vertical wind component are neglected, while in my formulation (Stohl, 1996), furtheron referred to as S96, low-frequency fluctuations are allowed for all three wind components and also above the boundary layer.

In this paper, three different model configurations are used: (i) neglect of low-frequency wind variance (base case), (ii) the method of M98, and (iii) the original FLEXPART parameterization (S96) (iii). Concentrations in the lowest 100 m are calculated with a uniform kernel of $0.5\,^\circ\times 0.5\,^\circ$ bandwidth and are compared with surface measurement data. During the ATMES-II model intercomparison (Klug *et al.,* 1997), a parabolic kernel with bandwidths increasing with time was used within FLEXPART to account for the uncertainty in trajectory calculations. This method is not used here, since it smoothes the concentration distribution density function (DDF) and hence artificially creates plume growth.

MEASUREMENT DATA AND MODEL EVALUATION METHOD

To assess the performance of the different model configurations, measurement data from three large-scale perfluorocarbon tracer experiments, the Cross-Appalachian Tracer Experiment (CAPTEX), the Across North America Tracer Experiment (ANATEX) and the European Tracer Experiment (ETEX), are used. During CAPTEX, seven tracer releases (six of which are used here) were made and tracer samples were taken every six hours at distances of up to 1000 km from the release location and until approximately 48 h after the release. During ANATEX, 66 releases of two different perfluorocarbon tracers were made from two locations. Samples were collected at distances of up to 3000 km, but both the spatial and the temporal resolution (24 h) of the sampling network were rather coarse and it was often not possible to separate individual tracer plumes, since releases were made every 2.5 days. Although a special procedure, described in Stohl *et al.* (1998), is employed to separate the individual plumes, half of the releases must be excluded from this study and those remaining are not entirely free from possible contamination by other releases. The most recent experiment was ETEX. Only two releases were made, but the tracer plumes were recorded across Europe for 90 h by a dense sampling network and at a high temporal resolution (3 h). Operational ECMWF analyses every 6 h, supplemented by 3 h forecasts every other three hours are used to simulate the ETEX cases, and ECMWF re-analyses every 6 h are used for the CAPTEX and ANATEX cases. The spatial resolution is 1 ° in all simulations. 100000, 50000 and 20000 particles are released in the model simulations for ETEX, CAPTEX and ANATEX, respectively.

A sensible method to evaluate the model results with respect to the size of the plume is to compare the simulated concentration DDFs with the observed ones, since this is not sensitive to errors in the plume position. An underestimation of plume width results in more zero concentration values, less frequent medium concentrations and higher maximum concentrations than observed, while an overestimation of plume width yields just the opposite result. A difficulty with this approach is that a model bias, likely not caused by low-frequency wind variations, also affects the DDF. Therefore, the concentration values are normalized to their respective mean concentration. The resulting DDFs cannot be compared in terms of concentration levels, but only in terms of their spread around the mean concentration. Since most of the concentration values are zero, relatively few data are available to characterize the rest of the DDFs. Therefore, the data is binned into just three classes: (1) $C_i < 0.2$, (2) $0.2 \leq C_i \leq 5$, (3) $C_i > 5$, where C_i are the normalized concentrations. Data from the first 24 h after the start of a release are not used. All other data obtained during an experiment are jointly analyzed. No attempt is made to characterize the temporal evolution of plume growth.

RESULTS

Figure 1 shows the frequency of the concentrations in the three classes for the two ETEX cases. For both experiments, the highest frequency is in the low-concentration class, which indicates the large number of zero concentrations. All model versions clearly overpredict the frequency of zero concentrations. On the contrary, the frequency in the medium-concentration class is underpredicted by a factor of two and more. The frequency of the concentrations in the high-concentration class is again slightly overpredicted by the model. This signature clearly shows that all three model versions

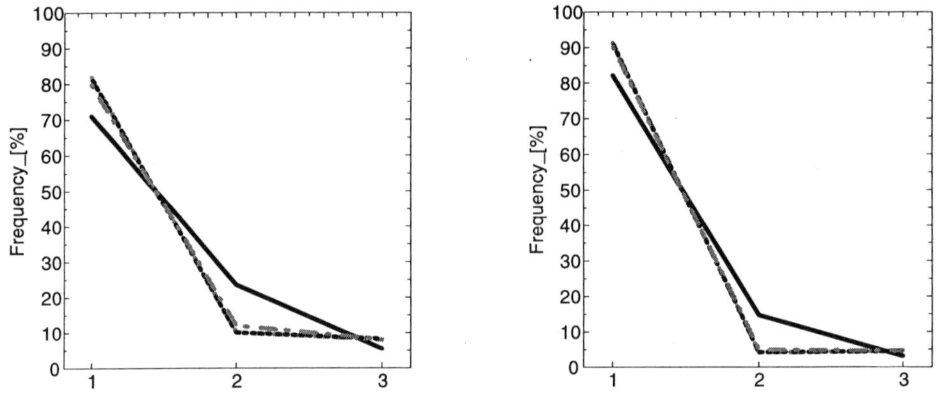

Figure 1. Frequency of data within the concentration classes (1), (2) and (3) for the first (left) and second (right) ETEX case. The black solid line shows the measurements, the grey dashed line the base case model results, the black dotted line the M98 simulation, and the gray dash-dotted line the S96 simulation.

underestimate the plume sizes during ETEX. Both the M98 and the S96 formulations of low-frequency wind variations make the model slightly more dispersive, but the improvement is rather small. This underestimation of the sizes of the plumes occurs even though the model results are smoothed by calculating cell-average concentrations.

More or less the same behavior is found for the CAPTEX cases (Figure 2). Again, all model versions are not dispersive enough on the average, but the improvement due to the addition of the low-frequency wind variance is quite large in some cases. For instance, for releases 1 and 3, the S96 simulation yields almost the correct frequency distribution. For release 7, both the M98 and the S96 simulations are too dispersive.

For the ANATEX cases (Figure 3), the discrepancy between the modeled and the measured DDFs is much larger than for the ETEX and CAPTEX cases. In the base case, the frequency of the medium concentrations is underestimated by a factor of six and four for the PDCH and the PTCH releases, respectively. The M98 parameterization improves this just a little; the S96 parameterization improves further but is also not sufficiently dispersive. The high frequency of the medium concentrations in the measurements is partly caused by contamination from previous or subsequent releases of the same tracer during ANATEX. To some part, however, this is also a result of the larger scale of this experiment. Data have been considered here up to five days after each release, so that at the end of this period the tracer is more dispersed than during CAPTEX and ETEX.

The underestimation of the frequency of the medium concentration class is the most striking feature of nearly all the model simulations presented here. Therefore, this information is repeated in Table 1. The M98 parameterization has a small positive effect on the frequency of the medium concentrations in nearly all cases, but increases dispersion not sufficiently. Although the S96 parameterization is more dispersive, its effect on the frequency of the medium concentrations is still too small. At least, both parameterizations improve the model results.

To determine whether the M98 and S96 parameterizations do not only have a positive effect on the DDFs, but also improve the model performance with respect to other statistical measures, global analyses of the concentrations paired in space and time are made. The Fractional Bias (FB), the Normalized Mean Square Error (NMSE), the Figure of Merit (FMS) (significance levels are 2 fl l^{-1} for CAPTEX and ANATEX and

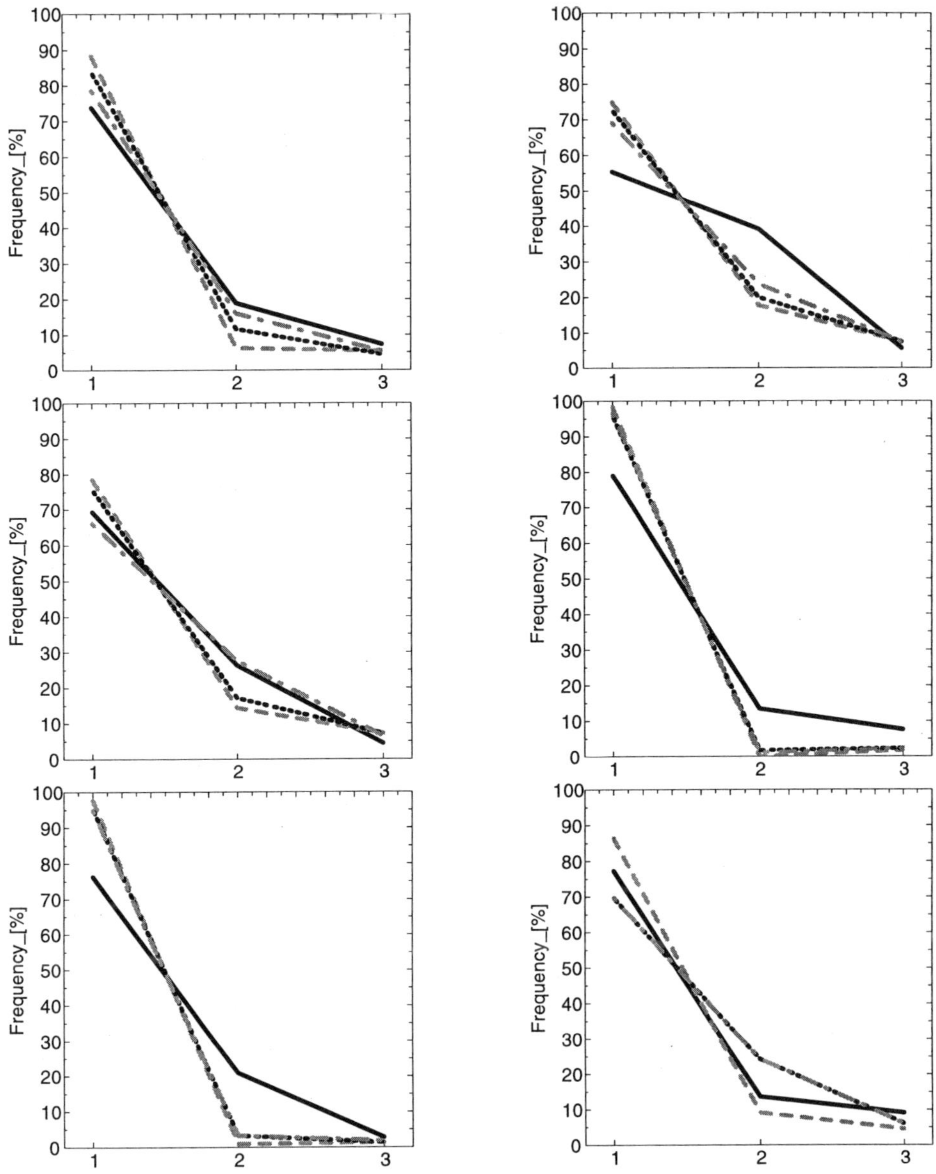

Figure 2. Same as Figure 1, but for the first (top left), second (top right), third (middle left), fourth (middle right), fifth (bottom left) and seventh (bottom right) CAPTEX case.

0.1 ng m^{-3} for ETEX), and the Pearson correlation coefficient (r) are calculated for all data available from later than 24 h after the start of the releases. Definitions of these measures can be found, for instance, in Stohl *et al.* (1998).

Tables 2, 3 and 4 show the results of this evaluation. They confirm that the introduction of low-frequency wind variance improves the results with respect to all parameters except FB which remains almost unaffected as is expected. Consistently with the previous results, the improvements are larger for the S96 method than for the M98 method. The largest improvements are found for FMS; for the seventh CAPTEX release, for instance, it is 56% for the S96 method over 26% for the base case.

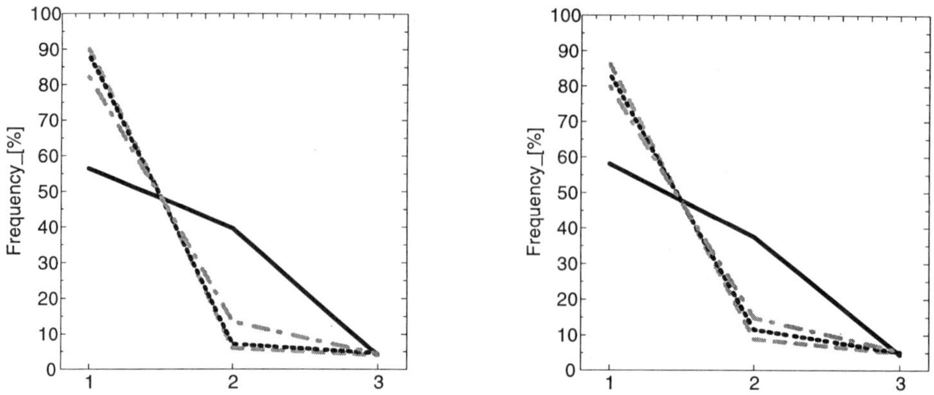

Figure 3. Same as Figure 1, but for all the PDCH (left) and all the PTCH (right) releases made during ANATEX.

DISCUSSION

The above evaluation has shown that the FLEXPART model severely underestimates the sizes of the tracer plumes. The introduction of two simple parameterizations for low-frequency wind fluctuations increased the plume spread and also clearly improved the agreement of the model results with the measurements with respect to other statistical measures. However, both methods failed to produce a plume growth as fast as observed in reality.

Underestimation of plume growth is not a problem specific to the FLEXPART model only. Similar behavior was found with the NAME model (Maryon, 1998, personal communication) and was also reported by several other authors for the ETEX cases (Nodop, 1997). Furthermore, the duration of the passage of the plume at specific receptor locations during the first ETEX case was almost always shorter for the model simulations than for the measurements (Mosca *et al.*, 1997). This supports the conclusion that the different models, Eulerian as well as Lagrangian, share the problem of too slow plume growth.

It seems unlikely that low-frequency wind fluctuations are so much stronger than expected. Therefore, additional processes, also not represented in the FLEXPART model, are likely to enhance the growth of the plumes. To exclude the possibility that an underestimation of small-scale turbulence is the cause for the difficulties, sensitivity runs with doubled standard deviations of v_t' were made. But the resulting increase in

Table 1. The frequency (%) of data in the medium concentration class according to the measurements and the model calculations for the different tracer experiments. N is the number of data points.

Release	N	Observations	Base case	M98	S96
ETEX1	2375	23.6	10.1	10.1	12.0
ETEX2	1659	14.7	4.2	4.3	4.9
CAPTEX1	206	18.9	6.3	11.6	16.0
CAPTEX2	271	39.1	17.7	19.9	23.6
CAPTEX3	251	26.3	14.3	17.1	27.5
CAPTEX4	171	13.4	0.0	1.8	1.2
CAPTEX5	215	20.9	0.9	3.3	3.3
CAPTEX7	66	13.6	9.1	24.2	24.2
All PDCH releases	2287	39.6	6.0	7.2	13.4
All PTCH releases	1828	37.5	8.9	11.6	14.8

Table 2. FB, NMSE, FMS and r obtained with the base case simulation

Release	N	FB	NMSE	FMS (%)	r
ETEX1	2375	-0.41	14.4	48	0.44
ETEX2	1659	1.22	132.2	0	-0.02
CAPTEX1	206	0.08	19.3	34	0.32
CAPTEX2	271	-0.04	6.7	50	0.51
CAPTEX3	251	0.33	25.7	42	0.23
CAPTEX4	171	-1.26	505.6	14	0.14
CAPTEX5	215	0.41	69.6	14	0.63
CAPTEX7	66	-0.76	39.0	26	0.22
All PDCH releases	2287	0.45	114.6	20	0.08
All PTCH releases	1828	0.32	45.2	26	0.61

Table 3. FB, NMSE, FMS (%) and r obtained with the M98 parameterization

Release	N	FB	NMSE	FMS	r
ETEX1	2375	-0.41	13.7	48	0.46
ETEX2	1659	1.22	127.5	0	-0.02
CAPTEX1	206	0.07	16.3	41	0.33
CAPTEX2	271	-0.03	5.4	53	0.53
CAPTEX3	251	0.33	25.2	41	0.25
CAPTEX4	171	-1.41	529.7	14	0.17
CAPTEX5	215	0.37	49.9	27	0.72
CAPTEX7	66	-0.80	20.8	54	0.32
All PDCH releases	2287	0.46	99.9	21	0.14
All PTCH releases	1828	0.27	33.8	28	0.62

Table 4. FB, NMSE, FMS (%) and r obtained with the S96 parameterization

Release	N	FB	NMSE	FMS	r
ETEX1	2375	-0.39	13.4	49	0.47
ETEX2	1659	1.22	119.7	0	-0.02
CAPTEX1	206	0.09	12.3	50	0.40
CAPTEX2	271	-0.03	4.1	56	0.57
CAPTEX3	251	0.36	23.6	34	0.28
CAPTEX4	171	-1.31	435.1	19	0.16
CAPTEX5	215	0.38	47.8	29	0.72
CAPTEX7	66	-0.79	19.3	56	0.36
All PDCH releases	2287	0.48	94.3	24	0.14
All PTCH releases	1828	0.31	33.2	29	0.64

plume width is less than half of that obtained with the M98 parameterization.

Visual qualitative comparisons of tracer concentration maps show that the observed tracer distributions are much more patchy than the modeled ones, which usually stay rather compact for quite a long time. The M98 and S96 parameterizations just widen these compact plumes. In reality, however, the plumes are often also split into several parts, separated by regions without tracer material. This splitting appears to happen regularly at the rear side of the tracer cloud, where small pockets of tracer material lag behind, giving the plume a ragged appearance. The small tracer pockets may subsequently experience fast initial growth, enhancing the dispersion of the tracer material. Consequently, the agreement between the measurements and the model simulations seems to be worse at the backside of a plume than at its front. The travel speed of the leading edge of the plume is mostly well predicted, while the trailing edge travels too fast in the model.

One important process that may explain this behavior is the captivation of tracer material below shallow ground-level inversions during the nighttime and its later release during the daytime. In depressions or in front of blocking (subgrid) topography, the wind speeds below the inversion may be close to zero. Under these conditions, the tracer material below the inversion may be completely separated from the material aloft which may explain the delayed tracer pockets found in the measurement data.

In this light, it is desirable to use high-resolution mesoscale model results that explicitly resolve many of the processes that are subgrid to the ECMWF and similar data to simulate plume dispersion. The fact that dispersion models based on mesoscale models did not perform better than dispersion models based on ECMWF data for the first ETEX case in the ATMES-II study, however, indicates that improved mesoscale *analyses*, possibly of still higher resolution, are needed to better understand and simulate the relevant processes.

Acknowledgments: I thank ECMWF for giving me access to their data and support (N. Kreitz) within the special project 'Validation of trajectory calculations' and Roy Maryon for helpful discussions. I am strongly indebted to all the people who organized CAPTEX, ANATEX and ETEX and to those who did the valuable measurements.

REFERENCES

Gupta, S., McNider, R.T., Trainer, M., Zamora, R., Knupp, K., and Singh, M.P., 1997, Nocturnal wind structure and plume growth rates due to inertial oscillations. *J. Appl. Meteor.* **36**, 1050-1063.

Hanna, S.R., 1982, Applications in air pollution modeling. In: Nieuwstadt F.T.M. and H. van Dop (ed.): *Atmospheric Turbulence and Air Pollution Modelling.* D. Reidel Publishing Company, Dordrecht, Holland.

Klug, W., Graziani, G., and Grippa, G., 1997, The European long range tracer experiment ETEX: Evaluation of the model intercomparison of the second release. In: Nodop, K. (editor): *ETEX Symposium on Long-Range Atmospheric Transport, Model Verification and Emergency Response*, European Commission EUR 17346, Brussels, 117-120.

Maryon, R.H., 1998, Determining cross-wind variance for low frequency wind meander. *Atmos. Environ.* **32**: 115-121.

Mosca, S., Graziani, G., Klug, W., Bellasio, R. and Bianconi, R. (1997) ATMES-II – Evaluation of Long-Range Dispersion Models Using 1st ETEX release data. Volume I. Draft report.

Nodop, K., ed., 1997, *ETEX Symposium on Long-Range Atmospheric Transport, Model Verification and Emergency Response*, European Commission EUR 17346.

Stohl, A., 1996, The FLEXPART Particle Diffusion Model – Scientific Description and Technical Documentation. Report to the Central Institute of Meteorology and Geodynamics, Hohe Warte 38, 1190 Wien, Austria. Available as an updated user guide via the internet from http://www.forst.uni-muenchen.de/LST/METEOR/stohl/flexpart.html.

Stohl, A., Hittenberger, M., and Wotawa, G., 1998, Validation of the Lagrangian particle dispersion model FLEXPART against large scale tracer experiments. *Atmos. Environ.* In press.

Stohl, A. and Thomson, D.J., 1998, A density correction for Lagrangian particle dispersion models. Submitted to *Boundary-Layer Meteorol.*

Stull, R.B., 1988, *An Introduction to Boundary Layer Meteorology*, Kluwer Academic Publishers, Dordrecht.

Thomson, D.J., 1987, Criteria for the selection of stochastic models of particle trajectories in turbulent flows. *J. Fluid Mech.* **180**, 529-556.

DISCUSSION

E. GENIKHOVICH: Have you analyzed meteorological conditions corresponding to high concentrations behind your diffusion front? Could they be due to convective bursts, local circulations and so on?

A. STOHL: In principle, convective bursts or local recirculation could be an explanation for localized high concentrations behind the main front. However, at least for the ETEX cases, this is not a likely explanation, since there was no large-scale recirculation at higher levels. With these processes one would also expect that sometimes tracer material appears in front of the main cloud, but the leading edge of the tracer cloud was usually well predicted by the model.

D. SYRAKOV: We try to introduce parameterization for the events lying in the meso-meteorological minimum of the energy spectrum. How do you do this, keeping in mind that this minimum is due to structures with high intensity but with small scale, which are rare and short-lasting?

A. STOHL: Although there is a minimum in the energy spectrum just in that range that is not grid-resolved and also not covered by the turbulence parameterization, the study of Maryon (1998) showed that there is still sufficient energy to enhance plume growth. Also it was a posteriori shown that the model results improved with the parameterization. The examples I have presented, however, have also shown that this paramterization is not sufficient. The observed tracer patches behind the main cloud are due just to the rare but intense mesoscale effects you mentioned.

M. KAASIK: Maybe the disappearance of the plume during the second ETEX experiment (some tracer remained near mountains) was due to passing over mountains (Alps), effects not included in your model?

A. STOHL: You are perfectly right. The ETEX-2 release was problematic also for other reasons (much of the tracer escaped into the free troposphere with the warm conveyor belt ahead of a cold front), but this specific tracer patch in the Alps was probably captured below inversions in complex terrain.

RTMOD: A WEB-BASED REAL-TIME STATISTICAL EVALUATION PACKAGE FOR LONG-RANGE DISPERSION MODELS

Roberto Bellasio[1], Roberto Bianconi[1], Stefano Galmarini[2], Giovanni Graziani[2], and Sonia Mosca[1]

[1]Enviroware, C.D.Colleoni - Andromeda 1, 20041 Agrate Brianza (MI) Italy
[2]EC Joint Research Centre Ispra, 21020 Ispra (VA), Italy

INTRODUCTION

RTMOD is an automated statistical evaluation package for the inter-comparison of the predictions of long-range atmospheric dispersion models. Its background is the five-year project ETEX that involved about 50 models run by several institutes around the world to simulate two real tracer releases over a large portion of the European continent.

In the preliminary phase of ETEX, three dry runs (i.e. simulations in real-time of fictitious releases) were carried out, and JRC-Ispra performed the statistical evaluation of model predictions. At that time, the World Wide Web was not so easily available at any location in the world. Thus, predicted data were sent to JRC-Ispra by fax and regular mail, and subsequently processed. Later, the growth of the World Wide Web and the experience gained during ETEX suggested the development of a tool including a friendly interface for data submission, and program modules running on a server, producing and displaying through any browser the numerical and graphical results of the statistical analysis.

This suggestion was also prompted by the main conclusions and recommendations of ETEX (Nodop, 1997), stressing the importance to maintain and improve the ETEX network of national institutes and international organizations, in order to continue model development and to demonstrate the technical capability to support emergency management in real cases. For this reason, it was decided by JRC-Ispra to organize and co-ordinate a concerted action to develop and manage a tool that could meet these recommendations.

The following paragraphs describe the RTMOD system and present its first applications. The system can be accessed through the web, at the address http://www.ei.jrc.it/rtmod.

Air Pollution Modeling and Its Application XIII, edited by S.-E. Gryning and E. Batchvarova, Kluwer Academic/Plenum Publishers, 2000

THE RTMOD SYSTEM

RTMOD is structured as a website interfacing the modeler on the client side, and the statistical analysis system on the server, at JRC Ispra. The modelers participating to the RTMOD exercises are informed by fax when a dry run (a fictitious release) takes place, and they access a web page where the release information is stored, starting the simulation using their meteorological data (forecast and analyzed) and long-range dispersion models. The simulation domain of these fictitious accidental releases into the atmosphere is currently the European region encompassing 5° W to 40°E in longitude and 40°N to 65° N in latitude.

After the completion of the simulation, the modeler can upload the results obtained to the RTMOD site, in a standard format. RTMOD processes the new results arrived, by comparing them with the results of other models available at the moment. When new model results arrive, the statistical intercomparison is updated where needed.

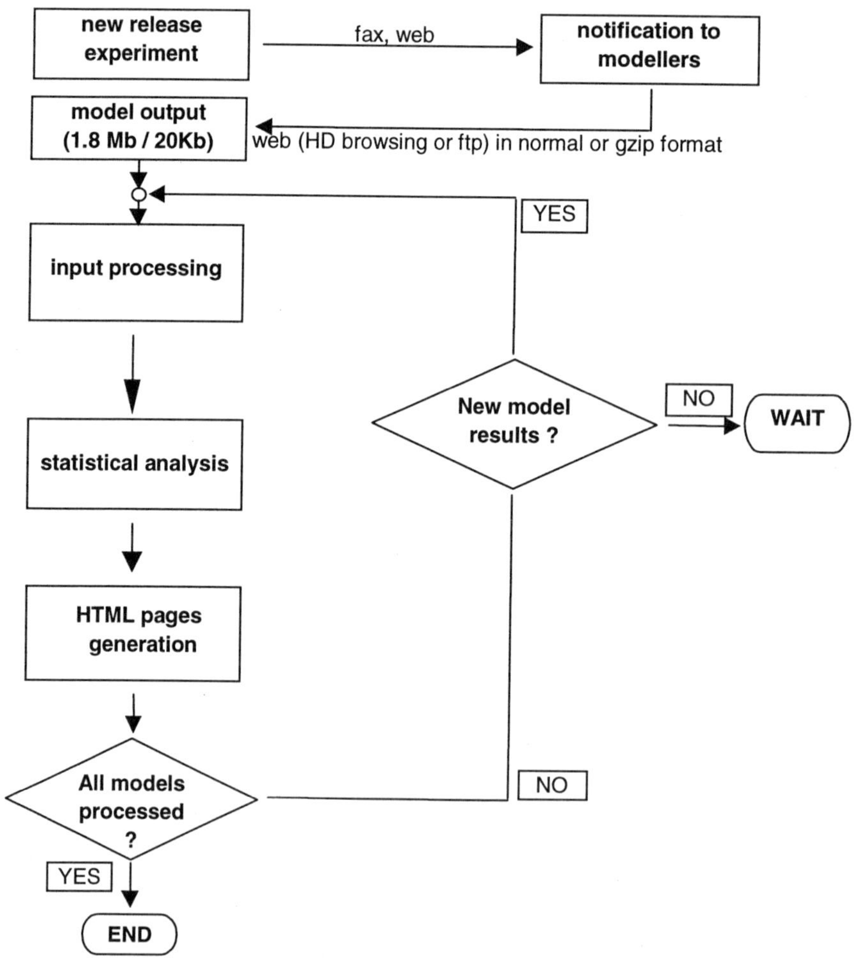

Figure 1. Main structure of the RTMOD system for the intercomparison of long-range dispersion models.

Three main types of analyses are performed:

Time analysis, where the concentrations at a fixed location are considered for the whole duration of the simulation. Discrepancies between predictions in time analysis can be due to over- or under-prediction of the concentration values at the considered location, to over- or under-prediction in the duration as well as to a time shifting between two predictions.

Space analysis, where the concentrations at a fixed time are considered all over the domain. Discrepancies in this analysis can be due to over- or under-prediction of the area involved by the tracer as well as to space shifting between different model predictions.

Global analysis, where all the concentration values at any time and location are considered. For this analysis the distribution of the values is important, as well as the overall tendency of a model to underestimate or overestimate concentrations respect to the other models.

These analyses produce results in numerical and graphical format, as described hereafter. It is only mentioned that the theory described can be also applied to the statistical comparison of model predictions against measurements. The additional complications that arise in this situation, due for example to the treatment of missing values or to the interpretation of null concentrations, are illustrated by Mosca et al. (1998). Table 1 lists the numerical analyses carried out in RTMOD.

Table 1. The numerical analyses performed by RTMOD statistical module

Time analysis	• FMT • Pearson's correlation coefficient • bias and NMSE • time of arrival and duration • time of peak and peak value • integrated concentration
Space analysis	• FMS • Confidence in Contamination Level
Global analysis	• FA2, FA5, FOEX • parameters of data distributions (percentiles, Kolmogorov-Smirnov, Cramer-VonMises-Smirnov, chi-square) • Pearson's correlation coefficient • bias and NMSE, geom. mean bias, geom. mean variance

Also, a number of graphical analyses and representations are produced by RTMOD when comparing any two models. A list is given in the following Table 2.

Table 2. The graphical analyses performed by RTMOD statistical module

Time analysis	• plot of time-series • integrated concentration plot
Space analysis	• contour maps • Confidence in Contamination Level (CCL) maps
Global analysis	• scatter diagrams • box plots

Most of these statistical parameters provide a comparison between any two models. However, a few parameters were computed aiming at determining the average prediction of all the models, to be used by decision makers in case of an emergency situation. For instance, the Confidence in Contamination Level map (CCL) determines for any domain location the percentage of models that predict concentrations above a certain threshold. This allows to individuate the areas that are most probably involved by high concentrations.

Following the interest of the modeling community after the first meeting held at Risoe (Denmark) in January 1998, the participation in RTMOD was enlarged to non-EU participants, from other European and non-European countries such as Japan, USA and Canada. A total of 28 participants was finally reached (Table 3).

Table 3. RTMOD participants by country

Austria	CMG	Japan	JAERI
Belgium	IAMB	Norway	NMI
Bulgaria	NIMH	Poland	IAE
Canada	CMC	Rumania	NIMH
Czech Republic	CHI	Russia	SPA
Denmark	DMI	Slovak Republic	SHI
	NERI	Sweden	SMHI
	RISOE	Switzerland	ISM
Finland	FMI	The Netherlands	KNMI
			RIVM
France	CEA-IPSN	United Kingdom	Meteorological Office
	MeteoFrance	USA	LLNL/ARAC-2
Germany	DWD		LLNL/ARAC-3
Italy	ANPA		NOAA
			Westinghouse S.R. Co.

The following paragraph presents some results obtained in the first two RTMOD experiments performed up to July 1998. Details are given on the timing of the experiments and on the functioning of the hardware and software components of the system.

THE FIRST TWO FICTITIOUS RELEASES

Chernobyl

The first hypothetical release, aimed at simulating a nuclear accident, was performed on the 28th of April 1998 from the location of Chernobyl (Ukraine). The release started at 0900 UTC and lasted 6 hours with constant emission rate of 10 g/s. At the same time the release start was notified to modelers by fax and e-mail.

The choice of the location was made on the basis of the meteorological circulation of that day, and to commemorate the 12th anniversary of the Chernobyl accident.

This first experiment was mainly planned to test the system, and to check the transmission protocols and the procedures. It was planned to receive the earliest replies from participating modelers as faxed maps showing the prediction of cloud location 24 hours after the release start time. At the same time the upload of the data files to be processed by the statistical analysis module of RTMOD were expected.

Eighteen models sent their results by fax to JRC Ispra, within a short time (Table 4) from the release beginning.

Table 4. Time of participants' response by fax in experiment 1 (Chernobyl). Release notification at 0900 UTC.

Participant	Fax date and time (UTC)	Participant	Fax date and time (UTC)
ANPA (I)	28-04-98 – 0954	LLNL/ARAC3 (USA)	(*)
CMC (CDN)	28-04-98 – 1330	Meteorological.Office (UK)	28-04-98 – 1058
CHMI (CR)	28-04-98 – 1449	MeteoFrance (F)	28-04-98 – 1014
DMI (DK)	28-04-98 – 1209	NIMH (BG)	28-04-98 – 1127
DNMI (DK)	28-04-98 – 1400	RHMI (RO)	29-04-98 – 0745
DWD (D)	28-04-98 – 1127	RISOE (DK)	(*)
ISM (CH)	28-04-98 – 1134	RIVM (NL)	29-04-98 – 1556
JAERI (J)	28-04-98 – 0904	SMHI (S)	28-04-98 – 0930
KNMI (NL)	28-04-98 – 1305	SPA (RU)	28-04-98 – 1100
LLNL/ARAC2 (USA)	28-04-98 – 1132	Westinghouse S.R.Co (USA)	28-04-98 – 1607

(*) results sent only on file

The transmission of files and their format showed some problems. Five output files arrived in the correct format within few hours from the release start. Other three correct files arrived the day after. The processing of the remaining models was delayed due to errors and necessary resubmission.

Figure 2 shows the Confidence in Contamination Level (CCL) at 12, 24, 36 and 48 hours after the release start. This is defined as the area contaminated above a certain level for a given fraction of the participating models: the higher is the consensus, the darker is the area on the map.

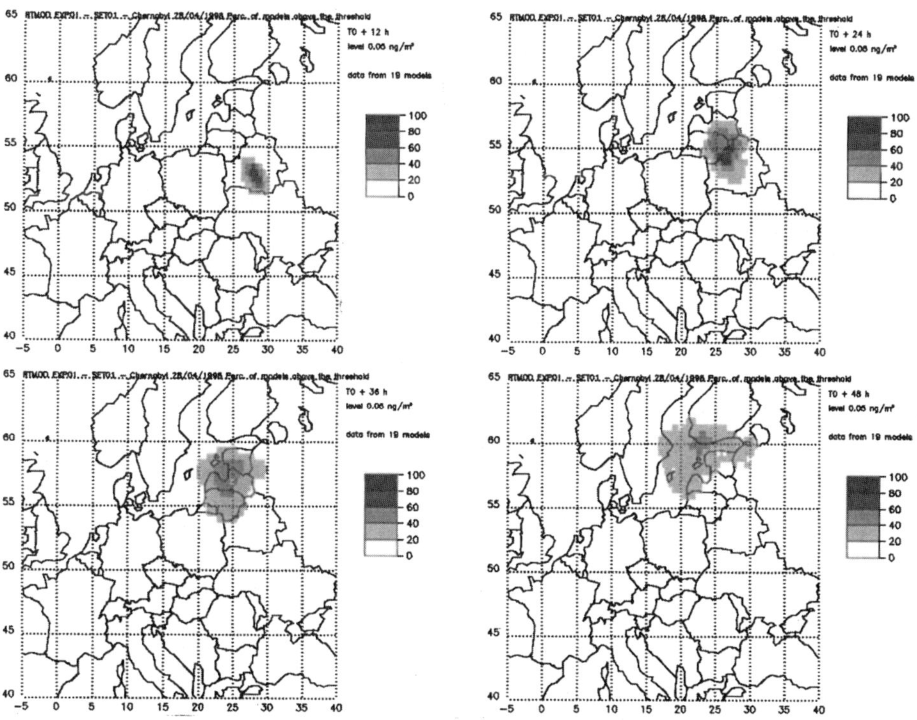

Figure 2. Experiment 1 (Chernobyl). Confidence on Contamination Level (CCL) maps at 12, 24, 36 and 48 hours after the release start. Dark areas have the highest confidence.

Figure 2 indicates that the horizontal cloud extension increases with time and the level of confidence decreases with time. This means that the general consensus on the contaminated area reduces with time. It seems however that there is a small group of participants whose forecasts were similar. Even for those, however, differences in concentration evolution at a given location can be relatively large, as it can be seen in Figure 3.

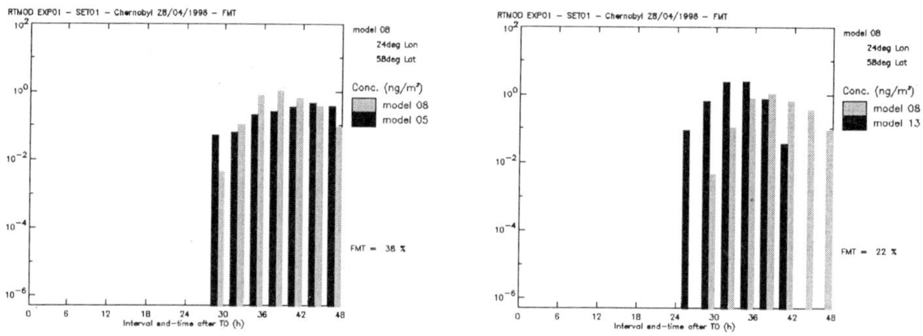

Figure 3. Experiment 1 (Chernobyl). Time series plot at (24E;58N). Left: agreement in time and concentration values. Right: agreement in concentration values but shift in time.

London

The second experiment was performed the 9th of July 1998. The fictitious release happened in London (UK) at 0900 UTC. Again, a release of 10 g/s intensity and lasting 6 hours was decided.

21 modelers sent the plots of their predictions by fax (Table 5).

Table 5. Time of participants' response by fax in experiment 2 (London). Release notification at 0900 UTC.

Participant	Fax date and time (UTC)	Participant	Fax date and time (UTC)
ANPA (I)	09-07-98 – 0957	KNMI (NL)	09-07-98 – 1206
CEA (F)	09-07-98 – 1240	LLNL-ARAC2 (USA)	09-07-98 – 2233
CMC (CDN)	09-07-98 – 1039	LLNL-ARAC3 (USA)	09-07-98 – 1749
CMG (A)	09-07-98 – 1314	Meteorological.Office (UK)	09-07-98 – 0958
DMI (DK)	09-07-98 – 1008	MeteoFrance (F)	09-07-98 – 0935
DNMI (DK)	09-07-98 – 1108	NIMH (BG)	09-07-98 – 1034
DWD (D)	09-07-98 – 1117	RIVM (NL)	09-07-98 – 1057
FMI (SU)	09-07-98 – 1127	SMHI (S)	09-07-98 – 1001
ISM (CH)	09-07-98 – 1022	SPA (RU)	09-07-98 – 1120
JAERI (J)	09-07-98 – 1028	Westinghouse S.R.Co (USA)	09-07-98 – 1246
KMI (B)	09-07-98 – 1305		

This time all the procedures went smooth, thanks to the testing and experience gained with the first preliminary experiment. No major problems were noticed in the transmission of output files and in the processing and publication of the results on the web, that was completed immediately after the data reception.

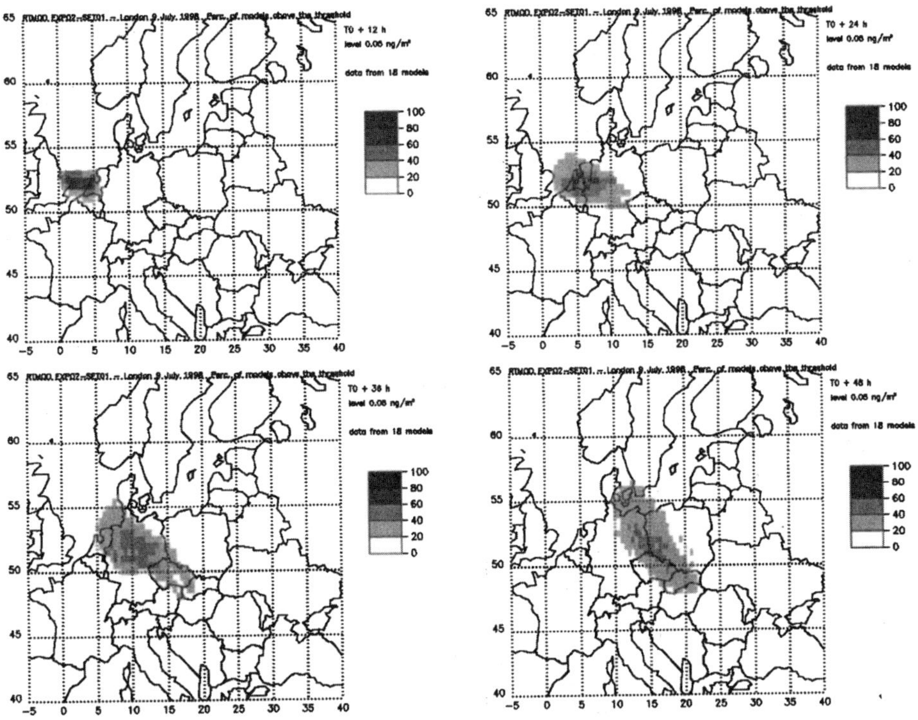

Figure 4. Experiment 2 (London). Confidence on Contamination Level (CCL) maps at 12, 24, 36 and 48 hours after the release start. Dark areas have the highest confidence.

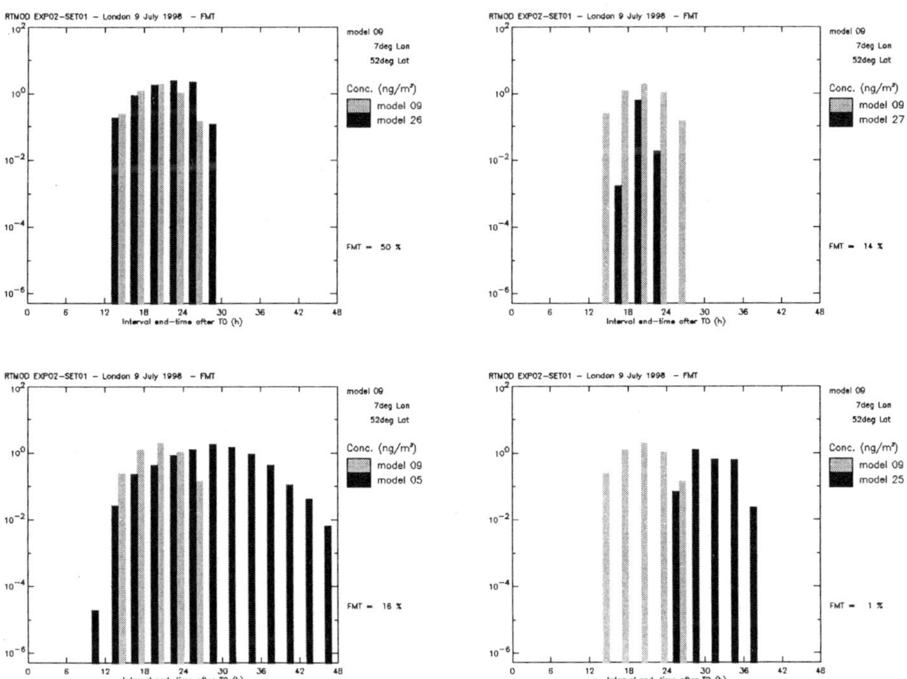

Figure 5. Experiment 2 (London). Time serie ɔt at (7E;52N). Upper left: agreement in time and concentration values. Upper right: disɛ ement in concentration values. Lower left: disagreement in duration. Lower right: shift in time.

In this experiment, the selected threshold contour level of the CCL map (Figure 4), is narrow at 12 hours and does not show significant changes in the transport direction even after hours. On the contrary, the contour stretches in the direction normal to the transport. This indicates that discrepancies in transport direction, more than on transport velocity, influence the predictions.

The time series plots for this release (Figure 5) are typical examples of how the over- or under-prediction of concentration values and/or of duration and the time shift can influence the performance of model predictions in the time analysis (FMT index).

CONCLUSIONS

The generally enthusiastic participation in the exercise and the expectancy among modelers to examine the results of the intercomparison are already proofs of the success of RTMOD.

Some necessary changes and improvements in the automatic internet procedures are foreseen for future exercises, including the analysis of predicted dry and wet deposition.

Despite the fact that no real data are available to compare with model results, RTMOD can quickly provide a vast information on long-range transport models reliability in forecasting atmospheric dispersion of harmful substances in many different meteorological conditions.

The variability in model results can be certainly reduced by making frequent use of the system. This may help in finding errors, and in simulation improvement. At the same time, the comparative analysis of model results variability and weather conditions allows to identify the situations that produce larger spread in calculations, and thus higher uncertainty.

RTMOD is useful to decision makers for acquiring the necessary knowledge and understanding in the variability of model results. Also, in case of a real accident many institutes would provide results in real-time to support the decision process. As an example, in late May 1998, an accidental release of radionuclides from an iron-melting factory in Algeciras (Spain) caused a very low but diffuse contamination over whole Europe. The release was not notified to the authorities, and only days after the accident it was detected by monitoring networks in European countries. If episodes like this were disclosed just after they happen, it would be possible and useful to activate the long-range modeling groups all over the world to evaluate the situation and suggest possible interventions to safeguard people, based on a number of estimates.

REFERENCES

Klug W., Graziani G., Grippa G., Pierce D., Tassone C., 1992,: *Evaluation of Long-Range Atmospheric Transport Models Using Environmental Radioactivity Data from Chernobyl Accident. The ATMES Report*, Elsevier Applied Science, London.

Mosca S. Graziani G., Klug W., Bellasio R. and Bianconi R., 1998, A statistical methodology for the evaluation of long-range dispersion models: an application to the ETEX exercise. Accepted for publication in *Atmospheric Environment.*

Nodop K. editor (1997): Proceedings of the ETEX Symposium on Long-Range Atmospheric Transport, Model Verification and Emergency Response, 13-16.5.1997. EUR 17346 EN, Office for Official Publications of EC, Luxembourg

NEW DEVELOPMENTS

chairmen:	E. Batchvarova
	B. Fisher
	T. Iversen
	M. Rotach
	R. Yamartino
rapporteurs:	P. de Haan
	M. Kolarova
	J. H. Sørensen
	J. Wilkinson

STUDYING OZONE EPISODES IN EUROPE WITH THE DANISH EULERIAN MODEL

Carsten Ambelas Skjøth, Annemarie Bastrup-Birk,
Jørgen Brandt and Zahari Zlatev

National Environmental Research Institute
Department of Atmospheric Environment
Frederiksborgvej 399, P. O. Box 358
DK-4000 Roskilde, Denmark

The Danish Eulerian Model is a large-scale air pollution model, which has been developed to study pollution levels in Europe and transport of air pollutants to Denmark and to the Southern part of Scandinavia. The model is described by a system of partial differential equations. The number of equations is equal to the number of chemical compounds involved in the model. In the present operational version there are 35 species, but experiments with chemical schemes containing 56 and 168 species are also carried out. All important pollutants (sulphur pollutants, nitrogen pollutants, ammonia-ammonium, ozone, PAN, many hydrocarbons and some radicals) can be studied by the model. Variations of the ozone concentrations in the period 1989-1995 as well as episodes during the summers in this period will be discussed in this paper. The output results have been stored in huge files. An advanced animation program has been used to visualise the ozone levels during the summer months and, thus, to determine both the time-periods in which episodes appear and the countries where the episodes take place. A video-tape will be used to illustrate the efficiency of the animation program. The results indicate that the critical levels accepted in EU (the European Union) are very often exceeded in many European countries. Sometimes the exceedances are very large. The main conclusion is that it is necessary to use reliable air pollution models for forecasting high ozone concentrations during the summer months.

1. INTRODUCTION

Ozone is one of the most harmful air pollutants. High ozone concentrations may have different damaging effects on plants animals and humans. This is why the ozone

Air Pollution Modeling and Its Application XIII, edited by S.-E. Gryning
and E. Batchvarova, Kluwer Academic/Plenum Publishers, 2000

concentration levels should be carefully studied. This is stated as a major requirement in the newest EU ozone directive. In this directive some critical levels, concerning health requirements for certain groups of humans and based on recommendations of WHO (the World Health Organisation), are clearly stated. Furthermore, certain critical levels concerning plants (and first and foremost crops and forest trees) are also proposed. Finally, there is a clear requirement in the new EU ozone directive for producing, after year 2000, ozone forecasts during the summer periods in all regions in Europe where there is a danger that the critical levels will be exceeded.

The introduction of new and more stringent critical levels as well as the requirements for achieving acceptable ozone concentrations in Europe lead in a natural way to an increased need for development of new more reliable mathematical models for studying the ozone concentrations in different parts in Europe and for forecasting situations where the critical levels will be exceeded in certain parts of Europe. Some results obtained in this direction at the National Environmental Research Institute (Roskilde, Denmark) will be reported in this paper.

The paper is organised in the following way. A short description of the mathematical model used in this study, the Danish Eulerian Model, will be given in the Section 2. The validation of the calculated by the model results, by comparing them with measurements, will be discussed in Section 3. Results, which shows that the critical levels are often exceeded in many European countries, will be given in Section 4 (these results either are direct snapshots from a video tape animating the distribution of the ozone concentrations in Europe during the summer of 1997 or are derived by using the data from this video tape). The possibilities of calculating reliable ozone forecasts will shortly be outlined in Section 5. Some comments and concluding remarks will be listed in Section 6.

2. MATHEMATICAL DESCRIPTION OF THE DANISH EULERIAN MODEL

The Danish Eulerian Model is described, as many other large-scale air pollution models, by a system of q partial differential equations (PDE's):

$$(1)\quad \partial c_s/\partial t = -\,\partial(uc_s)/\partial x - \partial(vc_s)/\partial y - \partial(wc_s)/\partial z$$

$$+\,\partial(K_x\partial c_s/\partial x)/\partial x + \partial(K_y\partial c_s/\partial y)/\partial y + \partial(K_z\partial c_s/\partial z)/\partial z$$

$$-\,(k_{1s} + k_{2s})c_s(x,y,z,t) + E_s(x,y,z,t) + Q_s(c_1,c_2,\dots,c_q),$$

where (a) the concentrations of the chemical species are denoted by c_s, (b) u, v and w are wind velocities, (c) K_x, K_y and K_z are diffusion coefficients, (d) the emission sources are described by E_s, (e) k_{1s} and k_{2s} are deposition coefficients, (f) the chemical reactions are represented by the non-linear functions $Q_s(c_1,c_2,\dots,c_q)$. The chemical scheme actually used in the Danish Eulerian Model at present is a CBM IV scheme proposed by Gery et al. (1989) with 35 chemical species, but the some experiments have also been carried out by using chemical schemes with 56 and 168 compounds. The model is fully described in Zlatev (1995) and Zlatev et al. (1996a).

It is difficult to treat the PDE system (1) directly. Therefore, different splitting procedures are normally used. Moreover, it is essential to select good (both sufficiently accurate and sufficiently fast) numerical algorithms. Finally, the algorithms chosen should be carefully optimised for runs on modern parallel and vector architectures, which is by far not an easy task. The complexity of the large-scale Eulerian models is discussed in detail in Peters et al. (1995).

Following ideas described in Marchuk (1985) and McRae et al. (1984), one can split, according to the involved physical processes, the model defined by (1) into five sub-models: (a) a horizontal advection sub-model, (b) a horizontal diffusion sub-model, (c) a deposition sub-model, (d) a chemistry sub-model and (e) a sub-model for the treatment of the vertical exchange. These sub-models have to be discretized, by using different discretization algorithms, and treated successively at each time-step (see, for example, Zlatev, 1995 or Zlatev et al., 1994). If the model is discretized on a *(96x96x10)* grid, then a system of ordinary differential equations, ODE's, containing 3225600 equations is to be treated in each sub-model at each time-step when the number of chemical species that are involved in the Danish Eulerian Model is $q = 35$. If the model has to be run with meteorological data covering one month + five days to start-up the model (Zlatev, 1995), then *3456* time-steps are needed; assuming here that the time-stepsize is *900 s* (the number of chemical time-steps is six times greater because the chemical sub-model is stiff and, therefore, has to be run with a much smaller time-stepsize; a time-stepsize of *150 s* is actually used). Many hundreds or even several thousands of such runs are to be performed in a long simulation process, see, for example, Bastrup-Birk et al. (1997) and Zlatev et al. (1996b). It is clear, from the above discussion, that the numerical treatment of model (1) results in huge computational tasks and, therefore, causes great difficulties when a long series of runs have to be carried out, which is true even when big and fast modern supercomputers are available. This explains why the two dimensional version of the Danish Eulerian Model, which has been used in Harrison et al. (1994), Zlatev (1995) and Zlatev et al. (1992, 1993), has also been applied in all experiments the results of which will be discussed in the following sections.

The number of sub-models is reduced from five to four and the number of equations in each of the discretized sub-models is reduced from *3225600* to *322560* when the two-dimensional version is used. The computational tasks are still very big and it is, also in this situation, necessary to apply different vectorization and parallelization algorithms in order to be able to perform the jobs (such algorithms for different types of supercomputers are discussed in Bendtsen and Zlatev, 1997, Brown et al., 1995, Zlatev, 1995 and Zlatev et al., 1994, 1996). The numerical algorithms used in the treatment of the model on computers will not be discussed here. However, these algorithms are discussed in many publications; see, for example, Alexandrov et al. (1997), Zlatev (1995) and Zlatev et al. (1994,1996a). The treatment of the chemistry is important. An improved QSSA (Quasi-Steady-State-Approximation) algorithm, see Alexandrov et al. (1997) is used. This algorithm is better than the original QSSA;; see Hestvedt et al. (1978) or Hov et al. (1988). One can also use methods similar to that used in Hertel et al. (1993) or classical methods for solving ordinary differential equations as in Alexandrov et al. (1997). The parallel architecture actually used in the runs at present is an IBM SP computer with at most 32 processors available. The exploitation of the parallelism is based on the Message Passing Interface (MPI). This is a standard tool for exploiting the parallelism (see, for example, Gropp et al., 1994). Therefore one should expect to get good results (with not too many modifications) also when other parallel computers are used.

3. VALIDATION OF THE MODEL RESULTS

Comparisons of calculated by the model concentrations with measurements taken at many EMEP stations located in different European countries have been used in the attempt to show that reliable results can be obtained by using the Danish Eulerian Model. Results from such comparisons have been discussed in many papers (see Bastrup-Birk et al., 1997, Harrison et al., 1994, Zlatev, 1995 and Zlatev et al, 1992, 1993, 1996b). Some new results, comparing the time variation of calculated and measured ozone concentrations over a long period of seven years, 1989-1995, are given in Fig. 1 - Fig. 4. The comparisons are at four measurement stations which are located in four different European countries (Denmark, Sweden, Germany and Great Britain).

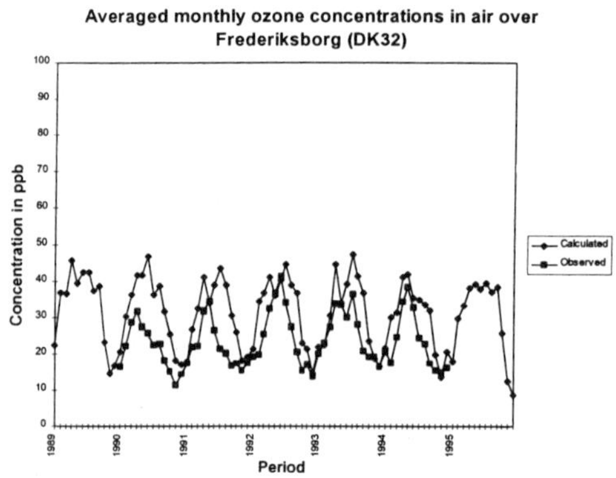

Figure 1. Comparison of averaged monthly calculated and measured ozone concentrations at Frederiksborg (Denmark) over a seven year period.

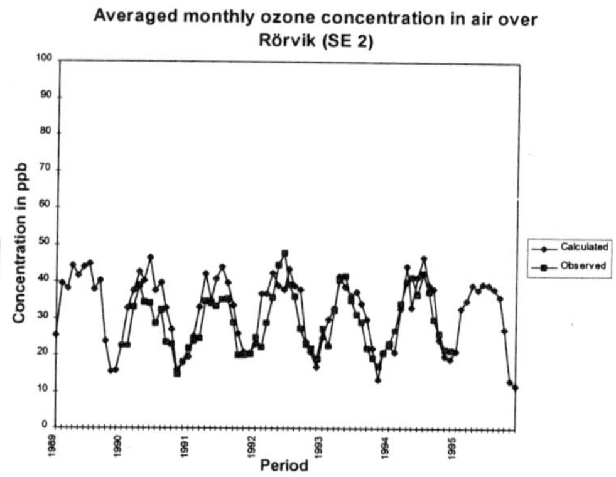

Figure 2. Comparison of averaged monthly calculated and measured ozone concentrations at Rörvik (Sweden) over a seven year period.

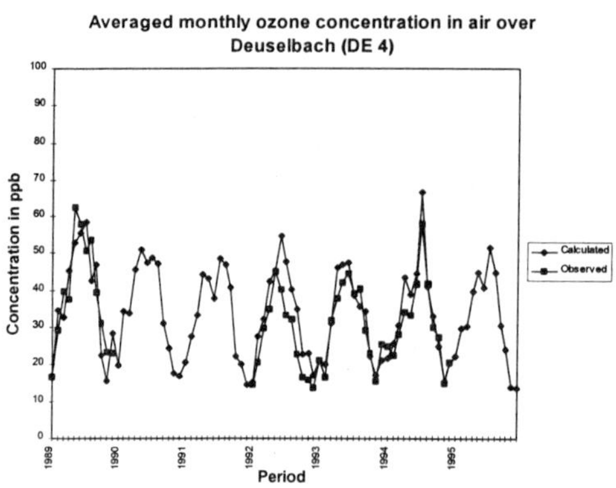

Figure 3. Comparison of averaged monthly calculated and measured ozone concentrations at Deuselbach (Germany) over a seven year period.

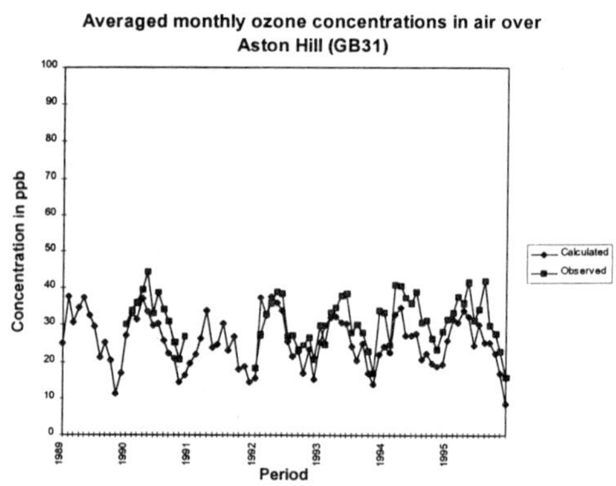

Figure 4. Comparison of averaged monthly calculated and measured ozone concentrations at Aston Hill (Great Britain) over a seven year period.

4. EXCEEDANCE OF THE CRITICAL LEVELS FOR OZONE IN EUROPE

Several critical levels are discussed in the drafts for a new *EU Ozone Directive*. Results concerning the following four ozone critical levels will be presented in this section. The numbers of days in which the ozone concentrations are exceeding *90 ppb* are given in Fig. 5 (according to the last draft of the *EU Ozone Directive* the authorities must inform the population when this happens). The numbers of days in which the ozone concentrations are over *120 ppb* are given in Fig. 6 (according to the last draft of the *EU Ozone Directive* the authorities must send an warning when this happens). The results in Fig. 6 show clearly the big ozone episode over Western Europe in the summer of 1995. Periods of 8 hours in which the ozone concentrations are over *60 ppb* are also causing some damages. The numbers of days in which such periods occur in different European countries during the summer of 1995 are given in Fig. 7. The time period (September 1997) has not been randomly chosen. High ozone levels were measured at this period in the Northern France. The French authorities offered free tickets for the public transportation in Paris in a couple of days in an attempt to reduce temporarily the emissions from the traffic. The model results indicate that the critical levels were indeed exceeded in many days of September.

The AOT40 values (see Bastrup-Birk et al., 1997) are important in connection with damages on plants. The results in Fig. 8 indicate that the critical level for crops (3000 ppb.hours) is greatly exceeded in many European countries. In some countries the actual AOT40 values are greater than the critical levels by a factor of seven (see Bastrup-Birk et al., 1997 and Zlatev et al., 1997).

The results presented in this section demonstrate the fact that the critical levels are often exceeded in many European countries and, thus, the ozone levels should be carefully studied. It should also be mentioned that such periods were found in many summer months in the period 1989-1997.

5. CALCULATING OZONE FORECASTS

Many ozone critical levels are often exceeded (some examples are given in the previous section). Therefore it is important to be able to tell the population when the occurrence of high ozone levels is likely. This is why ozone forecasts are needed. Several conditions must be satisfied in order to be able to calculate a reliable ozone forecast. It is necessary to have (i) a reliable meteorological forecast, (ii) to evaluate in a reliable way the emission fields in the area under consideration, (iii) to have access to some reliable measurements (which have to be used in the calculation of the initial concentration fields by using some kind of data assimilation), (iv) to have a fast and high resolution air pollution model, (v) to have fast and high quality visualisation routines by which the enormous output data files prepared by the air pollution model should be represented so that it could easily be understand also by non-specialists. This means that the calculations are to be carried out in three stages:

Stage 1 - Pre-processing (calculate or receive a meteorological forecast, prepare reliable emissions, use the available observations to obtain good initial concentration fields).

Stage 2 - Actual forecast calculations (perform the calculations to obtain the air pollution forecast).

Stage 3 - Post-processing (prepare the data which should be used by other specialists; for example in the TV news).

Data transmission, data manipulation (as, for example, changing formats) and data mining are needed in Stage 1 and Stage 3. Comparing results obtained by several different large scale air pollution models may be useful (and perhaps necessary) in Stage 3. High speed computers are needed in all stages, but their use will be crucial when high resolution is required if one wants to avoid the introduction of unnecessary simplifying assumptions which may affect the reliability of the results. The use of modern graphic tools and networking is most important for Stage 3, but it is also important for the other stages.

The work on producing air pollution forecasts has recently be initialised at the National Environmental Research Institute (Denmark).

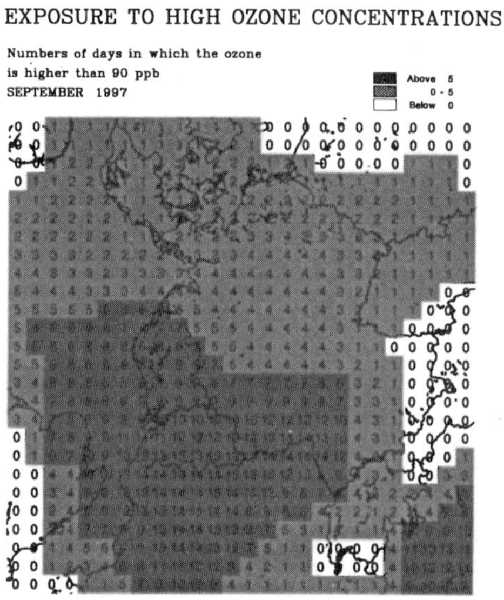

Figure 5. Numbers of days in which ozone concentrations greater than 90 pbb. were found in one or more hours.

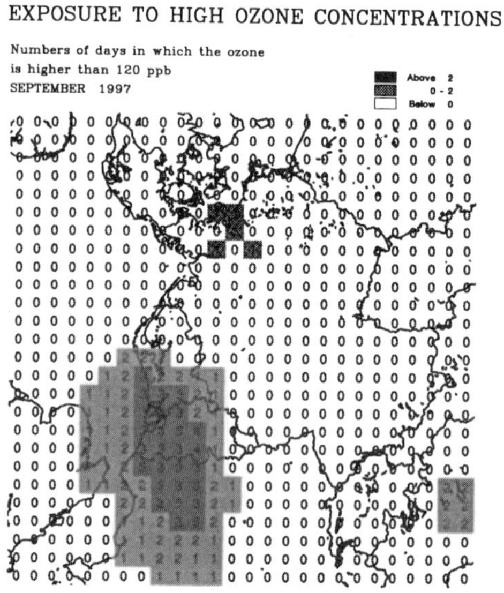

Figure 6. Numbers of days in which ozone concentrations greater than 120 pbb. were found in one or more hours.

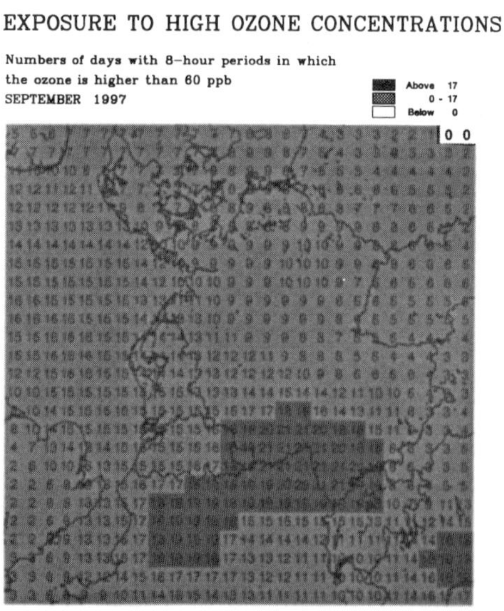

Figure 7. Numbers of days in which one or more 8-hour-periods were found with ozone concentrations more than 60 ppb.

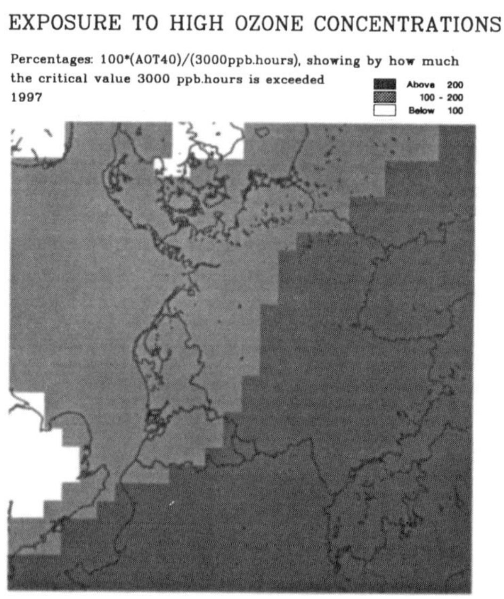

Figure 8. The AOT40 values in 1997 (multiplied by 100 and divided by the critical value of 3000 ppb.hours).

6. CONCLUDING REMARKS

The new and stringent requirements for more reliable information about the pollution levels and, what is even more important, for more reliable information about the situations where certain critical levels are exceeded (or will be exceeded in the next dew days) could be satisfied only if:

1. closer connections between specialists in all related areas (as, for example, specialists in preparing weather forecasts and meteorological data, specialists in preparing and analysing measurements, specialists in computer visualisation and animation techniques and specialists in air pollution modelling) are established,
2. better air pollution models, in which all important physical and chemical processes are adequately described, are developed,
3. all advantages of using the potentially powerful computer technology are efficiently exploited for achieving both high speed data communications and high speed computations (here one must understand that very often the computer technology is indeed only potentially powerful, and some hard efforts are often needed when high efficiency is to be achieved; this is especially true when the problems are very large, which is the case in air pollution modelling).

The importance of these three requirements (and some other requirements can also be added to the list) will be even more increased in the 21st century when the demands for a clean and healthy environment as well as for a sustainable development of the economy in all European countries will become much greater.

Acknowledgements

This research work was partially supported by NATO (ENVIR.CRG.930449, OUTS.CRG.960312 and ENVIR.ARW.971731), EU (ESPRIT, #22727 and # 24618) and NMR (the Nordic Council of Ministers) in a project in which teams from Denmark, Finland, Norway and Sweden are collaborating. A grant from the Danish Natural Sciences Research Council gave us access to all Danish supercomputers.

REFERENCES

Alexandrov, V., Sameh, A., Siddique, Y. and Zlatev, Z.,1997, Numerical integration of chemical ODE problems arising in air pollution models. Environmental Modelling and Assessment, Vol. 2, pp. 365-377.

Bastrup-Birk, A, Brandt, J., Uria, I and Zlatev, Z., 1997, Studying cumulative ozone exposures in Europe during a seven-year period. Journal of Geophysical Research, Vol. 102, pp. 23917-23935.

Bendtsen, C. and Zlatev, Z., 1997, Running air pollution models on message passing machines. In: *Parallel Virtual Machine and Message Passing Interface* (M. Bubak, J. Dongarra and J. Wasniewski, eds.), pp. 417-426. Springer-Verlag, Berlin.

Brown, J., Wasniewski, J. and Zlatev, Z., 1995, Running air pollution models on massively parallel machines. Parallel Computing, Vol. 21, pp. 971-991.

Gery, M. W., Whitten, G. Z., Killus, J P. and Dodge, M. C., 1989, A photochemical kinetics mechanism for urban and regional computer modeling. Journal of Geophysical Research, Vol. 94, pp. 12925-12956.

Gropp, W., Lusk, E. and Skjellum, A., 1994, *Using MPI: Portable Programming with the Message Passing Interface*. MIT Press, Cambridge, Massachusetts.

Harrison, R. M., Zlatev, Z. and Ottley, C. J., 1994, A comparison of the predictions of an Eulerian atmospheric transport chemistry model with measurements over the North Sea. Atmospheric Environment, Vol. 28, pp. 497-516.

Hertel, O., Berkowicz, R., Christensen, J. and Hov, Ø., 1993, Test of two numerical schemes for use in atmospheric transport-chemistry models. Atmospheric Environment, Vol. 27A, pp. 2591-2611.

Hesstvedt, E., Hov, Ø., and Isaksen, I. A., 1978, Quasi-steady-state approximations in air pollution modelling: comparison of two numerical schemes for oxidant prediction. International Journal of Chemical Kinetics, Vol. 10, pp. 971-994.

Hov, Ø., Zlatev, Z., Berkowicz, R., Eliassen, A. and Prahm, L. P., 1988, Comparison of numerical techniques for use in air pollution models with non-linear chemical reactions. Atmospheric Environment, Vol. 23, pp. 967-983.

Marchuk, G. I., 1985, *Mathematical Modeling for the Problem of the Environment.* Studies in Mathematics and Applications, No. 16, North-Holland, Amsterdam.

McRae, G. J., Goodin, W. R. and Seinfeld, J. H., 1984, Numerical solution of the atmospheric diffusion equations for chemically reacting flows. Journal of Computational Physics, Vol. 45, pp. 1-42.

Peters, L. K., Berkowitz, C. M., Carmichael, G. R., Easter, R. C., Fairweather, G., Ghan, S. J., Hales, J. M., Leung, L. R., Pennell, W. R., Potra, F. A., Saylor, R. D. and Tsang, T. T., 1995, The current state and future direction of Eulerian models in simulating the tropospherical chemistry and transport of trace species: A review. Atmospheric Environment, Vol. 29, pp. 189-221.

Zlatev, Z., 1995, *Computer Treatment of Large Air Pollution Models.* Kluwer Academic Publishers, Dordrecht-Boston-London.

Zlatev, Z., Christensen, J. and Eliassen, A., 1993, Studying high ozone concentrations by using the Danish Eulerian Model. Atmospheric Environment, Vol. 27A, pp. 845-865.

Zlatev, Z., Christensen, J. and Hov, Ø., 1992, An Eulerian air pollution model for Europe with non-linear chemistry, Journal of Atmospheric Chemistry, Vol. 15, pp. 1-37.

Zlatev, Z., Dimov, I. and Georgiev, K., 1994, Studying long-range transport of air pollutants. Computational Science and Engineering, Vol. 1, No. 3, pp. 45-52.

Zlatev, Z., Dimov, I. and Georgiev, K., 1996a, Three-dimensional version of the Danish Eulerian Model. Zeitschrift für Angewandte Mathematik und Mechanik, Vol. 76, S4, pp. 473-476.

Zlatev, Z., Fenger, J. and Mortensen, L., 1996b, Relationships between emission sources and excess ozone concentrations. Computers and Mathematics with Applications, Vol. 32, No. 11, pp. 101-123.

EDDY DIFFUSIVITY PARAMETERIZATION IN THE DECAYING CONVECTIVE RESIDUAL LAYER

G. Degrazia[1], D. Anfossi[2], O.L.L. Moraes[1], A. Goulart[1], S. Trini Castelli[2]

[1]Universidade Federal de Santa Maria (RS), Brasil
[2]CNR-Istituto di Cosmogeofisica, Turin, Italy

INTRODUCTION

In the last decades, important studies on the airborne pollutant dispersion in Convective (CBL) and Stable Boundary Layers (SBL) have been carried out. However, less attention has been paid to the dispersion in the Residual Layer (RL), the nearly adiabatic remnant of the daytime boundary layer, where the dispersion of contaminants occurs in condition of decaying turbulence.

The decay of energy-containing eddies in the CBL is the physical mechanism that can maintain efficient the dispersion process in the RL. Besides this mechanism, a second turbulence mechanism, the shear-driven turbulence, may be present in the RL (Kim and Mahrt, 1992). However it will not be considered in this paper.

It is the aim of the present paper to propose a general semi-empirical method to derive eddy diffusivities in a decaying turbulence and to apply it to the case of the decay of convective turbulence in the RL.

The expressions for the eddy diffusivity derived in this paper may be used in mesoscale transport and dispersion simulations.

DYNAMICAL EQUATION FOR THE ENERGY DENSITY SPECTRUM

From the Fourier transform of 3-D turbulence kinetic eq. (TKE) for an isotropic turbulence of an incompressible fluid, when buoyant and shear productions are not important, the dynamical equation for the energy density spectrum (EDS) is obtained (Hinze, 1975; Stanisic, 1988), namely:

$$\frac{\partial E(k,t)}{\partial t} = - W(k,t) - 2\nu k^2 E(k,t) \qquad (1)$$

Air Pollution Modeling and Its Application XIII, edited by S.-E. Gryning and E. Batchvarova, Kluwer Academic/Plenum Publishers, 2000

where k is the wavenumber, $E(k,t)$ is the EDS function; $W(k,t)$ is referred to as the energy-transfer-spectrum function and represents the contribution due to the inertial transfer of energy among different wavenumbers (Chandrasekhar, 1949; Batchelor, 1949; Stull, 1988; Stanisic, 1988); the second term on the r.h.s. of eq. (1) is the energy loss due to ordinary viscous dissipation. Following the tradition in micrometeorology (see, for instance, Kaimal et al., 1972) space spectra can be replaced by frequency spectra, since frequency n, not wavenumber, is measured. Consequently, eq. (1) becomes:

$$\frac{\partial S(n,t)}{\partial t} = T(n,t) - \frac{8\pi^2 n^2 \nu}{U^2} S(n,t) \qquad (2)$$

where $n = kU/2\pi$, U is the mean wind speed, $T(n,t) = W(k,t)\ 2\pi/U$ represents the inertial energy transfer among different frequencies and $S(n,t) = E(k,t)\ 2\pi/U$.

Heisenberg (ref. from Stanisic, 1988) explained the mechanism of inertial transfer of energy from large to small eddies in terms of an additional eddy viscosity, called kinematic turbulence viscosity (KTV). Thus, the effect of the inertia term can be regarded as equivalent to a virtual turbulent friction, ν_T, produced by the smaller eddies and acting on the larger eddies. ν_T represents the kinematic turbulence viscosity caused by the eddies with frequency ranging from n to infinity. Therefore $T(n,t)$ can be represented as:

$$T(n,t) = -\frac{8\pi^2 n^2}{U^2} \nu_T\ S(n,t),$$

so that eq.(2) becomes:

$$\frac{\partial S(n,t)}{\partial t} = -\frac{8\pi^2}{U^2} [\nu_T + \nu]\ n^2\ S(n,t) \qquad (3)$$

The correct choice of KTV is the major difficulty in Heisenberg's approach. Indeed, from a physical point of view, the introduction of KTV to account for the energy transfer from the larger to the smaller eddies, can be correct only if the small eddies, responsible for the existence of such a turbulence viscosity, are statistically independent of the large eddies (Batchelor, 1953).

DECAYING EDDY TURBULENCE IN THE PBL

As it is wellknown, the 1-D spectrum $S_i(n)$ in the PBL can be subdivided in three major spectral regions: energy-containing, inertial and dissipation subranges (Kaimal and Finnigan, 1994). In the energy-containing subrange, where eddies make the main contribution to the TKE, EDS shows its maximum, so it is possible to choose its corresponding frequency n_e to characterise the size of these eddies. In the dissipation subrange, where the viscosity plays its major role, it is also possible to associate a frequency n_d with the size of the eddies that provide the main contribution to the dissipation. Following the second Kolmogorov hypothesis (Hinze, 1975), if the Reynolds number is infinitely large, the dissipation in the region of frequencies very far below the region of maximum dissipation will be negligibly small compared with the flux of energy transferred by inertial effects. In such an inertial subrange, the effect of molecular viscosity would then vanish ($\nu_T >> \nu$,). Consequently we may associate a frequency n_I with the size of the eddies that provide the main contribution to this inertial energy flux, that is:

$$n_e \ll n_I \ll n_d \qquad (4)$$

With this assumption, the turbulence, in this subrange, is statistically independent of the range of energy-containing eddies and a relationship for KTV can be obtained. ν_T can be calculated directly from Taylor's statistical diffusion theory for large travel times τ (Batchelor, 1949; Hinze, 1975; Hanna, 1981; Degrazia and Moraes, 1992; Degrazia et al., 1988) as:

$$\nu_T = \frac{\beta_i}{6} \frac{\sigma^2}{n_I} \qquad (5)$$

where σ^2 is the turbulent variance calculated in the inertial subrange and β_i is defined as the ratio of the Lagrangian to the Eulerian time scales. Eq. (3) can thus be solved (setting $S_0(n) = S(n,t=0)$), obtaining:

$$S(n,t) = S_0(n) \exp\left(-8\pi^2 \nu_T \frac{n^2}{U^2} t\right) \qquad (6)$$

Having obtained an expression for the time evolution of EDS in decaying turbulence, an equation for the eddy diffusivity in decaying turbulence (K_α, α=x,y,z) may be easily derived by using the same theoretical framework used to derive $\nu_T(n)$. Let us assume that in the initial stage of the decay process, this last is predominantly determined by the decay of the energy-containing eddies (Hinze, 1975; Nieuwstadt and Brost, 1986; Sorbjan, 1997), so that these large eddies decay according to their intrinsic time scales. Therefore, as the energy-containing spectral range is characterised by n_e, an equation like (5) can be written for K_α, namely:

$$K_\alpha = \frac{\beta_i}{6} \frac{\sigma^2}{n_e} \qquad (7)$$

where:

$$\sigma^2 = \int_{n_e}^{\infty} S_0(n) \exp\left(-8\pi^2 \nu_T \frac{n^2}{U^2} t\right) dn.$$

As in the case of eq. (5), for large travel time, K_α can be expressed as a function of local turbulence properties and in terms of the characteristic discrete spectral frequencies n_e (energy-containing eddies) and n_I (inertial-transferring eddies). Eq. (7) represents the general semi-empirical method to derive eddy diffusivities in a decaying turbulence that we propose. Looking at this equation, suggests that the decrease of eddy diffusivity K_α with time depends on n_e. As n_e refers to the large eddies, this means that a certain level of turbulence can be sustained for a long time. On the other hand, from eq. (7) it can be seen that the decrease of K_α with time depends also on ν_T. This, in turn, depends on n_I, that describes the dimension of eddies providing the major contribution to the inertial energy flux. Consequently, this additional viscosity may be estimated from observed spectra. Since the inertial transfer of energy is the dominant factor in the inertial subrange, n_i will be considered here as the initial frequency of this range (Kaimal et al., 1976).

DERIVATION OF KTV AND K_z IN THE RL

As the 3-D EDS for an isotropic and homogeneous turbulent flow can be related to the 1-D EDS, we can calculate KTV and K_z in the residual layer. Our starting point is that in the first stage the turbulence structure of the RL is the same as that of the previously existing CBL. This means that the turbulence energy spectra are assumed to have the same form as they had in the recently decayed mixed layer.

The vertical turbulence energy spectrum in the inertial subrange, derived by Kaimal et al. (1976) on the basis of direct observation, can be written as:

$$S_w(n) = 0.36 \left(\frac{\kappa U \Psi_\varepsilon}{h}\right)^{\frac{2}{3}} w_*^2 \; n^{-\frac{5}{3}} \tag{8}$$

where $\Psi_\varepsilon = (\varepsilon h)/(w_*^3)$ is the non dimensional molecular dissipation rate function, w_* is the convective velocity scale, $\kappa = 0.4$ is the von Karman constant and h is the CBL height. In convective conditions $\Psi_\varepsilon \approx 0.65$ (Caughey and Palmer, 1979; Kaimal and Finnigan, 1994).

By setting $\beta_w = 0.55\ U/\sigma_w$ (Degrazia and Anfossi, in press), computing σ_w from the integration from n_i to infinity of eq. (8), the following expression for KTV is obtained (from eq. 5):

$$\nu_T = 0.067\ w_* \left(\frac{U}{n_i}\right)^{\frac{4}{3}} \left(\frac{\kappa\ \Psi_\varepsilon}{h}\right)^{\frac{1}{3}} \tag{9}$$

Concerning eq. (6), let us remember that, according to the convective similarity theory (Kaimal and Finnigan, 1994; Degrazia et al., 1996), the Eulerian velocity spectra under unstable conditions can be expressed by the following relationships:

$$S_{w,0}(n) = \frac{0.38\ \dfrac{z}{U}\left(\dfrac{\Psi_\varepsilon\ z}{h}\right)^{\frac{2}{3}} w_*^2}{\left(f_m^*\right)_w^{\frac{5}{3}} \left[1 + \dfrac{1.5\ n\ z}{U\left(f_m^*\right)_w}\right]^{\frac{5}{3}}} \tag{10}$$

Since the spectral peak for the vertical component can be approximated (Kaimal et al., 1976; Caughey and Palmer, 1979) by $(\lambda_m)_w = a_w\ h\ q_w$, the following relationships hold:

$$\left(f_m^*\right)_w = \frac{z}{(\lambda_m)_w} = \frac{z}{a_w\ h\ q_w} \quad \text{and} \quad n_e = \frac{U}{a_w\ h\ q_w}$$

and, consequently, eq. (10) becomes:

$$S_{w,0}(n) = \frac{0.38\,\frac{h}{U}\,(a_w q_w)^{\frac{5}{3}}\,\Psi_{\varepsilon 3}^{\frac{2}{3}}\,w_*^2}{\left(1+\frac{1.5\ n\ h\ a_w\ q_w}{U}\right)^{\frac{5}{3}}} \qquad (11).$$

We remind that, according to Caughey and Palmer (1979) and Kaimal and Finnigan (1994), for the vertical wind component $a_w = 1.8$ and $q_w = 1-\exp\left(-4\frac{z}{h}\right) - 0.0003\exp\left(8\frac{z}{h}\right)$ q_w is thus related to the vertical profile of more energetic eddies and accounts for the level of turbulence at each height.

To derive explicit relationships to be used in practical applications, it is necessary to have a good experimental estimation of n_I to be inserted in eq. (9). Following Kaimal et al. (1976), who found that the onset of the inertial subrange in the CBL occurs at wavelength $\lambda_w \approx 0.1\ h$, we assume in this work $n_I \approx (10\ U)/h$. It is of interest to point out that, using n_e values estimated by Kaimal et al. (1976) in a well mixed layer and Sorbjan (1977) in large eddy simulations, the ratio n_I/n_e ranges from 15 to 20. Consequently the condition $n_e << n_I$ is justified and the inertial subrange can be considered statistically independent of the subrange energy-containing eddies. With this assumption eqs. (9) becomes:

$$\nu_T = 1.98\text{x}10^{-3}\ h\ w_* \qquad (12)$$

Considering typical CBL values ($w_* = 2.0\ ms^{-1}$ and $h = 1500$ m), it results $\nu_T \approx 6.0$ m^2s^{-1} for the w components. It is worth noting that this value verifies the above condition $\nu_T >> \nu$.

By considering eq. (7) and by integrating eq. (6) (valid here for the 1-D turbulent energy spectrum) from n_e to infinity to obtain σ_w^2, and accounting for eq. (11) and (12), it results:

$$\frac{K_z}{w_* h} = 0.15 q_w^{11/6}\left[\int_{(1.8q_w)^{-1}}^{\infty} \frac{\exp\left(-0.16 f_h^2 w_* t/h\right)}{\left(1+2.7 f_h q_w\right)^{5/3}} df_h\right]^{\frac{1}{2}} \qquad (13)$$

where $f_h = n\ h\ /\ U$. Regarding eq.(13), it may be worthwhile to recall that a formally similar expression for the eddy diffusivity in decaying turbulence was obtained by Degrazia et al. (1997) on a completely empirical ground.

The vertical profile of RL eddy diffusivity (as given by eq. 13), obtained by inserting these typical CBL values, and normalised by dividing by $(h\ w_*)$, versus z/h, is illustrated in Fig. 1. Six profiles, evaluated at different times (t = 0, 1, 2, 4, 8 and 11 hrs), are shown in a log-linear scale. In this graph, curves begin at z/h = 0.2 in order to exclude the lower part of the PBL, in which the nocturnal ground-based stable layer is likely to develop. The K_z temporal evolution shown in this Figure is in a good qualitative agreement with the corresponding evolution presented by Freedman and Bornstein (1997). In particular, also in our case, it appears that RL eddy diffusivity is higher than that typical of the SBL by

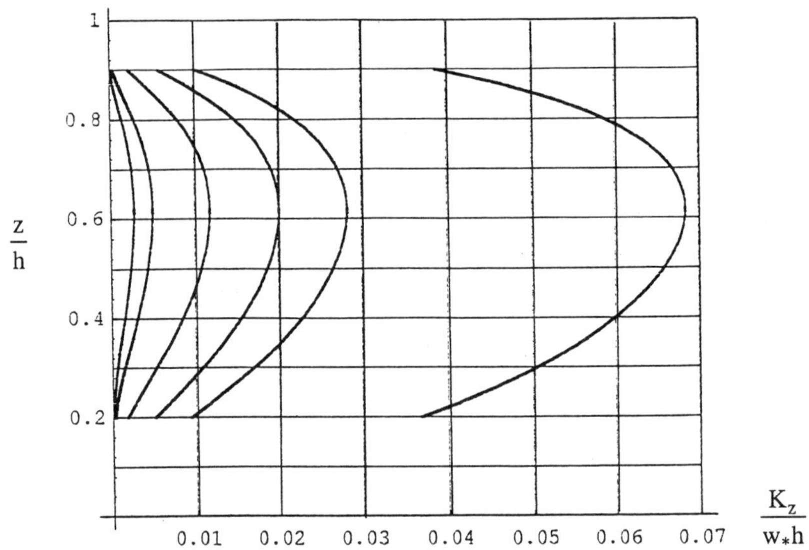

Fig. 1 Temporal evolution of the vertical profiles of normalised RL eddy diffusivity $K_z/(h\ w_*)$ as a function of dimensionless height z/h. The profiles are evaluated at six different times, t = 0, 1, 2, 4, 8 and 11 hrs

about an order of magnitude. This means, in particular, that effluents, either released directly into the RL during night-time from elevated sources or emitted during daytime and trapped in the RL because of the time evolution of the PBL structure, may be effectively diluted.

CONCLUSIONS

The problem of the theoretical derivation of a parameterization for the eddy diffusivity in decaying turbulence is addressed. This derivation makes use of the three-dimensional dynamical equation for the energy spectrum density and the classical statistical diffusion theory. The starting point is the Heisenberg's elementary decaying turbulence theory. The main assumption is related to the identification of a frequency, lying in the inertial subrange, characterising the inertial energy transfer among eddies of different size. The resulting eddy diffusivity parameterization is then applied to the decay of convective turbulence in the residual layer. Besides the intrinsic scientific interest, this topic has relevance for mesoscale transport and diffusion simulations.

REFERENCES

Batchelor G.K.: 1949, 'The role of big eddies in homogeneous turbulence', *Proc. Roy. Soc.* **A195**, 513-532

Batchelor G.K.: 1959, 'The theory of homogeneous turbulence', Cambridge monograph of mechanics and applied mathematics, Cambridge University Press, 197 pp

Chandrasekhar F.R.S.: 1949, 'On Heisemberg's elementary theory of turbulence', *Proc. Roy. Soc.* **A200**, 20-35

Caughey S.J.and Palmer S.G.: 1979, 'Some aspects of turbulence structure throught the depth of the convective boundary layer' *Quart. J. Roy. Met. Soc.*, **105**, 811-827

Degrazia G.A. and Moraes O.L.L.: 1992, 'A model for eddy diffusivity in a stable boundary layer'. *Boundary-Layer Meteorol.*, **58**, 205-214

Degrazia G.A., Moraes O.L.L., and Oliveira A.P.: 1996, 'An analytical method to evaluate mixing length scales for the Planetary Boundary Layer, *J. Appl. Meteor.*, **35**, 974-977

Degrazia G., Anfossi D., Moraes O.L.L. and Trini Castelli S.: 1997, 'A model for the turbulence parameterization in the residual layer". AIR POLLUTION V, H.Power, T. Tirabassi and C.A. Brebbia editors, 101-107

Degrazia G.A., Anfossi D., Fraga De Campos Velho H. and Ferrero E. : 1998 'A Lagrangian decorrelation time scale for non-homogeneous turbulence' *Boundary-Layer Meteorol.*, **86**, 525-534

Degrazia G. and Anfossi D.: in press, 'Estimation of the Kolmogorov constant C_0 from classical statistical diffusion theory' in press on*Atmospheric Environment.*

Freedman F.K. and Bornstein R.D.: 1997, 'Study of the spacial and temporal structure of turbulence in the nocturnal residual layer", Proceedings of 22nd NATO/CCMS International technical Meeting on Air Pollution Modelling and its Application, Clermont-Ferrand, France, June 2-6, 1997

Hanna S.R. : 1981 'Lagrangian and Eulerian time-scale i the daytime boundary layer'. J. Appl. Meteor., **20**, 242-249

Hinze J.O.: 1975, 'Turbulence', Mc Graw Hill, 790 pp

Kaimal J.C., Wyngaard J.C., Izumi Y. and Cote' O.R.: 1972, 'Spectral characteristics of surface layer turbulence' *Quart.J.Roy. Meteorol. Soc.*, **98**, 563-589

Kaimal J.C. and Finnigan J.J.: 1994, 'Atmospheric boundary layer flows', Oxford University Press, 289 pp

Kaimal J.C., Wyngaard J.C., Haugen D.A., Cote' O.R., Izumi Y., Caughey S.J. and Readings C.J.: 1976, 'Turbulence structure in the convective boundary layer'. *J. Atmos. Sci.*, **33**, 2152-2169

Kim J. and Mahrt L.: 1992, 'Simple formulation of turbulent mixing in the stable free atmosphere and nocturnal boundary layer' *Tellus* 44A, 381-394

Nieuwstadt F.T.M. and Brost R.A.: 1986, 'The decay of convective turbulence', *J. Atmos. Sci.* **43**, 532-546

Sorbjan Z.:1997, 'Decay of convective turbulence revisited', *Boundary-Layer Meteorol.* **82**, 501-515

Stanisic, M.M.: 1988, 'The mathematical theory of turbulence', Springer Verlag, Berlin and New-York, 501 pp

Stull, R.B.: 1988, 'An introduction to Boundary Layer Meteorology', Kluwer Academic Publishers, Boston, 666 pp

DISCUSSION

A. STOHL: How do your results depend on the choice of w_* and at what time do you take your w_*?

D. ANFOSSI: One of our assumptions, equal to the assumptions made by the other authors quoted in our paper, is that in the first stage of the decay process, the turbulence structure is the same as that of the previously existing CBL. This means that we consider the w_* values appropriate to the final stage of the CBL development. The idea is that w_* and z_i completely define and characterize that turbulence structure of the previously existing CBL.

D. STEYN: As the R.L. forms, turbulence decays by viscous dissipation which removes small scale turbulent motions first, leaving large scales to dominate the R.L. K theory is known to be inapplicable to cases of large scale turbulence where transfer is non-local. How can you apply K theory here?

D. ANFOSSI: By using Taylor statistical diffusion theory or Large Eddy Simulation with the incorporation of the counter-gradient correction it is possible to obtain well behaved eddy diffusivities for a CBL (t=0 from our figure). In this case the eddy diffusivities are expressed in terms of the energy-containing eddies, which are responsible for the main turbulent transport process in the CBL.

E. GENIKHOVICH: From the picture shown, one can conclude that the characteristic time for decaying turbulence is about 10 hours. Over such time period the temperature profile and other quantities will be significantly changed, gravitational waves will be generated, and the wind flow will be influenced by all these factors. Did you take them into account?

D. ANFOSSI: You are right that the factors you mention could influence the decay process. However in this paper we did not addressed these problem. For the moment, we decided to consider the decay process only. To consider also the shear-driven turbulence will be the task of future work.

ON CHEMISTRY AFFECTING THE TURBULENT FLUX AND TURBULENCE AFFECTING CHEMISTRY

Gé Verver [1]

Royal Netherlands Meteorological Institute (KNMI)
P.O.Box 201
3730AE De Bilt
The Netherlands

INTRODUCTION

In the atmospheric boundary layer (BL) simultaneous mixing and reacting of gases occur. Mixing is done by turbulence, which effectively brings the reactants together, but which also dilutes the mixture. To describe transport and chemical transformation in a model, two effects may become important: -1- the diffusion coefficient will depend on the chemical properties of the transported scalar, and therefore the conventional gradient diffusion approach to turbulent transport may fail; -2- in the case of bimolecular reactions, the mean transformation rates can significantly change due to correlated concentration fluctuations.

These interaction of mixing and chemistry can be best illustrated with the Reynolds averaged equations. As an example we consider only one bimolecular reaction ($i + j \rightarrow$ *product*). The mean concentration of species i in horizontal homogeneous conditions is given by:

$$\partial C_i / \partial t = -\partial \overline{c_i w} / \partial z - k_{ij}(C_i C_j + \overline{c_i c_j}), \quad (1)$$

where C_i and c_i are the mean and fluctuating parts of the instantaneous concentration of species i and k_{ij} is the reaction constant. If either species i or j has a chemical lifetime much longer than the timescale of turbulence, $\overline{c_i c_j}$ can usually be neglected. A relevant dimensionless number is the Damköhler number (Da), defined as the ratio of the turbulent timescale and the chemistry timescale. The importance of $\overline{c_i c_j}$ relative to the mean chemistry term is reflected in the intensity of segregation I_s, defined as:

$$I_s \equiv \frac{\overline{c_i c_j}}{C_i C_j}. \quad (2)$$

The lower limit of I_s is -1, when species i and j are fully segregated. When $I_s = 0$ the concentration fluctuations are uncorrelated, as in a well-mixed situation. When I_s is much larger than 1, both gases are concentrated in the same air parcels.

[1] I like to thank Han van Dop for presenting this paper

For a single bimolecular reaction the equation for the covariance $\overline{c_i c_j}$ reads:

$$\frac{\partial \overline{c_i c_j}}{\partial t} = -\overline{c_i w}\frac{\partial C_j}{\partial z} - \overline{c_j w}\frac{\partial C_i}{\partial z} - \frac{\partial \overline{c_i c_j w}}{\partial z} - 2\nu_c \overline{(\frac{\partial c_i}{\partial x_k})(\frac{\partial c_j}{\partial x_k})} + R_{ij}, \tag{3}$$

where the chemical term, R_{ij}, is defined as

$$R_{ij} = -k(C_i\overline{c_i c_j} + C_j\overline{c_i c_j} + C_i\overline{c_j^2} + C_j\overline{c_i^2} + \overline{c_i^2 c_j} + \overline{c_i c_j^2}). \tag{4}$$

For R_{ij} we have two effects. The correlation, either positive or negative, of concentration fluctuations are destroyed by chemistry due to the first two terms of equation (4), and will be forced to zero. Terms 3 and 4 are always negative, showing that concentration variance of one species may induce a negative covariance due to chemistry. The shape of the triple correlation is unknown, but when we assume symmetric distributions around the mean concentrations these terms are zero. Schumann (1993) showed with large-eddy simulations that due to the negative correlation of two tracers that were released near the top respectively near the bottom of a convective boundary layer, I_s could reach a value as low as -0.7, thus diminishing the effective (or "well-mixed") reaction rate by 70%.

The failure of conventional K-theory for chemical reactive gases can be illustrated by the equation for the mean flux:

$$\frac{\partial \overline{c_i w}}{\partial t} = -\overline{w^2}\frac{\partial C_i}{\partial z} + \beta\overline{c_i \theta_v} - \frac{\partial \overline{c_i w^2}}{\partial z} - \frac{1}{\rho}\overline{c_i \frac{\partial p}{\partial z}} + R_{wi}, \tag{5}$$

with, for one bimolecular reaction,

$$R_{wi} = -k_{ij}(C_i\overline{c_j w} + C_j\overline{c_i w} + \overline{c_i c_j w}). \tag{6}$$

where β is g/Θ_{v0}. Θ_{v0} and θ_v are mean and fluctuating parts of the virtual potential temperature. The chemistry term specified in equation (6), shows that the turbulent flux now depends on the chemical properties of the reactive scalar itself, as well as on the flux of the gases with which it reacts. It suggests that K, defined as the ratio of the flux and the gradient of the mean concentration, must depend on the reaction rate k_{ij}. The effect of chemistry terms in the flux equation is studied by *Vilá Guerau de Arellano and Duynkerke* (1992) and *Gao and Wesely* (1994).

Here we study the higher order chemistry effects on the fluxes and the effect of segregation on a comprehensive chemical scheme that describes the oxidation of isoprene. Isoprene is one of the most important non-methane hydrocarbons determining the background tropospheric ozone concentration. It has been suggested (*Davis*, 1992) that segregation effects of OH and isoprene may influence its rate of destruction. *Gao and Wesely* (1994) used an similar set of chemical reactions, but this study was limited to a neutral non-entraining boundary layer, in which segregation effects were neglected.

The approach consists of two steps: 1) an integration over several days of a simple two-layer model, and 2) one-hour integrations of a one-dimensional second-order closure model. We compare the results of the simple model with observations. The two-layer model provides us with a consistent and realistic set of initial and boundary conditions for the second-order closure model. The latter model solves the Reynolds averaged equations for the mean concentration, turbulent flux and covariance of concentration and temperature fluctuations (*Verver et al.*, 1997). The chemistry terms of all these equations are taken into account explicitly. To assess the impact of segregation and the chemistry terms in the second-order equations, we compare runs that include these terms with runs in which they are neglected.

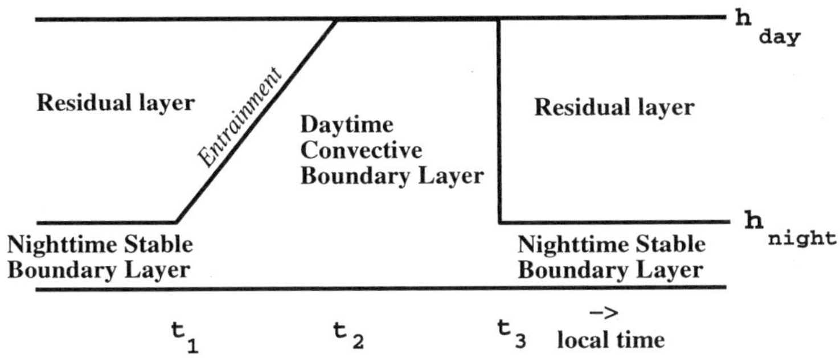

Figure 1. Prescribed boundary layer evolution during the day for the two-layer model.

In this study we focus on the Amazon Boundary Layer Experiment (ABLE-2A). This field experiment characterised chemistry and dynamics of the dry atmospheric boundary layer over the Amazon Basin during July and August 1985 (*Harriss et al.*, 1988).

Model descriptions

A hierarchy of two models is used to study the impact of higher order chemistry terms in the case of a fairly extensive chemistry scheme. Both models incorporate the same set of chemical reactions, based on work of *Atkinson et al.* (1982) and *Trainer et al.* (1987), and applied by *Gao and Wesely* (1994). Photolysis rates are calculated with an algorithm described in *Kuhn et al.* (1998). In the first model, which consists of two layers (next section), chemical transformations are calculated using mean concentrations only. The second-order closure model (*Verver et al.*, 1997) includes higher order chemistry terms that are specified further on.

The Two-Layer Model

The two-layer model incorporates the whole boundary layer (BL) as a well-mixed box with entrainment fluxes, surface fluxes and a prescribed boundary layer evolution. The equation for the layer-average concentration (indicated by $<>$) of species i in the BL is:

$$\partial < C_i >_{BL} / \partial t = \frac{Surface\ Flux(t) - Entrainment\ Flux(t)}{h(t)} + Chemistry(t), \quad (7)$$

where h is the BL height, that varies during the day. The diurnal variation of h is prescribed as depicted in figure . The entrainment flux between times t_1 and t_2 of species i is calculated by (Nappo and vDopzoeken):

$$Entrainment\ flux = \frac{dh}{dt} * (< C_i >_{BL} - < C_i >_{RL}), \quad (8)$$

1D Model description

For a full description of the one dimensional boundary layer model we refer to *Verver et al.* (1997); here we only give a brief description. The model solves the equations for scalars i in horizontally homogeneous conditions:

$$\partial C_i/\partial t = -\partial \overline{c_i w}/\partial z + R_i, \tag{9}$$

where R_i represents chemical production or destruction of the reactive species i. The model contains budgets for the second moments $\overline{c_i c_j}$, $\overline{c_i w}$ and $\overline{c_i \theta_v}$, all incorporating chemical terms that are specified in next section. Furthermore, the model uses diagnostic equations expressing third moments of concentration ($\overline{c_i c_j c_k}$, $\overline{c_i c_j w}$, $\overline{c_i w^2}$ and $\overline{c_i w \theta_v}$) in terms of second moments and their vertical derivatives.

We prescribe the profiles of the physical quantities that characterise the boundary layer and determine the turbulent transport of scalars (i.e. profiles of $\overline{w^2}$, Θ, $\overline{w\theta}$, $\overline{\theta^2}$ and the lengthscale l). These profiles are constructed using the convective velocity scale ($w_* \equiv ((g/\Theta_v)h\overline{w\theta_v})^{1/3}$) and the friction velocity (u_*) that represent the local situation at a specific time. These profiles are independent of the small mixing ratios of trace gases and are kept constant during the one-hour simulations. Boundary and initial conditions are taken from the two-layer model described in the previous section.

Chemistry Scheme

The oxidation of isoprene consists a complex chain of reactions that starts with O_3 or OH. During daytime the reaction with OH dominates (*Zimmerman et al.*, 1988), producing an isoprene peroxy radical ($RISOO2$). This is quickly transformed into stable products methyl vinyl ketone (MVK), metacrolein ($MACR$) and formaldehyde. These stable species react again with OH to form the radicals $MVKOO$, MRO_2, $MAOO_2$ and CH_3CO_3. Eventually all species are oxidised to CO and CO_2 consuming OH and transforming NO to NO_2, which leads to photochemical production of O_3.

The generic form of the chemistry terms in the flux and covariance equations are given (*Vilá Guerau de Arellano and Lelieveld*, 1998), and are explicitly taken into account in the second-order closure model.

RESULTS

In figures 2 and 3 BL concentrations of some tracers are shown during the last day of simulation with the two-layer model. The diurnal variation is a result of boundary layer dynamics, emission, chemistry and deposition. The diurnal variation of isoprene is dominated by a large surface emission during daytime, and a deposition flux during the night. BL concentration in the morning hours are also affected by dilution due to entrainment of air from the residual layer. The reaction with OH is the main chemical sink of isoprene. The model underestimates isoprene concentrations during the day by roughly 1ppb. Concentrations of ozone in the BL and RL are plotted in figure 2. They closely agree with the BL observations during ABLE2A (*Gregory et al.*, 1988; *Jacob and Wofsy*, 1988). The sharp increase of BL ozone concentration in the morning is caused by entrainment from the RL.

Although there is a continuous emission of NO, concentrations of NO_x increase more rapidly in the early morning, due to the photolyses of NO_2 which is otherwise efficiently deposited. Concentrations of NO_x above the BL are lower, so that entrainment dilutes BL concentrations of NO_x. The peak concentration of OH in the morning results in a sink of NO_x due to the formation of HNO_3. However, during daytime, the formation of PAN is the dominant chemical sink.

The intermediate products of the breakdown of isoprene, methacrolein ($MACR$) and methyl vinyl keton (MVK) show a strong diurnal cycle. Early morning is dominated by an influx from above the BL, which causes the sharp rise of concentration,

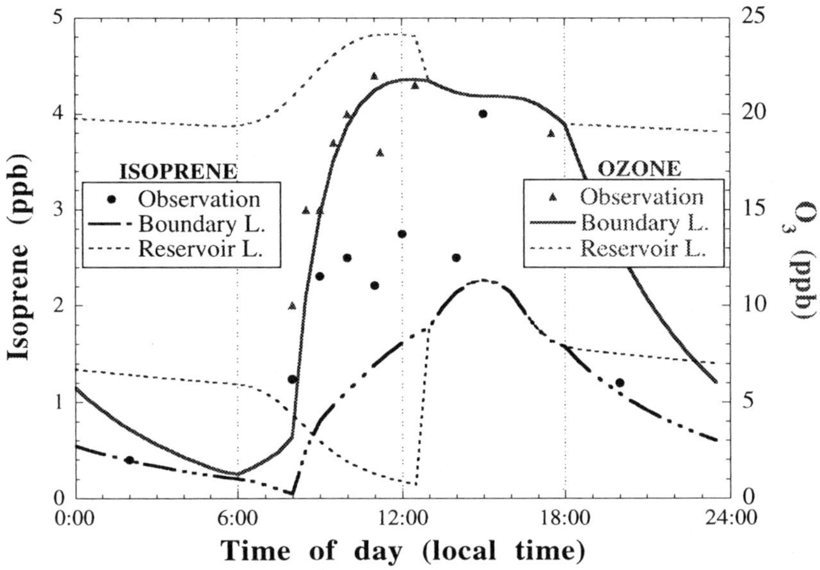

Figure 2. Observed and modelled concentrations of O_3 (*Gregory et al.*, 1988; *Jacob and Wofsy*, 1988) and isoprene (*Zimmerman et al.*, 1988; *Jacob and Wofsy*, 1988) during ABLE-2A. Model values are taken from day 5 of the simulation with the two-layer model.

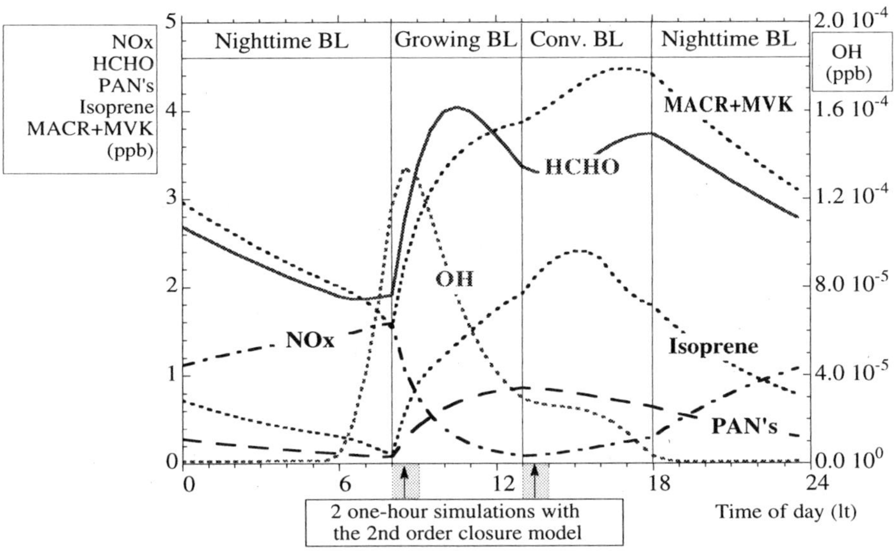

Figure 3. Modelled BL concentrations during ABLE-2A. Model values are taken from day 5 of the simulation with the two-layer model.

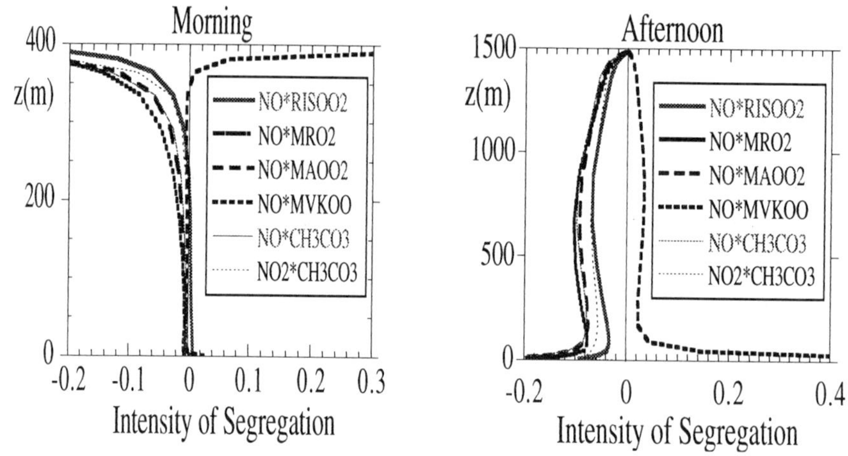

Figure 4. Intensity of segregation $I_s \equiv \overline{c_i c_j}/(C_i C_j)$ for NO and radical species or O_3.

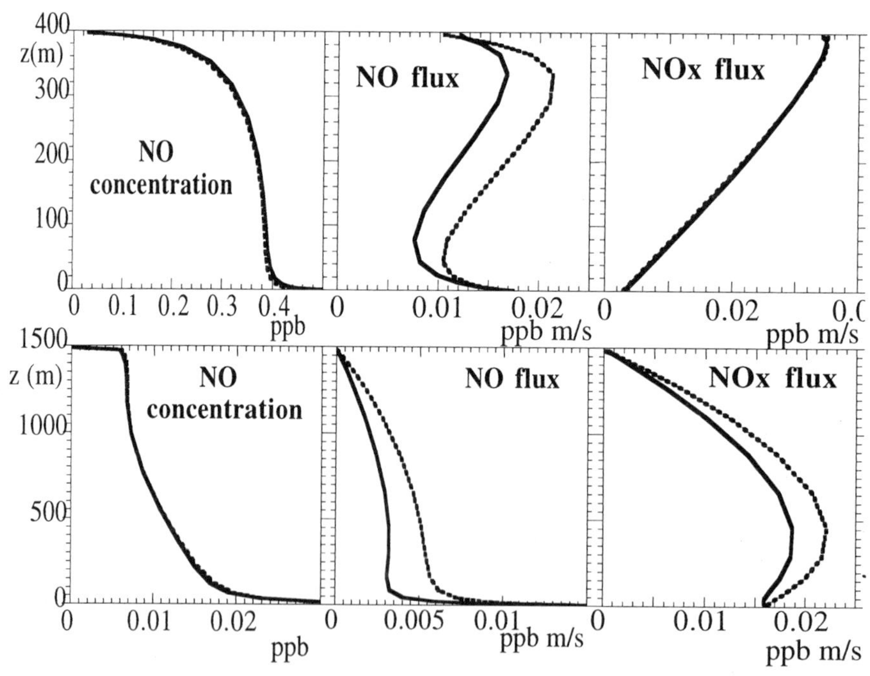

Figure 5. Concentration profiles for NO and flux profiles for NO and NO_x , in the morning (upper row) and afternoon (lower row). The solid line refers to the full 2nd-order closure model, the dashed line refers to a run where higher order chemistry terms as well as I_s are neglected.

damped a little by chemical destruction through the reaction with OH which is at its maximum concentration during these hours. Both $MACR$ and MVK are effectively removed by deposition at night. The simple model with two layers is able to represent the basic features of boundary-layer dynamics during one day. The diurnal variation of isoprene, O_3, NO_x, PAN and OH is in reasonable agreement with observations and previous model studies (*Zimmerman et al.*, 1988; *Gregory et al.*, 1988; *Jacob and Wofsy*, 1988).

Concentration, flux and segregation profiles obtained from the 1D second-order closure model are presented in figures 4 and 5. The 2nd order model is initialised with values obtained from the simple two layer model at two selected times: 8:00 LT and 13:00 LT (see arrows in figure 3). The 2nd-order model is run for one hour, which is long enough to produce linear flux profiles for the stable species such as O_3 and isoprene.

In figure 4 profiles of I_s of NO with some radical species are depicted. It is seen that in the afternoon run for most radicals roughly a 10 % reduction of the effective reaction rate is found in the middle of the boundary layer. For the morning case we find such effects in the upper quarter of the boundary layer, which is due to the entrainment fluxes that are present during this time of day. Eventually this segregation caused a small increase of radical concentrations, which resulted in similar transformation rates. The concentration profiles of NO and NO_2 remain unchanged in both cases (see figure 5). We do not find a significant effect of segregation of the reactions of OH with isoprene, MVK, $MACR$, $MGLY$ and formaldehyde.

A significant effect of the chemical terms in the flux equation for NO and NO_2 is found (figure 5). In the morning simulation the NO_x flux is unchanged due to compensating effects of the NO-flux and the NO_2-flux.

During the afternoon this is not the case, and a 10-20% decrease of the flux of NO_x is found in the middle of the boundary layer.

CONCLUSIONS

- The simple 2-layer model of the ABL gives a fair representation of the dynamics and chemistry during ABLE-2A (*Harriss et al.* (1988)).
- The intensity of segregation of OH with the stable intermediates is negligible for both cases.
- The intensity of segregation of NO with the radical species is significant.
- Fluxes of NO and NO_2 (and their flux-gradient relationships) are significantly changed due to the incorporation of chemistry terms in the flux equation.
- Higher-order chemistry terms impact on radical concentration profiles, but not on the concentrations of stables species.
- The results do not encourage the incorporation of higher order chemistry terms in regional and global scale models.

References

Atkinson, R., A. Lloyd, and L. Winges, An updated chemical mechanism for hydrocarbon/no_x/so_2 photooxidations suitable for inclusion in atmospheric simulation models, *Atmos. Environ.*, *16*, 1341–1355, 1982.

Davis, K., Surface fluxes of trace gases derived from convective profiles, Ph.D. thesis, University of Colorado, 1992.

Gao, W., and M. Wesely, Numerical modelling of the turbulent fluxes of chemically reactive trace gasses in the atmopheric boundary layer, *J. Appl. Meteorol.*, *33*, 835–847, 1994.

Gregory, G., E.V.Browell, and L. Warren, Boundary layer ozone: An airborne survey above the amazon basin, *J. Geophys. Res.*, *93*, 1452–1468, 1988.

Harriss, R., et al., The amazon boundary layer experiment (able 2a): Dry season 1985, *J. Geophys. Res.*, *93*, 1351–1360, 1988.

Jacob, D., and S. Wofsy, Photochemistry of biogenic emissions over the amazon forest, *J. Geophys. Res.*, *93*, 1477–1486, 1988.

Kuhn, M., et al., Intercomparison of the gas-phase chemistry in several chemistry and transport models, *Atmos. Environ.*, *32*, 693–709, 1998.

Trainer, M., E. Hsie, S. McKeen, R. Tallamraju, D. Parrish, F. Fehsenfeld, and S. Liu, Impact of natural hydrocarbons on hydroxyl and peroxy radicals at a remote site, *J. Geophys. Res.*, *92*, 11879–11894, 1987.

Verver, G., H. V. Dop, and A. Holtslag, Turbulent mixing of reactive gases in the convective boundary layer, *Boundary-Layer Meteorol.*, *85*, 197–222, 1997.

Vilá Guerau de Arellano, J., and P. G. Duynkerke, Influence of chemistry on the flux-gradient relationships for the no-o3-no2 system, *Boundary-Layer Meteorol.*, *61*, 375–387, 1992.

Vilá Guerau de Arellano, J., and J. Lelieveld, *Clear and Cloudy Boundary Layers*, chap. Chemistry in the atmospheric boundary layer, p. 360, Netherland Acedemy of Sciences, Amsterdam, 1998.

Zimmerman, P., J. Greenberg, and C. Westberg, Measurements of atmospheric hydrocarbons and biogenic emission fluxes in the amazon boundary layer, *J. Geophys. Res.*, *93*, 1407–1416, 1988.

DISCUSSION

The paper was presented by H. van Dop

W. GONG: Can you comment on the uncertainties of your 2nd order closure model. How much do your full solution results depend on the particular parameterisation you used to close your 2nd-order equations?

H. van DOP: We compared fluxes and covariances calculated by the model with observations of inert tracers with fairly good results. However, it is likely that the terms that are parameterised are affected by chemistry, which we did not take into account. For fast chemistry the errors may be significant. We found for instance a strange behaviour of our third moments close to the surface for OH and RISOO2. However, for moderately and slowly reacting species we expect these effects to be small.

B. FISHER: Does your conclusion that mixing can be neglected relative to chemistry in global and regional scale models still apply when one considers plumes from point sources?

H. van DOP: We do not find a strong impact of covariances of concentration fluctuations on the overall transformation rates. We defined horizontal homogeneous emissions, deposition and entrainment fluxes. In the case of inhomogeneous sources (like point emissions), other authors do find significant effects.

AN ANALYTICAL MODEL OF THERMAL INTERNAL BOUNDARY LAYER GROWTH IN NEAR-NEUTRAL ONSHORE FLOWS

Ashok K. Luhar

CSIRO Atmospheric Research
PMB 1, Aspendale, Victoria 3195, Australia

INTRODUCTION

Most analytical models of the height (h) of the thermal internal boundary layer (TIBL) employed in operational coastal fumigation modelling are based on the slab approach using a well-mixed potential temperature distribution in the vertical (e.g., Venkatram, 1977). They give a parabolic growth of h with distance inland, and are derived by assuming that the onshore flow is stable (i.e., $\gamma > 0$, where γ is the potential-temperature gradient or the lapse rate over water). While this assumption may often be valid, occurrences of near-neutral onshore flows (i.e., $\gamma \approx 0$) are not uncommon (Rayner et al., 1990). Under such conditions, these models overpredict h due to their singular behaviour at $\gamma = 0$.

The singularity problem can be overcome by including the Zilitinkevich (1975) correction (see next section), also called the 'spin-up' term, in the parameterisation of the entrainment heat flux used in the model formulation (Venkatram, 1977; Rayner et al., 1990), but only at the expense of a numerical solution. In this paper, a generalised analytical model for h is derived that is applicable under neutral to strongly stable onshore flow conditions (i.e., $\gamma \gtrsim 0$). The model formulation is based on the current framework for the slab modelling approach involving the spin-up term. The model predictions are also compared with some field measurements taken under near-neutral onshore flow conditions A detailed account of this work is given by Luhar (1998).

MODELLING THE TIBL HEIGHT USING THE SLAB APPROACH

The governing slab model equations for the TIBL growth are (e.g., Melas and Kambezidis, 1992)

$$\overline{u}\,\frac{d\theta_m}{dx} = \frac{H_o - H_i}{h}, \tag{1}$$

$$H_i = -\Delta\theta\,\overline{u}\,\frac{dh}{dx}, \tag{2}$$

$$\frac{d(\Delta\theta)}{dx} = \gamma\frac{dh}{dx} - \frac{d\theta_m}{dx}, \tag{3}$$

Air Pollution Modeling and Its Application XIII, edited by S.-E. Gryning and E. Batchvarova, Kluwer Academic/Plenum Publishers, 2000

where $\overline{u}$ is the mean wind speed, x the downwind distance, θ_m the mean boundary-layer virtual potential temperature, $H_o[= Q_o/(\rho c_p)]$ the kinematic virtual heat flux, Q_o the virtual heat flux at the surface, ρ the air density, and c_p the specific heat at constant pressure. The parameters H_i and $\Delta\theta$ are the kinematic entrainment heat flux and the potential temperature jump, respectively, at the top of the boundary layer.

To close the above set of equations, it is commonly assumed that $H_i = -C_K(w_m^3/h)/(g/T)$ (Driedonks, 1982), where g/T is the buoyancy parameter and $w_m^3(= w_*^3 + C_N^3 u_*^3)$ is the combined rate of surface mixing energy input from mechanical and buoyancy contributions (Rayner and Watson, 1991). The quantity $w_*[= (gH_oh/T)^{1/3}]$ is the convective velocity scale, u_* the friction velocity, g the acceleration due to gravity, and C_K and C_N are constants. When convective mixing dominates over mechanical turbulence with γ assumed to be constant with height and H_o independent of x, this closure yields

$$h^2 = h_o^2 + \left[\frac{2(1+2C_K)H_o}{\gamma}\frac{x}{\overline{u}}\right], \tag{4}$$

where h_o is the TIBL height at $x = 0$. The parabolic form (4) is one of the most commonly used parameterisations for the TIBL height in coastal dispersion models. A similar expression for the mechanical internal boundary layer (MIBL) can be obtained when mechanical turbulence dominates convection (i.e. when solar radiation is low and the winds are strong). There is, however, no simple means of deriving a model that combines both convective and mechanical contributions and at the same time carries an analytical (or algebraic) form. The substitution of the above closure for H_i in Eqs. (1)–(3) results in a rather complex formula for $\Delta\theta$ that does not lead to an explicit expression for h. In view of this, Gryning and Batchvarova (1990) suggested the following interpolative approximation for $\Delta\theta$ (termed $\Delta\theta_{GB}$) with the correct limiting behaviours for convective and mechanical turbulence, neglecting the influence of the initial conditions

$$\Delta\theta_{GB} = \frac{\gamma h(C_K h - C_K C_N^3 kL)}{(1+2C_K)h - 2C_K C_N^3 kL}, \tag{5}$$

where L is the Obukhov length $[= -u_*^3 T/(kgH_o)]$ and k the von Kármán constant $(= 0.4)$. An implicit expression for h can be obtained by inserting this approximation in Eq. (2).

The main problem with the above models, however, is that they approach a singular behaviour when $\gamma \to 0$. This is because the closure for H_i does not admit a dependence of H_i upon γ. Our aim here is to develop a counterpart of Eq. (4) that can be used for all γ values.

Derivation of the New TIBL Model

Zilitinkevich (1975) derived a non-stationary energy budget equation at the entrainment interface, which gives

$$\frac{g}{T}H_i = -C_K\frac{w_m^3}{h} + C_T\overline{u}\frac{w_m^2}{h}\frac{dh}{dx} \tag{6}$$

where C_T is a constant. The second term on the right hand side of Eq. (6) is the spin-up term which accounts for a finite entrainment velocity when γ is small. Substituting Eq. (6) in Eqs. (1)–(3), we obtain

$$\overline{u}\frac{dh}{dx} = \frac{C_K w_m^3}{C_T w_m^2 + g\,h\,\Delta\theta/T}, \tag{7a}$$

$$h\,\overline{u}\frac{d\theta_m}{dx} = \Delta\theta\,\overline{u}\frac{dh}{dx} + H_o. \tag{7b}$$

Although the set of Eqs. (7a,b) can account for $\gamma \approx 0$ as well as a height-varying γ, it can only be solved numerically. Here we use $C_K = 0.18$, $C_N = 1.33$ and $C_T = 0.8$ as suggested by Rayner et al. (1990) who successfully tested this model against field TIBL data. It should be noted that, while $C_K \sim 0.2$ is widely used, there is considerable uncertainty in the values of the constants C_N and C_T.

In Eqs. (7a,b) the parameters $\Delta\theta$ and h are interdependent and are both influenced by the spin-up term. Gryning and Batchvarova (1990) have attempted to include the spin-up term in deriving an algebraic expression for h. However, their model is not complete in that it does not consider the effect of the spin-up term on $\Delta\theta$ and instead substitutes expression (5) directly in Eqs. (7a) to calculate h. Furthermore, their model neglects the convective contribution in the spin-up term, which leads to a singular solution when $\gamma = 0$ under convective mixing conditions.

Assuming γ to be constant with height [as in Eq. (5)], we obtain an analytical expression for $\Delta\theta$ when $\gamma = 0$ by solving Eqs. (1)–(3) and (6) using the initial condition $\Delta\theta = 0$ at $h = h_o$; thus

$$(\Delta\theta)_{\gamma=0} = \frac{3\delta C_K}{(3+2C_K)}\left[-\frac{(h+\eta)^{2/3}}{h} + \frac{(h_o+\eta)^{\frac{2}{3}+\frac{1}{C_K}}}{h(h+\eta)^{\frac{1}{C_K}}}\right], \tag{8a}$$

where $\delta = (H_o^2 T/g)^{1/3} C_T/C_K$ and $\eta = C_N^3\, u_*^3/(g\, H_o/T)$.

We now formulate as follows an approximate expression for $\Delta\theta$ that also includes the correct asymptotic limit for the $\gamma = 0$ case

$$\Delta\theta = (\Delta\theta)_{GB} + (\Delta\theta)_{\gamma=0}, \tag{8b}$$

where $(\Delta\theta)_{GB}$ and $(\Delta\theta)_{\gamma=0}$ are given by Eqs. (5) and (8a) respectively. Under the three limiting cases of $u_* = 0$, $H_o = 0$ and $\gamma = 0$, the $\Delta\theta$ expression (8b) reduces to the same exact analytical forms as obtained from Eqs. (7a,b).

Substituting $\Delta\theta$ from Eq. (8b) in Eq. (7a) and integrating, we obtain the following implicit solution for h,

$$\frac{H_o\gamma(3+2C_K)}{3C_K\delta}\left[-\frac{\beta}{\alpha^2}(h-h_o) + \frac{\beta^2}{\alpha^3}\ln\left(\frac{\alpha h+\beta}{\alpha h_o+\beta}\right) + \frac{1}{2\alpha}(h^2-h_o^2)\right] + \left[(h+\eta)^{\frac{2}{3}} - \frac{(h_o+\eta)^{\frac{2}{3}+\frac{1}{C_K}}}{(h+\eta)^{\frac{1}{C_K}}}\right] = \left(\frac{3+2C_K}{3C_T}\right)\left(\frac{g}{T}H_o\right)^{\frac{1}{3}}\frac{x}{\overline{u}}, \tag{9}$$

where $\alpha = (1+2C_K)H_o$ and $\beta = 2C_K\, C_N^3\, u_*^3\, T/g$.

In most TIBL cases relevant to shoreline fumigation, the convection can be assumed to dominate over mechanical mixing. Under such cases Eq. (9) gives

$$\psi\,(h^2 - h_o^2) + \left[h^{2/3} - h_o^{2/3}\left(\frac{h_o}{h}\right)^{1/C_K}\right] - \chi = 0, \tag{10}$$

where $\psi = H_o\,\gamma(3+2\,C_K)/(6\,C_K\,\delta\,\alpha)$ and $\chi = [(3+2\,C_K)/(3\,C_T)](g\,H_o/T)^{1/3}(x/\overline{u})$.

If the spin-up term in Eq. (10) is neglected we obtain Eq. (4). On the other hand, if $\gamma = 0$ and $h_o = 0$ we have $\psi = 0$ and the spin-up term in Eq. (10) leads to $h = \chi^{3/2}$. When $h_o = 0$, Eq. (10) can be solved analytically to give

$$h = \frac{\left\{(2/3)^{1/3}\,\phi^{2/3} - 2\right\}^{3/2}}{2\sqrt{3}\;\psi^{3/4}\phi^{1/2}}, \tag{11}$$

where $\phi = 9\,\chi\,\psi^{1/2} + [3\,(4 + 27\,\chi^2\,\psi)]^{1/2}$. Expression (11) is the desired counterpart of Model (4) when $h_o = 0$.

When mechanical turbulence dominates, we obtain $\Delta\theta = \gamma h/2$ which, when substituted in Eq. (7a), results in an analytical solution for h [see Luhar (1998) for details].

Figure 1a compares predictions by the commonly used model (4), the numerical model (7) and the new analytical model (11) for an example case with $H_o = 0.3$ m K s^{-1} (i.e., $Q_o \approx 350$ W m^{-2}), $u_* = 0.6$ m s^{-1}, $\overline{u} = 10$ m s^{-1}, $T = 295$ K and $h_o = 0$. This comparison effectively tests the validity of the approximation given by Eq. (8b). When $\gamma = 0.05$ K m^{-1} all models estimate almost the same TIBL variations. However, as γ gets smaller, Eq. (4) progressively overpredicts h compared to the other models. For very small γ values, the spin-up term in Models (7) and (11) becomes more influential, preventing them from giving near-singular h estimates. It is clear that the new model (11) gives nearly the same predictions as Eqs. (7a,b) for all γ cases.

When γ is close to zero, the calculated temperature jump ($\Delta\theta$) in Models (7) and (11) can be slightly negative (e.g., -0.2 K), especially during the initial phase of the TIBL development, causing an accelerated boundary-layer growth. A negative $\Delta\theta$ is indicative of instability which, in the present context, corresponds to the fact that the heating of the land increases the air temperature within the mixed layer while the temperature of the (near-neutral) air above remains at a lower value. Under such conditions the inclusion of mechanical mixing in the model slows down the TIBL growth somewhat by reducing the influence of the instability. This can be observed in Figure 1a for $\gamma = 0.001$ K m^{-1} where the predictions by Model (11) (which does not include u_* effects) are larger than those by Model (7) (which includes u_* effects) in the initial phase of the TIBL development.

When h_o is non-zero, Eq. (10) can only be solved numerically. It is possible, however, to use the following method that leads to an approximate analytical solution. With $C_K \approx 0.2$ the term $(h_o/h)^{1/C_K}$ in Eq. (10) is significant only during the initial growth of the TIBL and its influence diminishes rapidly as h grows. We approximate this term as

$$\left(\frac{h_o}{h}\right)^{1/C_K} \approx \left(\frac{h_o}{h_o + c\,.\,(h)_{h_o=0}}\right)^{1/C_K}, \tag{12}$$

where $(h)_{h_o=0}$ is the TIBL height given by Eq. (11) involving $h_o = 0$ and c is a constant to be determined. Hence, the approximate analytical solution is basically the same as Eq. (11) but with χ replaced by

$$\chi_{\text{new}} = \chi + \psi\,h_o^2 + h_o^{2/3}\left(\frac{h_o}{h_o + c\,.\,(h)_{h_o=0}}\right)^{1/C_K} \tag{13}$$

When $\gamma = 0$ we obtain $h = \chi_{\text{new}}^{3/2}$.

Figure 1b presents the TIBL curves obtained using Models (4), (7) and (11) [with Eq. (13)] for the same input conditions as for Figure 1a but with $h_o = 50$ m. A value of 0.25 for c was used. As indicated earlier, the solution is sensitive to c only during the initial phase of the boundary-layer growth. The agreement between the analytical Models (11) and the numerical model (7) is again very good.

Figures 1a and 1b also indicate that the influence of h_o diminishes rapidly with distance, and that for the input H_o value considered, the influence of u_* on h predictions is practically negligible.

COMPARISON OF THE NEW ANALYTICAL MODEL WITH FIELD DATA

The model Eq. (11) is applicable for all γ values; however, we are more interested in testing it with data for near-neutral onshore flow conditions since under stable conditions

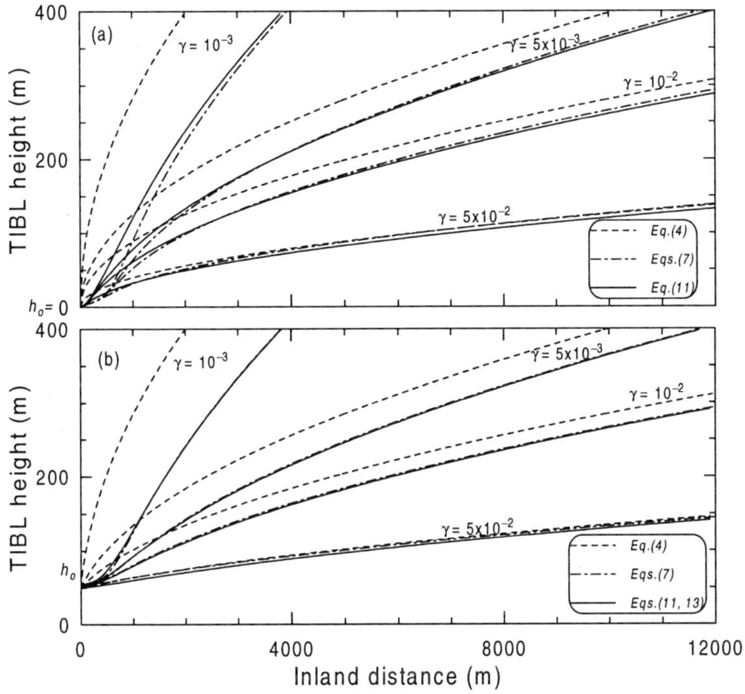

Figure 1. The TIBL predictions of Models (4), (7) and (11) for various γ values (in K m^{-1}): (a) $h_o = 0$ and (b) $h_o = 50$ m.

Table 1. Data from Kwinana experiments on 20 March 1989 (Rayner et al., 1990; Rayner, pers. comm., 1997), used in the TIBL height calculations.

Time (h)	Q_o (W m^{-2})	u_* (m s^{-1})	T (K)	$\overline{u}$ (m s^{-1})	Wind dir. (deg)	γ K m^{-1}
1425-1500	360	0.61	300	6.0	250	0.003
1510-1530	310	0.55	300	5.7	245	0.003
1725-1740	95	0.47	298	5.5	210	0.001

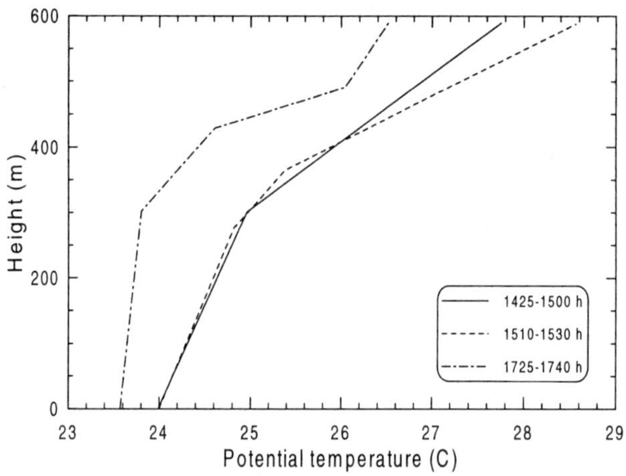

Figure 2. Observed profiles of the overwater potential temperature on 20 March, 1989.

our model behaviour is similar to that obtained by existing parameterisations that have been tested elsewhere in the literature (e.g., Stunder and SethuRaman, 1985).

Of five days aircraft TIBL data reported by Rayner et al. (1990) for the summers of 1987 and 1989 at Kwinana in Western Australian, near-neutral onshore flows were encountered on one day (20 March, 1989) in the lowest few hundred metres. Table 1 presents values of input meteorological parameters averaged over the time period of individual experiments on this day. The u_* and T values, and the data for the last two periods, not originally published, have been obtained from Rayner (pers. comm., 1997). Figure 2 presents observed profiles of the overwater potential temperature for the three experimental periods of 20 March, 1989. These profiles have been obtained from aircraft spiral data, and are presented as simplified piecewise linear profiles in the form used by our models. The spirals started at a minimum height of 60 m. The lowest sections of the profiles were produced by downward extrapolation of the measured profile.

Each temperature profile in Figure 2 shows a near-neutral layer about 300 m deep above the ground, capped by more stably stratified air. The γ values given in Table 1 correspond to the near-neutral regions in Figure 2.

Model Results and Discussion

The TIBL predictions made by various models (with $h_o = 0$ at $x = 0$) and the data are shown in Figures 3a–c for the three experimental periods of 20 March, 1989. The TIBL variations represented by the dash-dot lines in Figure 3 have been calculated using the numerical model Eqs. (7a,b) with u_* effects included. This model can account for a height varying γ and, therefore, the temperature profiles shown in Figure 2 were used directly in the computations. Since Model (4) does not account for a variable γ, it was used separately for the successive layers of the onshore flow each having a different γ value, and h_o was taken as the initial TIBL height equivalent to the bottom level of the onshore flow within which the TIBL is growing. The TIBL predictions by this model are represented by the dotted lines. The new model Eq. (11), together with Eq. (13), was applied in the same manner as Eq. (4) for the successive onshore flow layers, and the resulting TIBL variations are shown by the solid lines.

Figure 3a indicates that the TIBL growth into the near-neutral region obtained by the new analytical Model (11) closely follows the numerical model variation. The small

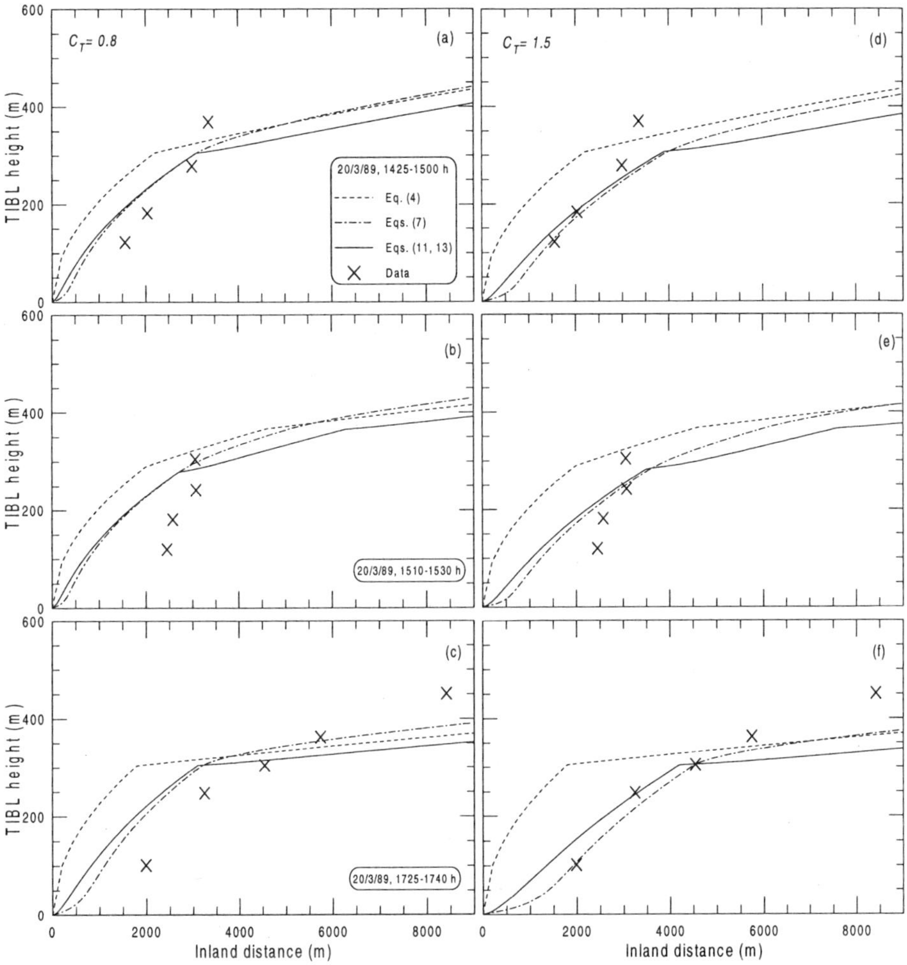

Figure 3. Comparison of predicted TIBL variations with the field data for various periods on 20 March, 1989: (a–c) $C_T = 0.8$ and (d–f) $C_T = 1.5$. Dashed lines - predictions obtained by the commonly used Eq. (4) which incorporates a parabolic TIBL variation, dash-dot lines - direct numerical solutions to Eqs. (7a,b), and solid lines - predictions obtained by the new analytical model Eq. (11) incorporating the h_o influence through Eq. (13).

differences between the two model curves in this region are solely due to the omission of the mechanical turbulence in the former. The subsequent TIBL growth into the region above the near-neutral environment, predicted by the new model, is also not much different from that obtained by the numerical model; most of this deviation is due to the implicit difference in the initial condition on $\Delta\theta$ between the two models for an onshore flow layer above the first layer. Figure 3a shows that although Models (7) and (11) overestimate the observed TIBL heights in the near-neutral region, they certainly provide improved predictions compared to Model (4). In Figure 3b, the observed TIBL growth is more rapid, but similar conclusions as given above can be drawn. For the case represented by Figure 3c, the value of γ in the near-neutral layer is smaller by a factor of three compared to the previous two cases; however, in the TIBL height calculations this is compensated by a factor of about four decrease in the value of the surface heat flux (see Table 1). Hence, both the observed and the predicted TIBL variations in Figure 3c are very similar to the corresponding ones in Figures 3a and 3b.

As mentioned earlier, while $C_K = 0.2$ is commonly used, the value of C_T is less certain. The latter becomes important when the spin-up term dominates under the near-neutral onshore flow conditions. So far we have used $C_T = 0.8$ based on an analysis by Rayner and Watson (1991). While discussing the problem of uncertainty, Driedonks (1982) suggests $C_T = 1.5$ with the emphasis that the value of C_T must follow from experiments in which the spin-up term is important. There have not been many such experiments and the uncertainty remains. Figures 3d–e show the same TIBL curves as in Figures 3a–c but with $C_T = 1.5$. The TIBL growths obtained by Model (4), which corresponds to $C_T = 0$, are unaffected but there is a significant improvement in the predictions by the numerical model and the new analytical model. It should be noted, however, that the observed data also have some limitations; for example, there is inadequate sampling time, and uncertainty in the downwind distance calculations exists due to the non-uniformity of the coastline and to the wind shear. Other limitations include the simplifying assumptions involved in the slab approach itself. Nevertheless, the above model comparison does suggest that the inclusion of the spin-up term in Eqs. (7a,b) or Eq. (11) leads to better TIBL estimates compared to Model (4).

CONCLUSIONS

In this paper, a generalised analytical model [Eq. (11)] for the TIBL height, valid for neutral to strongly stable lapse rate cases, has been derived based on the existing framework for the slab approach involving the Zilitinkevich correction. The height variation of the lapse rate (γ) was accounted for in the model by including the initial TIBL height (h_o). A form of the model, given by Eq. (9), can account for both mechanical and convective contributions to the TIBL growth and is, therefore, appropriate for use when H_o is small and/or u_* large.

The new model (11) was compared with some field data obtained under near-neutral onshore flow conditions. The performance of the model was observed to be better than the commonly used parameterisation (4). The comparison also indicated that a value of 1.5 for the model constant C_T gives better results; however, a much larger data set under near-neutral conditions would be required to obtain a definitive value.

REFERENCES

Driedonks, A. G. M., 1982, Models and observations of the growth of the atmospheric boundary layer, *Boundary-Layer Meteorol.* 23:283.

Gryning, S. E., and Batchvarova, E., 1990, Analytical model for the growth of the coastal internal boundary layer during onshore flow, *Quart. J. Roy. Meteorol. Soc.* 116:187.

Luhar, A. K., 1998, An analytical slab model for the growth of the coastal thermal internal boundary layer under near-neutral onshore flow conditions, *Boundary-Layer Meteorol.* In press.

Melas, D. and Kambezidis, H. D., 1992, The depth of the internal boundary layer over an urban area under sea-breeze conditions, *Boundary-Layer Meteorol.* 61:247.

Rayner, K. N., Bell, B. P., and Watson, I. D., 1990, Coastal internal boundary layers and chimney plume dispersion, Preprints, *International Clean Air Conference*, Clean Air Society of Australia and New Zealand, Auckland.

Rayner, K. N. and Watson, I. D., 1991, Operational prediction of daytime mixed layer heights for dispersion modelling, *Atmos. Environ.* 25A:1427.

Stunder, M. J. and SethuRaman, S., 1985, A comparative evaluation of the coastal internal boundary layer height, *Boundary-Layer Meteorol.* 32:177.

Venkatram, A., 1977, A model of internal boundary layer development, *Boundary-Layer Meteorol.* 11:419.

Zilitinkevich, S. S., 1975, Comments on 'A model for the dynamics of the inversion above a convective boundary layer', *J. Atmos. Sci.* 32:991.

DISCUSSION

J. LEVITIN: From your experimental results it leads that the TIBL growth has linear character in opposite to standard parabolic form, is it real situation?

A. LUHAR: The parabolic variation of the TIBL often relates to the situation that involves a stably stratified onshore flow. The data presented in this paper correspond to near-neutral onshore flows and, therefore, such a TIBL variation is not explicit in the data. In fact the model used in the paper gives a $x^{3/2}$ growth of the TIBL for neutral flows, so under near-neutral conditions it would be close to linear.

SECONDARY POLLUTANTS FORMATION: VARIATION OF EXPOSURE DUE TO QUANTITATIVE CHANGE IN ENERGY EMISSIONS

Alexander V. Keiko

Melentev Institute of Energy Systems
Russian Academy of Science
130 Lermontov Rd.
Irkutsk 664033 Russia

INTRODUCTION

The reactions leading to formation of secondary air pollutants such as atmospheric oxidants, aldehydes and others are now studied rather well. Nitrogen oxides and unsaturated hydrocarbons the dominant fraction of which is emitted by stationary and mobile energy sources are known to be chemical precursors for the majority of secondary products in the atmosphere.

A number of investigations have been devoted to the combined influence of NOx and hydrocarbons emissions on concentrations of relevant atmospheric species (e.g. Bowman F.M. 1994, 1994a, Poppe D. 1992, Rovinskii F.Ya. 1986). Ozone isopleth diagrams have been used for prediction of its dangerous levels on a base of its precursors concentrations and later on for testing chemical mechanisms (Tonnesen S. 1994). The endup of ozone isopleth diagrams was predefined as the actual ozone concentrations are influenced by a vast variety of other chemical, meteorological and geophysical factors. Moreover, the usefulness of correlations between concentrations of secondary pollutants and their health effects stays doubtful until we consider the exposure. Fortunately this has been understood nowadays almost everywhere.

As the energy production and the related emissions are inevitable so far it seems reasonable to clarify the span of possible variations of exposure to secondary pollutants over the whole range of primary pollution scenarios. Besides, it can be easily done with the help of a simple kinetic model.

THE MODEL SETUP

In order to disclose the target dependencies a box photochemical model has been built up, using the software **NICK 2.2** (Keiko A.V. 1996) developed at the Institute of Energy

Air Pollution Modeling and Its Application XIII, edited by S.-E. Gryning
and E. Batchvarova, Kluwer Academic/Plenum Publishers, 2000

Table 1. The subset of olefin reactions

No.	Reaction	k at 298K *)
1	C2H4+OH=HOCH2CH2	$9.00\cdot10^{-12}$
2	HOCH2CH2+O2=HOCH2CH2OO	$4.80\cdot10^{-12}$
3	HOCH2CH2OO+NO=HOCH2CH2O+NO2	$7.68\cdot10^{-12}$
4	2HOCH2CH2OO=2HOCH2CH2O+O2	$1.00\cdot10^{-13}$
5	HOCH2CH2OO+CH3OO=HOCH2CH2O+CH3O+O2	$1.00\cdot10^{-13}$
6	HOCH2CH2O=CH2OH+HCHO	24
7	C3H6+OH=CH3CH(OH)CH2	$1.05\cdot10^{-11}$
8	C3H6+OH=CH3CHCH2OH	$1.95\cdot10^{-11}$
9	CH3CH(OH)CH2+O2=CH3CH(OH)CH2OO	$4.80\cdot10^{-12}$
10	CH3CHCH2OH+O2=CH3CH(OO)CH2OH	$4.80\cdot10^{-12}$
11	2CH3CH(OH)CH2OO=2CH3CH(OH)CH2O+O2	$1.00\cdot10^{-13}$
12	CH3CH(OH)CH2OO+CH3OO=CH3CH(OH)CH2O+CH3O+O2	$1.00\cdot10^{-13}$
13	2CH3CH(OO)CH2OH=2CH3CH(O)CH2OH+O2	$1.00\cdot10^{-13}$
14	CH3CH(OO)CH2OH+CH3OO=CH3CH(O)CH2OH+CH3O+O2	$1.00\cdot10^{-13}$
15	CH3CH(O)CH2OH=CH3CHO+CH2OH	28
16	CH3CH(OH)CH2O=CH3CHOH+HCHO	28
17	CH3CHOH+O2=CH3CHO+HO2	$1.30\cdot10^{-11}$
18	CH3CHO+OH=CH3CO+H2O	$1.59\cdot10^{-11}$
19	CH3CHO+NO3=HNO3+CH3CO	$2.73\cdot10^{-15}$
20	CH3CO+O2=CH3C(O)OO	$5.00\cdot10^{-12}$
21	CH3CO+NO2=CH3COO+NO	$2.50\cdot10^{-11}$
22	CH3COO+OH=CH3OH+CO2	$7.35\cdot10^{-13}$
23	CH3COO=CH3+CO2	$1.00\cdot10^{7}$
24	CH3CHO+$h\nu$=CH3+CHO	**)
25	CH3C(O)OONO2=CH3C(O)OO+NO2	$5.80\cdot10^{-4}$
26	CH3C(O)OO+NO2=CH3C(O)OONO2	$8.40\cdot10^{-12}$
27	CH3C(O)OO+NO=CH3COO+NO2	$9.98\cdot10^{-12}$
28	CH3C(O)OO+CH3OO=CH3O+CH3COO+O2	$2.60\cdot10^{-12}$
*)	$(cm^3/molecule)^{n-1}/s$, where n is the formal order of the reaction	
**)	the function of solar irradiation	

Systems (formerly Siberian Energy Institute), RAS. The software allows to formulate and study kinetic models of photochemical processes under the conditions of lower troposphere.

The model is intended for simulation of detailed atmospheric chemistry of olefins. For that purpose the basic model describing transformations of the key atmospheric species (Keiko A.V. 1996) has been expanded with the subset of reactions listed in the table 1.

While nitrogen oxides are among the most important constituents of the background atmosphere, the consideration of unsaturated hydrocarbons has been limited by lowest olefins represented by ethene and propene as the most abundant in the energy emissions. The resulting kinetic scheme comprises 152 reactions and 52 species. Rate constant coefficients are mostly taken according to the recommendations of IUPAC (Atkinson R. 1992, Atkinson R. 1989).

The model accounts for dry deposition of nitric acid, ozone and hydrogen peroxide as well as the upward flux of fixed nirogen from the surface. These material flows have been well balanced, following the results of earlier works (e.g. Keiko A.V. 1997). The chosen diurnal variations of temperature and solar flux are typical for June 21 in Irkutsk (52°N, 104°E). The calculations have been done for cloudless conditions.

THE RESULTS

The initial air composition assumed in the base case is a typical one for the continental background atmosphere in summer. In comparison to this state energy emissions are known to cause an increase in concentrations of both the olefins and nitrogen oxides within four orders of magnitude. That is why the 48 other scenarios imply a rise in concentrations of NOx and/or either alkene to cover the whole range of their possible tropospheric concentrations upto those of the worst smog incidents ever recorded.

For each the scenario a computation over the period of 24 hours have been done starting with the sunrise. The value of exposure are calculated for the following five secondary pollutants -- ozone, hydrogen peroxide, peroxy acetyl nitrate, formaldehyde and acetaldehyde. In order to liken the results obtained both for different scenarios and for different secondary products the relative daily production (RDP) is calculated as well.

The RDP index of a pollutant originates from the general theory of sensitivity and has the form:

$$P_i = \frac{B_i - A_i}{A_i},$$

where P_i is the RDP of i-th pollutant, A_i and B_i are calculated as the integrals of concentration over time

$$A_i = \int_t C_i^\circ(t)dt \text{ and } B_i = \int_t C_i(t)dt,$$

$C_i(t)$ is the diurnal course of concentration, and the superscript zero indicates background atmosphere (base case). Thus, RDP shows the increase in exposure to a pollutant in comparison with those in clean atmospheric environment and cannot be less than -1.

The figures present RDP values of the considered pollutants as the function of initial precursors concentrations. The results show that

1) the dependence of exposure on the concentrations of secondary products precursors is highly nonlinear;
2) the nonlinearity is similar to those known for ozone productivity of hydrocarbons, and that seems to be quite natural;

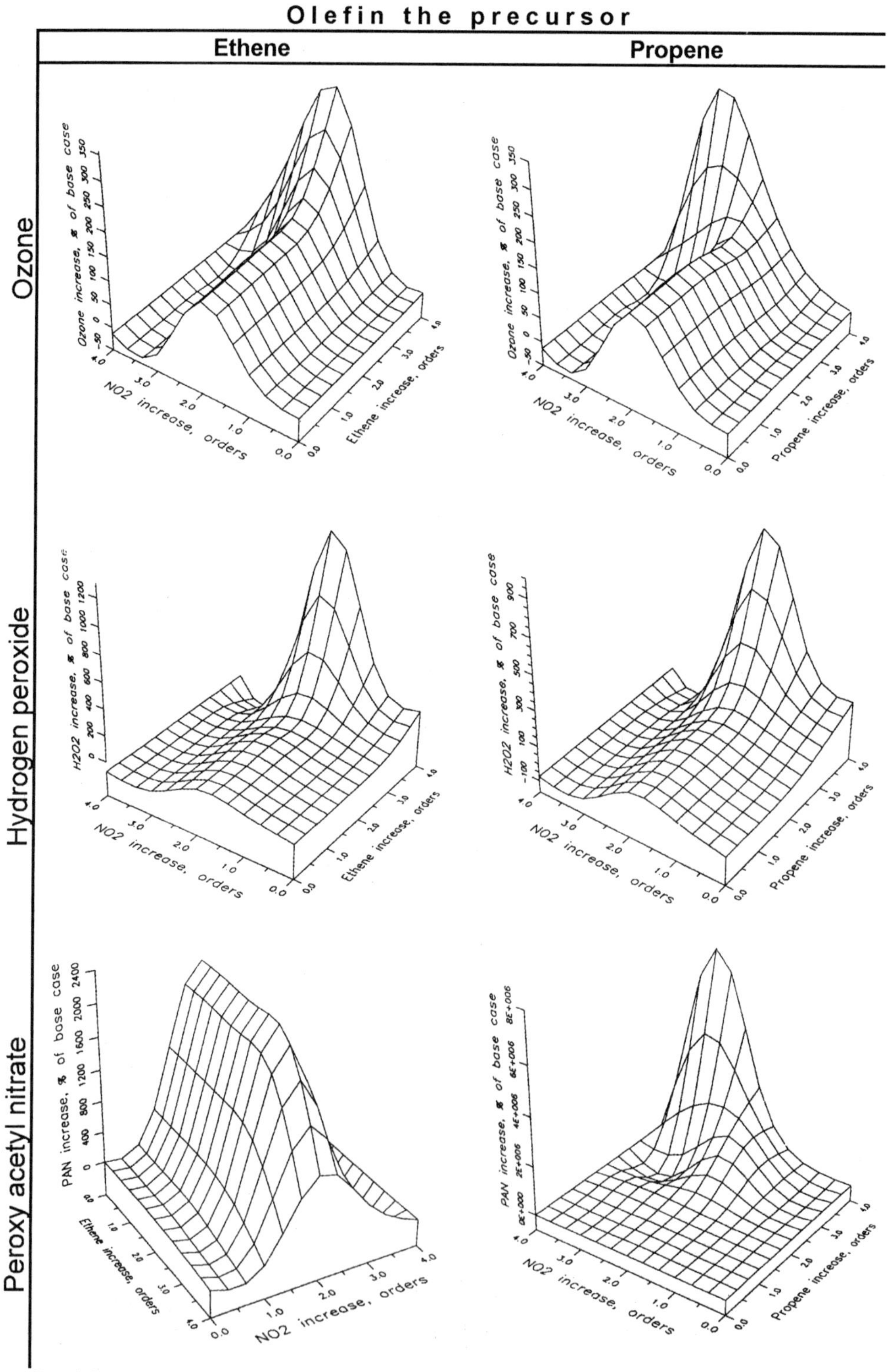

The daily exposure to secondary pollutants as a function of precursor concentrations

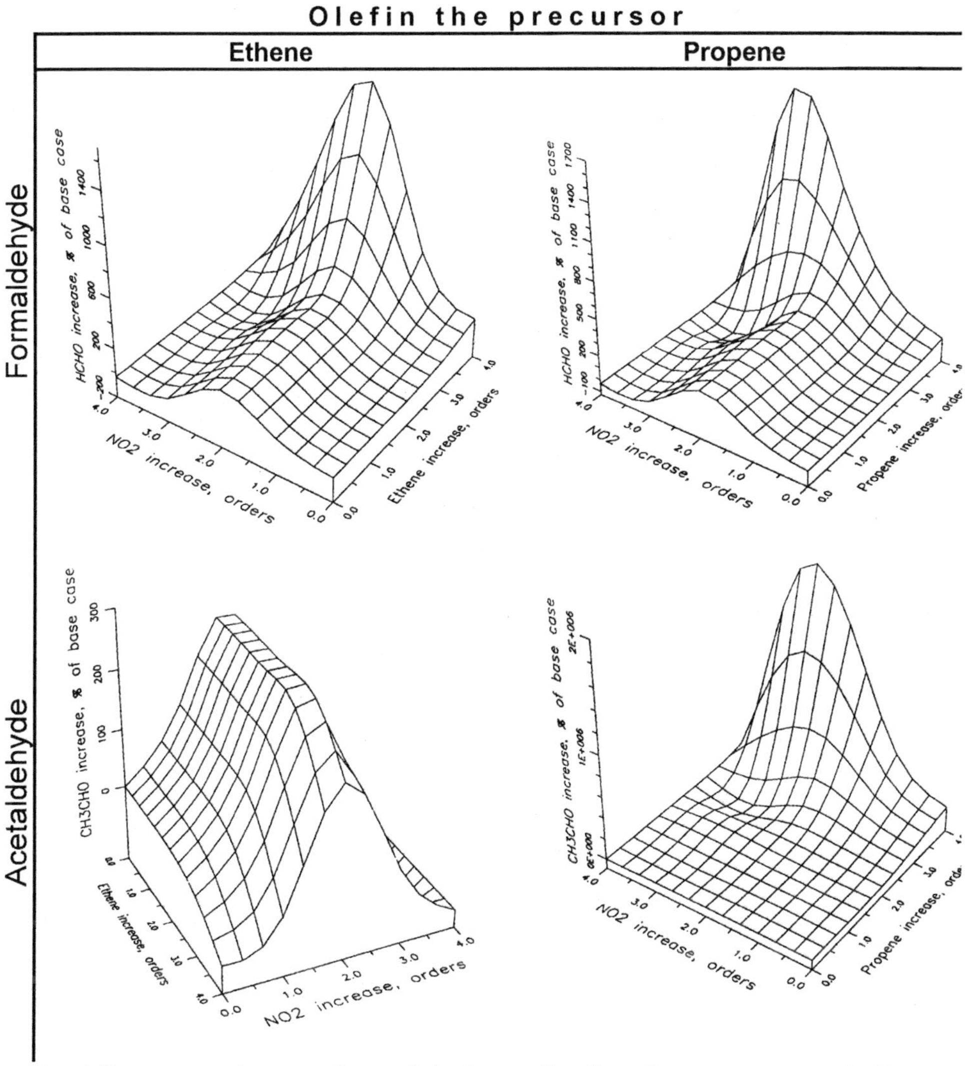

The daily exposure to secondary pollutants as a function of precursor concentrations (continued)

3) the dependence on NOx concentration has a maximum at about its hundredfold rise (ca. ten ppb);
4) the rise in olefin concentration causes monotonous increase in exposure;
5) the emission of propene leads to approximately the same change in exposure to inorganic secondary pollutants and formaldehyde (O_3 and H_2O_2) as those of ethene;
6) the exposure to C2 organic compounds is much higher in the scenarios with raised propene emissions;
7) low exposure to secondary pollutants in the cases of highest pollution by both the precursors results from higher rates of secondary products destruction;
8) the increase in exposure to PAN and acetaldehyde exceeds the increase in concentrations of precursors in the adverse ranges;
9) the exposure to C2 secondary air pollutants is nearly independent on the emission of ethene and has even reciprocal correlation at higher its concentrations.

The likeness of the shown exposure functions for ozone, hydrogen peroxide, aldehydes and PAN indicates the uniformity of chemical routs controlling secondary air pollution in general. Some of them are known, the other are yet to be understood. Anyway, the likeness means that the technique being widely developed to provide ozone control strategies is to be not less suitable in respect of other secondary pollutants.

A correlation between RDP values and short-term health statistics is expected to exist. However before investigating this correlation the dependencies of RDP on seasonal variation of solar flux and temerature as well as on cloud cover are to be clarified. Besides, the RDP value depends obviously on the time of day it is calculated over. When revealed, these functions may offer a useful tool for understanding health effects of secondary pollution.

REFERENCES

Atkinson R., Baulch D.L., Cox R.A. et al., 1992, Evaluated kinetic and photochemical data for atmospheric chemistry. *J. Phys. Chem. Ref. Data* 21(6):1125.

Atkinson R., Baulch D.L., Cox R.A. et al. , 1989, Evaluated kinetic and photochemical data for atmospheric chemistry. *J. Phys. Chem. Ref. Data* 18(2):881.

Bowman F.M., Seinfeld J.H., 1994, Fundamental basis of incremental reactivities of organics in ozone formation in VOC/NOx mixtures. *Atmospheric Environment* 28(20):3359.

Bowman F.M., Seinfeld J.H., 1994a, Ozone productivity of atmospheric organics. *J. Geophys. Res.* 99(D3):5309.

Keiko A.V., 1996, The software package for kinetic analysis of thermodynamic processes. Siberian Energy Institute, Russian Academy of Science, Irkutsk, Russia.

Keiko A.V., 1997, The dependence of OH concentration on the intensity of atmosphere-surface exchange with nitrogen species. Proc. of the 4th Int. Conf. Chem. Kinet., NIST, Gaithersburg, USA. Rep. L15.

Poppe D., 1992, Photochemical production and measured diurnal cycles of ozone, peroxides, HCHO, and PAN in the planetary boundary layer: A comparison of field data and model calculations, in: *Photo-oxidants: Precursors and Products*. SPB Academic Publishing, The Hague.

Rovinskii F.Ya., Egorov V.I., 1986, *Ozone, Nitrogen and Sulphur Oxides in Lower Atmosphere.* Gidrometeoizdat, Leningrad. (in Russian)

Tonnesen S., Jeffries H.E., 1994, Inhibition of odd oxygen production in the Carbon Bond Four and generic reaction set mechanisms. *Atmospheric Environment* 28(7):1339.

SENSITIVITY ANALYSIS OF URBAN PHOTOCHEMICAL SMOG BY USING A REDUCED VERSION OF AN AIR QUALITY MODEL

Josep Calbó

Dpt. Enginyeria Industrial
Universitat de Girona
17071 GIRONA, Spain

INTRODUCTION

A reduced version of a comprehensive urban-scale photochemistry model has recently been developed in order to make it possible the inclusion of urban-scale processes as sub-grid scale phenomena in global-scale models (Calbó et al., 1998). More specifically, this reduced model is presently being linked to the MIT Integrated Global System Model for Climate Policy Assessment (Prinn et al., 1998). The reduced model is in fact a parameterization consisting of a set of analytical expressions that approximate the predictions by the CIT Urban Airshed Model as functions of fourteen input parameters. For each species, effective emissions are a function of actual urban emissions of this and other species and of other properties such as meteorology. The probabilistic collocation method was applied to develop the parameterization. This method uses the probability density functions of the inputs to generate a set of orthogonal polynomials. These polynomials are then used as the basis for a polynomial expansion that approximates the actual response of the CIT model to its inputs. Seasonal variation is assumed to be represented by sinusoidal functions. Since the parameterization provides a computationally very efficient simulation of the actual model behavior, it is applicable to uncertainty and sensitivity analysis of the urban pollution physics and photochemistry.

Urban pollution is an intensively studied system, from a broad range of points of view, including measurements, basic physics and chemistry, and modeling. The modeling approach is a very important tool to study air pollution in an urban airshed, and in particular, to assess air pollution control policies (Zannetti, 1990). There are several studies about the sensitivity of urban pollution, i.e., about the system's response to changes in some environmental conditions, including meteorology, pollutants' emissions, and background air quality. Although there are some experimental works (e.g., Kelly, 1992), most sensitivity studies are based upon photochemical models. For example, Liu et al. (1976) studied the effect of several atmospheric parameters on concentration levels of both primary and secondary pollutants. Kleinman (1991) analyzed the seasonal dependence, which was simulated by varying some meteorological inputs, of several photochemically generated species using a zero-dimensio-

Air Pollution Modeling and Its Application XIII, edited by S.-E. Gryning
and E. Batchvarova, Kluwer Academic/Plenum Publishers, 2000

nal model. Similarly, Walcek (1997) used a photochemical box model to study the sensitivity of ozone formation rate to changes in meteorologically controlled parameters. Sillman et al. (1990) studied the sensitivity of ozone to NO_x and hydrocarbons by running several models of different complexity. Schneider et al. (1996) showed the effect of emission input data on ozone level predictions, using the EUMAC Zooming Model as the basis for their analyses. Jiang et al. (1997) analyzed the sensitivity of ozone concentrations to VOC and NO_x emissions, using the SARPC90 chemical mechanism implemented in the CALGRID model.

All the above mentioned sensitivity studies were performed by running the models several times, and changing one or more input values in each run. Since urban photochemical models are usually complex, that can be a quite time-consuming task. Nevertheless, there are several methodologies, which are, at least in theory, more attractive for sensitivity analyses. We can cite, for example, perturbation theory (Uliasz, 1983); Fourier sensitivity test (McRae et al., 1982a) and automatic differentiation (Carmichael et al., 1997). These methodologies, however, have been scarcely used in air pollution sensitivity studies.

In this paper, a sensitivity analysis of some pollutant concentrations to meteorological conditions, pollutant emissions, and background air quality is performed. The basis of the work is a reduced version of the CIT model. Both this model and the methodology used to obtain its reduced version are presented in next section. Some details about the inputs and outputs of the reduced model are then given. Subsequently, results of the sensitivity study and some discussion are presented. Finally, conclusions of the work are summarized in the last section.

BASIC TOOLS

The CIT Urban Airshed Model

The California Institute of Technology - Carnegie Institute of Technology (CIT) Urban Airshed Model is a photochemical air quality model, designed for predicting the distribution of both inert and chemically active pollutants (McRae et al., 1982b; Harley et al., 1993). The model solves the species continuity equations using the operator splitting method. Horizontal Cartesian, and vertical terrain-following coordinates are used to discretize the domain. Diffusion is handled with explicit second-order finite differences. Horizontal advection is treated using implicit fourth-order finite element integration with spatial filtering, and vertical advection is solved by means of a first-order upstream finite difference scheme. Turbulent mixing is parameterized by applying the K theory, and turbulent diffusion coefficients are functions of the Monin-Obukhov length L and surface roughness height. Dry deposition is modeled through pollutant deposition velocities, which are also functions of L.

The photochemical reaction formulation employed at present in the CIT model is based on the work by Lurmann et al. (1987). This formulation is focused on fast reactions between nitrogen oxides and volatile organic compounds that drive ozone production. Transformations from NO_x to nitrous and nitric acids, and from SO_2 to SO_3 (i.e., from oxides to aerosol precursors) are also considered. Thirty-five species (single species or lumped classes) and 106 reactions are included. Sixteen of these species are treated as directly emitted species, and fourteen of them are used to set initial and boundary conditions. Hourly, grid-resolved emission data must be provided to the model. Boundary conditions, given by the combination of surface and vertical profiles of concentrations, can also have an hourly resolution if needed. Hourly 3D fields of several meteorological variables must be also provided: wind, humidity, temperature, mixing layer height, and two radiation scaling factors (one for the whole solar radiation spectrum, the other for the UV light). These fields may either be derived from observed surface and sounding data or supplied by a meteorological model.

The collocation method

The probabilistic collocation method (Tatang et al., 1997) is a tool designed originally to simplify uncertainty analysis of computationally expensive models, and is one of several mathematical techniques for reducing a model to a simpler form. The method approximates the response of the model by a polynomial function of the inputs. Suppose we have:

$$y=f(x_1,\dots,x_k,\dots,x_K) \tag{1}$$

where x_k is one of the K input parameters to a "black-box" model f, and y is the output (or "response" to the inputs) of the model. Both input parameters and outputs of the model are considered as random variables. The goal of the collocation method is to approximate the K-dimensional "response surface," y, by a simple polynomial function of x_k. The polynomials used by the collocation technique are orthogonal polynomials $P_k^j\ (x_k)$ relative to the probability density functions (PDF) of each independent random variable x_k.

Using these polynomials, the dependent variable y can be approximated by a polynomial chaos expansion (PCE):

$$y = y_0 + \sum_{i_1,i_2,\dots,i_K}^{1\le i_1+i_2+\dots+i_K\le N} y^{i_1 i_2 \dots i_K}\ P_1^{i_1}(x_1)\,P_2^{i_2}(x_2)\dots P_K^{i_K}(x_K) \tag{2}$$

where y_0, $y^{i1i2\dots iK}$ are coefficients to be determined, and N is the order of the approximation. By running the actual model a number of times we obtain the required number of sets of (x_k, y) values to solve for y_0 and all $y^{i1i2\dots iK}$.

The orthogonal polynomials are used in the polynomial chaos approximation of the model, but they are also used to generate the sets of parameter values x_k (the so-called collocation points) for solving for the approximation through (2). Specifically, the N+1 roots of the (N+1)th order polynomial corresponding to each parameter x_k are used to define the collocation points. Because these roots help define the high probability region of each input parameter, we obtain an approximation of y that is particularly good within the most probable range of values of the input parameters. The approximation defined by (2) is therefore designed to perform better in the region of highest probability of the inputs. Versions of equation (2), i.e. the "reduced model," are developed for each output variable (y) of interest.

Table 1. Reduced model inputs

Name	Short description	Range of values (units)
DATE	Day in the year, in Julian days	[0,364] (days)
LATI	Latitude of the urban area	[0,67.5] (°N)
DTEM	Difference between actual daily mean temperature and climatic prediction	[-8,+8] (°C)
CLOU	Fraction of sky covered by clouds	[0,10] (tenths)
MIXI	Maximum mixing layer height during the day	[200,2000] (m)
TIME	Residence time, i.e. ratio between domain size and wind velocity	[6,72] (h)
SO2	Emissions of sulfur oxides SO_2 and SO_3	[0,50] (kg km^{-2} day^{-1})
CO	Emissions of carbon monoxide CO; this input drives also emissions of both VOC and NO_x	[0-500] (kg km^{-2} day^{-1})
DVOC	Deviation of actual VOC emission from those derived after input CO	[-0.7,0.7]
DNOX	Deviation of actual NO_x emission from those derived after input CO	[-0.7,0.7]
AQINOX	Determines initial and boundary concentration of NO and NO_2	[0,1] [(a)]
AQISO2	Determines initial and boundary concentration of SO_2	[0,1] [(a)]
AQIVOC	Determines initial and boundary concentration of several VOC and CO	[0,1] [(a)]
AQIOZO	Determines initial and boundary concentration of O_3 and PAN	[0,1] [(a)]

[(a)] AQIs = 0 means clean atmosphere, AQIs = 1 means rather polluted atmosphere (e.g., in the city center, [NO_2] = 300 ppb; [SO_2] = 100 ppb; [formaldehyde] = 40 ppb; [O_3] = 250 ppb)

THE REDUCED MODEL: INPUTS AND OUTPUTS

The reduced model that we derived is, specifically, a second order polynomial expansion for 13 input variables multiplied by a function of the input DATE that takes account of the annual cycle in each output y. Since a second order polynomial expression does not have a periodic behavior, we augmented the expansion by including a cosine function:

$$y = y_0 + \left[A(x_2,\dots,x_{14})\cos(\omega\, x_1 + \delta(x_2,\dots,x_{14}))\right] PCE(x_2,\dots,x_{14}) \tag{3}$$

$$PCE = \sum_{i=2}^{14} y_1^i\, P_i^1(x_i) + \sum_{i=2}^{14} y_2^i\, P_i^2(x_i) + \sum_{i=2}^{14}\sum_{j=i+1}^{14} y_{11}^{ij}\, P_i^1(x_i)\, P_i^1(x_j) \tag{4}$$

Here x_1 is the input DATE, $x_2,\dots,x_{14}$ are the other 13 inputs, A and δ are the amplitude and the phase angle to be found for each set of input parameters, and y_0 is the mean value for the output. Note again that there is one expansion (i.e., a set of coefficients) for each output.

Name and description for the inputs, as well as the assumed ranges of values are shown in Table 1. The first six inputs describe meteorological conditions, since they define temperature, incoming solar radiation, wind and turbulence. Next, there are four parameters defining pollutant emissions in the urban domain, and other four values to account for the background air quality. From the inputs, four-dimensional (time and space) meteorological fields are derived by making several simplification assumptions. Similarly, emission parameters are mapped to simulate space and time variation of actual emissions of pollutants in a city. Among other hypothesis, flat terrain, circular-shaped city, and typical cycles of temperature, mixing layer height and emissions are assumed. Emissions of VOC are speciated according to Los Angeles emission inventory used by Harley et al. (1993).

The outputs of interest for the present work, whose response surfaces have been parameterized through an expansion as above explained, are the concentrations of several pollutants. Specifically, reduced models for domain-averaged and 8 h-averaged concentrations for SO_2, NO_x, CO and O_3 are available and will be used in our sensitivity analysis.

RESULTS AND DISCUSSION

A set of reference values for the input parameters was chosen as a base case of the present analysis. This values are: DATE=171, i.e., June, 21; LATI=40 °N, i.e, a "Mediterranean" city; DTEM=0, i.e, average daily temperature, 21 °C; CLOU=2, i.e., almost clear sky; MIXI =1000 m, i.e., moderate convective activity; TIME = 1 day, i.e., 2.5 m/s wind speed in a 200 km domain length; SO2=10 kg km^{-2} day^{-1}; CO=100 kg km^{-2} day^{-1}; DNOX=0, i.e., 33 kg km^{-2} day^{-1} of NO_x; DVOC=0, i.e., 23 kg km^{-2} day^{-1} of VOC; AQINOX=0.1; AQISOX=0.1; AQIVOC=0.1; and AQIOZO=0.1. Subsequently, some of these values (one at a time or simultaneously) were changed to obtain the results shown below.

In Figure 1 averaged concentrations for the period 9 to 16 hours (solar time) are shown for SO_2, CO, NO_x and O_3 as a function of changes in mixing layer height and residence time. SO_2 and CO behave similarly, since they are almost passive on the time-scales assumed. Obviously, mixing ratios for CO are much higher than those for SO_2, since so are the respective emission rates. Note that although the pattern is simple and reasonable, it does not follow strictly the single box model behavior, represented by dotted lines. Specially for low mixing layers and long residence times (i.e., stagnant and stable conditions), a box model would predict higher concentrations. This is probably due to the effect of deposition, which is considered by the CIT model, and therefore it is also accounted for in the reduced version. NO_x apparently behaves like a passive pollutant, although a close analysis reveals some differences. This is explained by the fact that NO and NO_2 are both included in this plot. In addition, the rates of transformation of NO_2 into the final or temporal sinks (HNO_3 and PAN, basically) are relatively slow. Behavior of ozone is rather different and it has to be explained considering both dynamics and chemistry. For short residence times (i.e., windy conditions) ozone concentrations keep low despite of the mixing layer height (MLH). It might be that, when the MLH is low, a short supply of OH radicals prevent ozone from rapidly forming. In contrast, in calm conditions higher mixing layers result in lower ozone concentrations, as expected from having a larger volume of air where pollutants can be mixed.

Figure 2 is devoted to show O_3 concentrations as a function of emission rates of VOC and NO_x, and to demonstrate how the relation among these three species changes when other conditions are modified. Note that $[O_3]$ refers to domain and 8-h averages (between 9 and 16 h), thus resulting in apparently low values. After some tests, it turns out that peak $[O_3]$ in the domain is usually 10 to 20 times higher than the averages shown here. "Basic isopleths" refers to results obtained by changing only VOC and NO_x emissions, but keeping the other parameters equal to the above "base case" values. VOC emissions range between 15 and 50 kg km^{-2} day^{-1}, which correspond to 0.7 and 2.4×10^{12} C-atoms cm^{-2} s^{-1} given the assumed speciation. NO_x emissions range between 10 and 35 kg km^{-2} day^{-1}, which correspond to 2.3 and 8.0×10^{11} molecules cm^{-2} s^{-1} assuming 95% of NO_x emissions to be NO. The maximum and minimum values of the ratio VOC/NO_x emissions are 10 and 1 approximately, which correspond to typical urban areas (Gillani and Pleim, 1996). We can see in the "basic isopleths" picture that $[O_3]$ increases as VOC emissions increases or NO_x emissions decreases. Therefore, it seems that the whole region defined by the ranges of emissions lays in the so-called "VOC-limited region."

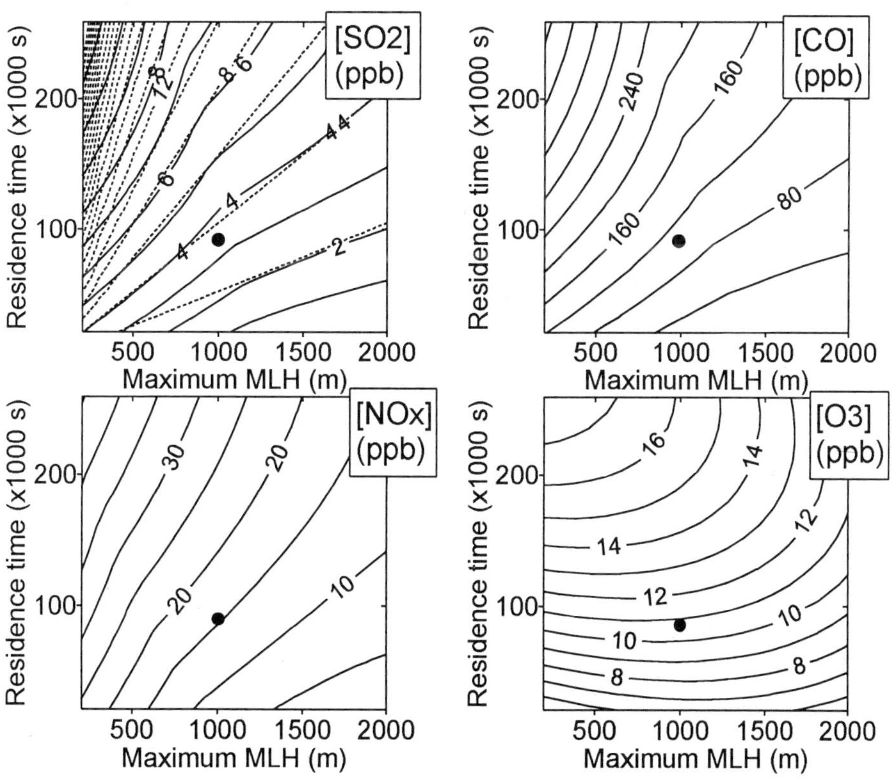

Figure 1. Domain averaged and 8 h-averaged (from 9 to 16 h) model's first layer concentrations of SO_2, CO, NO_x and O_3 as function of input parameters TIME and MIXI. Dotted lines in the upper left picture correspond to a simple box-model (i.e., [SO_2] is proportional to TIME/MIXI). The black dot is the base case.

The effect of atmospheric factors is explored in the next 4 plots of Figure 2. Note that DATE and LATI define solar radiation, temperature, and daily cycle of mixing layer height. We see that the effect of changing the parameter DATE to spring (DATE=81) or winter (DATE= =355) is to dramatically reduce [O_3]. In addition, in winter conditions, the increase of VOC emissions does not enhance so much ozone formation. Moving the city towards the tropics (LATI=15°) or to higher latitudes (LATI=60°) results in an increase and a reduction of [O_3] respectively. While the reduction is substantial (~ 40%), the increase in the tropics is less significant (~ 20%). Actually, with low VOC/NO_x emissions ratio, predicted [O_3] is almost the same (8 ppb) as at 40°. Reducing mixing layer height (MLH) has little effect on [O_3] for the base case. However, lower MLH results in lower [O_3] when the VOC/NOx emissions ratio is low, and results in higher [O_3] when the VOC/NOx emissions ratio is high. Augmenting MLH has little effect in the base case, although in general [O_3] is lowered. In these highly convective conditions, [O_3] tends to be independent of VOC emissions, particularly for high NO_x emissions. Finally, the effect of wind, which is accounted for through the parameter TIME, is as expected: [O_3] reduction in windy conditions and increase in clam conditions.

The three next plots correspond to changing characteristics of the background air quality (i.e., to changing initial and boundary conditions). If initial and incoming air is almost NO_x-free (AQINOX=0.02) predicted [O_3] concentrations are somewhat higher than in the base case. The effect of a high AQINOX value is in general to reduce the ozone outcome, although [O_3] may augment in such conditions if the VOC/NO_x emission ratio is very low. As it corresponds to a VOC limited atmosphere, a decrease of VOC in background air results in lower [O_3], while increasing VOC air quality results in a large increase of [O_3]. Finally, substantially decreasing the ozone concentration in initial and incoming air does not have significant effect on [O_3] after several hours. This means that initial ozone is almost completely removed at night even in the base case conditions. In contrast, if background air has high ozone content, it results in higher [O_3] in the domain during the morning and afternoon hours.

CONCLUSIONS

Several thousands of runs of a reduced version of the CIT urban airshed model have been used to analyze urban air pollution response to changes of some environmental conditions. The reduced model is analytical, although it still requires a computer to be run. The reduced model was obtained through the probabilistic collocation method, originally designated to perform uncertainty and sensitivity studies of complex "black-box" models. The reduced model needs 14 input parameters, including some values that affect meteorological conditions, some values that describe emissions and some values that define initial and boundary air quality conditions. Outputs of reduced model are, among others, spatial and time averages of concentrations of the most typical pollutants. The present study is an example of the ability of the methodology to perform sensitivity analysis.

Since the number of possible sensitivity analyses is enormous, given the number of inputs and outputs and the probable interactions among them, only a small fraction of results from a sensitivity study has been shown in this paper as example. In a first step, the effect of two meteorological-related parameters on concentrations of SO_2, CO, NO_x and O_3 has been investigated. Differences between our results and a simple box-model have been set, and the distinct behavior of chemically active (NO_x, O_3) and passive (SO_2, CO) pollutants has been described.

Furthermore, the sensitivity of O_3 concentration to several input parameters has been investigated. After this analysis, we can say that under the described conditions, high ozone concentrations have to be expected in summer, being high mixing layer heights unable of significantly reducing ozone. If air flowing into the urban area has high VOC content, ozone concentrations in the city will rise. The same effect will result if the incoming air does not carry any nitrogen oxides. For the city assumed here, VOC control would be required with the goal of ozone reduction, since NO_x control would result in higher ozone concentrations.

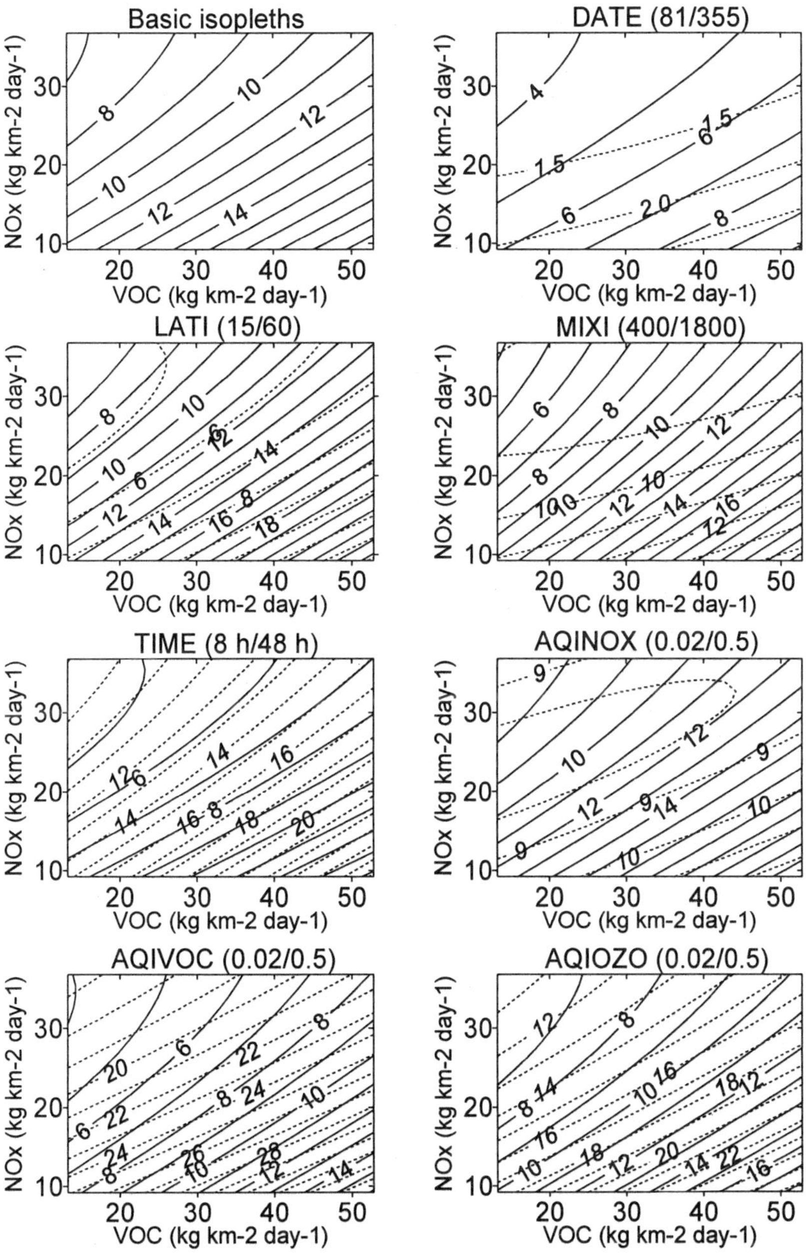

Figure 2. Ozone concentrations, in ppb, as a function of VOC and NO_x emissions. The upper left plot shows the isopleths corresponding to the base case for all parameters but DVOC and DNOX. In the other plots, the effect of giving a low (solid line) or high (dotted line) value to some parameters is shown.

It is quite difficult to compare the results presented here with other previously published results, since both input parameters and responses have different definitions. However, present results agree satisfactorily, from a qualitative point of view, with results presented for example by Milford et al. (1989), Sillman et al. (1990), Kleinman (1991) or Jiang et al. (1997). Further work is required to deeply investigate the results presented here, to perform other sensitivity analysis and to apply the methodology presented here to a real city.

REFERENCES

Calbó, J., Pan, W., Webster, M., Prinn, R.G., and McRae, G.J., 1998, Parameterization of urban sub-grid scale processes in global atmospheric chemistry models, *J. Geophys. Res.*, 103, D3: 3437-3451.

Carmichael, G.R., Sandu, A., and F.A. Potra, 1997, Sensitivity analysis for atmospheric chemistry models via automatic differentiation, *Atmos. Environ.*, 31, 3: 475-489.

Gillani, N.V., and Pleim, J.E., 1996, Sub-grid-scale features of anthropogenic emissions of NO_x and VOC in the context of regional eulerian models, *Atmos. Environ.*, 12: 2043-3059.

Harley, R.A., Russell, A.G., McRae, G.J., Cass, G.R., and Seinfeld, J.H., 1993, Photochemical modeling of the southern California air quality study, *Environ. Sci. and Technol.*, 27: 376-388.

Jiang, W., Singleton, D.L., Hedley, M., and McLaren, R., 1997, Sensitivity of ozone concentrations to VOC and NO_x emissions in the Canadian Lower Fraser valley, *Atmos. Environ.*, 31, 4: 627-638.

Kelly, N.A., 1992, A captive-air irradiation study of the response of nitric acid and peroxyacetyl nitrate to ozone control strategies in Los Angeles, *Atmos. Environ.*, 26B, 4: 463-472.

Kleinman, L.I., 1991, Seasonal dependence of boundary layer peroxide concentration: the low and high NO_x regimes, *J. Geophys. Res.*, 96, D11: 20721-20733.

Liu, M-K., Whitney, D.C., and P.M. Roth, 1976, Effects of atmospheric parameters on the concentration of photochemical air pollutants, *J. Appl. Meteorol.*, 15: 829-835.

Lurmann, F.W., Carter, W.P., and. Coyner, L.A., 1987, A surrogate species chemical reaction Mechanism for urban-scale air quality simulation models, Volume 1- Adaptation of the mechanism, Final report to the U.S. EPA under contract No. 68-02-4104, ERT Inc. and Statewide Air Pollution Research Center, University of California.

McRae, G.J., Tilden, J.W., and Seinfeld, J.H., 1982a, Global sensitivity analysis -a computational implementation of the Fourier Sensitivity Test (FAST), *Computers and Chem. Eng.*, 6: 15-25.

McRae, G.J., Goodin, W.R., and Seinfeld, J.H., 1982b, Development of a Second Generation Mathematical Model for Urban Air Pollution: I. Model Formulation, *Atmos. Environ.*, 16: 679-696.

Milford, J.B., Russell, A.G. and McRae, G.J., 1989, A new approach to photochemical pollution control: implications of spatial patterns in pollutant responses to reductions in nitrogen oxides and reactive organic gas emissions, *Environ. Sci. and Technol.*, 23: 12901301.

Prinn, R., Jacoby, H., Sokolov, H., Wang, C., Xiao, X., Yang, Z., Eckaus, R., Stone, P., Ellerman, D., Melillo, J., Fitzmaurice, J., Kicklighter, D., Holian. G., and Liu, Y., 1998, Integrated global system model for climate policy assessment: feedbacks and sensitivity studies, *Climatic Change*, in press.

Schneider, Ch., Kessler, Ch., and Moussiopoulos, N., (1996), Influence of emission input data on ozone level predictions for the Upper rhine valley, in: *Urban Air Pollution, Vol. 2*, H. Power and N. Moussiopoulos, eds., Computational Mechanics Publications, Southampton, UK.

Sillman, S., Logan, J.A., and Wofsy, S.C., 1990a, The sensitivity of ozone to nitrogen oxides and hydrocarbons in regional ozone episodes, *J. Geophys. Res.*, 95, D2: 1837-1851.

Tatang, M.A., Pan, W., Prinn, R.G., and McRae, G.J., 1997, An efficient method for parametric uncertainty analysis of numerical geophysical models, *J. Geophys. Res.*, 102: 21925-21932.

Uliasz, M., 1983, Application of the perturbation theory to sensitivity analysis of an air pollution model. *Zeits. Meteorol.*, 33: 355-362.

Walcek, C.J., 1997, Temperature-related factors affecting ozone formation in polluted areas, presented at the Air & Waste Management Association's 90th Annual Meeting & Exhibition, Toronto, Ontario, Canada, online acces at http://www.awma.org/.

Zannetti, P., 1990, *Air Pollution Modeling. Theories, Computational Methods and Available Software*, Computational Mechanics Publications, Southampton, UK.

DISCUSSION

C. MENSINK: Did you compare the results of the simplified model with the results of the original model for certain (test) cases?

J. CALBÓ: Yes, I did. Actually, I was waiting this question. I checked the results of the reduced model with those obtained by the true model. The comparison showed a good agreement in general. Only some outputs, not used in the present paper, showed poor performance. I have not shown the comparison here because of time restrictions, but detailed information about it can be found at our paper in the Journal of Geophysical Research, 1998, Vol. 103, D3, 3437-3451.

C. MENSINK: Is the simplified model able to simulate the non-linear phenomena described by the original model?

J. CALBÓ: Yes, it is in some way. We have used second order polynomials, and, as shown in the expression (4) in the paper, cross-products among the input parameters are included in the polynomial expansion.

STUDY ON AIR POLLUTANT DISPERSION IN COASTAL THERMAL INTERNAL BOUNDARY LAYER

Akinori Kouchi[1], Ryoji Ohba[1], Yanping Shao[2]

[1]Nagasaki Research & Development Center, Mitsubishi Heavy Industries, LTD., Nagasaki 851-0392, Japan
[2]Center for Advanced Numerical Computation in Engineering and Science, University of New South Wales, Sydney, Australia

INTRODUCTION

Complicated terrain and atmospheric stability are important factors in the prediction of gas diffusion in the atmosphere where a stack is located near a shoreline. In coastal onshore air flows, a thermal internal boundary layer (TIBL) often develops as a result of abrupt change in temperature between cool sea surface and warm land surface. In TIBL, turbulent properties are considerably different from those of in stable or neutral boundary layers, and these differences have a significant influence on dispersion of air pollutants in coastal area, and causes fumigation phenomenon which brings high ground level concentration (GLC) as shown in Fig.1.

In this paper, we describe a wind tunnel experiment conducted in thermally stratified wind tunnel in Nagasaki Research & Development Center, MHI, Japan for the purpose of investigating the effect of TIBL on GLCs and turbulent statistical properties. The GLCs obtained by wind tunnel experiment were compared with field observations carried out in Tokai area, Japan, 1984[1]. The both GLCs agree well with each other and it indicates validity of wind tunnel experiment.

On the other hand, numerical calculation simulating the same condition as wind tunnel experiment was also mentioned. The numerical model used in this simulation is Lagrangian stochastic model which was originally developed by CANCES (Center for Advanced Numerical Computation in Engineering and Science), University of New South Wales, Australia. This model is able to predict GLCs taking account of TIBL effect on turbulent dispersion both in stable layer (homogeneous turbulence) and unstable convective layer (inhomogeneous turbulence). The well mixed condition and Kolmogorov's local similarity theory are applied to determine the advection and diffusion coefficients in the model. The

Air Pollution Modeling and Its Application XIII, edited by S.-E. Gryning and E. Batchvarova, Kluwer Academic/Plenum Publishers, 2000

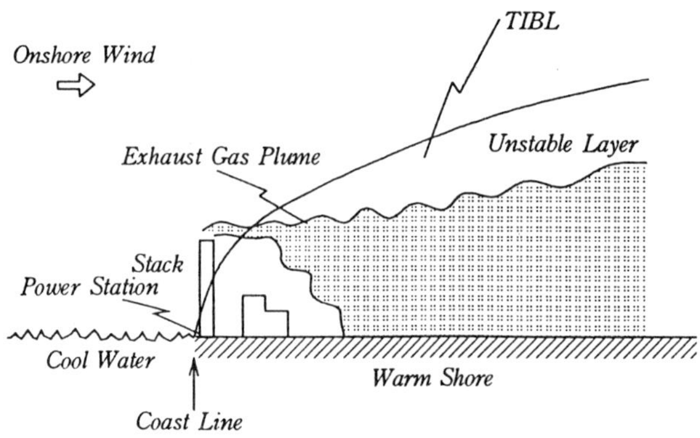

Figure 1. Schematic Diagram of TIBL Fumigation.

turbulent parameters required in the model such as velocity variance, skewness, and dissipation rate of turbulent kinetic energy are determined by the similarity theory of convective layer based on airbone observations.

This study provides quantitative comparisons between these results.

WIND TUNNEL EXPERIMENT

Experimental facilities

The thermally stratified wind tunnel shown in Fig.2 is used to simulate the sea-breeze TIBL. This wind tunnel is suction-type and has a test section with 15.5m length, 2.5m width and 1.0m height. A space-heating unit is placed upstream of the test section. This unit is horizontally divided into 10 layers and each layer has electric heater and temperature of air-flow passing in each layer can be controlled separately to generate continuous temperature stratification. The floor of wind tunnel is also divided into several panels and temperature of each panel can be controlled independently by using electrical floor heating unit and floor cooling unit. In the wind tunnel experiments simulating coastal TIBL, the air-flow is stably stratified by the space-heating unit and a floor panel corresponding to sea surface was cooled to simulate a stable layer, and then floor panel corresponding to land surface was heated to generate the unstable convective boundary layer.

Methane gas (CH_4) is used as tracer gas and released from Γ-shape stack model as a point source which simulates stack of power plant or chemical plant near the coast. To investigate the ground level concentration, arrays of sampling tubes are mounted on the wind tunnel floor, from which air samples are taken. The samples are later analyzed in the laboratory using a Hydro Carbon gas analyzer (Shimazu, Japan). Air samples are also taken from a mobile sampling tube at different places in the airflow, and from these samples the tracer gas concentration are analyzed using a fast response hydrocarbon meter, HFR-400 (Cambustion, U.K.).

The velocity components of flow are measured with a FLV (Fiber Laser Velocimeter, Nihon Kagaku Kougyo, Japan). The measurements are taken at various X (downwind distance) and Z (height) in the center of the test section. Flow temperature is also measured with a platinum resistance sensor (Cold-wire). The sampling rate for temperature measurements is 100 Hz.

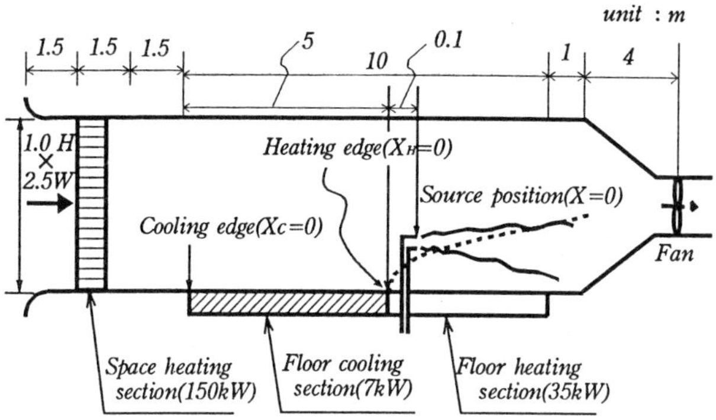

Figure 2. Schematic of wind tunnel experiment in thermally stratified wind tunnel.

Experimental method

The similarity rule of bulk Richardson number, Ri_b was applied to determine the wind tunnel flow speed and temperature difference between cooling panel and heating panel. The bulk Richardson number is defined by next equation,

$$Ri_b = \frac{g \cdot L \cdot \Delta T}{U^2 \cdot T} \qquad (1)$$

where g is the gravity acceleration, L is reference length (chosen as the height of pollutant souse), U is wind speed and ΔT is the temperature difference between two floor panels.

The detail reason of using the bulk Richardson number as the similarity rule in non-neutral atmospheric flows has been discussed by Ohba et al[2].

In this study, five cases experiments were carried out as summarized in Table 1. The results were compared with results of numerical calculations using Lagrangian stochastic dispersion model described later, and especially with regard to Case5, its result was also compared with field observation carried out in Tokai area, Japan, 1984[1].

Table 1. Experimental parameters; where x_s is the onshore distance of the source from coastline, z_s is source height, U_m mean flow speed, H_0 surface heat flux and γ temperature gradient in the stable layer. Parameters are in given in equivalent field scale values.

Case	x_s (m)	Z_s (m)	U_m (m/s)	ΔT (K)	H_0 (W/m^2)	γ (K/m)
Case1	-1000	200	6	4.5	350	0.0075
Case2	-1000	50	6	4.5	350	0.0075
Case3	100	200	6	4.5	350	0.0075
Case4	100	50	6	4.5	350	0.0075
Case5	100	120	6	4.5	350	0.0075

The parameters in Table 1 are converted from corresponding model-scaled values in the wind tunnel. The spatial scaling between wind tunnel and field observation (Case5 is corresponding to Tokai area field observation.) is 1/2000. In this scaling, according to the bulk Richardson number similarity theory, a flow speed of 0.42 m/s in the wind tunnel corresponds to a 6m/s in field scale, and a temperature difference of 45K corresponds 4.5K in field scale.

Surface sensible heat flux H_0 and temperature gradient in the stable layer are estimated from the temperature profile. The technique used to estimate the surface heat flux is as described by Ohba et al[2], where H_g is expressed as

$$H_0 = \int_0^{z_i} \frac{dT}{dx} U(z) dz \qquad (2)$$

where z_i is the TIBL height.

NUMERICAL SIMULATION

Lagrangian stochastic dispersion model

The Lagrangian stochastic dispersion model[3] was applied to simulate pollutant dispersion in TIBL.

In Lagrangian stochatic model, the movement of passive particle in a turbulent flow is described by a equation system below

$$dU_i = a_i dt + \sqrt{C_0 \varepsilon}\, d\xi_i \quad (3) \qquad\qquad dX_i = U_i dt \quad (4)$$

where U_i and X_i are the velocity and position of particle respectively, t is time and $d\xi_i$ is a random acceleration, C_0 is the Kolmogorov constant, and ε is the dissipation rate for turbulent kinetic energy. The drift coefficient a_i is determined by solving the Fokker-Planck equation assuming the well mixed condition. To apply the Lagrangian stochastic model to coastal areas, the probability density function of Eulerian velocity fluctuations, P_E, needs to be specified.

In this study, for the stable region of the coastal atmospheric boundary layer, P_E of each velocity component is assumed to have Gaussian distribution. On the other hand, within the TIBL region, P_E of u and v component are assumed to be Gaussian but no longer Gaussian for the case of w component because of influence of convection. Instead of normal Gaussian distribution, P_E of w component in TIBL is modeled using a combination of bi-Gaussian distributions, namely updraft component and downdraft component, i.e.

$$P_E(w) = pN(m_1, \sigma_1) + (1-p)N(m_2, \sigma_2) \qquad (5)$$

$$N(m, \sigma) = \frac{1}{\sqrt{2\pi}\sigma} exp\left[-\frac{(w-m)^2}{2\sigma^2}\right] \qquad (6)$$

where m_1 and σ_1 refer to mean and standard deviation of updrafts while m_2 and σ_2 to those of downdrafts; p represents the probability of updrafts; and (1-p) that of downdrafts. The first three moments are constrained by the corresponding statistic moments of the Eulerian velocity field. This leads to

$$\left.\begin{aligned} &pm_1 + (1-p)m_2 = 0 \\ &p(m_1^2 + \sigma_1^2) + (1-p)(m_2^2 + \sigma_2^2) = \sigma^2 \\ &p(m_1^3 + 3m_1\sigma_1^2) + (1-p)(m_2^3 + 3m_2\sigma_2^2) = S_k\sigma^3 \end{aligned}\right\} \qquad (7)$$

where S_k is the skewness. These equations in the five unknowns p, m_1, m_2, σ_1 and σ_2 have two degrees of freedom, so two other requirements are chosen to simplify the arithmetic by imposing the conditions $m_1 = \sigma_1$ and $m_2 = \sigma_2$. It then follows from Eq.7 that p, m_1, and m_2 must be

$$p = 0.5 - 0.5\sqrt{1 - 8/(8 + S_k^2)} \qquad (8)$$

$$m_1^2 = 0.5\sigma^2(1-p)/p \qquad (9)$$

$$m_2 = -m_1 p/(1-p) \qquad (10)$$

As can be seen, to determine the parameters in the Gaussian or bi-Gaussian distribution, turbulent properties including variances and skewness of turbulent fluctuations and the dissipation rate of turbulent kinetic energy are required.

For the stable region of the coastal boundary layer, the variance of velocity components as well as the dissipation rate for turbulent kinetic energy can be assumed to be constant and skewness is zero. In the numerical simulation presented later, these turbulence variables are specified according to the wind tunnel observations.

For the convective region of the coastal boundary layer, turbulence properties vary in space. Following Shao et al.[4] the turbulent quantities required by the Lagrangian stochastic model can be expressed as functions depending on z/z_i, when properly scaled by convective scaling velocity w_{*x}, where z_i is TIBL depth. The scaling velocity and z_i[5] are expressed as below

$$w_{*x} = \left(\frac{g}{T}\frac{H_0}{\rho C_p} z_i\right)^{1/3} \qquad (11) \qquad\qquad z_i = \left(\frac{2H_0 x}{\rho C_P \gamma U}\right)^{1/2} \qquad (12)$$

where ρ is air density and C_p is the specific heat of atmosphere. H_0 is assumed independent of the onshore distance.

The similarity relationships relevant to the present study are expressed as follows

$$\sigma_w = \sqrt{2}\, w_{*x}\left(\frac{z}{z_i}\right)^{1/3}\left(1 - 0.8\frac{z}{z_i}\right)^{3/4} \qquad (13)$$

$$S_K = 0.42\left(1 - \frac{z}{z_i}\right)\left(1 - 0.8\frac{z}{z_i}\right)^{-2} \qquad (14)$$

$$\varepsilon = \frac{1}{\sqrt{8}}\left(1.3 + 0.1\frac{z}{z_i}\right)^{3/2}\frac{w_{*x}^3}{z_i} \qquad (15)$$

where σ_w and S_k are the standard deviation and skewness of the vertical velocity component respectively.

Fig.3 is a comparison of σ_w^2 normalized with w_{*x}^2 between the wind tunnel observations and the field observations[4]. The solid line in the Fig.3 represents Eq.13 used in the numerical model. They show good agreement and this means that the assumptions of similarity relationships used in the model is valid.

Fig.4 also shows the comparison of TIBL height between the model (Eq.12), wind tunnel measurement and field observation of Tokai area. As can be seen, the Eq.12 fits well to the observed data and it indicates the reasonableness of the approximation using Eq.12.

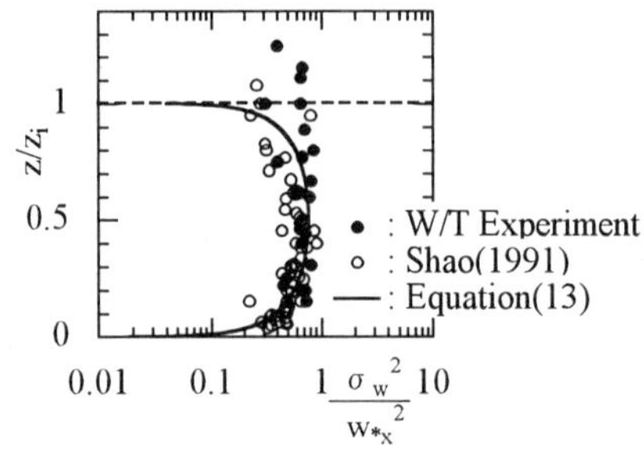

Figure 3. Comparison of Turbulent Intensity.

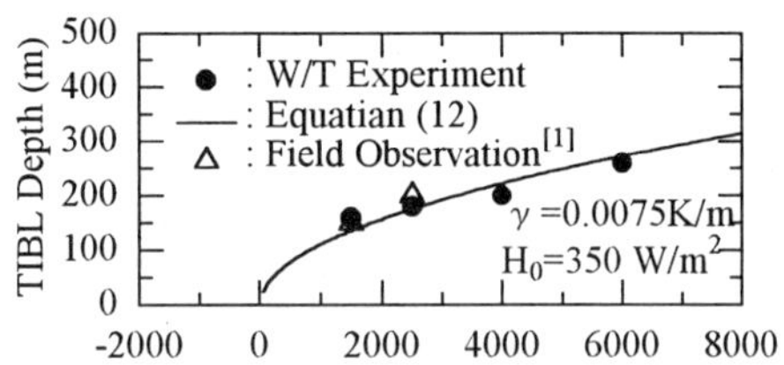

Figure 4. Comparison of TIBL Depth.

Table 2. Summary of parameters used in the numerical simulation. The velocity variance and dissipation rate of turbulent kinetic energy ε were measured in the stable region of the wind tunnel flow.

H_0 (W/m^2)	γ (K/m)	U_m (m/s)	σ_u (m/s)	σ_v (m/s)	σ_w (m/s)	ε (m^2/s^3)
350	0.0075	6	0.25	0.25	0.25	0.001

RESULTS

Comparison of GLC

Ground level concentration (GLC) is the major concern for practical environmental assessment. In this section, comparisons of axial GLC are shown.

The parameters used in the numerical simulation are estimated from the wind tunnel measurements, as summarized in Table 2. In Table 2, the turbulence statistics are measured for the stable region of the coastal boundary layer, while for the convective region, the turbulence properties are specified according to the similarity laws described in the previous section.

Numerical simulations for five cases shown in Table 1 were carried out and compared with the wind tunnel observations. Fig.5 gives a calculated result of the X-Z cross-section of concentration contours. The basic features of the simulated concentration fields are as expected. In the stable region the dispersion of the pollutant are relatively small and show typical characteristics of Gaussian plumes. On the other hand in the convective region particles are rapidly dispersed to lower layers, increasing the ground level concentration.

Comparison of axial GLC between the numerical simulation, the wind tunnel experiments and field observation(for Case 5 only) is shown in Fig.6. The overall agreement between the numerical simulation and the wind tunnel observations is very good.

In Case 5, comparison with field observation is given and the agreement is good. The best agreement appears in Case 1 and 3, for which the source height is 200m. The worst agreement is Case 4, for which the source height is 50m. For this case, maximum GLC is at a smaller distance than observed. Possible reason of this discrepancy is as follows. In the numerical model, turbulence properties are determined according to the similarity theory of convective layer, and therefore may not properly describe the features of turbulence near the surface. In this case, the discrepancy between wind tunnel measurements and numerical model results will be most obvious.

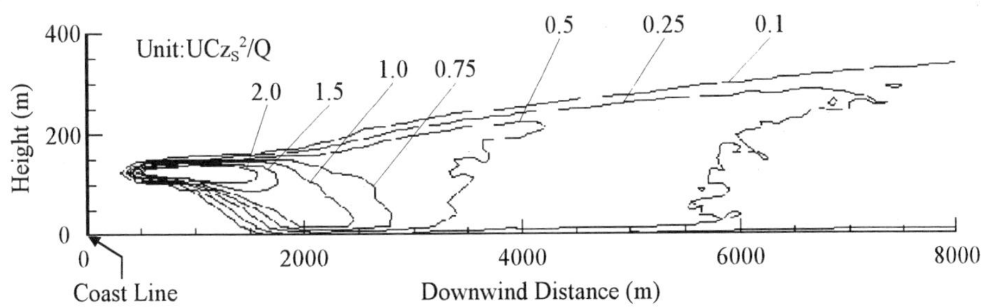

Figure 5. Contour of Concentration in Vertical Cross Section.

CONCLUSIONS

In this study, numerical simulation and wind tunnel experiment are conducted for the purpose of investigating the fumigation in the coastal atmospheric boundary layers.

Lagrangian stochastic dispersion model was applied to this problem and its results were compared with wind tunnel results and filed observation. As far as the axial ground level concentration is concerned, they showed good agreement and this confirms the usefulness of Lagrangian stochastic model in practical application to predict GLC in coastal areas.

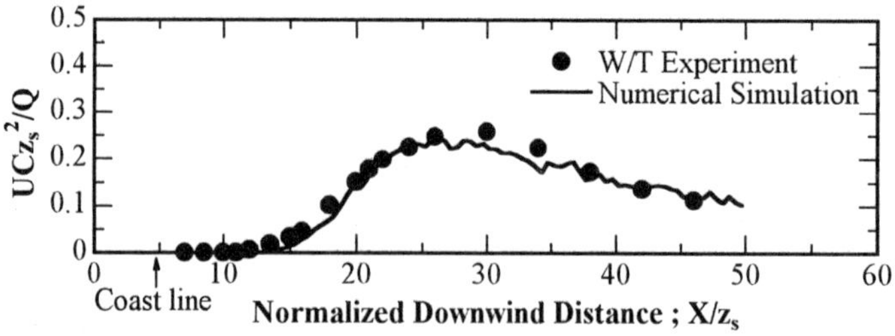

(a) CASE1 (z_s=200m, x_s=-1000m)

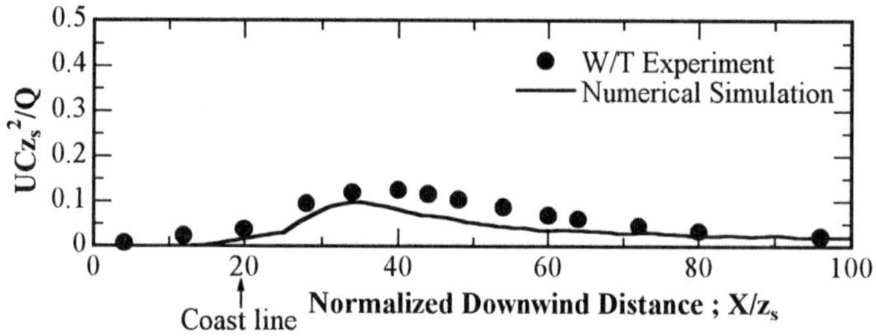

(b) CASE2 (z_s=50m, x_s=-1000m)

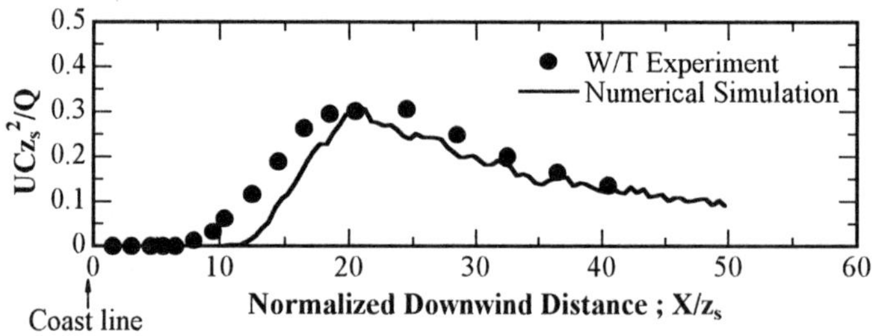

(c) CASE3 (z_s=200m, x_s=100m)

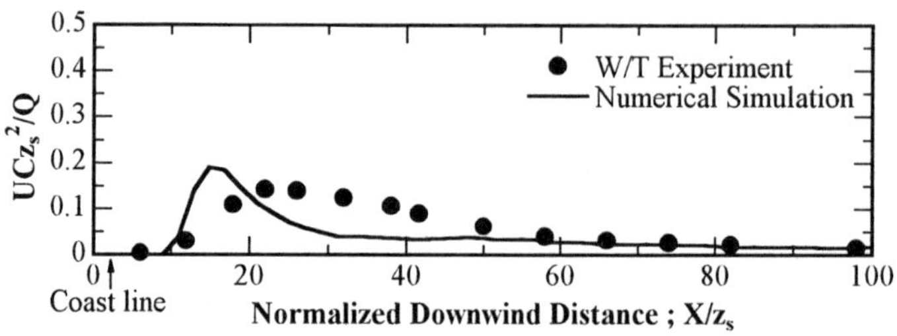

(d) CASE4 (z_s=50m, x_s=100m)

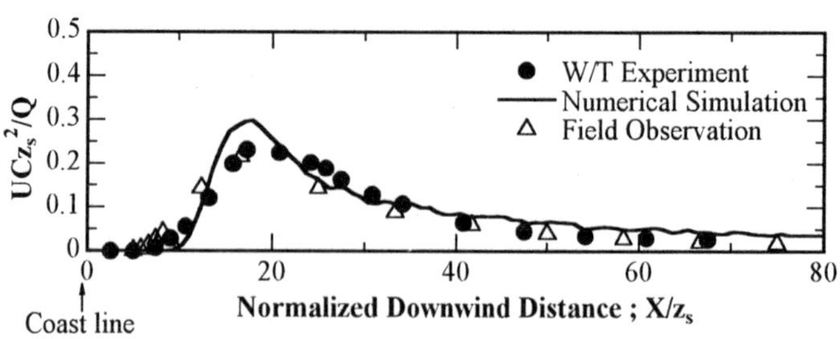

(e) CASE5 (z_s=120m, x_s=100m)

Fig.6. Comparison of axial GLC. X-axis means the downwind distance from source normalized with source height z_s, Y-axis means axial GLC normalized with U, z_s and source strength Q.

REFERENCES

[1] M. Kakuta, T. Hayashi : Results of atmospheric diffusion experiments Vol.2 TOKAI82,TOKAI83, JAERI-M 86-097, Vol.2, 1986.

[2] R. Ohba, S. Kakishima and S. Ito : Water tank experiment of gas diffusion from a stack in stably and unstably stratified layers under calm conditions, Atmos. Environ., Vol.25, pp.2063-2076, 1991.

[3] Y. Shao: Turbulent dispersion in coastal atmospheric boundary layers: an application of Lagrangian model., Boundary Layer Meteorology, Vol.59., pp.363-385, 1992.

[4] Y. Shao, J. M. Hacker and P. Schwer
dtfeger: The structure of turbulence in a coastal atmospheric boundary layer, Quart. J. Roy. Meteorol. Soc., Vol.117, pp.1299-1324, 1991.

[5] B. Weisman: On the criteria for the occurrence of fumigation inland from a large lake – A reply., Atmospheric Environment, Vol.12, pp.172-173, 1976.

DISCUSSION

D. WILSON: Wind tunnel simulation of heated surface has a serious problem in getting the correct heat flux from the surface because of the large difference in Reynolds numbers between wind tunnel and full scale. Please tell me how you addressed this problem in your model that simulated surface temperature by a bulk Richardson number matching.

A. KOUCHI: In the wind tunnel experiment, we did not directly adjust the surface heat flux. Because there are considerable difficulties in adjusting and measuring the surface heat flux from the wind tunnel floor and now we have no strictly correct opinion regarding similarity rule to be satisfied. Therefore in this study we satisfied the bulk Richardson number similarity rule concerning the surface temperature. However the surface heat flux estimated from vertical temperature profile in the wind tunnel agreed in almost with the value observed in the field experiment.

D. WILSON: You might try to adjust heaters in the wind tunnel to produce the expected surface heat flux directly, (rather than the surface temperature).

FOOTPRINT ANALYSIS FROM RANDOM WALK MODELS – SENSITIVITY TO BOUNDARY CONDITIONS

Anne Katrine Vinther Falk and Sven-Erik Gryning

Risø National Laboratory
DK-4000 Roskilde
Denmark

INTRODUCTION

Footprint analysis is the discipline of determining the relative contribution from an upwind source to a measured vertical flux of a passive scalar. This relationship between source and receptor is of interest for calculation of short-range dispersion from surface area sources, e.g. a chemical spill, as well as for aggregation of subscale fluxes used for input in numerical weather prediction models. The latter is often used to provide wind fields for long-range atmospheric transport and diffusion models.

In this paper the analysis is based on backward trajectories from a random walk model and it is the aim to develop a method of footprint analysis which applies to measurement points above the surface layer. Under convective conditions the turbulence in the mixed layer is skewed, therefore "perfect reflection" distorts the velocity probability density function (PDF) at the boundaries. Since the footprint depends on the near-ground concentrations it is interesting to investigate how sensitive it is to the boundary condition. We have calculated the footprint using three different reflection algorithms namely "perfect reflection", the scheme proposed by Weil (1990) and the scheme proposed by Thomson and Montgomery (1994).

In order to exclude effects from height-inhomogeneities the investigation was carried out in height-homogeneous turbulence. Theoretically the use of an incorrect reflection scheme gives rise to a systematical error in the contribution to the footprint. Our numerical experiments confirm this theory since cumulative footprint curves obtained from simulations where "perfect reflection" was used systematically take lower values than those obtained from simulations where the algorithm of Thomson and Montgomery and the algorithm of Weil were used. The scheme of Thomson and Montgomery and the scheme of Weil perform equally good and it is hard to tell which one of them is doing the best job.

REFLECTION SCHEMES

Boundary conditions in atmospheric flows are complicated, since to an example the ground almost never acts as a smooth wall, in the vicinity of which the velocity statistics can be specified. Instead there is a layer adjacent to the ground called the unresolved basal layer (UBL) where the velocity statistics are unknown. When a flow close to this boundary is modelled it is thus not possible to track particles in the UBL and other methods for handling these situations must be used.

In a random walk model the boundary can be treated in two ways: Either by preventing particles from crossing it at all (nonattainable boundaries (Wilson and Flesch, 1993)) or by reflection of particles which actually cross the boundary. In the following we will concentrate on reflection and envisage the problem from an "applied point of view" in the context of footprint analysis.

Perfect reflection. It is a well-known problem in random walk models that when the turbulence is skewed (i.e. the PDF of the velocity fluctuations is skewed) then the simulated velocity distribution is distorted near the boundaries if the particles are reflected with their incident velocities. The problem is illustrated in fig. 1, where the skewed PDF used for the simulations in this paper is shown (solid line). The dashed line is the distribution of particles reflected at the lower boundary. Since the PDF is not symmetric the "reflected distribution" does not coincide with the desired PDF and there will be a deficit of particles with small positive velocities (where the reflected distribution takes lower values than the desired) and a surplus of particles rising rapidly. A reflection scheme that retains the PDF at the lower boundary should in this case map the incident velocity (w_i) into a reflected velocity (w_r) that is smaller than the absolute value of w_i, $|w_r| \leq |w_i|$. To retain the PDF at the upper boundary the reflection scheme should provide $|w_r| \geq |w_i|$.

The scheme of Weil. Weil (1990) proposed a reflection scheme for skewed turbulence that provides a bijective mapping of the cumulative probability for the negative velocities onto the cumulative probability for the positive velocities. For the lower boundary this scheme reads

$$\frac{\int_{-\infty}^{w_i} P(z_b, w)dw}{\int_{-\infty}^{0} P(z_b, w)dw} = \frac{\int_{w_r}^{\infty} P(z_b, w)dw}{\int_{0}^{\infty} P(z_b, w)dw} \tag{1}$$

where $P(z, w)$ is the PDF of the vertical velocity fluctuations w at the height z and z_b is the height of the reflecting boundary. Generalization to the upper boundary is obvious.

The condition of Thomson and Montgomery. Thomson and Montgomery (1994) argues that since the net flux at the boundary must equal zero, then a reflection scheme (at the bottom) must fulfill the condition

$$\int_{-\infty}^{0} w_i P(z_b, w_i)dw_i = -\int_{0}^{\infty} w_r P(z_b, w_r)dw_r \tag{2}$$

This is a very weak condition, which can be implemented in more than one way. Thomson and Montgomery (1994) give two possibilities: either to choose the reflected velocity at random from a distribution proportional to $w_r P(z_b, w_r)$ or to use a bijective map of the cumultive incoming flux onto the cumulative outgoing flux. The latter can again be done in more than one way, Thomson and Montgomery themselves suggest

$$\int_{-\infty}^{w_i} wP(z_b, w)dw = -\int_{w_r}^{\infty} wP(z_b, w)dw \tag{3}$$

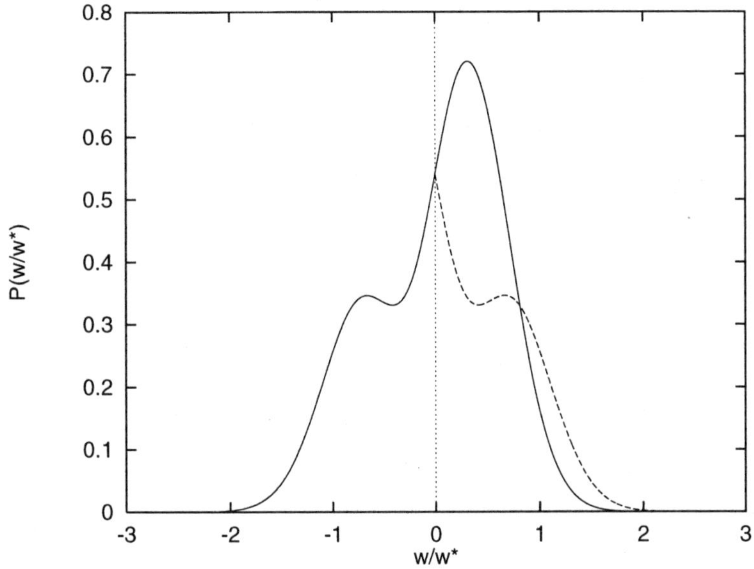

Figure 1. The solid line is the skewed PDF used for the simulations in this paper. The dashed line is the distribution of particles that are reflected at the lower boundary with their incident velocities. Velocities are scaled by the convective velocity scale w_*.

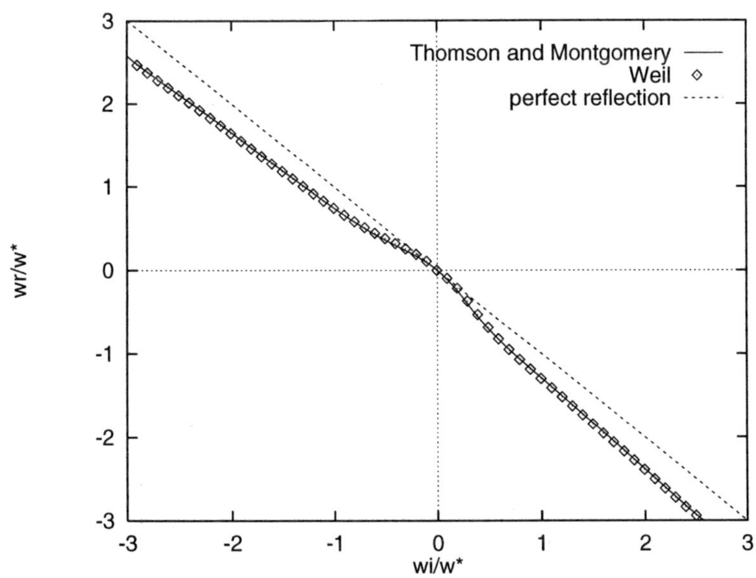

Figure 2. Reflected velocity, w_r, as function of incident velocity, w_i. The left half of the figure is reflection at the ground (negative w_i) and the right half is reflection at the top.

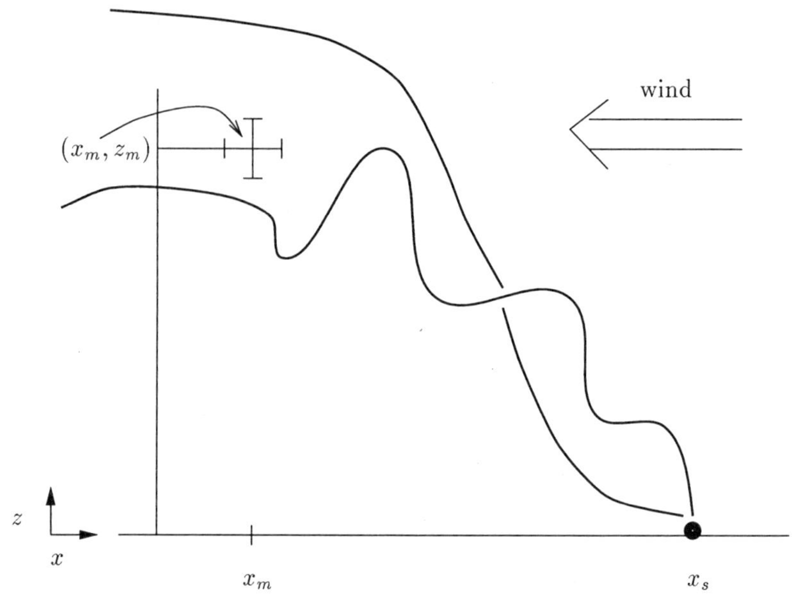

Figure 3. Measurement point and point source.

which is the scheme that in the sequel will be referred to as "the scheme of Thomson and Montgomery".

In fig. 2 the reflected velocity versus incident velocity for the PDF in fig. 1 is shown for the three reflection schemes. Reflection at the ground is drawn in the left half of the graph (negative incident velocities) while reflection at the top of the boundary layer is drawn in the right half (positive incident velocities). Perfect reflection gives a straight line of slope -1 since $w_r = -w_i$. Neither of the two other schemes can be solved explicitely for w_r, instead a table is calculated for both the lefthand and righthand side and the values compared in order to find w_r. It turns out that the scheme of Thomson and Montgomery and the scheme of Weil predict almost identical reflected velocitites and both of them posess the property of reflecting particles with higher speed at the top and lower speed at the ground.

BASIC CONCEPTS OF FOOTPRINTS

The analysis in the following section is carried out in a crosswind integrated (CWI) frame under the assumptions listed below

- Horizontal and vertical transport are uncorrelated
- The vertical transport is only due to turbulent velocity fluctuations (no vertical mean wind)
- The horizontal transport is only due to the mean wind, i.e. streamwise diffusion is neglected.

Consider the situation in fig. 3 where a sensor is placed in the point (x_m, z_m) above an emitting surface. The total vertical flux in (x_m, z_m) consists of contributions from all upwind sources ($x \geq x_m$). We assume that emission takes place at the roughness length z_0 as did Horst and Weil (1992). The flux footprint at position (x, z_0) with respect to the measurement point (x_m, z_m) is called $\overline{f}^y(x, z_0|x_m, z_m)$, where the bar denotes that the quantity is crosswind integrated. It is a "source weight" that relates the vertical flux in the measurement point $\overline{F}^y(x_m, z_m)$ to the spatial distribution of surface emission fluxes $\overline{F}^y(x, z_0)$. In the sequel we supress z_0 in the notation by renaming the surface flux and the footprint: $\overline{S}^y(x) = \overline{F}^y(x, z_0)$ and $\overline{f}^y(x, z_0|x_m, z_m) = \overline{f}^y(x|x_m, z_m)$. Using this notation the integral equation of diffusion reads

$$\overline{F}^y(x_m, z_m) = \int_{x=x_m}^{\infty} \overline{S}^y(x)\overline{f}^y(x|x_m, z_m)dx \tag{4}$$

If the flow is horizontally homogeneous the footprint depends only on the separation between the source and the measurement point, i.e. $\overline{f}^y(x|x_m, z_m) = \overline{f}^y(x - x_m|z_m)$.

The footprint in (x_s, z_0) can be determined by placing a point source of strength Q [kg/s] at the location, $\overline{S}^y(x) = Q\delta(x - x_s)$, as shown in fig. 3. The flux in (x_m, z_m) due to the point source in (x_s, z_0) is then

$$\begin{aligned} &\overline{F}^y(x_m, z_m|x_s) = \\ &\int_{x=x_m}^{\infty} Q\delta(x - x_s)\overline{f}^y(x|x_m, z_m)dx = \\ &Q\overline{f}^y(x_s|x_m, z_m) \end{aligned} \tag{5}$$

from which the footprint can be isolated

$$\overline{f}^y(x_s|x_m, z_m) = \overline{F}^y(x_m, z_m|x_s)/Q \tag{6}$$

i.e. the footprint in (x_s, z_0) is equal to the flux in (x_m, z_m) due to a point source in (x_s, z_0), divided by its strength.

FOOTPRINT CALCULATION FROM A BACKWARD MODEL

In backward Lagrangian models the particles are tracked backward in time, from the sensor to the surface source. A desirable property of backward models is that we actually consider trajectories that coincide exactly in the measurement point - thus the quantities calculated from this type of models are not averaged over a sensor volume as in forward models. However, the real advantage shows up when the terrain is horizontally inhomogeneous because in a backward model only a single release of particles (from the sensor point) is necessary for calculation of flux and concentration in the sensor point, whereas a forward model needs to track particles from each and every upwind element. These are the main reasons that this paper concerns backward trajectories. A minor backdraw of backward models is that e.g. a footprint analysis is confined to the specific (measurement) height from which the particles were released, this means that for footprint calculations for different measurement heights, a release of particles must be made for each height.

The particle trajectories are simulated by the Langevin equation. The turbulence is chosen to be height-homogeneous for two reasons: no effects are introduced due to height-inhomogeneities and the PDF is so skewed at the boundary that distortion by a reflection scheme becomes obvious. The skewed PDF is constructed as a fourth order expansion in Hermite polynomials as described by Jensen and Gryning (1997). A version that includes a horizontal meanwind profile and where the turbulence is considered height-homogeneous can be found in Falk (1998). In the present paper the same parameterizations for turbulence and meanwind profile as in Falk (1998) are used, but the dispersion is reversed in order to track the particles backward in time as described by Flesch et al. (1995) and Thomson (1987). How to calculate concentrations and fluxes from backward trajectories is derived by Flesch et al. (1995) (concentrations) and Flesch (1996) (fluxes). Here we will use a modified version of eq.(1) in Flesch (1996) to calculate the footprint. The modification enables the original formula to account for a reflected velocity that does not equal the incident velocity and it reads

$$\frac{\overline{F}^y(x_m, z_m|(x,z) \in V_{src})}{q} = \frac{1}{N} \sum_{\{k|(x,z)\in V_{src}\}} \left(w_{k,ini} \left(\frac{1}{|w_{k,i}|} + \frac{1}{|w_{k,r}|} \right) \right) \tag{7}$$

where $w_{k,ini}$ is the initial velocity of a particle (labelled k) which subsequently touches down in V_{src}, a sourcevolume next to the ground of infinitesimal height dz and finite length Δx_s. When a particle touches down its incident and reflected velocities $w_{k,i}$ and $w_{k,r}$ are recorded together with the horizontal coordinate and $w_{k,ini}$. The summation runs over all particles that touches down within the specific V_{src} and N is the total number of released particles. The surface emission rate is q. If the footprint is assumed constant over V_{src} it can be calculated as

$$\overline{f}^y(x_s|x_m, z_m)\Delta x_s = \frac{\overline{F}^y(x_m, z_m|(x,z) \in V_{src})}{q} \tag{8}$$

As is evident from equations (7) and (8) the footprint calculated from backward trajectories depends on the near ground velocities $w_{k,i}$ and $w_{k,r}$. This indicates that the simulated footprint might be sensitive to how the particles are reflected at the boundaries. This is examined by running simulations that are identical except for the reflection scheme and subsequently calculating the footprints. The problem is now to evaluate the "quality" of the simulated footprints. The following argument is valid if the surface emission rate is constant and equal to q all over the upwind area, i.e. for a homogeneous surface of infinite extent. Then according to eq. (4)

$$\frac{\overline{F}^y(x_m, z_m)}{q} = \int_{x=x_m}^{\infty} \overline{f}^y(x|x_m, z_m)dx \tag{9}$$

Eq. (9) says that the fraction of the vertical surface flux that is preserved at height z_m equals the streamwise integral of the crosswind integrated footprint. Falk (1998) showed that this fraction is related to the horizontal advective velocity through

$$\int_{X=X_m}^{\infty} \overline{f}^y(X|X_m, Z_m)dX = 1 - \int_{Z=Z_0}^{Z_m} U(Z)dZ \quad . \tag{10}$$

This relation is given in dimensionless quantities which are obtained from the dimensional as follows:

The vertical coordinate is scaled by the inversion height (z_i), $Z = z/z_i$, and vertical velocities are scaled by the convective velocity scale (w_*), $W = w/w_*$. The dimensionless horizontal coordinate is given by $X = x\frac{w_*}{z_i\overline{u}}$ where $\overline{u} = \frac{1}{z_i}\int_{z=z_0}^{z_i} u(z)dz$ is the average dimensional velocity over the boundary layer depth. The dimensionless horizontal velocity is chosen to be $U(Z) = u(Z)/\overline{u}$ which gives $\overline{U} = \int_{Z=Z_0}^{1} U(Z)dZ = 1$. The footprint as a function of the dimensionless coordinates corresponds to that of the dimensional coordinates through $\overline{f}^y(X|X_m, Z_m) = \frac{\overline{u}z_i}{w_*}\overline{f}^y(x|x_m, z_m)$ which gives $\int_{X=X_m}^{\infty} \overline{f}^y(X|X_m, Z_m)dX = \int_{x=x_m}^{\infty} \overline{f}^y(x|x_m, z_m)dx$.

Eq. (10) gives means to an evaluation of the simulated footprints: If the footprint systematically takes too low or too high values, the asymptotic value of its streamwise integral will not tend to the value predicted by eq. (10).

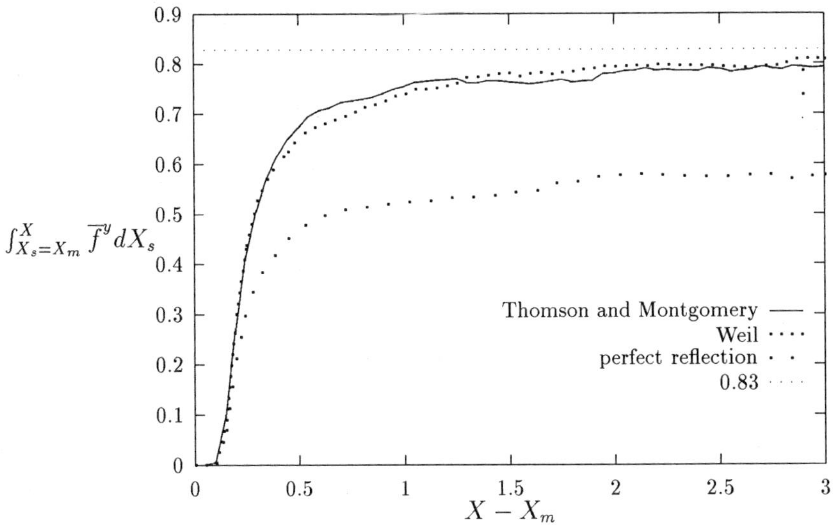

Figure 4. cumulative footprints calculated for $Z_m = 0.195$ for the three reflection schemes under consideration.

RESULTS

Fig. 4 shows the cumulative footprint calculated for a sensorheight $Z_m = 0.195$. For this height the asymptotic value of the cumulative footprint should be 0.83 according to eq. (10); this is shown as a horizontal line in fig. 4. Each of the calculations are based on 20000 backward trajectories and the only difference in the simulations is in the reflection scheme. The horizontal size of the source volume is taken to be $\Delta X_s = 0.05$. As expected perfect reflection performs worst, giving an asymptotic value of the cumulative footprint less than 0.6 which should be compared to the predicted value 0.83. This behaviour can be explained from eq.(7) and fig. 1: For reflection at the ground there will be a surplus of particles reflected with a high speed and a deficit of particles reflected with small speed. Since the reflected velocity w_r occurs in the denominator of eq. (7) this expression will generally take too low values and so will the footprint. The scheme of Thomson and Montgomery and the scheme of Weil turns out to behave very similar to each other, which should also be expected from fig.2. With both schemes an asymptotic value around 0.8 is reached, which is a satisfactory result.

In fig. 5 the asymptotic value of the cumulative footprint for a number of heights in the entire boundary layer is shown together with the values predicted by eq. (10). Here the effect of using different reflection schemes becomes a little clearer. Perfect reflection gives a gross underestimation of the flux simulated at the sensor and the problem becomes larger as the sensorheight approaches the ground.The scheme of Weil gives results that are very close to the theoretical values but with a slight underestimation. When using the scheme of Thomson and Montgomery the results also get close to the theoretical values but the discrepancies from these are both positive and negative, which means that there is no tendency of neither underestimation nor overestimation. It is hard to say whether the scheme of Weil or the scheme of Thomson and Montgomery does the best job since the former gives a systematical (but small) underestimation and the latter gives larger deviations from the theoretical value of $\int_{X_m}^{\infty} \overline{f}^y(X|X_m, Z_m)dX$. The conclusion is that both schemes are suitable for footprint

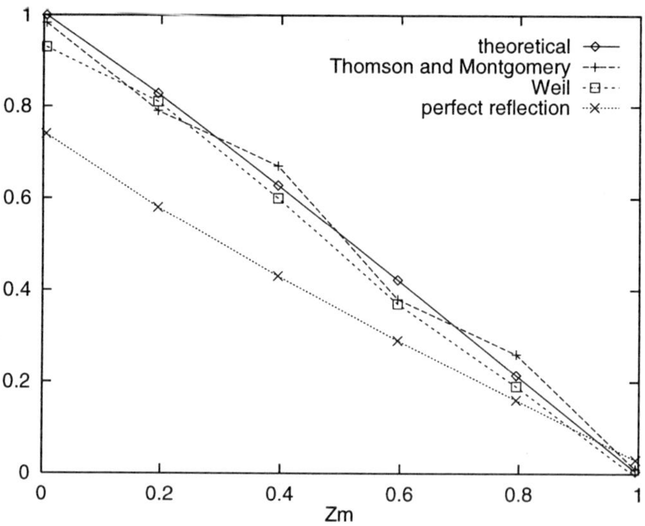

Figure 5. Simulated values of $\int_{X_m}^{\infty} \overline{f}^y(X|X_m, Z_m)dX$ obtained by using the three reflection schemes. These are compared to the theoretical value for the heights Z_m = 0.005, 0.195, 0.395, 0.595, 0.795, 0.995.

analysis made from backward trajectories whereas perfect reflection gives an insufficient description when the turbulence is skewed at the ground.

REFERENCES

Falk,A.K.V. (1998). On the area under the footprint curve. *in preparation*

Flesch,T.K. (1996). The footprint for flux measurements, from backward Lagrangian stochastic models. *Boundary-Layer Meteorol.* 78:399-404.

Flesch,T.K., Wilson,J.D. and Yee,E. (1995). Backward-time Lagrangian stochastic dispersion models and their application to gaseous emmisions. *J. Appl. Meteor.* 34:1320-1322.

Horst,T.W., and Weil,J.C. (1992). Footprint estimation for scalar flux measurements in the atmospheric surface layer. *Boundary-Layer Meteorol.* 59:379-296

Jensen,A.K.V., and Gryning,S.-E.(1997). A new formulation of the probability density function in random walk models for atmospheric dispersion. In *Air Pollution Modelling and its Application XII*, pp.429-438, Plenum Press, New York.

Thomson,D.J (1987). Criteria for the selection of stochastic models of particle trajectories in turbulent flows. *J. Fluid Mech.* 180:529-556.

Thomson,D.J. and Montgomery,M.R. (1994). Reflection boundary conditions for random walk models of dispersion in non-Gaussian turbulence. *Atmos. Environ.* 28:1981-1987.

Weil,J.C. (1990). A diagnosis of the assymetry in top-down and bottom-up diffusion using a Lagrangian stochastic model. *J. Atmos. Sci.*47:501-515.

Wilson,J.D., and Flesch,T.K. (1993). Flow Boundaries in random-flight dispersion models: Enforcing the well-mixed condition. J. Appl. Meteorol. 32:1695-1707.

DISCUSSION

B. FISHER: Does the method allow one to estimate the minimum extent of the uniform surface which should exist upwind of a turbulence flux measurement?

A. K. V. FALK: Footprint analysis is developed exactly for that purpose. However the present study cannot be used directly for estimation of how long uniform fetch that is needed. It is a sensitivity study of the model and no validation against measurements has been carried out yet.

TOWARDS REVISION OF CONVENTIONAL FLUX-PROFILE RELATIONSHIPS FOR THE STABLY STRATIFIED ATMOSPHERIC SURFACE LAYER

Sergej Zilitinkevich[1,2]

[1]Department of Earth Sciences - Meteorology, Uppsala University
Villavägen 16, S-752 36 Uppsala, Sweden
[2]Department of Wind Energy and Atmospheric Physics
RISØ National Laboratory P.O. Box 49, DK 4000 Roskilde, Denmark

INTRODUCTION

Conventional flux-profile relationships for the stably stratified atmospheric surface layer are modified accounting for possible effects of the free-flow stability on the boundary layer and the surface layer. In the revised log-linear profiles, slope factors traditionally considered as constants become functions of $S = NL/u_*$, where u_* is the friction velocity, L is the Monin-Obukhov (MO) length, and N is the Brunt-Väisälä frequency in the free flow. The proposed model allows well-developed turbulence at very strong static stability, up to Richardson numbers, Ri, of order 10, and results in a reasonable dependence of the turbulent Prandtl number, Pr_T, on Ri in the z-less stratification layer (both features are disregarded in the traditional similarity theory). The model provides simple tools for improved flux parameterization.

MEAN PROFILES

In the proposed formulation, two types of the stable-static-stability boundary layer (SBL) are distinguished, namely, the mid-latitude nocturnal SBL developing against the neutrally stratified residual layer (the surface-flux-dominated SBL), and the wintertime high-latitude SBL developing against pronounced static stability in the free-flow (the imposed-stability-affected SBL). The latter SBLs are shown to be essentially dependent on the length scales u_*/N and $u_*/\sqrt{fN}$, where f is the Coriolis parameter (Zilitinkevich and Mironov, 1996). Typically they are much more shallow than the surface-flux-dominated SBLs (Handorf, 1996). It is reasonable to assume that the imposed stability affects not only the SBL depth but also the vertical structure of the SBL including the surface layer.

Air Pollution Modeling and Its Application XIII, edited by S.-E. Gryning and E. Batchvarova, Kluwer Academic/Plenum Publishers, 2000

To develop an extended similarity-theory formulation, we consider firstly the atmospheric surface layer (ASL) affected by the free-flow stability only. Here, taking negligible the surface flux of potential temperature, $F_{\theta s} \to 0$, and consequently the reciprocal of the MO length, $L^{-1} \equiv -\beta F_{\theta s} / u_*^3 \to 0$ (where $\beta = g / T$ is the buoyancy parameter), the basic turbulence length scale in the z-less stratification layer is $l \sim u_* / N$, so that the eddy-viscosity scale is $K_M \sim u_*^2 / N$. Hence the velocity gradient is

$$\frac{\partial u}{\partial z} = \frac{u_*^2}{K_M} \sim N = \frac{u_*}{z}\left(\frac{Nz}{u_*}\right) \tag{1}$$

(cf. Zilitinkevich et al., 1998b). Similarly, taking negligible the conventional M-O theory scales for potential temperature, $\theta_* = -F_{\theta s} / u_* \to 0$, and for its gradient, $\theta_* / L \to 0$, the only available scale for the potential temperature gradient in the z-less stratification layer is

$$\frac{\partial \Theta}{\partial z} \sim \frac{N^2}{\beta} = \frac{\theta_*}{z}\left(\frac{N^2 L z}{u_*^2}\right). \tag{2}$$

Taking the eddy conductivity, K_H, proportional to the eddy viscosity, $K_H \sim u_*^2 / N$, Eq. (2) implies a non-zero potential-temperature flux in the z-less stratification layer, $F_\theta = K_H \partial\Theta / \partial z \sim N u_* / \beta$, even in the case that the above flux diminishes at the surface, $F_\theta |_{z=0} \equiv F_{\theta s} = 0$.

Interpolation between the above asymptotic expressions, Eqs. (1) and (2), for the mean gradients and the conventional MO-theory log-linear formulation yields

$$\frac{\partial u}{\partial z} = \frac{u_*}{kz}\left[1 + \left(C_{uL} + C_{uN}\mathrm{S}\right)\frac{z}{L}\right], \quad u = \frac{u_*}{k}\left[\ln\frac{z}{z_{0u}} + \left(C_{uL} + C_{uN}\mathrm{S}\right)\frac{z}{L}\right], \tag{3}$$

$$\frac{\partial \Theta}{\partial z} = \frac{\theta_*}{k_T z}\left[1 + \left(C_{\theta L}^{1/2} + C_{\theta N}^{1/2}\mathrm{S}\right)^2 \frac{z}{L}\right], \quad \Theta = \Theta_s + \frac{\theta_*}{k_T}\left[\ln\frac{z}{z_{0T}} + \left(C_{\theta L}^{1/2} + C_{\theta N}^{1/2}\mathrm{S}\right)^2 \frac{z}{L}\right]. \tag{4}$$

Here, z_{0u} and z_{0T} are the surface roughness lengths for wind and temperature, respectively, Θ_s is the potential temperature at the surface, $\mathrm{S} = NL / u_*$ is a dimensionless number characterising the role of the free-flow stability, k and k_T are the von Karman constants characterising the logarithmic layer, C_{uL}, C_{uN}, $C_{\theta L}$ and $C_{\theta N}$ are the same type dimensionless constants characterising the z-less stratification layer.

Empirical values of the constants that appear in the traditional formulation are $k \approx 0.40$, $k_T \approx 0.47$, $C_{uL} \approx 2.1$ and $C_{\theta L} \approx 3.2$ (e.g., Yaglom, 1979; Högström, 1995). The new wind-profile constant $C_{uN} \sim 0.4$ has been roughly estimated by Zilitinkevich at al. (1998b). An order of magnitude estimate of $C_{\theta N} \sim 3$ follows from experimental data on the dependence of turbulent Prandtl on Richardson number (see next Section).

In Eqs. (3) and (4), the slop factors $\left(C_{uL} + C_{uN}\mathrm{S}\right)$ and $\left(C_{\theta L}^{1/2} + C_{\theta N}^{1/2}\mathrm{S}\right)^2$ are no longer constants, as is the convention. Moreover, their dependence on S is linear for wind and quadratic for temperature, reflecting higher sensitivity of thc temperature profile to the free-flow stability.

TURBULENT PRANDTL NUMBER

The conventional MO similarity theory allows no developed turbulence at Richardson numbers, $\mathrm{Ri} \equiv \beta(\partial\Theta / \partial z) / (\partial u / \partial z)^2$, higher than $k^2 C_{\theta L} / k_T C_{uL}^2 \approx 0.24$. It also predicts that the turbulent Prandtl number, Pr_T, turns into a constant in the z-less stratification layer, $\mathrm{Pr}_T \equiv K_M / K_H \to k C_{\theta L} / k_T C_{uL} \approx 1.3$.

By contrast, well-developed turbulence is often observed in the ASL at Richardson numbers much higher than 0.3 (Larsen et al., 1990; Sempreviva et al., 1992).

Moreover, a well pronounced dependence of Pr_T^{-1} on Ri was obtained from marine measurements (Taylor, 1960), laboratory experiments (Ellison and Turner, 1960), and atmospheric measurements (Kondo et al., 1978). Empirical plots of Pr_T^{-1} versus Ri suggest that Pr_T^{-1} decreases from 1.2 to 0.06 as Ri increases from 10^{-2} to 8.

Eqs. (3) and (4) yield a single-valued dependence of Pr_T^{-1} on Ri,

$$\mathrm{Pr}_T = \frac{u_* \partial\Theta / \partial z}{\theta_* \partial u / \partial z} = \frac{k\left(C_{\theta L}^{1/2} + C_{\theta N}^{1/2} \mathrm{S}\right)^2}{k_T\left(C_{uL} + C_{uN} \mathrm{S}\right)}, \quad \mathrm{Ri} = \frac{k^2\left(C_{\theta L}^{1/2} + C_{\theta N}^{1/2} \mathrm{S}\right)^2}{k_T\left(C_{uL} + C_{uN} \mathrm{S}\right)^2}, \tag{5}$$

which is consistent with the above experimental evidence. Comparing Eq. (5) with empirical plots of Pr_T^{-1} versus Ri provides an order of magnitude estimate of $C_{\theta N} \sim 3$.

CONCLUDING REMARKS

Comprehensive empirical validation of the proposed formulation is needed. In particular the wind and temperature slop factors in the z-less stratification layer, $(kL / u_*)\partial u / \partial z$ and $(k_T L / \theta_*)\partial\Theta / \partial z$, should be examined for their dependence on S, and the constants, C_{uN} and $C_{\theta N}$ should be specified.

Then parameterization schemes for turbulent fluxes currently used in numerical models are to be changed, to a first approximation by substituting effective values of the wind velocity, $u(z_1) - (C_N / k) N z_1$, and the potential temperature increment, $\Theta(z_1) - \Theta_s - (C_{\theta N} / k_T)(N^2 / \beta) z_1$, for conventionally employed $u(z_1)$ and $\Theta(z_1) - \Theta_s$, where z_1 is the lowest model level.

In the conditions of well pronounced free-flow stability throughout the lower atmosphere, an improved flux parameterization scheme based on Eqs. (3)-(5) suggests much stronger turbulent mixing than was the case in earlier schemes based on the MO theory.

In the light of recent evidence from field experiments (Larsen et al., 1990; Sempreviva et al., 1992), climate modelling (Roeckner and Jacob, 1998), and pollution dispersion modelling (Gryning, 1998), this is the requested result.

ACKNOWLEDGEMENTS

This work is a contribution to co-operative studies of the stable-boundary-layer turbulence and dispersion, carried out at Uppsala University and National Defence Research Establishment, Umeå, Sweden. It has been partially supported from EU Project "Surface Fluxes in Climate System" (SFINCS) - EC Contract ENV4-CT97 0573.

REFERENCES

Ellison, T.H., and Turner, J.S., 1960, Mixing of dense fluid in a turbulent pipe flow, *J. Fluid. Mech.*, 8: 514-544.

Gryning, S.-E., 1998, Some aspects of atmospheric dispersion in the stratified atmospheric boundary layer over homogeneous terrain, submitted to: *Boundary-layer Meteorol.*

Handorf, D., 1996, Zur Parametrisierung der stabilen atmosphärischen Grenzschicht über einem antarktischen Schelfeis, *Berichte zur Polarforschung*, 204: 133 S.

Högström, U., 1995, Review of some basic characteristics of the atmospheric surface layer, *Boundary-Layer Metorol.*, 78: 215-246.

Kondo, J., Kanechika, O., and Yasuda, N., 1978, Heat and momentum transfer under strong stability in the atmospheric surface layer, *J. Atmos. Sci.*, 35: 1012-1021.

Larsen, S.E., Courtney, M., and Mahrt, L., 1990, Low frequency behaviour of horizontal power spectra in stable surface layers, in: *Proc. 9th AMS Symposium on Turbulence and Diffusion*, 401-404.

Roeckner, E., and Jacob, D., 1998, Requirements from climate modelling to improved flux parameterization, *RISØ I-1245 (EN).*

Sempreviva, A., Larsen, S.E., Mortensen, N.G., 1992, Experimental study of flow modification inland from a cost for non-neutral conditions, *RISØ M-2924.*

Taylor, G.I., 1960, *Scientific Papers*, Vol. 2, Cambridge University Press.

Yaglom, A.M., 1979, Similarity laws for constant-pressure and pressure-gradient turbulent wall flows, *Ann. Rew. Fluid Mech.*, 11: 505-540.

Zilitinkevich, S., Johansson, P.-E., Mironov, D.V., and Baklanov, A., 1998, A similarity-theory model for wind profile and resistance law in stably stratified planetary boundary layers, *Journal of Wind Engineering and Industrial Aerodynamics*, 74-76: 209-218.

Zilitinkevich, S. S., and Mironov, D. V. 1996 A multi-limit formulation for the equilibrium depth of a stably stratified boundary layer. *Boundary-Layer Meteorol.*, 81: 325-351.

DISCUSSION

G. SCHAYES: You mentioned the use of Antarctic data. On the sloped ice shelf these include frequently strong katabatic winds in stable conditions. On the other extreme, we know that stable situations occur frequently with very weak winds with the result of intermittent turbulence. Can you comment on the validity of your scheme in these extreme situations.

S. ZILITINKEVICH: Of these two extremes, the scheme is applicable to katabatic winds, hardly to weak intermittent turbulence.

X. CAI: In the proposed momentum/heat flux relationships, namely, $\frac{kz}{u_*}\frac{\partial u}{\partial z} = 1 + C_{uL}\frac{z}{L} + C_{uN}\frac{zN}{u_*}$, the new term is expressed in a linear fashion. In my opinion, the relationship should be in a more general form, i.e., $\frac{kz}{u_*}\frac{\partial u}{\partial z} = F\left(\frac{z}{L}, \frac{Nz}{u_*}\right)$, or $1 + g\left(\frac{z}{L}, \frac{Nz}{u_*}\right)$. Could you comment on this?

S. ZILITINKEVICH: Generally speaking, you are right. However, a more general form looks neither promising nor practically useful. Simple reasoning is given in my paper in support of the proposed formulation. To my mind it deserves empirical validation.

DOUBLE-FLUX DESCRIPTION OF THE TRANSPORT OF PASSIVE SCALARS IN THE CONVECTIVE ATMOSPHERIC BOUNDARY LAYER

Eugene Genikhovich

Air Pollution Modeling and Forecasting Laboratory
Main Geophysical Observatory
194021 St. Petersburg, Russia

INTRODUCTION

Convective diffusion was a "hot issue" during the last two decades. Numerous papers were devoted to analyses of the main physical features of this process. In particular, it was recognized that the pattern of the concentration field from the point source in the convective boundary layer (CBL) is different from those predicted with use of the advection-diffusion equation and corresponding local eddy diffusivities (Willis and Deardorf, 1976; Deardorf and Willis, 1984). It is generally accepted nowadays that the most effective theoretical description of the convective flow and diffusion can be obtained using numerical techniques like LES and DNS. These techniques, however, can be implemented only if powerful computational platforms are available, and even in this case they are rather time-consuming. More simple parametrizations are needed for many practical applications including climatic and dispersion modeling.

Since publishing the pioneering works by J. Wyngaard and his co-authors, the idea of the double-flux description of the convective dispersion due to the "bottom-up" and "top-down" diffusion has been widely used by many authors. In 1984 Wyngaard and Brost (this paper will be referred as WB) and Moeng and Wyngaard introduced a corresponding eddy diffusivity and explained that this parameter in their scheme should be determined as non-local one. It was recognized (Baerensten and Berkowicz, 1984; Wyngaard, 1987; Weil, 1990) that the skeweness of the probability distribution function (PDF) of turbulent fluctuations of the vertical component of the wind velocity, w, is one of the most important governing parameters responsible, in particular, for the striking difference in bottom-up and top-down eddy diffusivities. The concept of the contragradient diffusion as a tool for description of the convective mass transfer was also discussed in numerous publications dealing, for example, with high-order turbulent closure for convective flows (*e.g.*, Randall et al., 1992; Abdella and McFarlane (1997)). Modeling results seem to be in reasonable agreement with data of field and laboratory experiments (see for example Briggs, 1985; Piper et al., 1995; Weil at al., 1998).

When developing a stochastic model of plume dispersion in CBL, Baerensten and

Berkowicz (1984) decomposed PDF into a sum of two normal distributions corresponding to updrafts and downdrafts, respectively. Similarly, the concentration field from a point source was decomposed into a sum of two Gaussian plumes in so-called PDF models (Venkatram, 1983; Weil *et al.*, 1997); the plume axes in these models were tilted with slopes W_u/U and $-W_d/U$ where U is the mean wind speed (assumed to be constant with height), and W_u and $-W_d$ are mean vertical velocities in updrafts and downdrafts, respectively. From these papers one can suppose that updrafts and downdrafts, being considered separately, can be described with use of the local eddy diffusivity. The mix of the updrafts and downdrafts can be considered as an "intermittent" flow consisting of two superimposing regimes. In such a case, the method of conditional sampling is an effective tool for description of effects in the complex flow. This method was earlier used in practical applications (see for examlpe Genikhovich and Snyder, 1994).

In the presented paper, the Wyngaard's concept of the double-flux description of the convective dispersion is modified in order to describe explicitly the vertical transport of scalars in the convective atmospheric boundary layer. The simplest model of the local positive eddy diffusivity is used here. To demonstrate the main features of the proposed approach, two problems are considered. The first of them, concerning with the horizontally homogeneous convective boundary layer, has been investigated by Wyngaard and co-authors using LES and experimental data. The second one, dealing with dispersion from the point source, had been initially studied in laboratory conditions by Deardorf and Willis and then was investigated by many other authors. For the sake of simplicity, only the "steady-state" case with constant height and other properties of CBL is discussed in this paper; entrainment processes and structure of the interfacial layer at the top of CBL will not be considered here.

GOVERNING EQUATIONS

Following to Baerensten and Berkowicz (1984), let us decompose PDF of turbulent fluctuations of the vertical component of the wind velocity into a sum of two Gaussian distributions corresponding to positive and negative mean values. Accordingly, the fluctuatuions will be classified as corresponding to updrafts and downdrafts. They are characterized by the probabilities (weights in this decomposition) p_u and p_d, respectively (obviously, $p_u + p_d = 1$).

Let us consider CBL as an "intermittent" flow which is a mixture of updrafts and downdrafts. Following to simplifications introduced in WB, let us assume that CBL is quasi-steady and do not describe directly the entrainment effects at the bottom of the capping temperature inversion. With such an approach, variations in time of the height of CBL, L, can be taken into account parametrically. For the sake of simplicity, mean vertical velocities in updrafts and downdrafts, W_u and $-W_d$ are assumed constant. It is assumed also that the horizontal component of the wind velocity, U, is also constant with height and small enough to neglect the dynamically generated turbulence, and that the rotation of the wind direction inside CBL is negligible.

The advection-diffusion equation for a conservative scalar written separately in updrafts and downdrafts with use of the conditional averaging results in the following two equations:

$$\begin{aligned}\frac{\partial C_u}{\partial t}+U\frac{\partial C_u}{\partial x}+W_u\frac{\partial C_u}{\partial z}&=-\frac{\partial cw_u}{\partial z},\\ \frac{\partial C_d}{\partial t}+U\frac{\partial C_d}{\partial x}-W_d\frac{\partial C_d}{\partial z}&=-\frac{\partial cw_d}{\partial z},\end{aligned}\tag{1}$$

where (x, y, z) are Cartesian coordinates (with the x-axis oriented along wind direction), C_u

and C_d, and cw_u and cw_d are mean concentrations and turbulent fluxes in updrafts and downdrafts, respectively. Obviously, the resulting mean value of concentration, C, is represented as follows:

$$C = p_u C_u + p_d C_d . \tag{2}$$

Similarly, the turbulent flux, wc, can be written as a linear combination of the bottom-up and top-down fluxes with weights equal, respectively, to p_u and p_d. In this work cw_u and cw_d are parametrized with use of the first-order closure:

$$cw_u = -K_u \frac{\partial C_u}{\partial z}, \quad cw_d = -K_d \frac{\partial C_d}{\partial z}, \tag{3}$$

where K_u and K_d are bottom-up and top-down eddy diffusivities.

Accordingly to Baerensten and Berkowicz (1984) and Weil *et al.* (1997), the following parametrizations are used in this paper

$$\begin{aligned} p_u &= \frac{W_d}{W_u + W_d}, \quad p_d = \frac{W_u}{W_u + W_d}, \\ \frac{\sigma_{wu}}{\sigma_w} &= 0.5\left(\sqrt{\gamma_1^2 S^2 + 4/\gamma_2} + \gamma_1 S\right), \\ \frac{\sigma_{wd}}{\sigma_w} &= 0.5\left(\sqrt{\gamma_1^2 S^2 + 4/\gamma_2} - \gamma_1 S\right), \end{aligned} \tag{4}$$

where σ_{wu} and σ_{wd} are standard deviations of the vertical velocity in updrafts and downdrafts, respectively, σ_w is the "total" standard deviation of the vertical velocity, and S is the vertical velocity scewness. Parameters γ_1 and γ_2 are determined as

$$\gamma_1 = \frac{1 + r^2}{1 + 3r^2}, \quad \gamma_2 = 1 + r^2, \tag{5}$$

where $r = \sigma_{wu}/W_u = \sigma_{wd}/W_d$. It was noted by Weil *et al.* (1997) that there is little difference between the computational results corresponding to r = 1 and r = 2 and that the value r = 2 could be preferable for comparisons with field measurements. The value r = 1 which is used in this paper provides slightly better agreement with laboratory results. In the convective limit considered in this paper S = 0.6 and probabilities p_u and p_d are close to 0.4 and 0.6, respectively. The vertical velocity variance is given as follows:

$$\sigma_w^2 = 1.2u_*^2 + 0.31w_*^2, \tag{6}$$

where u_* is the friction velocity and w_* is the convective velocity scale (in the convective limit, therefore, $w_* = 1.8\,\sigma_w$), and the scewness is parametrized as $S=0.104\,(w_*/\sigma_w)^3$.

VERTICAL SCALAR TRANSPORT: HORIZONTALLY HOMOGENEOUS CASE AND DISPERSION FROM A POINT SOURCE

When applying the approach introduced in WB to Eqs. (1) and (3), the following expressions describe the vertical fluxes in the horizontally homogeneous CBL:

$$-K_u \frac{\partial C_u}{\partial z} + W_u(C_u - C_{u0}) = \frac{P_0}{p_u}(1 - \frac{z}{L}),$$

$$K_d \frac{\partial C_d}{\partial z} + W_d(C_d - C_{dL}) = R\frac{P_0}{p_d}\frac{z}{L}, \tag{7}$$

where $P_0 = -cw_{u0}$, $R = cw_{dL}/P_0$, C_{u0} and cw_{u0} are values of C_u and cw_u at $z = z_0$, and C_{dL} and cw_{dL} are analogous values of C_d and cw_d at $z = L-z_0'$ (the scalar C can be understood here, for example, as the potential temperature). Roughness parameters z_0 and z_0' are characteristics of the length scale of the eddies responsible for energy exchange with the ground surface and the atmosphere above the boundary layer. The probabilities p_u and p_d are introduced in the right-hand sides of Eq. (7) to ensure that the "final distribution", C, determined from Eq. (2) corresponds to the given flux, P_0, at the underlying surface and the counterflux, $-RP_0$, at the top of the boundary layer. Eddy diffusivities are parametrized with use a simplified version of the expressions suggested by Moeng and Wyngaard (1984):

$$K_u = 2.5 w_* L(\frac{z}{L})^{3/2},$$

$$K_d = 1.43 w_* L(1 - \frac{z}{L})^2. \tag{8}$$

The problem (7)-(8) can be easily solved either analytically or numerically. An example of the vertical distribution of the dimensionless deviation, T', of the air temperature from its reference value is given on Fig. 1 as a function of the dimensionless height Z = z/L.

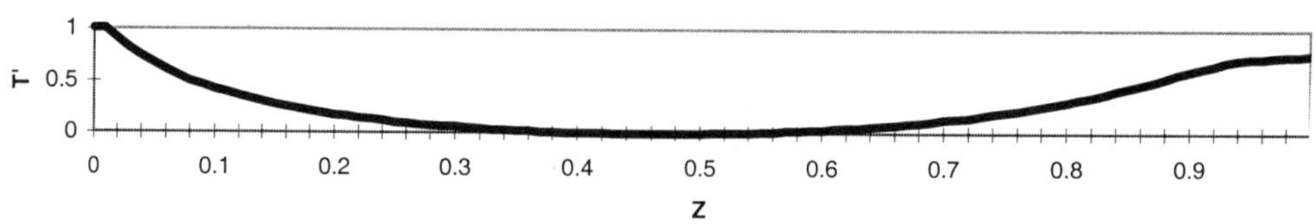

Fig. 1 Calculated temperature distribution in CBL.

The deviation is normalized with the temperature scale $T_* = P_0/w_*$ and additionally multiplied by $5{\cong}10^{-6}$. This profile corresponds to the values of the dimensionless (normalized with L) roughness parameters $z_0 = 10^{-3}$ and $z_0' = 1.8{\cong}10^{-2}$. It looks more-or-less similarly to the known measurement results. Therefore, one can assume that Eq. (8) can be used at least as a rough approximation of the eddy diffusivities profiles.

The same eddy diffusivities (8) are used with Eq. (1) when calculating the stationary field of the crosswind integrated concentrations of a passive scalar from the point source. In such a case, Eq. (1) is supplemented with boundary conditions which correspond to the point source located at X = 0; $Z = Z_s$ and to complete reflection of the scalar from the bottom and the top of the domain. The computed fields corresponding to $Z_s = 0.067$, 0.24, and 0.49 are shown on Figs. 2 - 4. One can see that these fields obtained from numerical solutions of Eq. (1) closely resemble those measured by Willis and Deardorf in the

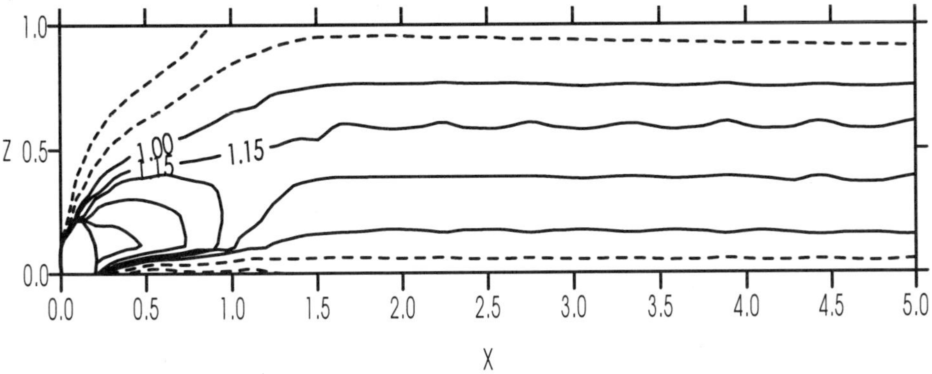

Fig.2 Normalized concentrations field (Zs = 0.067); dashed isolines correspond to C = 0.5 and 0.75, and solid ones to C = 1.0, 1.15, 1.25, 1.5, 2.0, 3.0, and 5.0.

convective tank or estimated with use of more sophisticated numerical models (see for example Baerensten and Berkowicz, 1984). Similarly, variations with the distance from the source of the mean height Z_c of the particles (Fig. 5) and their overall vertical standard deviations SIGMA (Fig. 6) estimated from the same numerical solutions resemble those measured in the convective tank and calculated with use of Monte-Carlo techniques (Baerensten and Berkowicz, 1984) or LES (Nieuwstadt, 1998).

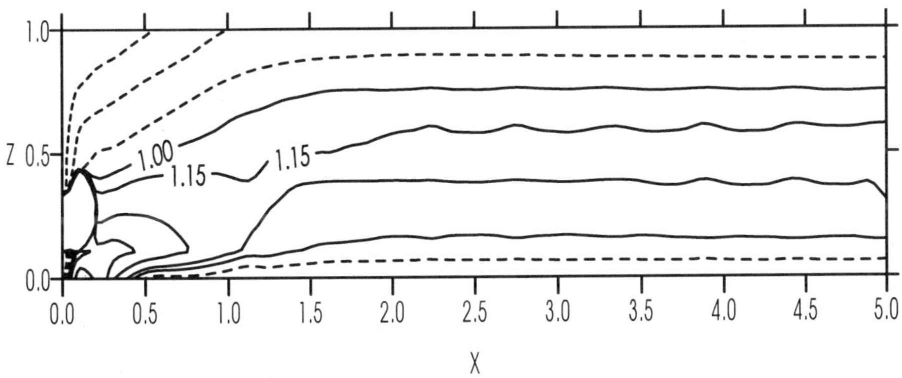

Fig.3 Normalized concentrations field (Zs = 0.24); dashed isolines correspond to C = 0.2, 0.5 and 0.8, and solid ones to C = 1.0, 1.15, 1.5, 2.0, 3.0, and 4.0.

DISCUSSION AND CONCLUSION

It is widely recognized nowadays that convective turbulent flows can be considered as "a-mix" of small-scale eddies and large-scale coherent structures. Turbulent fluxes in these flows cannot be related directly to local gradients. That is why several versions of the non-local description of these fluxes were introduced in last years. The main goal of this work is to demonstrate that, after decomposition of this multi-scale convective flow into bottom-up and top-down components, the resulting fluxes can be described with use of "conventional" gradient approximation plus terms describing systematic transport in either updrafts or downdrafts. The results obtained indicate that the solution of Eq. (1) can reproduce some characteristic features of "convective" fields studied in laboratory and numerical experiments. If this conclusion is correct, it can simplify the problem of parametrization of the convective planetary boundary layer in climatic models. Moreover,

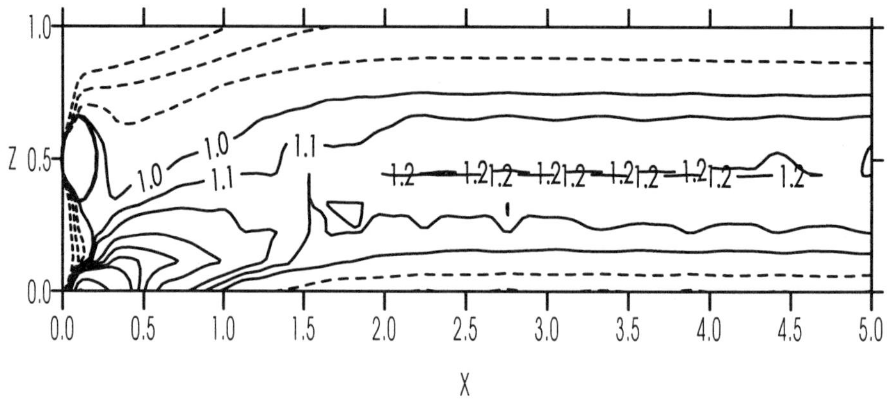

Fig.4 Normalized concentrations field (Zs = 0.49); dashed isolines correspond to C = 0.4, 0.6 and 0.8, and solid ones to C = 1.0, 1.1, 1.2, 1.4, 1.6, 1.8, 2.0, 3.0, and 4.0.

this idea of decomposition of a complex multi-scale flow into a set of simpler flows each consisting of a large-scale motion superimposed with the smaller-scale turbulence, seems to be applicable not only to convective flows. The problem of the turbulent closure could be addressed in this case with use of well-known approaches (see for example Abdella and McFarlane, 1997).

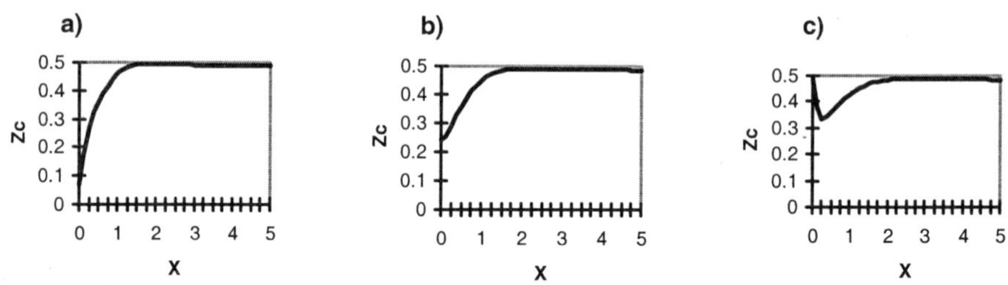

Fig. 5 Mean height of the particles corresponding to Z_s = 0.067 (a), 0.24 (b) and 0.49 (c).

The goal of the presented paper is only to demonstrate that this approach could be explored in future studies. There are also some additional technical problems to be resolved in these studies. In particular, this approach should be universal working not only in convective but, for example, in neutral conditions. It means that Eq. (4) - (6) should be generalized to provide proper limits of corresponding quantities at S = 0. More realistic models of the wind flow should also take into account variations of W_u and W_d with height and corresponding variations of the horizontal component of the wind speed to satisfy the continuity equation. Nevertheless, even the results presented in this paper show that decomposition ("splitting") of the flow into a set of more "simple" flows with use of the local-gradient description of turbulent fluxes can be considered as an alternative to the non-local description of these fluxes and even more sophisticated numerical techniques.

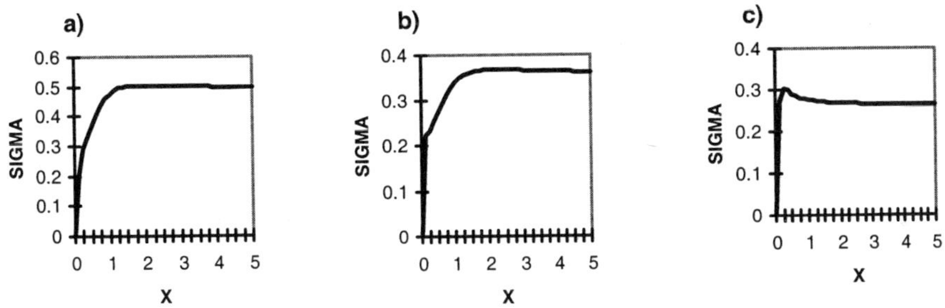

Fig. 6 Overall standard deviation of the particles corresponding to Z_s=0.067 (a), 0.24 (b) and 0.49 (c).

Acknowledgments

The author wishes to acknowledge R. Berkowicz for helpful discussions and W. Snyder and D. Mironov for their help in collecting published papers on convection. Preliminary results of this work were obtained during the visit to the National Environmental Research Institute, Roskilde, Denmark, which was supported by the Danish National Research Council.

REFERENCES

Abdella, K., and McFarlane, N., 1997 A new second-order closure scheme for the planetary boundary layer, *Journ. Atmos. Sci.* 54:1850

Baerensten, J.H., and Berkowicz, R., 1984 Monte Carlo simulation of plume dispersion in the convective boundary layer, *Atmos. Environ.*, 18:701

Briggs, G.A., 1985 Analytical parametrizations of diffusion: the convective boundary layer, *Journal Climate Appl. Meteorol.*, 24:1167

Deardorf, J.W., and Willis, G.E., 1984 Groundlevel concentration fluctuations from a buoyant and a non-buoyant source within a laboratory convectively mixed layer, *Atmos. Environ.*, 18:1297

Genikhovich, E.L., and Snyder, W.H., 1994 A new mathematical model of pollutant dispersion near a building, *Eight Joint Conf. on Applications of Air Pollut. Meteorology With AWMA,* AMS: 254

Moeng, and C.-H., Wyngaard, J.C., 1984 Statistics of conservative scalars in the conservative boundary layer, *Journ. Atmos. Sci.* 41:3161

Nieuwstadt, F.T.M., 1998 Review of diffusion processes in the convective boundary layer, *Buoyant Convection in Geophysical Flows*. E.J. Plate, E.E. Fedorovich, D.X. Viegas, and J.C. Wyngaard, ed., Kluver, Dordrecht.

Piper, M., Wyngaard, J.C., Snyder, W.H., and Lawson, R.E., 1995 Top-down, bottom-up diffusion experiments in a water convective tank, *Journ. Atmos. Sci.* 52:3607

Randall, D.A., Shao, Q., and Moeng, C.-H., 1992 A second order bulk boundary-layer model, *Journ. Atmos. Sci.*, 49:1903

Venkatram A., 1983 On dispersion in the convective boundary layer, *Atmos. Environ.*, 17:529

Weil, J.C., 1990 A diagnosis of the asymmetry in top-down and bottom-up diffusion using a Lagrangian stochastic model, *Journ. Atmos. Sci.,* 47:501

Weil, J.C., Corio, L.A., and Brower, R.P., 1997 A PDF dispersion model for buoyant plumes in the convective boundary layer, 1997, *Journ. Appl. Meteorol.* 36:982

Weil, J.C., Snyder, W.H., Lawson, R.E., and Shipman, M.S., 1998 Recent experiments on buoyant dispersion in a laboratory convective tank, *Tenth Joint Conf. on Applications of Air Pollut. Meteorology With AWMA,* AMS:9B3

Willis, G.E., and Deardorf, J.W., 1976 A laboratory model of diffusion into the convective planetary boundary layer, *Quart. Journ. Roy. Meteor. Soc.,* 102:427

Wyngaard, J.C., and Brost, R.A., 1984 Top-down and bottom-up diffusion of a scalar in the convective boundary layer, *Journ. Atmos. Sci.* 41:102

DISCUSSION

S. ZILITINKEVICH: What about the heat flux in your example, is it prescribed or modeled?

E. GENIKHOVICH: It is prescribed in the same way as in Wyngaard's model.

S. ZILITINKEVICH: The model presented is in fact an essentially non-local model. It distinguishes between the large-scale structures, namely updrafts and downdrafts, and small-scale chaotic turbulence. It is just the small-scale turbulence that is modelled with the use if a local turbulence closure. Contrastingly, the contribution to the vertical transport from large-scale structures is due to a non-local advection.

E. GENIKHOVICH: I agree, but this advection is local in terms of the decomposed quantities.

AN INTERSCALE APPROACH TO ASSESS NO_2 AND OZONE LEVELS IN URBAN AREAS

N. Moussiopoulos, P. Sahm and G. Theodoridis

Laboratory of Heat Transfer and Environmental Engineering
Aristotle University Thessaloniki, 54006 Thessaloniki, Greece

ABSTRACT

Urban air quality depends on both regional and local scale transport and transformation phenomena. Therefore an interscale approach is needed for reliable assessments of air pollution levels in conurbations. Microscale models taking into account chemical processes can provide mesoscale models with more accurate emission data. This is demonstrated by the aid of simulations with a CFD model. Furthermore, a suitable combination of a regional and local scale photochemical model is needed for properly describing ozone formation at the urban scale. An efficient combination of that type is proposed which may be used for reliable assessments of longer term air pollution levels and the effectiveness of control strategies.

INTRODUCTION

Air pollutants are known to have significant effects on agricultural crops and cause severe health problems. Most of the latter occur at places of higher population density, i.e. in conurbations. According to air quality measurements in urban stations, in the year 1995 80% of the EU urban population was exposed to exceedances of the EU ozone threshold value for the protection of human health (Berge et al., 1997). Despite the downward trend detected over the last decade in the case of SO_2 and NO_2, the corresponding air quality standards are still exceeded in numerous European cities. Hence, emission reductions will still be needed in the future, and if the corresponding interventions are to be effect-oriented and cost-effective, any air pollution abatement strategy in conurbations should be properly assessed prior to be implemented.

Strategies to abate air pollution at any relevant scale may be assessed with already available models (Borell et al., 1997). Among other model systems, the European Zooming Model (EZM; previous name: EUMAC Zooming Model) was developed for air quality assessments in conurbations (Moussiopoulos, 1995). This model system was successfully utilised to optimise the air pollution abatement strategy for Athens (Moussiopoulos et al., 1994): Ozone measurements of the Athenian monitoring network for the period 1990-1995

Air Pollution Modeling and Its Application XIII, edited by S.-E. Gryning
and E. Batchvarova, Kluwer Academic/Plenum Publishers, 2000

confirm the trends derived from model simulations performed in 1990. Other examples of urban air quality analyses can be found elsewhere (Ebel and Singh, 1997; Ebel et al., 1997).

In spite of the usefulness of their results, previous urban air quality assessments generally did not go beyond a qualitative analysis. This handicap was caused by the poor understanding of processes governing air pollutant transport and transformation at local scales. Examples for scientific issues that have to be dealt with prior to proceeding to quantitative assessments are:

• Description of pollution levels in single streets as an example for zones where the atmospheric flow is heavily disturbed by buildings and other obstructions. This may affect local concentrations by more than one order of magnitude compared to unobstructed terrain. Consequently, considering the buildings not individually but only within the roughness parameterization may lead to significant errors in source-receptor relationships.

• Analysis of the multiscale character of pollutant dispersion in urban areas. Due to the combined effect of natural wind variability and obstacles, concentrations may differ substantially over relatively short distances. Besides, the receptor points are normally close to the sources. The presence of large concentration gradients makes it complicated to find representative locations for air quality monitoring stations.

• Identification of important chemical reactions occuring in urban areas. Different equally important situations need to be described depending on the pollutant residence time in the urban canopy. These situations range from the "fast chemistry" scenario (situation in street canyons where it is sufficient to treat the NO-NO_2-ozone system) to stagnant conditions over a city (situation in which VOC chemistry is important even within the urban canopy). In any case, chemical cycles in urban atmospheres are characterised by high pollutant concentration levels, thus differing considerably from cycles in rural and marine atmospheres.

The above scientific issues are examples for topics included in the work programme of the EUROTRAC-2 subproject SATURN (1997).

The present paper addresses the issue of adequately describing multiscale effects in urban air pollution models with emphasis on those affecting ozone formation at the urban scale. Towards this aim, we first discuss the need to combine several models in a multiscale cascade. Subsequently, we present simulation results for the air flow and pollutant transport and transformation in typical street canyon configurations. Finally, we address the issue of ozone formation in an urban plume taking into account the findings of microscale model applications.

MULTISCALE AIR POLLUTION MODELLING

Recent model intercomparison studies (an example for such an intercomparison study is given by Moussiopoulos and Papagrigoriou (1997)) show that contemporary mesoscale air pollution models are capable of reproducing most of the dispersion characteristics affecting air pollutant concentrations in complex airsheds (Moussiopoulos et al., 1996). Discrepancies between results of individual models are in general smaller than differences computed with any of these models for various emission scenarios. On the other hand, the results of air pollution models are found to be very sensitive to the larger scale information imposed at the domain boundaries.

Usually, larger scale processes are taken into account by coupling models of different scales via an appropriate nesting technique (Moussiopoulos et al., 1996). Model coupling is, however, recommendable not only for prescribing boundary conditions to the next smaller scale: Fluxes defined at domain boundaries generally depend on processes occuring at scales which are too small to be resolved in the given model. In principle, an appropriate smaller scale model should be applied for calculating such fluxes.

Special attention is necessary in the case of pollutant fluxes at the lower model boundary, i.e. the source terms associated with surface level emissions. In the rather usual case that a flux refers to a group of emitted pollutants (e.g., NO_x or VOC), it is common practice to assume that the composition merely depends on the source type regardless of the dispersion length scale. However, chemical reactions with small time scales (i.e. of the order of a few minutes) can affect decisively the composition of the emitted pollutants before scales are reached which are comparable to the resolution of mesoscale models. More accurate emission data can be provided to mesoscale models by microscale models taking into account chemical processes such as the NO-NO_2-ozone cycle.

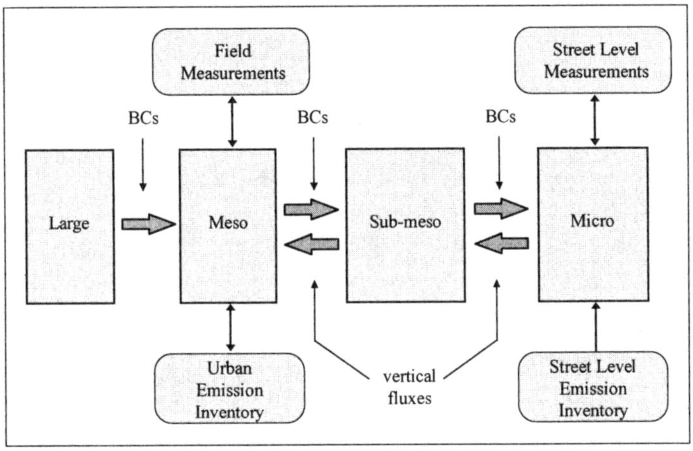

Figure 1. Multiscale model cascade for addressing urban air pollution.

In contrast to nested models, multiscale model systems which allow both prescribing refined boundary conditions to the next smaller scale and calculating accurate fluxes to the larger scales are at an early stage of development. As an example, Figure 1 shows a schematic representation of the multiscale cascade envisaged in the ZEUS model system which is under development at the Aristotle University Thessaloniki. This model system starts at the local-to-regional scale and goes down to the microscale. Within ZEUS, each scale provides the next smaller one with appropriate lateral boundary conditions for momentum, heat and pollutant concentrations; microscale parametric computations provide improved momentum, heat and pollutant vertical fluxes as input to the larger scales.

INFLUENCE OF BUILDINGS ON LOCAL SCALE AIR POLLUTION LEVELS

Placed within the atmospheric boundary layer, buildings act as artificial obstacles to the wind flow and cause stagnant conditions in an urban area, even for relatively high ambient wind conditions. A typical configuration is the so-called street canyon, formed along a street in densely built urban areas. At the bottom of a street canyon the vehicles emit atmospheric pollutants in the form of buoyant plumes, at the same time causing an initial dispersion of pollutants due to their motion. Atmospheric pollutants are subsequently dispersed by the wind flow within the street canyon.

Street canyons are recognised as important configurations with regard to urban air pollution and have been the subject of extensive experimental investigations (Hosker,

1985). The most characteristic feature of the street canyon is the formation of a wind vortex such that the direction of the wind at street level is opposite to the flow above roof level. The systematic study of the available experimental data resulted in a classification of flow regimes in urban street canyons (Oke, 1988). More specifically, three flow types are observed, depending mainly on two geometric ratios, the ratio of the average building height H over the canyon width W and the ratio of the length of the building L over the average height of the building H. For widely spaced buildings ($H/W<0.3$), the flow fields associated with the buildings do not interact. At closer spacing ($0.3<H/W<0.7$) the wake created by the upwind building is disturbed by the downwind building, giving rise to a downward flow along the windward face of the latter. Even closer spacing ($0.7<H/W$) results in a stable vortex in the canyon and the ambient flow is decoupled from the street flow. As the ratio H/W is further increased ($H/W>1.5$), an additional weak tertiary counter rotating vortex of lesser extent can develop at the bottom of the street canyon. The ratio L/H gets important if $L/H<20$. In such a configuration, the vortices generated at street intersections interact with the main vortex in the street canyon, resulting in higher pollutant concentration levels (Hoydysh and Dabberdt, 1998).

The improvement of methods that are used for assessing traffic pollution in streets is the subject of the TMR project TRAPOS (1998). The work programme of TRAPOS includes measurements in streets, wind tunnel experiments, advanced numerical modelling and data evaluation. The results from the studies will be incorporated into simpler, operational models.

MODELLING OF AIR FLOW AND POLLUTANT DISPERSION AND TRANSFORMATION IN STREET CANYONS

As a preparatory action towards formulating the microscale module of ZEUS (see section 2), 2-D computations were performed in typical street canyon configurations with the CFD code TASCflow (Raw et al., 1989).

Model description - TASCflow

TASCflow may be used for modelling fluid flow, heat and mass transfer and fast chemistry in complex geometries. The code uses arbitrary curvilinear, body-fitted, multi-block, structured, non-staggered grids, a strong conservative form of the governing differential equations, a first-order accurate backward fully implicit scheme in time, a second-order bounded scheme for the spatial discretisation of advection terms, a coupled solution procedure for momentum and continuity equations, an algebraic multi-grid method for the solution of the sets of the algebraic difference equations and the standard k-ε turbulence model with wall functions.

A module for the coupled treatment of fast chemical reactions (NO-NO_2-O_3 cycle) within street canyons has been developed. An implicit treatment of source terms due to chemical reactions is accomplished by incorporating negative source terms to the active part of the coefficients of the discretised equations. Velocity, turbulence and pollutant concentration fields are first computed as quasi-steady, treating pollutants as chemically inert. Velocity and turbulence fields are subsequently frozen and unsteady computations follow for the evolution of the concentration fields of NO, NO_2 and O_3. Background values for O_3, NO and NO_2 are provided at the inflow boundary, while NO and NO_2 sources at street level are used in order to simulate heavy NO_x traffic emissions.

Case specifications

The geometry of the domain studied and the boundary conditions are illustrated in Figure 2. Two street canyon configurations were used, a rectangular one with H/W=1 and a deep one with H/W=2 (H=28m). At the inflow boundary, a horizontal velocity profile $U=U_{ref}(z/\delta)^{\alpha}$ was applied (α=0.33, U_{ref}=5ms^{-1}, δ=140m) with a turbulence intensity of Tu=10%, a dissipation length scale of L=25m, zero background concentration for NO, NO_2 and a concentration of 70ppb for O_3. A uniform emission rate of 1,25μgm^{-1}s^{-1} for NO_x with NO_2/NO_x=10% was assumed at street level. Night- and day-time conditions were analysed.

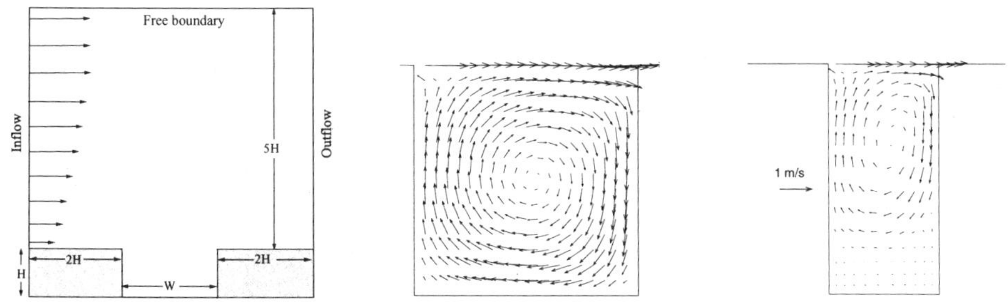

Figure 2. Sketch of the computational domain for TASCflow (left); Velocity fields for two street canyon configurations as predicted with TASCflow (middle and right).

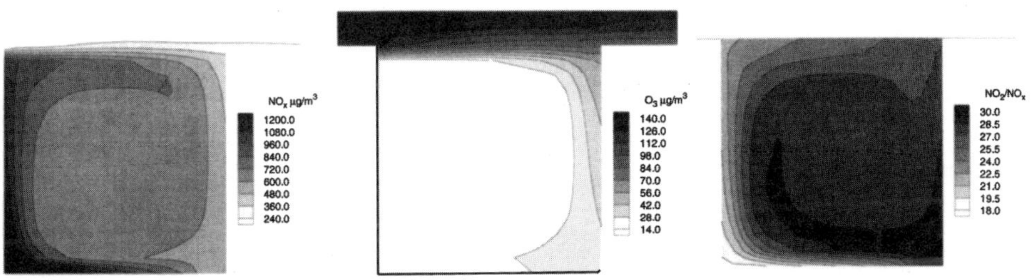

Figure 3. Pollutant concentrations as predicted with TASCflow. NO_x (left), O_3 (middle) and NO_2/NO_x (right).

Results

The main findings from the numerical investigation of the flow field and the dispersion and chemical transformation of pollutants in typical street canyon configurations are as follows:

- The flow fields resulting from the simulations correspond to the patterns observed in street canyons (Figure 2).
- In particular, and in good agreement with observations, a dual vortex system is predicted for the deep street canyon configuration (case with H/W=2, Figure 2 right panel).
- Steady state is reached at a time scale of the order of two to three minutes.
- High NO_x concentration levels are predicted on the leeward side of the street canyon, while O_3 is depleted within the street canyon (Figures 3 (left) and (middle)).
- The oxidation of NO to NO_2 leads to a significant increase of the NO_2 concentration (Figure 3 (right)). Thus, the NO_2/NO_x ratio in the air exiting the urban canopy may be much higher than the value following from the emission inventory.
- In particular, the NO_2/NO_x ratio was found to vary linearly with the background O_3 levels.

MODELLING OZONE FORMATION IN URBAN PLUMES

The impact of specific emission reductions to urban air quality can be addressed with suitable simulation models. In principle, such analyses can be performed with sophisticated mesoscale air pollution models. The computational effort of the latter limits, however, their applicability to episodic simulations only which can hardly be considered as representative for longer time periods. Much more efficient models are needed for applications of statistical nature, i.e. the analysis of long term air quality indicators (e.g. yearly average, 98th percentile values). In the past a higher model efficiency could only be achieved at the cost of a considerably lower modelling accuracy. At present, the development of novel modelling tools is envisaged which will combine high efficiency at an acceptable degree of reliability. This will be possible thanks to the experience gained over more than one decade from the application of sophisticated models.

Model description - OFIS

The Ozone Fine Structure (OFIS) model allows an adequate description of urban photochemistry at a low computational effort: This newly developed Eulerian model is capable of calculating ozone levels due to transport and photochemical transformation within an urban plume (Moussiopoulos and Sahm, 1998). Moreover, it can be used for obtaining exceedance statistics of ozone threshold values based on meteorological data and regional background concentrations for the period considered.

The conceptual basis of OFIS is a coupled 1D-2D approach:

• Background boundary layer concentrations are calculated with a three-layer box model representing the local-to-regional conditions in the surroundings of the city considered. This model uses at input non-urban emission rates, meteorological data and regional scale model results for pollutant concentrations (e.g. EMEP model results (Simpson, 1993; 1995)).

• Pollutant transport and transformation downwind of the city (along the prevailing wind direction) is calculated with a three-layer multibox model representing a substantially refined version of MARS-1D (Moussiopoulos, 1990).

The distinction of three individual layers of time depending thickness allows adequately describing the dynamics of the atmospheric boundary layer. At the same time, vertical transport is taken properly into account by considering the exchange between adjacent layers. For prescribing the thickness of the three layers, a 1D version of the non-hydrostatic mesoscale model MEMO (Moussiopoulos, 1995) is utilised: The vertical profiles of temperature, mean wind speed and turbulent exchange coefficient as well as the mixing height are calculated both for the city surroundings and the urban plume assuming Monin-Obukhov similarity at the lower boundary.

Case specifications

Simulations of the ozone levels in the area around Stuttgart were performed with OFIS for each day between 1 April and 30 September 1990. For this period, meteorological data and EMEP model results (regional background concentrations) were available for an area covering 250×250 km^2 around Stuttgart at a spatial resolution of 50 km and a temporal resolution of 6 hours. Emission data for Stuttgart, specifically the diurnal variation of the emission rates for urban, suburban and rural areas at a temporal resolution of 1 hour were supplied by the Institute of Energy Economics and the Rational Use of Energy (IER), Stuttgart. The emission inventory was compiled with a NO_2/NO_x ratio of 14.56 %.

Background boundary layer concentrations were computed with the three-layer box model embedded in OFIS for a domain of 150×150 km^2 which was assumed to be rural area. For each day in the period considered, pollutant transport and transformation downwind of Stuttgart was calculated in 5 km steps assuming the wind direction valid for the respective day, an initial plume width of 14 km and a plume widening angle of 30°. Dry deposition was accounted for by using a three-resistance model approach. Biogenic emissions were taken into account for rural areas.

As shown in Figure 4, prevailing wind directions at surface level in Stuttgart are those from W-SW. Also, winds from NE appear to occur rather frequently in the area.

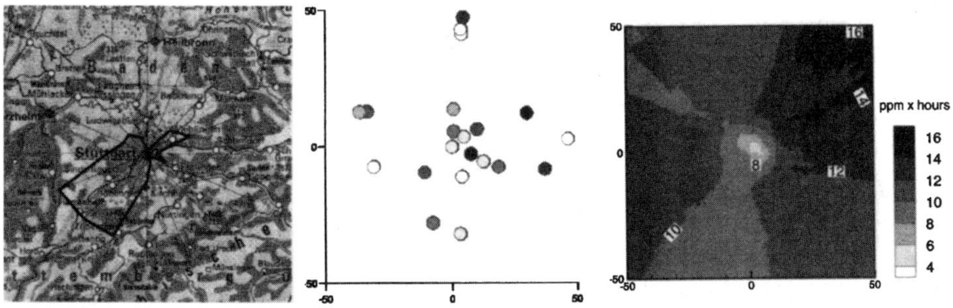

Figure 4. Study area around Stuttgart and assumed wind direction statistics in the period April-September 1990 (left); AOT60 values calculated from the results of OFIS (right) and those derived from the measurements (middle).

Results

The accumulated ozone exposure above a threshold value of 60 ppb (AOT60) is frequently used to quantify the exceedance of the WHO air quality guideline for protection of human health.

In Figure 4 AOT60 values calculated with OFIS for the period 1 April – 30 September 1990 are compared with corresponding data derived from observations. According to the OFIS results for AOT60 in the surroundings of Stuttgart, highest AOT60 levels are reached to the NE of Stuttgart, i.e. the area which happens to be most of the time downwind of the city. Values ranging up to 15 ppm×hours agree satisfactorily with available measurements in spite of the fact that the situation in the individual measuring stations is inevitably influenced by a variety of local influences not considered in OFIS (emission inhomogeneities, terrain effects etc.).

Influence of the NO_2/NO_x emission ratio on the ozone formation

The above presented TASCflow results reveal that the NO_2/NO_x ratio in a street canyon may deviate considerably from the value following from the emission inventory (Figure 3). In particular, the air exiting the urban canopy may contain more than the double of the actually emitted NO_2.

To investigate how the NO_2/NO_x ratio in the air exiting the urban canopy affects the formation of ozone in the urban plume, the simulation with OFIS described in the previous section was repeated for the NO_2/NO_x ratios 10%, 15%, 20% and 25%.

The results reveal that the ozone levels increase with the assumed NO_2/NO_x ratio. The AOT60 values exhibit a high sensitivity: At certain locations AOT60 increases by up to almost 3 ppm×hours for a NO_2/NO_x ratio change from 10% to 25%. The results furthermore reveal that the predicted maximum ozone concentration varies almost linearly with the NO_2/NO_x ratio.

CONCLUSIONS

Processes affecting urban air quality extend over several scales. Urban air pollution models should therefore allow adequately describing interscale relationships. In particular, the pollutant concentration load in the air exiting the urban canopy may deviate considerably from the emission composition; it should therefore be computed with a comprehensive CFD code accounting for fast chemical reactions which occur in street canyons. Moreover, both regional and urban scale effects play a decisive role regarding the ozone levels downwind a city; therefore, air pollutant dispersion and transformation in the urban plume should be described with an appropriate multiscale model cascade.

REFERENCES

Berge E., Beck J., Larssen S., Moussiopoulos N. and Pulles T., 1997, Air Pollution in Europe 1997, 107 pp., EEA Environmental Monographs No. 4, (Editors: A. Jol and G. Kielland), European Environment Agency, EU Publications, Copenhagen. (cf. http://www.eea.dk/ Document/Monogr/air97/).

Borrell P., Builtjes P.J.H., Grennfelt P., Hov P.Ø., van Aalst R., Fowler D., Mégie G., Moussiopoulos N., Warneck P., Volz-Thomas A. and Wayne R.P., 1997, Photo-oxidants, Acidification and Tools: Policy Applications of EUROTRAC Results, 216 pp., Transport and Chemical Transformation of Pollutants in the Troposphere, Vol. 10, Springer, Heidelberg.

Ebel A. and Singh H., eds, 1997, EUMAC: European Modelling of Atmospheric Constituents, Atmos. Environ. 31, 3117-3282.

Ebel A., Friedrich R. and Rodhe H., 440 pp., eds, 1997, Troposphere Modelling and Emission Estimation, Transport and Chemical Transformation of Pollutants in the Troposphere, Vol. 7, Springer, Heidelberg.

Hosker R.P., 1985, Flow Around Isolated Structures and Building Clusters, 1671-1692, A Review, ASHRAE Trans., Vol 91.

Hoydysh W.G. and Dabberdt W.F., 1988, Kinematics and dispersion characteristics of flows in asymmetric street canyons, 2677-2689, Atmos. Environ. 22.

Moussiopoulos N. and Papagrigoriou S., 1997, Athens 2004 Air Quality, 183 pp., eds; Proceedings of the International Scientific Workshop "Athens 2004 Air Quality Study", Athens, 18-19 February 1997, (electronic version cf. http://www.envirocomp.org/html/ publish/CDROM/Athens/flyer.pdf).

Moussiopoulos N., 1990, Influence of power plant emissions and industrial emissions on the leeward ozone levels, 1451-1460, Atmos. Env. 24A.

Moussiopoulos N., 1994, Air pollution in Athens, 3-41, Urban Air Pollution Vol. 1 (H. Power, N. Moussiopoulos and C.A. Brebbia, eds), Computational Mechanics Publications, Southampton.

Moussiopoulos N., 1995, The EUMAC Zooming Model, a tool for local-to-regional air quality studies, 115-133, Meteorol. Atmos. Phys. 57.

Moussiopoulos N., Berge E., Bøhler T., de Leeuw F., Grønskei K.E., Mylona S. and Tombrou M., 1996, Ambient Air Quality, Pollutant Dispersion and Transport Models, 94 pp., European Topic Centre on Air Quality, Topic Report 19, 1996, European Environment Agency, EU Publications, Copenhagen. (cf. http://www.eea.dk/ Document/Topicrep/air/9619.zip).

Moussiopoulos N. and Sahm P., 1998, The OFIS model: An efficient tool for assessing ozone exposure and evaluating air pollution abatement strategies, Proc. EUROTRAC-2 Symposium.

Oke T.R., 1988, Street Design and Urban Canopy Layer Climate, 103-113, Energy Building, Vol. 11.

Raw M.J., Galpin P.F. and Hutchinson B.R., 1989, A co-located finite-volume method for solving the Navier-Stokes equations for incompressible and compressible flows in turbomachinery, 189-196, Results and applications, Canadian Aeronautics and Space Journal, 35.

SATURN Project Description, 1997. (cf. http://aix. meng.auth.gr/lhtee/saturn.html).

Simpson D., 1993, Photochemical model calculations over Europe for two extended summer periods: 1985 and 1989. Model results and comparison with observations, 921-943, Atmos. Env. 27A.

Simpson D., 1995, Biogenic emissions in Europe 2: Implications for ozone control strategies, 22891-22906, J. Geophys. Res. 100 D11.

TRAPOS Project Description, 1998. (cf. http://www .dmu.dk/AtmosphericEnvironment/trapos/).

DISCUSSION

M. SOFIEV: Just a comment to AOT metric as a measure of ozone impact. It seems to me that this measure is not the best one, because it is: 1) Based on threshold approach 2) integrates the exceedances. It is known that such a characteristics can not be stable and, in particular, very sensitive to any fluctuations of conditions. Do you agree with this?

P. SAHM: As you say in 1) the values are based on the threshold approach and thus they are very sensitive statistics. This is bad for AOT40 (the effect parameter used by UNECE to assess the effects of ozone on agricultural crops and ecosystems) and much worse for AOT60 (the effect parameter used in this paper). However, the fact that the AOT integrates the exceedances is good. In fact, this is the best feature of AOTs as this integration gives a longer time-perspective to the results. Also, this is biologically sound - vegetation seems to care about itslong term exposure rather than short-time events.

The main defence of AOT40 is that it does indeed correlate very well with yield loss in species such as wheat. And it is much simpler to use and understand than the complex dose-response functions favoured in the States (e.g.sigmoidal functions).

For the above reasons we decided to express the exceedances of the World Health Organisation (WHO) guideline for the protection of human health (120 μgm^{-3}, 8h average) as an AOT60, using the 60 ppb level as a threshold.

J. LEVITIN: What time emissions (hourly or daily) do you use in this model?

P. SAHM: The diurnal variation of the emission rates for urban, suburban and rural areas at a temporal resolution of 1 hour are required in the OFIS model.

GRAVITY WAVE STRESS PARAMETERIZATION IN A MESOSCALE MODEL

Carmen J. Nappo [1] and William Physick [2]

[1]Atmospheric Turbulence and Diffusion Division, Air Resources Laboratory
National Oceanic and Atmospheric Administration
456 S. Illinois Ave., Oak Ridge, TN 37830, USA

[2]CSIRO, Division of Atmospheric Research
PMB1 Aspendale, Victoria 3195, Australia

INTRODUCTION

It is well known that terrain obstacles can launch vertically propagating gravity waves in a stably-stratified atmospheric flow. The generation of these mountain waves produces a form drag on the ground surface, and the reaction to this drag acts on the atmospheric flow at some height where wave dissipation occurs. Baines (1995) suggests that the form drag on the atmosphere generated by topography is approximately 50 % of the total drag from the ground, the other half being frictional drag generated by turbulence in the planetary boundary layer (PBL). The importance of accounting for form drag generated by subgrid scale terrain features in global-scale numerical weather prediction and general circulation models is also well known, and has been recently discussed by Lott and Miller (1997). Blumen and Raeder (1991) and Kim and Mahrt (1992) examined wave drag on scales comparable to the mesoscale, and Nappo and Chimonas (1992) and Grisogono (1994) examined gravity waves and wave drag in the stable (PBL). For example, Nappo and Chimonas (1992) showed that under typical nighttime conditions the wave stress over hills and ridges with height and width scales as small as 100 m and 1000 m respectively can be as great or greater than the friction stress at the ground surface. However, these mesoscale and PBL studies only evaluated the wave stress. In this paper, we describe a method whereby the wave stress is fed back into the numerical model.

For mesoscale numerical models such as the Regional Atmospheric Modeling System (RAMS) (Pielke *et al.*, 1992) or the Lagrangian Atmospheric Dispersion Model (LADM) (Physick *et al.*, 1994), horizontal grid resolutions of 10 km or less can be considered fine scale, yet many terrain features exist on smaller scales. In addition to increased drag on the atmosphere, breaking gravity waves result in increased vertical diffusion, and if this process is not considered in air quality models, significant error may result. This paper examines the effects on the LADM model of using a subgrid-scale parameterization of gravity-wave-generated turbulence in a simulation of a two-dimensional sea breeze over a coastal ridge. The work described here is a natural extension of a one-dimensional PBL study by Nappo and Andren (1995).

Air Pollution Modeling and Its Application XIII, edited by S.-E. Gryning and E. Batchvarova, Kluwer Academic/Plenum Publishers, 2000

WAVE STRESS PARAMETERIZATION

The wave stress parameterization is simple and straightforward. At each time step of LADM, linear wave theory is used to evaluate the wave stress $-\langle\rho\tilde{u}\tilde{w}\rangle$ where $\tilde{u}$ and $\tilde{w}$ are the horizontal and vertical wave-perturbation velocities respectively, and the braces indicate horizontal averaging over a wavelength for the case of a monochromatic wave or over a large horizontal distance for the case of a wave packet. The stress is calculated using the model wind and temperature profiles in the vertical grid column containing the unresolved terrain feature. The wave stress is added to the Reynolds stress, and the mean-flow momentum equation becomes,

$$\frac{D(\rho U)}{Dt} = -\frac{\partial p}{\partial x} + \frac{\partial}{\partial z}\left(\rho\overline{u'w'} + \rho\langle\tilde{u}\tilde{w}\rangle\right). \tag{1}$$

where U is the LADM-predicted x-component of velocity and $\rho\overline{u'w'}$ is the Reynolds stress. Equation (1) is then used to calculate a new mean-flow U, and the calculation proceeds to the next time step.

The wave drag, D, per unit crosswind distance over a two-dimensional ridge is given by (see, for example, Nappo and Chimonas, 1992)

$$D = \int_{-\infty}^{\infty} -\rho\tilde{u}\tilde{w}\,dx = \int_{0}^{\infty} -\frac{\rho}{\pi}\left[\frac{U_0\hat{h}(k)}{\hat{w}(k,0)}\right]^2 k \times Im\left[\hat{w}(k,z)\frac{d}{dz}\hat{w}^*(k,z)\right]dk \tag{2}$$

where $U_0 = U(0)$, $\hat{h}$ is the Fourier transform of the terrain height profile, $\hat{w}(k,z)$ is the Fourier transform of the wave-perturbation vertical velocity, k is the horizontal wavenumber, the asterisk denotes the complex conjugate, and Im indicates the imaginary part of the object in braces. The vertical velocity, $\hat{w}(k,z)$, is obtained from the Taylor-Goldstein equation (see, for example, Nappo and Chimonas, 1992). The wave stress over an isolated ridge is obtained by dividing D by a suitable horizontal length such as the width of the ridge; however, when considering flow over a two-dimensional surface corrugation the wave stress is given by averaging $-\rho\tilde{u}\tilde{w}$ over one wavelength of the corrugation.

Vertically-propagating gravity waves launched by terrain obstacles transport mean flow horizontal momentum at a constant rate unless wave dissipation occurs (see, for example, Eliassen and Palm, 1960; Baines, 1995). From (1) we see that if the wave stress is constant with height, then flow acceleration will not occur. The process of wave dissipation and subsequent exertion of wave drag on the mean flow is generally referred to as wave saturation (Fritts, 1984). When the background wind speed in the direction of wave propagation decreases with height, wave amplitudes will grow as the wave propagates upward. Eventually, the wave amplitude becomes so great that the wave breaks much as surface waves approaching a beach. Wave breaking marks the onset of convective instability when more dense fluid is forced above less dense fluid. From linear wave theory, the wave-perturbation temperature is

$$\theta(x,z) = -\frac{\partial\Theta(z)}{\partial z}\,\zeta(x,z) \tag{3}$$

where $\Theta(z)$ is the background potential temperature, and $\zeta(x,z)$ is the vertical displacement of a flow streamline from its upstream level. Convective stability occurs when

$$\frac{\partial(\Theta + \theta)}{\partial z} \geq 0. \tag{4}$$

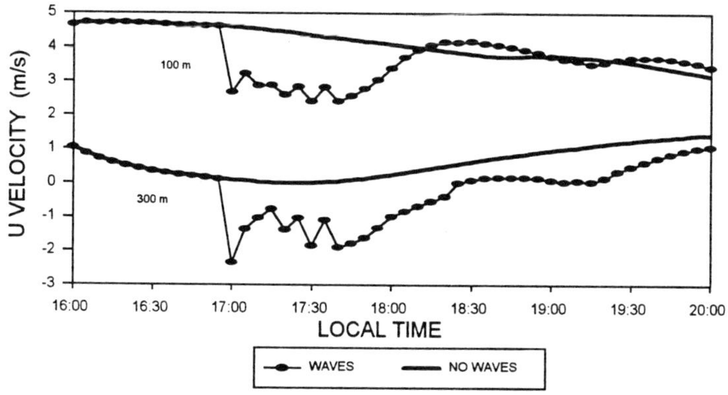

Figure 1. Time series of U velocity at 100 m and 300 m AGL at the point x=100 km, y=50 km. Results with wave stress parameterization indicated by WAVES, and results without parameterization indicated by NO WAVES.

Using (3) in (4) and assuming that $\partial^2\Theta/\partial z^2 << 1$, the condition for convective stability is $\partial\zeta/\partial z \leq 1$. Where $\partial\zeta/\partial z = 1$, the flow streamlines become vertical, and further increases in wave amplitude as the wave moves upward results in wave breaking, *i.e.*, dissipation. The saturation hypothesis is that dissipation is just balanced by wave growth so that wave amplitude either decreases or remains constant.

In the wave saturation parameterizations used by Schoeberl (1988) and Kim and Mahrt (1992), the values of the streamline displacements, $\zeta(x,z)$ are locally adjusted so that (4) holds throughout the wave field. The streamline adjustment decreases wave amplitude, and this results in decreased wave stress. Schoeberl (1988), Kim and Mahrt (1992) and others used the WKB method for solving the wave equation (see, for example, Laprise, 1993; Baines 1995). This method assumes that the vertical variations in wind speed and static stability are gradual enough so as not to induce reflections of the upward propagating waves. This is equivalent to requiring that the scale of the vertical variations of the background quantities

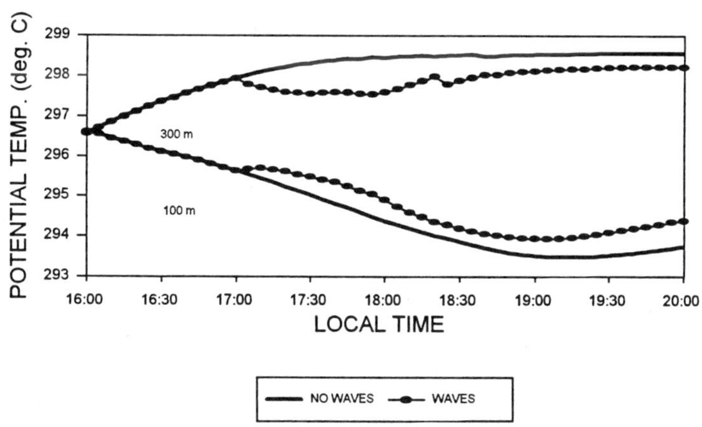

Figure 2. Same as Figure 1 but for potential temperature.

is large compared to the vertical wavelength of the gravity wave. An estimate of the vertical wavelength λ_z is $2\pi(N^2/U^2 - k^2)^{-1/2}$ where k is the dominant horizontal wavenumber of a terrain-generated gravity wave. If the width of the terrain feature is W, then $k = 2\pi/W$. For $U = 4$ ms^{-1}, $N = 0.03$ s^{-1}, and $W = 1000$ m, $\lambda_z = 1530$ m. Since the wind speed and static stability can change significantly within about the first

200 m of the stable PBL, it is clear that the WKB method is not applicable in the PBL, but this also means that the method of streamline adjustment used by Schoeberl (1988) and Kim and Mahrt (1992) is not applicable. In the initial stages of the work done by Nappo and Andren (1995), the streamline adjustment method was used, but instead of wave stress decreasing with height it increased. Because the flow streamlines are due to the additions of both upward and downward (reflected) propagating waves, the adjustment process reduced the amplitudes of upward propagating waves, but increased the amplitudes of the downward propagating waves. An alternative is to decrease amplitude of all the waves by gradually decreasing the terrain height until (4) is satisfied throughout the wave field. The original terrain height is restored at the start of each time step.

THE LAGRANGIAN ATMOSPHERIC DISPERSION MODEL (LADM)

LADM is an air quality dispersion model which consists of two main components: a hydrostatic mesoscale meteorological model and a Lagrangian particle dispersion model. The meteorological model employs the equations of motion, the hydrostatic equation, the compressible form of the continuity equation and the thermodynamic and moisture equations in a terrain-following coordinate system in which the vertical coordinate is pressure normalized by surface pressure. A semi-Lagrangian approach is used for advection calculations. The model can be run in a nested manner (one-way) with zero-gradient boundary conditions on the outermost grid. Turbulent fluxes are parameterised by K-theory, following Louis (1979),

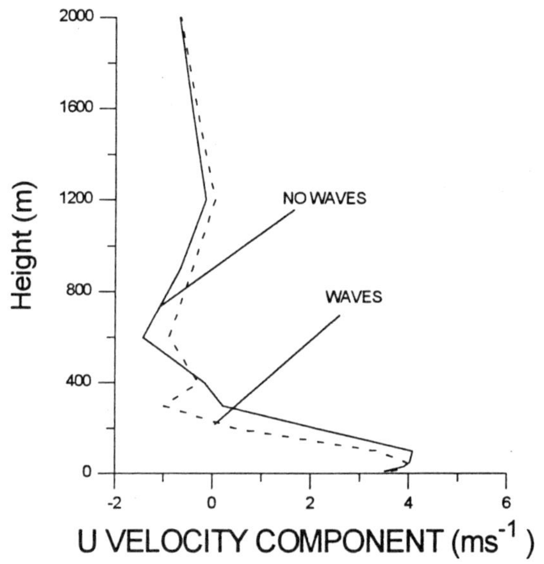

Figure 3. Wind speed profiles at x=100 km, y=50 km.

and a vegetative canopy scheme is employed at the lower boundary with a two-layer soil moisture scheme. Surface temperature is computed from an energy balance equation, and simple formulae are used to allow for the effect of cloud cover on surface net radiation.

The dispersion component of LADM uses a Lagrangian scheme in which marked particles released from sources, and representing neutrally buoyant gases, are advected and diffused according to the local wind and turbulence. The turbulent components are obtained from the general form of the Langevin equation derived by Thomson (1987). Skewed homogeneous turbulence is used for convective conditions (Hurley and Physick, 1993) and Gaussian inhomogeneous turbulence in stable conditions. Ground-level concentrations are calculated by counting the number of particles in small "boxes" over the region of interest. Plume rise is determined from a set of equations (based on those of Glendening *et al.*, 1984) valid for both zero and non-zero ambient winds.

THE NUMERICAL EXPERIMENT

We test the wave-stress parameterization in a two-dimensional sea breeze simulation over a coastal ridge. The transition height z_c, where for example on-shore flow switches to off-shore flow and U vanishes is a critical level for mountain waves. At a critical level z_c, the mountain waves are essentially absorbed; however, as the wave approaches z_c its amplitude grows in response to decreasing wind speed. Wave breaking begins at some distance below z_c, as described in Fritts and Geller (1976) and Dörnbrack and Nappo (1997). In our parameterization, we assume complete wave absorption at the critical level, and a linear decrease of wave stress between the wave-breaking level and z_c. The location of the simulation is the south-west coast of Australia near the city of Perth. The size of our single model domain in the x, y, and z directions is 200 × 100 × 19 km respectively. The horizontal grid resolution is 10 × 10 km, and there are 24 vertical levels. The coordinate system is terrain following, but for this simulation the resolved terrain is flat, and at sea level. The shore line is 80 km from the west boundary of the model, and the two-dimensional subgrid-scale ridge is located in the 10th row of grid cells, *i.e.*, 20 km inland from the shoreline. The time step size is 300 s. The simulation begins at 00 hrs on February 4, 1994, and lasts 24 hours. A steady geostrophic wind speed of 0.2 ms^{-1} from the east is imposed. This small geostrophic wind speed was used to get a well-defined sea breeze circulation.

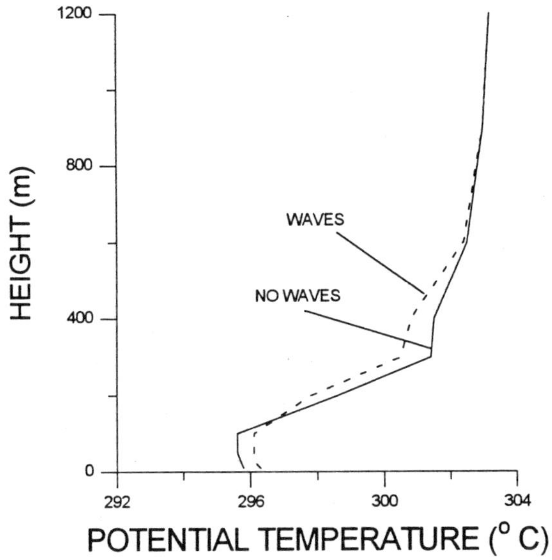

Figure 4. Same as Figure 3 but for potential temperature.

The Gaussian-shaped ridge is described by $h(x) = H_R \exp(-x^2/b^2)$ where H_R = 200 m, and b = 600 m. The wave equation must be solved for each wavenumber, k, of the Fourier transform of $h(x)$ for each grid cell containing the ridge. This would require considerable computation time. Instead, we choose to replace the ridge with a surface corrugation of amplitude H_C and single wavenumber k_C. The requirement in selecting H_C and k_C is that the stress over the corrugation is the same as that over the ridge. In the case of constant wind and stratification, and assuming that N^2 / U^2 is much greater than either k or k_C, we require $H_C = \sqrt{2}\, H_R$, and $k_C = 1/b$.

To test the effect of the wave-stress parameterization on vertical dispersion, we simulate a continuous emission from a 100 m high stack located at grid-cell coordinates i = 9, and j = 5, *i.e.*, 10 km inland from the shoreline and 50 km north of the southern boundary of the model. The Lagrangian dispersion model uses a 20-second time step and concentrations are calculated in boxes 1000 x 1000 x 25

m. The wind and turbulence fields are updated every 10 minutes. The source strength is 1000 gs^{-1}, and the plume is non-buoyant.

RESULTS AND DISCUSSION

Figure 1 shows time series of *U*-velocity component at heights of 100 m and 300 m at the grid cell located at x = 100 km, y = 50 km. Values calculated without the wave parameterization are marked NO WAVES, and values calculated with the parameterization are marked WAVES. Sunset is at 19:20, but wave breaking begins at 16:55 as seen by the sudden deceleration in the wind speed, and continues until about 23:00. Before this time, the wave stress has no effect on the flow. At the start of wave breaking,

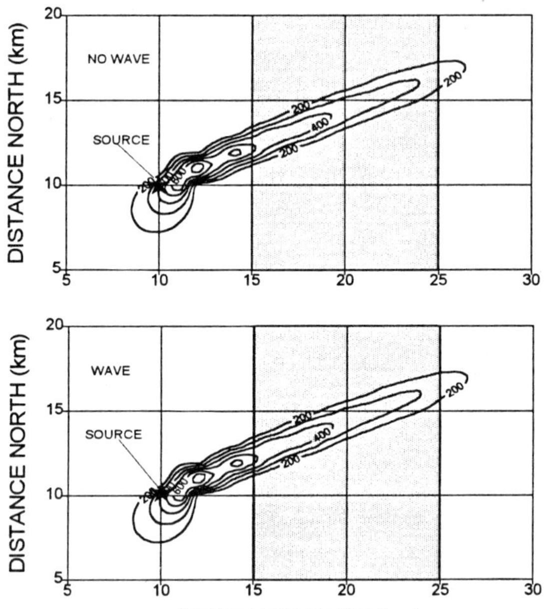

Figure 5 Average ground level concentration for the period 15:00 to 16:00.

wind speeds oscillate from time step to time step, but after about an hour this ends and a low frequency oscillation develops. These oscillations are most likely numerical, and are probably due to the sudden changes in the flow when wave breaking begins. From 17:00 to 00:00 z_c above this grid square rises from about 260 m to 915 m for the WAVE case, and from 440 to 960 for the NO WAVE case. These changes in z_c due to the dynamics of the sea breeze and not the parameterization scheme. Because of this rise, the effects of wave breaking at 100 and 300 m decreases through the night. Figure 2 shows time series of potential temperature at 100 m and 300 m. The parameterization results in warming at 100 m and cooling at 300 m which would indicate an enhanced downward heat flux.

Figure 3 shows profiles of the U- component of velocity at 18:00 hrs. The wave stress parameterization results in a lowering of the critical level from about 350 m to about 250 m, and a decrease in the magnitude of the U-velocity at all heights up to 2000 m (except at 300 m). The lowering of z_c has been observed in laboratory and numerical experiments (Dörnbrack and Nappo, 1997), and is consistent with linear wave theory. The decrease in velocity is a natural consequence of the wave stress, and the overall effect of the parameterization on the flow is to decrease the bulk wind shear. Potential temperature profiles are shown in Figure 4; the wave stress parameterization results in a warming below about 200 m and cooling above 200 m to about 900 m. This result suggests an increased downward heat flux over the NO WAVES case. Here also the overall effect is to decrease the temperature gradient.

The effects of the wave-stress parameterization on the dispersion calculations are illustrated in Figures 5 and 6 which show contours of hourly-averaged ground lcvel concentrations for the periods 15:00 - 16:00 and 16:00 - 17:00 respectively. The shaded area depicts the region containing the terrain. The concentration patterns shown in Figure 5 are identical for the WAVES and NO WAVES cases. The maximum concentration value is 730 μg m^3. Figure 6 shows no ground level concentration for the NO WAVE case, but considerable concentration for the WAVE case. The peak concentration is 850 μg m^3.

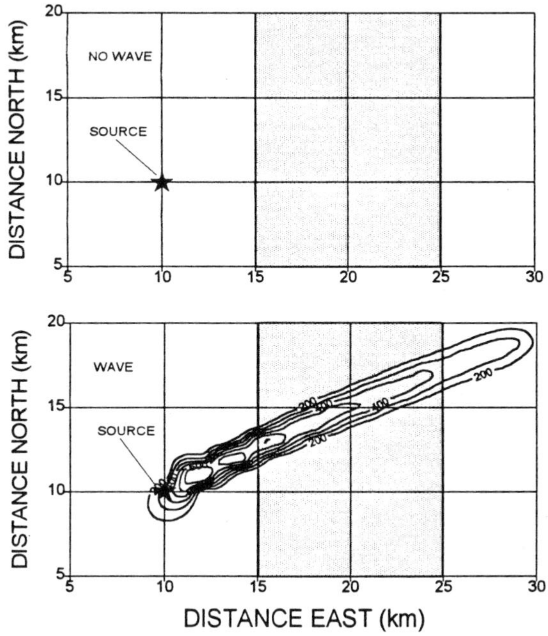

Figure 6 Average ground level concentration for the period 16:00 to 17:00.

The peak concentration is 1200 μg m^3 at 18:00, and 630 μg m^3 at 9:00. After 19:00, there is no ground level concentration although wave breaking continues until about 23:00, Since the only source of pollutant is at 100 m agl, vertical diffusion must be responsible for the presence of material in the surface layer. Before wave breaking, the parameterization has no effect on vertical diffusion. Wave breaking begins shortly after the PBL becomes statically stable. This marks the beginning of high-amplitude gravity waves. But this is also the time when vertical diffusion due to turbulence shuts down. It is clear that the wave stress parameterization results in increased downward transport of airborne material.

CONCLUSIONS

Gravity waves launched by terrain in the stable PBL can grow in amplitude and then break down resulting in a stress on the PBL flow. When a critical level for these waves is present, wave breaking will almost always occur. We have parameretized the effects of wave breaking using linear wave theory to calculate the wave stress, and linear saturation theory to calculate stress divergence. We have used this parameterization in the LADM model to simulate a sea breeze circulation over a coastal two-dimensional subgrid-scale ridge with a maximum height of 200 m and width of about 2000 m. The results indicate that the wave-stress parameterization affects all aspects of the flow including temperature and pollutant concentrations. Specifically, it results in significant increases in the downward fluxes of heat, momentum, and atmospheric pollutants. It is possible that failure to account for gravity-wave-generated turbulence can lead to significant under prediction of ground level concentrations due to elevated releases of material.

ACKNOWLEDGMENTS

This collaborative work was done during an extended visit by the first author to CSIRO, which was made possible under an informal agreement between NOAA's Air Resources Laboratory and the Division of Atmospheric Research of the CSIRO, Aspendale, Australia.

REFERENCES

Baines, P. G., 1995:*Topographic Effects in Stratified Flows*, Cambridge University Press, New York.

Blumen, W. and K. Raeder, 1991: Dissipation of linear mountain waves over a ridge. *Tellus*, **43A**, 226-234.

Dörnbrack, A. and C. J. Nappo, 1997: A note on the application of linear wave theory at a critical level. *Boundary-Layer Meteorology*, **82**, 399-416.

Eliassen, A. and E. Palm, 1960: On the transfer of energy in stationary waves. *Geofys. Publ.*, **22**, 1-23.

Fritts, D. C., 1984: Gravity wave saturation in the middle atmosphere. A review of theory and observations. *Rev. Geophys. Space Phys.*, **22**, 275-308.

Fritts, D. C. and Geller, M. A., 1976: Viscous stabilization of gravity critical level flows. *J. Atmos. Sci.*, **33**, 2276-2284.

Hurley P.J. and W.L. Physick, 1993. Lagrangian particle modelling of buoyant point sources: Plume rise and entrainment under convective conditions. *Atmospheric Environment*, **27A**, pp. 1579-1584.

Glendening J.W., Businger J.A., and R.J. Farber, 1984. Improving plume rise prediction accuracy for stable atmospheres with complex vertical structure. *J. Air Pollut. Control Ass.*, 34, pp. 1128-1133.

Gossard, E. E. and W. H. Hooke, 1975: *Waves in the Atmosphere*, Elsevier, 456 pp.

Grisogono, B. 1994: Dissipation of wave drag in the atmospheric boundary layer. *J. Atmos Sci.*, **51**, 1237-1243.

Kim, J. and L. Mahrt, 1992: Momentum transport by gravity waves. *J. Atmos. Sci.*, **49**, 735-748.

Laprise, J. P. R., 1993: An assessment of the WKBJ approximation to the vertical structure of linear mountain waves: Implications for gravity-wave drag parameterization., *J. Atmos. Sci.*, **50**, 1469-1487.

Lott, F. and M. J. Miller, 1997: A new subgrid-scale orographic drag parameterization: Its formulation and testing. *Q. J. R.. Meteorol. Soc.*, 123, 101-127.

Louis, J.F., 1979. A parametric model of vertical eddy fluxes in the atmosphere, *Boundary-Layer Meteorol.*, 17, 187-202.

Nappo, C. J. and A. Andren, 1995: A parameterization of subgrid scale gravity-wave-generated turbulence in a mesoscale boundary layer model. *11th Symposium on Boundary Layers and Turbulence, March 27-31, 1995, Charlotte, NC*, American Meteorological Society, Boston, MA.

Nappo, C. J. and G. Chimonas, 1992: Wave exchange between the ground surface and a boundary-layer critical level. *J. Atmos. Sci.*, **49**, 1075-1091.

Physick, W. L., J. A. Noonan, J. L. McGregor, P. J. Hurley, D. J. Abbs, and P. C. Manins, 1994: *LADM: A Lagrangian Atmospheric Dispersion Model*, CSIRO Division of Atmospheric Research Technuical paper; no. 24.

Pielke, R. A., W. R. Cotton, R. L. Walko, C. J. Tremback, W. A. Lyons, L. D. Grasso, M. E. Nicholls, M. D. Moran, D. A. Wesely, T. J. Lee, and J. H. Copeland, 1992: A comprehensive meteorological modeling system - RAMS, *Meteorol. Atmos Phys.*, 49, 69-91.

Schoeberl, M. R., 1988: A model of stationary gravity wave breakdown with convective adjustment. *J. Atrmos. Sci.*, **45**, 980-992.

Thomson, D. L., 1987: Criteria for the selection of stochastic models of particle trajectories in turbulent flows. *J. Fluid Mech.*, **180**, 529-556.

DISCUSSION
The paper was presented
by A. Luhar

B. WIMAN: Is there any significant relationship between the effects you describe and the state of the sea (ocean-atmosphere dynamics affecting the model you described)?

C. NAPPO: The model does not consider air-sea interactions other than the sea breeze, and it is unlikely that a significant relation exists between these small scale gravity waves and the sea state in our simulations.

A SKEWED PUFF MODEL

T. Tirabassi[1] , S. Malta[1], U. Rizza[2], and C. Mangia[2]

[1] Institute FISBAT of C.N.R., via Gobetti 101, I-40129 Bologna, Italy
[2] Institute ISIAtA of C.N.R., via Arnesano, I-73100 Lecce, Italy

ABSTRACT

A practical puff model based on the Monin-Obukhov similarity theory is presented. The model (named SPM: Skewed Puff Model) utilises approximate solutions proposed by van Ulden for the dispersion of a cloud of passive contaminant released from a source near the ground and where particular importance is given in describing the interaction between wind shear and vertical diffusion. The model can be applied routinely using as input simple ground-level meteorological data acquired by an automatic network.

The concentrations calculated with the model are compared with an experimental data set for a relatively tall source.

INTRODUCTION

A distorting effect of the variation with height on the mean wind, both in speed and direction, is often observed in the development of puffs or plume smoke. This effect is most prominent in stable stratified conditions. In fact, wind shear creates a variance in the wind direction, while vertical diffusion destroys this variance and tries to re-establish a non-skewed distribution. The interaction between vertical mixing and velocity shear is continuously effective.

In order to take into account the above phenomenon, we have developed practical puff model based on the Monin-Obukhov similarity theory where continuously emitting sources are represented by the superposition of a series of the above clouds. The model (written in FORTRAN and named SPM: Skewed Puff Model) utilises approximate solutions proposed by van Ulden (1992) for the dispersion of a cloud of passive contaminant released from an instantaneous source near the ground and where particular importance is given in describing the interaction between wind shear and vertical diffusion and in the process that

Air Pollution Modeling and Its Application XIII, edited by S.-E. Gryning and E. Batchvarova, Kluwer Academic/Plenum Publishers, 2000

transforms shear produced skewness into diffusive variance in the wind direction. In fact the longitudinal dispersion is given by three processes: i) direct diffusion, described in the same manner as lateral diffusion; ii) the effect of wind shear ; iii) the interaction of skewness and vertical diffusion.

A preliminary performance evaluation is shown in the case of continuous emission from an elevated source in a variable boundary layer.

THE SKEWED PUFF MODEL

SPM is a model written in FORTRAN that utilises approximate solutions proposed by van Ulden for the dispersion of a cloud of passive contaminant released from a source near the ground.

Following van Ulden (1992) the equation that describes the concentration field is the sum of puffs, where the shape of the concentration (χ) distribution of one puff is given by:

$$\chi(\xi, y, z) = \frac{AQ}{2\pi\Sigma_d\Sigma_\eta < z >}\exp\left\{-\frac{1}{2}\left(\frac{\xi-\xi_s}{\Sigma_d}\right)^2 - \frac{1}{2}\left(\frac{y}{\Sigma_\eta}\right)^2 - \left(\frac{Bz}{< z >}\right)^s\right\} \tag{1}$$

where Q is the emission flux, y is the crosswind co-ordinate, z the vertical co-ordinate and <z> is the mean cloud depth, which is defined as:

$$\langle z \rangle \equiv \frac{1}{Q}\int z\cdot\chi\cdot d\xi\cdot dy\cdot dz \tag{2}$$

ξ is a cloud following co-ordinate defined as:

$$\xi \equiv x - X(t)$$

where x is the along mean wind co-ordinate and X(t) is defined as:

$$X(t) \equiv \int x\cdot C\cdot dx\cdot dy\cdot dz$$

with C(x,y,z) the concentration function in the fixed co-ordinates, ξ_s is the co-ordinate of the cloud centroid, A and B are defined as:

$$A(s) = s\Gamma(2/s)/\{\Gamma(1/s)\}^2$$
$$B(s) = \Gamma(2/s)/\Gamma(1/s)$$

where Γ is the Gamma function and the shape factor s is the parameter that governs the shape of the vertical concentration profile (see Eq. (1)) and it is function of the mean cloud depth and of the Monin-Obukhov length L:

$$s = \frac{1+6\frac{\langle z \rangle}{L}}{1+3\frac{\langle z \rangle}{L}} \qquad \text{for } L > 0$$

$$s = \frac{1-5\frac{\langle z \rangle}{L}}{1-10\frac{\langle z \rangle}{L}} \qquad \text{for } L \leq 0$$

Σ_d is the standard deviation of the concentration distribution around the cloud centroid, while the total standard deviation take in to account also the stretching of the cloud due to wind shear (for more details see van Ulden (1992)):

$$\Sigma_d \equiv \langle(\xi - \xi_s)\rangle^{1/2}$$

where the symbols < > mean an averaging process like Eq.(2).
Σ_η is the standard deviation of the lateral concentrations in the cloud following co-ordinate:

$$\Sigma_\eta = \left\langle \eta^2 \right\rangle^{1/2}$$

where η is the cloud following co-ordinate in the y direction.
The variables of Eq. (1) are evaluated by means of the following equations (van Ulden, 1992) that describe the time evolution of the cloud depth, the lateral spread, the longitudinal skewness and the longitudinal diffuseness:

$$\frac{d < z >}{d\tau} = \frac{k}{\phi_h (E < z >)} \tag{3}$$

$$\frac{d\Sigma_\eta}{d\tau} = \frac{\alpha}{u_*}\sigma_v \tag{4}$$

$$\frac{d\Sigma_s}{d\tau} = \frac{D}{k}\phi_m(< z >) - \frac{k\Sigma_s}{D^2 \phi_h (< z >) < z >} \tag{5}$$

$$\frac{d\Sigma_d}{d\tau} = \frac{k\Sigma_s^2}{D^2 < z > \phi_h(< z >)\Sigma_d} + \frac{\alpha}{u_*}\sigma_u \tag{6}$$

D and E are function of the shape factor s and they are defined as:

$$D(s) = \left[\frac{\Gamma(1/s)\Gamma(3/s)}{\{\Gamma(2/s)\}^2} - 1\right]^{1/2}$$

$$E(s) = \left[s\{\Gamma(2/s)/\Gamma(1/s)\}^s\right]^{1/(1-s)}$$

τ is defined as $\tau \equiv u_* t$ with u_* the friction velocity and t the time; k is the von Karman constant (k=0.4), α is an empirical constant (α = 0.3; see van Ulden, 1992) and σ_u and σ_v are the standard deviation of two components of horizontal wind velocity and they are evaluated according the following formulae (Beljiars et al., 1983; Gryning et al., 1987):

$$\left(\frac{\sigma_u}{u_*}\right)^2 = 6 \qquad \text{for } L \geq 0$$

$$\left(\frac{\sigma_u}{u_*}\right)^2 = 6 + 0.35\left(-\frac{h}{kL}\right)^{2/3} \qquad \text{for } L < 0$$

$$\left(\frac{\sigma_v}{u_*}\right)^2 = 4 \qquad \text{for } L \geq 0$$

$$\left(\frac{\sigma_v}{u_*}\right)^2 = 4 + 0.35\left(-\frac{h}{kL}\right)^{2/3} \qquad \text{for } L < 0$$

where h is the depth of the boundary layer

ϕ_m and ϕ_h are the universal function for momentum and heat respectively. Following Dyer (1974) they are expressed as:

$$\begin{aligned} \phi_h &= 1 + 5z/L \\ \phi_m &= 1 + 5z/L \end{aligned} \qquad \text{for } L > 0$$

$$\begin{aligned} \phi_h &= (1 - 16z/L)^{-1/2} \\ \phi_m &= (1 - 16z/L)^{-1/4} \end{aligned} \qquad \text{for } L \leq 0$$

Following van Ulden (1992) we have used simple approximations (with an accuracy of few per cent) for the functions A(s), B(s), D(s) and E(s). The approximations are:

$$\begin{aligned} A(s) &= 1/(1 + 0.95 \ln s) \\ B(s) &= 1/(1 + 1.2 \ln s) \\ D(s) &= 1/(1 + 0.45 \ln s) \\ E(s) &= 1.54 \end{aligned}$$

The downwind position of a cloud X is evaluated by means an approximate equation obtained by van Ulden (1978) and van Ulden (1992):

$$\frac{dX}{dt} = \frac{u_*}{k}\left\{\ln\left(\frac{a<z>}{z_0} + 1\right) - \phi_m^{-1}\left(\frac{a<z>}{L}\right) + 1\right\} \qquad \text{for } L > 0 \tag{7}$$

$$\frac{dX}{dt} = \frac{u_*}{k}\left\{\ln\left(\frac{a<z>}{z_0} + 1\right) + \phi_m\left(\frac{<z>}{L}\right) - 1\right\} \qquad \text{for } L \leq 0 \tag{8}$$

where a = 0.6 (van Ulden, 1992) and z_0 is the roughness length.

The downwind position X together with ξ_s gives the position of the cloud centroid. A continuously emitting point source is represented by the superposition of a series of puffs.

MODEL VALIDATION AGAINST EXPERIMENTAL DATA

In Tirabassi and Rizza (1995) it is evaluated the performances of the SPM model in the cases of both an emission source near the ground using the data set of the Prairie Grass experiment (Olesen, 1995), as well as an elevated source, using the Copenhagen data set (Gryning and Lyck, 1984).

The Copenhagen data set is composed by tracer SF_6 data from dispersion experiments carried out in the northern part of Copenhagen. The site was mainly residential with a roughness length of 0.6 m. The tracer was released without buoyancy from a tower at a height of 115 m, and collected at the ground-level positions in up to three crosswind arcs of tracer sampling units. The sampling units were positioned 2-6 Km from the point of release.

In Tirabassi and Rizza (1995) the average sampling time was 1 hour, but in the model validation presented here, we used data with a greater time resolution kindly made available to us by Gryning. In particular, we used 20 minutes averaged measured concentrations and 10 minutes averaged values for meteorological data. Moreover, tracer releases typically started up 1 hour before the tracer sampling and stopped at the end of the sampling period, so we have considered 12 intervals (of 10 minutes) for meteorological data and 3 intervals (of 20 minutes) for concentration data.

Tables 1, 2 and 3 report the friction velocity, boundary layer height (only one value for each run) and the Monin-Obukhov length, respectively, used.

Table 1. Friction velocity (m/s) for the different runs and time steps. Every time step corresponds at 10 minutes

Run Time step	1	2	3	4	5	7	8	9
1	.36	.68	.46	.56	.58	.48	.65	.72
2	.37	.67	.45	.51	.52	.48	.79	.73
3	.40	.81	.47	.37	.51	.57	.67	.60
4	.43	.68	.39	.44	.58	.62	.67	.59
5	.35	.75	.39	.48	.59	.53	.68	.65
6	.34	.74	.40	.48	.52	.65	.65	.71
7	.42	.76	.40	.39	.52	.63	.68	.73
8	.43	.82	.41	.40	.45	.65	.67	.73
9	.40	.76	.31	.39	.44	.66	.73	.73
10	.37	.73	.34	.39	.44	.62	.73	.66
11	.35	.69	.39	.39	.44	.52	.75	.67
12	.36	.66	.40	.39	.43	.62	.69	.74

Table 2. Boundary layer height (m) for the different runs

Run	1	2	3	4	5	7	8	9
Height	1980	1920	1120	390	820	1850	810	2090

Table 3. Monin-Obukhov length (m) for the different runs and time steps. Every time step corresponds at 10 minutes

Run / Time step	1	2	3	4	5	7	8	9
1	-26	-178	-152	-75	-492	-71	-71	-793
2	-23	-227	-194	-42	-215	-80	-85	-471
3	-83	-311	-106	-23	-368	-64	-47	-202
4	-42	-160	-101	-32	-735	-111	-49	-366
5	-36	-203	-129	-71	-366	-177	-45	-633
6	-42	-286	-70	-80	-273	-67	-63	-13588
7	-47	-155	-83	-83	-273	-87	-41	-593
8	-38	-228	-60	-101	-262	-71	-47	-471
9	-83	-184	-106	-129	-395	-56	-70	-389
10	-21	-389	-42	-129	-395	-111	-64	-375
11	-32	-133	-101	-129	-395	-215	-52	-262
12	-29	-375	-70	-129	-759	-123	-39	-252

In table 4 the calculated and measured crosswind-integrated concentrations are shown for each time period (20 minutes) of the simulation. Moreover a statistical analysis of model performances has been evaluated. In table 5 we present some statistical indices: normalised mean square error (nmse), correlation coefficient (r), fractional bias (fb) and the factor of two (fa2). Statistical indices confirm the reliability of the model results.

Table 4. Observed (C_o) and predicted (C_p) crosswind-integrated concentrations, normalised with emission rate (10^{-4} s m^{-2}) at different distances from the source (m) for the three time periods (20 minute) of the simulation

Run	Dist.	I Period		II Period		III Period	
		C_o	C_p	C_o	C_p	C_o	C_p
1	1900.	5.60	5.89	8.27	5.40	5.51	6.34
1	3700.	1.74	3.37	2.25	4.94	3.02	5.20
2	2100.	4.36	7.38	5.14	5.84	6.73	6.33
2	4200.	2.72	4.19	1.96	2.80	4.20	2.98
3	1900.	6.00	9.38	9.26	7.95	9.32	8.59
3	3700.	4.70	6.54	6.53	5.19	7.62	6.28
3	5400.	3.93	6.66	5.24	5.27	4.01	5.77
4	4000.	6.26	17.64	9.97	13.48	17.37	11.25
5	2100.	5.78	11.16	8.62	11.42	5.89	12.43
5	4200.	5.09	6.48	6.55	6.39	5.91	7.31
5	6100.	5.07	6.11	5.37	5.47	4.65	5.69
7	2000.	2.72	4.52	12.74	4.60	5.25	5.39
7	4100.	2.31	2.39	1.34	2.59	2.42	2.51
7	5300.	2.45	2.38	0.64	2.56	1.49	2.16
8	1900.	4.00	6.13	4.84	5.55	3.65	5.34
8	3600.	2.31	6.02	1.34	5.52	2.42	5.21
8	5300.	2.45	5.82	0.64	5.62	1.49	5.19
9	2100.	3.98	5.19	3.93	5.94	5.90	6.75
9	4200.	3.46	2.52	2.44	3.07	3.40	3.68
9	6000.	3.96	1.81	2.04	2.05	1.76	2.52

Table 5. Statistical evaluation of model performances. The subscript "o" and "p" are for the observed and predicted concentrations respectively and σ is the standard deviation

Indices	Definition	Value
nmse	$\dfrac{\overline{(C_o - C_p)^2}}{\overline{C_o}\,\overline{C_p}}$	0.32
r	$\dfrac{\overline{(C_o - \overline{C_o})(C_p - \overline{C_p})}}{\sigma_o \sigma_p}$	0.58
fa2	fraction of data (%) for which $0.5 \le C_p / C_o \le 2$	78
fb	$2\dfrac{\overline{C_o} - \overline{C_p}}{\overline{C_o} + \overline{C_p}}$	-0.23

CONCLUSIONS

We have presented a model for puff dispersion that utilises an approach suggested by van Ulden (1992).

The model can be applied routinely using as input simple ground-level meteorological data acquired by an automatic network. In fact, the model needs diffusion parameterizations based on fundamental parameters to describe the characteristics of the atmospheric boundary layer that can be evaluated by ground measurements (Holtslag and Van Ulden, 1983; Weil and Brower, 1984; Van Ulden and Holtslag, 1985; Trombetti et al., 1986).

Model performances were evaluated using data from the Copenhagen data set, but with a time resolution greater than the data ones generally distributed (Olesen, 1995). In particular, we used 20 minute average concentrations of SF_6 and 10 minute average values for meteorological data. The validation has to be considered preliminary one. As a matter of fact, we have checked the model only with data referring to continuous emission (in variable meteorology) and in receptors points far from the source (2 - 6 Km).

ACKNOWLEDGEMENTS

We wish to thank Dr. Sven-Erik Gryning for making his data available to us.

REFERENCES

Beljaars A.C.M., Schotanus, P. and Nieuwstadt, F.T.M., 1983, Surface layer similarity under non uniform fetch conditions. *J. Clim. Appl. Meteor.* 22:1800.

Dyer A.J., 1974, A Review of flux-profile relationship. *Bound.-Layer Meteor.* 7:363.

Gryning, S.E. and Lyck, E., 1984, Atmospheric dispersion from elevated sources in an urban area: comparison between tracer experiments and model calculations, J. *Clim. Appl. Meteor.* 23: 651.

Gryning, S.E., Holtslag, A.A.M., Irwin, J.S. and Siversten, B., 1987, Applied dispersion modelling based on meteorological scaling parameters. *Atmos. Environ.* 21:79.

Holtslag, A.A.M. and van Ulden, A.P., 1983, A simple scheme for daytime estimation of surface fluxes from routine weather data. *J. Clim. Appl. Meteor.* 22: 517.

Olesen, H.R., 1995, Datasets and protocol for model validation, *Int. J. Environment and Pollution* 5: 693.

Tirabassi, T and Rizza, U., 1995, A practical model for the dispersion of skewed puffs. J. Appl. Meteor. 34: 989.

Trombetti, F., Tagliazucca, M., Tampieri, F. and Tirabassi, T. 1986, Evaluation of similarity scales in the stratified surface layer using wind speed and temperature gradient. *Atmos. Environ.* 20: 2465.

Van Ulden, A.P. and Holstlag, A.A.M., 1985, Estimation of atmospheric boundary layer parameters for diffusion applications, *J. Clim. Appl. Meteor.* 24:1196.

Van Ulden, A.P., 1978, Simple estimates for vertical diffusion from sources near the ground. *Atmos. Environ.* 12: 2125.

Van Ulden, A.P., 1992, A surface-layer similarity model for the dispersion of a skewed passive puff near the ground. *Atmos. Environ.* 26 (A): 681.

Weil, J.C. and Brower R.P., 1984, An update gaussian plume model for tall stacks, *JAPCA*, 34: 818.

DISCUSSION

R. YAMARTINO: Did you find significant differences or improvements in your modeling of Copenhagen tracer experiment by using the 10-minute meteorological data versus using the usually available, 60-minute meteorological data?

T. TIRABASSI: No, we did not. But you have to consider that during each run of the Copenhagen experiment the meteorological conditions were quasi steady.

D. WILSON: The key feature of your model is the vertical profile exponent, s. Did this model, using van Ulden's equation for s give significantly better agreement with measurements compared to giving s at s=2.0 (Gaussian) or s=1.5 the value often observed for ground level sources.

T. TIRABASSI: We can say that van Ulden 's equation give significantly different results compared to Gaussian approach (s=2) or to s=1.5 results for ground level sources. If our model works better than Gaussian approach it is an open question although we believe that as a whole it works better.

MEAN HEIGHT OF A PASSIVE PLUME IN AN URBAN CONVECTIVE BOUNDARY LAYER: A LARGE-EDDY SIMULATION

Xiaoming Cai

School of Geography, University of Birmingham, Edgbaston
Birmingham, B15 2TT, UK

INTRODUCTION

In a convective boundary layer (CBL) over an urban patchy surface pattern, turbulent structure may have a strong response to the surface forcing induced by the patchy pattern of sensible heat flux caused by different landuse type. It is not clear that the dispersion of a passive plume emitted into such a boundary layer will have different characteristics from that emitted into a CBL over a horizontally homogeneous surface such as a typical uniform rural area. The present study aims to investigate the case by using large-eddy simulation (LES) with a focus on the mean plume height.

There have been many previous LES studies of plume dispersion over a homogeneous surface. As early as in 1978, Lamb used the large-eddy field produced by Deardorff's model to drive a Lagrangian particle model to study plume dispersion over a homogeneous ground. Nieuwstadt and Valk (1987), however, adopted the Eulerian approach to solve a diffusion equation for a passive and a buoyant plume within a large-eddy model. They obtained very good results comparing with the laboratory experiments carried out by Willis & Deardorff (1976, 1981). Haren and Nieuwstadt (1989) used the same model to conduct a more careful study and compared with laboratory data as well as field data. Results showed that while the model can produce quite reasonable results for a passive plume, the plume motion caused by buoyancy does not obey Briggs' 2/3 law. Nieuwstadt (1992a) carried out another LES of a line source in a CBL and analysed the results with relation to two major contributions to dispersion, i.e., meandering and relative diffusion. The results demonstrated that contribution due to meandering is significant at the initial stage and when source height increases. For a buoyant line source, Nieuwstadt (1992b) illustrated that plume looping enhances entrainment and reduces plume rise. Sykes and Henn (1992), however, focused on the concentration fluctuations of a passive plume in a neutral atmospheric boundary layer, and comparison with wind-tunnel experiments is satisfactory for an elevated source. The simulation provided useful information of scalar fluctuation dissipation rate, the lateral spatial correlation function and the temporal autocorrelation for a passive plume. Henn and Sykes (1992) adopted a CBL over a homogeneous surface and the results agree reasonably well with the experiments and with the numerical results of Lamb (1978). Compared with the neutral case in Sykes & Henn (1992), they found that the convective turbulence is more effective than shear-driven turbulence in diffusing the plume concentration. Mason (1992) carried out a series of LESs for different stability regimes and the results demonstrated that for a small value of u_*/w_* (<0.3) the dispersion properties are close to free convection. In the study by Hadfield (1994), a Lagrangian particle model is used for a one-dimensional sinusoidal surface flux, and he found that the buoyancy perturbations diffuse like a passive scalar. Kemp & Thomson (1996) has recently conducted a particle model with LES of stable boundary layers and showed reasonable results for the vertical spread.

In the present study, dispersion of passive plume in the CBL over an idealised urban patchy surface is simulated by LES. Discussions are focused on the mean plume height and the vertical dispersion parameter. In the following sections, the model set-up of the LES of such a CBL will be introduced, followed by the description of the plume configuration; then discussion of the results will be given; and finally some conclusions will be provided.

LARGE-EDDY MODEL CONFIGURATION

Colorado State University's Regional Atmospheric Modeling System (CSU-RAMS) version 2a is used in the present study. More detailed description of the model specification can be seen in Cai et al. (1995) and Cai & Steyn (1996). In this study, Smogrinsky's subgrid-scale model is adopted and the model parameter C_s is taken as 0.1 for all cases. The results show that this value of C_s works very well and there is no grid-mode turbulence found in the power spectra of resolved velocities.

In this study, horizontal grid spacing is chosen as 60 m and number of grid is 64×64×52 covering a 3.84×3.84×1.22km, as shown in Table 1. In the vertical direction, the grid spacing is 2.4m immediately above the ground, is stretched with a ratio of 1.2 upwards, and has the maximum spacing of 30m. The top of the domain is at about 12km, and there are about 30 to 40 grid points within the CBL. In the present study, the CBL under the stability regime in which $Z_i/|L| << 1$ is investigated; therefore, there is no mean wind speed is given. This configuration is the same as that in the water tank experiment by Willis & Deardorff (1976, 1978, 1981, *etc.*).

At the bottom of the domain, a patchy pattern of an idealised urban surface is specified, as shown in Figure 1. In the figure, light grey colour represents built-up area and dark grey colour represents park area. Parks are of square with the width of p and surrounded by a rough built-up area with the width of b consisting of buildings, roads and other constructions. The distance between centres of two contiguous parks is denoted by d. There are three independent dimensionless parameters which may be important: (1) d/Z_i, the ratio of two length scales, the scale of the surface pattern and the boundary layer depth; (2) p/b, the ratio of the length scale of the "cold-surface" to the that of the "hot-surface"; and (3) $\overline{w'\theta'}_b / \overline{w'\theta'}_p$, the ratio of surface sensible heat flux over the "hot-surface", $\overline{w'\theta'}_b$, to that over the "cold-surface", $\overline{w'\theta'}_p$. In the current study, in order to make convincing comparisons, averaged surface sensible heat fluxes for all inhomogeneous cases are fixed as 0.0875 mT/s. An extra constraint is needed to specify the ratio $\overline{w'\theta'}_b / \overline{w'\theta'}_p$, which is 2.0 in the current study. The parameters adopted for all runs are indicated in Table 1. For the case with a homogeneous surface, roughness length is given as 1 m over the whole

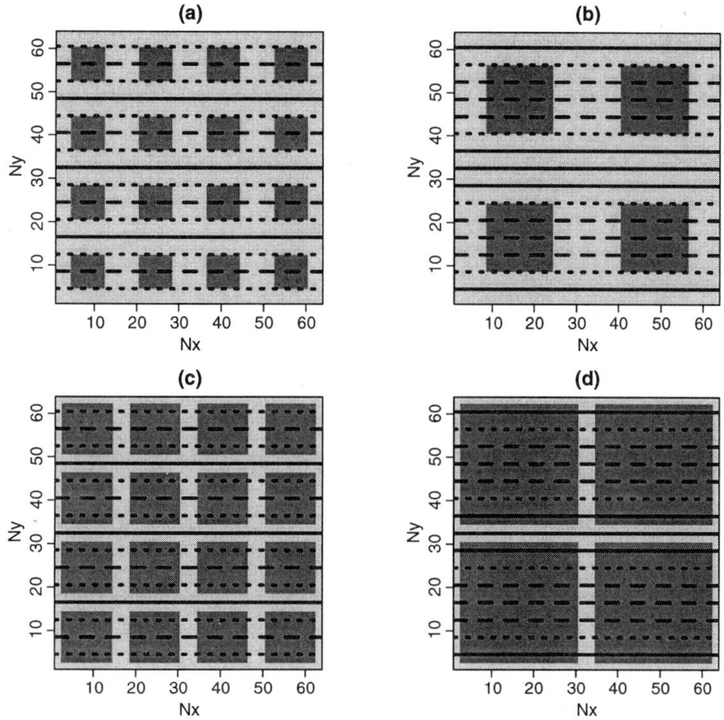

Figure 1. Surface pattern of heat flux for different runs listed in Table 1: (a) I64a; (b) I64b; (c) I64c; and (d) I64d. Light grey colour represents built-up area and dark grey colour represents park area. All thick lines represent the horizontal location of line source: solid thick line denotes the source above or near the centre of the built-up surface; dashed thick line denotes the source above or near the centre of the park area; and dotted thick line denotes the source near the half-way between the two surface types.

Table 1. List of simulations for the present study. "H" denotes "homogeneous surface", and "I" denotes "inhomogeneous surface". D is the domain size (in meter), d is the distance between centres of two parks (in meter, see Figure 1), b is the width of building area (in meter), p is the park width (in meter), $\overline{w'\theta'}_b$ and $\overline{w'\theta'}_p$ are surface sensible heat flux over the built-up and park area (in Km/s), respectively.

Run	D	d	b	p	$\overline{w'\theta'}_b$	$\overline{w'\theta'}_p$
H64	3840	-	-	-	-	-
I64a	3840	960	480	480	0.1	0.05
I64b	3840	1920	960	960	0.1	0.05
I64c	3840	960	240	720	0.122	0.061
I64d	3840	1920	240	1680	0.142	0.071

domain. For the cases with an inhomogeneous surface, roughness lengths are specified as 1 m over the built-up area and 0.01 m over the park area.

At t=0, potential temperature is given as 293K from the surface up to z=500 m, and above it, a strongly stable stratification of 9 *K/km* is specified. Each simulation runs for thirty minutes ($t/t_* \approx 4$, where $t_* = Z_i/w_*$) to allow the UCBL turbulence to be initially developed (Dornbrack & Schumann, 1993). After the initialisation, the model runs up to three hours for all cases, and then plume simulation starts. At this time, the CBL height, Z_i, is about 830m for all cases, and grows further during the period of plume simulation.

CONFIGURATION OF PASSIVE PLUME SIMULATION

Finite-difference scheme with second-order accuracy is used to calculate the transport and diffusion of passive plume in the domain. After convective eddies are fully developed, instantaneous line sources of passive plume are introduced in the domain at a specific height. Each line source is arranged similar to that in Nieuwstadt & Valk (1987), but with a cross-section of initial concentration in Gaussian distribution. This smooth distribution approximates the near-field dispersion close to the source, and significantly reduces amount of negative concentration caused by finite difference scheme. The standard deviations of the distribution are taken as 30m and 15m in the horizontal and the vertical directions, respectively. To obtain a good statistics, a total of sixteen line sources are used in the present study. These instantaneous line sources are located parallel to the x-axis and horizontal positions are shown in Figure 1. In order to compare with the laboratory data of Willis & Deardorff (1978, 1981), the initial heights of the line sources are at about $0.24Z_i$ and $0.49Z_i$. Using Taylor's hypothesis adopted by previous studies, the averaged results along the x-direction at the different times for this instantaneous line source can be interpreted as those at different downwind distances from a continuously released point source. Following the mixed-layer scaling proposed by Deardorff, the downwind distance x is scaled by $Ut_* = UZ_i/w_*$, where U is the mean wind speed at stack height, to make a dimensionless quantity, $X = xw_* /(UZ_i)$. When using Taylor's translation hypothesis to substitute x/U for travel time of the plume, this quantity, X, is equivalent to the dimensionless time, t/t_*.

Each simulation runs for 30 min, which is about $3t_*$, during which Z_i is approximately 850m (Cai, 1998). In order to achieve the aims of this study, statistics from the 16 line sources are selectively processed. The selective procedure are based on the line type shown in Figure 1. The results discussed in the next section may be derived only from one type of line source, e.g., solid lines or emissions above or near the centreline of the built-up area. The calculation of the mean plume height, Z_c, and the vertical dispersion parameters, S_z, is the same as that in Nieuwstadt & Valk (1987) and Henn & Sykes (1992).

RESULTS AND DISCUSSIONS

Figure 2 shows the nondimensionalised mean plume height, Z_c/Z_i, averaged over all 16 line source experiments for each individual case listed in the legend for two emission heights: $0.24Z_i$ and $0.49Z_i$. The solid lines represent the results from the homogeneous surface; this is the case investigated by all previous studies. For the emission height of $0.24Z_i$, Figure 2(a) shows that there are no significant differences for the mean plume height among the five cases, and all the results agree with the observation from the tank experiments. However, a clear indication is shown that variations across the 16 LES experiments from all cases with inhomogeneous surface are larger than that from H64, the

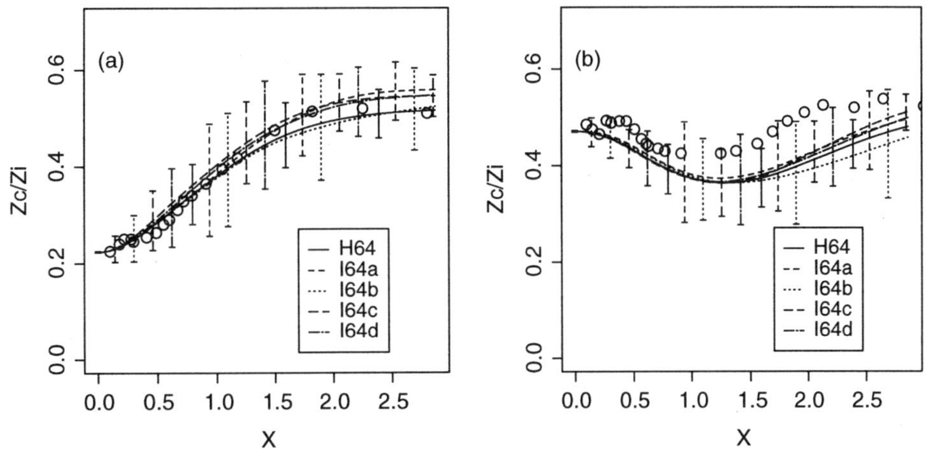

Figure 2. Dimensionless mean plume height Zc/Zi as a function of dimensionless downwind distance *X* for five cases list in Table 1. The mean plume height is obtained from the average of 16 line sources shown in Figure 1. The error bars show the standard deviation over the 16 experiments. Line types of curves and error bars are indicated in the legend. (a) Initial emission height is at $0.24Z_{i,}$ and circles indicate the data by Willis & Deardorff (1978), and (b) initial emission height at $0.49Z_i$; circles, data by Willis & Deardorff (1981).

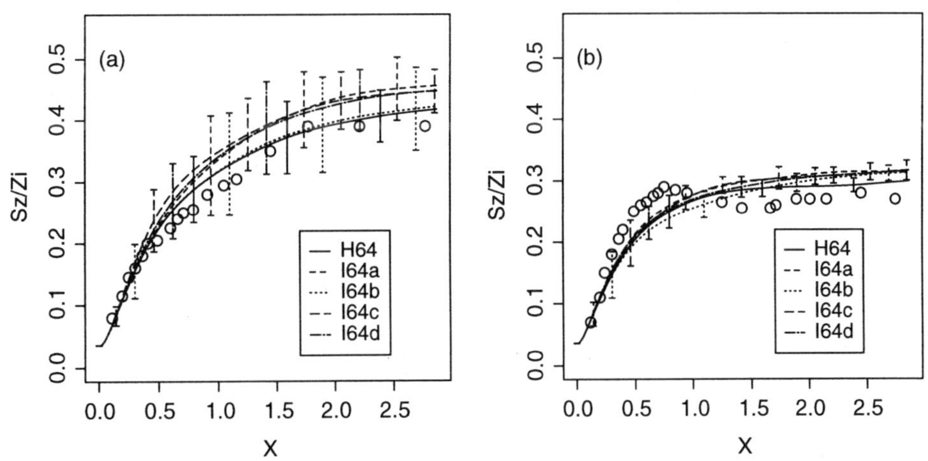

Figure 3 Dimensionless mean vertical dispersion parameter S_z/Zi as a function of dimensionless downwind distance *X* for five cases list in Table 1. All symbols are the same as those in Figure 2.

case with a homogeneous surface. The cases of I64b and I64d have the largest variation and this corresponds to their large value of d/Z_i shown in Figure 1. This is because when d/Z_i is large enough (>2), the surface-driven convective eddies can be fairly stable and maintain for several hours (Cai, 1998). These eddies have a very good correlation with the surface pattern (updrafts over the built-up area and downdrafts over the park area). For the emission height of $0.49Z_i$, the similar results are presented based on the comparison among the five cases. However, the mean plume heights from the present LES are lower than the observational data. This has also been shown in Henn & Sykes (1992).

For the dimensionless vertical dispersion parameter, S_z/Z_i, it is shown in Figure 3(a) that for a lower emission height, the results fit the laboratory data fairly well, and the mean values are about the same across the five cases. Variations across the different surface type for inhomogeneous cases are still larger than that for the homogeneous case. In Figure 3(b), for a high emission elevation, all cases slightly underestimate the observation for $X<1.0$, but overestimate for $X>1.0$. Variations for all cases are small so that the conclusion may be drawn that the surface pattern has a negligible effect on the vertical dispersion parameter for a elevated source in the middle of the CBL.

Figure 4 shows that the dimensionless mean plume heights for selective line sources for the four inhomogeneous cases in comparison with that for the homogeneous case. The mean plume height denoted by 'H-16' for case H64 is still obtained from the average of 16 line sources and plot of this case is for the purpose of comparison. In Figure 4(a) and 4(c), the result of case I64a averaged over 4 solid lines (see Figure 1) is denoted by 'I64a-*b*4', and the result averaged over 4 dashed is denoted by 'I64a-*p*4'. In Figure 4(b) and 4(d), the result averaged over 6 solid lines (see Figure 1) is denoted by 'I64b-*b*6', and the result

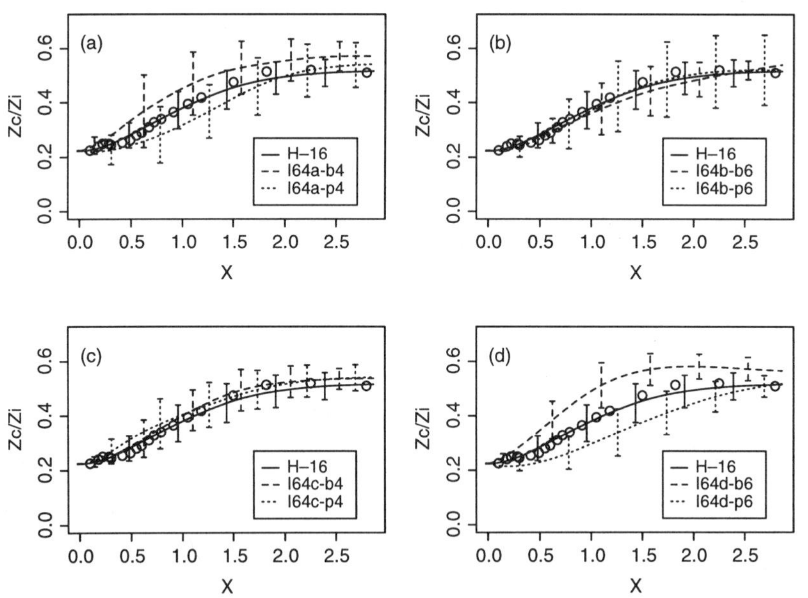

Figure 4 Dimensionless mean plume height *Zc/Zi* as a function of dimensionless downwind distance *X* for the emission height of z=0.24*Z_i. The mean plume height of 'H-16' for case H64 is obtained from the average of 16 line sources. (a) case I64a: 'I64a-*b*4' is averaged over 4 (solid line) sources in Figure 1(a) and 'I64a-*p*4' is from 4 dashed line sources; (b) case I64b: 'I64b-*b*6' is averaged over 6 (solid line) sources Figure 1(b) and 'I64b-*p*6' is from 6 (dashed line) sources; (c) case I64c: 'I64c-*b*4' is averaged over 4 (solid line) sources Figure 1(c) and 'I64b-*p*4' is from 4 (dashed line) sources; (d) case I64d: 'I64d-*b*6' is averaged over 6 (solid line) sources Figure 1(d) and 'I64b-*p*6' is from 6 (dashed line) sources. The error bars show the standard deviation over the corresponding experiments. Line types of curves and error bars are indicated in the legend.

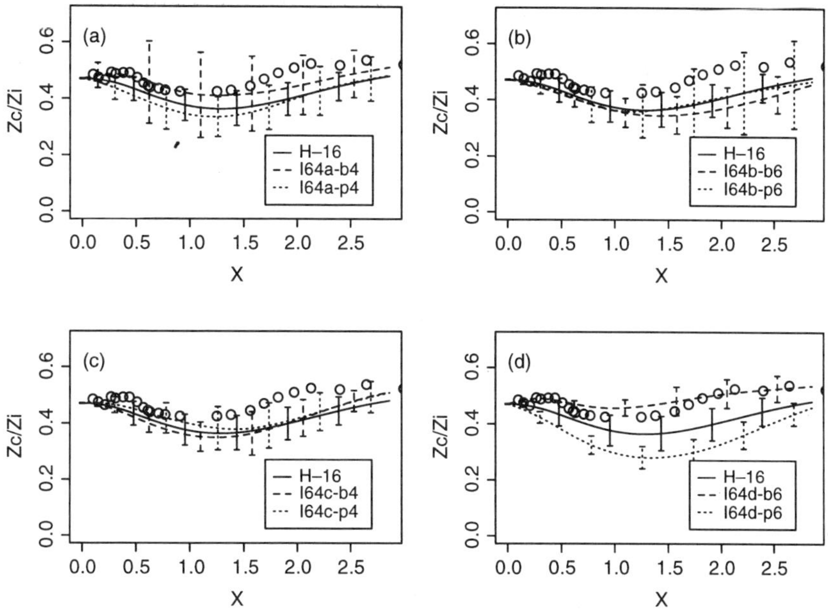

Figure 5 Dimensionless mean plume height *Zc/Zi* as a function of dimensionless downwind distance *X* for the emission height of z=0.49 Z_i. All symbols are the same as those in Figure 4.

averaged over 6 dashed is denoted by 'I64b-*p*6'. In the group of simulations with a small d/Z_i (case I64a & I64c), the mean plume height in case I64c (larger park size) is less affected by the surface pattern. When park size is small (case I64a), Z_c/Z_i is higher over the built-up area and lower over the park area. Also variation within the 4 experiments (*b*4 or *p*4) is much greater than that in I64c. In contrast, in the group of simulations with a large d/Z_i (I64b & I64d), the opposite situation is observed: larger park size (I64d) triggers stronger circulation so that the mean plume path over the built-up area is much higher than that over the park area. For a case with a smaller park size (I64b), the mean plume height is little influenced by the surface pattern, although variation from line to line is significant.

The above-discussed features are also seen in Figure 5 for a higher initial emission elevation of $0.49Z_i$. Because the variation across the line sources is relatively small, a clearer contrast for the mean plume height for case I64d is demonstrated.

CONCLUSIONS

The results show that turbulent structure induced by the inhomogeneous surface heat flux affects the dispersion of a passive plume emitted at elevated sources. The position of the point source with regards to built-up or park area affects the dispersion of passive plume in the downwind distances. When the point source is located above the central line across the park areas, the mean plume height is lower than that for a homogeneous case; if the point source is aligned with the built-up area, the mean plume height is slightly higher than that for a homogeneous case. The extent to which the surface heat flux pattern influences the mean plume height depends on two parameters: d/Z_i and b/p. However, the mean plume heights averaged over all line sources are about the same as those for a homogeneous case. This suggests that dispersion of a passive plume in a CBL over a patchy surface pattern has different behaviour from the previous studies for a homogeneous surface.

REFERENCES

Cai, X.-M., D. G. Steyn and I. S. Gartshore, 1995, Resolved scale turbulence in the atmospheric surface layer from large eddy simulation. *Boundary-Layer Meteorol.*, **75**, 301-314.

Cai, X.-M. and D. G. Steyn, 1996, The von Karman constant determined by large eddy simulation. *Boundary-Layer Meteorol.*, **78**, 143-164.

Cai, X.-M., Large-eddy simulation of the convective boundary layer over a patchy surface. Submitted to *Quart. J. Roy. Meteorol. Soc.*

Deardorff, J.W., 1985, Laboratory experiments on diffusion: the use of convective mixed-layer scaling. *J. Clim. Appl. Meteorol.*, **24**, 1143-1151.

Dornbrack, A. and U. Schumann, 1993, Numerical simulation of turbulent convective flow over wavy terrain. *Boundary-Layer Meteorol.*, **65**, 323-355.

Hadfield, M.G., 1994, Passive scalar diffusion from surface sources in the convective boundary-layer. *Boundary-Layer Meteorol.*, **69**, 417-448.

Haren, L. van and F.T.M. Nieuwstadt, 1989, The behaviour of passive and buoyant plumes in a convective boundary layer, as simulated with a large-eddy model. *J. Appl. Meteorol.*, **28**, 818-832.

Henn, D.S. and R.I. Sykes, 1992, Large-eddy simulation of a dispersion in the convective boundary layer. *Atmospheric Environment*, **26A**, 3145-3159.

Kemp, J.R. and D.J. Thomson, 1996, Dispersion in stable boundary-layers using large-eddy simulation. *Atmospheric Environment*, **30**, 2911-2923.

Lamb, R.G., 1978, A numerical simulation of dispersion from an elevated point source in the convective planetary boundary layer. *Atmospheric Environment*, **12**, 1297-1304.

Mason, P.J., 1992, Large-eddy simulation of dispersion in convective boundary layers with wind shear. *Atmospheric Environment*, **26A**, 1561-1571.

Nieuwstadt, F.T.M. and J.P.J.M de Valk, 1987, A large eddy simulation of buoyant and non-buoyant plume dispersion in the atmospheric boundary layer. *Atmospheric Environment*, **21**, 2573-2587.

Nieuwstadt, F.T.M., 1992a, A large eddy simulation of a line source in a convective atmospheric boundary layer. 1. Dispersion characteristics. *Atmospheric Environment*, **26A**, 485-495.

Nieuwstadt, F.T.M., 1992b, A large eddy simulation of a line source in a convective atmospheric boundary layer. 2. Dynamics of a buoyant line source. *Atmospheric Environment*, **26A**, 497-503.

Sykes, R.I. and D.S. Henn, 1992, Large-eddy simulation of concentration fluctuations in a dispersing plume. *Atmospheric Environment*, **26A**, 3127-3144.

Willis, G.E. and J.W. Deardorff, 1978, A laboratory study of dispersion from an elevated source within a modelled convective planetary boundary layer. *Atmospheric Environment*, **12**, 1305-1311.

Willis, G.E. and J.W. Deardorff, 1981, A laboratory study of dispersion from a source in the middle of the convectively mixed layer. *Atmospheric Environment*, **15**, 109-117.

DISCUSSION

P. BUILTJES: Do you have any plans to incorporate chemical transformations like NO-NO_2-O_3 in you LES-model?

X.-M. CAI: In this study, there is no any chemical transformation involved. The aim of this work is to investigate the effects of large convective eddies on dispersion of a passive plume. The time scale of these large eddies is more than ten minutes, which is much larger than the equilibrium time scale of NO-NO_2-O_3 reaction. There have been some studies that incorporated NO-NO_2-O_3 chemistry with a focus on the interaction of turbulent eddies and chemistry (e.g., Schumann, 1989, Atmos. Environ., Vol.23, 1713-1727; Sykes et al., 1992, J. Appl. Meteorol., Vol.33, 825-834.).

B. FISHER: Your calculations show the importance of surface inhomogeneities on plume height and plume spread. Do you have any plans for incorporating a surface inhomogeneity correction factor into simpler dispersion models?

X.-M. CAI: This study provides some interesting results that can be used for deriving a practical scheme. Before doing that, we probably need more systematic studies so that an empirical scheme can be firmly based. I am planning to do more careful LES investigations on the dispersion problem.

M. ROTACH: In a real urban environment the patches of different surface heat flux are associated with differences in surface roughness. Have you looked at the question which of the two inhomogeneities is actually dominating?

X.-M. CAI: In this study, I am dealing with a convective boundary layer in a free-convection regime. Effects due to surface roughness are not explicitly handled by the LES simulation, or they are parameterised by subgrid scale model and boundary conditions. Correct representation of turbulent structure near the surface by LES is still a challenging research subject. The present study cannot answer your question.

CLARIFYING THE RELATIONSHIP BETWEEN URBAN ROAD STRUCTURE AND AIR QUALITY EXCEEDENCES USING A TRAINING MODEL

B. E. A. Fisher[1] and A. G. Newlands[2]

[1]University of Greenwich
School of Earth & Environmental Sciences
Medway Campus
Chatham Maritime
Kent ME4 4TB U.K.

[2]67 Haslemere Avenue
London W13 9UH

INTRODUCTION

The European Union Framework Directive and daughter Directives on air quality (see Table 1), as well as individual national legislation, will require assessments of concentrations within urban agglomerations, together with a demonstration of how improvements may be brought about. There is therefore a need for 'easy to apply' roadside dispersion models. This paper describes a model which grew originally out of the need to train non-experts how to use dispersion models.

There is a requirement to be able to plan for reducing concentrations and therefore a need for urban and roadside dispersion models. The air quality limits are based on protection of human health. NO_2 and PM_{10} are likely to be the pollutants for which it will be most difficult to meet criteria. An earlier paper (Fisher and Sokhi, 1998) considered mainly NO_2 so that this paper pays most attention to PM_{10}.

A requirement for the successful implementation of legislation of this kind is that procedures should be easily understood. In order to train non-experts one needs to be able to explain how concentrations change when factors are varied. Of particular interest are the changes brought about by varying emission conditions. A contrasting approach is that of simulating actual conditions as closely as possible with the expectation that additional complexity brings greater accuracy. Errors can arise when a non-expert user tries to apply an advanced model. In both cases good agreement with measurements should be demonstrated before the model is applied to real situations. The GRAM model (Greenwich Review of Air Quality Method) was originally developed to help train people who were not experts in modelling with no intention to use the model for real assessments. However when tested against measurements its performance compares favourably with that of other more complex methods.

Air Pollution Modeling and Its Application XIII, edited by S.-E. Gryning and E. Batchvarova, Kluwer Academic/Plenum Publishers, 2000

Table 1 Some Limit Values for the Protection of Human Health in EU Daughter Directives

Pollutant	Criteria	Averaging Period	Limit Value	Date limit is to be met
NO_2	Hourly limit with reducing margin of tolerance from 1 January 2001	1 hour	$200\mu g/m^3$ not to be exceeded more than 18 times per year	1 January 2010
	Annual limit	1 year	$40\mu g/m^3$	1 January 2010
PM_{10}	Daily limit with reducing margin of tolerance from 1 January 2001	24 hours	$50\mu g/m^3$ not to be exceeded more than 35 times per year	1 January 2005
	Annual limit	1 year	$40\mu g/m^3$	1 January 2005

WORST CASE CONCENTRATIONS AND THE NEED FOR SCREENING MODELS

The form of the criteria shows that it is important for any model to be able to assess worst case situations when the highest concentrations are likely to occur. In addition the need in assessment is to be able to identify situations when high concentrations may occur in which case some degree of overprediction is not a disadvantage. In real situations most models and decisions are likely to be supported by measurements. Hence the concept of a simple screening model for use in practical situations is valid.

In the GRAM model used in this paper, vertical dispersion from a line source is a crucial parameter describing the mixing of pollution around a road and the more general mixing of material from ground-level area sources within the surrounding urban area. In GRAM vertical dispersion is based on the R91 model (Clarke, 1979). Highest concentrations are likely to occur in situations of light winds described by dispersion categories F and G. When comparing model performance it is vital to compare algorithms within models for these situations. The GRAM dispersion algorithm has been compared with that within the CALINE4 model (Benson, 1984) as an example of testing behaviour. In CALINE4 there are three distance zones around a road within which dispersion takes place. Closest to the road is a mixing zone also assumed within GRAM in which thermal and mechanical turbulence are assumed to maintain uniform concentrations. Then there is a zone of influence caused by the heat flux generated by each vehicle. Finally there is dilution at greater distances, where dispersion follows that of a passive release. Figure 1 shows a comparison of vertical dispersion in the GRAM and CALINE4 models in categories F and G for a typical six lane motorway. (The width of the road is a factor determining the size of the mixing zone in CALINE4. The roughness length is taken to be 1m but no vehicle heat flux is assumed.)

At distances close to the road typically less than 500m the GRAM dispersion is less than dispersion calculated from the CALINE4 model in both F and G dispersion categories. However at larger distances there is seen to be reasonable agreement between the GRAM

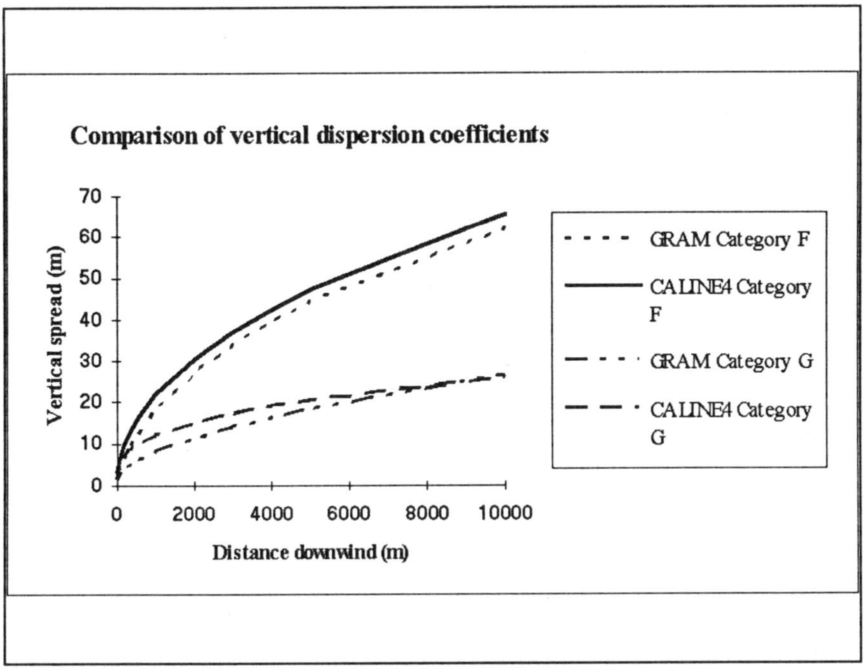

Figure 1 Example of a comparison between algorithms in two models.

results and the category G predictions from CALINE 4. This agreement is not surprising since both models have their origin in the same method developed by F B Smith. This shows the need to compare algorithms when comparing models and not just predicted values. It also shows that to determine when an air quality exceedence occurs one needs to be able to agree whether extremely stable conditions are likely to occur. Category G, not surprisingly, is not well defined theoretically. It corresponds roughly to an upward heat flux of $20W/m^2$ and a ten metre wind speed of 1m/s. Whether such extreme conditions occur within the urban atmosphere is unclear. There is a further difficulty if one decides to allocate all meteorological conditions into categories or bands, as the extreme cases may never be included in the analysis. It is also difficult to decide what should be the height of the mixing layer in very stable conditions. Somewhat arbitrarily this is taken to be 100m in GRAM. The assumption that category G will arise has been adopted as a representation of the worst case. These assumptions can only be tested by comparisons with roadside measurements.

PARALLEL VERSUS PERPENDICULAR WINDS

A related issue is whether highest concentrations occur when the wind is blowing parallel or perpendicular to the road. Ratio R of the concentration in a wind parallel to a road to the concentration when the wind is perpendicular to the road at a distance x from the road is given by

$$R(x)=\frac{\sigma_z(x)}{\sqrt{2\pi}}\int_0^l \frac{dy}{\sigma_z(y)\sigma_y(y)}\exp-\frac{x^2}{2\sigma_y(y)^2} \quad (1)$$

where the vertical dispersion (in m) is described by $\sigma_z(x)=1+0.151x^{0.557}$ and the horizontal dispersion (in m) is described by $\sigma_y(x)^2=\sigma_z(x)^2+0.065^2(7/u)T$, where x is distance downwind, T is the sampling time in hours and u is the ten metre wind speed in m/s and l is length of the road in metres. These formulae are either taken directly from the R91 model or fitted to the G dispersion curve in the R91 model. Taking x to be a receptor 10m from the side of the road, the length of the road to be 1000m and taking T=24h which would correspond to considering daily averages of PM_{10} , it turns out that in light winds of 1m/s that R=1.2. In light winds the lateral dispersion is high due to large wind direction fluctuations. Hence in worst case light wind conditions, although the parallel wind gives slightly higher concentrations, one may effectively assume that the wind is perpendicular to the road. This conveniently avoids the need to introduce another parameter l, the length of the road, into the model, which is another advantage of a training model. The number of parameters should be kept to a minimum.

COMPARISON WITH MEASUREMENTS AND OTHER MODELS

Sokhi et al (1998) have described some comparisons of GRAM with measurements and other models. The comparison has been based on statistical quantities that are relevant to the UK National Air Quality Strategy (namely the 99%ile of daily values of PM_{10}). Table 2 shows that if anything GRAM has a tendency to overpredict daily PM_{10} concentrations. Further comparisons were conducted for PM_{10} measurements at predominantly roadside locations. During 1997 seven kerbside and roadside sites within the National Network at which PM_{10} is measured, have become operational in the UK (in the London Boroughs of

Table 2 Comparison of Roadside Measurements with Predictions from Three Models

Pollutant	Criterion	Site	Measured	GRAM	ADMS	CALINE
CO	Maximum	M25	17	14	9	5
ppm	8h	Victoria	6.6	13.4		2.9
		Cambridge	6.4	5.6		
NO_2	Max 1h	M25	129	116		170
ppb	Annual mean	M25	31	40		54
PM_{10}	99% 24h mean	M25	101	169	105	111
$\mu g/m^3$	Annual mean	M25	30	21	33	51

Sutton, Haringey and Camden, in Glasgow and Bury, and at two further London sites near the major highways: the A3 and the Marylebone Road). If seven urban background sites associated with these roadside sites, two motorway sites along the M4 and M25, the Cambridge site and two of the Westminster City Council sites are included, one has a data set of 19 sites with 1997 PM_{10} concentrations which may be used to compare predictions from a dispersion model with measurements. Using the model GRAM (Fisher and Sokhi, 1998) the annual average and the 99 percentile of daily PM_{10} concentrations may be predicted. These concentrations have been compared with measured annual and 99 percentile daily PM_{10} concentrations.

Table 3 Comparison of PM_{10} Measurements at Roadside and Background Locations

Roadside Site		Background Site		Road Contribution	
Name	$\mu g/m^3$	Name	$\mu g/m^3$	AADF	$\mu g/m^3$
Sutton 1	24	Bexley/ Eltham	22	20000	2
Haringey	26	Brent	22	17100	4
Camden 1	32	Brent/ Kensington	23	45100	9
Glasgow Kerbside	31	Glasgow Central	21	15000	10
Bury	30	Bolton/ Manchester Piccadilly	22	26100	8
A3	29	Bexley	23	104000	6
Marylebone Road	39	Kensington	24	71200	15
M25	31	Bexley/ Eltham Brent/ Kensington	22.5	184000	8.5

AADF= annual average daily flow (vehicles/day)

The relationship between predicted and measured annual average PM_{10} concentrations takes the following form:

$$Mean_{measured} = 14.2 + 1.09 Mean_{calculated}$$

where the correlation coefficient r equals 0.823. The relationship between the predicted and measured 99 percentile of daily PM_{10} concentrations was found to be:

$$Peak_{measured}=41.4+0.28Peak_{calculated}$$

where the correlation coefficient r equals 0.783. The model is able to explain much of the variation in annual average concentrations and the 99 percentile of daily PM_{10} concentrations. However a sizeable extra PM_{10} concentration arising from regional sources mainly of secondary origin has to be included. For peak daily PM_{10} concentrations, represented by the 99 percentile of daily concentrations, the GRAM model overestimates the contribution from road and urban background sources by a large margin. The GRAM model may be considering situations which describe dispersion in a way which is unduly pessimistic. One suspects that the assumption that very stable conditions will persist for 24 hours is unrealistic. The large intercept of 41.4 $\mu g/m^3$, which is close in value to the UK National Air Quality Strategy Objective of 50 $\mu g/m^3$, is somehow related to the contribution from regional, predominantly secondary sources. However an extra contribution cannot be added directly to the predicted 99 percentile concentration because it is not known how well correlated daily PM_{10} concentrations from roadside and urban background concentrations are to regional PM_{10} concentrations. There is no simple way of deriving the combined frequency distribution from local and regional sources. By contrast annual average concentrations can be added together.

SECONDARY SULPHATE FORMATION

Measurements of SO_4 over the UK suggest that concentrations range from about $1\mu g/m^3$ S in the south to about $1\mu g/m^3$ S in the north. These concentrations correspond to 3 to 6 $\mu g/m^3$ sulphate. Models suggest that the sum of the NO_3 and ammonium contributions is about 1.5 to 2 times that of the sulphate and that 30 to 45% of the sulphate originates from sulphur dioxide sources outside the UK. An overall secondary contribution of up to $15\mu g/m^3$ may arise from sources within the UK and Europe. A contribution of this size can explain a substantial fraction of the intercept found in equation (2).

In order to satisfy the requirement of a screening model one does not wish to have to include complex factors outside the immediate scope of the model. However for PM_{10} one faces the dilemma of the need to be able to identify two different kinds of conditions which can cause high PM_{10} concentrations: regional particle episodes and urban particle episodes.

SURROGATE METHODS FOR IDENTIFYING PROBLEM AREAS

One simple way around the dilemma is just to focus attention on those episodes that are affected by local control measures. The model is used to predict what kinds of conditions will lead to local exceedences. The permitted number of occurrences of air quality standards during local episodes should be set more strictly since one expects additional exceedences during regional episodes. In other words one may for example need to consider the second highest local episode rather than the fourth highest, because of the occurrence of two additional regional episodes. The regional episodes will not be expected to vary much with local conditions i.e. the size of the urban area or the traffic flow.

For applications by non-experts what is required is an indication of the types of road traffic flows and size of urban areas which give local air quality exceedences. One is again looking at realistic worst case concentrations. By applying the model GRAM, estimates of peak PM_{10} concentrations in the traffic flow conditions representative of typical "busy roads" and "congested street canyons" may be made, with the estimated contribution from a surrounding urban area also included. GRAM considers street canyon effects by assuming that the wind direction in a street canyon is constrained to be parallel with the street axis. The background concentration depends on the size of the urban area. One ends up with the curves shown in Figure 2. The upper dotted curve corresponds to a traffic flow of 25,000 vehicles per day at 16km/h through in a street canyon typical of a busy urban two lane road. The lower curve corresponds to a traffic flow of 50,000 vehicles per day at 70km/h without any street canyon effect. A width of 0 km represents a road through a rural area.

One see that in rural areas exceedences of the $50\mu g/m^3$ standard may occur with traffic flows in excess of 50,000 vehicles per day, whereas in urban areas in small towns exceedences may occur when the traffic is 15,000 vehicles per day. The model therefore places a limit on the kinds of the conditions that need to be considered in more detail and screens out the circumstances when further analysis is not necessary.

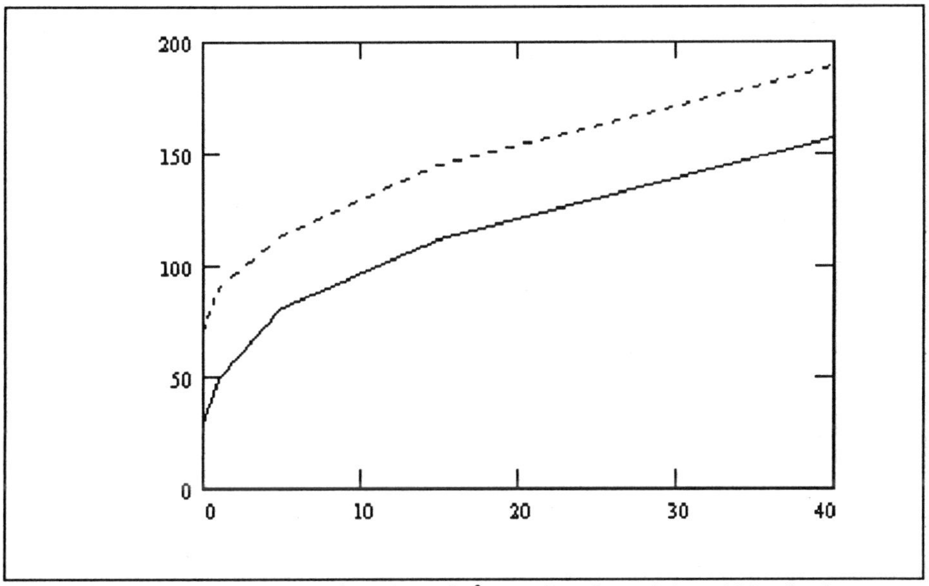

Figure 2 Peak PM_{10} concentrations in $\mu g/m^3$ as a function of the width of urban areas in km. Upper curve 50,000 veh/day no canyon, lower curve 25,000 veh/day in street canyon.

CONCLUSIONS

Simple models to assist air quality assessments are possible to formulate, which give satisfactory agreement with measurements. They are sensitive to the description of unusual meteorological events which are not well described. The same feature is true for more complex models. Attention should be paid to worst case conditions in a screening model. PM_{10} is particularly difficult because of the regional contribution from secondary particle formation. It is still possible in a simple model to broadly outline conditions which will lead to exceedences in terms of surrogate parameters, such as traffic flow and city size. Implementation of air quality legislation requires extrapolation to future conditions and this will introduce additional uncertainty to any model regardless of the extent to which it has been tested against measurements under current conditions.

REFERENCES

Airborne Particles Expert Group, 1998, Source apportionment of airborne particulate matter in the United Kingdom, to be published

Benson P E, 1984, CALINE4 - A dispersion model for predicting air pollutant concentrations near roadways, Caltrans Report Number FHWA-CA-TL-7080-77-25

Clarke R H (Chairman of Working Group on Atmospheric Dispersion), 1979, A model for short and medium range dispersion of radionuclides released to the atmosphere, NRPB Report R91

Fisher B E A and Sokhi R S, 1998, Investigation of roadside concentrations in busy streets using the model GRAM - Conditions leading to high short-term concentrations, 5th International Conference on Harmonisation within Atmospheric Dispersion Modelling for Regulatory Purposes, 18-21 May 1998, Rhodes, Greece

R S Sokhi, Fisher B, Lester A, McCrae I, Bualert S and Sootornstit N, 1998, Modelling air quality around roads, 5th International Conference on Harmonisation within Atmospheric Dispersion Modelling for regulatory Purposes, 18-21 May 1998, Rhodes, Greece

DISCUSSION

J. KUKKONEN: In modelling urban fine particles evaluation of emissions from traffic is problematic. How did you obtain your traffic emissions?

B. E. A. FISHER: Emissions are built into the model GRAM based on the best available commonly accepted information. It is accepted that for fine particles the emissions are rather uncertain. For average British vehicles under different driving conditions, emission factors for different years can be found on the London Research Centre Web Site: www.london-research.gov.uk/emission/webhtm.htm

J. LEVITIN: What is the difference between GRAM and CALINE4?

B. E. A. FISHER: GRAM and CALINE4 are based on similar algorithms describing dispersion. However whereas CALINE4 is used to produce concentrations under specified meteorological conditions, the results from GRAM are in the form of percentiles or annual averages which may be compared directly with air quality standards.

J. LEVITIN: Can the GRAM model be used for street canyons?

B. E. A. FISHER: There is an option in GRAM which allows one to estimate concentrations in street canyons. It is based on the assumption that the worst case corresponds to a wind blowing along a street in which pollution from vehicles in the street is well mixed.

DECOMPOSITION OF THE EMISSIONS AND METEOROLOGICAL COMPONENTS OF THE HOURLY NO_X CONCENTRATIONS IN CENTRAL LONDON USING ARTIFICIAL NEURAL NETWORKS

M.W.Gardner and S.R.Dorling

School of Environmental Sciences
University of East Anglia
Norwich, NR4 7TJ
United Kingdom

INTRODUCTION

Atmospheric concentrations of nitrogen oxides (NO_X=NO+NO_2) are determined by both the magnitude of NO_X emissions and the dispersive capabilty of the atmosphere. Emissions of NO_X are primarily in the form of NO which rapidly reacts with ozone (O_3) to produce NO_2. Only 2% of NO_X emissions are in the form of NO_2 (Derwent et al., 1995). In order to accurately model the hourly atmospheric concentrations of such primary pollutants it is important to resolve both the diurnal cycle of emissions and the effects of meteorology.

Atmospheric diffusion models attempt to solve the fundamental physical and chemical equations controlling pollutant concentrations. These models rely on detailed emissions and meteorological data for the domain under study. Collet and Oduyemi (1997) provide a review of such models. Alternatively statistical models can be developed from concurrent observations of pollutant concentrations and meteorological and emission predictor variables. Traditionally linear regression has been applied to develop such statistical models (Shi and Harrison, 1997; Ziomass et al., 1995). Typically the relationship being modelled is non-linear and involves the complex interaction of many predictor variables. Linear regression has been shown to underperform in such circumstances (Gardner and Dorling, 1998). More recently artificial neural networks, in particular the multilayer perceptron (MLP), have been successfully applied to model the behaviour of both primary and secondary pollutants Gardner and Dorling (In Press). MLP neural networks have the advantage of being able to model highly non-linear relationships between input (predictor) and output (predictand) data with no requirement to specify the nature of the non-linearity. MLP neural networks learn the input-output mapping during a supervised training process.

Air Pollution Modeling and Its Application XIII, edited by S.-E. Gryning
and E. Batchvarova, Kluwer Academic/Plenum Publishers, 2000

Table 1. The hourly meteorological data used to train the models

Variable (Abbreviation)	Units
Windspeed (WS)	knots
Low Cloud Amount (LOW)	oktas
Visibility (VIS)	metres
Dry Bulb Temperature (DRY)	degrees Celsius
Vapour Pressure (VP)	millibars
Base of Low Cloud (BASE)	metres

Table 2. The input data for the models. Abbreviations are taken from table 1

Model Number	Input Data	Architecture
01	WS, LOW, VIS, DRY, VP, BASE	6:20:20:1
02	WS, LOW, VIS, DRY, VP, BASE, sin(hour), cos(hour)	8:20:20:1
03	WS, LOW, VIS, DRY, VP, BASE, sin(hour), cos(hour), 7 binary day of week inputs	15:30:30:1

METHODOLOGY

In this work MLP neural network models were developed following the methodology described in Gardner and Dorling (In Press). The Stuttgart Neural Network Simulator, which is freely available via the internet (`ftp://ftp.informatik.uni-stuttgart.de`), was used to develop the MLP models. The MLP neural networks were trained using the scaled conjugate gradient algorithm (Moller, 1991) which is efficient compared to other training algorithms and has a high probability of finding a good solution. Unlike many neural network training algorithms scaled conjugate gradient is not sensitive to the selection of initial training parameters. The models were trained to learn the relationship between the meteorological input variables listed in table 1 and ambient NO_X and NO_2 concentrations. The hourly pollutant data was obtained from the Department of the Environment, Transport and the Regions automatic monitoring network for two Central London sites (`http://www.aeat.co.uk/netcen/aqarchive/auto.html`). Both sites were very close to each other and the data were combined to produce a longer dataset. The hourly meteorological data was obtained for London Weather Centre from the British Atmospheric Data Centre (`http://www.badc.rl.ac.uk/`). In this paper the analysis will be restricted to the NO_X models. Results concerning the NO_2 models, alongside a comparison between regression and neural network models, can be found in Gardner and Dorling (In Press).

All the MLP models were trained on hourly data from 1990 and subsequently tested on data from 1991. Ideally an independant validation data set would be used to gauge the network performance during training to avoid overfitting. Unfortunately data was limited to a two year period. Bootstrap estimates of standard error were used to gauge the degree of bias introduced into the model performance statistics (Wilks, 1995).

Three NO_X models were developed as described in table 2. Model 01 was trained to model the NO_X variability from purely meteorological data. The subsequent models were trained to capture the emissions related component of the NO_X variability. This component of the signal would be expected to be strongly linked to time of day and day of the week, reflecting the dominant traffic related source of NO_X in the London urban environment (Derwent et al., 1995). In order to resolve the diurnal variation in source strength model 02 had an additional pair of inputs, the sine and cosine of the hour of the day normalised between 0 and 2π. The two inputs are required in order to realistically represent the cyclical nature of time of day such that the network is aware that 23:00 hours is close to 00:00 hours. Presenting the network with the time of day directly would result in a discontinuity around midnight. Model 03 was designed with seven binary inputs each representing a day of the week. These inputs were either "on" or "off" depending on the day of the week, and should allow the network to learn the daily variation in source strength, for example between weekdays and weekends.

This approach to resolving the diurnal variation in source strength contrasts with that of Shi and Harrison (1997) who derived an hourly emissions factor for each day of the week. They used a very simple box model to estimate emissions from ambient concentrations. The approach adopted here removes the requirement to calculate the emissions rate and reduces the uncertainties inherent in their estimation. More importantly the method adopted in this paper will allow the meteorological and emissions related components of the NO_X signal to be identified.

To discover what the various models had learnt comparisons between the models were made to assess the effect of adding the additional inputs. For example to determine what model 02 learns, compared to model 01, it is necessary to look at the difference between the predicted concentrations from model 02 and model 01. In this way the component of the relationship learnt by model 02 due to the addition of a new input becomes apparent. Such comparisons were made between models 01 and 02 and models 01 and 03 over the test data period (1991). This comparison goes some way to overcoming the "black box" reputation of neural network models.

RESULTS

Various measures of network performance are presented in table 3 for all models. Figure 1a shows the observed mean diurnal NO_X concentrations by day of the week for 1991 whilst figure 1b shows model 01 predicted diurnal NO_X concentrations. Model 01 had no means of resolving the diurnal variation in emissions since NO_X emissions are generally unrelated to meteorology. Model 01 predicts higher concentrations in the morning, peaking around 07:00, followed by a decrease in concentrations during the afternoon until 17:00. This pattern of predicted concentrations does not match the observed pattern of concentrations and is largely related to atmospheric stability derived from the meteorological predictor data. Model 01 explains 45% of the variability in hourly NO_X concentrations. Interestingly model 01 predicts mean Sunday morning concentrations to be on average 30ppb higher than other days. Initial analysis of the 1991 hourly wind and temperature data by day of the week (not shown here) show that Sunday morning median windspeeds are lower than at any other time during the week and also that the median temperatures recorded on Sunday mornings are amongst the lowest. These two factors suggest that on average the Sunday morning boundary layer is more stable and results in the model predicting higher concentrations. This could explain why Sunday morning observed NO_X concentrations, figure 1a, are relatively high even though the emissions from traffic sources are probably at their lowest. The cause of this relative Sunday morning stability could be related to the weakening of the urban heat island effect (with lower urban to rural temperature and humidity diferences) and also less turbulent mixing in the urban boundary layer.

Model 01 performs poorly with respect to predicting a typical diurnal cycle. The addition of the time of day inputs (model 02) improves the predictions such that the general pattern of diurnal variation in emissions is resolved. Model 02 explains 54%

Table 3. Performance statistics for the models when tested on 1991 hourly data. $\bar{O}$,$\bar{P}$ - mean observed and predicted concentrations. s_o,s_p - observed and predicted standard deviation. MAE - mean absolute error. RMSE - root mean squared error. R^2 - coefficient of determination. a,b - straight line fit through observed versus predicted concentrations, intercept and gradient. Bootstrap estimates of standard error or given in brackets.

	Model 01	Model 02	Model 03
$\bar{O}$	101.49(2.40)	101.49(2.40)	101.49(2.40)
$\bar{P}$	96.70(1.75)	95.58(1.79)	95.26(1.95)
s_o	114.85(7.66)	114.85(7.66)	114.85(7.66)
s_p	82.65(3.94)	80.63(4.25)	86.54(5.77)
MAE	46.97(1.55)	38.47(1.46)	38.24(1.38)
RMSE	85.54(5.00)	78.57(5.96)	77.63(5.34)
R^2	0.45(0.04)	0.54(0.03)	0.55(0.04)
a	47.73(2.68)	43.45(3.45)	38.74(4.00)
b	0.48(0.03)	0.51(0.04)	0.56(0.04)

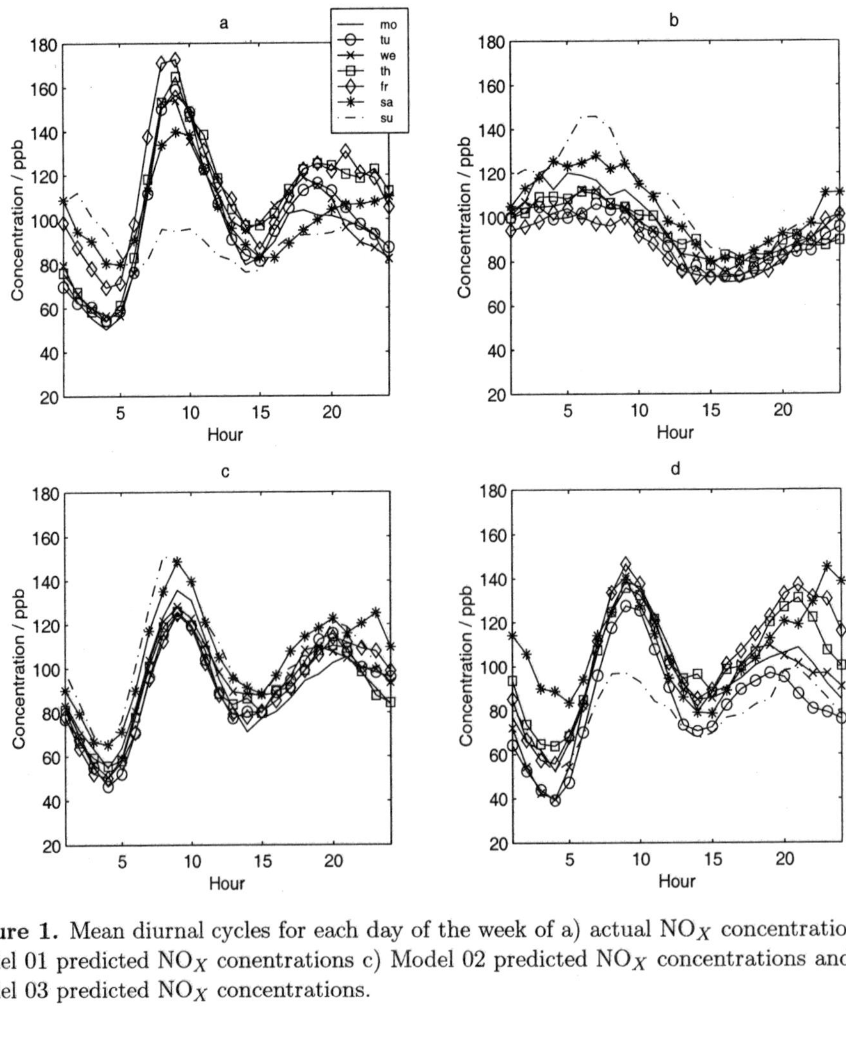

Figure 1. Mean diurnal cycles for each day of the week of a) actual NO_X concentrations b) Model 01 predicted NO_X conentrations c) Model 02 predicted NO_X concentrations and d) Model 03 predicted NO_X concentrations.

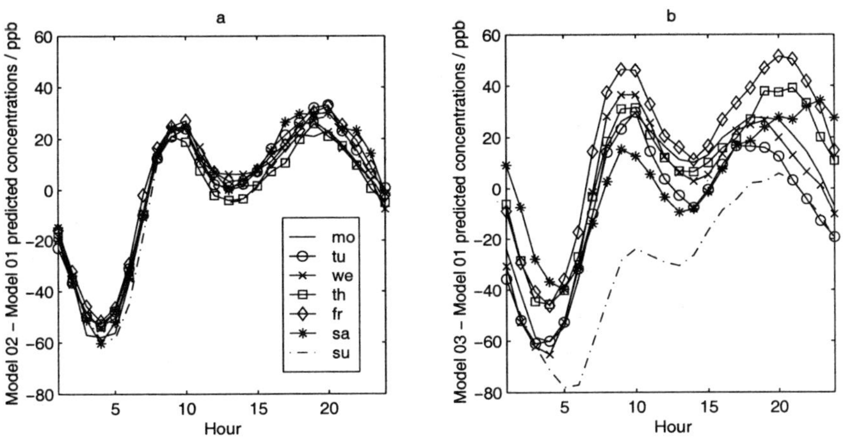

Figure 2. Mean diurnal profiles for each day of the week of a) (model 02)-(model01) predicted concentrations and b) (model 03)-(model 01) predicted concentrations.

of the NO_X variability which is considerably better than the performance achieved by Shi and Harrison (1997) with their multiple linear regression models. Gardner and Dorling (In Press) developed a multiple linear regression model that could only explain 32% of the hourly NO_X variability using exactly the same data as was used to train model 02. Clearly MLP neural networks, and their non-linear modelling capabilities, are better suited to resolving the complex interaction between the predictor variables that determine NO_X concentrations.

Figure 2a shows the difference between model 02 and model 01 predictions. This plot illustrates the improvements made by model 02 due to the addition of the time of day inputs. The profiles shown in figure 2a all cluster closely together and illustrate that the model is only picking up the mean daily pattern of emissions and has no mechanism by which to adjust the predictions for day of the week. This is evident in figure 1c which shows that Saturday and Sunday concentrations are overpredicted between 09:00 and 10:00. The model is applying the general relationship, between time of day and concentrations, even though emissions are much lower on these days and concentrations do not generally reach such high levels.

The addition of seven binary day of the week inputs (model 03) is designed to allow the network the freedom to adjust the learnt relationship for each day of the week. Model 03 shows only a slight improvement over model 02 in terms of the performance statistics calculated over a whole year of hourly data (table 3). This is because the input data is now less densely distributed over the 15-dimensional input space due to the differentiation between days of the week. The model is required to generalise over a much larger input space than was the case for models 01 and 02. Increasing the amount of training data would help alleviate this problem.

Although the improvement in hourly performance is only slight, model 03 has begun to learn the differences between the days of the week with mean annual diurnal profiles of model predictions more closely resembling the observed concentrations (figure 1d). Comparison with model 01, figure 2b, shows that the model is resolving the emissions related component of the NO_X concentrations as would be expected. Figure 2b represents the component of the atmospheric concentration that the model has learnt to be related to the time of day and day of the week. This could be interpreted as the atmospheric response to NO_X emissions with the estimated meteorological signal removed. This curve differs from a NO_X emission rate since there is an element of transport of background NO_X from the suburbs and outside of London, as well as a temporal lag before emissions become evident in the measured atmospheric concentrations.

Figure 2b shows that the predicted Friday curve peaks higher than any other day, a recognised phenomenon due to increased commuter traffic leaving London on Friday afternoon. This also tends to extend the length of the Friday evening peak. The Sunday curve is the lowest curve with the morning response begining almost two hours later than on other days. The very early Saturday morning signal is higher than on any other day probably reflecting the emissions from traffic associated with Friday evening recreational activity. Interestingly the Saturday evening curve continues to increase until around 23:00, again possibly due to the effect of increased recreational traffic in central London throughout Saturday evening.

CONCLUSIONS

It has been demonstrated that MLP neural network models can model the relationship between local meteorological data and NO_X concentrations in an urban environment. The models can be designed to learn the day to day variation in the stength of NO_X emissions without any external guidance. This allows the models to be easily constructed without any requirement to have detailed emissions data.

Comparisons between models has shown that it is possible to decompose the measured NO_X concentration signal into a meteorological and an emissions related component. Analysis of the emissions related component appears to correlate with the expected pattern of the predominantly traffic related emissions in central London.

REFERENCES

Collet, R. S. and Oduyemi, K. (1997), 'Air quality modelling : a technical review of mathematical approaches', *Meteorological Applications* **4**, 235–246.

Derwent, R. G., Middleton, D. R., Field, R. A., Goldstone, M. E., Lester, J. N. and Perry, R. (1995), 'Analysis and interpretation of air quality data from an urban roadside location in central London over the period from July 1991 to July 1992', *Atmospheric Environment* **29**(8), 923–946.

Gardner, M. and Dorling, S. (In Press), Neural network modelling of hourly NO_X and NO_2 concentrations in urban air in London. To appear in Atmospheric Environment.

Gardner, M. W. and Dorling, S. R. (1998), 'Artificial neural networks (the multilayer perceptron) - a review of applications in the atmospheric sciences', *Atmospheric Environment* **32**(14–15), 2627–2636.

Moller, M. F. (1991), 'A scaled conjugate gradient algorithm for fast supervised learning', *IEEE Transactions on Systems, Man and Cybernetics* **21**, 272–280.

Shi, J. P. and Harrison, R. M. (1997), 'Regression modelling of hourly NO_X and NO_2 concentrations in urban air in London', *Atmospheric Environment* **31**(24), 4081–4094.

Wilks, D. S. (1995), *Statistical methods in the atmospheric sciences : an introduction*, International Geophysics Series, Academic Press, San Diego, CA.

Ziomass, I. C., Melas, D., Zerefos, C. S. and Bais, A. F. (1995), 'Forecasting peak pollutant levels from meteorological variables', *Atmospheric Environment* **29**(24), 3703–3711.

DISCUSSION

D. SYRAKOV: Why, introducing the time angle h as additional factor, do you introduce the two dependant on each other parameters (predictors), sin(h) and cos(h)?

M. GARDNER: By including only one of the predictors the network will not be able to resolve the time of day exactly. For example sin(h) will take the same value at 0300h and 0900h. Inclusion of both the sine and cosine components overcomes this problem.

S. RAFAILIDIS: The measurements show that the morning rush hour NO_x peak is greater than during the evening NO_x rush hour period. Does your neural network model give any insight to explaining this phenomenon? Judging from the figures the neural network does not reproduce this effect too well.

M. GARDNER: Model 02 (meteorological and time of day inputs) captures this effect the best, however the very high morning concentrations are generally underpredicted. This is a characteristic of all statistical models.

Insight into reasons for the higher concentrations in the morning can be obtained from model 01 (meteorological inputs only) which suggests that given no information about sources of NO_x the best predictions are made by considering atmospheric stability. Atmospheric stability is greater in the morning and hence the morning rush hour emissions result in higher atmospheric concentrations. In the afternoon stability is reduced as windspeeds and temperatures increase and there is greater dispersion of the NO_x emissions resulting in lower atmospheric concentrations.

B. FISHER: Have you got any "feeling" on how much the performance of your model could have been improved if fairly detailed emission inventory data were also input? There are now certain such datasets becoming available.

M. GARDNER: This work has indicated that emissions vary greatly from day to day, hence using an average emissions inventory for an idealised day of the week would not offer much advantage. If emissions data were available real-time at an hourly resolution then I think this would produce a considerable improvement in the model performance.

A NEW MODEL FOR FLOW AND DISPERSION IN A STREET-CANYON

Lionel Soulhac and Richard J. Perkins

Laboratoire de Mécanique des Fluides et d'Acoustique,
CNRS UMR 5509, École Centrale de Lyon & Université Lyon I
36, avenue Guy de Collongue, 69 131 Ecully - France

INTRODUCTION

It is necessary to understand the transport and dispersion of pollutants in city streets, for many reasons. Vehicle emissions form a large proportion of the atmospheric pollutants in urban environments, and these emissions occur within streets, at street level; we need to know how these pollutants are transported between streets, and into the overlying urban canopy. Pollution monitors are usually located within streets, often at street level (see fig. 1), and in order to exploit their measurements fully, we need to understand the relationship between what is measured at a particular location, in very specific circumstances, and the more general state of atmospheric pollution in different areas of the city. Most of the urban population will be exposed to atmospheric pollution within city streets at street level, or within the buildings that border and define the streets. We need to know how this exposure depends on the emissions, the flow and the dispersion within the street.

There have already been many studies of flow and dispersion in a street-canyon (see, for example, JOHNSON *et al.*, 1973; YAMARTINO and WIEGAND, 1986; BERKOWICZ *et al.*, 1997). Most of these studies have involved a two-dimensional configuration in which the street is perpendicular to the wind. In general, two approaches have been adopted to model the flow in a street-canyon; numerical simulations such as those of SINI *et al.* (1996) are capable of reproducing many of the features of the velocity and concentration fields inside the street canyon, but they require large computing resources, and their use is therefore restricted to rather simple, isolated cases. Much simpler models have been developed (e.g. HOTCHKISS and HARLOW, 1973), for practical applications, but they are generally limited to rather simple configurations (e.g. symmetrical streets), they depend on empirical coefficients, and they do not permit detailed calculation of the concentration field in the streets. But there are many problems for which it is necessary to compute the structure of the concentration field inside the street, and such calculations may be needed for many different configurations and circumstances. So

Air Pollution Modeling and Its Application XIII, edited by S.-E. Gryning
and E. Batchvarova, Kluwer Academic/Plenum Publishers, 2000

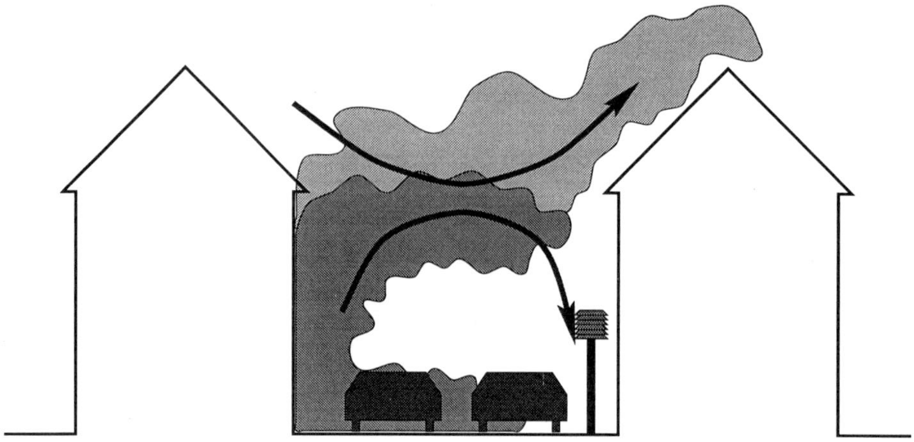

Figure 1. Dispersion in a street-canyon.

there is a need for a practical model that is intermediate between the two approaches described above. This paper describes such a model. The model itself is presented in §2, and then, in §3, we compare some preliminary results from the model with those from other models. Finally, in §4, we discuss possible further developments of the model.

DESCRIPTION OF THE MODEL

The model is based on an analytical solution of the advection-diffusion equation, using an approach similar to that adopted by HUNT and MULHEARN (1973) for modelling dispersion around a cylinder. The flow within the street canyon is modelled assuming potential flow, and a conformal transformation of this flow field then enables us to use an analytical solution of the advection-diffusion equation. Calculation of the velocity and concentration fields is therefore rapid enough for the model to be useful as a practical tool.

The Velocity Field

We assume that the flow outside and within the cavity can be modelled as a stationary, 2-D potential flow. (This is clearly not the case in practice, but the results from the model show that this is a reasonable simplification for the advection terms in the advection-diffusion equation.) Because the fluid is inviscid, there can be no recirculating flow within the cavity, so we create a recirculation by adding an irrotational vortex within the cavity. The strength of the vortex is determined so as to ensure separation of the flow at a specified location.

The boundary of the velocity field consists of straight line segments, so the velocity field can be mapped onto the Φ-Ψ plane, using the Schwarz-Christoffel transform. Analytical solutions are only available for rather simple geometries, so we use a general numerical technique, as described by DRISCOLL and VAVASIS (1996).

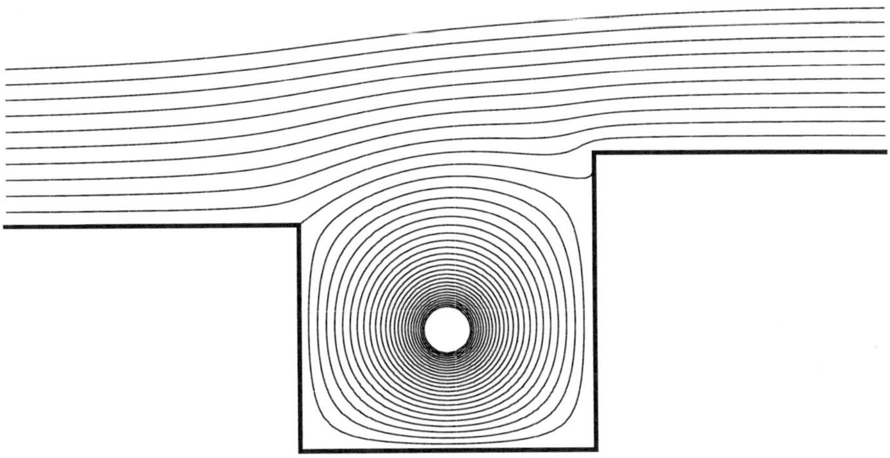

Figure 2. Flow streamlines in an asymmetric street.

Figure 2 shows an example of the resulting streamlines, for flow in an asymmetric cavity. The vortex is located at the centre of the cavity, and the imposed separation point occurs at the corner of the upstream building.

The Concentration Field

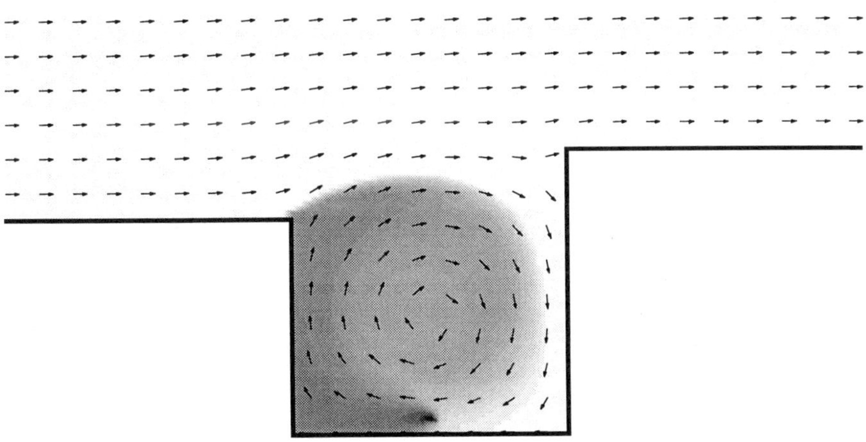

Figure 3. Concentration field in an asymmetric street.

The conformal mapping of the velocity field from the real x-z plane to the complex potential plane (Φ-Ψ) transforms the velocity fields into uniform flow. Since the advection-diffusion equation is a form of Laplaces' equation, its solution is not modified by the conformal mapping, and so the concentration field in the Φ-Ψ plane becomes the standard Gaussian plume. In order to satisfy the condition of zero flux on the solid boundaries of the flow domain, the transport and dispersion within the cavity has to be simulated separately from that outside the cavity. The two calculations are coupled through the flux of material across the separation streamline which divides the two regions. We model this as a turbulent mixing layer created by two parallel streams travelling at slightly different velocities, using a solution analogous to that proposed by GOERTLER (1942). This enables us to compute the flux analytically, as a function of the velocity and the turbulence in the external flow.

As an example of this type of calculation, figure 3 shows the concentration field generated by a source at street level in the cavity.

RESULTS

Comparison with other Models

We have computed the concentration field in a square street-canyon, with a point source at street level, and we compare the results with those obtained from a k-ε calcu-

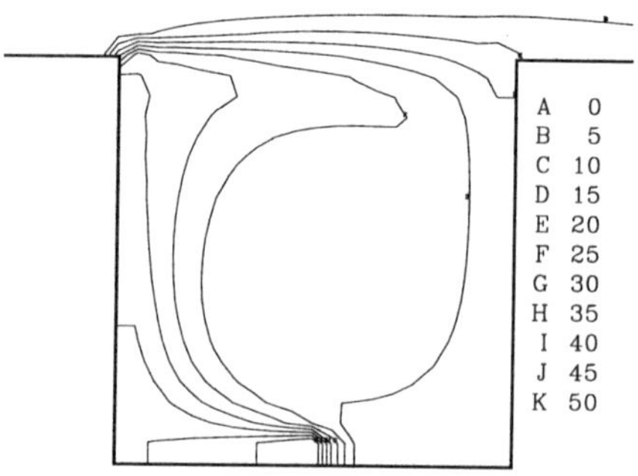

MERCURE k-ε model
($\simeq$ 10 hrs CPU)

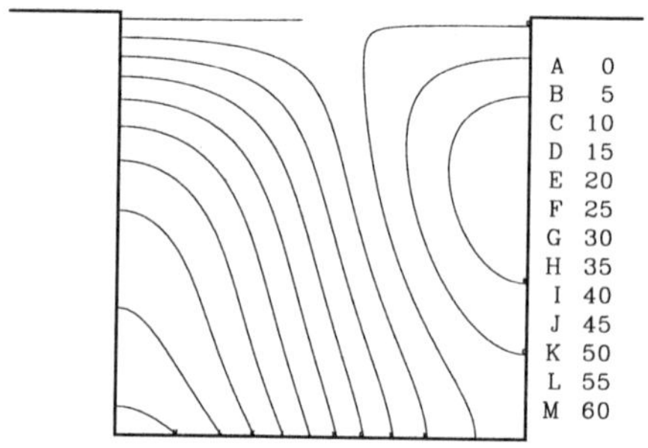

Analytical model (HOTCHKISS & HARLOW, 1973)

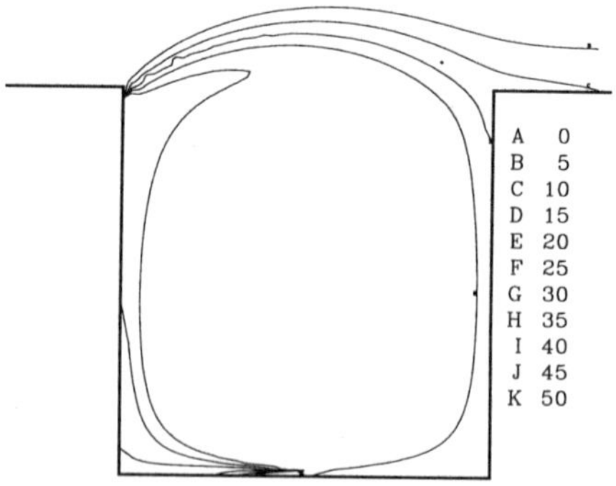

SOULHAC & PERKINS *model ($\simeq$ 1 min CPU)*

Figure 4. Dimensionless concentration $C^{\star}$: comparison between different models.

lation (the code *Mercure* , see Carissimo *et al.*, 1995 for further details) and from the model of Hotchkiss and Harlow (1973). Figure 4 shows the dimensionless concentration $C^{\star} = CUW/Q$ (where U is the velocity at roof level, W is the width of the street and Q is the mass flux of pollutant per unit length at the source) for the three models. All three models predict higher concentrations on the upwind side of the cavity, but only *Mercure* and the current model show any evidence of the existence of a "plume" within the cavity. (This is probably due, at least partly, to the fact that the model of Hotchkiss and Harlow assumes a uniformly distributed line source on the floor of the cavity, rather than a point source.) The general concentration levels predicted by our model agree reasonably well with those computed using *Mercure*. The calculation with *Mercure* took about 10 hrs cpu time on a work station; the calculation with the new model took about 1 min cpu time on the same machine.

Influence of the Position of an Air Pollution Monitor

In order to illustrate one of the possible practical applications of the model, we have used it to compute the concentration that would be measured by a stationary air pollution monitor, as a function of its position in the street canyon. We consider a square cavity (height H=20 m, width W=20 m) with a point source of pollutant placed in the cavity, at ground level, and at varying distances from the upwind face of the cavity. The monitor is located 2 m from the upwind face of the cavity, and we consider three different heights above the street (z=1 m, 2 m and 3 m).

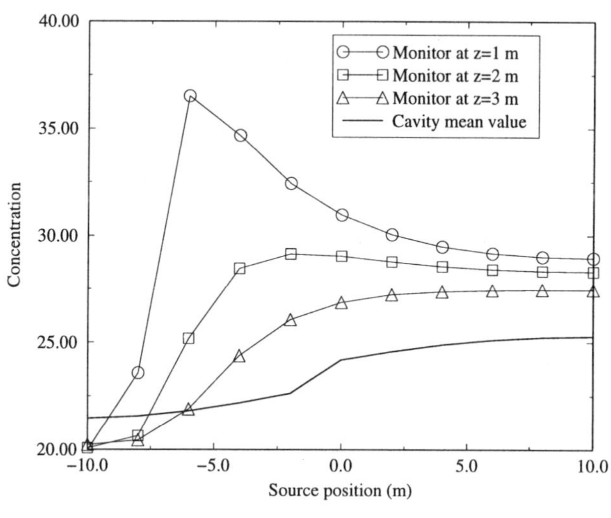

Figure 5. Dimensionless concentration $C^{\star}$: influence of the position of the monitor in the street.

Figure 5 shows the concentrations measured at each of the three monitor locations, as a function of the position of the source. These results show that, in almost all configurations, the concentrations measured by the monitor exceed the concentration at the centre of the cavity (which is almost equal to the mean concentration in the cavity), and that the monitor closest to the ground is the most sensitive to the position of the source. Even when the source is as far as possible from the monitors, the three monitor positions yield different concentration measurements

DISCUSSION

The new model has several advantages over existing approaches. It reproduces the details of the concentration field within the street canyon rather well, but it requires much less computer time and memory than a full scale CFD calculation. Thus it is feasible to use this type of model in problems where it is necessary to study many different cases.

The model can easily be applied to complex geometries, including features such as asymmetric cavities, series of streets and irregularly-shaped cavities. Figure 6 shows a typical example of such a calculation, involving several streets.

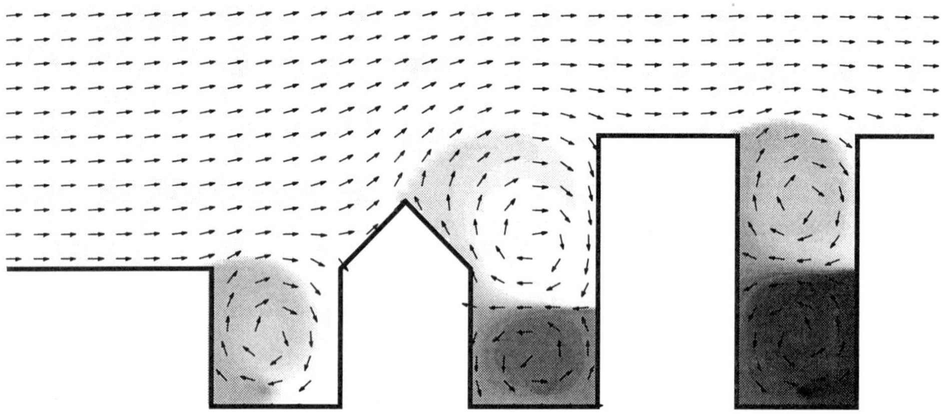

Figure 6. Flow and dispersion over a complex geometry.

Finally, the use of a mixing layer model to compute the flux between the cavity and the external flow enables us to account specifically for the turbulence in the external flow.

The model also suffers from several disadvantages. Firstly, it is two-dimensional, with an infinitely long street; this is imposed by the use of conformal mapping to provide an analytical solution of the advection-diffusion equation. Secondly, in order to construct the velocity field, it is necessary to know where to place the centre of the vortex in the cavity, and where to place the separation point. At the moment, these parameters can only be determined empirically. It is also necessary to use empirical estimates for the diffusion coefficients for the flow within the cavity and outside it.

Future research will be devoted to improving our estimates for these parameters, and developing the model further, to take three-dimensional effects into account, including situations in which the street is no longer perpendicular to the wind. It will also be necessary to investigate the interactions between street canyons at intersections.

REFERENCES

BERKOWICZ, R., HERTEL, O., LARSEN, S. E., SORENSEN, N. N., and NIELSEN, M. 1997. *Modelling traffic pollution in streets.* National Environmental Research Institute.

CARISSIMO, B., DUPONT, E., MUSSON-GENON, L., and MARCHAND, O. 1995. *Note de principe du code MERCURE version 3.1.* EDF-DER, HE-3395007B.

DRISCOLL, T. A., and VAVASIS, S. A. 1996 (January). *Numerical conformal mapping using cross-ratios and Delaunay triangulation.* Cornell University.

GOERTLER, H. 1942. Berechnung von aufgaben der freien turbulenz auf grund eines neuen naherungsansatzes. *Z.A.M.M.*, **22**, 244–254.

HOTCHKISS, R. S., and HARLOW, F. H. 1973. Air pollution transport in street canyons. *Environmental Protection Agency*, 1–128.

HUNT, J. C. R., and MULHEARN, P. J. 1973. Turbulent dispersion from sources near two-dimensional obstacles. *J. Fluid Mech.*, **61**, 245–274.

JOHNSON, W. B., LUDWIG, F. L., DABBERDT, W. F., and ALLEN, R. J. 1973. An urban diffusion simulation model for carbon monoxide. *Journal of Air Pollution Control Association*, **23**, 490–498.

SINI, J-F., ANQUETIN, S., and MESTAYER, P. 1996. Pollutant dispersion and thermal effects in urban street canyons. *Atmos. Env.*, **30** - **15**, 2659–2677.

YAMARTINO, R. J., and WIEGAND, G. 1986. Development and evaluation of simple models for the flow, turbulence and pollutant concentration fields within an urban street canyon. *Atmos. Env.*, **20** - **11**, 2137–2156.

DISCUSSION

C. MENSINK: From the potential flow solution you obtain a steady state solution. How would you therefore make a comparison with measured data or observations in the urban canopy which have a more transient nature?

L. SOULHAC: Firstly, we should notice that most of the street-canyon models only provide steady state solution. Then, to validate this model, we will make a comparison with wind tunnel measurements, in which data will be averaged on a sufficient time to get steady state values. For a comparison with field measurements, we have to average meteorological and emission conditions on a sufficient time, and to compare the calculation for the mean conditions with the averaged measured concentrations.

E. GENIKHOVICH: At the corners of your domain and at the center of the vortex you have singularities in the wind velocity field where the wind speed is infinite. How does it influence applicability of your scheme?

L. SOULHAC: The velocity field is not completely realistic, but it provide a first good approximation of the advection characteristics of the flow. At the corners of the domain and at the center of the vortex, the wind speed is infinite but on a small area. Because the pollutant source is located far from this area, and because the concentration gradient is weak where the velocity is infinite, it does not affect so much the results of the model.

R. YAMARTINO: I understand that a double vortex can arise in potential flow and that this second vortex is seen in water channel studies, but I have not seen this double vortex appear in the lower half of a deep, narrow street canyon in either full-scale studies or wind tunnel studies. Do you have any comment on the use of this second vortex in your model ?

L. SOULHAC: The number and the location of vortices are input parameters of the model, which must be adjusted with experimental data. I wanted only to show that it is possible to take into account several vortices in this model. To justify the use of two vortices, I have used the numerical results of Lee and Park (1984) who show that a double-vortex structure can arise if the aspect ratio H/W is greater than 2.1.

S. RAFAILIDIS: Comment on the question of R. Yamartino on why one doesn't observe two vortices in narrow street canyons both in the field or other models (e.g. MISKAM).

Insofar as field measurements are concerned, these are not observed because of highly insufficient averaging times. The narrower the street canyon is, the slower the rotations in it will be for a given roof wind. The averaging times required to observe two vortices in actual street canyons are thus much longer than that allowed by wind meandering and/or diurnal meteorology changes.

No comparison with other numerical models makes sense, since also the other models are steady-state, whereas the phenomenon itself is intermittent. Therefore, no basis exists for realistic comparison.

CHAOTIC ADVECTION AND GEOPHYSICAL APPLICATIONS (A REVIEW)

S. Panchev

Dept. of Meteorology and Geophysics, Faculty of Physics
Sofia University, 1126 Sofia, Bulgaria

INTRODUCTION

Historically two types of motion of fluids are distinguished - LAMINAR and TURBULENT. Laminar is the motion with regular velocity in space and time, which can be described by well behaved functions of $(\vec{x}, t)=(x, y, z, t)$, e.g. trigonometric ones or represented graphically by smooth curves. Turbulent is the motion with highly irregular velocity in the $(\vec{x}, t)$ space-time occupied by the flow and mathematically is described by random functions, requiring probabilistic (stochastic) methods for studying.

These definitions are based on the so-called Eulerian approach to the motion. In the alternative approach known as Lagrangian one, trajectories of the fluid particles or passive tracers are monitored. Then, in laminar flows the trajectories are smooth curves, while in turbulent one they are highly irregular.

During the last two decades a third (intermediate) kind of flow regime, independent of the previous two was discovered. The Eulerian velocity is fully deterministic, prescribed by smooth nonlinear function of (x, y, z, t), but nevertheless, the trajectories are highly irregular - the particle coordinates

$$\vec{x}(\mathrm{t}, \vec{x}_0)=(x(t), y(t), z(t))$$

are random-like functions of the time t.

Meanwhile, the modern theory of chaos was developed (e.g. Hilborn,1994; Panchev,1996; Strogatz,1996) and this new kind of motion was labeled LAGRANGIAN CHAOS (LC) or LAGRANGIAN TURBULENCE (LT). The term CHAOTIC ADVECTION(CA), first proposed by Aref (1984), is now widely accepted. In this paper the problem of advection of a passive scalar substance $\theta(\vec{x}, t)$ in a fluid flow - a process described by the equation

$$\frac{d\theta}{dt}=\frac{\partial\theta}{\partial t}+\vec{\mathrm{v}}.\nabla\theta=K\nabla^2\theta \tag{1}$$

Air Pollution Modeling and Its Application XIII, edited by S.-E. Gryning
and E. Batchvarova, Kluwer Academic/Plenum Publishers, 2000

was studied, where $\vec{v}(\vec{x},t)$ is a prescribed Eulerian velocity field, K is diffusivity coefficient. Many people still think that the field $\theta(\vec{x}, t)$ is at worst as complicated as that of $\vec{v}(\vec{x},t)$. However, as it was definitely shown by Aref (1984) and subsequent authors, this would be a wrong conclusion. Very simple velocity field $\vec{v}(\vec{x},t)$ can induce highly complex advection patterns.

In the next sections an attempt for a brief review of the problem is made, based on the available to me publications. In an Addendum some of our preliminary results in this direction are presented.

THE NATURE OF THE CA

The above assertion about the possibility for extreme complexity of the θ- field against a simple structure of the $\vec{v}$-field follows explicitly, if the so called Lagrangian representation of the problem is used - the trajectories of the individual advected material particles are followed:

$$\dot{\vec{v}} = d\vec{x}/dt = \vec{v}(\vec{x},t) \tag{2}$$

or in scalar notations

$$\dot{x} = \mathrm{u}(x,y,z,t), \qquad \dot{y} = \mathrm{v}(x,y,z,t), \qquad \dot{z} = \mathrm{w}(x,y,z,t). \tag{3}$$

Here u, v, w are prescribed functions, at least one of which has same kind of nonlinearity, so that the system of ordinary differential equations (ODEs) as a whole is nonlinear one.

Actually, being equations of motion of individual particles (2), (3) can be viewed also as 3-dimensional (3-D) nonauthonomous (t enters explicitly) nonlinear dynamical system (NDS) and the general theory of such systems and of chaos (Hilborn,1994; Panchev, 1996; Strogatz, 1996) can be applied. According to this theory, the system (3) can have random-like solution for $\vec{x}(t)$ even at stationary velocity field $\vec{v}(\vec{x})$ and thus highly irregular trajectories in the space. This is CA! Three classical examples will illustrate the phenomenon.

Example 1. The "*ABC*"- Flow (Grisanti et al., 1991)
The equations of notation are prescribed as follows:

$$\dot{x} = \mathrm{u} = A\sin z + C\cos y$$

$$\dot{y} = \mathrm{v} = B\sin x + A\cos z \tag{4}$$

$$\dot{z} = \mathrm{w} = C\sin y + B\cos x$$

with A, B, C - real parameters. This velocity field is a special (steady-state) solution of the Euler equations of motion of inviscid incompressible fluid. At $A \geq B > 0$, C=0 the system (4) is integrable and the trajectories are smooth predictable curves. However, for other values of A,

B, C, for example $A=\sqrt{3}$, $B=\sqrt{2}$, $C=1$ the solution to (4) is chaotic and thus unpredictable.

Example 2. The Lorenz system:

$$\dot{x} = -\sigma x + \sigma y \equiv \mathrm{u}, \quad \dot{y} = r x - y - x z \equiv \mathrm{v}, \quad \dot{z} = -b z + x y \equiv \mathrm{w} \tag{5}$$

with (σ, b, r) >0 – parameters. Obviously, the "velocity" field in (5) is deterministic, quadratically nonlinear and quite simple. However, as it is widely known, for some values of the parameters, e.g. σ =10, b=8/3, r =28, the numerical solution of (5) exhibits chaotic behaviour.

Example 3. The Rosler system:

$$\dot{x} = -y - z \equiv \mathrm{u}, \quad \dot{y} = x + a y \equiv \mathrm{v}, \quad \dot{z} = b - c z + x z \equiv \mathrm{w} \tag{6}$$

The "velocity" field (u, v, w) is now even simpler than the previous example, but nevertheless, for some values of the parameters, e.g. a=b=0.2, c=4.2 the solution is chaotic and the corresponding trajectories - irregular.

At present, countless number of similar 3-D NDSs with chaotic behaviour of their solutions are known (Panchev, 1996; Strogatz, 1996). Those of them which can be verified experimentally (the "velocity" field $\vec{\mathrm{v}}(\vec{x})$ can be reproduced in laboratory) confirm the theoretical conclusion that LC can be observed in the absence of Eulerian chaos (EC). However, the relation between LC and EC is a very complicated issue. It may look very natural that EC implies a chaotic motion of particles. Indeed, it appears to hold in many systems, but nevertheless there are exceptions - EC may not yield LC (Lupini et al., 1989).

For example, let

$$\dot{x} = \mathrm{u} = \xi(t)\sin x \cos z, \qquad \dot{z} = \mathrm{w} = -\xi(t)\cos x \sin z \tag{7}$$

where $\xi(t)$ is a purely random or random-like function of time, i.e. Eulerian velocity is chaotic. However, because of the invariant following from (7)

$$\sin x(t) . \sin z(t) = \text{const} \tag{8}$$

the trajectories in the $x0z$ plane are closed curves (orbits) and LC does not exist. The angular velocity however depends on $\xi(t)$. This example originates from the Lorenz system (5).

GEOPHYSICAL IMPLICATIONS OF GENERAL CONCEPTS ABOUT CA

The above examples are not so interesting from point of view of application in the Geophysical Fluid Dynamics (GFD). The reason is that the large-scale flows in the atmosphere and ocean are quasi-two-dimensional (vertical scales much less than horizontal ones) and nonstationary - e.g. periodically modulated in time. Before going to description of the underlying general basis of this case, few comments of terminological nature, related to the phenomenon CA are necessary:

* Laminar advection corresponds to deterministic velocity field $\vec{\mathrm{v}}(\vec{x},t)$ (linear or nonlinear) and regular trajectories. It realizes at low Reynolds numbers ($Re\to 0$) in the fluid flows.

* Turbulent advection and diffusion corresponds to fully disordered, random velocity fields $\vec{v}(\vec{x},t)$, i.e. stochastic equations of motion (3) and hence irregular particles trajectories and high level of diffusivity. It realizes at $Re \to \infty$ in the fluid flows.

* Chaotic advection corresponds to deterministic velocity field $\vec{v}(\vec{x},t)$ and irregular (chaotic-sensitively dependent on the initial conditions) trajectories. It is not necessary the Reynolds number (*Re*) to be large. The CA can exist quite independently of any "Eulerian turbulence".

In all three cases in real fluids molecular diffusion is also acting, but it is a slow process and can be neglected. Compared to it, turbulent diffusion is much stronger process and, as a result, turbulent flows are well mixed. In other words, the MIXING is a diffusion related process, by means of which substances released in the flow are mixed across intermaterial surfaces. CA is inherently related to stretching and folding of these surfaces (lines in 2-D flows) and, as a result, to STIRRING of the flow. Hence, STIRRING is of kinematic nature and is not related to balance of forces in the momentum equation governing the flow. This justifies one more name for the CA - "Kinematic chaos". That is why, in principle, a two-phase fluid system can be "stirred but not mixed" as well as "mixed but not stirred". In real fluid flows stirring and mixing act simultaneously. But there may have situations when one of them dominates-natural or man made in some technological processes. For example, it can be very costly to mix two highly viscous fluids by turbulence and very cheep and enough to stir them. In other cases, turbulent mixing if possible and cheep, can be undesirable, i.e. in biofluids containing long-chain molecules. Then, the gentle stirring is preferable. It produces chaotic particle trajectories in a deterministic flow field by passive advection.

Finally, a phenomenon similar to the CA is well known in plasma physics and high-energy physics (particle accelerators). The motion of the accelerated particles can be chaotic in simple electromagnetic field configuration and it has been observed in the corresponding devices.

In view of the geophysical applications the idea about the CA is clearer in the case of 2-D unsteady, incompressible flow. The general equations (3) take the form

$$\dot{x} = u(x,y,t) = -\frac{\partial}{\partial y}\psi(x,y,t), \qquad \dot{y} = v(x,y,t) = \frac{\partial}{\partial x}\psi(x,y,t). \tag{9}$$

In hydromechanical interpretation $\psi(x,y,t)$ is a nonstationary stream function. It is well known that the corresponding stream lines do not generally coincide with the particle paths. In alternative interpretation $\psi(x,y,t)$ is nonauthonomous Hamilton's canonical equations with conjugate variables just the particle coordinates x, y. Hence, the theory of Hamiltonian systems can be applied. If $\psi = \psi(x,y)$ is independent of time, the system (9) has one degree of freedom and is integrable, so that chaotic solutions about $x(t)$, $y(t)$ are impossible - the stream lines are also particle trajectories. However, if $\psi(x,y,t)$ is nonlinear and nonauthonomous, (9) has two degrees of freedom and can be rewritten as 3-D NDS:

$$\dot{x} = u(x,y,z), \qquad \dot{y} = v(x,y,z), \qquad \dot{z} = w \equiv 1 \tag{10}$$

which is generally nonintegrable and can generate chaos. An important particular case of (10) is a periodic time flow

$$\psi(x,y,t) = \psi(x,y,t+T) \tag{11}$$

The Hamiltonian chaos, unlike the nonhamiltonian (dissipative) one, e.g. (5), is not confined in attractors and can coexist with areas of regular motion. In a poetic language this means that in the time periodic case, the fluid domain decomposes into "islands of order" immersed in a "chaotic sea". The particles moving according to (9) can neither penetrate nor leave these islands. They remain as "prisoners" in either of the two areas (chaotic and nonchaotic). Nonperiodic flows (e.g. 2-D turbulence) are free of such islands and complete mixing is observed.

THEORETICAL MODELS OF CA IN GEOPHYSICAL FLOWS

a) *CA in 3-D Steady Ocean Circulation.* In a recently published paper (Yang et al., 1997) a model of CA in a simple 3-D laminar, steady, basin-scale model of ocean flow consisting of the gyre and the thermohaline circulation mode is presented. The model flow resembles the *ABC* flow (4) and is confined in a box

$$0 \le x \le 2L, \quad -L \le y \le L, \quad -H \le z \le 0 \tag{12}$$

L=2000 km, H = 4 km, (x, y, z) denote eastward, northward and upward coordinate variables. The flow is constructed by two parts:

1. The long time mean horizontal double-gyre wind-driven ocean circulation - Figure 1a), with a stream function

$$\psi_w(x,y) = (x-2)(1-e^{-x/\delta_s})\sin \pi y \tag{13}$$

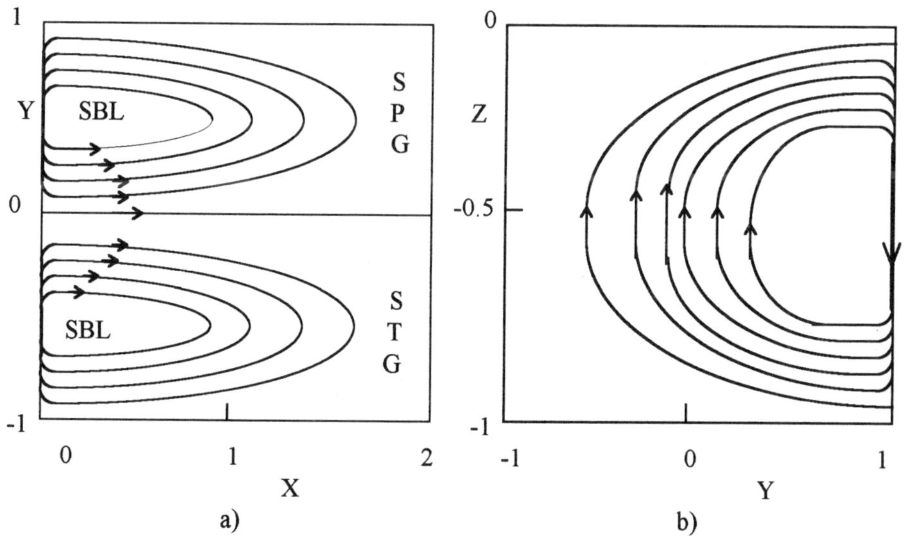

Figure 1. The structure of the model flow (13), (14) (sketched after Yang et al., (1997)): SPG - subpolar gyre, STG - subtropical gyre, SBL- Strommel boundary Layer.

2. The long-time mean meridional overturning buoyancy driven thermohaline cell - Figure1.b), with a stream function

$$\psi_\beta(y,z) = (y+1)(1-e^{-\frac{y-1}{\delta_\beta}}).\varepsilon \sin\frac{\pi z}{b} \tag{14}$$

$\varepsilon > 0$, $b = H / L$ - aspect ratio.

Then, the total 3-D long time averaged oceanic flow field is given by

$$(\mathrm{u}, \mathrm{v}, \mathrm{w}) = \left(-\frac{\partial \psi_w}{\partial y}, \frac{\partial \psi_w}{\partial x} + \frac{\partial \psi_\beta}{\partial z}, -\frac{\partial \psi_\beta}{\partial y}\right), \tag{15}$$

$$\partial \mathrm{u} / \partial x + \partial \mathrm{v} / \partial y + \partial \mathrm{w} / \partial z = 0 \tag{16}$$

and equations for the Lagrangian trajectories are

$$\dot{x} = \mathrm{u}(x,y,z), \qquad \dot{y} = \mathrm{v}(x,y,z), \qquad \dot{z} = \mathrm{w}(x,y,z) \tag{17}$$

In this model the flow defined by (13)-(16) is laminar, incompressible and in analytical form, so that the equations (17) are easy for numerical integration. According to (16), the system (17) is conservative, the ocean domain (here the phase space) can have both chaotic and nonchaotic regions (trajectories will not be attracted by a strange attractor). Actually, the numerical results presented in Yang et al.,(1997) show that the trajectories could be regular, "almost" chaotic or fully chaotic, depending on the area of starting. the Lyapunov exponents have been calculated to classify the trajectories in one of the categories and to quantify the degree of chaoticity. Several colour pictures (Yang et al., 1997) illustrates the author's results.

b) *CA in 2-D Unsteady Ocean Circulation.* In an earlier paper (Yang et al., 1994) the horizontal Lagrangian trajectories and the concentration of a passive tracer in a double gyre ocean as sketched in Fig.1a, driven by time-dependent wind, is studied. The stream function used in the model is constructed on the base of (13):

$$\psi(x,y,t) = (x - 2L)(1 - e^{-x/\delta_s}) . \frac{\partial \tau}{\partial y} \tag{18}$$

$$\tau = \frac{\tau_0}{\pi} \cos\left[\pi(y - y_0(t))\right], \qquad y_0(t) = a \sin \omega t \tag{19}$$

where τ is wind stress allowed to seasonally migrate according to (19). Because of these, the intergyre boundary line will also periodically migrate with respect to the climatological boundary y=0. The authors argue that in the North Atlantic Ocean this line corresponds to the Gulf Stream.

With $\psi(x,y,t)$ as defined above, the numerical solution of the equation $\dot{x} = -\psi_y$, $\dot{y} = \psi_x$ clearly demonstrates the chaotic character of the trajectories for different $a \neq 0$ and ω.

c) *CA in 2-D Wave Atmospheric Systems.* Many kinds of waves are observed both in the real atmosphere or in its theoretical models. Some of the "theoretical" waves are exact explicit solutions to the governing linearized equations and some of them even to the nonlinear ones. Since the flows under consideration are assumed 2-D and nondivergent, equations (9) will hold, where ψ is periodic in time (11). in Yang (1993, 1993a) the following expression for ψ is constructed

$$\psi(x,y,t) = A \sin(K_1(x - c_1 t)) \sin L_1 y + \eta \sin(K_2(x - c_2 t)) \sin L_2 y \tag{20}$$

The various parameters in (20) have clear meaning. This expression is intended to model any type of waves in geophysical fluids and particularly in the atmosphere, such as Rosby waves, gravity and tidal waves as well as combinations of them.

Suppose after Yang (1993, 1993a), that some amount θ of passive and conservative admixture is released into the model flow and diffusion is neglected. (θ could be potential temperature, concentration of long lived trace gas such as CO_2, N_2O, CH_4, O_3 and other pollutants). Then θ will be advected according to (1):

$$\frac{d\theta}{dt}=\frac{\partial\theta}{\partial t}+u\frac{\partial\theta}{\partial x}+v\frac{\partial\theta}{\partial y}=0 \tag{21}$$

where (u, v) and the particle coordinates $x(t)$, $y(t)$ are determined by (9) and (20). Fascinating colour pictures illustrating the phenomenon CA for various combinations of the parameters' values in (20) are published in Yang (1993, 1993a). It is very worth mentioning how a smooth, well behaved large-scale pattern can produce almost infinitely small-scale patterns.

d) *CA in Atmospheric Nonperiodic Flows*. In the previous three cases (a, b, c) the stream functions (13), (14) and (18)-(20) were constructed on the basis of some physical arguments. Another approach is $\psi(x,y,t)$ to be obtained as solution to some equation governing the motion in Eulerian formulation. In Ipatova et al. (1997) this is the so called barotropic vorticity equation of atmospheric dynamics

$$\frac{\partial}{\partial t}\nabla^2\psi+J(\psi,\nabla^2\psi)+\beta\frac{\partial\psi}{\partial x}+\sigma\nabla^2\psi-\nu\nabla^4\psi=f(x,y), \tag{22}$$

integrated in a squared area with L=1200 km, $\psi=\nabla^2\psi=0$ on the boundaries, f=$A\sin(2\pi y/L)$ and for some typical values of the parameters. Having determined $\psi(x, y, t)$ from (19), then under initial conditions (x_0, y_0) the equations $\dot{x}=-\psi_y$, $\dot{y}=\psi_x$ are integrated numerically. The results clearly show the random-like behaviour of $x(t)$, $y(t)$ and consequently the same about the trajectories $y = y(x)$ of individual particles marked by their initial coordinates (x_0, y_0).

Much more impressive in this direction are the results presented in Yang (1993a). The global winds (u, v) are taken from a 100-year run of a particular (GFDL-Princeton) general circulation model. Then the equations of motion on a sphere are solved

$$\dot{\lambda}=\frac{u(\lambda,\theta,t)}{a\cos\theta},\qquad \dot{\theta}=\frac{1}{a}v(\lambda,\theta,t) \tag{23}$$

Two numerical experiments have been conducted to simulate the difference between zonal and meridional mixing. Spectacular pictures in color illustrate the numerical results.

DOES CA REALLY EXIST IN GEOPHYSICAL FLOWS?

This question can be answered on the basis of real observations and correct interpretation in the framework of the existing theoretical schemes. During the past two decades several experiments had been designed and conducted in the atmosphere and ocean.

a) *Experiments in the Atmosphere*. The most direct observation on the dispersal of Lagrangian particles in the atmosphere comes from tracking high-altitude balloons. Two field experiments are known:

- The EOL-experiment in which a set of 483 balloons released from the ground in several clusters to fly at the 200 mb height was tracked. The results were analyzed in Morel (1970), Morel et al. (1974). The main finding was that during the first 5 days of the balloons' flight, their r.m.s. separation grew exponentially with *e*-folding time $\tau \approx 2.7$ days. This result can be interpreted as a signature of CA. According to Paldor et al. (1992), the main reason for such behaviour are the pressure planetary tidal wave fluctuations.

- The second source of similar data comes from the TWERL-Experiment (Twerle team, 1977). The dispersion rate is closed to that found in the EOL (Brown et al., 1991). It has been attributed mainly to the Rossby waves.

b) *Experiments in the Ocean*. Similarly to the atmospheric constant-level balloons, so called submerged floats have been used to collect data for Lagrangian trajectories.. A collection known as "SOFAR-trajectories" is analyzed in Brown et al. (1990, 1991). The main conclusion of the authors is that their analysis does not unambiguously identify SOFAR float trajectories as being chaotic, but it provides no evidence that they are not chaotic, i.e. inconclusive result.

CONCLUSION

Whether the CA is, or it is not a dominant factor in the real atmosphere and if it is, then when and where remains to be established by current and future studies. However, since the large-scale atmospheric motions are quasi-2D and quasi-nondivergent, one can still expect to see some portion of the spatial patterns being similar to those in the Hamiltonian systems. A practical implication of this will be that if a pollutant (e.g. radioactive particles) is released instantaneously at a point in a chaotic region, then there will have a mild pollution by fallout over a large area. Otherwise, if the pollutant is released at a nonchaotic area, then there will have intense pollution over a small area. An important implication is seen also in understanding of the earth's climate variability and particularly the role of the World ocean climate in this respect.

From a fundamental view point another important fact is that the equations (9) provide a visualization of a Hamiltonian system thereby providing an experimentally observable picture of phase space. This possibility had been first noted 100 years ago by J.W. Gibbs, who introduced in physics the notion of "phase space".

Until recently, the turbulent diffusion paradigm was leading in construction of transboundaries pollution models. The CA paradigm is another approach to the problem, which can supplement the first one.

ADDENDUM *

Lagrangian Trajectories in the Shallow Water Geophysical Flows with Uniform Horizontal Shear

A great number of theoretical models in the geophysical fluid dynamics are based on the so called shallow water equations (SWE) on the rotating Earth

* Some results from the author's current research with a collaborator.

$$U_t + UU_x + VU_y - fV = -H_x$$

$$V_t + UV_x + VV_y + fU = -H_y \qquad \text{(A.1)}$$

$$H_t + UH_x + VH_y + H(U_x + V_y) = 0$$

where f is the Coriolis parameter, $U_t = \partial U / \partial t$ etc. It is known from long ago (see Panchev, 1996 and Young, 1986 and the references there in) that the system (A,1) can be exactly converted into a system of ordinary differential equations by the substitutions

$$U(x, y, t) = U_1(t)x + U_2(t)y,$$

$$V(x, y, t) = V_1(t)x + V_2(t)y, \qquad \text{(A.2)}$$

$$H(x, y, t) = H_{00}(t) + \frac{1}{2}H_{11}(t)x^2 + \frac{1}{2}H_{22}(t)y^2 + H_{12}(t)xy.$$

Instead of U_i, V_i, H_{ik} more informative characteristics are introduced as follows:

$$D(t) = U_x + V_y = U_1(t) + V_2(t) \qquad \text{- divergence}$$

$$\zeta(t) = V_x - U_y = V_1(t) - U_2(t) \qquad \text{- vorticity}$$

$$F(t) = U_x - V_y = U_1(t) - V_2(t) \qquad \text{- distortion (deformation)} \qquad \text{(A.3)}$$

$$M(t) = V_x + U_y = V_1(t) + U_2(t) \qquad \text{- distortion (deformation)}$$

$$S(t) = H_{11}(t) + H_{22}(t) \qquad \text{- Laplacian}$$

$$R(t) = H_{11}(t) - H_{22}(t), \qquad Q(t) = 2H_{12}(t). \qquad \text{(A.4)}$$

The structure of the solution (A.2) is a generic property of the system (A.1). After substitution (A.2) into (A.1) one derives the following coupled systems of ODEs for seven unknown functions (A.3), (A.4):

$$\dot{\zeta} = -D(\zeta + f), \qquad \dot{F} = -DF + fM - R$$

$$\dot{M} = -DM - fF - Q, \qquad \dot{D} = f\zeta - S - \frac{1}{2}(D^2 + F^2 + M^2 - \zeta^2) \qquad \text{(A.5)}$$

and

$$\dot{S} = -2DS - FR - MQ, \quad \dot{R} = -2DR - FS - \zeta Q, \quad \dot{Q} = -2DQ - MS + \zeta R \tag{A.6}$$

They have to be solved under initial conditions $\zeta_0, F_0, \ldots, Q_0$ at $t = 0$. Having determined $\zeta, F, \ldots, Q$ as functions of t, then the coefficients U_i, V_i, H_{ik} in (A.2) can be found through (A.3), (A.4). Finally, the equations for the trajectories are to be solved

$$\dot{x} = U_1(t)x + U_2(t)y, \qquad \dot{y} = V_1(t)x + V_2(t)y. \tag{A.7}$$

Analytical solution to (A.5), (A.6) in the general case is impossible. However, if we consider a nondivergent flow

$$D(t) = U_1(t) + V_2(t) \equiv 0 \tag{A.8}$$

such solution is possible in some particular cases:

a) One ignores the H-equation in (A.1) and assumes $H=H(x, y)$ with $R=Q=0$ and $S\neq 0$, i.e. circular isolines of the H-field on the (x, y)-plane. It immediately follows from (A.5) that

$$\begin{aligned} &F(t) = r_0 \sin(f\,t + \delta_0), \qquad M(t) = r_0 \cos(f\,t + \delta_0), \\ &\zeta(t) = \zeta_0, \qquad r_0^2 = F_0^2 + M_0^2, \qquad tg\delta_0 = F_0 / M_0, \\ &\zeta_0^2 + 2f\zeta_0 - 2S = r_0^2. \end{aligned} \tag{A.9}$$

In the Northern hemisphere $f > 0$, but ζ_0 and S can be positive or negative.

b) The H-field is allowed to evolve in time (S, R, Q depend on t) and (A.6) holds. It can be shown that the simplified ($D = 0$) system (A.5), (A.6) admits five invariants of motion and those parts of the complete solution we need to insert into (A.7) are

$$\begin{aligned} &\zeta(t) = \zeta_0 = \text{const}, \qquad S(t) = S_0 = \text{const}, \\ &F(t) = r_0 \sin(\omega_0 t + \delta_0), \qquad M(t) = r_0 \cos(\omega_0 t + \delta_0) \end{aligned} \tag{A.10}$$

where

$$\omega_0 = f + \sqrt{\frac{Q_0^2 + R_0^2}{F_0^2 + M_0^2}} \tag{A.11}$$

and $\zeta_0^2 + 2f\zeta_0 - 2S_0 = r_0^2$. Therefore, in both cases $\zeta = \zeta_0$ while $F(t)$ and $M(t)$ are periodic functions with frequencies f and $\omega_0 > f$. The corresponding stream function

$$\psi(x, y, t) = \frac{1}{4}(M + \zeta)x^2 - \frac{1}{4}(M - \zeta)y^2 - \frac{1}{2}F(t)x\,y \tag{A.12}$$

is also periodic - (11). Hence, one can expect the system $\dot{x} = -\psi_y$, $\dot{y} = \psi_x$, i.e. (A.7), to have chaotic solutions. In the case of irrotational (potential) flow $\zeta = \zeta_0 = 0$ and (A.12) simplifies.

Let now consider a passive scalar advected by the flow (A.1), (A.2) according to (21). The latter equation admits solution of polynomial type

$$\theta(x,y,t) = \sum_{m,k} T^*_{mk}(t)\, x^m y^k \tag{A.13}$$

the simplest approximation being the quadratic one

$$\theta(x,y,t) = T_{00} + \frac{1}{2} T_{11} x^2 + \frac{1}{2} T_{22} y^2 + T_{12} x\, y \tag{A.14}$$

With this expression inserted into (21) together with (A.2), the following system similar to (A.6) can be derived (Spassova, 1992)

$$\dot{S}_T = -DS_T - F R_T - MQ_T$$

$$\dot{R}_T = -DR_T - F S_T - \zeta Q_T \tag{A.15}$$

$$\dot{Q}_T = -DQ_T - M S_T + \zeta R_T$$

where $S_T(t) = T_{11} + T_{22}$, $R_T(t) = T_{11} - T_{22}$, $Q_T(t) = 2\,T_{12}$ and D, F, M, ζ as determined before. For the previous two particular cases $D \equiv 0$, $\zeta = \zeta_0 =$ const and $F(t)$, $M(t)$ periodic functions, (A.15) represent a 3-D nonauthonomous dynamical system and is likely to have chaotic solutions. The principle difference between the general system (3) and (A.15) is that in the latter case S_T, R_T, Q_T are phase variables which characterize the local time evaluation of θ.

Solutions to the system (A.1) of the type (A.2) are known in hydrodynamics as Kirchoff vortex. Thus, we have shown here that the problem for Lagrangian trajectories and advection in the plane occupied by such vortex can be formulated and worked out by comparatively simple analytical and numerical tools. For specified vortices the above general results have to be adjusted accordingly.

ACKNOWLEDGEMENTS

This work was supported by the Sofia University Research Fund under No 258/97.

REFERENCES

Aref H., 1984, Stirring by chaotic advection, *J. Fluid Mech.* 143, 1-21

Aref H., Balachandar S., 1986, Chaotic advection in a Stokes flow, *Phys. Fluids*, 29(11), 3315-21

Aref H., Jones S.W., Mofina S., Zawadski I., 1989, Vortices, kinematics and chaos., *Physica* D37, 423-440

Aref H., 1991, Stochastic particle motion in laminar flows., *Phys. Fluids*, A3, 1009-1016

Brown M.G., Smith K.B., 1990, Are SOFAR float trajectories chaotic?, *J. Phys. Ocean*, 20, 1, 139-149

Brown M.G., Smith K.B., 1991, Ocean stirring and chaotic low-order dynamics, *Phys. Fluids*, A3(5), 1186-1192

Cox S.M., Drazin P.G., Ryrie S.C., Slater K., 1990, Chaotic advection of irrotational flows and of waves in fluids, *J. Fluid Mech.*, 214, 517-534

Er-El J., Peskin R. L., 1981, Relative diffusion of constant-level balloons in the Southern hemisphere., *J. Atm. Sci.*, 38, 2264-2274

Grisanti A., Falconi M., Vulpiani A., 1991, Lagrangian chaos: Transport, mixing and diffusion in fluids, *Rivista Del Nuovo Cimento*, 14, 12, 1-84

Hillborn R.C., 1994, *Chaos and Nonlinear Dynamics*, Oxford Univ. Press, pp.654

Ipatova V.M., Filatov A.N.,1997, Lagrangian and Eulerian chaos in atmospheric models *Izv. Acad. Nauk, Phys. Atm. Ocean*, 33, 4, 443-451

Lupini R., Siboni R., 1989, Mixing in laminar flows by chaotic advection, *Nouvo Cimento*, 103, 3, 237-257

Morel P., 1970, Large scale dispersion of constant-level balloons in the Southern general circulation, *Ann. Geophys.*, 26, 815-828

Morel P., Larcheveque M., 1974, Relative dispersion of constant - level balloons in 200- mb general circulation, *J. Atm. Sci.*, 31, 2189-2196

Ottino J.M., 1988, *The Kinematics of Mixing Stretching, Chaos and Transport*, Cambridge Univ Press

Paldor N., Boss E., 1992, Chaotic trajectories of tidally perturbed internal oscillations, *J. Atm. Sci.*, 49, 23, 936-948

Panchev S., 1996, *Theory of Chaos*, Bulgarian Acad. Press, pp. 382

Pierrehumbert R.T., 1991, Chaotic mixing of tracers and vorticity by modulated travelling Rossby waves, *Geophys. Astrophys. Fluid Dyn.*, 58, 285-319

Provenzale A., Babiano A., Villone B., 1995, Single - particle trajectories in two-dimensional turbulence, *Chaos, Solitons, Fractals*, 5, 10, 2055-2071

Spassova T., 1992, Temperature advection by gradient wind in the atmosphere, *Bulgarian J. Meteorol. and Hydrol.*, 3, 3, 136-140

Strogadz S.H., 1996, *Nonlinear Dynamics and Chaos*, Addison-Wesley Publ. Co, pp. 498

TWERLE Team, 1977, The TWERL - Experiment, *Bull. Amer. Meteor. Soc.*, 58, 936-948

Wiggins S., 1992, *Chaotic Transport in Dynamical Systems*, Springer-Verlag

Yang H., 1993, Dependence of Hamiltonian chaos on perturbation structure, *Int. J. Bifurcation and Chaos*, 3, 4, 1013-1028

Yang H., 1993a, Chaotic mixing and transport in wave systems and the atmosphere, *Int. J. Bifurcation and Chaos*, 3, 6, 1423-1445

Yang H., Liu Z., 1994, Chaotic transport in a double gyre ocean, *Geophys. Res. Letters*, 21, 7, 545-548

Yang H., Liu Z., 1997, The three - dimensional chaotic transport and the great ocean barrier, *J. Phys. Oceanography*, 27, 1258-1273

Young W.R., 1986, Elliptic vortices in shallow water, *J. Fluid Mech.*, 171, 101-119

DISCUSSION

D. KAASIK: Is it possible to use the chaotic advection scheme in real atmospheric modelling. What is your opinion?

S. PANCHEV: Yes, it is! This possibility follows from the fact that the large (synoptic) scale atmospheric motions are quasi-two-dimensional, quasi-nondivergent, i.e. quasi-Hamiltonian and very often periodically (wavy) modulated. It has been realised in a number of publications (see the reference to our review).

H. van DOP: In response to the question of M. Kaasik: In a special issue of Atmospheric Environment, J. Pudykiewicz (Pudykiewicz and Koziol, 1998: An Application of the Theory of Kinematics of Mixing to the Study of Tropospheric Dispersion, **32**, 4227-4244) presents an application of chaotic advection to the global dispersion of an inert tracer released in France, simulating a nuclear accident (ETEX, a European Tracer Experiment)

S. PANCHEV: This is a comment, not a question. I have not seen the paper you are talking about and I could not say anything about it.

G. DJOLOV: At present the modellers are using trajectories obtained numerically or experimentally without having knowledge if the weather systems are in chaotic regime. The trajectories are used to transport (advect) air masses with different concentration content. Would you tell us if the "stirring" process defined in the theory of chaos is identical or covers the transport and diffusion process used by the modellers today?

S. PANCHEV: The "stirring" process cannot be identical to the diffusion one used by modellers. It has much in common with the transport of air volumes along the trajectories, if the latter have not been much smoothed.

A NEW HIGH PERFORMANCE VERSION OF THE LAGRANGIAN PARTICLE DISPERSION MODEL SPRAY, SOME CASE STUDIES

G. Tinarelli,[1] D. Anfossi,[2] M. Bider,[3] E. Ferrero,[4] S. Trini Castelli [2]

[1]ENEL/SRI/Polo Ambiente, via Reggio Emilia 39, 20090 Segrate (MI), Italy
[2]CNR, Istituto di Cosmogeofisica, Corso Fiume 4, 10133 Torino, Italy
[3]Picodata S.R.L., via Alserio 22, 20100 Milano, Italy
[4]Dipartimento di Scienze e Tecnologie Avanzate, Corso Borsalino 54 Alessandria, Italy

INTRODUCTION

Lagrangian particle models are powerful tools to simulate the atmospheric dispersion of gaseous releases. Although having a quite complex mathematical basis (Thomson, 1987, Rodean, 1994), their practical implementation is generally simple and intuitive, allowing to easily take into account complex situations such as the presence of the topography or terrain inhomogeneities, low wind speeds, spatial and temporal variations of meteorological fields. In these models the atmospheric dispersion is simulated by the motion of fictitious particles splitted in a mean part due to the mean wind, and a stochastic fluctuation related to the statistical characteristics of the turbulent flow. It is quite clear that the model accuracy is strongly dependent on the number of emitted particles and the computer time often limits the kind of simulations that can be performed. For this reason, the earlier version of these models were mainly devoted to reproduce the dispersion of a limited number of emissions at local scale. The recent wide and rapid diffusion of very fast computational tools lead to the development of more sophisticated codes, able to take into account more general situations. SPRAY (Tinarelli et al., 1992) is a Lagrangian stochastic particle model designed to perform dispersion simulations in complex terrain. The version 1 of the code, based on a three dimensional form of the Langevin equation for the random velocity with coupled non-gaussian random forcing following Thomson (1984, T84 in the following) and subsequently improved (Tinarelli et al., 1992), was able to satisfactorily reproduce local to regional scale dispersion both over flat (Brusasca et al., 1989, 1992) and complex terrain (Brusasca et al., 1995, Nanni et al., 1996) taking into account the emission from single or multiple sources. The development of a better based theory (Thomson, 1987) and the further demand of more complex regional scale simulations able to cover longer periods with a variety of emissions of different kinds (i.e. main roads, industrial or urban area) called for a new version of the code. The new version 2 of SPRAY code contains some improvements regarding the

Air Pollution Modeling and Its Application XIII, edited by S.-E. Gryning and E. Batchvarova, Kluwer Academic/Plenum Publishers, 2000

theoretical approach, turbulence parametrizations and time response characteristics. In this paper we describe these new developments, comparing model performances with those of the previous version through simulations performed both in theoretical and real cases.

NEW DEVELOPMENTS

The version 2 of the code contains many new algorithms. They can be resumed into the following three main classes.

Langevin Equations and PDFs

A generalized form of the Ito's type Langevin equation has been added as new alternative option to the previous formulation. The equation for the vertical displacement z and for the vertical velocity w of each particle take the following form:

$$dz = w dt \tag{1}$$

$$dw = a(z,w)\, dt + \sqrt{B_0(z)dt}\, d\mu \tag{2}$$

in eq. 2 $d\mu$ is a normally distributed random variable, $B_0 = (C_0\varepsilon)/2$, being C_0 a constant whose value is not yet definitely established, ranging from 2 to 7 (Rodean, 1994), ε the ensemble average dissipation rate of the turbulent kinetic energy, and $a(z,w)$ is a function depending on the form of the atmospheric Eulerian PDF $P(z,w)$ of the turbulent velocity. Following Thomson (1987), the form of $a(z,w)$ can be deduced imposing the so called 'well mixed condition' leading to a solution of the following Fokker-Planck equation for stationary conditions

$$w\frac{\partial P}{\partial z} = -\frac{\partial aP}{\partial w} + B_0\frac{\partial^2 P}{\partial w^2} \tag{3}$$

To deal with non-uniform or convective turbulence, the PDF of the vertical motion can be non-Gaussian. In this case, two different approaches have been adopted and are both present in the model, in order to give a solution of the Fokker-Planck equation with different degrees of approximation and different time responses. The first one, already presented by different authors (Baerentsen and Berkowicz, 1984, Luhar and Britter, 1989) defines P as a linear combination of two Gaussian PDFs (BG in the following)

$$P(z,w) = \alpha P_u(m_u,\sigma_u) + \beta P_d(m_d,\sigma_d) \tag{4}$$

where α and β are the weights of the two Gaussian distributions P_u and P_d whose mean and standard deviations are respectively m_u, σ_u and m_d, σ_d. The six unknown parameters in (4) are determined as a function of the first four moments of $P(z,w)$ defined as follows

$$\overline{w^n} = \int w^n P(z,w)dw \qquad n=0,1,2,3 \tag{5}$$

Substituting (4) into (3) it is possible to obtain an analytical form of $a(z,w)$ (Luhar and Britter, 1989). The second method, proposed by Ferrero and Anfossi (1998) consists in the adoption of a Gram-Charlier expansion, truncated as option to the third (GC3) or fourth (GC4) order, to approximate the PDF form with given moments (Kendall and Stuart, 1977)

$$P(x,z)=\frac{e^{-x^2/2}}{2\pi}(1+C_3H_3+C_4H_4) \tag{6}$$

where H_3 and H_4 are Hermite polynomials and C_3 and C_4 are their coefficients, whose expressions are

$$H_3 = x^3 - 3x \tag{7}$$

$$H_4 = x^4 - 6x^2 + 3 \tag{8}$$

$$C_3 = \overline{\mu^3}/6 \tag{9}$$

$$C_4 = (\overline{\mu^4} - 3)/24 \tag{10}$$

where $x = w/\sigma_w$, $\overline{\mu^3}$ and $\overline{\mu^4}$ are the standardized moments of w. Substituting (6) into (3) the following expression for the coefficient $a(z,w)$ is found

$$a=\sigma_w\frac{\frac{1}{\tau}(T_1)+\frac{\partial\sigma_w}{\partial z}(T_2)}{T_3} \tag{11}$$

where $\tau = \sigma_w^2/B_0$ is the Lagrangian decorrelation time scale and

$$T_1 = -3C_3 - x(15C_4+1) + 6C_3x^2 + 10C_4x^3 + C_3x^4 - C_4x^5 \tag{12}$$

$$T_2 = 1 - C_4 + x^2(1+C_4) - 2C_3x^3 - 5C_4x^4 + C_3x^5 + C_4x^6 \tag{13}$$

$$T_3 = 1 + 3C_4 - 3C_3x - 6C_4x^2 + C_3x^3 + C_4x^4 \tag{14}$$

Variable time step discretization

In the version 1 of the code, the T84 Lagrangian equations corresponding to (1) and (2) are numerically integrated by means of a discrete constant time step Δt. Its value is prescribed setting

$$\Delta t = \frac{\tau_{min}}{c} \tag{15}$$

where τ_{min} is the minimum value of τ_i, i=x,y,z which represent the horizontal and vertical Lagrangian decorrelation time scales. The empirical coefficient c is generally set equal or greater than 10. Obviously, the larger is c and the lesser are the integration errors due to the discretization. On the other hand, the lesser is c and the greater is the computational time and the value 10 for c represents a good compromise between the two needs. Looking at Fig.1 showing a typical CBL τ_z profile computed according to the Weil (1990) parametrization, it is evident that, far from the domain boundaries, τ_z is much larger than τ_{min}. For the particles lying into a large portion of the domain, the restriction imposed in (15) leads to a number of time steps significantly superior to what strictly necessary to obtain an equally correct simulation.

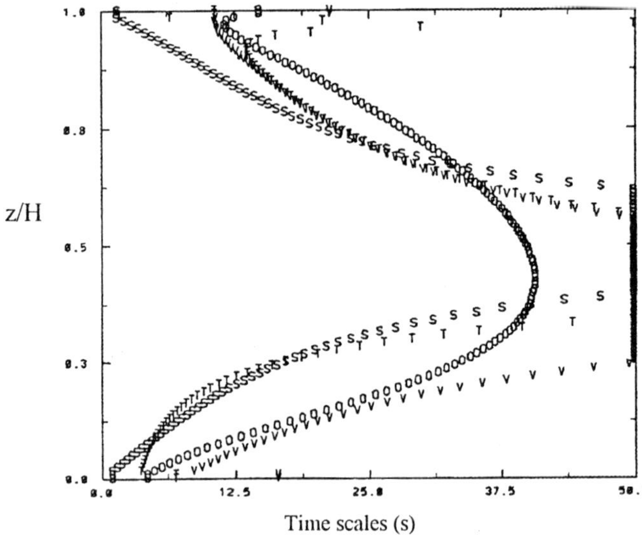

Fig. 1 Vertical profiles of τ_z (0 symbols), τ_w (V), τ_s (S) and $\tau_{\tau z}$ (T) as a function of z/H in a CBL of height H.

It is worth noting that the Lagrangian decorrelation time scale cannot be the only limiting parameter to take into account. Even into the region where the condition $\tau_i >> \tau_{min}$ take place, Δt cannot become too big if strong gradients of the turbulent characteristics are present. Because of these inhomogeneities, particularly evident in the vertical direction, excessive long paths performed by the particles in a single time step could bring them into regions characterized by completely different turbulent properties. This would tend to distribute the particles in a non-uniform way thus violating the well-mixed condition. To solve this problem, three further "vertical inhomogeneity time scales" are taken into account

$$\tau_w(x,y,z) = \left(\frac{\sigma_w}{\overline{w^2}} \frac{\partial \overline{w^2}}{\partial z} \right)^{-1} \tag{16}$$

$$\tau_s(x,y,z) = \left(\frac{\sigma_w}{\overline{w^3}} \frac{\partial \overline{w^3}}{\partial z} \right)^{-1} \tag{17}$$

$$\tau_{\tau_z}(x,y,z) = \left(\frac{\sigma_w}{\tau_z} \frac{\partial \tau_z}{\partial z} \right)^{-1} \tag{18}$$

where τ_w, τ_s, and $\tau_{\tau z}$ represent the inhomogeneity time scales due to the vertical variations of variance, skewness and Lagrangian decorrelation time of the vertical velocity fluctuations respectively. The first two time scale have been proposed by Wilson and Flesch (1993) whereas the third one is proposed in this work. The model selects the time step Δt as follow

$$\Delta t = \min\left(\frac{\tau_x, \tau_y, \tau_z, \tau_w, \tau_s, \tau_{\tau z}}{c_{10}} \right) \tag{19}$$

Where c_{10} is a constant of the order 10 given by the user. Vertical profiles of τ_z, τ_w, τ_s, $\tau_{\tau z}$ are represented in Fig.1 for the Weil's CBL parametrization. It is evident that, particularly close to the boundaries, many intersections between the different curves occur. τ_s and $\tau_{\tau z}$ are also consistently greater than τ_z inside non- negligible vertical layers.

Pre-processing codes

Two meteorological pre-processors are now present in SPRAY model in order to give the turbulence variables needed by the code. The first one, already present in version 1, generates vertical profiles of turbulent variables on the basis of the wind fields coming from a diagnostic mass consistent tool, land-use characteristics and surface meteorological data. In addition, a new pre-processor is operational to interface SPRAY with the mesoscale prognostic code RAMS (Pielke at al. 1992). This new code, named MIRS (Method for Interfacing Rams and Spray) and already presented in a preliminar version (Trini Castelli and Anfossi, 1997) allows SPRAY to take as input the turbulence fields generated by the different closure schemes present in the 3b version of RAMS. In particular, TKE fields coming from the Mellor-Yamada 2.5 scheme can be used to estimate both the PBL height and σ_w profiles. Many optional turbulence parametrization models are implemented in MIRS to define all the input parameters to SPRAY code not directly given by RAMS, such as the Lagrangian decorrelation time scales and the skewness of the vertical velocity PDF.

CASE STUDIES

Some case studies have been performed in order to test the new code and compare it against the previous version. In particular, laboratory and real field simulations have been done, to check both the model consistency in controlled conditions and the performances in a real complex scenario. As regards the first case, the well-mixed condition has been tested in the convective boundary layer parametrized as in Fig. 1.

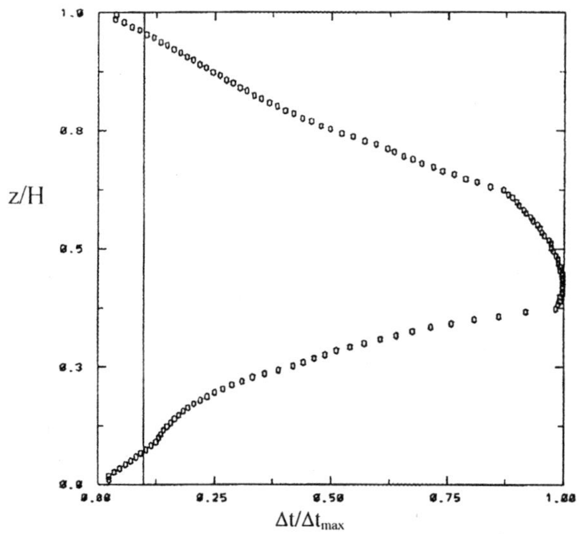

Fig.2 $\Delta t/\Delta t_{max}$ ratio as a function of z/H in the CBL indicated in Fig. 1 (O symbols). The continuous vertical line represents the $\Delta t/\Delta t_{max}$ ratio corresponding to a fixed Δt with c=10.

Fig. 2 shows the comparison between the variable Δt (normalized with the maximum value) determined by the model using the limiting rule indicated in (19) and the fixed time step following (15) with c=10, at each vertical level in the simulated CBL. The variable time step method allows an improvement into almost all the vertical domain but the two layers close to the boundaries, where vertical gradients are stronger. The actual gain depends on the vertical distribution of particles during the simulation, but an overall better time response could be expected in these conditions. The BG and GC schemes showed to be able to satisfactorily mantain well mixed an initially uniform vertical particle distribution (not shown here) both using a fixed and a variable Δt. Table 1 illustrates the speedup factor with respect to the previous T84 formulation obtained in this test case using the different models and the values of 10 and 20 for the c_{10} constant (the c constant of the fixed time step scheme was always set to 10).

Table 1 Speedup factors of the new schemes with respect to the T84 version

Δt scheme	PDF scheme	c_{10}	speedup factor
constant	T84	-	1
constant	BG	-	0.93
constant	GC4	-	1.17
variable	BG	10	2.95
variable	BG	20	1.76
variable	GC4	10	3.72
variable	GC4	20	3.00

The T84 method shows to be slightly more efficient than the BG method, but its capability to maintain the well mixed condition is only approximate. The GC method (only the GC4 results are shown in Table 1) is more efficient than the BG one, as one could expect due to its lesser mathematical complexity. In general, the GC method permits a limited speedup factor, of the order of about 20%, whereas larger factors can be obtainable using variable time steps. In this last case, better physical results have been obtained using c_{10}=20 but a reasonable performance is still reachable using c_{10}=5.

The model has been applied in a more complex situation, in order to verify the capabilities offered by the new algorithms in a real scenario. Simulations have been performed in the Valle D'Aosta regional area (Manzi et al., 1998), over a 100x80 km^2 computational domain located in Italy, adjacent to the northwestern boundary. This is a very complex topographical site where the higher summits of the Alps (more than 4400m a.s.l.) are present together with the bottom of the main valley (about 500m a.s.l.). Mean wind three dimensional fields have been reconstructed using a mass-consistent model and then used as input to the dispersion code. About 1000 emitting sources of different type have been considered, to simulate a complex network of heating systems, industrial and urban areas and main roads. Fig. 3 shows an horizontal view of the domain, with an example of the particle horizontal distribution in a typical summer diurnal condition. Two periods have been simulated, the first one lasting 11 days during summer 1996 and the second one lasting 10 days during winter of 1996. This choice was made in order to extrapolate from the obtained daily mean concentration fields some climatological aspects of the regional impact due to the considered emissions. These severe run were performed on a DEC-Alpha 600 S/333 Workstation with 256 Mbytes of central memory using the version 1 of SPRAY code. Huge peaks of the CPU time demanding were observed during the simulation and the total 21 days run took about 8 days to be completed. A 5 s fixed time step was used to disperse a mean quantity of 40000 particles with a peak number of about 85000 particles during some low wind nocturnal periods. The comparison with daily and climatological average concentrations collected by a regional network of samplers demonstrated a rather satisfactory performance of the model, but a better time response is required to realize longer simulations of this type enhancing the climatological significance of the results.

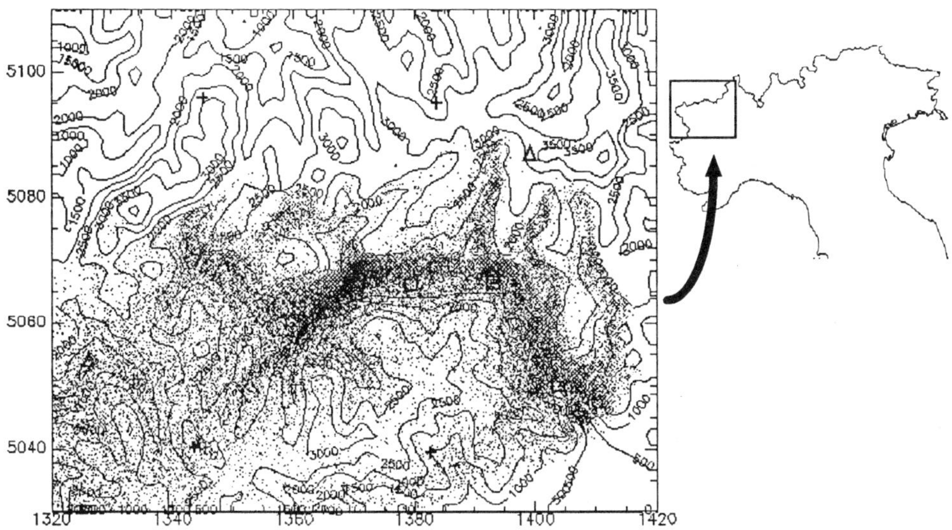

Fig.3 Computational domain used for the dispersion simulations of the Valle D'Aosta regional area (on the right side, the position with respect to the north Italian region is indicated). Contour labels indicate topography levels in meters. About 40000 particles are represented.

For this reason, some comparative simulations have been performed with the version 2 of the code. Preliminar results shows that the speedup in these conditions is, on the average, about 1.7 using variable time steps with $c_{10}=5$ and the BG scheme. The overall quality of the simulation results remains the same even if some differencies during peak episodes took place with respect to the T84 version. The presence of a large number of emitting points close to the ground probably does not permit the same performances listed in Table 1 due to strong gradients limiting the time step magnitude for a great number of particles. Better results can be obtained using the GC method, therefore allowing the realization of longer simulations, of the order of one month, in a reasonable time.

CONCLUSIONS

New algorithms has been added to the version 1 of the Lagrangian Particles dispersion model SPRAY in order to get better performances. These improvements involve the mathematical basis of the Langevin stochastic differential equations, the insertion of variable time stepping and the connection with a prognostic code to get input parameters. Some of these developments has been checked through controlled simulations, demonstrating an enhanced efficiency of this new version of the code and improving the physical correctness with respect to the previous one. Severe simulations in real field have also been performed and compared with those of the previous version, obtaining sensible speedup factors and showing good results, even if better performances can be expected in more simple situations such as the presence of a limited number of elevated or buoyant emissions. This new version 2 of the SPRAY code should now be able to deal with problems requiring a large amount of CPU time, such as climatological reconstructions or real time applications, with a satisfactory degree of correctness in more than acceptable response times.

NOTICE

SPRAY is a commercial code jointly marketed by ENEL SpA/PAM, Via Reggio Emilia 39, 20090 Segrate (Milano, Italy) and Aria Technologies, le Charlebourg, 14/30 rue de Mantes, 92700 Colombes (France).

REFERENCES

Baerentsen J.H. and Berkowicz R., Monte-Carlo simulation of plume diffusion in the convective boundary layer, *Atmospheric Environment*, 18, 701-712 (1984).

Brusasca G., Tinarelli G., Anfossi D., Particle model simulation of diffusion in low windspeed stable conditions, *Atmospheric Environment* 26, 707-723 (1989).

Brusasca G., Tinarelli G., Anfossi D., Comparison between the results of a Monte Carlo atmospheric diffusion model and tracer experiments, *Atmospheric Environment* 23, 1263-1280 (1992).

Brusasca G., Ferrero E., Anfossi D., Desiato F., Tinarelli G., Morselli M.G., Finardi S., Sacchetti D., Intercomparison of 3-D flow and particle models with Transalp 1989 meteorological and tracer data", Proc. of the 21st CCMS-NATO meeting, Baltimore, 6-10 November, 1995, 386-394.

Ferrero E. and Anfossi D., Sensitivity analysis of Lagrangian Stochastic models for CBL with different PDF's and turbulence parameterizations. Air Pollution Modelling and its Applications XI, S.E. Gryning and N. Chaumerliac eds., Plenum Press, New York, 22, in press.

Kendall M. and Stuart A., The advanced theory of statistics, MacMillan, New York (1977).

Luhar A.K., and Britter R.E., A random walk model for dispersion in inhomogeneous turbulence in a convective boundary layer, *Atmospheric Environment*, 23, 1191-1924 (1989).

Manzi G., Brusasca G., Morselli M.G., Tinarelli G., Indagine generale per lo studio pluridisciplinare del deperimento del pino silvestre ed altre specie botaniche in Val D'Aosta, simulazione della dispersione in atmosfera degli inquinanti emessi dal traffico veicolare e dal riscaldamento, ENEL/CRAM report 1998-0022 (1998).

Nanni A., Riva M., Tinarelli G., Brusasca G., Particle model simulation of pollutants dispersion from a line source in complex terrain, *The Science of the Total Environment*, 189-190, 301-309 (1996)

Rodean H.C., Notes on the Langevin model for turbulent diffusion of "marked" particles, UCRL-ID-115869 Report of Lawrence Livermore National Laboratory (1994).

Thomson D.J., Random walk modelling of diffusion in inhomogeneous turbulence. *Quart. J. Roy. Meteor. Soc.*, 110, 1107-1120 (1984).

Thomson D.J., Criteria for the selection of stochastic models of particle trajectories in turbulent flows. *J. Fluid Mech*, 180, 529-556 (1987).

Tinarelli G., Anfossi D., Brusasca G., Ferrero E., Giostra U., Morselli M.G., Moussafir J. Tampieri F., Trombetti F., Lagrangian particle simulation of tracer dispersion in the lee of a schematic two-dimensional hill, *Journal of Applied Meteorology*, 33, 744-756 (1994).

Trini Castelli S., Anfossi D., Intercomparison of 3D turbulence parametrizations for dispersion models in complex terrain derived from a circulation model, Il Nuovo Cimento C, 20, 287-313 (1997).

Weil J.C., A diagnosis of the asymmetry in top-down and bottom-up diffusion using a Lagrangian stochastic model, *J. Atmos. Sci*, 47, 501-515 (1990).

Wilson J.D. and Flesch T.K., Flow boundaries in random-flight dispersion models: enforcing the well-mixed condition., *Journal of Applied Meteorology*, 32, 1695-1707 (1993).

DISCUSSION

R. YAMARTINO: How many particles do you need to use in your applications and do you use point particles or kernel estimators?

G. TINARELLI: The peak number of particles moved in the 'Valle d'Aosta' simulation in a single time step was about 150000. We did not use kernel estimators to compute concentrations but we used instead boxes. This is the main reason for which we considered such a large number of particles.

A. STOHL: It is possible to pre-calculate a large set of random numbers and then randomly select subset out of this. Although not strictly correct, it works in practice and speeds up the computations considerably.

G. TINARELLI: We will try to experiment this method in the near future.

A METHOD TO SPEED UP A LAGRANGIAN STOCHASTIC PARTICLE DISPERSION MODEL

Mathias W. Rotach and Stefan Schwere

Swiss Federal Institute of Technology, Zürich, Switzerland

INTRODUCTION

Lagrangian stochastic particle dispersion models are widely accepted to be the most powerful tools for the simulation of passive tracers in the planetary boundary layer (Wilson and Sawford, 1996). However, one of their major disadvantages is their enormous requirement of computing time (cpu). This is mainly due to the large number of particle required in order to obtain a reliable estimate of the concentration distribution in combination with the sometimes (i.e., for certain conditions of stability) small time steps that is mandatory in order to keep the simulated acceleration of the particles within certain limits (see, e.g., Thomson, 1987 or Rotach et al., 1996 for time step criteria). In a theoretical estimate van Dop (1992) found $O(10^4)$ particles to be necessary for a 'good' accuracy of the resulting concentration fields if a *perfect* random number generator is used and correspondingly more for a 'real' random number generator. Consequently, de Haan (1998) uses $5 \cdot 10^5$ particles as a 'true reference' when trying to assess the effect of reducing the number of particles on the results. Nevertheless, the number of particles actually used in particle-model simulations is generally on the order of 10^4 particles, i.e. about an order of magnitude less than actually required. And this is simply due to the wish to minimize the 'computing costs' for each simulation.

If the number of required particles is $O(10^5)$ and the number of required time steps for a particular situation is, say, $O(10^3)$ and if we further assume the number of operations necessary to compute the acceleration of one particle to be $O(10^2)$, what is probably a conservative assumption for most models, this means that the total number of operations required for 'one run' will be $O(10^{10})$ only to determine the accelerations (no I/O statements and not counted the determination of the random numbers). It follows that any procedure to reduce the number of required operations can effectively contribute to reduce the required cpu time. The most effective approach will certainly be to reduce the number of particles. de Haan (1998) has shown that when using a kernel method to determine the concentration fields rather than employing the widely used 'box counting' method, the number of required particles is reduced by about two orders of magnitude. In the present contribution a method is presented to further reduce the number of operations for the determination of the acceleration of each particle during each time step, resulting in a speed-up of about a factor of eight. In Section 2, the necessary background for he particle models is briefly restated and the model under consideration is introduced. The actual speed-up procedure is described in Section 3, while in Section 4 the results are presented concerning the ability of the approximate model to reproduce the characteristics of the full model.

LAGRANGIAN STOCHASTIC PARTICLE MODELS

In Lagrangian stochastic particle dispersion models the evolution of the particle's velocities (u_i) and positions (x_i) are generally assumed to be Markovian and hence

$$du_i = a_i(\mathbf{x},\mathbf{u},t)dt + b_{ij}(\mathbf{x},\mathbf{u},t)d\zeta_j$$
$$dx_i = u_i dt \qquad (1)$$

where dt is the time step, $d\zeta_j$ are the increments of a Wiener process with zero mean and variance dt and the a_i and b_{ij} are functions, which have to obey certain restrictions. Thomson (1987) has shown that the most general of these restrictions is the 'well-mixed condition', which in short can be invoked by requiring that $P=P_a$ obeys the Fokker-Planck equation:

$$\frac{\partial P}{\partial t} = -\frac{\partial}{\partial x_i}(u_i P) - \frac{\partial}{\partial u_i}(a_i P) + \frac{\partial^2}{\partial u_i \partial u_j}(B_{ij}P) \qquad (2)$$

Here, P is the probability density function (pdf) of the particles and P_a is that of the air (i.e. fluid elements). Furthermore, in (2) and in the following we use small letters to denote the fluctuating wind speed components $u_i = U_i - \overline{u}_i$, where the capital letters refer to the actual total wind speed components and the overbar denotes an ensemble average. From (2) and with the assumption of stationarity the functions a_i can be expressed as

$$a_i = \frac{1}{P_a}\left[\frac{\partial}{\partial u_i}(B_{ij}P_a) + \Phi_i\right] \qquad (3)$$

with

$$\frac{\partial \Phi_i}{\partial u_i} = -\frac{\partial}{\partial x_i}(u_i P_a) \qquad (4)$$

and the condition $\Phi_i \to 0$ for $|u| \to \infty$. In (3), $B_{ij} = 1/2\, b_{ik} b_{jk}$ is simply an abbreviation and the Einstein summation convention applies. The random contribution to the accelerations is often modeled based on an approach which takes into account the Lagrangian velocity structure function in the inertial subrange, yielding $B_{ij} = \delta_{ij}\, 1/2\, C_o \varepsilon =: B_*$, where C_o is a universal constant and ε is the dissipation rate of turbulent kinetic energy (Thomson, 1987). This formulation for B_{ij} is convenient since it does not depend on the particle velocities. However, (3) reveals that it is sufficient to specify $P=P_a$ for a certain situation (e.g., scaling regime) to be modeled in order to derive the appropriate particle model (if it exists).

Thomsons model for Gaussian inhomogeneous turbulence

For Gaussian inhomogeneous turbulence the pdf P_a can be written as

$$P_a = \frac{1}{(2\pi)^{3/2}(\det V)^{1/2}} \exp\left\{-\frac{1}{2}u_i V_{ij}^{-1} u_j\right\} \tag{5}$$

where V_{ij} is the Reynolds stress tensor and V_{ij}^{-1} its inverse. For this pdf Thomson (1987) derived what he calls the 'simplest solution' of (3) [note that in more than one dimension there is no unequivocal solution of (3), see, e.g. Sawford and Guest, 1988] to yield the functions a_i:

$$\begin{aligned} a_i = -B_{ij}V_{jk}^{-1}u_k + \frac{1}{2}\frac{\partial V_{il}}{\partial x_l} + \frac{\partial \bar{u}_i}{\partial t} + \bar{u}_l \frac{\partial \bar{u}_i}{\partial x_l} \\ + \left[\frac{1}{2}V_{lj}^{-1}\left(\frac{\partial V_{il}}{\partial t} + \bar{u}_m \frac{\partial V_{il}}{\partial x_m}\right) + \frac{\partial \bar{u}_i}{\partial x_j}\right]u_j \\ + \frac{1}{2}V_{lj}^{-1}\frac{\partial V_{il}}{\partial x_k}u_j u_k \end{aligned} \tag{6}$$

Here, we consider an idealized situation of stationary, horizontally homogeneous turbulence with vanishing correlation between the lateral (u_2) and the vertical (u_3) velocity components and we get for Φ_i/P_a in eq. (3) the following relations:

$$\begin{aligned} \frac{\Phi_1}{P_a} = \frac{1}{2}\frac{\partial V_{13}}{\partial z} - B_* V_{11}^{-1}u_1 + \left[\frac{\partial \bar{u}_1}{\partial z} - B_* V_{13}^{-1}\right]u_3 \\ + \left[\frac{1}{2}V_{11}^{-1}\frac{\partial V_{11}}{\partial z} + \frac{1}{2}V_{31}^{-1}\frac{\partial V_{13}}{\partial z}\right]u_1 u_3 \\ + \left[\frac{1}{2}V_{13}^{-1}\frac{\partial V_{11}}{\partial z} + \frac{1}{2}V_{33}^{-1}\frac{\partial V_{13}}{\partial z}\right]u_3 u_3 \end{aligned} \tag{7a}$$

$$\frac{\Phi_2}{P_a} = -B_* V_{22}^{-1}u_2 + \frac{1}{2}V_{22}^{-1}\frac{\partial V_{22}}{\partial z}u_2 u_3 \tag{7b}$$

$$\begin{aligned} \frac{\Phi_1}{P_a} = \frac{1}{2}\frac{\partial V_{33}}{\partial z} - B_* V_{33}^{-1}u_1 - B_* V_{33}^{-1}u_3 \\ + \left[\frac{1}{2}V_{11}^{-1}\frac{\partial V_{31}}{\partial z} + \frac{1}{2}V_{31}^{-1}\frac{\partial V_{33}}{\partial z}\right]u_1 u_3 \\ + \left[\frac{1}{2}V_{13}^{-1}\frac{\partial V_{31}}{\partial z} + \frac{1}{2}V_{33}^{-1}\frac{\partial V_{33}}{\partial z}\right]u_3 u_3 \end{aligned} \tag{7c}$$

THE SPEED-UP PROCEDURE

Looking at (6) or (7) it becomes evident that the solution for the functions a_i is composed of a sum of products of certain functions of height only, $f_{ij}(z)$ and the instantaneous fluctuating velocity components, u_i. Specifically, under the assumptions leading to (7), we have

$$a_1 = f_{11}(z) + f_{12}(z)u_1 + f_{13}(z)u_3 + f_{14}(z)u_1u_3 + f_{15}(z)u_3^2 \tag{8a}$$

$$a_2 = f_{21}(z)u_2 + f_{22}(z)u_2u_3 \tag{8b}$$

$$a_3 = f_{31}(z) + f_{32}(z)u_1 + f_{33}(z)u_3 + f_{34}(z)u_1u_3 + f_{35}(z)u_3^2 \tag{8c}$$

In a 'full' particle model simulation, the functions f_{ij} are evaluated for each particle and time step (i.e. height) separately, leading to a large number of required operations. Much computing time can therefore be saved if the functions f_{ij} (z) are evaluated 'once for all' prior to the actual simulation, with the result that only a limited number of multiplications and additions is required to finally determine the a_i for each particle and time step. In principle, the functions f_{ij} can be pre-determined to any desired accuracy using, e.g. spline interpolation procedures. Here, we consider the much simpler (and more effective in terms of cpu time savings) approach of determining the f_{ij} at fixed heights with a given, pre-selected spacing of Δz. If, for example, Δz=1m is chosen, the actual particle height will directly yield the appropriate array element (index), where the required functional values of the f_{ij} can be obtained from during the simulation.

RESULTS

In this section the results from the original model are compared to those of its ‘speeded up version’ in order to see what the effect of this speed-up procedure is on the cpu consumption on the one hand and on the accuracy of the resulting concentration distributions on the other hand. The simulations will be denoted ‘full model’ and ‘approximate model’, respectively. As an example, one of the tracer release experiments from Copenhagen (Gryning and Lyck, 1984) is chosen - simply to assign reasonable input variables (stability, mixing height, etc.) for the simulations. A total number of 5000 particles is released, i.e. the minimum required number as recommended by de Haan (1998) if the kernel method is used to determine the concentration fields. From Table I, it can be seen that the approximate model is roughly eight times faster in determining the accelerations if a height interval of Δz=1m is used to pre-calculate the functions f_{ij} prior to the actual simulation. For the approximate model, the random number generator now spends two thirds of the total cpu time, indicating that a further speed up of a Lagrangian particle dispersion model will have to deal with speeding up the random number ‘production’ (note, that in order to demonstrate the improvement due to the approximate model, one of the quickest random number generators according to Press et al., 1992 was used). All together, i.e. for a complete model run including the calculation of the random numbers and the random

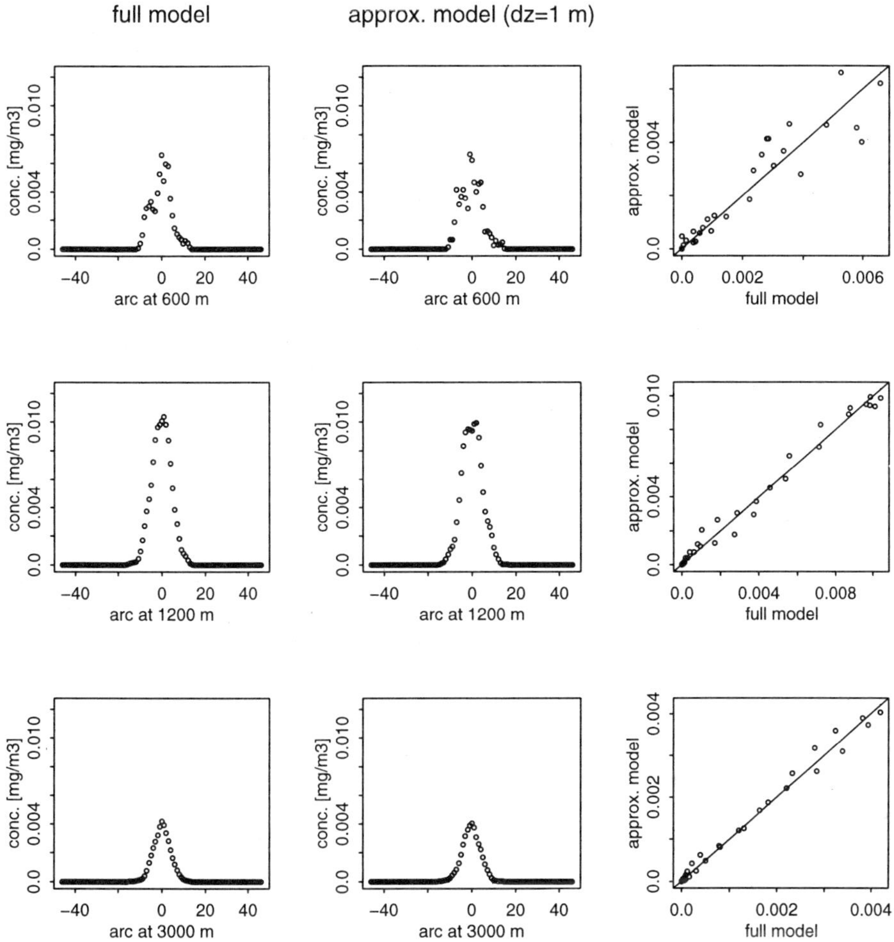

Figure 1 Concentration distributions on three arcs (at the indicated distance from the source) for the full model (left panels) and the approximate model (middle panels). In the right panels, a scatter plot of the full model and approximate model results is shown.

Table I cpu time requirements for the simulation of 5000 particles with models according to eqs. 3/7 ('full model') or its approximate version according to eq. 8. The cpu requirements for calculating the accelerations and those for picking the random numbers are given separately.

		cpu time [s]			
		1. timestep	10^1 timesteps	10^2 timesteps	10^3 timesteps
full model	total	0.070	0.540	5.080	50.520
	accelerations	0.047	0.411	4.083	40.702
	random numbers	0.010	0.097	0.962	9.705
approx. model (Δz=1m)	total	0.040	0.190	1.500	14.750
	accelerations	0.006	0.052	0.500	5.110
	random numbers	0.010	0.097	0.959	9.660

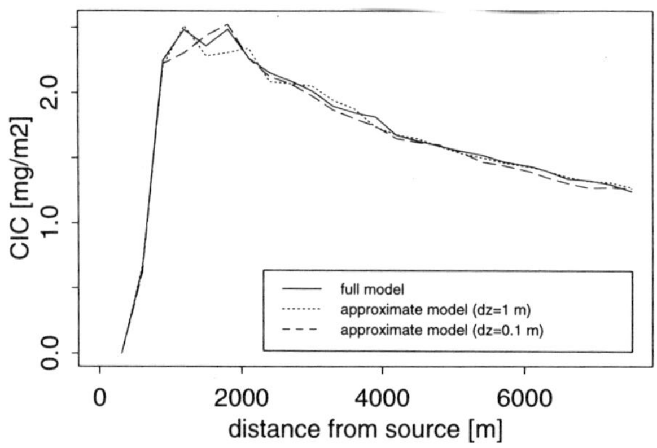

Figure 2 CIC as a function of distance from the source for the full model (solid line), the approximate model with Δz=1m (dotted line) and the approximate model with Δz=0.1m (dashed line) for one of the Copenhagen tracer release experiments (Nov. 9, 1978).

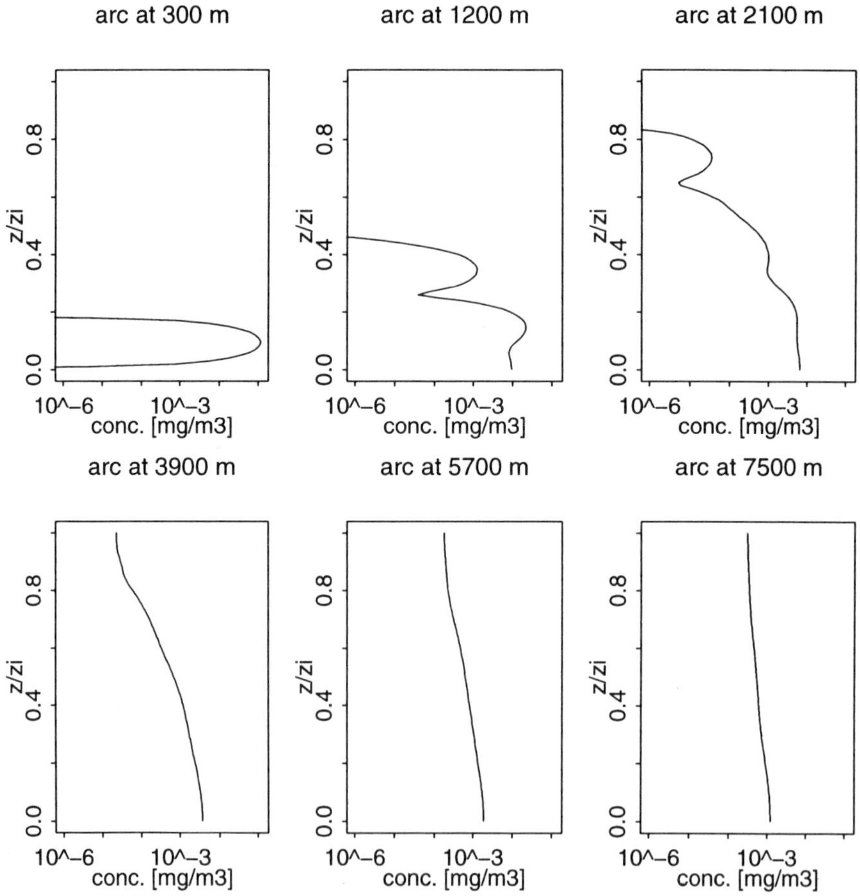

Figure 3 Concentration profiles at various distances downwind of the source for the simulation of Fig. 2.

Table II Statistical comparison of the full model to the approximate model. Presented arc the Normalized Mean Square Error (NMSE), the Fractional Bias (FB) and the correlation coefficient (cor) for the cross-wind integrated concentration (CIC), the arcwise maximum (ARCMAX) and the lateral spread (σ_y). Example for the data set from Copenhagen, exp. No 5. Evaluated are N=25 arcs.

		NMSE	FB	cor
CIC	Δz=1m	0.00094	0.0050	0.956
	Δz=0.1m	0.00095	0.016	0.957
ARCMAX	Δz=1m	0.0035	-0.011	0.957
	Δz=0.1m	0.0100	-0.007	0.955
σ_y	Δz=1m	0.00068	0.0135	0.960
	Δz=0.1m	0.00073	0.0060	0.959

number generation, the effective speed up of the approximate model is about a factor of three. Table I also reveals that the cpu time required to pre-calculate the f_{ij} is essentially negligible as soon as some hundred time steps are integrated. In Fig. 1, the lateral concentration distribution on certain arcs downwind of the source are compared for run #5 (Nov. 11, 1978, see Gryning and Lyck) of the Copenhagen data set. Close to the source some scatter is observable which is probably due to the relatively small number of particles rather than due to the approximation. For arcs further downwind the concentration distributions can be seen to become very similar between the full and the approximate model. This is even more evident when looking at the downwind evolution of the cross-wind integrated concentration (CIC) in Fig. 2. It is essentially the maximum concentration, which exhibits some differences to the 'full model' and these differences appear not to be sensitive to the discretization interval. In Table II, the statistical measures (Normalized Mean Square Error, NMSE; Fractional Bias, FB and the correlation coefficient, cor) are given for the simulation of Figs. 1 and 2. The finer discretization of Δz=0.1m does not seem to improve the correspondence to the 'full model', thus indicating that a vertical resolution of Δz=1m is sufficient in the 'approximate model' for the present purpose. Finally, Fig. 3 shows that the 'well-mixed property' of the model is not affected through the approximation in the functions f_{ij}, a result that can be anticipated, given the well-mixed property of the full model (Thomson, 1987) and the straight forward approximation in this speed-up version.

SUMMARY

An efficient procedure is presented which allows to speed up the calculation of the accelerations in a Lagrangian stochastic particle dispersion model by a factor of about eight. The speed-up procedure essentially consists of an additional subroutine, in which all those parts of the acceleration terms, which are only a function of height (or, more general, of position) are pre-calculated prior to the actual simulation. Having invoked this speed-up procedure the cpu requirements for a simulation are essentially limited by the generation of random numbers. Therefore, the total speed-up factor (accelerations and random numbers) is about three. The procedure is shown not to significantly change the resulting concentration distributions, and a discretization interval of Δz=1m is found to be sufficient.

The model, for which the performance of the speed-up procedure was evaluated in the present study is that for Gaussian inhomogeneous turbulence of Thomson (1987). Due to its simple structure it is perfectly suited for this exercise. For other models, which contain more (computationally) expensive functions such as the error function (e.g., the one-dimensional model of Luhar and Britter, 1989 for convective turbulence) the speed-up factor can be expected to become smaller than the value cited here.

REFERENCES

van Dop, H.: 1992, 'Buoyant plume rise in a Lagrangian framework', *Atmos. Environ.*, **26A**, 1335-11346.

Gryning, S. E. and Lyck, E.: 1984, 'Atmospheric dispersion from elevated sources in an urban area: comparisons between tracer experiments and model calculations', *J. Clim. Appl. Meteorol.*, **23**, 651-660.

de Haan, P.: 1998, 'On the use of density kernels for concentration estimates within particle and puff dispersion models', submitted to *Atmos. Environ.*

Luhar, A.L. and Britter, R.E.: 1989, 'A random walk model for dispersion in inhomogeneous turbulence in a convective boundary layer', *Atmos. Environ.*, **23**, 1911-1924.

Press, W.H.; Teukolsky, S.A.; Vetterling, W.T. and Flannery, B.P.: 1992, 'Numerical Recipes. The Art of Scientific Computing', 2nd Edition, Cambridge University Press, 963pp.

Rotach M.W., Gryning, S.E. and Tassone C. 1996: 'A two-dimensional stochastic Lagrangian dispersion model for daytime conditions', *Quart. J. Roy. Meteorol. Soc.*, **122**, 367–389.

Sawford, B. L., and Guest, F. M.: 1988, 'Uniqueness and universality of Lagrangian stochastic models of turbulent dispersion', AMS, *Preprints, 8th Symposium on Turbulence and Diffusion*, San Diego, CA, April 25-29, 1988, Amer. Meteorol. Soc., Boston, MASS, 96-99.

Thomson, D.J., 1987: 'Criteria for the selection of stochastic models of particle trajectories in turbulent flows', *J. Fluid Mech.*, **180**, 529-556.

Wilson, J.D. and Sawford, B.L: 1996, 'Review of Lagrangian stochastic models for trajectories in the turbulent atmosphere', *Boundary-Layer Meteorol.*, **78**, 191-210.

DISCUSSION

A. LUHAR: Do you get similar speed improvement for the skewed (and inhomogeneous) turbulence case?

M. ROTACH: We have indeed implemented this speed-up procedure to a more complicated model which is well-mixed also for skewed (in the vertical velocity component), inhomogeneous turbulence. The speed-up in this case is smaller (due to the more complicated model structure) but it still amounts to almost a factor of two.

D. ANFOSSI: This is a comment, not a question. Even in the computations of the SPRAY model, just presented by Dr. Tinarelli, all parameters depending on position only (not on particle's velocities) are pre-computed, so as to speed up the calculations. This means that the speed-ups shown in the previous presentation did not include the above pre-computation (already included in the old SPRAY version).

M. ROTACH: Thank you.

MODELLING OF CHEMICAL PROCESSES IN CLOUDS: SCAVENGING AND PARTITIONING OF SPECIES AMONG GAS AND LIQUID PHASES

N. Chaumerliac, N. Audiffren, and S. Cautenet

LaMP/OPGC, CNRS, Université Blaise Pascal
24 Avenue des Landais 63177 AUBIERE Cedex FRANCE
Tel. 4 73 407372, e-mail: chaumerl@opgc.univ-bpclermont.fr

INTRODUCTION

More than 50% of the earth's surface is covered by clouds and theoretical calculations of Ravishankara (1997) have shown that clouds can alter the composition on a global scale. Clouds interact in many ways with chemicals and on a wide range of scales from micrometers up to thousand of kilometers.

At large scale (thousands of kilometers), clouds are organized in broad and complex systems which are responsible for the transport of species from the boundary layer to the free troposphere (Renard et al., 1994; Edy et al., 1996). Tracer redistribution can be greatly changed in case of precipitating clouds systems due to their efficient scavenging. Photochemical processes are modified through cloud/radiation interactions (Thompson, 1984). Within these systems, each individual cloud is the host of complex microphysical processes that influence the partitioning of species among the air, the cloud and the precipitation (Grégoire et al. , 1994). Finally, at microscale, gas absorption, chemical reactions greatly depend on the microstructure of the cloud such as the droplet spectrum, the crystal shape and one has to consider complex interfacial transfer between gaseous, liquid and solid phases. Moreover, these small scale features cannot be ignored at larger scale because removal processes, radiative properties of clouds that perturb photochemistry depend on the microphysical characteristics of the clouds. In order to simulate such complex interactions on the whole range of scales at which they are efficient, it is necessary to use several types of models from box chemical model, to cloud model and mesoscale model.

In this paper, scavenging processes that occur in clouds and their dependency on the fine microphysical features such as droplet size, liquid water content, phase of the cloud (liquid or solid) will be discussed for one particular chemical species, the hydrogen peroxide which is both a soluble and a reactive compound in clouds. The way clouds interact with this particular species will be described in details. In particular, deviations from Henry's law can occur in clouds for this species at the sudden apparition of the aqueous phase or in a cloud presenting different types of granulometry (cloud droplets versus raindrops).

Air Pollution Modeling and Its Application XIII, edited by S.-E. Gryning
and E. Batchvarova, Kluwer Academic/Plenum Publishers, 2000

DEVIATIONS FROM HENRY'S LAW FOR H_2O_2 DURING A CLOUD LIFETIME

The evolution of hydrogen peroxide has been followed during a cloud event by means of a chemical box model (Madronich and Calvert, 1990), which interprets any chemical mechanism, including aqueous phase chemistry based upon Grégoire et al. (1994). The chemical system is a standard mechanism including methane and CO in the presence of NO_x and sulfur dioxide. Table 1 lists the initial concentrations of the species.

O_3	H_2O_2	SO_2	CH_3O_2H	CH_2O	CO	HNO_3	HO_2	OH	NO_2	NO	CH_3O_2	HCOOH
25ppb	2ppb	1ppb	2 ppb	0.5ppb	140ppb	200ppt	9ppt	0.08ppt	12ppt	6ppt	9ppt	0.3ppt

Table 1: Initial Values of the Chemical Species

In order to simulate a cloud event, a variable liquid water content has been introduced in the chemical model. The liquid water content evolution (Figure 1) corresponds to observations at Kleiner Feldberg on October 31th 1990 for a stratocumulus cloud, during the 1990's Cloud Experiment (EUROTRAC Subproject Ground-based Cloud Experiment, Fuzzi, 1994). The time duration of the simulation is four hours with an initial one hour of solely gas phase chemistry. The cloud is essentially maintained during one hour and then it evaporates. The last hour period is also pure gas phase chemistry. The maximum water content was 0.2g/kg.

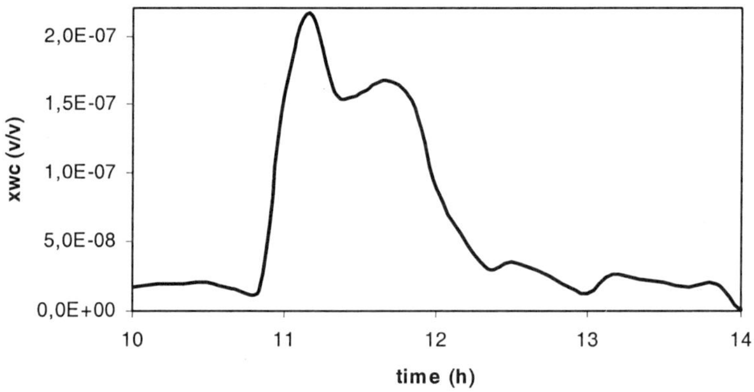

Figure 1. Time evolution of the liquid water content (in v/v) during the Kleiner Feldberg Experiment (Fuzzi et al., 1994).

On this simple cloud scenario, three sensitivity runs are performed. The first one only includes gas-phase chemical reactions and mass transfer between gas and aqueous phases. The two others consider gas and aqueous phase chemistry , mass transfer and include or exclude sulfur dioxide chemistry. When SO_2 is taken into account, higher reactivity is obtained in aqueous phase for H_2O_2. The sensitivity of the drop size (10μ versus 100 μ) is also examined in each case. Then, the uptake of the gases by the drops can be either described using instantaneous Henry's law equilibrium or full kinetic mass transfer following Schwartz (1986). Figure 2 shows the time evolution of hydrogen peroxide in every runs.

In Figure 2a, the assumption of instantaneous Henry's law equilibrium is valid for droplets of 10μ radius but gives slightly different result for larger drops. This is due to the fact that for soluble species, equilibrium between gas and aqueous phases is established more slowly for larger drops (Iribarne and Cho, 1989; Huret et al ., 1994).

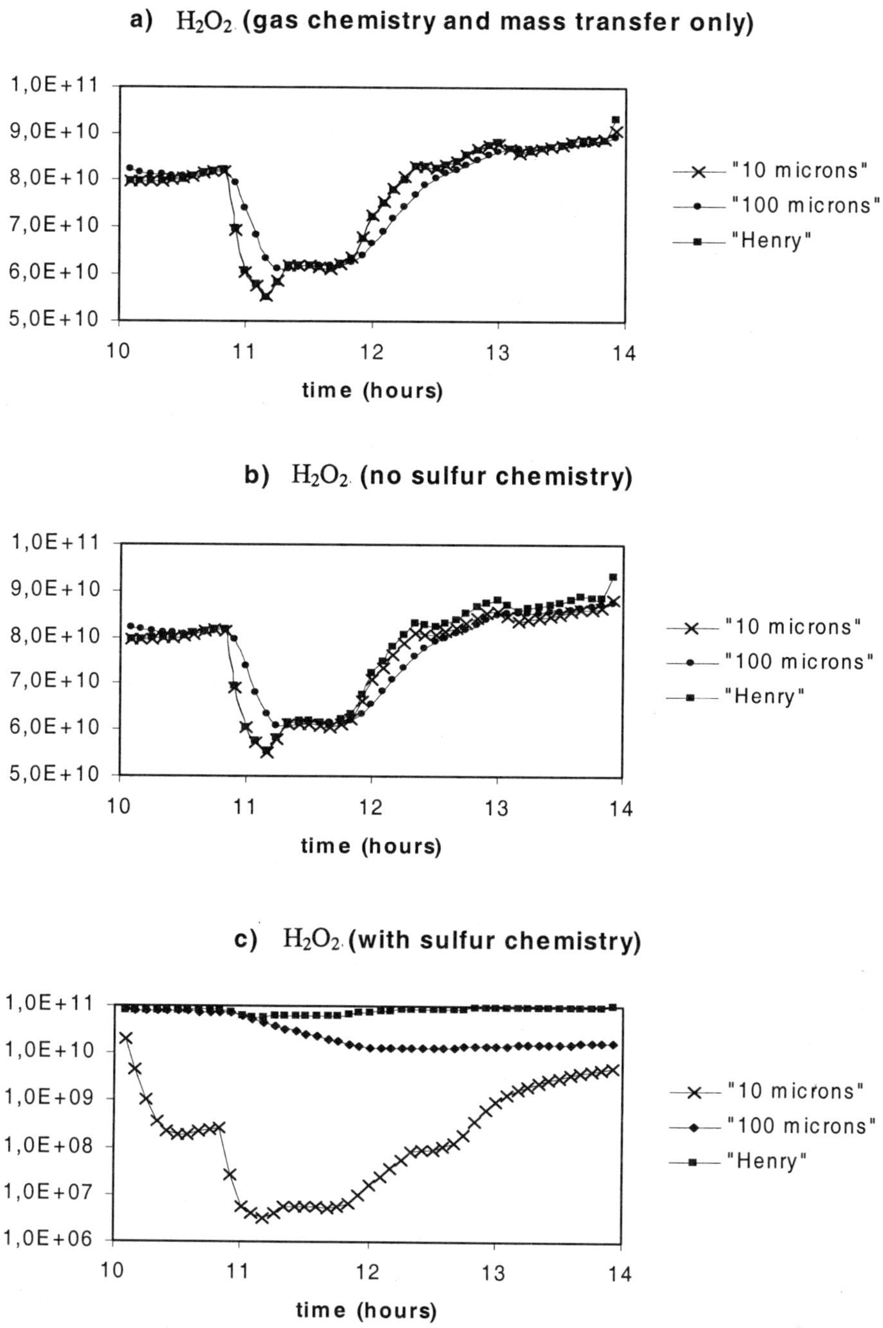

Figure 2. Time evolution of the peroxide hydrogen concentration in molec/cm3, for various hypothesis (lines with squares, Henry's law equilibrium is assumed, with crosses and diamonds, mass transfer is considered for small and large drops respectively) and for the three sensitivity runs, which are described as follows:

a) gas phase chemistry and mass transfer are considered,
b) same as a) plus aqueous phase chemistry
c) same as b) plus SO_2 chemistry.

If aqueous phase reactions are now added (Figure 2b), some deviations arise between the run assuming Henry's law equilibrium and the run including mass transfer. This is essentially true for larger drops as explained before. There is 16% discrepancy for large drops at the entrance of the cloud. During cloud dissipation, differences between the two assumptions are of 13% for the smaller droplets and 8% in the case of larger ones.

The more drastic discrepancies are observed if furthermore SO_2 chemistry is added (Figure 2c). For hydrogen peroxide, by considering Henry's law equilibrium, gas phase concentration is greatly overestimated whatever the drop size is. Especially, for low water contents (from 10 to 11h), the two assumptions (Henry's law equilibrium and mass transfer) severely deviate for smaller drops. The first assumption leads to higher amount of gas uptake when entering the cloud because of higher speed of transfer. For bigger drops, the speed of transfer is limited and deviations from Henry's law are less important.

Hence, soluble and reactive species such as H_2O_2 may not be considered in an instantaneous equilibrium between the gas and aqueous phases during the formation stage or the evaporation of clouds.

A simple interpretation of the previous results can be deduced from the equilibrium times towards chemical and phase equilibrium.The rate equation in gaseous phase for a particular species may be written as:

$$\frac{dC_g}{dt} = P_g - D_g C_g \tag{1}$$

where P_g and D_g are respectively the production and destruction terms and C_g the gaseous concentration of the species.

With the introduction of cloud water, this equation becomes:

$$\frac{dC_g}{dt} = P_g - D_g C_g - Lk_t C_g + \frac{k_t C_{aq}}{H_{eff} RT} \tag{2}$$

where k_t describes the mass transfer between gas and aqueous phases, L the liquid water content, H_{eff} the Henry's law effective constant of the species (Schwartz, 1986) and C_{aq} the aqueous concentration.

As soon as cloud water is present, additional rate equation for the aqueous concentration C_{aq} has to be accounted for:

$$\frac{dC_{aq}}{dt} = P_{aq} - D_{aq} C_{aq} + Lk_t C_g - \frac{k_t C_{aq}}{H_{eff} RT} \tag{3}$$

In pure gaseous chemistry, the characteristic time scales (lifetimes) are expressed by the inverse of the term D_g , and in gas-aqueous chemistry two different terms are involved:

$(D_g + Lk_t)^{-1}$ and $(D_{aq} + \frac{k_t}{H_{eff} RT})^{-1}$.

The equilibration time is often taken as the characteristic lifetime of the species. In fact, the equilibration time τ_{eq} is just proportional to this lifetime for pure gaseous chemistry, and it is given by:

$$\left|\frac{C_{g_{eq}} - C_g}{\tau_{eq}}\right| = D_g C_g \tag{4}$$

where D_g is the inverse of the lifetime of the species (chemical destruction rate) and $C_{g_{eq}}$ the equilibrium concentration.

With transfer to aqueous phase, this equation becomes:

$$\left|\frac{C_{g_{eq}} - C_g}{\tau_{eq}}\right| = D_g C_g + Lk_t C_g (1-q) \tag{5}$$

where $q = \dfrac{C_{aq}}{LH_{eff} RTC_g}$ expresses the deviation from Henry's law equilibrium (q equals 1 in Henry's law equilibrium).

On Figure 3, the factor q has been drawn for small and large drops in the three runs described previously: the first run only considers gas chemistry plus mass transfer, while the two others take into account both gas and aqueous chemistries and either exclude or include SO_2 chemistry. The curves of Figure 3 just reflect what we have already seen on Figure 2 but will be helpful in the following section to compare the chemical behaviour of H_2O_2 in a cloud simulated by a mesoscale model.

At a sudden introduction of aqueous phase the term $Lk_tC_g(1\text{-}q)$ with q equals to zero, becomes the driving term of the equation for most of the species. When Henry's law equilibrium is established, the relaxation time will be driven by D_g again. Any deviation from Henry's law can therefore significantly increases the equilibration time in each phase. For the aqueous phase, the equilibration time can be derived in the same manner:

$$\left|\frac{C_{aq_{eq}} - C_{aq}}{\tau_{eq\,aq}}\right| = D_{aq} C_{aq} + \frac{k_t}{H_{eff} RT} C_{aq} (1-q)/q. \tag{6}$$

Hence, three types of equilibrium determine the equilibration time of each species: the physical equilibrium between the two phases and the chemical equilibrium of each phase.

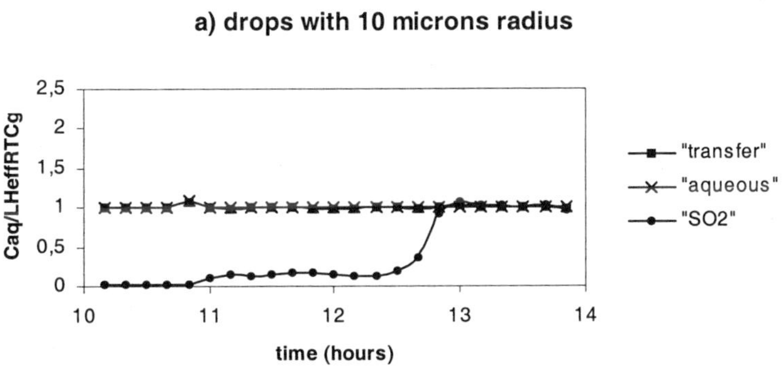

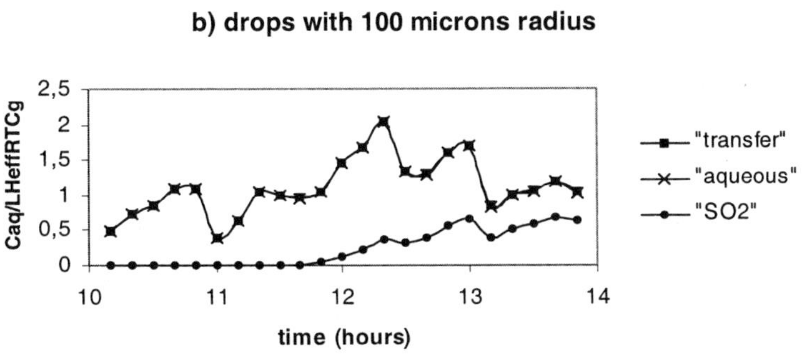

Figure 3. Time evolution of the factor $q=C_{aq}/LH_{eff}RTC_g$, for the three sensitivity runs and two drop sizes:

1) gas phase chemistry and mass transfer are considered (squares),
2) same as 1) plus aqueous phase chemistry (crosses)
3) same as 2) plus SO_2 chemistry (circles).

Assuming that the equilibration time is mainly due to chemical reactions, this leads to consider that the partition between aqueous phase and gaseous phase always follows the Henry's law (instantaneous Henry's law equilibrium). This assumption requires that for most species the fluxes at the interface are faster than gas or aqueous chemistry which means that the characteristic times of chemical reactions D_g^{-1} and D_{aq}^{-1} are respectively higher than $(Lk_t)^{-1}$ and $(k_t/H_{eff}RT)^{-1}$. This is observed for H_2O_2 scavenged by small drops of 10µm radius when no SO_2 chemistry is involved as can be seen on Figure 3a.

For big drops, the total content of H_2O_2 in a single drop may become so important in phase equilibrium, that the resulting equilibration time seriously exceeds the chemical equilibration times and the model timestep), even with fast exchanges at the interface of the drops (Lk_t term). In such case, deviations from Henry's law is observed with $q<1$ at the onset of the cloud and $q>1$ during the evaporation stage of the cloud. Fast exchanges are also revealed through the high variations in the q factor.

In the second run (without SO_2) the value of D_g of the hydrogen peroxide is small compared to Lk_t for both 10 µm and 100 µm cases, as D_{aq} is small versus $(k_t/H_{eff}RT)^{-1}$. It is noteworthy that both $(D_{gas} + Lk_t)$ and $(D_{aq} + (k_t/H_{eff}RT)^{-1})$ decreases by a factor of 100 between the 10µm and 100µm cases, meaning that the equilibration time is much longer for big drops. When sulfur dioxide is added, Lk_t is still higher than D_g while D_{aq} becomes higher than $(k_t/H_{eff}RT)^{-1}$ by a factor of 2000 in the case of small drops and only by a factor of 2 (up to 8) in the big drop case. The driving time scale in the aqueous phase equilibrium is then $(k_t/H_{eff}RT)^{-1}$. Hence, the method assuming instantaneous Henry's law equilibrium is not applicable anymore, and the q factor is always less than 1 (Figure 3).

DEVIATIONS FROM HENRY'S LAW FOR H_2O_2 IN A POLYDISPERSE CLOUD

The same type of study (Audiffren et al., 1996) has been performed in the framework of a mesoscale model simulating orographic clouds formed from different air masses (continental versus maritime). This mesoscale meteorological model is coupled with the chemical module described in the previous section, except that chemical equation system is solved with a QSSA type solver more appropriate (see Audiffren et al., 1998a for more details). A complete description of the meteorological mountain wave scenario can be found in Chaumerliac et al. (1987,1990). The half width of the mountain is 25km and its height is 1km. The initial wind is horizontal and homogeneous with a velocity of 20m/s. The initial atmosphere consists in a layer of constant lapse rate with 80% relative humidity up to 250mb which is topped by a dry isothermal layer. The mesoscale model (Nickerson et al., 1986) covers an horizontal domain of 300km with a grid spacing of 10km. The timestep is 10s and there are 16 vertical levels spaced in a terrain-following coordinate. The dynamical, microphysical and chemical processes are fully interactive in the model. The simulation lasts three hours and is two-dimensional. The microphysical parameterization is based upon the work from Berry and Reinhardt (1974).

Two runs are performed with different types of orographic clouds, originated from two different air masses (Figure 4). A continental air mass gives rise to a cloud composed with many small droplets, non-precipitating, while the maritime cloud formed on a maritime air mass contains less but larger drops that can precipitate. The total liquid water contents are identical and only the partitioning between cloud water and rainwater, due to the autoconversion process differs. This provide us with an idealized situation for comparing chemical contents of such clouds. Chemical species dissolve both in cloud and rain phases and once included in the drops they can undergo transfer through microphysical processes. This last characteristic complexifies the previous situation where only the time evolution of a cloud was considered and make the cloud chemistry more realistic.

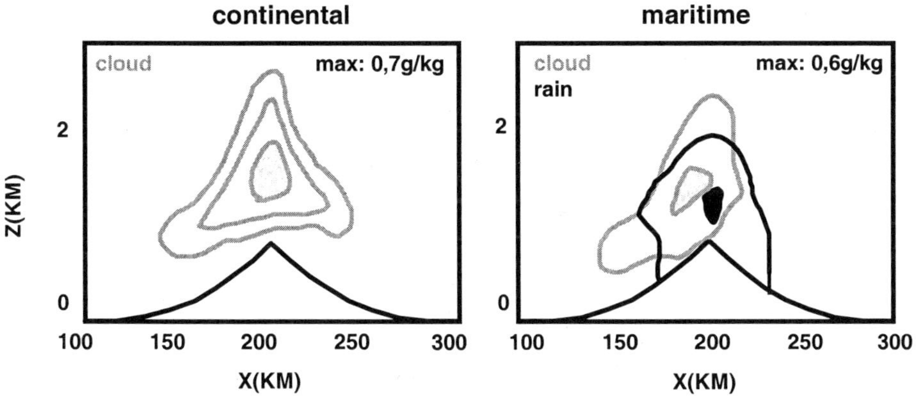

Figure 4. Vertical cross-sections of the cloud water and rainwater mixing ratios for a continental non-precipitating cloud and a maritime precipitating cloud.

As seen before, the ratio $q = C_{aq} / LH_{eff}RTC_g$ gives a measure of the deviation from Henry's law. C_{aq} can be either the cloud water concentration or the rainwater concentration (molec per cubic centimeter air) and L the corresponding liquid water content. This ratio has been plotted for both the soluble species H_2O_2 (Figure 5). Figures 5a and 5b show the evolution of this q factor along the mountain slope for the continental and maritime cloud, while Figure 5c refers to the maritime rainwater.

We clearly observe that the cloud droplets are always subsaturated upwind of the mountain top and supersaturated downwind where cloud evaporation occurs. Henry's law equilibrium is attained in the middle of the cloud and not on cloud edges. This can be explained by the fact that surrounding air acts as an important gaseous source for hydrogen peroxide on cloud edges. In the case of the maritime cloud also, deviations from Henry's law occur in parts of the cloud where rain is present. Vertically, the cloud water content decreases not only because of evaporation but mainly through conversion of cloud into rain. This can be related to an open system as studied by Pandis and Seinfeld (1992), where a decrease in cloud water leads to an increase in the q factor. In contrast, in maritime rainwater, drops on the upwind side are supersaturated in the upper part of the cloud. In all remaining parts of the precipitating cloud, raindrop concentrations of H_2O_2 are very low under Henry's law value (q is less than 1 even equal to 0.2 near the ground). This net deviation from Henry's law is likely due to the raindrop size which increases drastically on that side of the mountain.

In conclusion, dealing with a mesoscale meteorological model simulating orographic cloud has highlighted some more features about chemical species scavenging, compared to what we already showed with a box model. In particular, the box model did not include effects of microphysical processes that link a small drop population to a large drop population in the maritime case. It seems that microphysical processes that transport in fact H_2O_2 from the cloud to the rain phase act in the mesoscale model as the reagent SO_2 was acting in the box model, displacing the equilibrium below Henry's law value. Also for big drops (like the ones in the maritime case), we retrieve a high variability of the q factor, related to fast exchanges as in the box model. Finally, precipitation efficiently removes H_2O_2, while in the box model, a decrease in the liquid water content allows a complete release of the chemical species into the gas phase.

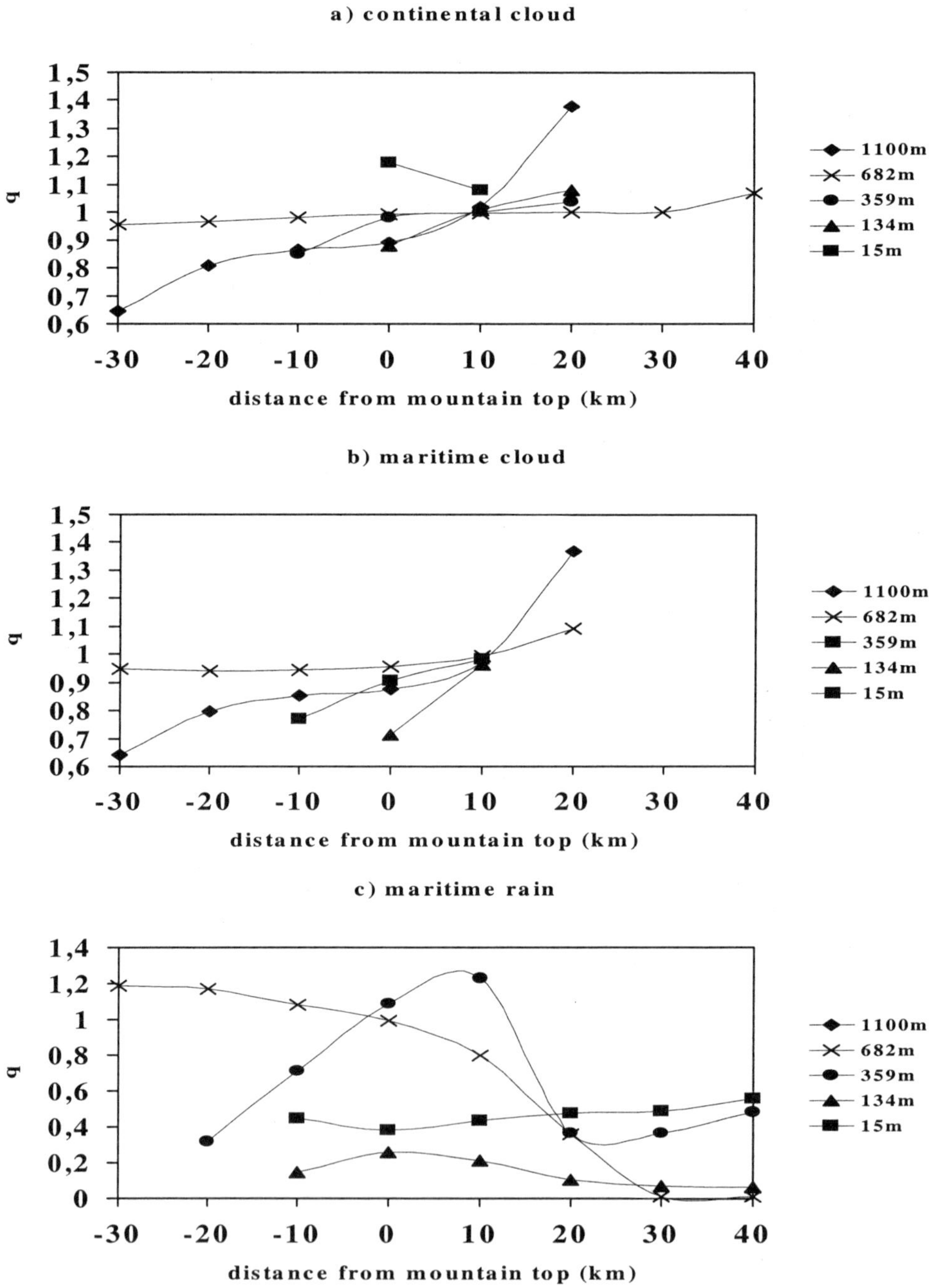

Figure 5. Evolution as a function of the distance from the mountain top of the factor $q=C_{aq}/LH_{eff}RTC_g$, showing the deviations from Henry's law in case of a) a continental cloud, b) a maritime cloud and c) the maritime rain field.

CONCLUSION

In this study, several examples of liquid clouds have been simulated by various types of models: a process model and a mesoscale model, including the multiphase chemistry. Results demonstrate the ability of all these modeling studies to complete lacking information upon the partitioning of species among the different phases of a cloud. Experimentally, it is still very difficult to isolate concentrations of chemical species in one particular phase. Moreover, most of the time, samplings of cloud chemistry are either measurements from pluviometers or impactors that provide time integrated results, or airborne measurements that often suffer from detection limits of trace quantities. Within the models, it is possible to redistribute soluble and reactive gas species, such as hydrogen peroxide among the gas and liquid parts of a cloud and to follow each component during the whole cloud time evolution and spatial evolution.

We have shown that deviations from Henry's law equilibrim depends on drop size, but also on chemical reactivity in both gas and aqueous. So, for hydrogen peroxide in the presence of sulfur dioxide, the deviation is large even for small droplets. The partitioning of gases is a function of cloud development conditions like the air mass on which the cloud has been formed (continental versus maritime), the microphysical properties (cloudwater, rainwater content). For an orographic cloud, the deviations are also contrasted on the upwind side, on the downwind side or at top of the mountain.

In conclusion, several causes to deviations from Henry's law have been examined and have been estimated through the ratio q, equal to 1 in case the equilibrium between phases is reached. If more measurements of this quantity were available in real clouds, this could be a way of parameterizing the multiphase chemistry in large scale models.

ACKNOWLEDGMENTS

The work reported in this paper was supported by the INSU/CNRS (Institut National des Sciences de l'Univers/Centre National de La Recherche Scientifique) Program PNCA (Programme National de Chimie Atmosphérique). Computer resources were provided by I.D.R.I.S. (Institut du Développement et Des Ressources en Informatique Scientifique), project n°187.

REFERENCES

Audiffren N., N. Chaumerliac and M. Renard: Effects of a polydisperse cloud on tropospheric chemistry, *J. Geophys Res*, 101, D20, 25,949-25,966, 1996.

Audiffren N., M. Renard, E. Buisson and N. Chaumerliac: Deviations from the Henry's law equilibrium during cloud events: a numerical approach of the mass transfer between phases and its specific numerical effects, accepted for *Atm. Res.*, 1998a.

Berry, E.X. and R.L. Reinhardt, An analysis of cloud drops growth by collection, II. Single initial distributions, *J. Atmos. Sci.*, *31*, 1825-1831, 1974.

Chaumerliac N., E. Richard, J.P. Pinty, et E.C. Nickerson: Sulfur scavenging in a mesoscale model with quasi-spectral microphysics: Two-dimensionnal results for continental and maritime clouds. *J. Geophys. Res.*, **92**, 3114-3126, 1987.

Chaumerliac N., E. Richard, et R. Rosset: Acidity production in orographic clouds and rain in a mesoscale model with semi-spectral microphysics. *Atm. Env.*, **24A**, 1573-1584, 1990.

De More W.B. et al., Chemical kinetics and photochemical data for use in stratospheric modeling, *JPL Publ N°*. 94-26 (Jet Propulsion Laboratory, Pasadena, CA), 1994.

Edy J., S. Cautenet and Ph. Brémaud, Modeling ozone and carbon monoxide redistribution by shallow convection over the amazonian rain forest, , *J. Geophys. Res*, 101, D22, 28671-28682, 1996.

Fuzzi S., The Kleiner Feldberg Cloud Experiment 1990: Introduction, EUROTRAC Subproject, Ground-based Cloud Experiment (GCE), Kluwer Academic Publishers, S. Fuzzi Ed., 1994.

Grégoire P. J., N. Chaumerliac and E. C. Nickerson: Impact of cloud dynamics on tropospheric chemistry: advances in modeling the interactions between microphysical and chemical processes. *J. Atm. Chem.*, **18**, 247-266, 1994.

Huret N., N. Chaumerliac , H. Isaka, and E.C. Nickerson: Impact of different microphysical schemes on the dissolution of highly and less soluble non-reactive gases by cloud droplets and raindrops. *J. Appl. Meteor.*, **33**, n°9, 1096-1109, 1994.

Iribarne J. V. and H. R. Cho, Models of cloud chemistry, *Tellus*, 41B, 2-23, 1989.

Lelieveld J and P.J. Crutzen, The role of clouds in tropospheric photochemistry, *J. Atmos. Chem.*, 12, 229-267, 1991.

Madronich S., and J.G. Calvert, Permutation reactions of organic peroxy radicals in the troposphere, *J. Geophys. Res.*, 95, 5697-5715, 1990.

Nickerson E.C., E. Richard, R. Rosset and D.R. Smith, The numerical simulation of clouds, rain and airflow over the Vosges and Black Forest mountains: a meso-β model with parameterized microphysics, *Mon. Wea. Rev.*, *23*, 477-487, 1986.

Pandis S.N. and J.H. Seinfeld, Should bulk cloudwater or fogwater samples obey Henry's law?, *J. Geophys. Res.*, *96*, D6, 10,791-10,798, 1991.

Ravishankara A.R., Heterogeneous and multiphase chemistry in the trposphere, *Science*, **276**, 1058-1065, 1997.

Renard M., Chaumerliac N., S. Cautenet, N. Audiffren and E.C. Nickerson, A numerical redistribution of the seasonal variations for tracer redistribution by clouds over West Africa, *J. Atmos. Chem.*, 1995.

Schwartz S., Mass-transport considerations pertinent to aqueous phase reactions of gases in liquid water clouds, *Chemistry of mulitphase systems*, W.Jaeschke, Ed. Springer, 415-471,1986.

Taylor, G.R.: Sulfate production and deposition in midlatitude continental cumuls clouds. Part I: Cloud Model formulation and base run analysis. *J. Atmos. Sci.*, **46,** 1989.

Thompson A. M., The effect of clouds on photolysis rates and ozone formation in the unpolluted troposphere, *J. Geophys. Res.*, 89, 1341-1349, 1984.

Wang, C., and J.S. Chang : A three-dimensional numerical model of cloud dynamics, microphysics and chemistry. I. Concept and formulation. *J.Geophys.Res.*, **98**, 14827-14844, 1993.

ACCIDENTAL RELEASES

chairman: A. Graff

rapporteur: J. H. Sørensen

A NEW MODEL FOR ACCIDENTAL RELEASES OF SO_3 AND OLEUM (CLOUD BEHAVIOUR)

T. Kapias and R.F. Griffiths

Environmental Technology Centre - Department of Chemical Engineering
UMIST, PO Box 88
M60 1QD, Manchester, UK

INTRODUCTION

SO_3 and oleums are highly reactive and aggressive materials that are used widely in the process industries. Their main feature is the violent and highly exothermic reaction of SO_3 and water that occurs in both the liquid and the vapour phase. On escape to the environment they create liquid pools that can boil or evaporate or even solidify. The pool behaviour is governed by the amount of water available for reaction. SO_3 and possibly H_2SO_4 will evolve from the pool. These vapours react with atmospheric moisture yielding sulphuric acid aerosol that presents a major hazard to humans and to the environment. The behaviour of the cloud (or plume) is very complicated as several interrelated physical and chemical processes are involved. The cloud initially behaves as a dense-gas cloud and only after some distance downwind will it become passive.

Two models have been developed, one describing the pool behaviour and the processes that occur in the liquid phase and the other describing the dispersion and the interactions that take place in the vapour phase. A detailed description of the model describing the pool behaviour can be found elsewhere[1]. Its main characteristic is that it describes the pool behaviour in a realistic way taking into account the numerous phenomena occurring in the pool. The same model can also be used for pools created from spills of other reactive substances. Model results indicate that the pool behaviour is mainly determined by the amount of water available for reaction. Another parameter that has an important influence is the friction velocity of the airflow[2,3].

CLOUD BEHAVIOUR

Previous modelling attempts assumed that the cloud generated from pools of SO_3 or oleum behaves as a passive one that has reached its chemical end point almost instantaneously. However it has been shown[4,5,6] that there is not usually enough atmospheric moisture for complete and almost instantaneous reaction of SO_3 to H_2SO_4 aerosol. A new model is being developed that describes the dispersion of SO_3/ H_2SO_4 clouds. It consists of two parts: the description of the vapour cloud dispersion behaviour and the thermodynamic description.

Dispersion model

A simple approach has been adopted in order to calculate the cloud geometrical size, the distribution of chemical concentration within the cloud etc., as a function of its downwind distance from the release point. It is based on a numerical box model[7,8] and the cloud characteristics are estimated by a process of iteration by solving the lateral expansion velocity, mass continuity (top and edge air entrainment) and momentum continuity equations. The main assumption is that the source evolution rate, temperature and radius are constant. It should be noted that although these properties are not constant during the evolution period, they are not highly variable for the pool spreading time[2]. The mean values are calculated by the pool model and they are used as input parameters in the source window uptake model[7]. The Richardson number Ri used in the model is given from the following equation:

$$Ri = \frac{g\Delta' L}{u^2_1}$$

where Δ' is the fractional density excess, u_l the longitudinal turbulent velocity (m s^{-1}), g is the acceleration due to gravity (m s^{-2}) and L is the ambient turbulent eddy size (or turbulent length scale) (m). The criterion[7] that is used for transition to passive behaviour is: $Ri < 1$ or $\Delta' < 0.001$.

Thermodynamic model

Although there is much work in the literature on heavy gas dispersion models that include thermodynamic models for reactive chemicals, none of them is judged to be satisfactory in describing the processes that occur in a SO_3/ H_2SO_4 cloud. In order to have a better understanding of the processes that occur in such a cloud two different stages are needed.

First stage. During this stage SO_3 vapour, H_2SO_4 vapour and H_2SO_4 aerosol will be present and the cloud will be denser than air. For this period the processes that occur simultaneously are:

- SO_3 vapour reacts homogeneously with atmospheric water yielding H_2SO_4 vapour.
- SO_3 vapour reacts heterogeneously with water on the ground yielding H_2SO_4 liquid.
- H_2SO_4 vapour and water vapour nucleate to sulphuric acid aerosols (nucleation).
- H_2SO_4 vapour diffuses within the cloud and some amount is lost on the ground.
- H_2SO_4 aerosols are formed and interact with atmospheric water producing droplets of greater size (aerosol growth).

SO_3 and H_2SO_4 will evolve from the pool and both will react with the atmospheric water vapour. The first will yield H_2SO_4 vapour and the latter H_2SO_4 aerosol. The water fraction being used by each of these two processes will be proportional to the relative molar fractions of the respective reactants.

The homogeneous reaction of SO_3 vapour with H_2O vapour has a long history of investigation. Even so, only very recently has it been found that the reaction mechanism involves one SO_3 and two H_2O molecules. Jayne et al.[9] have reported that the gas-phase reaction has a strong negative temperature dependence and no significant pressure dependence. The first-order rate coefficient for the SO_3 loss was reported to be equal to:

$$k_l = 3.9 \times 10^{-41} \exp(6830.6/T)[H_2O]^2 \text{ (s}^{-1}\text{)}$$

where $[H_2O]$ is in units of molecule cm^{-3} and T is in Kelvin.

The same authors also reported on the heterogeneous reaction of SO_3 with wet surfaces[9]. Because of the high SO_3 concentration this reaction will be diffusion limited. The SO_3-N_2 diffusion coefficient was reported to be equal to 94.6 (± 3) Torr cm^2 s^{-1} at 300 K. Some quantity of H_2SO_4 vapour will also be lost on the ground. The H_2SO_4-N_2 diffusion coefficient was reported[10] to be 66.8 (± 1.1) Torr cm^2 s^{-1}.

Unfortunately, all the theoretical and experimental data on the H_2SO_4 aerosol nucleation and growth correspond to ambient atmospheric (or tropospheric and stratospheric) H_2SO_4 aerosol concentrations which are extremely low, compared to those likely to be encountered in the cloud of interest here. Furthermore, there is a large discrepancy between different theories and experimental results. For example the H_2SO_4 mass accommodation coefficients are variously reported to be in the range 0.02-1. From accidents and experiments that have occurred in the past[11, 12] it seems that H_2SO_4 aerosols are usually small (particle radius 0.01-1 μm).

In order to describe aerosol nucleation the approach followed by Easter and Peters[13] has been adopted. The nucleation rate is given by the following equation modified by a factor of 10^{-6} to account for hydrate effects[14]:

$$J = \alpha \times S_a^{\beta} \times 10^{-6} \text{ (particles cm}^{-3}\text{ s}^{-1}\text{)}$$

where α and β are constants dependent on the atmospheric relative humidity and S_α is the saturation ratio for the acid, given as a ratio to its value over pure sulphuric acid[14]. The value for equilibrium pressure can be found in Ayers et al[15].

It should be noted that the nucleation rate calculated from the above equation is usually higher than the actual aerosol concentration would allow. Thus it is judged that the aerosol nucleation is a water availability limited process.

The aerosol vapour phase will condense if the temperature drops to the dew point which is usually above 393 K[16]. The H_2SO_4 aerosol generated will grow, but unfortunately no existing model can describe this phenomenon when the aerosol concentration is higher than the atmospheric water concentration. In the model presented here it is assumed that the nucleation process yields $H_2SO_4.H_2O$ aerosols that do not grow.

At a certain distance downwind, all the SO_3 vapour and the H_2SO_4 vapour will be consumed and only liquid aerosol will be present in the cloud.

Second stage. In this stage only H_2SO_4 aerosol will be present in the cloud. The processes that occur are:

- H_2SO_4 aerosols condense and interact with atmospheric water producing droplets of larger size (aerosol growth).
- Some of the H_2SO_4 aerosol will be lost on the ground.

After dilution with air the cloud density will continuously decrease and transition to passive behaviour will occur. A description of the processes that occur in the cloud is depicted in figure 1.

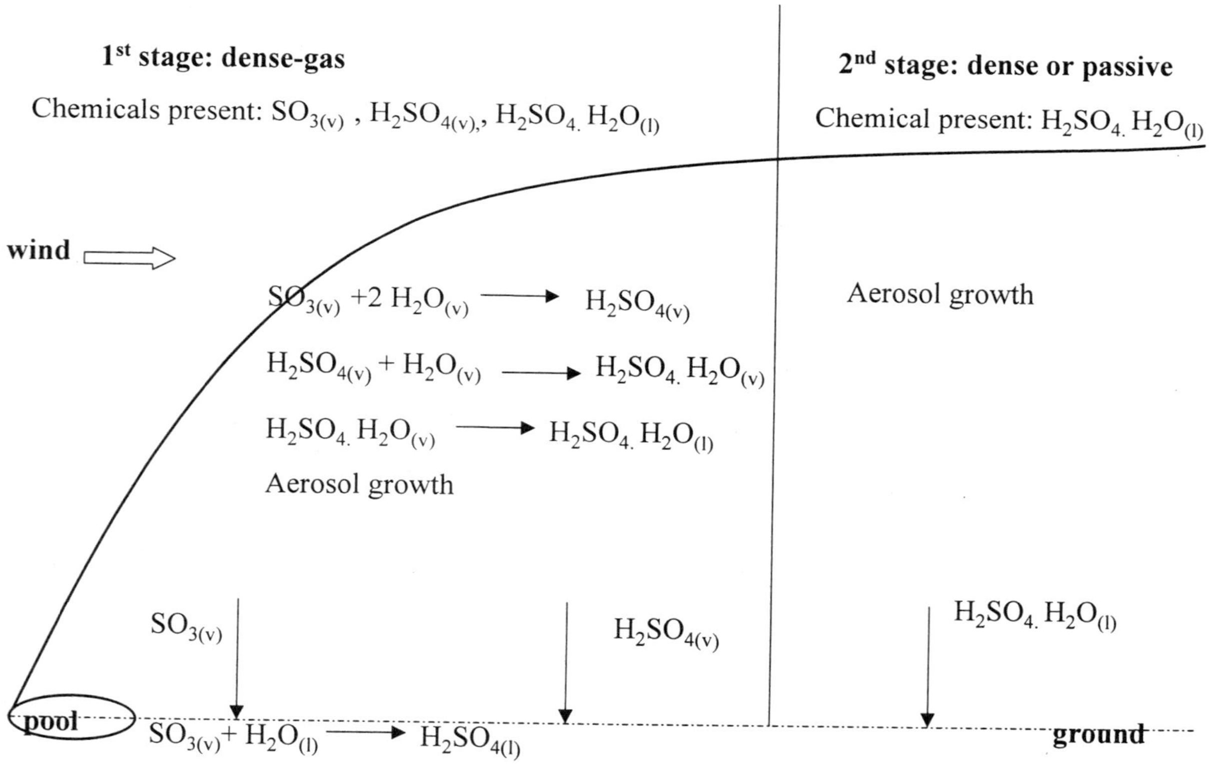

Figure 1. Behaviour and processes that occur in a cloud generated from pools of SO_3 and oleum.

MODEL CALCULATIONS AND RESULTS

A new model has been developed that describes the dispersion behaviour and the processes that occur in the cloud. Calculations are performed at a distance increment equal to dx = 0.01m. At every increment an energy balance is conducted taking into account the thermochemistry of all the reactions and processes (all of them are exothermic or thermoneutral) and of the energy exchange with the substrate. The heat flux transferred from the substrate into the cloud takes into account both the forced heat convection and the natural heat convection as described by Holman[17] and by McAdams[18] respectively. It should be noted that the ground temperature is not constant as the exothermic heterogeneous reaction of SO_3 vapour with liquid water occurs on the ground surface. All the properties of the vapours and liquids involved have been parameterised. The cloud density is calculated by assuming that the perfect gas law holds. The presence of the aerosol liquid phase increases significantly the total cloud mass but has a trivial effect on the total volume.

The inputs to the model are the σ_θ (standard deviation of the horizontal wind direction, degrees), the wind speed at a height of 10 m U_{10} (m s^{-1}), the roughness length of the substrate z_o (m), the air temperature T_a (K), the air mass mixing ratio f_l (kg of water vapour/kg of total air) or the relative humidity (%RH), the initial ground temperature T_{grd} (K), the free water film thickness on the ground w_g (m). The SO_3 and H_2SO_4 mean evolution rates M_{SO3} and M_{H2SO4} (kg s^{-1}), the mean pool radius R_p (m) and temperature T_p (K) are calculated by the pool model. The main outputs are the cloud width W (m) and height H (m), the mass rates of the substances in the cloud C_i (kg s^{-1}), the cloud temperature T_c (K), the cloud density d_c (kg m^{-3}) and the Richardson number Ri, as functions of the distance from the source x (m).

Five different release scenarios are investigated. All of them correspond to continuous releases of 16 kg s^{-1} of 65% oleum for 600 s under different values of free ground water film thickness and air mass mixing ratio. These release scenarios were first run by the pool model in order to calculate the mean evolution rates, pool radius and temperature. The results of these runs are summarised in table 1.

Table 1. Results given by the pool model

Scenario	w_g (m)	M_{SO3} (kg s^{-1})	M_{H2SO4} (kg s^{-1})	T_p (K)	R_p (m)	f_l (kg of water /kg of total air)
1	0.001	3.51833	0.0004	385.41	7.396	0.001 (RH≈10%)
2	0.001	3.51833	0.0004	385.41	7.396	0.005 (RH≈50%)
3	0.001	3.51833	0.0004	385.41	7.396	0.01 ((RH≈100%)
4	0.002	4.4512	0.2105	479.37	7.46	0.005 (RH≈50%)
5	0.0005	1.8468	0	343.5	7.755	0.005 (RH≈50%)

The rest of the input parameters have the same values for all the scenarios and are equal to: σ_θ = 10 degrees, U_{10} = 5 m s^{-1}, z_o = 0.1 m, T_a = 288 K and T_{grd} = 283 K. The new model has been implemented in a computer programme in Microsoft Visual Basic Professional Edition 5.0. Results of the runs are shown in figures 2 and 3. It should be noted that the model is still under development. The main difficulty is the lack of any theoretical or experimental data on the aerosol nucleation and growth under high acid concentrations.

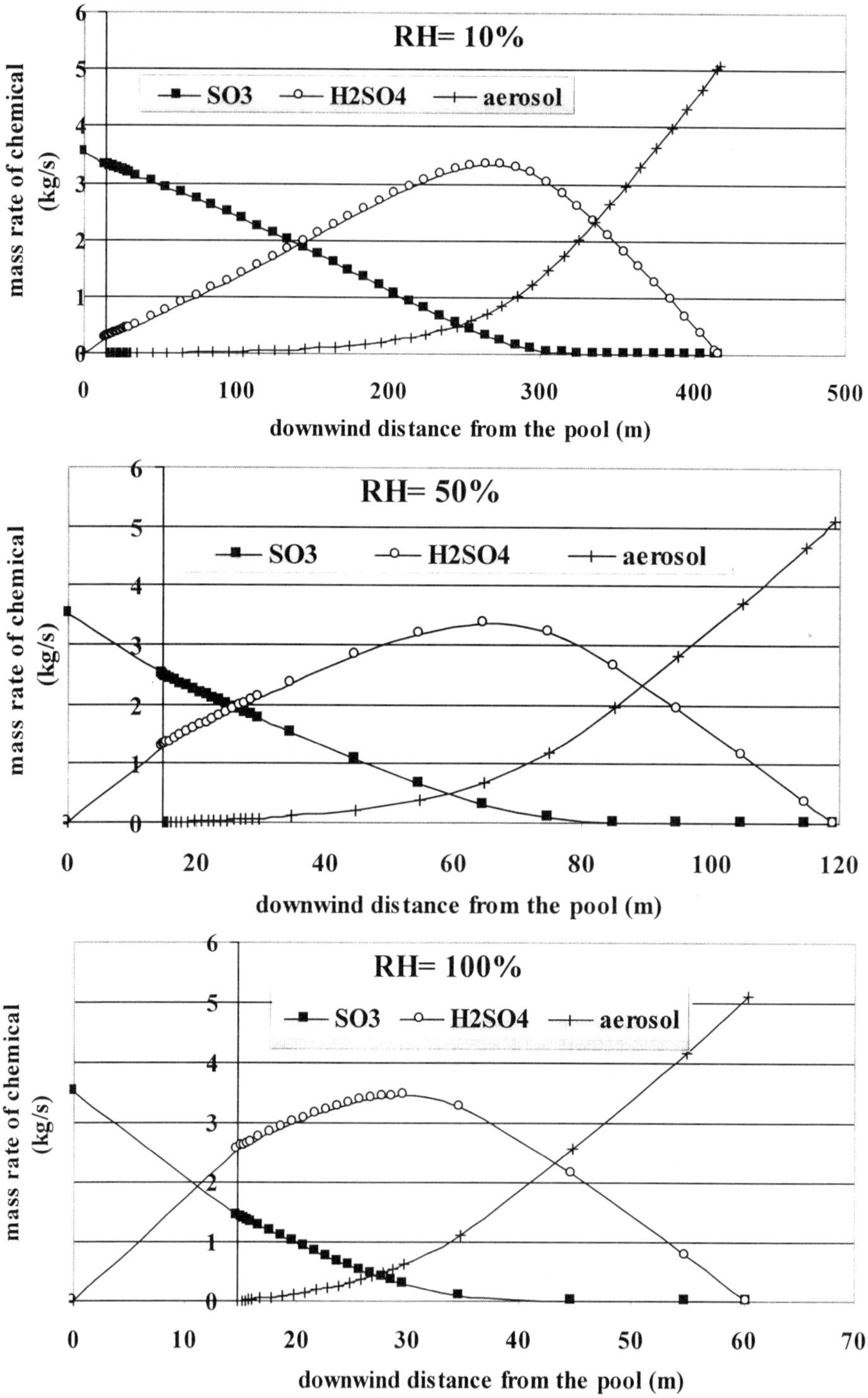

Figure 2. Cloud composition for scenarios 1, 2 and 3 (RH ≈ 10%, 50% and 100% respectively).

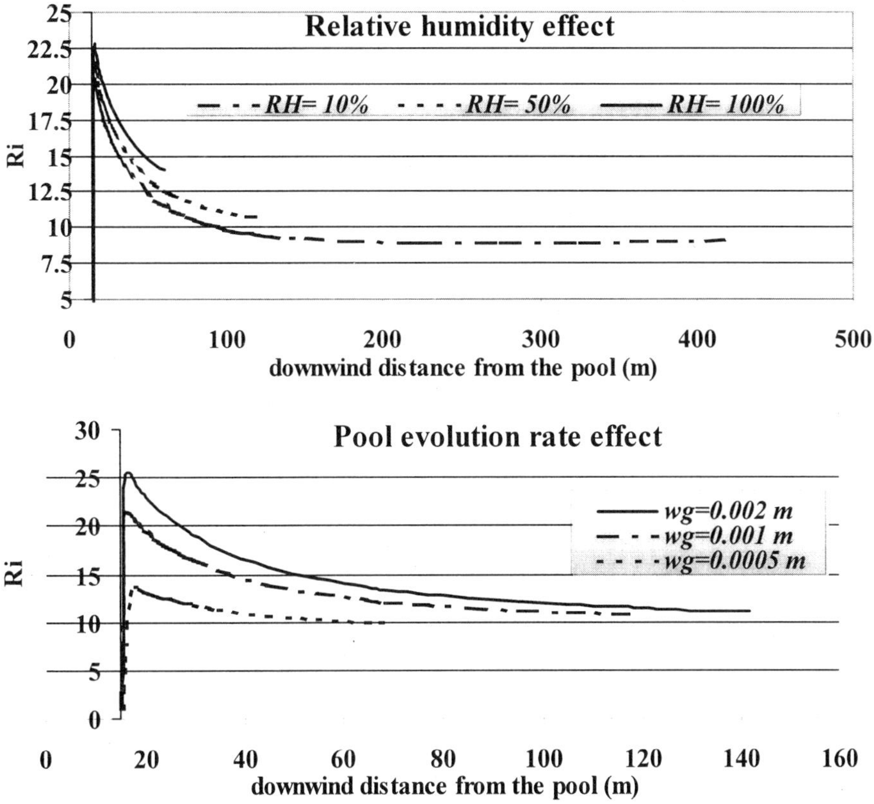

Figure 3. Relative humidity (Scenarios 1, 2 and 3) and pool evolution rate effect (Scenarios 2, 4 and 5) on the Richardson number.

DISCUSSION AND CONCLUSIONS

The behaviour of the cloud generated from pools of SO_3 and oleum is very complicated. There are six physical and chemical processes occurring in the cloud, the characteristics of which are not well established. Generally it will initially behave as a dense gas cloud. It should be noted that the calculated Richardson numbers are not extremely high (figure 3), indicating that the cloud will not be very dense. In the first stage SO_3, H_2SO_4 and H_2SO_4 aerosol are present in the cloud (figure 2) (the end of the curves in figure 3 corresponds to the point where the second stage starts). In the second stage only H_2SO_4 aerosol is present and due to the dilution with air and the aerosol loss on the ground, the cloud density drops. Eventually, at a certain distance, transition to passive behaviour will occur. The cloud behaviour is strongly affected by the relative humidity (figures 2 and 3) and by the pool evolution rate (figure 3). The higher the relative humidity, the higher the Richardson number is and the lower are the lifetimes of SO_3 and H_2SO_4 in the cloud. The higher the SO_3 and H_2SO_4 pool evolution rates, the higher is the Richardson number. It should be noted that in figure 2, the x-axis corresponds to the centreline masses of the chemicals in the cloud. In both figures (2 and 3) the y-axis encounters the x-axis at a value equal to the pool diameter. The model is not yet complete, as the aerosol growth and loss to the ground have not yet been modelled. Further improvement of the presented model depends on the availability of either theoretical or experimental data.

REFERENCES

1. T. Kapias and R.F. Griffiths, A model for spills of SO_3 and oleum, part I: model description, *J. Haz. Mat.* to be published.
2. T. Kapias and R.F. Griffiths, A model for spills of SO_3 and oleum, part II: results, conclusions and discussion, *J. Haz. Mat.* to be published.
3. T. Kapias and R.F. Griffiths, SO_3 and H_2SO_4 releases from accidental spills of SO_3 and oleum, *International Conference on Protection and Restoration of the Environment IV*, Halkidiki-Greece (1998).
4. T. Kapias and R.F. Griffiths, Modelling accidental releases of SO_3 and oleum, *9th International Symposium on Loss Prevention and Safety Promotion In the Process Industries*, Barcelona-Spain (1998).
5. T. Kapias, *Modelling the Behaviour of Pools of Sulphur trioxide and Oleum*, MSc dissertation, UMIST, Manchester (1996).
6. R.F. Griffiths (president of working party), *Sulphur trioxide, Oleum and Sulphuric acid mist*, Major hazards monograph, IChem^E (1996).
7. P.K. Raj and J.A. Morris, *Source Characterization and Heavy Gas Dispersion Models for Reactive Chemicals*, Technology and Management Systems, Inc., Burlington (1987).
8. Manju Mohan, T.S. Panwar and M.P. Singh, Development of dense gas dispersion model for emergency preparedness, *Atm. Env.* 29 (16): 2075-2087 (1994).
9. J.T. Jayne, U. Pöschl, Y. Chen, D. Dai, L.T. Molina, D.R. Worsnop, C.E. Kolb and M.J. Molina, Pressure and temperature dependence of the gas-phase reaction of SO_3 with H_2O and the heterogeneous reaction of SO_3 with H_2O/H_2SO_4 surfaces, *J. Phys. Chem. A* 101: 10000-10011 (1997).
10. U. Pöschl, M. Canagaratna, J.T. Jayne, L.T. Molina, D.R. Worsnop, C.E. Kolb and M.J. Molina, The mass accommodation coefficient of H_2SO_4 vapor on aqueous sulfuric acid surfaces and the gaseous diffusion coefficient of H_2SO_4 in N_2/H_2O, *J. Phys. Chem. A* to be published.
11. R.L. Basket, P.L. Vogt, W.W. Schalk, B.M. Pobanz, *ARAC Dispersion Modeling of the July 26, 1993 Oleum Tank Car Spill in Richmond, California*, Lawrence Livermore National Laboratory, Pleasanton (1994).
12. B.E. Wyslouzil, J.H. Seinfeld and R.C. Flagan, Binary nucleation in acid-water systems. II. Sulfuric acid-water and a comparison with methanesulfonic acid-water, *J. Chem. Phys.* 94 (10): 6842-6850 (1991).
13. R.C. Easter and L.K. Peaters, Binary homogeneous nucleation: temperature and relative humidity fluctuations, nonlinearity, and aspects of new particle production in the atmosphere, *J. App. Met.* 33: 775-784 (1994).
14. Jaecker-Voirol and P. Mirabel, Nucleation rate in a binary mixture of sulfuric acid and water vapor, *J. Phys. Chem.* 92: 3518-3521 (1988).
15. G.P. Ayers, R.W. Gillet and J.L. Gras, On the vapor pressure of sulfuric acid, *Atm. Env.* 15: 1221-1225 (1980).
16. Kirk-Othmer's Encyclopedia of Chemical Technology, *Sulfuric acid and Sulfur trioxide*, 3rd ed., John Wiley and Sons, (1983).
17. J.I. Holman, *Heat Transfer*, 5th ed., McGraw-Hill, New York (1981).
18. W.H. McAdams, *Heat Transmission*, McGraw-Hill, New York (1954).

DISPERSION FROM STRONGLY BUOYANT SOURCES

Jaakko Kukkonen[1], Juha Nikmo[1], Sue A. Ramsdale[2], David Martin[2], David M. Webber[3], Michael Schatzmann[4] and Joachim Liedtke[4]

[1]Finnish Meteorological Institute, Air Quality Research
Sahaajankatu 20 E, FIN-00810 Helsinki, Finland
[2]AEA Technology plc
Thomson House, Risley, Warrington, WA3 6AT, United Kingdom
[3]Integral Science and Software
484 Warrington Road, Culcheth, Warrington, WA3 5RA, United Kingdom
[4]University of Hamburg, Meteorological Institute
Bundesstraße 55, D-20146 Hamburg, Germany

INTRODUCTION

This paper reviews the project "Dispersion from strongly buoyant sources - BUOYANT" (1994-1997), which has addressed the atmospheric dispersion of pollutants emitted from typical fires in warehouses and chemical stores (Ramsdale et al., 1997). Such fires may represent a major hazard to people and the environment, and the fire plumes may contain a variety of harmful or toxic chemical compounds.

The BUOYANT project is related to earlier CEC research programme projects dealing with warehouse fires and the combustion of chemical substances, namely "Major Hazards arising from Fires in Warehouses and Chemical Stores" (Jagger, 1994), "Combustion of Chemical Substances and the Impact on the Environment of the Fire Products - COMBUSTION" and the TOXFIRE project (Smith-Hansen, 1994). Results of wind tunnel experiments on warehouse fires have been previously presented by, e.g., Hall et al. (1993). The previous projects have addressed in detail the combustion and fire development and the chemical quantification of the toxic products, whilst this project is focused on the dispersion of the toxic products in the atmosphere.

Integral models predicting the downstream evolution of steady plumes have been developed, and successfully applied in hazard analysis, for a number of physical situations, including passive plumes, plumes from stacks, high momentum jets, and steady continuous releases of a heavy gas cloud. The success of such models is based largely on their simplicity, which makes them readily available to the hazard analyst. The derivation of the models has relied heavily on insight gained from experiment, and their accuracy and credibility rests on validation against good quality, comprehensive experimental data.

Air Pollution Modeling and Its Application XIII, edited by S.-E. Gryning
and E. Batchvarova, Kluwer Academic/Plenum Publishers, 2000

Strongly buoyant plumes such as plumes above large fires have in the past received rather less attention than the other flows mentioned, both from the theoretical and experimental point of view. All the flow models share a number of common features, but the physics of dispersion is quite different in each case. This arises from the overall buoyancy of the flow and the very different turbulence properties which are strongly dependent on stratification, buoyancy and momentum.

For a more detailed discussion of this work, the reader is referred to the final reports of the individual partners, see Martin et al. (1997), Liedtke and Schatzmann (1997), and Nikmo et al. (1997).

OBJECTIVES

The overall objectives of the project have been to (1) develop a simple mathematical model of a plume designed for conditions of very high buoyancy, as are found in toxic plumes from fires (2) generate a carefully designed set of experimental data for the high buoyancy, near field region, covering appropriate ranges of the important source variables (3) write a PC-based computer program to solve the model equations in a way which is readily accessible to hazard analysts and (4) validate the model against this and other data.

Whilst AEA Technology (AEAT) has concentrated on the near and intermediate field of the plume, the Finnish Meteorological Institute (FMI) describes a model for the far-field dispersion (Figure 1). Far field dispersion is of particular importance for highly toxic substances. The models developed by AEAT and FMI have been encapsulated in a computer code called BUOYANT, which can be used by hazard analysts to predict the concentration of toxics at different distances from a highly buoyant source such as a large fire. The model has been validated against experimental data generated by the University of Hamburg (UH) in their wind tunnel facility, and also the Kincaid experimental field data.

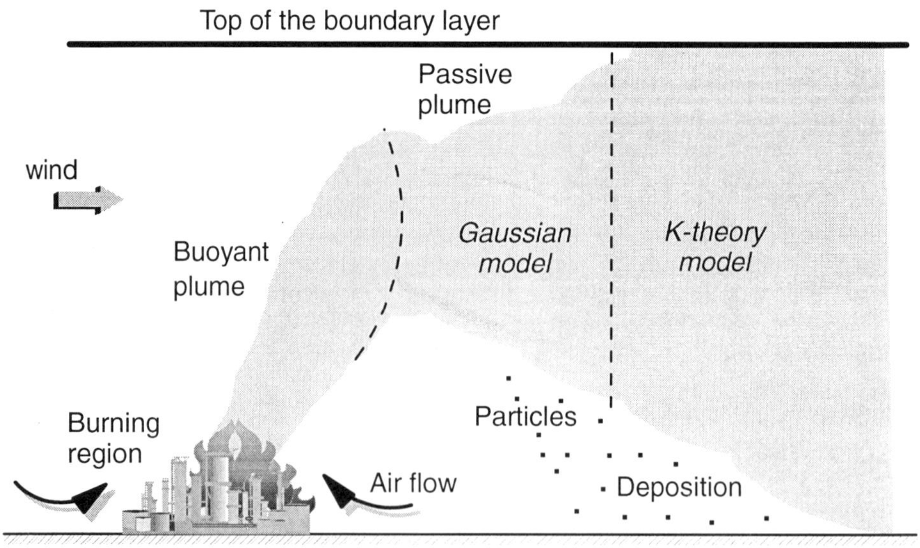

Figure 1. A schematic presentation of the dispersion of a fire plume and the modelling regimes.

Figure 1 shows schematically a plume formed in a major fire (Nikmo et al., 1996 and 1997). Horizontally, the scale of interest is a few tens of kilometres, and the vertical scale is in general that of the height of the atmospheric boundary layer (ABL). The source is assumed to represent a continuous, steady-state release over homogeneous terrain.

THE NEAR- AND INTERMEDIATE-FIELD DISPERSION

A complete analytical solution has been derived for a one dimensional model of a buoyant plume in still air (Martin et al., 1997). The complete solution reveals that there may be a neck (point of minimum radius) in the plume and a point of maximum velocity. If there is a neck, then the point of maximum velocity is always above it.

The three dimensional Reynolds equations (reduced to two dimensions by the assumption of cylindrical symmetry) have been examined to see whether a self similar solution is possible within and below the neck region. It is shown that, although it is impossible for all the important physical fields to be self similar simultaneously, a limited form of self similarity is possible in principle. This consists of the concentration and the vertical component of the plume velocity being self similar, but not the horizontal component of velocity or the density. The full equations can then be reduced to the one dimensional case already solved. The solution predicts the linearity of the 5/2 power of the momentum flux with the square of the mass flux, which provides a precise means of testing the model against experimental data.

The "self similar" solution has been tested against helium plume data of Liedtke and Schatzmann (1997). The data include radial profiles of concentration and vertical velocity at different heights through the neck region, providing an excellent test of the model. Although it is known that a self similar solution cannot extend all the way to the source, the model is found to be a good approximation down through the neck region of the plume.

A model has been developed for a buoyant plume in the presence of a wind. Instead of the plume rising vertically, as for the case when the atmosphere is calm, the plume is bent over by the wind. First of all, a model is described for the isothermal case, where only the wind velocity varies with height. Secondly, a model is described for the non-isothermal case, where other atmospheric properties vary with height.

A model has been described for what happens when a buoyant plume encounters a temperature inversion layer above the atmospheric boundary layer. The plume is assumed to be bent over, so that its upper side hits the inversion layer first. Buoyancy is gradually depleted as the plume interacts with the inversion layer, and the plume may run out of buoyancy while some material is still within the mixing layer. Alternatively, the plume may be sufficiently buoyant to fully penetrate the inversion layer, in which case its contents are effectively isolated from the ground over the distances and timescales of interest. A tentative expression is presented for the fraction of material available for dispersion to ground level after plume rise termination.

Experimental data corresponding to isothermal conditions (no inversion layer) with wind have been analysed, and compared with the predictions of the model. In most of the isothermal experiments, the plume either touched the ground (which lies outside the scope of the model) or bifurcated. However, one of the tests is used to determine best estimates for the entrainment coefficient and the added mass term. Further calculations are then performed for two other suitable tests, using the best estimate values. The calculations give some confidence in the use of the plume model under these conditions. The trajectory of the plume can be modelled quite well, and there is reasonable agreement between the calculated concentrations and relevant experimental results.

Experimental data have been analysed for a plume rising into an inversion layer in the absence of wind, and compared to the model. The final heights of the plumes (as determined experimentally by the maximum radius, and when the buoyancy goes to zero in the model) show reasonable agreement. However, the predicted concentrations are higher than those measured in the experiments and the predicted plume radii are lower than that measured. This appears to indicate that there is less entrainment occurring in the model than in the experiment. However, the model was not intended to be applied to cases where the plume was rising vertically, as in these tests, so good agreement with these data could not be expected. Experimental data for non-isothermal conditions with wind would be required to properly validate the simple model which has been developed.

THE FAR FIELD DISPERSION

The concept 'far field dispersion' refers here to the passive plume regime, in which the influence of plume buoyancy is no longer important.

A model has been developed for the local-scale dispersion of a continuous plume Nikmo et al. (1996 and 1997). The model is based on atmospheric boundary-layer scaling theory. In the vicinity of the source (~ 100 m – 1 km), Gaussian equations are used in both the horizontal and vertical directions. After a specified transition distance, gradient transfer (K-) theory is applied in the vertical direction, while the horizontal dispersion is still assumed to be Gaussian. The model also includes a treatment of the dry deposition of gases and particles, but wet deposition has been neglected.

Particular attention has been given to the description of the atmospheric boundary layer scaling (BLS), and the related problems of modelling the eddy diffusivity and the dispersion coefficients. The eddy diffusivity profiles have been modelled according to the ABL parametrization of Holtslag and Nieuwstadt (1986). We have presented a new model of the vertical eddy diffusivity (K_z) which is a continuous function of the height in all applied atmospheric scaling regions.

The computer program applies a numerical solver for the atmospheric diffusion equation (ADE). The numerical deviations of the model predictions from two analytic solutions of the ADE were less than two per cent for the computational regime.

The model has been tested against the Kincaid experimental field data (Olesen, 1995a). Figures 2a-b present scatter and quantile-quantile (cumulative distribution) plots of observed versus predicted concentrations respectively. The figures show normalised concentrations, defined as C / Q, where Q is the emission strength.

The average agreement of the predictions and the data is reasonably good, but the internal variation of the scatter plot is substantial. In particular, there is an excessive number of zero concentrations in the predicted data, compared with the measured values. This feature is also evident in the comparisons of other local scale models and the Kincaid data (Olesen, 1995b).

The model version presented here has certain limitations, as follows. The influence of the inversion layers and the dry deposition of gases and particles is included in the gradient transfer modelling, but neglected in the near field, in which the Gaussian dispersion model is applied in the vertical direction. The influence of these factors is usually more important at larger downwind distances. The model does not allow for wet deposition nor for the influence of inhomogeneous terrain or downwash effects.

The numerical testing of the various submodels showed that their behaviour is theoretically correct and produces numerically reasonable values. The functioning of the program is robust, and it has been thoroughly tested numerically. The accuracy of the numerical model was examined, by comparing its predictions against two analytic solutions of the atmospheric diffusion equation. For the selected example cases, the maximum error caused by the numerical method was approximately 2 %.

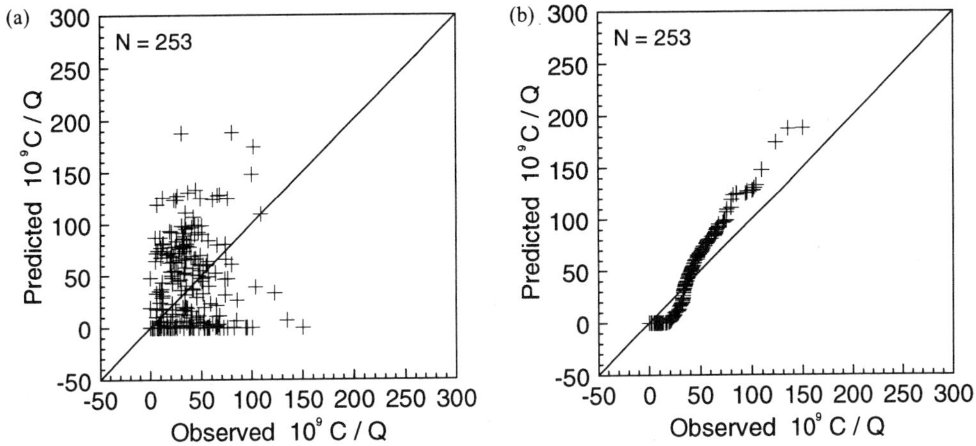

Figures 2a-b. Scatter and quantile-quantile (cumulative distribution) plots of observed versus predicted normalised concentrations C / Q (10^{-9} s/m^3), where C is the concentration and Q is the source strength. N is the total number of data points.

THE EXPERIMENTAL RESULTS IN A WIND TUNNEL

The experimental wind tunnel simulation of fire plume behaviour was conducted in unstratified boundary layers and in the presence of an elevated inversion (Liedtke and Schatzmann, 1997).

Experimental programme

The measurements were made in the Multi – Layer – Windtunnel (MLW) of the Meteorological Institute of University of Hamburg.

Due to the short (approximately 2 min) time intervals over which the inversion layers remained stationary, it was necessary to apply a measurement system which allows fast, complete and reliable determination of time-mean concentration fields. This was achieved by a Video Imaging System (VIS). With this instrument it was possible to collect and monitor the concentrations inside the fire plume during the time when temperature profile of the inversion layer was approximately constant.

The large density defect of the fire plumes was simulated with a mixture of helium and air (ρ_{air} = 1.293 kg/m^3, ρ_{He} = 0.1786 kg/m^3). Area sorces with diameters 16 mm < D_s < 150 mm were used for the measurements.

In order to achieve more information about the behaviour of the plumes, measurements of the vertical and lateral velocities inside the plume were made. The velocity measurements inside the plumes were done with a two-dimensional Laser-Doppler-Anemometer (LDA). To simulate the boundary layer, vorticity generators and roughness elements on the floor of the MLW were used. The boundary layer profiles were measured by a two-dimensional LDA system. The dispersion of the fire plumes was investigated at three oncoming free stream velocities at Froude numbers $0.031 < Fr_s < 1.058$, velocity ratios $0.057 < W_s / U_\infty < 2.0$ and density ratios $0.347 < \Delta\rho / \rho_s < 1.86$. The diameters of the source were 34 mm < D_s < 90 mm.

To study the behaviour of the fire plume beneath an elevated inversion layer without wind, two inversion layers with temperature gradients of approximately 3 °C / 100 m (in the large scale) were generated. With the geometrical scale of 1:1875 the heights of the inversion layers (in large scale) is $z_{i,P} = 640$ m ($z_{i,m} = 340$ mm) and $z_{i,P} = 450$ m ($z_{i,m} = 240$ mm).

Results

Experiments in an unstratified environment in zero wind speed conditions indicate that:

- the z-dependence of the vertical velocity is qualitatively the same as measured by Benech (1976) in a large scale experiment,
- the experiments also showed that the vertical velocity can be described by an equation given by Chen and Rodi (1975) for an unstratified environment under zero wind speed conditions.
- both the concentration and velocity profiles are approximately of Gaussian shape,
- the concentration decay at the plume centerline can also be described by the relationship given by Chen and Rodi (1975), but the value of the constant is again different.

From the investigations in an unstratified, turbulent boundary layer the following parameters of the boundary layer were determined: (1) a boundary layer thickness $\delta = 500$ mm, (2) a roughness length $z_0 = 0.7$ mm, (3) an exponent of the velocity profile $n = 0.22$, and the ratio of friction velocity u_* to free stream velocity U_∞, $u_* / U_\infty = 0.059$. We also found a good agreement of the measured turbulence intensities in the simulated boundary layers, compared with the turbulence intensities in the atmospheric boundary layer (ESDU - data).

The concentration measurements in the experiments with unstratified, turbulent boundary layers show:

- that the concentration profiles separate in two parts as described by Fanaki (1975) for the plume from a chimney,
- a good agreement of the plume rise with the equation given by Arya and Lape (1990),
- consistency with an entrainment coefficient $\beta_b = 0.6$ as also given by Arya and Lape (1990) for a large number of experiments.

For the simulation of plumes under an elevated inversion layer and under zero wind conditions:

- inversion layers with base heights of $z_{i,P} = 690$ m (IS1) and of $z_{i,P} = 450$ m (IS2) (at full scale with a geometric scale factor of M = 1:1875) were realised; the temperature gradients within the inversion layers were $\Delta T / \Delta z = 3.17$ and 2.99 °C / 100 m,
- the stratification below the inversion layer was neutral or slightly stable.

The concentration measurements in these inversion layers confirm:

- the power laws of the concentrations decay along the plume centerline is the same as published by Chen et al.,
- the value of the constant C_c was found to be $C_c = 4.42$ independent on the height of the inversion layer,
- the equilibrium heights of the plumes appeared to be independent of the Froude number (over the narrow range of Froude numbers investigated in these experiments),
- the final height of rise was mainly dependent on the height of the inversion layer base.

CONCLUSIONS

The near- and intermediate-field dispersion

A complete analytical solution has been derived for a one dimensional model of a buoyant plume in still air. The complete solution reveals that there may be a neck (point of minimum radius) in the plume and a point of maximum velocity. If there is a neck, then the point of maximum velocity is always above it.

The three dimensional Reynolds equations (reduced to two dimensions by the assumption of cylindrical symmetry) have been examined to see whether a self similar solution is possible within and below the neck region. It is shown that, although it is impossible for all the important physical fields to be self similar simultaneously, a limited form of self similarity is possible in principle.

A model has been developed for a buoyant plume in the presence of a wind. Instead of the plume rising vertically, as for the case when the atmosphere is calm, the plume is bent over by the wind. First of all, a model is described for the isothermal case, where only the wind velocity varies with height. Secondly, a model is described for the non-isothermal case, where other atmospheric properties vary with height.

A model has been developed for what happens when a buoyant plume encounters a temperature inversion layer above the atmospheric boundary layer. The plume is assumed to be bent over, so that its upper side hits the inversion layer first. A tentative expression is presented for the fraction of material available for dispersion to ground level after plume rise termination.

The far-field dispersion

We have developed a mathematical model for the local-scale dispersion of a continuous plume, emitted from typical fires in warehouses and chemical stores. We have focused on the passive plume regime, in which the influence of plume buoyancy is no longer important.

We have also presented a new model for the vertical eddy diffusivity, which is a continuous function of height in the various atmospheric scaling regions. Further, we have developed a numerical solver for the atmospheric diffusion equation. Numerical results are presented for the vertical profiles of the eddy diffusivity and the dispersion parameters, and spatial concentration distributions in various atmospheric conditions. We have also tested the model against the Kincaid experimental field data.

The experimental results in a wind tunnel

We have developed an experimental setup to measure the concentration profiles beneath an elevated inversion layer. For this purpose it was necessary to built a video imaging system for the on-line measurement of the concentrations. Moreover, a new kind of area source was developed to obtain a turbulent outflow out of the source at low or moderate Reynolds numbers.

In unstratified turbulent boundary layers experiments were made at Froude numbers lower than those covered in the past. The plume rise formula developed by Arya and Lape (1990) was found to be valid even under these conditions.

ACKNOWLEDGEMENTS

The project "Dispersion from strongly buoyant sources" was financed within the Environment programme of the European Union. The authors are grateful also for the financial support of the Academy of Finland, the Ministry of the Environment in Finland and the Landesumweltamt NRW in Essen, Germany. We wish to thank our coworkers in this project, Dr. Brian Underwood (AEAT), Mr. Steve Jones (AEAT), Mr. Juha-Pekka Tuovinen (FMI) and Mr. Ilkka Valkama (FMI) for successful cooperation. We are also grateful to Prof. Sylvain Joffre (FMI), Dr. Göran Nordlund (FMI), Dr. Gunnar Omstedt

(Swedish Meteorological and Hydrological Institute) and Mr. Erkki Rantakrans (FMI) for their valuable comments.

REFERENCES

Arya, R.S. and Lape, jr. J.F., 1990, A comparative study of different criteria for the modelling of buoyant plume rise in a neutral atmosphere. *Atmos. Environ.* 24a, pp. 289-295.

Benech, B., 1976, Experimental study of an artificial convective plume initiated from the ground. *J. Appl. Meteor.* 15, pp. 127-137.

Chen, C.J. and Rodi, W., 1975, A review of experimental data of vertical turbulent buoyant jets. Sonderforschungsbereich 80, Ausbreitungs- und Transportvorgaenge in Stroemungen, University of Karlsruhe.

Fanaki, F.H., 1975, Experimental observations of a bifurcated buoyant plume. *Bound.-Layer Meteor.* 9, pp. 479-495.

Hall., D.J., Kukadia, V. and Marsland, G.W., 1993, Dispersion of plumes from warehouse fires. Proceedings of the Industrial Fires Workshop, 11.-12.3.1993, Apeldoorn, Netherlands. Commission of the European Commission, EUR 15340 EN, pp. 203-222.

Holtslag, A.A.M. and Nieuwstadt, T.M., 1986, Scaling the atmospheric boundary layer. *Bound.-Layer Meteor.* 36, pp. 201-209.

Jagger, S.F., 1994, Major hazards arising from fires in warehouses and chemical stores: a summary of project STEP CT-90-0096. Proceedings of the Industrial Fires II Workshop, 17.-18.5.1994, Cadarache, France. European Commission, EUR 15967 EN, pp. 75-79.

Liedtke, J. and Schatzmann, J., 1997, Dispersion of plumes from strongly buoyant sources, Final Report.

Martin, D., Webber, D.M., Jones, S.J., Underwood, B.Y., Tickle, G.A. and Ramsdale, S.A., 1997, Near- and intermediate-field dispersion from strongly buoyant sources, AEAT/1388, Final Report.

Nikmo, J., Tuovinen, J.-P., Kukkonen, J., Valkama, I. and Rantakrans, E., 1996, Modelling local scale plume dispersion from industrial fires. Proceedings of the Industrial Fires III Workshop, 17.-18.9.1996, Risø, Denmark. European Commission, EUR 17477 EN, pp. 321-332.

Nikmo, J., Tuovinen, J.-P., Kukkonen, J. and Valkama, I., 1997, A hybrid plume model for local-scale dispersion. Finnish Meteorological Institute, Publications on Air Quality 27. Helsinki, 65 p.

Olesen, H.R., 1995a, Datasets and protocol for model validation. *Int. J. Environment and Pollution* 5, pp. 693-701.

Olesen, H.R., 1995b, The model validation exercise at Mol: overview of results. *Int. J. Environment and Pollution* 5, pp. 761-784.

Ramsdale, S.A., Martin, D., Nikmo, J., Kukkonen, J., Liedtke, J. and Schatzmann, M., 1997. Dispersion from strongly buoyant sources – Overall executive summary. AEA Technology Report AEAT/1408, Warrington, 16 p.

Smith-Hansen, L., 1994, The COMBUSTION and TOXIFIRE projects. Proceedings of the Industrial Fires II Workshop, 17.-18.5.1994, Cadarache, France. European Commission, EUR 15967 EN, pp. 43-58.

Webber, D.M., Jones, S.J. and Tickle, G.A., 1997, Dispersion in very buoyant plumes, Presented at IChemE Major Hazards Conference XIII, UMIST.

DISCUSSION

E. GENIKHOVICH: What was the reason to have two separate regions corresponding to the Gaussian model and the K-theory model? Actually, the Gaussian equation can be obtained as a solution of the K-equation.

J. KUKKONEN: It is known from basic diffusion theory that K-theory is not valid, if the characteristic length and time scales for changes in the mean concentration field are large, compared with the corresponding scales for turbulent transport. We therefore included an option to use Gaussian modelling near the source, if the source dimensions are small. However, the computer program can easily bypass the Gaussian modelling regime, if the user considers this to be appropriate.

M. KAASIK: Can this model be applied to the stack gas emissions?

J. KUKKONEN: The model has originally been developed to describe dispersion from strongly buoyant area sources. But in principle, it should be applicable also for moderately buoyant small area sources, like stack gases.

D. ANFOSSI: Which are the differences in your treatment of plume rise from the models, either simple analytical ones or based on the complete set of conservation equations, present in the literature?

J. KUKKONEN: The basic conservation laws are the same, which have been utilised, for instance, by Briggs. However, there are also substantial differences, for instance, in describing the penetration of inversion layers and in allowing for atmospheric boundary layer structure. There is a more detailed description in the project final report, published by AEA Technology.

A MODEL OF TRANSPORT, DEPOSITION AND RE-SUSPENDING OF HAZARD POLLUTION RELEASED BY POINT LARGE SOURCE AND CARRIED BY THE POLYDISPERSIONAL PARTICLES

Michael V.Galperin,[1] Olga G.Scotnikova[2], Michael P.Panin[2]

[1]Independent researcher, Studencheskaya str. 38-31, Moscow, 121165 Russia
[2]Moscow Physical - Engineering Institute, Kashirskoe sh. 31, Moscow 115230 Russia

INTRODUCTION

A model is intended for calculation of accidental release and continuous source of dangerous substances carried by particles (radioactive materials, hazard «heavy» metals, persistent organic species). The model includes two sub - models:

1.Advection and deposition of original release. The inherent time scale in this sub - model is 100 - 100000 sec..

2.Re-suspending (re-emission) of particles from surface by wind and extension of primary polluted area. The time scale of this sub - model is up to 10000 days.

ADVECTION, SEDIMENTATION, WET DEPOSITION AND PARTICLE SIZE DISTRIBUTION OF ORIGINAL RELEASE

The used advection scheme SCD is fully considered in the paper (*Galperin, 1997)* and its short description is included in the present proceedings and omitted here. It is a pseudo-Lagrangian 3D scheme with uniform step intracell concentration distribution and total time splitting of co-ordinates and processes.

The sedimentation, coagulation and wet deposition of the particles - carriers of hazard substances had been comprehensively studied at 50-60-th years during the nuclear weapon tests in atmosphere and results of this investigations are unique up to the present (e. g. *Frankiel and Sheppard, 1959; Stade, 1968; Junge, 1977; Makhonko, 1968*).

The velocity of sedimentation (gravitational deposition) V_S and washout efficiency (capture factor) γ are also governed by Stokes law which is well fulfilled within the range 0.01-1000 μm^2 g/cm^3 (*Wark and Warner, 1977*). At less value of ρD^2 (ρ is density and D is mean size of particles) the particle behaviour is similar to that of «weightless» gas. V_S and γ strong depend on particle size (Fig. 1).

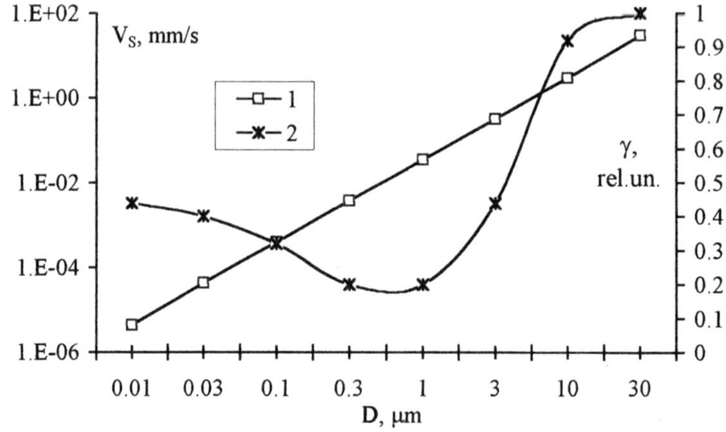

Fig. 1 Sedimentation velocity V_S (1) and washout capture factor γ (2) versus particle size $\boldsymbol{D}$ for particle density 1g/cm^3.

While the particles have been removed from the atmosphere their air distribution is distorted (Fig. 2) and mean values of V_S and γ are changed in their turn. The usual method of attack the problem is to split the particle distribution into ranges and to calculate each of ones independently. This approach requires the large computer resources for getting a good accuracy.

However the knowledge of mean values of V_S and γ (these are W and $\bar{\gamma}$) is sufficiently for calculation and the problem is to determine them at each time step. Due to one-to-one connection between V_S and γ it is reasonably to store W only. Let K_z be a vertical diffusion coefficient and z be a vertical co-ordinate, then

$$\partial C/\partial t = \partial (CW)/\partial z + \partial [K_z(\partial C/\partial z)]/\partial z \tag{1}$$

is equation for vertical concentration $C(z,t)$ and

$$\partial (CW)/\partial t = \partial (C\sigma_v^2)/\partial z + \partial (CW^2)/\partial z + \partial \{K_z[\partial (CW)/\partial z]\}/\partial z \tag{2}$$

is equation for vertical impulse $CW(z,t)$, where σ_v^2 is a variance (second central moment of distribution) of sedimentation velocity $V_S(z,t)$ (*Galperin and Masljaev, 1996; Galperin and Syrakov, 1997*).

The same equation should be constructed for $C\sigma_v^2$ with third moment in right part and so on. This is a problem of closure similar to that in turbulence theory. For important real case the situation is more simply. Let assume that distribution density of V_S can be approximated by Pearson curve with only parameter W:

$$f(V_S) = V_S^{p-1} exp(-pV_S/W)/[(W/p)^p \Gamma(p)], \tag{3}$$

where $p \geq 1$ is a factor of the shape and $\Gamma(p)$ is a gamma-function. It is easy to show that shape factor is a constant ($p = const.$) during sedimentation and washout. The equation for vertical impulse (2) gets the view:

$$\partial (CW) / \partial t = (1+1/p)\partial (CW^2) / \partial z + \partial \{K_z[\partial (CW)/\partial z]\} / \partial z \tag{4}$$

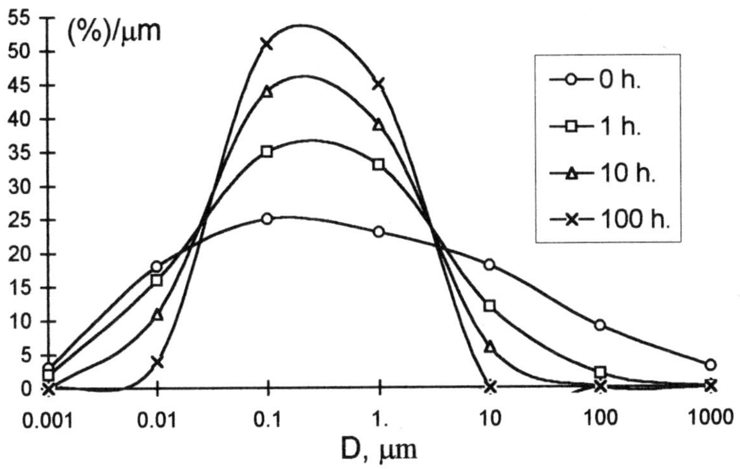

Fig. 2 Particle size distribution versus residence time in the atmosphere.

Arguments in favour of choice $p \approx 1$ can be advanced under maximum uncertainty conditions, that is inherent in practice. Under this assumption the particles have a Rayleigh distribution with size

$$f(D) = (D/\sigma^2)\, exp(-D^2/2\sigma^2), \qquad (5)$$

where $\sigma^2 = 2D_m^2/\pi$ and D_m is mathematical expectation of D.

It is assumed above that the hazard substance resides on particle surface, substance mass is proportional to D^2 and consequently $f(V_S)$ of substance is expressed in relative particle area. When the substance itself forms the particles, its mass is proportional to D^3 and its $f(V_S)$ should be expressed in relative mass (or volume). For example, if (3) has a view $f_S(V_S) = exp(-V_S/W)/W$ for surface, then for volume we have:

$$f_V(V_S) = 2(3/2)^{-3/2}\, V_S^{1/2}\, exp(-3V_S/2W)/\pi^{1/2}(2W/3)^{3/2}$$

with mathematical expectation $3W/2$and standard deviation $W(3/2)^{1/2}$.

The surface dry deposition is described by usual way *(Sehmel, 1980; Williams, 1982; Faith and Atkisson, 1972)*.

It is supposed that washout of particles is described by linear equation:

$$dC/dt = -\Lambda C, \qquad (6)$$

where Λ is Langmuir factor, which depend on precipitation intensity I (*Engelmann, 1963, Galperin, 1989)*:

$$\Lambda = \gamma S_0 I/V_P \approx 1.5\, \gamma I/d. \qquad (7)$$

Here S_0 is a great circle area of spherical drop (or an effective surface of a snow flake), V_p is drop volume, d is its diameter.

Drop sizes are varied within a wide range and their spectrum depends on precipitation intensities (*Kelkar, 1959; Engelmann, 1965; Makhonko, 1968, etc.*). Summarised experimental data on washout velocities (*Avramenko and Makhonko, 1971; Tseselev et al, 1970*) indicate that an excess of drops with small d at low precipitation intensity leads to shift of dependence (7) and the approximation

$$\Lambda = 1.5\,\gamma\,(I + I_0)/d_{max}\,, \tag{8}$$

where I_0 = 0.4mm/h for rain and I_0 = 0 for snow, d_{max} = 1mm, may be accepted.

The efficiency of capture is proportional to and $\bar{W}$ within range 1 - 20 μm (*Chemberlain, 1953, 1960*). Taking into account the dependence of γ on drop size, shown on Fig.1, it is possible to express $\bar{\gamma}$ by sum: $\bar{\gamma} = \gamma_0 + \bar{\gamma}_W$, where $\gamma_0 \approx 0.15$, $\bar{\gamma}_W = 0.85W/(W_{ref} + W)$ and $W_{ref} \approx$ 0.5cm/s.

The particle spectrum variations during washout are much faster than during dry deposition. Turning back to the multilayer description of the deposition, we may write the following equations for vertical impulse in case of precipitation's:

$$d(CW)/dt = -\,1.5\,(\gamma_0 + 2\bar{\gamma}_W)\,(I + I_0)(CW)/d_{max}. \tag{9}$$

In (9) multiplier 2 upstream of $\bar{\gamma}_W$ is arisen due to simultaneous decrease of the mass and sedimentation velocity. At the same time rapid leap-wise decrease of W and consequently $\bar{\gamma}_W$ takes place. It is just the reason for the fact that the application of rough models leads to big errors.

Processes of particle washout in clouds are very complicated. Particles themselves can be the centres of coagulation of drops and enter the air again at the evaporation. In a rough approximation it may be supposed that in clouds $\bar{\gamma}$ increases as much as 4 since the whole surface of drops and ice crystals takes part in particles capture.

SOURCE HEIGHT AND VERTICAL PARTICLE CONCENTRATION DISTRIBUTION

The study of vertical particle distributions was carried out with the two goal: the testing of multilayer SCD model and investigation the distribution under different conditions. The features of 25-layers SCD scheme with unlimited upper layer give the possibilities to eliminate of numerical effects and to obtain the realistic pattern of phenomenon's. The dispersion of point puff is studied. The example of vertical distribution of particles and the K_z shape are shown in the Fig. 3. This is pattern, integrated over whole horizontal domain, and in single point the shapes may differ. There are two concentration maximum after ten hours. The first one is close to surface and associated with gravitational sedimentation. The second one is close to boundary layer height (***BLH***). The minimum of concentration coincides with maximum of K_z. To explain this phenomena the equation (1) should be transformed:

$$\partial C/\partial t = \partial\,(CW)/\partial z + (\partial K_z/\partial z)(\partial C/\partial z) + K_z(z)\,\partial^2 C/\partial z^2. \tag{10}$$

The third summand in right member of (!0) is the diffusion term as such and the second one includes a directional «velocity factor» (with velocity dimensionality) $\partial K_z/\partial z$. This factor should push the particles (and other substances) out of the K_z maximum and attract them to its minimum. Consequently, the maximum of K_z should produce the barrier (Fig. 4) for the substances. The height of the barrier layer does not equal to ***BLH***, which is determined by wind or K_z profile. When unstable conditions are take place bimodal shape is smoothed and the concentration profile does not depend on source height after one hour. In case surface source (H_s=1m) an elevation of puff up to ***BLH*** takes about one hour, then it is retarded near ***BLH*** and the pattern becomes similar to high source.

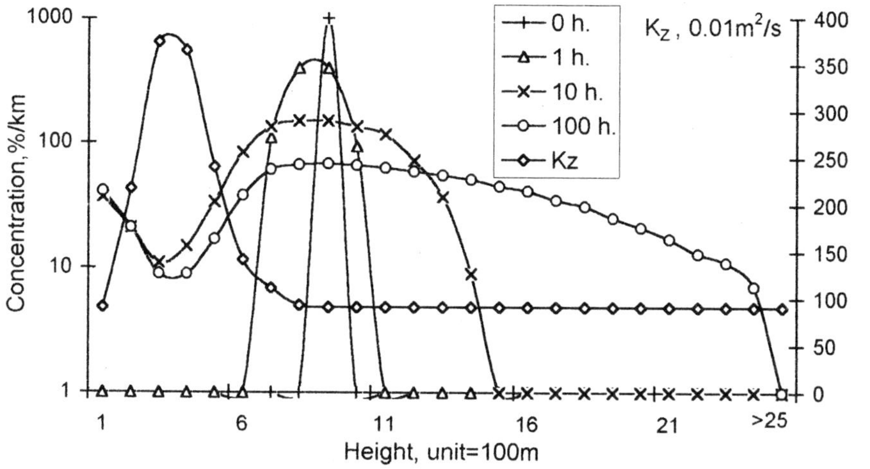

Fig. 3. Vertical profile of particle concentration versus residence time at stable atmosphere and instantaneous point source placed over a boundary layer height (***H_S>BLH***). The assumed profile of K_Z is shown also.

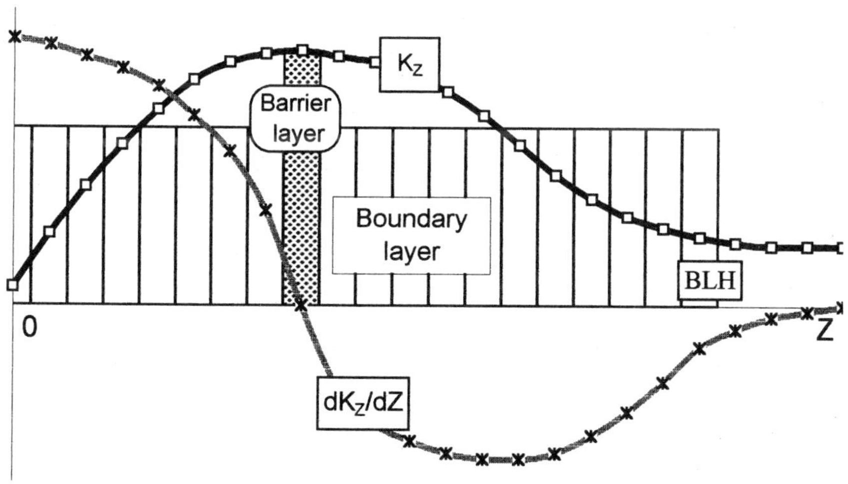

Fig. 4. The barrier layer formation.

RE-SUSPENDING OF PARTICLES FROM SURFACE BY WIND

The re-suspending of hazard particles from surface to atmosphere is governed by two processes taken into account in the model. On the one hand the hazard substances are chemically bonded or degraded step by step in the soil, for another the remained part of them may be escalated in air by strong winds. For description these interrelated processes the following equation set are used

$$\Delta Q_{free}(t)/\Delta t = F_{dep}(t) - F_{rem}(t) - \Delta Q_{fix}(t)/\Delta t; \tag{11.1}$$

$$F_{rem}(t) = \lambda_{rem}(\tau, I, u, \psi)\, Q_{free}(t); \tag{11.2}$$

$$\Delta Q_{fix}(t)/\Delta t = \lambda_{decay}(\tau, \psi)\, Q_{free}(t), \tag{11.3}$$

where $Q_{free}(t)$ is the deposited quantity of substance capable of re-suspending, $F_{dep}(t)$ is deposition flux, $F_{rem}(t)$ is re-suspending flux, $Q_{fix}(t)$ is the mass of decayed substance, $\lambda_{rem}(\tau, I, u, \psi)$ and $\lambda_{decay}(\tau, \psi)$ correspondingly re-suspending and decay factors depending on age of substance τ, precipitation I, wind velocity u and surface conditions ψ.

The difficult problem in framework of K-theory (in fact there is not other one for modelling practice) is true surface source ($H_s = 0$). We deal with it in case of re-suspending. It may be suggested that re-suspending starts, if the turbulence penetrates into roughness layer. It can occur, if Reynolds number exceeds the critical value:

$$Re = u_{*crit}\, z_0 / \nu \geq Re_{crit}, \tag{12}$$

where u_{*crit} is the threshold friction (shear) velocity, z_0 is roughness length and ν is kinematic viscosity of air.

Threshold velocities were estimated by a number authors (see *Gillette, 1979;Gillette and Passi, 1988*) for soil erosion, but the erosion is significantly differed from considered problem. During the erosion a part of wind energy may be spent for detachment particles from soil, because elevated mass is a material of surface itself, and secondary knocking out (splash) of particles is of basic importance. The splash does not influence on re-suspending of particles - carriers due to their small concentration under usual conditions. An erosion also is absent from many of the surfaces (for example, the forest or grass), but re-suspending is possible for each ones. Therefore the values of u_{*crit} estimated by equation (12) should quite differ for some surfaces. The estimations of u_{*crit} obtained under usual values of $Re_{crit} \approx 1200$ and $\nu \approx 15 \cdot 10^{-6}\ m^2/s$ are given in Table 1.

Table 1. Estimations of threshold friction velocity of different surface

Type of surface	*Equation (12)*	*Gillette's estimation*
Ploughed sand soil	0.2 - 0.5	0.45
Ploughed black earth	0.2 - 0.5	0.7
Forest	0.01 - 0.05	no data
City, mountains	0.005 - 0.01	up to 10.0
Salt marshes	0.4 - 4.0	1.0
High grass	0.05 - 0.15	no data
Sand dunes	0.2 - 0.8	0.25

The erosion is usually described by power function of 3 - 5-th order, for example according to the formula obtained by Gillettte, the re-suspending factor is:

$$\lambda_{rem}(\tau, I, u, \psi) = \lambda_{rem\,0}(\tau) u_*^4 (1 - u_{*crit}/u_*),\ u_* \geq u_{*crit}. \qquad (13)$$

It is conventionally assumed, that $\lambda_{rem\,0}(\tau)$ depended on substance age due to partially binding of substance: $\lambda_{rem\,0}(\tau) = \lambda_{rem\,00}(\tau) exp(-\tau/\tau_0)$.

Since the re-suspending exists only with high wind, the small particles are subjected to long-range transport (LRT). At the same time the large heavy particles can be ejected from a surface, but quickly fall backward (the saltation, see *Anderson and Haff, 1988*). As a result the particles - carriers separate into two populations depending on friction velocity and their size and density.

Friction stress is $\rho_{air} u_*^2$ ($\rho_{air} \approx 1.25 \cdot 10^3$ g cm^3 is a air density) and a force of shift

$$F = \rho_{air} u_*^2 \pi D^2 / 4, \qquad (14)$$

acts on a particle. A static friction force of particle is

$$F_{fr} = k_{fr} \rho g \pi D^3 / 6, \qquad (15)$$

where k_f is a friction coefficient and g is gravitational acceleration.

Taking into account the Stokes law for spherical particle

$$V_S = \rho g D^2 / 18 \rho_{air} \nu,$$

we obtain value of u_*, which results in particle ejection from surface

$$u_{*\,eject} = (8 k_{fr}^2 \rho g \nu V_S / \rho_{air})^{1/4} = (\alpha V_S)^{1/4}. \qquad (16)$$

If a particle distribution with V_S is (3) and $p = 1$, then distribution of ejected (shifted) particles with u_* is

$$f(u_*) = [(4 u_*^3 / \alpha W) exp(-u_*^4 / \alpha W)]. \qquad (17)$$

A total mass of ejected particles is

$$M = \int_0^{u_*} f(u_*)\, d u_*$$

and it is proportional to u_*^4. This result well agrees with dependence (13) obtained in (*Gillette D.A., 1979; Gillette D.A., Passi R., 1988*).

When inequality $u_* \geq u_{*\,eject}$ is fulfilled, the particle is implicated into saltation and not necessarily into LRT. The ejected heavy particles can not be subject to LRT and slowly move jumping lengthways the surface.

The vertical diffusion velocity of particle escalation is (see equation (10))

$$\partial K_z / \partial z \approx æ u_*, \qquad (18)$$

where æ is a Karman constant, and wind escalation of particle is possible only if $\partial K_z/\partial z \geq V_S$.

CONCLUSION

The number of questions in considered problem is more than number of known answers. However a practice can not wait until the all scientific problems will be solved.

Therefore two air transport models for real time predictions of consequences of accidental release of particles, carrying the hazard radioactive substances, were developed on the presented ideas. The first one has been designed for solution of local problems and a grid 50×50 cells with grid size 1km was used in it. The second one is regional scale model with a grid 60×60 cells and grid size 20km. It is aimed for evaluation the worst consequences of large emergency in far receiver (for example big city). The models can include up to 50 isotopes. The 21 isotopes are included in the second model at present.

REFERENCES

Anderson R.S., Haff P.K., 1988, Simulation of eolian saltation, *Science*, 241, pp.821-823.

Avramenko A.S., Makhonko K.P., 1971, A study of washout of radioactive products from the atmosphere in the mountains of the central Caucasus, *Proc. of the Institute of Experimental Meteorology*, 21, Obninsk. [Авраменко А.С., Махонько К.П., 1971, Исследование вымывания радиоактивных примесей в горах центрального Кавказа, *Труды ИЭМ*, 21, Обнинск.]

Chamberlain A.C., 1953, *Aspects of travel and deposition of aerosol and vapour clouds*, British Report AERE-HP/R-1261.

Chamberlain A.C., 1960, Aspects of deposition of radioactive and other gases and particles, *Int. J. Air Pollution*, vol. 3, pp.63-88.

Engelmann R.J., 1963, *Rain scavenging of particulates*, USA EC Report HW-79382, Hanford Atomic Products Operation.

Engelmann R.J., 1965, Rain scavenging of zinc sulfide particles, *J.Atmosph.Sci.*, v.22, 6, pp.719-729.

Faith W.L., Atkisson A.A., 1972, *Air pollution*, Second ed., John Wiley and Sons, NY.

Frankiel F.N. and Sheppard P.A. (eds), 1959, *Atmospheric diffusion and air pollution*, Academic Press, NY, London.

Galperin M.V., 1989, Adsorption-kinetic nonlinear washout model of sulphur and nitrogen compounds from the atmosphere, in: *Air Pollution Modeling and Its Application VII*, Plenum Press, N.Y. & London.

Galperin M.V., 1997, Two numerical advection schemes for atmospheric models: development and comparison testing, in: *Proc. of Swedish - Bulgarian Workshop on Long-Range Air Pollution*, Bulgarian Academy of Science, Sozopol.

Galperin M.V., Masljaev A.M., 1996, *Pilot version of the EMEP/MSC-E model for transport of airborne persistent organic pollutants. Description and preliminary sensitivity analysis*, EMEP/MSC-E Technical report 3/96, June, 1996, MSC-E, Moscow

Galperin M.V. Syrakov D.E., 1997: A model for airborne poly-dispersive particle transport and deposition, in: *Proc. of the XXII NATO/CCMS Int. Technical Meeting on Air Pollution Modeling and Its Application*, Clermont - Ferrand, France, pp.111-118.

Gillette D.A., 1979: Environmental factors affecting dust emission by wind erosion, in: Saharah dust, ed. by C.Morales, SCOPE 14, Wiley and Sons, NY.

Gillette D.A., Passi R., 1988: Modelling emission caused by wind erosion, *J. of Geoph.Res.*, 93, pp.14233-14242.

Gusev N.G., Belyaev V.A., 1991: Radioactive release in biosphere (Handbook), Energoatomizdat, Moscow. [Гусев Н.Г., Беляев В.А., 1991: Радиоактивные выбросы в биосфере (Справочник), Энергоатомиздат, М.]

Junge C.E. 1977: Basis consideration about trace constituents in the atmosphere is related to the fate of global pollutants, in: Fate of pollutants in the air and water environment, Part I, I.H. Suffet (ed.) (Advances in environmental science and technology, vol. 8), Wiley-Interscience, NY.

Kelkar V.N., 1959: Size distribution of raindrops, *Indian J. Meteorol. Geophys.*, v.10, 2, pp.125-136.

Makhonko K.P., 1968: Sedimentation of radioactive dust and its washout from the atmosphere with precipitation's, Atomizdat, Moscow. [Махонько К.П., 1968: Оседание радиоактивной пыли и её удаление из атмосферы осадками, Атомиздат, М.]

Sehmel G.A., 1980: Particle and gaz dry deposition: a review, *Atmospheric environment*, vol. 14, No 9, pp.983-1011.

Stade D.H. (ed.), 1968: Meteorology and atomic energy, Pergamon Press, NY.

Tsevelev M.A., Chalov P.I., Makhonko K.P., 1970: A study of regularities of washout of radioactive fission products in mountains and foothills, Proceedings of the Institute of Experimental Meteorology, 5, Obninsk.[Цевелев М.А., Чалов П.И., Махонько К.П. (1970) Изучение закономерностей вымывания радиоактивных продуктов деления в горах и предгорьях, Труды ИЭМ, 5, Обнинск.]

Wark K., Warner C.F., 1977: Air pollution, its origin and control, Harper & Row, NY.

Williams R.M., 1982: A model for the dry deposition of particles to natural water surfaces, *Atmospheric environment*, vol. 16, 1933-1938.

DISCUSSION

D. WILSON: Please explain the special conditions under which the shape factor p of Gamma distribution remains constant during a deposition and sedimentation process.

M. GALPERIN: There are no such special conditions. Only parameter of Gamma distribution (that is mathematical expectation) is changing during a sedimentation process. I should note that Gamma distribution corresponds the maximum of Shannon's entropy and it is only distribution finite at the left, which is stable in its family and has a limited moments.

CLIMATOLOGICAL TRAJECTORY ANALYSIS FOR POTENTIAL ACCIDENTS AT THE KOLA NUCLEAR POWER PLANT

Jørgen Saltbones, Anstein Foss, and Jerzy Bartnicki

Norwegian Meteorological Institute (DNMI)
PO Box 43 Blindern 0313 Oslo, Norway

INTRODUCTION

Norwegian Authorities have for a long time been concerned with different aspects of the safety at nuclear installations in former Soviet Union. As a result, Report to Norwegian Parliament, nu 34 for 1993 - 1994 was worked out: Nuclear Activity and Chemical Weapons in our near Northern Surroundings[1]. A plan for follow-up activities was worked out by the Ministry of Foreign Affairs[2].

The unsatisfactory safety of the Kola Nuclear Power Plant (Kola NPP) was the focus of attention in these official reports. As a response to the 'follow-up' plan[2], DNMI has together with two Norwegian institutions; Norwegian Radiation Protection Authority (NRPA) and Institute for Energy Technology (IFE), been involved in a joint project: Consequence Analysis of Potential Accidents at the Kola Nuclear Power Plant[3,4].

In an ongoing review of the Norwegian preparedness organization against nuclear accidents, the results from this project were described as very useful[5]. Working out 'worst case scenarios' for the threat to the country from nuclear power plants are useful examples of what the preparedness organization might come up against. This also give background information on the level of preparedness the organization should aim for.

CLIMATOLOGICAL TRAJECTORY ANALYSIS

Some longer time scale aspects of the atmospheric transport of radioactive air pollutants can tentatively be formulated as the following question: What is the probability for a radioactive cloud, emitted during an accident, to 'hit' a certain location? We have approached this problem by a methodology we have denoted as 'climatological trajectory analysis'. By the word 'climatological' we here mean the 'long time scale' aspects. Similar methodology has been used earlier by other authors[6,7].

Air Pollution Modeling and Its Application XIII, edited by S.-E. Gryning
and E. Batchvarova, Kluwer Academic/Plenum Publishers, 2000

An important by-product of the trajectory analysis was a screening procedure for the selection of suitable meteorological transport/deposition episodes for 'worst case scenarios'. These episode studies will hopefully give better insight into the meteorological variability of the atmospheric transport for the emitted pollutants, then can be obtained by statistical analysis of trajectories alone.

To highlight the episodic nature of the atmospheric transport of the accidental releases, it is important to stress that: It is the weather at the time of the release and in the following period which will determine the consequences for specific locations. For a given weather situation, a specific location may get the full impact from the accidental releases, even though the statistical probability for this outcome may be low. This means that the consequences from one nuclear accident can have quite another distribution, then the probability assessment of the consequences from hypothetical accidents. We must remember that consequences and risks can be large even though the probability may be low. This was one of the lessons learned from the Chernobyl accident. In other words, the analysis of the 'long time scale' aspects formulated in terms of probability concepts, must not be 'over-interpreted'.

With this precaution in mind, we have formulated questions that can best be answered in 'probabilistic terms', such as: - What is the probability for a radioactive cloud from a given source to arrive (or pass over) a specified location of interest? We have limited our application to the following setup:

- Assume that the accident can happen at any time with equal probability. What will then be the probability for a specific location to be effected (or 'hit') by such accidental releases?

The climatological trajectory analysis was performed in two steps:

Step 1: We compiled meteorological data for a period of 11 years - (1985 - 1995) - with assured homogeneous characteristics in time. The temporal resolution of the meteorological fields was 6 hours. This data set has previously been used at DNMI for computing transboundary transport of air pollutants in Europe for the European Monitoring and Evaluation Programme (EMEP)[8]. The methods and procedures for setting up this data set, are described in details in ref. (9). The EMEP data used in our study includes: Precipitation fields (accumulated 6-hourly values), fields for the mixing height (which is updated every 12 hours) and instantaneous wind fields (every 6 hours from the level σ=0.925 in the LAM150 model). LAM150 is abbreviation for DNMI's previous numerical weather prediction model: Limited Area Model - 150 km grid squares[10,11]. This level (σ=0.925) corresponds roughly to ~ 600 m height above the terrain and represents reasonably well the level for the bulk transport of pollutants in the atmospheric boundary layer (ABL).

Step 2: The compiled wind fields were used to compute forward trajectories (10 days long) for air parcels starting from the location of the Kola NPP. The trajectories are computed in a 50 km grid square system and according to a method described by S. Pettersen[12]. One air parcel was released every hour in these 11 years: (1985 - 1995). The time-step between the points along each trajectory is 15 minutes to assure that the distance between sequential points during one time-step is shorter than the grid square size (50 km). All together - 96 162 - air parcels were released during the 11-years period.

In our analysis, we have assumed that the probability of arrival - PA_{ij} - of the radioactive cloud to a given grid square (i,j) in case of an accident during these 11 years, is equal to the number of trajectories crossing this grid square - N_{ij} - divided by the total number of air parcels/trajectories released - N_{tot}:

$$PA_{ij} = \frac{N_{ij}}{N_{tot}} \tag{1}$$

This formulation of the problem and the use of two dimensional trajectories involves an implicit assumption that the transport of radioactive material released during a nuclear accident mainly takes place in the atmospheric boundary layer. This is not true in general and for the episode studies performed as part of this project, the transport at higher levels of the atmosphere is be taken into account.

RESULTS

The main body of results from the climatological trajectory analysis described in the previous paragraph consists of:

1: Map of probability of arrival - or more precisely - the probability for a grid square (i,j) to be hit (or crossed) by the trajectories of air parcels starting from the Kola NPP.
2: Map of average travel time.
3: Map of minimum travel time.
4: Distribution of the travel time for selected sites in Norway.
5: Calendar for selection of episodes for closer study.

Probability of being 'hit'

A map of the probability of being hit by the emitted material is shown in Figure 1 in a grid system with 50 km resolution. The probability of the arrival of 'a radioactive cloud' from the Kola NPP to each grid square (i,j), was calculated according to Equation (1). The computed probability is dependent on the length of the trajectories. In general, the probability of arrival increases with the length of the trajectories, but for trajectories longer than 8 days it does not change much. Therefore, we have used 10 days trajectories in our analysis.

The probability distribution presented in Figure 1 is rather smooth and surprisingly isotropic (note the logarithmic scale for the different probability values). The pattern for the different probability values, and especially for the larger values, resemble flattened circles extending slightly in the north-east direction. A clear blocking effect of Greenland as well as a less pronounced blocking effect of the Scandinavian mountains are visible in Figure 1.

Looking more closely at Figure 1 for areas in Norway, the probability of arrival is as expected highest for eastern part of Finnmark, ~ 5%. For Trøndelag (central section of the country) the probability is ~ 1% and for the southern part ~ 0.5%.

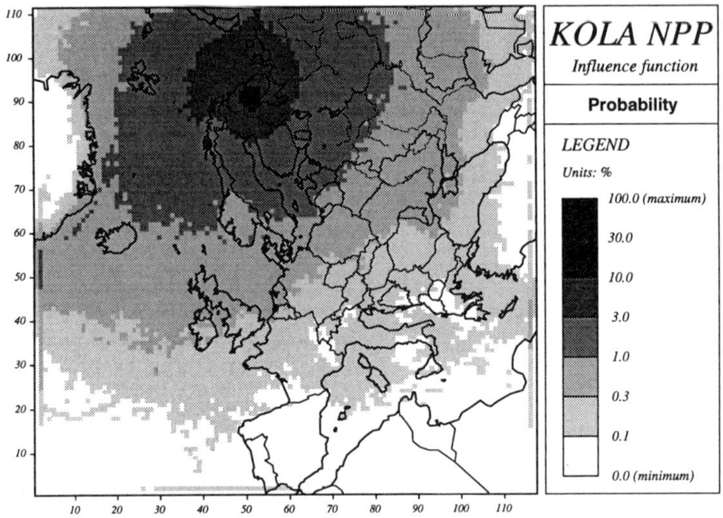

Figure 1. Map for the probability of arrival based on Equation (1) and calculated for the trajectories of air parcels released every hour from Kola NPP for a period of 11 years; 1985 - 1995.

Average travel time

Travel time from the Kola NPP to each 50 km grid square was computed in the following way: The number of points along the trajectories (marked every 15 min.) between the Kola NPP and a given grid square (i,j) was calculated. On each trajectory, the counting started from the second point (the location of Kola NPP was omitted) and was continued until the first point in the grid square (i,j) was reached. The number of points was multiplied by the time step for the trajectories ($\Delta t = 15\ min.$) to calculate the travel time. In this procedure, we did not take into account other trajectory points in the same grid square (i,j).

A map of the average travel time from the Kola NPP to each 50 km grid square (i,j) is shown in Figure 2. As in the case for the probability of arrival, computations were based on the same 11 years of meteorological data. The maximum of average travel time in Figure 2 is 239.75 hours, which is almost the same as 10 days; the chosen trajectory length. It means that trajectories 10 days long are still not long enough for the climatological trajectory analysis.

The pattern on Figure 2 is also quite regular. However, as seen from Figure 2, the transport in north-east direction is slightly faster then in other directions, indicated from the slightly elongated pattern in this direction. Among the Nordic countries, Finland can be reached relatively fast. Typical values for the average travel time is 1-2 days. However, we must bear in mind that the radioactive cloud in case of an accident, seldom travels according to average meteorological conditions.

Average travel time to Norway is between 2 and 6 days, depending on the location. The influence of mountains in the south-western part of the country is visible. For example, average travel time to Oslo is shorter than to Bergen.

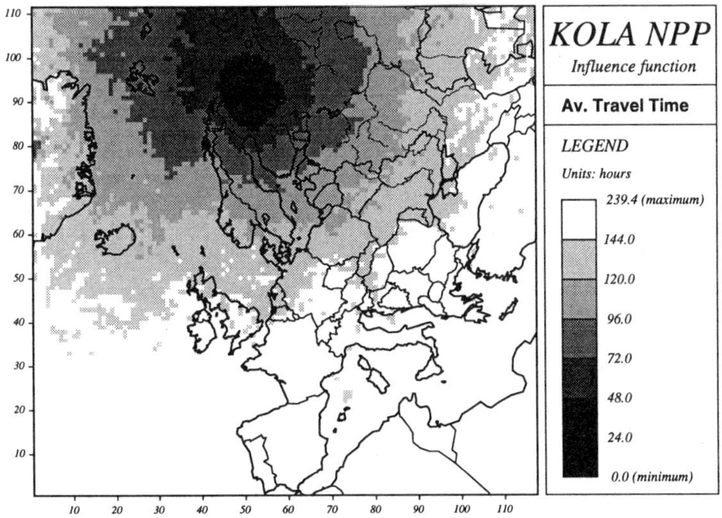

Figure 2: Map of the average travel time from the Kola NPP to any given grid square (i,j) based on 11 years of meteorological data; 1985 - 1995.

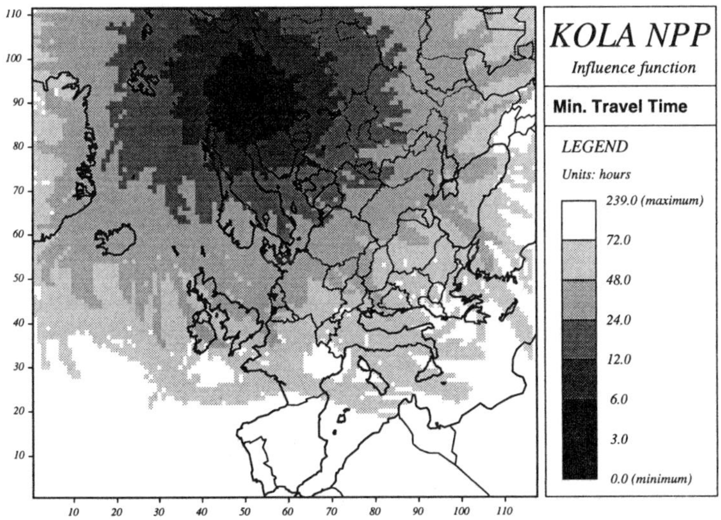

Figure 3. Map of the minimum (shortest) travel time from the Kola NPP to any given grid square (i,j) based on 11 years of meteorological data; 1985 - 1995.

Minimum travel time

Of special interest is perhaps the map for minimum travel time from Kola NPP to any grid square (i,j), as shown in Figure 3. It is somewhat surprising to see how isotropic this map looks like. No direction show markedly stronger winds then any other direction. We do not claim that this map is the absolute extreme values for the minimum travel time, but it is the minimum found from quite a large data set - covering a period of 11 years from 1985 to 1996. We can see from Figure 3 that a 'radioactive cloud' from Kola NPP can reach the Norwegian border in Eastern Finnmark in about 3 hours, Bodø in less then 12 hours and Trondheim within 24 hours.

Distributions of the travel time for selected sites in Norway

A statistical analysis of the travel time has been performed for some selected receptors: Kirkenes, Tromsø, Bodø, Trondheim, Oslo, Bergen and Kristiansand. In addition, we have included one grid square in Russia, next to the grid square containing the Kola NPP. On Figure 4 in shown the distributions of the travel time for each of these selected receptors.

We can make the following statement for Kirkenes: If the cloud 'hits' this location, the probability of arrival is in the range ~ (3 - 10)% as seen from Figure 1, there is ~ 45% probability that the cloud will arrive within 24 hours after the release, as seen from Figure 4. Similar statements can be given for other locations.

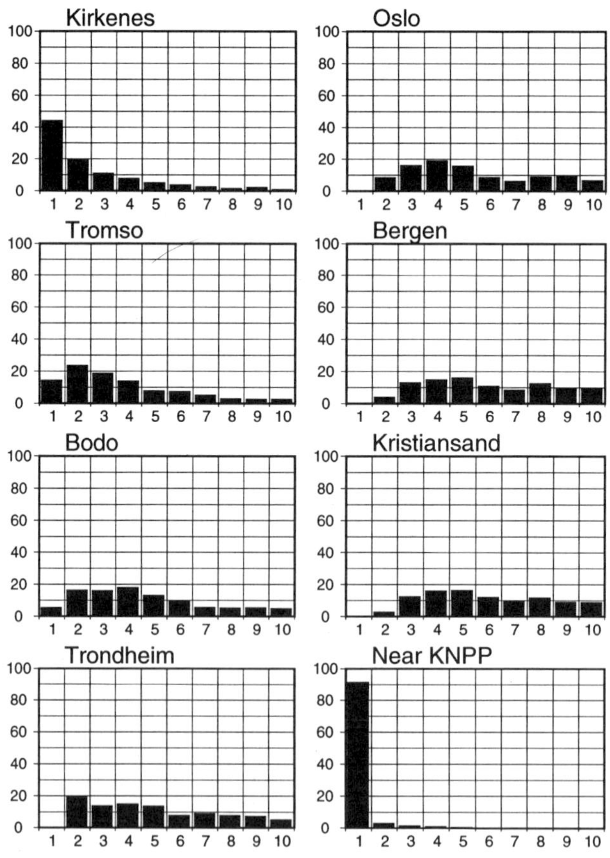

Figure 4. Distribution (in percent) of the travel time from the Kola NPP to the selected receptors in ten classes. Class 1,2, . . ,10 corresponds to 1, 2, . . . ,10 days of transport time, respectively.

For Bergen, as for other far away points, we observe from the distribution of travel time on Figure 4, no clear decrease in the frequency distribution representing the longer travel times, class (5-10). This means: If the cloud hits Bergen (~0.5% probability from Figure 1), it is almost the same probability that it will arrive on day no 10 (~10% probability) as on any of the days from 3 to 10, (maximum probability ~15% for day 5).

It may be interesting to see these statements in connection with Table 1, where we have listed the number of arriving air parcels from the Kola NPP, the probability of arrival and the minimum travel time - for the selected receptors. Of the - 96 162 - hourly releases of air parcels from the Kola NPP through the 11 years period (1985 - 1995), - 4 212 had trajectories which passed over the 50 by 50 km grid square where Kirkenes is located. This gives a probability of arrival of 4.38%. Among all these released air parcels, the one with the shortest travel time, will make its presence felt within 3 hours after its release from Kola NPP. As far away as Oslo, we still have 921 'crossings', representing about 1% probability of arrival and the shortest travel time is 26.5 hours. For Bodø the numbers are - 1560 crossings, 1.6% probability of arrival and 9 hours as the shortest travel time.

Table 1. Number of arriving trajectories, probability of arrival and minimum travel time from the Kola NPP to selected receptors

Receptor	No. of arriving trajectories	Probability of arrival (in %)	Minimum travel time (hours)
Kirkenes	4212	4.38	3.00
Tromsø	2030	2.11	7.50
Bodø	1560	1.62	9.00
Trondheim	867	0.90	22.75
Oslo	921	0.96	26.50
Bergen	777	0.81	35.75
Kristiansand	935	0.97	35.50
Near KNPP	21343	22.19	0.50

SUMMARY AND CONCLUSIONS

This paper focus mainly on aspects of air pollution transport seen from the longer time scales. The cause of concern is the threat of accidental releases from the Kola NPP. The formulation of the problem can be summarized in the following statement: Assume that an accident (with releases) could happen at any moment with a given probability. What will then be the probability for a specific location to be harmfully effected by these accidental releases?

'Climatological trajectory analysis' as a method, was chosen to fit this problem formulation. The word 'climatological' have been used in the following meaning; - covering a reasonable long period of time; (11 years: 1986 - 1995). From the available meteorological data was computed forward trajectories for air parcels released from the Kola NPP every hour in 11 years; - 96 162 releases.

These trajectories enabled us to construct maps and tables for: Probability of arrival, average travel time, minimum travel time and statistical distribution of travel time.

Examples of conclutions that can be drawn from these Figures and Tables:
The Kirkenes area has been chosen as an example ~ 250 km from the potential source:
- If the 'cloud' hits this location, (the - If - has been calculated to imply a probability ~ 5% from Figure 1 or 4.38% from Table 1), there is ~ 45% probability that the 'cloud' will arrive within the first 24 hours after the release. From Table 1, the minimum travel time from Kola NPP to this location is computed to be 3 hours. Similar statements can be given for the other locations.

At a distance as far away as Bergen, ~ 1600 km from the potential source, we observe from the distribution of travel time as presented in Figure 4; no clear decrease in the number of trajectory crossings on the last day (day no. 10) compared with earlier days. This indicates that: If the cloud hits this location (with ~ 0.8% probability from Table 1), there is almost the same probability that it will arrive on day no. 10 (~10% probability) as any of the days from 3 to 10, (maximum probability ~15% for day 5). This also points to a limitation in our calculations; too short trajectories. On the other hand, the harmful effect transmitted after such a long transport time/distance, would probably have diminished.

REFERENCES

1. - Report to Norwegian Parliment, no. 34 for 1993-1994: Nuclear Activity and Chemical Weapons in our near Northern Surroundings.
2. - Ministry of Foreign Affairs (1995): Plan of Action for Follow-up Activities to Report no. 34 to Norwegian Parliment (1993-1994). Project Description.
3. J. Bartnicki and J. Saltbones (1997) Analysis of Atmospheric Transport and Deposition of Radioactive Material Released During a Potential Accident at Kola Nuclear Power Plant. Research Report no. 43, DNMI, Oslo, January 27 1997. ISSN 0332-9879.
4. J. Saltbones, J. Bartnicki and A. Foss (1997) Atmospheric Transport and Deposition from Potential Accidents at Kola Nuclear Power Plant. Part II: Worst Case Scenarios. Research Report no. 56, DNMI, Oslo, 5 November 1997. ISSN 0332-9879.
5. Scandpower: Evaluation of the Norwegian Preparedness against Nuclear Accidents, Pr. nr. 24.31.01, September 1997.
6. G. Nordlund, J. Rossi, I. Valkama and V. Seppo (1988) Probabilistic trajectory and dose analysis for Finland due to hypothetical radioactive releases at Sosnovyy Bor. Research Note 847. Technical Research Centre of Finland. Espoo, Finland. ISBN 951-38-3106-X.
7. - OECD (1977) The OECD Programme on Long Range Transport of Air Pollutants. Measurements and Findings. Organization for Co-operation and Development. Paris.
8. K. Barrett and E. Berge (1996) Transboundary Air Pollution in Europe. Part 1 and Part 2. EMEP/MSC-W Status Report 1/96. Meteorological Synthesizing Centre - West, Oslo. ISSN 0332-9879.
9. S. Grønås and K. H. Midtbø, (1987) Operational multivariate analysis by successive corrections. *J. Meteo. Soc.*, Japan, **65**, 61 - 74.
10. T.E. Nordeng, (1986) Parameterization of physical processes in a three dimensional numerical weather prediction model. DNMI, Techn. Rep. No. 65.
11. H.A. Jakobsen, (1996) Preparation of meteorological input data to the 150 km Lagrangian air-pollution models. Research Report No. 40. Norwegian Meteorological Institute, Oslo, Norway. ISSN 0332-9879.
12. S. Pettersen, (1956) Weather analysis and forecasting. McGraw-Hill, New York.

DISCUSSION

The paper was presented by S. Mylona

D. STEYN: It would be very helpful if maps of 'probability of hit', travel time etc were accompanied by maps indicating the confidence we can place in the estimated statistics.

S. MYLONA: Right, it would be helpful. In fact, other studies of similar type have been carried out, such as a study by the Swedish Defence Research Establishment (February 1998) and a NATO/CCMS Pilot Study (March 1998), which show results that differ somewhat from our work. Part of the difference can be due to the different spatial resolution applied (it is easier to hit a large grid square than a small one), as well as to the different set-up and assumptions. We would have liked to follow up this study with a sensitivity analysis, as well as statistics showing seasonal variations.

M. HONGISTO: The probability distribution of the trajectories seems to be very even; in Kola Peninsula there are high emission source areas surrounded by a network of air concentration monitoring stations. They might measure (at Finnish side) twice-three times during a month a short peak (the concentration distribution based on this could be more oriented towards east).

S. MYLONA: One of the surprises we found in our study was the isotropic distribution of 'probability of hit', travel time etc. We had expected to see the effect of the westerly wind being more pronounced. The isotropic patterns must be due to the fact that the study shows mean values for the whole period. Part of the year (winter) this region (Kola) is under the influence of the outflow of arctic air from the north. Other periods can be dominated by the westerlies. We would have liked to follow up this study by dividing the period into seasons and studying them separately. It should also be stressed that our statistics is based on the wind at σ-level 0.925 (ca. 500-600 m above the terrain). The forcing effect of the topography is much reduced at this level and cannot very easily be compared with measurements at the surface.

M. HONGISTO: The precipitation depends on wind direction at least during the wintertime, thus the consequences of an accident might not be the same at the south-western and north-eastern side of this nuclear PP.

S. MYLONA: The seasonal variation is one of the points we would have liked to follow up. When it comes to the consequences of a nuclear accident, we must remember that the environmental impact will be depending on the weather conditions at the time of the accident, not on the statistics of the weather elements.

STOCHASTIC RECONSTRUCTION OF INTERMITTENT ZERO CONCENTRATION PERIODS IN PLUMES FOR ACCIDENTAL TOXIC AND FLAMMABLE RELEASES

David J. Wilson and Trevor L. Hilderman

Combustion and Environment Group
Department of Mechanical Engineering
University of Alberta
Edmonton, AB, Canada, T6G 2G8

ABSTRACT

Most plumes in the atmosphere are highly intermittent, with varying intervals of zero concentration separating bursts of high intensity concentration fluctuations. Experiments in the atmosphere and in laboratory scale models have shown that these concentration fluctuations are not a simple first order Markov process with a purely random time derivative of concentration, but instead exhibit a strong correlation of the variance of the time derivative with the instantaneous concentration level. A method is presented for generating realistic concentration fluctuation time series using stochastic methods with specified levels of mean concentration, fluctuation intensity, zero concentration intermittency fraction and fluctuation integral time scale. Comparisons between simulated and measured concentration time series are made using data from water channel experiments of point-source plumes dispersing in a rough-wall boundary layer designed to simulate neutral atmospheric conditions. The stochastic simulation gave reasonable estimates for the probability of observing zero concentration intervals shorter than the average zero interval, but overpredicted the duration of zero periods longer than the average.

INTRODUCTION

Toxicity and odor effects have a strong non-linear dependence on concentration. For example, ten Berge, Zwart and Appleman (1986) analyzed animal experiments for 20 different acutely toxic gases and found that fatalities are a function of concentration c^n where $n = 1.0$ to 3.5, with most toxic gases in the range of 2.0 to 3.0. The Center for Chemical Process Safety of the American Institute of Chemical Engineers (CCPS, 1989) recommends similar non-linear models for predicting acute toxicity from common industrial chemicals.

If the exponent $n = 1$, then high concentrations are no more important than low concentrations and the only variables determining toxicity are mean concentration C and exposure duration t_e. However, if $n > 1$, concentration fluctuations increase the toxicity of the exposure because high concentrations with $c > C$ will dominate the toxic load $\langle c^n \rangle t_e$. In a typical point source gas plume, where the fluctuation standard deviation, $c' = \sqrt{\langle (c-C)^2 \rangle}$, is often several times larger than the mean concentration, C, fluctuations have a significant effect on the predicted toxic hazard.

Advanced toxicity models include both a biological uptake and detoxification rate. The influence of the detoxification (recovery) rate on the final toxic load is strongly dependent on whether the zero concentration periods in an intermittent plume are sufficiently long to allow the biological system to partially recover.

In the present study, we generate ensembles of reconstructed random time series of intermittent concentration fluctuations that can be applied directly to a hazard model. We will examine the probability distribution function for the length of these zero concentration time gaps as a measure of whether our reconstructed time series can be used to estimate biological recovery during a toxic event. The intermittent concentration fluctuation time series are generated directly as first order "memoryless" Markov processes using a stochastic model discussed in more detail in Hilderman and Wilson (1998). This stochastic model is an extension of the non-intermittent model by Du, Wilson and Yee (1998).

INTERMITTENT TIME SERIES PROBABILITY DISTRIBUTIONS

Probability distributions functions (pdfs) of intermittent concentration fluctuations usually focus on the non-zero "in-plume" concentrations with the intermittent periods described by a delta function at zero concentration. Wilson (1995, chap. 5) examined several different distributions and recommended the lognormal as the best fit to a wide variety of data for the conditional (non-zero) concentrations. The choice of the lognormal pdf is supported by the water channel concentration data analyzed by Yee, Wilson and Zelt (1993). Zelt (1992) found that the shape of the pdf evolves with plume travel time, and that lognormal, gamma, and clipped normal pdfs may be used as approximations as the plume evolves.

The stochastic simulation is expressed in pseudo-concentration coordinates c_+ where the subscript "+" denotes parameters related to these pseudo-concentrations. In c_+ coordinates, the concentration fluctuations are represented by a lognormal distribution with only positive concentrations and an intermittency factor $\gamma_+ = 1.0$. There are no zero concentration intermittent periods in the c_+ time series.

After a simulated time series is generated, the c_+ concentrations are shifted down by an amount c_{base} to give positive and negative concentrations $\tilde{c} = c_+ - c_{\text{base}}$ as shown in Figure 1. The probability of obtaining a positive concentration is the intermittency factor γ, and the probability of obtaining a negative concentration is $(1 - \gamma)$. The magnitude of the negative concentration is interpreted physically as being inversely proportional to the likelihood of obtaining a positive concentration in the next time step.

The assumption of a clipped lognormal pdf was confirmed by comparison with experimental water channel concentration pdfs. The shifted and clipped lognormal performs slightly better than a standard lognormal in describing concentrations less than the conditional mean.

STOCHASTIC MODEL FOR INTERMITTENT TIME SERIES

Experimental evidence from Yee et al. (1993) and water channel data presented later in this paper shows that the root mean square concentration derivative $\dot{c}'$ increases with the concentration level c at which it is measured. This requires that our stochastic model for the time series of concentration fluctuations have a deterministic component that changes dc_+/dt as concentration c_+ increases along with the usual random component. We assume

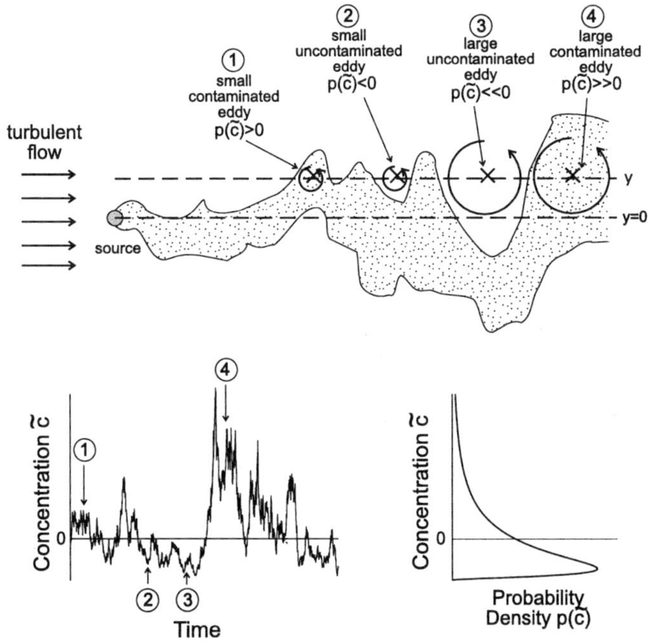

Figure 1. Example of a dispersing plume and corresponding time series and probability density generated by the stochastic model. Negative concentrations $\tilde{c}$ are interpreted as intermittent periods of clean air (zero concentration).

that the concentration time series at a fixed point in a dispersing plume can be modelled as a continuous first-order Markov (memoryless) process described by the one-dimensional stochastic differential Langevin equation:

$$\frac{dc_+}{dt} = a(c_+, t) + b(c_+, t)\frac{d\zeta}{dt} \tag{1}$$

where $a(c_+, t)$ is the deterministic portion of the time derivative, dependent on concentration c_+ and time t; and $b(c_+, t)d\zeta$ is a random forcing function where $d\zeta$ is a Gaussian random number with a mean of zero, variance dt and no dependence on the instantaneous level of c_+. The Langevin equation is discussed in detail by Gardiner (1983, pp. 80–83) and Durbin (1983). Here, we will apply the Langevin equation in a fixed Eulerian frame of reference.

If we require the pdf of concentration c_+ to be statistically stationary, a useful constraint between the a and b terms can be derived by writing a differential equation for the time evolution of the pdf $p(c_+)$, and then setting $\partial p/\partial t = 0$. The one dimensional Fokker-Planck equation is a deterministic equation describing the time evolution of the pdf in a Markov process, see Durbin (1983)

$$\frac{\partial p}{\partial t} = -\frac{\partial(ap)}{\partial c_+} + \frac{1}{2}\frac{\partial^2(b^2 p)}{\partial c_+^2} \tag{2}$$

Stationarity requires that a, b and the pdf p do not change with time. With this assumption $\partial p/\partial t = 0$ and Equation (2) can be integrated and solved for b^2 in terms of a. Following Du, Wilson and Yee (1998)

$$b^2 = \frac{2}{p}\int_{c_+}^{\infty} -ap \; dc_+ \tag{3}$$

Equation (3) relates the pseudo concentration time series generation parameters a and b in Equation (1) and the probability density function p specified by the user.

The model proposed by Du, Wilson and Yee (1998) for the non-zero (conditional) part of the concentration time series showed that for a stationary process the a term can only be a linear return to mean of the form

$$a = \frac{C_+ - c_+}{T_{c+}} \tag{4}$$

where T_{c+} is the integral autocorrelation time scale of the pseudo-concentration fluctuations.

The pdf $p(c_+)$ must be specified to complete the model. In this study, $p(c_+)$ is a log-normal distribution. The value of the b term is calculated by substituting the pdf and the definition of a from Equation (4) into Equation (3). After the reconstructed time series are generated they are shifted down by an amount c_{base} and the result is a clipped lognormal distribution for the intermittent time series. The c_+ lognormal distribution must be chosen so that after it is clipped the desired intermittency factor γ and the conditional fluctuation intensity i_p^2 are obtained. The conditional fluctuation intensity is $i_p^2 = c_p'^2/C_p^2$, where $c_p'^2$ is the conditional variance of the non-zero concentrations (i.e. excluding the time periods when $c = 0$)and C_p is the conditional mean excluding the zero periods.

CONCENTRATION FLUCTUATION MEASUREMENTS

The data were collected by Wilson, Zelt and Pittman (1991) in the water channel shown schematically in Figure 2. Measurements were taken at three downstream positions $x/h_s = 9$, 19 and 29 and various positions y across the width of the plume at each of these downstream locations. In the present study, only data taken at the source height $z = h_s$ and at the position $x/h_s = 19$ were used.

For short range dispersion ($\leq$ 1 km) from an elevated source in a neutrally stable atmosphere, the crosswind plume spread, σ_y, is well approximated by a linear function of downwind distance x. In the water channel we found that the crosswind plume spread was also linear, so the water channel to atmosphere length scale factor could be chosen from a range of possibilities because a linear spread is the same in any scale. In the present study, we chose a scale factor of 1000:1 based on the thickness of the surface log-law velocity profile layer. In the water channel this log-law layer is about 100 mm thick while in the full scale atmosphere it is about 100 m.

The mean velocity profile of the shear layer in the water channel followed the log law profile, $U/u_* = (1/\kappa)\ln((z - d_0)/z_0)$ where U is the mean flow velocity, $u_* = 14.6$ mm/s is the friction velocity, $\kappa = 0.4$ is the von Karman constant, $d_0 = 1.5$ mm is the displacement height of the surface roughness, $z_0 = 0.15$ mm is the roughness length scale and z is the height

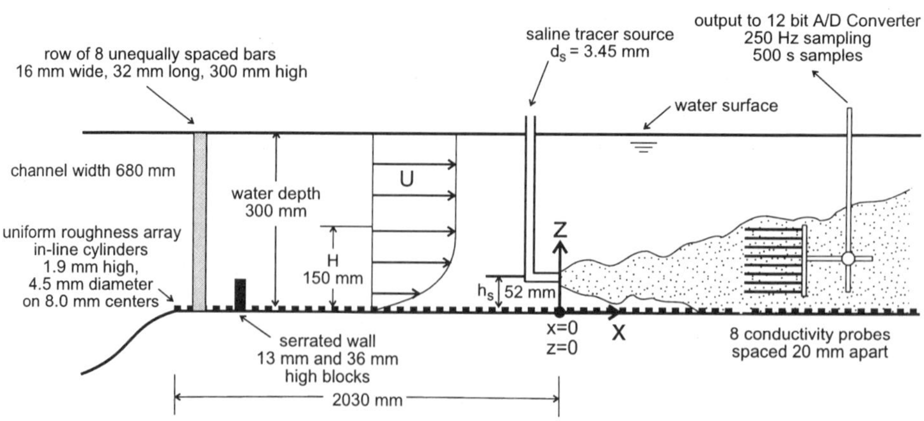

Figure 2. Schematic diagram of the water channel shear layer generator test section.

above ground level. At a 1000:1 length scale factor this corresponds to a full scale roughness of $z_0 = 0.15$m.

A downstream facing isokinetic source from a 3.45 mm ID tube 2030 mm downstream of the beginning of the channel emitted saline tracer solution. At the source height, $h_s = 52$ mm, above the bottom of the channel the mean velocity U_{h_s} was 210mm/s. The tracer was a neutrally buoyant mixture of water, ethanol and 50g/l salt emitted with a flow rate of 2.0 ml/s.

The concentration fluctuation data were obtained with a rake of 8 electrical conductivity probes mounted with the probe tips 20 mm apart. Each probe had a spatial resolution of about 1 mm. Throughout the experiments the probe response was corrected for background buildup and electronic drift in the output. Each fluctuation time series consisted of 125,000 points sampled at 250 points per second for a total sample time of 500 seconds. At 1000:1 scale this corresponded to samples spaced about 1 m apart in the wind direction.

The data were processed to correct for the probe response time constant that produced a -3dB rolloff frequency of 35 Hz. Deconvolution of the digital signals from the probes with the inverse of a first order impulse response function enhanced the effective frequency response to 105 Hz.

The "zero" concentration threshold was set at 8.0 times the background noise level to eliminate the effect of electronic noise on the measured intermittency. This threshold corresponds to about $0.09C_p$ and was determined empirically by analyzing time series with no source emission and adjusting the threshold to produce the required $\gamma = 0$ intermittency.

STOCHASTIC MODEL PERFORMANCE

The correct distribution of concentration values in the reconstructed time series is guaranteed by the input clipped lognormal pdf with its measured values of C_p, i_p and γ that match the statistics of the data set. In order to compare the reconstructed and experimental time series, the reconstructed time series were filtered to match the 105 Hz frequency response of the conductivity probes and a zero cutoff was applied at $0.09C_p$ to match the experimental data. As a measure of our success in stochastic reconstruction we will look at the derivative of concentration and the zero concentration gap lengths. These recurrence statistics agree with the experimental data only if the simulation produces the correct fluctuations.

Time Derivative of Concentration

One important assumption in the stochastic model is that the magnitude of the derivative of concentration is dependent on the concentration level through the deterministic term $a(c_+, t)$ in Equation (1). Figure 3 shows that the water channel measurements confirm the field observations of Yee et al. (1993), with the rms (root mean square) concentration derivation $\dot{c}'$ exhibiting a strong dependence on concentration level c/C_p.

The linear return to mean stochastic model coefficient a in Equation (4) will produce a non-zero root mean square concentration as c goes to zero. Figure 3 also shows that this form gives realistic results for the derivative standard deviation at least down to the cutoff limit of 10% of the conditional mean concentration C_p.

Zero Concentration Gap Durations

The probability distribution of the zero period gap intervals in the measured and reconstructed time series provides a sensitive measure of the accuracy of the stochastic model in simulating intermittency. The pdfs of gap duration in Figure 4 show that the stochastic model is capable of reconstructing zero periods as long as 10 times the integral time scale T_c.

The stochastic model showed no consistent trend of underpredicting or overpredicting the gap duration. The average gap duration $\overline{t_g}$ of the stochastic model was 0.82 of the water

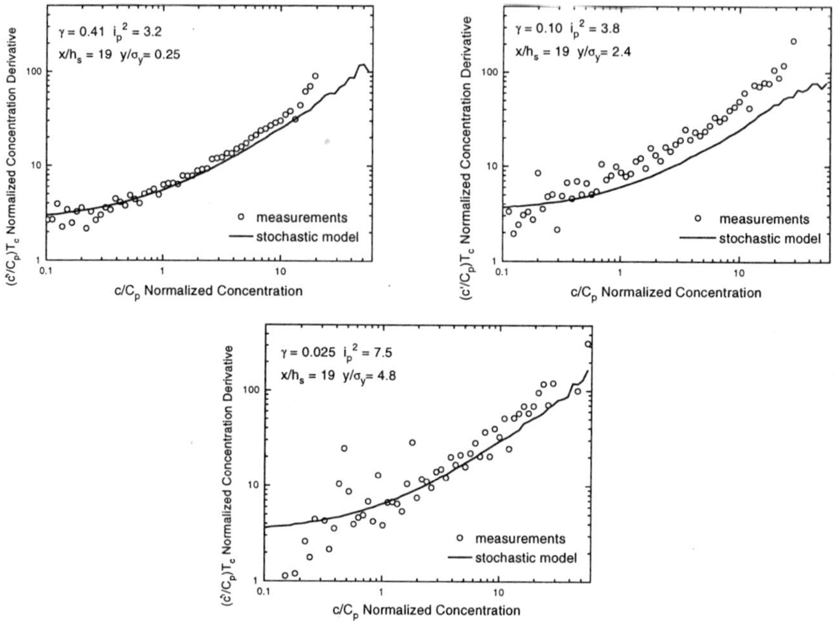

Figure 3. Normalized root mean square time derivative of concentration $(\dot{c}'/C_p)T_c$ at the normalized concentration c/C_p at downstream distance $x/h_s = 19$ from the source compared to the stochastic simulation. Cross stream distance from centreline of source (a) $y/\sigma_y = 0.25$ (b) $y/\sigma_y = 2.4$ (c) $y/\sigma_y = 4.8$

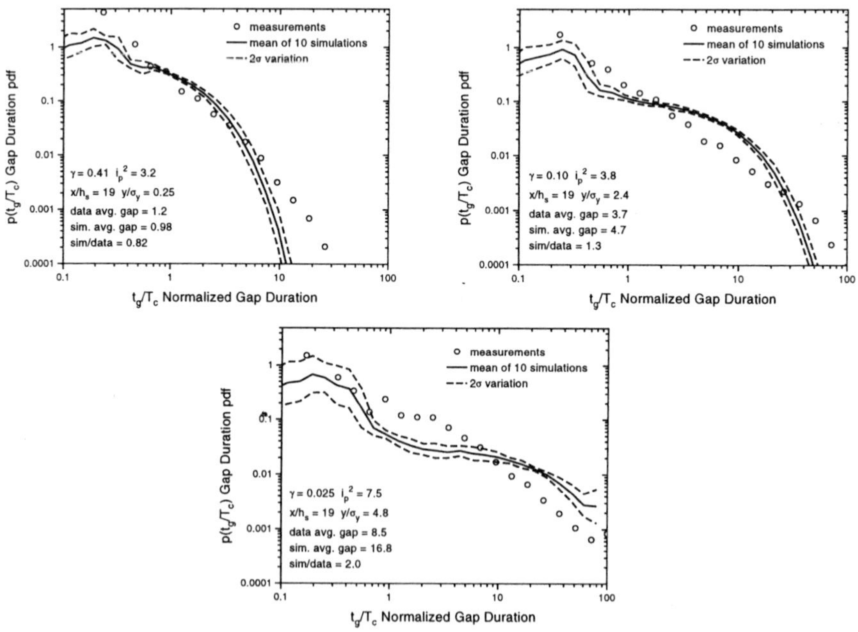

Figure 4. Probability density p of normalized zero concentration gap duration t_g/T_c at downstream distance $x/h_s = 19$ from the source compared to the stochastic simulation. Cross stream distance from centreline of source (a) $y/\sigma_y = 0.25$ (b) $y/\sigma_y = 2.4$ (c) $y/\sigma_y = 4.8$

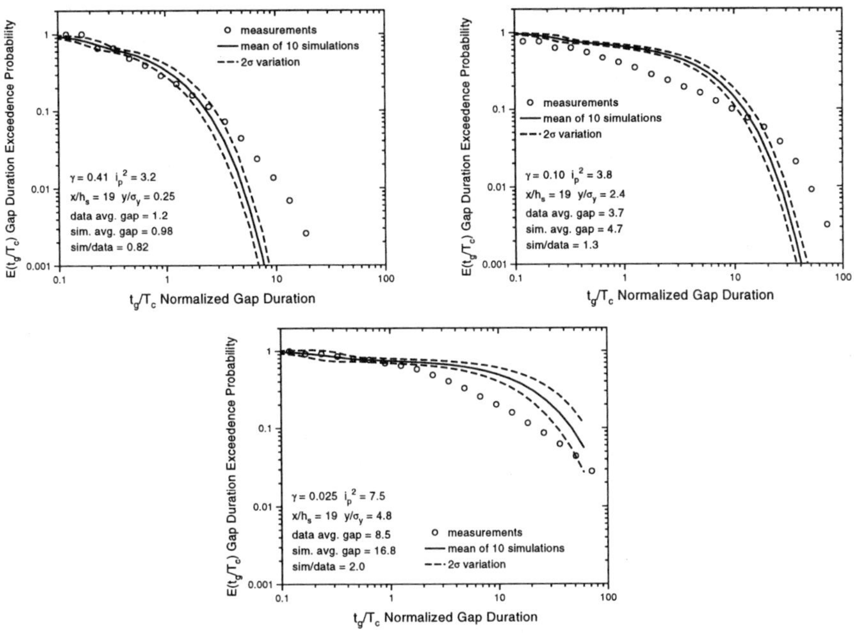

Figure 5. Exceedance probability E of zero concentration gap duration t_g/T_c at downstream distance $x/h_s = 19$ from the source compared to the stochastic simulation. Cross stream distance from centreline of source (a) $y/\sigma_y = 0.25$ (b) $y/\sigma_y = 2.4$ (c) $y/\sigma_y = 4.8$

channel measurement average gap at $y/\sigma_y = 0.25$ near the plume centreline, and up to a factor of 2.0 at $y/\sigma_y = 4.8$ at the outer edge of the plume. The probability of observing a specified zero-concentration gap duration t_g was predicted within a factor of 2 over a 50:1 range of t_g, from $0.2\overline{t_g}$ to $10\overline{t_g}$. This difference may be caused partly by the different frequency spectra of fluctuations in the water channel and in the stochastic simulation.

Wilson, Zelt and Pittman (1991) note that for salt tracer diffusing into water, the mass diffusivity D is much smaller than the molecular viscosity of the fluid ν and the Schmidt number $Sc = \nu/D \gg 1$. The effect of this large Schmidt number is that velocity-driven straining of the concentration field decays before molecular diffusion smears out the concentration fluctuations. We found that water channel data follow the viscous-convective Batchelor spectrum with $F_c \propto f^{-1}$ at high frequency f. In contrast, the shape of the concentration spectrum of the stochastic simulation is determined by the assumption of a memoryless Markov process that produces $F_c \propto f^{-2}$ at high frequency. The spectrum of concentration fluctuations in the atmosphere where $D \approx \nu$ so $\text{Sc} \approx 1$ has $F_c \propto f^{-5/3}$ at high frequencies, see Wilson (1995, pp. 143-146).

For recovery from toxic effects, the long zero concentration gap times are more important than the short ones. Therefore, the most useful information is contained in exceedance probability plots rather than pdfs or cumulative probability plots. Figure 5 shows the normalized conditional exceedance probability $E(t_g/T_c)$ of finding a gap time greater than t_g/T_c. The agreement is satisfactory only up to gap times as long as the integral time scale. The shapes of the curves are also quite different with position across the plume. However, if the curves were replotted with the gap time normalized by the average measured gap time (listed on the plots) the curve shapes would be quite similar at all three positions across the plume.

CONCLUSIONS

The advantage of a stochastic model is that it can be used to reconstruct time series with any realistic combination of intermittency and fluctuation intensity to generate large ensembles of random time series with identical means, variances, and intermittencies. Each simulated time series represents an individual realization of the event and complex hazard models can be time stepped through simulated releases to observe the effects of realization to realization variability as well as large ensemble averages.

The key assumption in the reconstruction is that intermittent periods of zero concentration are part of the same physical mixing process as the non-zero fluctuations. Positive concentrations are actual fluctuations while negative concentrations are interpreted as periods of zero concentration where the magnitude of the negative concentration represents the likelihood of obtaining a non-zero concentration in the next step.

The accuracy of the stochastic time series reconstruction was measured by two factors: its prediction of the conditional rms derivative $\dot{c}'$, and its estimate for the probability density function for the time gaps of zero concentration periods. The stochastic model accurately tracked the 50:1 increase in $\dot{c}'$ over a 120:1 concentration range of the fluctuations. The probability of observing zero-concentration gap times t_g was reasonably well estimated within a factor of 3 over a 50:1 range from 0.2 to 10 times the average gap time.

In its present form, the stochastic simulation is useful for generating time series to evaluate hazardous effects. However, some additional work is required to make this model more realistic near ground level. All data used in the present study were taken at the source emission height, but near ground level where most receptor exposures occur, the mixing is complicated by high velocity shear levels.

REFERENCES

CCPS, 1989, *Guidelines for Chemical Process Quantitative Risk Analysis*, Center for Chemical Process Safety of the American Institute of Chemical Engineers.

Du, S., Wilson, D. J., and Yee, E., 1998, A stochastic time series model for threshold crossing statistics of concentration fluctuations in non-intermittent plumes, manuscript submitted to *Boundary-Layer Meteorology*, April 1998.

Durbin, P. A., 1983, Stochastic Differential Equations and Turbulent Dispersion, Technical report, National Aeronautics and Space Administration, NASA Reference Publication 1103.

Gardiner, C. W., 1983, *Handbook of Stochastic Methods*, Springer-Verlag.

Hilderman, T. L. and Wilson, D. J., 1998, Simulating concentration fluctuation time series with intermittent zero periods and level dependent derivatives, manuscript submitted to *Boundary-Layer Meteorology*, April 1998.

ten Berge, W. F., Zwart, A., and Appelman, L. M., 1986, Concentration-time mortality response relationship of irritant and systemically acting vapours and gases, *Journal of Hazardous Materials*, 13:301–309.

Wilson, D. J., 1995, *Concentration Fluctuations and Averaging Time in Vapor Clouds*, Center for Chemical Process Safety of the American Institute of Chemical Engineers.

Wilson, D. J., Zelt, B. W., and Pittman, W. E., 1991, Statistics of Turbulent Fluctuation of Scalars in a Water Channel, Technical report, Department of Mechanical Engineering, University of Alberta, Edmonton, Alberta.

Wilson, J. D. and Sawford, B. L., 1996, Review of Lagrangian stochastic models for trajectories in the turbulent atmosphere, *Boundary-Layer Meteorology*, 78:191–210.

Yee, E., Kosteniuk, P. R., Chandler, G. M., Biltoft, C. A., and Bowers, J. F., 1993a, Recurrence statistics of concentration fluctuations in plumes within a near-neutral atmospheric surface layer, *Boundary-Layer Meteorology*, 66:127–153.

Yee, E., Wilson, D. J., and Zelt, B. W., 1993b, Probability distributions of concentration fluctuations of a weakly diffusive passive plume in a turbulent boundary layer, *Boundary-Layer Meteorology*, 64:321–354.

Zelt, B. W., 1992, *Concentration Fluctuations and their Probability Distributions in Laboratory Plumes*, PhD thesis, University of Alberta, Edmonton, Alberta.

DISCUSSION

P. de HAAN: Stochastic Markov processes are commonly used in Lagrangian "random - walk" particle models to describe ensemble-averaged dispersion, i.e. particles at the same position are not correlated. Using such a process to emulate in-plume fluctuations caused by eddies might yield a correct pdf, since the first and second moments are an input from elsewhere. But, what is the significance of the time series of the stochastic fluctuations itself?

D. WILSON: A memoryless Markov process will produce an exponential autocorrelation for the concentration fluctuations, and a frequency spectrum that follows a -6/3 slope at high frequencies. Real concentration fluctuation time series in atmospheric gas plumes have a - 5/3 spectrum at high frequencies, so the simulated time series tends to produce more high frequency fluctuations than the actual time series. We assessed the physical realism of the simulation by measuring the upcrossing rates of the concentration through specified concentration levels, above or below the mean C. The data presented here shows that even for the extreme value of a zero concentration threshold, the pdf of gap time (zero concentration) durations was in reasonable agreement with the measured gap times, a fairly severe test of the physical realism of the time series simulation.

MODEL ASSESSMENT AND VERIFICATION

chairmen: C. Borrego
H. van Dop
D. Syrakov

rapporteurs: A. K. V. Falk
S. Rafailidis

MAESTRO SYSTEM 1.0 – A NEW MODELLING SYSTEM FOR IMPACT STUDIES OVER COMPLEX TERRAIN

A. Dutrieux [1] and G. Schayes [2]

[1] ATM-Pro s.p.r.l, Rue St-André 5, B-1400 Nivelles, Belgium

[2] Institut d'Astronomie et de Géophysique, Univ. cathol. de Louvain, B-1348 Louvain-la –Neuve, Belgium

INTRODUCTION

To perform air dispersion exercises numerous software exist world-wide. Most of them are based on the classical Gaussian dispersion model. Although they are the most commonly used and available for short term and long term studies, this kind of model is not able to treat correctly complex terrain situations. Their inputs are the classical meteorological data from synoptic weather stations and the emission rates. They calculate averaged concentration, integrated deposition, concentration value for specified percentile and average period. They run on PCs and the CPU time required is very low. In general they are implemented in attractive and user-friendly interfaces.

More sophisticated software are also found based on 3-D meso-scale meteorological models. These scientific software are often under continuous development in research centres. They intend to be as realistic as possible. Therefore they take into account a lot of physical processes. Different types of associated 3-D dispersion models exist (Lagrangian or Eulerian). They can be diagnostic or prognostic models. In general their inputs are much more complicated than those used in Gaussian models. Moreover the required information is not always available. These models demand much computational power and run on workstations. They produce the same outputs as Gaussian models, as well as other relevant information for dispersion, like winds and turbulence. They are not suited for long term studies, but are commonly used for short term episode analysis or specific research level studies.

For conducting impact studies over moderately complex terrain and performing long term statistical analysis, we did not find any operational scientific software.

ATM-PRO s.p.r.l. decided to provide such a tool with its *MAESTRO* system 1.0*. It includes a new 3-D dispersion model that allows both short term and long term integration over complex terrain. Moreover this module runs on PCs and requires low CPU time with the present available computers.

This document presents the main features of *MAESTRO* and its validation. Typical application examples will be shown.

SHORT MODEL DESCRIPTION

The modelling system includes a 3D mesoscale wind field generator coupled to a Lagrangian dispersion model.

The meso-meteorological model is a wind field generator based on a simplified version of a mesoscale model namely the TVM model (Schayes et al. 1997, Thunis et al. 1998). solving the the approximated Navier-Stokes equations in the vorticity formulation. This formulation takes into account the terrain characteristics, i.e. topography and land type roughness. The simplifications adopted in the meteorology module include : (i) a reduced grid size, typically a grid of 20 x 20 km^2 with 1 km horizontal resolution is used, centered around the site of interest ; (ii) a reduced formulation of the model physics in which the vertical structure of the atmosphere (turbulence and temperature) are computed from similarity theories and observations; (iii) searching for stationary solutions.

The model inputs are similar to those needed by classical Gaussian models, and consist in meteorological data from synoptic weather stations. A meteorological preprocessor included in the system derives the needed input data. The main outputs are the wind components and turbulence parameters necessary to compute the dispersion.

The Lagrangian model used in *MAESTRO* is based on the so-called « kernel » method which associates a certain quantity of substance to particles. These are dispersed in the 3-D wind-field produced by the meso sub-model. The dispersion model takes into account the dry and wet depositions, the gravitational settling and a plume rise formulation. It can consider various types of sources and emitted substances for which physical characteristics and emission rates must be provided

Integration of this coupled 3D meso dispersion model can be obtained at reasonable computer cost (CPU) in comparison to Gaussian models or sophisticated 3-D models. It is thus the best compromise between realism and cost.

MAESTRO VALIDATION EXERCISES

The validation of *MAESTRO* consists in a series of classical tests (Thomson, 1987 and Zanetti, 1990). Purely theoretical tests have been made to verify that the dispersion performs well in stationary and homogeneous turbulence where analytical solutions exists. Comparison with Pasquill and Briggs Gaussian formulations are performed in case of flat and homogenous terrain for different air stability conditions..

---

* MAESTRO system 1.0 is a proprietary air modelling system of ATM-Pro s.p.r.l. Protection of rights and trade marks under application.

Figure 1 presents the comparisons for the urban case for stable situations, showing good agreement between the *MAESTRO* and Gaussian approaches.

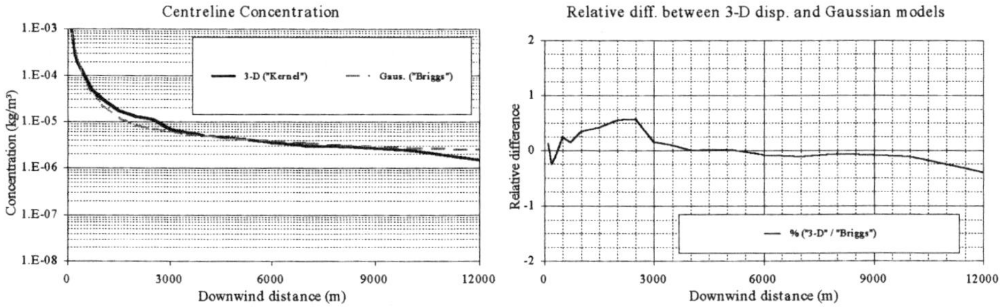

Figure 1. Comparison to standard Gaussian model in Flat Terrain (urban case). Results shown correspond to the stable case concentrations at 1 m.

The Copenhagen case of the Olesen (1994) validation kit has been used to confront the system with a more real situation The quality of the results depends of the available input data. In the case when only standard data are available 56 % of the results lie within a factor of 2 from the observations. When one takes also into account the available mixing heights, this score is still better with 91 % (fig 2).

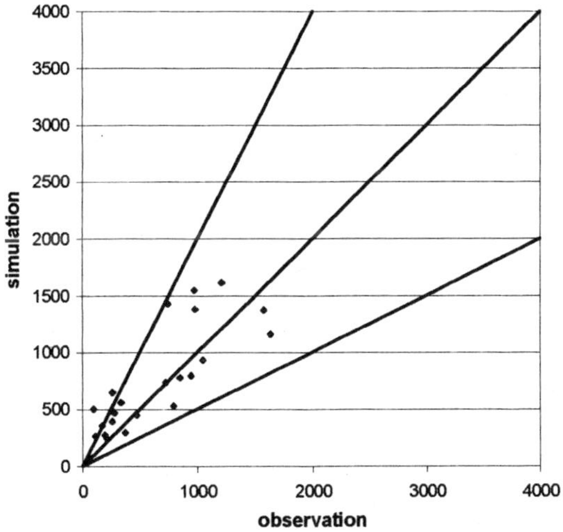

Figure 2. MAESTRO results for the Copenhaguen experiment obtained with input of the standard wind speed and boundary layer height. The lateral lines represent a factor of 2 from the observations.

Unfortunately, very few examples exist right now over complex terrain to validate the model for non flat topography. In this latter case, we rely on the fact that the wind field generator used here derives from a mesoscale model (TVM) that has passed many tests and realistic applications over complex topography (e.g. Bornstein et al., 1996). Validations in case of complex topography are planned in the near future.

Indeed, one has to remind that the validation of any model is a continuous process for which each new study case enables to validate and improve it.

In the case of flat topography, the 3-D dispersion model uses only twice as much CPU time as a Gaussian model for the same kind of computation. This low computer cost shows the effectiveness of the 3-D dispersion model. Nevertheless, CPU time increases with terrain complexity but remains still reasonable in all circumstances. So *MAESTRO* may be considered as an effective software for easy and realistic air dispersion studies at low computer cost.

APPLICATION EXAMPLES

Three cases will be shown here : (i) flow and dispersion around an idealised hill; (ii) an example over real topography : "belgian Ardennes" case; (iii) reconstruction of a potential 1-year data series.

Idealised Hill case

The figure 3 shows the result of an idealised case using a synoptic forcing of 2 m/s wind speed coming from West (260°), a 300 m high gaussian shaped hill, air having the E stability class. As expected, the air flow is well deflected around the hill and the contours show a possible concentration pattern due to this type of flow.

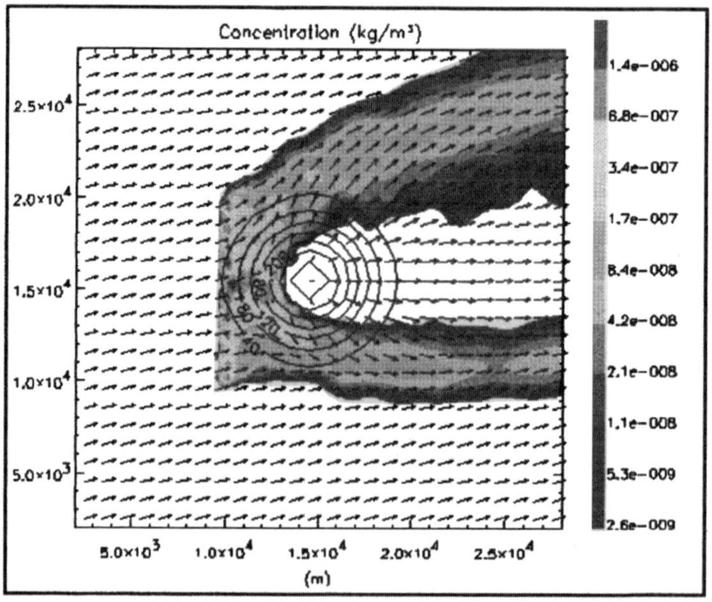

Figure 3. *MAESTRO* results of 3-D dispersion around an idealised gaussian 250 m high hill from a linear source in westerly conditions. Arrows indicate direction and intensity of winds. Contour lines represent the topography. Contours give concentration fields 1 m above ground.

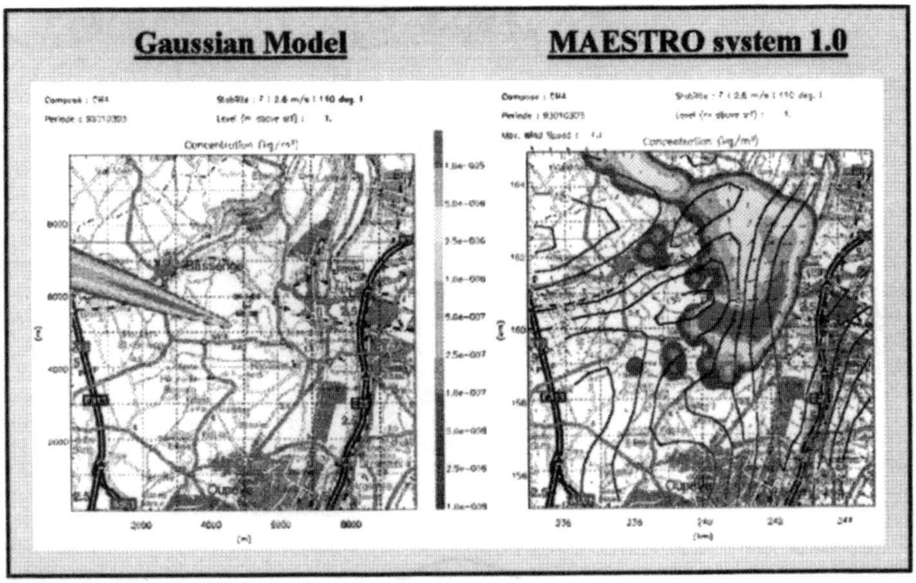

Figure 4. Results obtained with (a) a Gaussian dispersion module in the belgian Ardennes from a hypothetical surface source, (b) *MAESTRO* wind field generator and LPD module for the same situation. The forcing is a synoptic wind direction of 110°. Atmospheric stability is Pasquill class F. Contours give concentration fields at 1 m above ground. Arrows represents the wind field and bold contour lines the topography.

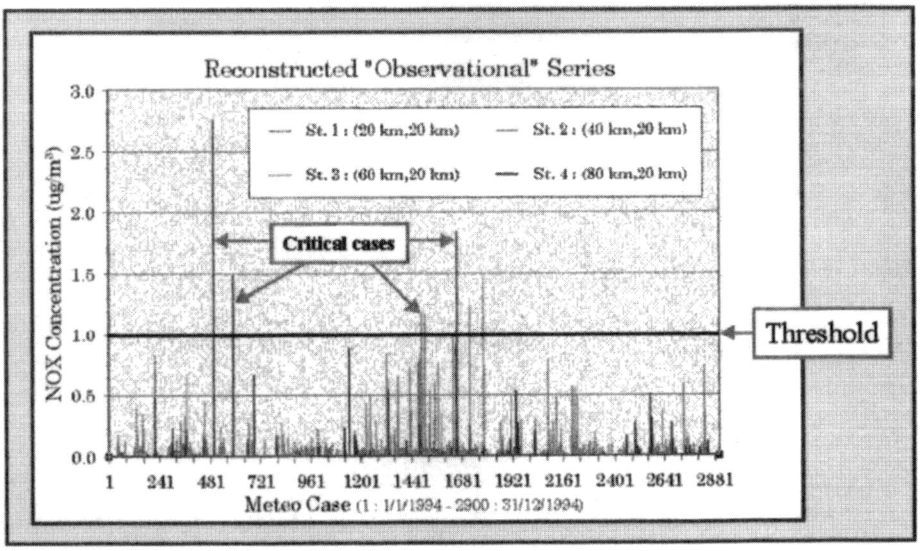

Figure 5. Reconstruction of potential concentration series over 1 year using real meteorological data over the region of Geneva. Four point sources have been taken into account. Four receptor sites have been selected in the region.

Belgian Ardennes Case

Figure 4 presents the results obtained (a) with standard Gaussian dispersion module (i.e. no topography, homogeneous and constant wind field) in the belgian Ardennes from a hypothetical surface source (e.g. from waste disposal). The forcing is based on a synoptic wind direction of 110°. The atmospheric stability is strong (Pasquill class F). Contours show concentration fields at 1 m above ground, and (b) *MAESTRO* results for the same situation using the 3-D wind field generator and the LPD module, taking into account the terrain complexity. Arrows represents the wind field and bold contour lines the topography. This example shows that very different things are produced by both models!

Reconstruction of a 1-Year Data Series

This last example shows that due to the low CPU requirements, analysis of long series of meteorological data is possible with *MAESTRO*. A one-year series of meteorological data (3-hourly) has been used to compute the daily potential concentrations due to a four point-source configuration in the region of Geneva. Four receptor sites have been selected over the 100 x 100 mesh simulated by the dispersion model. This helps to select and analyse critical days due to unfavorable meteorological situations or with respect to the emissions.

CONCLUSIONS

The *MAESTRO* modelling system shows great potential for treating impacts of pollution sources in complex terrain in a realistic way while avoiding high CPU.

It uses currently available data bases :

- site description (topography, land use),
- emission inventory (types, substances emission rates),
- meteorological inputs (standard synoptic weather station measurements)

and generates data bases for effective impact assessment and strategy planning as :

- reconstructed meteorological fields in 3-D around the site,
- reconstructed series of concentration fields over dense networks
- general statistics (annual mean concentrations, peaks, percentiles with respect to some threshold concentration, ...).

It is an effective tool to perform impact studies or air quality assessment and to identify environmental strategies and solutions. These informations are the starting point for environmental strategy development in order to comply with environmental regulations.

REFERENCES

Bornstein R., Thunis P., Grossi P., Schayes G., 1996. Topographic vorticity-mode mesoscale-β (TVM) model. Part II: Evaluation. Journ. Appl. Meteor. **35**, 1824-1834.

Olesen (1994). "Model validation kit for the workshop on Operational Short-Range Atmospheric Dispersion Models for Environmental Impact Assessments in Europe", *National Environment Research Institute*, Denmark, juin 1994.

Thunis, Ph. (1995). Formulation and evaluation of a nonhydrostatic vorticity-mode mesoscale model. Ph. D. Thesis, Univ. cathol. de Louvain ; also *EU Report* EUR 16141 EN, 151 p.

Schayes, G., Thunis, Ph., and R. Bornstein. (1996). "Topographic vorticity-mode mesoscale-beta (tvm) model. part I : formulation". *J. Appl. Meteor.*, N°10, 1815-1823, 1996.

Thomson D.J. (1987). "Criteria for the selection of stochastic models of particle trajectories in turbulent flows", *J. Fluid Mech.*, vol. 180, p. 529-556, 1987.

Zanetti (1990). *Air Pollution Modelling : Theories, Computational Methods and available Software. Computational Mechanics Publications*, Southampton Boston, (Ed.) Van Nostrand Reinhold, N.Y., 1990 .

DISCUSSION

Z. ZLATEV: What is the influence of the boundary conditions on the results? Did you make any sensitivity tests?

G. SCHAYES: The boundary conditions used in the wind field model are open i.e. with all horizontal derivatives put equal to zero. As the model is intended to compute the effect of local pollution sources, these are placed inside the modelling domain (no background pollution is considered).

However, while using this model in moderately complex topography, one must take care of avoiding disturbances created by a too steep topography close to the borders. This can be done by selecting the most adequate study domain.

A VERIFICATION OF A NEW COASTAL AREA DISPERSION MODEL

Joseph Levitin

Israel Meteorological Service
P.O.Box 25, Bet Dagan, 50250 Israel

INTRODUCTION

Theoretical and experimental research of internal boundary layers (IBLs) have received much attention during the last two decades. A typical and most interesting situation involving the IBL takes place when a daytime onshore wind (sea breeze) is coupled with a significant land-sea temperature difference. In this case an intensive thermal convection within the surface atmospheric layer leads to the intensification of the turbulent mixing of air pollution. In case of a tall smoke stack close to a shoreline, pollution are at first released into a stable or neutral stratified onshore air flow. After the plume penetration into the surface mixing layer, also called the thermal internal boundary layer (TIBL), the pollutant plume is rapidly dissipated (fumigation effect); this results in high ground-level pollutant concentrations, often exceeding the accepted air quality standards. The growing tendency to locate new industrial units in coastal areas associated with increase of population densities has stimulated a great practical interest in TIBL research. Starting with the Lyons and Cole's (1973) model, a simple scheme (Figure 1) is used to describe the TIBL phenomena as a convective ground-based underlayer with depth, H, that increases with fetch, x, from the shoreline and is defined by the following expression

$$H = A\,x^{b}$$

where A is an empirical constant or an analytical expression containing different physical

Air Pollution Modeling and Its Application XIII, edited by S.-E. Gryning and E. Batchvarova, Kluwer Academic/Plenum Publishers, 2000

parameters such as wind speed, land-sea temperature difference or sensible heat flux, the lapse rate over a sea. Exponent, b, is usually an empirical constant of the order of 0.5 (Stander and Sethu Raman, 1985; Gryning and Batchvarova, 1990; Melas and Kambezidis, 1992 etc.). In particular, the U. S. EPA recommended Shoreline Diffusion Model (1988) bases on the above simple TIBL conception and Weisman's formula (Weisman, 1976) for the TIBL depth. Pollution diffusion is described by the Gaussian function above the TIBL and instantly and uniformly mixed down to the ground after penetration into the TIBL.

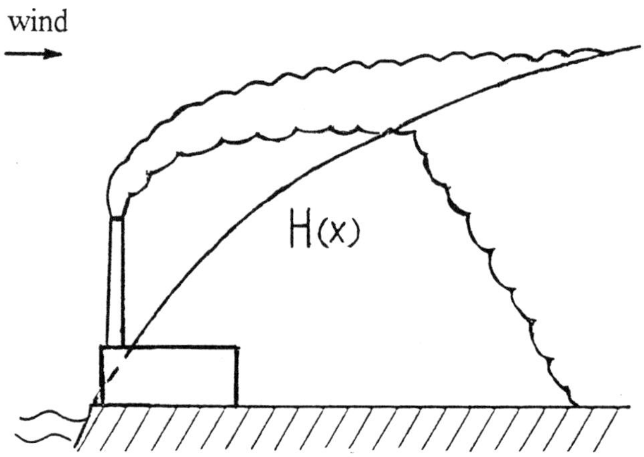

Figure 1. Representation of plume fumigation within the TIBL.

SHORT MODELS DESCRIPTION

In the present paper a series of observed high ground-level pollution concentrations are compared with the estimated values computed using a new Boundary Layer Transformation Model (BLTM) (Levitin, 1994; Levitin and Kambezidis, 1997). It is based on the semi-empirical theory (K-theory) of the ABL (Monin and Obukhov, 1954; Bobyleva et al., 1965, Laykhtman, 1970) and represents a system of equations for the ABL above areas with different surface temperatures. In the model equations the Boussinesq and hydrostatic approximations and one-and-a-half order closure are used. The BLTM does not use the concept of TIBL depth as such. The TIBL is interpreted as a surface atmospheric layer with parameters, which change in comparison to those over the sea surface (during an onshore wind). Namely, during the air flow movement inland the lapse rate transforms from the neutral or slightly stable, over-sea stratification to an unstable one in the surface layer leading to formation of the convective, TIBL. Thus the TIBL is related to the surface layer with intensive turbulence initiated by a thermal convection mechanism. It is to be expected that the depth of this layer grows with increased distance from the shoreline.

Wind profiles are a more conservative characteristic and in case of tall stack emission their variation with fetch from a shoreline inland may be neglected.

Unlike other models, a separate equation is not needed in the BLTM to calculate the TIBL height. The TIBL height is taken as the level at which the eddy diffusivity value returns to the background, over-sea, value. It can be seen, for instance, from Figure 2, where the eddy diffusivity profiles at different distances from a shoreline are shown for one run. In this calculation wind speed is 3 ms^{-1}, sea and land temperatures are 29C and 45C, respectively, that is relevant to daytime summer conditions.

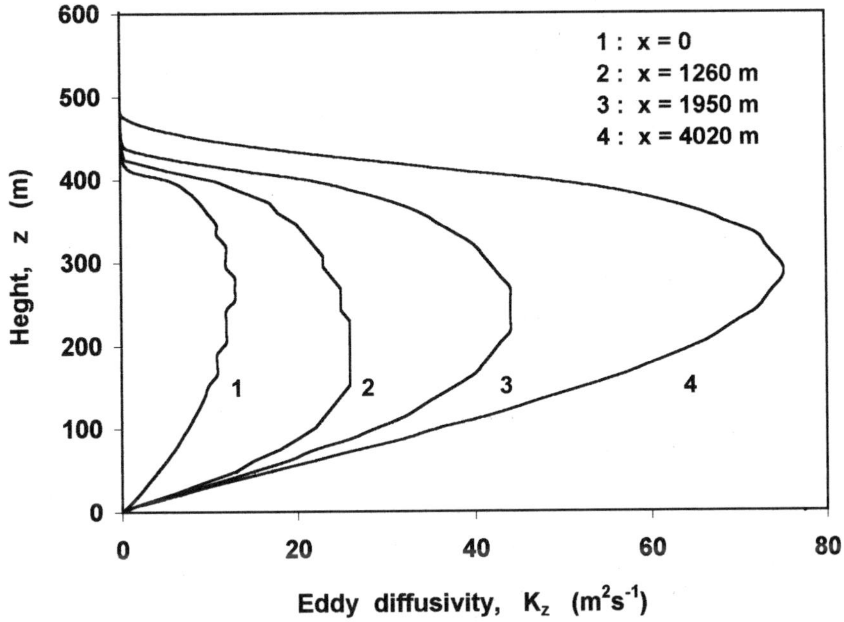

Figure 2. Transformation of K_z profiles with downwind fetch, x, from a shoreline.

Transport of pollution emitted from tall smoke stacks is simulated by the turbulent diffusion equation with the wind component and vertical eddy diffusivity, K_z, obtained from the BLTM. The horizontal crosswind pollutant distribution is described by the generally accepted Gaussian function. Therefore it is not necessary to use the unrealistic hypothesis of uniform vertical plume mixing within the TIBL. The main parameter, which defines the plume dispersion, is the vertical eddy diffusivity, K_z, and its transformation determines the scenario of the plume diffusion. In particular, under daytime breeze conditions an abrupt change of surface temperature under a sea-land transaction leads to TIBL formation and smoke plume fumigation.

INPUT DATA AND RESULTS OF CALCULATIONS

For model verification the Ashdod area was chosen as a typical example of a flat coastal region with an elevated pollution sources located near the Mediterranean shoreline. The Ashdod fuel-fired power plant generates at a total capacity of 1200 MW and is equipped with three 50 m and two 150 m stacks. Concentrations of SO_2 (used as a tracer) as well as a number of other pollutants and a set of standard meteorological parameters are continuously registered by the network of automatic monitoring stations and are transmitted to the Computer Centre in IMS.

The 44 highest SO_2 concentration episodes registered during the summers 1993-1997 were used for the model test. All these episodes took place during the typical sea breeze meteorological conditions, i.e. summer daytime, cloudless sky, gentle onshore wind of 3-5.5 ms^{-1}. The summer sea temperature near the Israel Mediterranean beach and the midday land surface temperature are about 28-29C and 40-45C, respectively, average terrain roughness is about 0.03 m (farmland). The analysed data set included 8 observations at a 3-km distance (the nearest monitoring station to the power plant) and 36 observations at 6-8 km distances (other monitoring stations).

For each episode calculations were carried out for two different thermal stratifications of onshore flow, i.e. for stable and neutral ones, since we have not the corresponding observation data. These assumptions concerning the atmospheric stability ranges during a sea breeze agree,

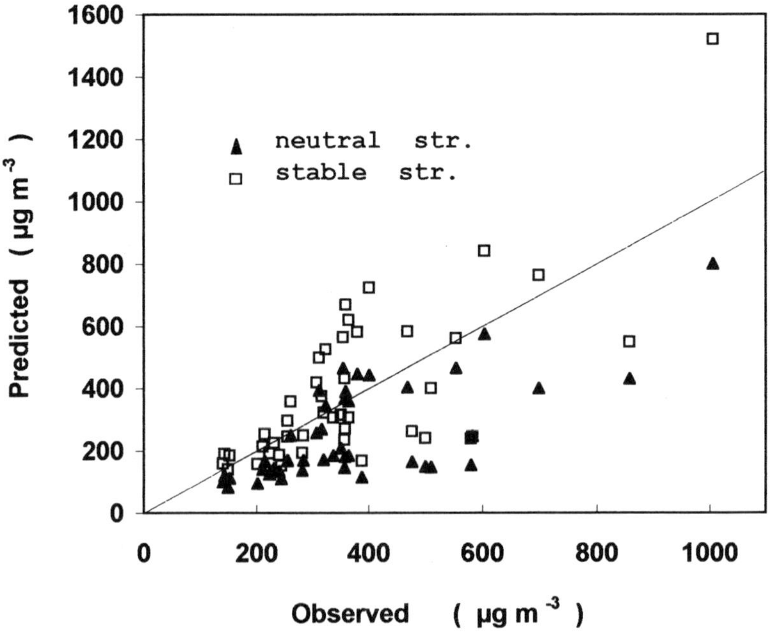

Figure 3. Scatterplots of observed vs predicted concentrations for two initial atmospheric stratification.

for instance, with the field experimental data observed during ATHens Internal Boundary Layer EXperiment (ATHIBLEX) (Kambezidis et al., 1995).

The scatterplots of the predicted vs observed values are shown in Figure 3. The corresponding root mean square errors (RMSE) and correlation coefficients, R, are given in Table 1. As can be seen from Table 1, the RMSE and R are close each other for both initial (over-sea) stratifications. It confirms the fact that the plume dispersion mainly occurs in the formed surface convective layer (TIBL) with a large diffusivity, K_z.

Table 1. Quantitative measures of model performance

Neutral stratification		Stable stratification	
RMSE	*R*	RMSE	*R*
177	0.70	175	0.72

As an example, the calculated ground-level concentrations and corresponding observed values for two episodes are shown in Figure 4.

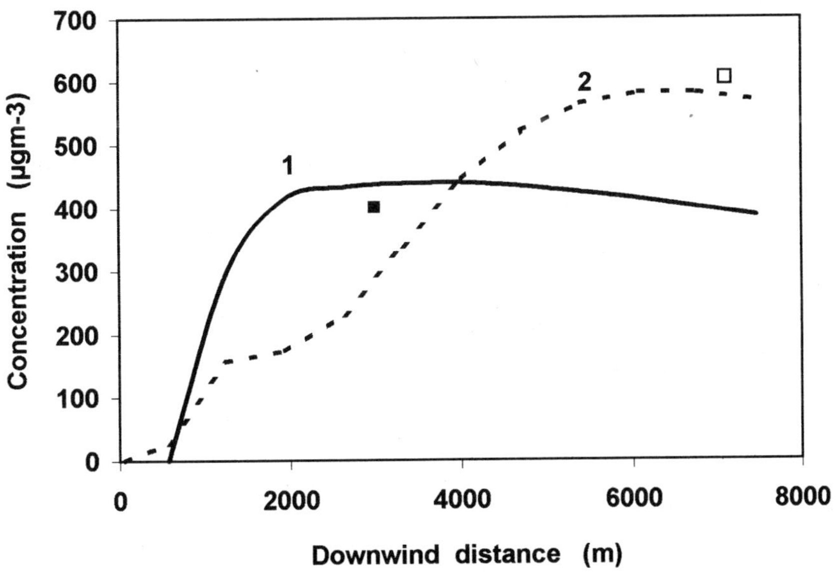

Figure 4. Predicted and observed ground-level concentrations for two episodes:
1 – 26 July, 1994 / —— predicted; ■ observed /;
2 – 30 June, 1997 / .... predicted; □ observed /

CONCLUSIONS

Interest in the TIBL phenomena and coastal area diffusion models is caused by serious air quality problems connected with industrial activity in the densely populated coastal areas.

The new coastal area diffusion model agrees quantitatively with the observed high SO_2 concentrations during fumigation conditions.

The model gives a more realistic distribution of ground-level concentrations, as a result of gradual turbulent diffusion of the plume with different spread rate about and within the TIBL in contrast to the empirical models. Transport of pollution plumes from industrial sources is also simulated without requiring any special TIBL height formulations and additional hypotheses about diffusion processes within the TIBL.

The BLTM itself appears to be more universally applicable in comparison with the above empirical models. In the framework of this model inserted IBLs over a patchwork of surface features, such as a coastal area, lakes, rural regions, forests etc. may be simulated. Addition the turbulent diffusion equation to the BLTM allows to simulate pollution transport within these IBLs inserted.

REFERENCES

Bobyleva, I.M., Zilitinkevich, S.S. and Laykhtman, D.L., 1965, A turbulent regime in the thermal stratified planetary boundary layer of the atmosphere. In: *Proc. Int. Seminar on Atmospheric Turbulence and Spreading of Radiowaves* (in Russian), Moscow.

Gryning, S.E. and Batchvarova, E., 1990, Analytical model for the growth of the coastal IBL during onshore flow, *Quarter Journal Royal Meteorological Society*. 116:187.

Kambezidis, H.D., Peppes, A.A. and Melas, D., 1995, An environmental experiment over Athens urban area under sea breeze conditions, *Atmospheric* Research. 36:139.

Laykhtman, D.L., 1970, *Physics of the Atmospheric Boundary Layer* (in Russian), Gidrometeorizdat, Leningrad.

Levitin, J., 1994, Modelling of air pollution transport over a patchwork of surface features, In: Proc. 2nd International Conference on Air Pollution, Barcelona, Spain, Wessex Institute of Technology, Southampton.

Levitin, J. and Kambezidis, H.D., 1997, Numerical modelling of the thermal internal boundary-layer evolution using Athens field experimental data, *Boundary-Layer Meteorology*. 84:207.

Lyons, W.A.and Cole, H.S., 1973, Fumigation and plume trapping on the shore of lake Michigan during stable onshore flow, *Journal Applied Meteorol*ogy, 12:494.

Melas, D. and Kambezidis, H.D., 1992, The depth of the internal boundary layer over an urban area under sea-breeze conditions, *Boundary-Layer Meteorology*, 61:247.

Monin, A.S. and Obukhov, A.M., 1954, The basic laws of turbulent mixing in the atmosphere near the ground, *Tr.Geophiz. Ins. Akad. Nauk. SSSR,* 24(151):1963.

Stander, M. and Sethu Raman, S., 1985, A comparative evaluation of the coastal IBL height Equations, *Boundary-Layer Meteorol*ogy,32:177.

U.S. Environment Protection Agency, 1988, User's Guide to SDM - A Shoreline Dispersion Model. EPA-450/4-88-017, Office of Air Quality Planning and Standards, Research Triangle Park, NC.

Weisman, B., 1976, On the criteria for the occurrence of fumigation inland from a large lake – a Reply, *Atmospheric Environment*, 12:172.

DISCUSSION

A. LUHAR: The observations show, the turbulence within the TIBL is skewed and is mostly due to large scale convective eddies that leads to counter-gradient scalar fluxes within the boundary layer. Under such conditions the K-theory models usually become inapplicable. I was wondering how you could justify your K-theory model for describing fumigation within the TIBL?

J. LEVITIN: The standpoint that the K-theory models are inapplicable for the convective internal boundary layer is not unquestionable one. As regards our model, it is verified by two series of field measurements. Namely, the TIBL model was tested against data on the observed TIBL depth during the Athens Field Experiment (referenced in the paper). The plume diffusion model, basing on the TIBL model solutions, was tested against pollutants concentration data registered by the monitoring stations network in the Israel coastal area. These results were reported in this paper.

E. GENIKHOVICH: What kind of scaling do you use in your model, Monin-Obukhov's or Deardorf's one?

J. LEVITIN: The Monin-Obukhov's scaling is used in the model.

VALIDATION OF TWO PHOTOCHEMICAL NUMERICAL SYSTEMS UNDER COMPLEX MESOSCALE CIRCULATIONS

C. Borrego, N. Barros, A. I. Miranda, A. C. Carvalho, and M. J. Valinhas

Department of Environment and Planning
University of Aveiro
3810 AVEIRO - PORTUGAL

INTRODUCTION

Portugal has a quite extensive coastline associated with significant terrain features and sea/land breeze circulation, which results in a complex wind field with strong implications for the production and transport patterns of photochemical pollutants with a typical time-scale larger than 24 hours. Since the coastal zone is a major component in global budgets and global resources availability and utilisation, effective sustainable management strategies in such areas require a special ability to think beyond divisions between different types of resources and human activities. Portuguese littoral is the region with the highest population density and economic development. The resulting pollutant emissions place a high pressure on the environment and natural resources of the Portuguese urban coastal areas.

The study of mesoscale circulation is rather important to the assessment of photo-smog formation and the transport of ozone (O_3) at urban coastal areas. The Great Lisbon Area (GLA) is an example of the influence of mesoscale circulation on production and transport of photo-oxidants like O_3 because it is placed in a region of complex topography, dominated by two large estuaries and multiple hills. The University of Aveiro has been studying the problems related with photochemical air pollution in the urban coastal areas of Portugal, particularly in GLA because of its industrial and urban importance and high emissions level.

For the numerical simulation of photochemical production and transport, is crucial an accurate prediction of complex mesoscale circulation's occurring in this area. In order to overcome this limitation two prognostic air quality systems were prepared for a temporal large application, both including a meteorological and a photochemical model. The information acquired through the application of these models would be validated with measurements from intensive field campaigns. Within this context, a field measurement campaign was organised at the GLA, during the summer of 1996.

The main purpose of this work is to present the validation of the two photochemical numerical systems by comparison with measurements from the meteorological and air quality campaign that occurred in Lisbon region from 8 to 17 July 1996.

Air Pollution Modeling and Its Application XIII, edited by S.-E. Gryning and E. Batchvarova, Kluwer Academic/Plenum Publishers, 2000

THE PHOTOCHEMICAL NUMERICAL SYSTEMS

The numerical tools that have been used to analyse the atmospheric circulations and the air quality patterns in the Lisbon region during summer were MEMO/MARS and MAR models systems. Both systems were the subject of a validation using data from a meteorological and air quality campaign.

MEMO/MARS Models

The mesoscale model MEMO is a non-hydrostatic prognostic model developed to simulate the wind flow over complex terrain (Moussiopoulos, 1989). Within the model, the conservation equations for mass, momentum and scalar quantities are solved numerically, in terrain following co-ordinates. Radiative transfer is calculated with an efficient scheme based on the emissivity method for longwave radiation and an implicit multilayer method for short-wave radiation. For turbulence parameterization k-theory is applied. Recently, the University of Aveiro has implemented in MEMO an advanced scheme for the simulation of turbulence (Coutinho and Borrego, 1995), using the transilient theory, a non-local closure for turbulence.

MARS model describes the dispersion and chemical transformation of air pollutants in a three dimensional region (Moussiopoulos *et al.*, 1995). This model is directed towards the photo-oxidants simulation, from which O_3 is the major component. It is a fully vectorised model that solves the concentration parabolic equation system for known meteorological variables. KOREM and RADM2 can be used on MARS as chemical mechanisms. The first one, which has been chosen to this work, includes 39 chemical reactions with 20 reactive pollutants.

MEMO/MARS system is one of the core models of the EUMAC Zooming Model, being currently applied in several international research institutions.

MAR System

A version of the Colorado State University Meso-meteorological Model (CSUMM) (Kessler and Douglas, 1992), a prognostic meteorological model was used to generate the meteorological inputs needed by the Urban Airshed Model (UAM CB IV) (Morris *et al.*, 1990). This integrated system, specially developed to be applied to coastal regions, has been named MAR (Modelo Atmosférico Regional).

The CSUMM model used in this work is a version of the hydrostatic, incompressible, primitive-equation model originally developed by Pielke (1974). It includes a surface heat budget and parameterizations of the atmospheric surface and planetary boundary layers. A terrain following co-ordinate system simplifies the application of the model to domains with an irregular lower boundary.

UAM is a tridimensional Eulerian model developed for the simulation of the physico-chemical processes associated to the production, transport and dispersion of reactive and inert atmospheric pollutants. The model is based on the resolution of the continuity equation, which expresses the mass conservation of each pollutant in a turbulent flow where chemical reactions occur. The chemical mechanism used on UAM ("Carbon Bond Mechanism", CB - IV) includes 80 chemical reactions with 30 reactive and inert pollutants.

THE LISBON ATMOSPHERIC FIELD EXPERIMENT

The knowledge and characterisation of mesoscale atmospheric flow patterns, as well as, the description, by mathematical models, of dispersion and transformation mechanisms of photo-oxidants in the GLA are fundamental. Models also have to be validated by appropriate field measurements. In this context, a field measurement campaign (Lisbon Atmospheric Field Experiment - LisbEx 96) was organised in GLA, from 8 to 17 July 1996, in order to: (i) characterise meteorology and air quality during the Summer; (ii) study the formation and transport of photochemical pollution; and (iii) validate models.

The campaign area (200 km x 200 km) was structured in order to integrate all ground based monitoring stations (public or private) located in the study domain. Mobile monitoring stations were also used, location based on previous numerical simulations. Temperature, relative humidity, wind speed and direction were measured at every meteorological station (see figure 1).

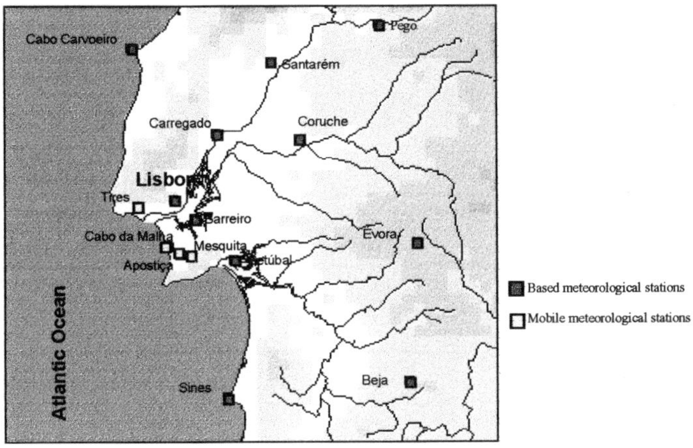

Figure 1. Location of meteorological stations at GLA.

Because of their locations, the stations can be divided in coastal (Cabo da Malha, Cabo Carvoeiro and Sines), inland (Beja, Évora, Pego, Santarém and Coruche) and urban (Lisbon, Barreiro, Setúbal and Tires). Pollutant's concentrations, including O_3, were measured at air quality stations located in the areas of Sines, Setúbal, Mesquita, Tires and Pego. The diurnal O_3 concentrations at measuring stations showed the typical pattern of photochemical smog formation: a steep rise in the first morning hours with a maximum in the first afternoon hours around.

Besides the ground-based information, radiosondes and tether-balloon soundings have been done. The tether balloon sounding system included classic meteorological data acquisition and an O_3 sensor.

NUMERICAL SYSTEMS VALIDATION

Numerical models quality assurance should include at least three steps: (i) sensitivity analysis; (ii) evaluation of model performance by comparison of predicted results with available measurements; and (iii) for a given problem, inter-comparison of results from different models. The first step is a particular task, dependent of the model formulation and specific sub-modules and is not considered in this work. Concerning the other two validation tasks, both have been taken into account, by the comparison between model results and measured data and by the inter-comparison of model estimates, respectively. For the considered GLA domain resolution, both hidrostatic and non-hidrostatic approaches can be applied and the two model systems performance compared.

Evaluation methodology

Wind direction and velocity, temperature and O_3 values predicted by MEMO/MARS and MAR systems were compared with measured data from LisbEx 96 experimental campaign. An index of proximity (Martins *et al.*, 1996) was applied for the quantitative evaluation of wind speed and temperature results and the cosine of the difference between the simulated and the estimated wind direction for the study of the wind direction. The index of proximity is defined as 100 (1 - | S - R | / (S + R)), where S and R represent, respectively, the obtained and the measured values of wind speed or temperature. A value of 1 for the cosine of the difference angle between estimated and measured wind direction indicates the agreement of the estimated values with the real data and a value of -1 indicates that there is a shift of 180°.

The evaluation of the predicted O_3 by both systems were done following the Environmental Protection Agency (EPA) performance measure formulations (EPA, 1991):

- Unpaired highest-prediction accuracy (A_u), which quantifies the difference between the magnitude of the highest 1-hour observed value and the highest 1-hour predicted value:

$$A_u = \frac{c_0(.,.) - c_p(.,.)}{c_0(.,.)} x100\%$$

- Normalised bias test (D):

$$D = \frac{1}{N_T} \sum_{i=1}^{N} \sum_{j=1}^{Hi} \frac{c_0(i,j) - c_p(i,j)}{c_0(i,j)}$$

- Gross error of all pairs > 180 μg.m^{-3} (E_d)

$$E_d = \frac{1}{N_T} \sum_{i=1}^{N} \sum_{j=1}^{Hi} \frac{c_0(i,j) - c_p(i,j)}{c_0(i,j)}$$

Where:

A_u - unpaired highest-prediction accuracy;

D - normalised bias obtained from all hourly prediction-observation pairs;

E_d - normalised gross error for all hourly prediction-observation pairs for hourly observed values > 180 μg.m^{-3};

c_0 (.,.) - Maximum 1-hour observed concentration over all hours and monitoring stations;

c_p (.,.) - Maximum 1-hour predicted concentration over all hours and monitoring stations;

c_0 (i,j) - observed value at monitoring station i for hour j;

c_p (i,j) - predicted value at monitoring station i for hour j;

N - number of monitoring stations;

H_i - number of hourly prediction-observation pairs for monitoring station i;

N_T - total number of station-hours.

The predicted value c_p should be derived from bilinear interpolation of the predicted values at the four grid cells nearest to station i for the given hour.

Simulation conditions

MAR and MEMO/MARS systems were applied to a modelling domain of 200 km x 200 km, equivalent to the previously referred LisbEx96 domain, with a horizontal grid resolution of 4 km. In the vertical direction 28 non-equidistant layers until 8 km were considered by CSUMM and 20 non-equidistant layers until 6000 km were considered by MEMO. MARS used the same vertical structure than MEMO and the photochemical model that integrates the MAR system considers 5 layers up to 3 km.

In order to specify the boundary concentrations, a clean air approach was assumed. The strong influence of the clean air advected from the Atlantic Ocean over the Lisbon air-shed, as well as the results of field campaigns performed in rural and forest atmospheres, were considered in the definition of typical trace constituents concentrations.

Simulations results

The systems meteorological output, wind and temperature, were compared with measured data recorded during days 2 and 3 (9 and 10 of July) of the Lisbex96 campaign, which were considered as the most representative summer days. The lower troposphere pressure field for these days was characterised by the presence of the Azores anticyclone extended in ridge over the northern part of the Iberian Peninsula, promoting a continental dry and very hot circulation over Portugal. Strong insulation allows the formation of mesoscale circulation and photochemical production.

Figure 2 presents the observed and predicted hourly data of temperature, wind velocity and direction at Lisbon station, as well as the indexes of proximity for the first two variables and the cosine of the difference angles for the last one.

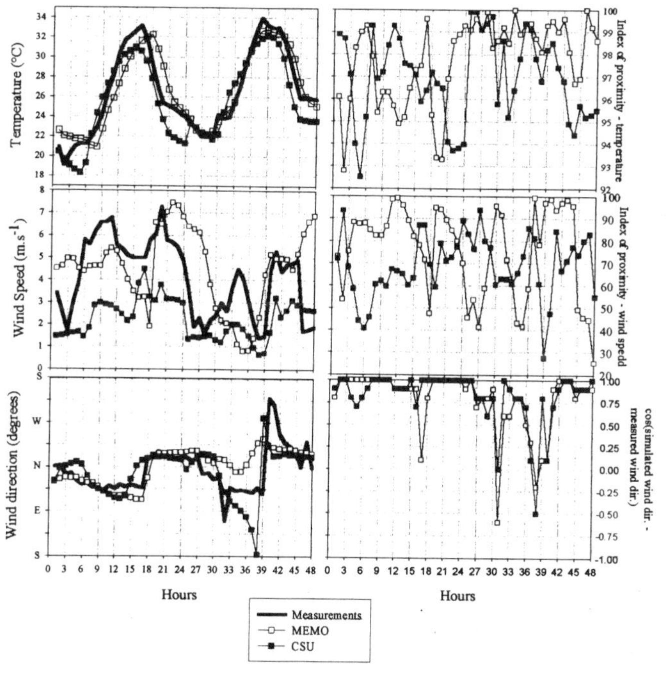

Figure 2. Observed and predicted hourly meteorological data and indexes of proximity for wind speed and temperature and the cosine of the difference angles for wind direction, obtained at Lisbon station.

Temperature results from both systems were quite reasonable, following the measurements behaviour along the two simulated days. This is confirmed through the analysis of the estimated proximity index, always greater than 92%. However, it should be noticed that CSUMM results have shown a better agreement with measurements on the 1st simulated day and MEMO results were closer to temperature data on the 2nd one. Both models performed worse during night periods.

Results are not so good concerning wind speed. The index of proximity varies from circa 30 to 100%. CSU under-predict wind speed and MEMO seems to have some difficulties to simulate the lowest winds. Differences between wind direction estimates and measurements are in general lesser than 45°, which is a quite good model performance.

Inter-comparison between model's results allowed verify that usually they do not agree, exceptions occurring when both results were close to measurements.

Table I present averaged values of 48 hours for each meteorological station, allowing to evaluate the global behaviour of both systems. Five of the total meteorological stations were not included in the evaluation process due to the lack of reliability of measured data.

Table I. Global evaluation of CSU and MEMO models

Meteorological Station	Index of Proximity Temperature		Index of Proximity Wind Speed		Cos (simulated-observed) Wind Direction	
	CSU	MEMO	CSU	MEMO	CSU	MEMO
Apostiça	91.9	91.2	61.1	67.8	0.7	0.6
Mesquita	95.0	93.9	75.4	77.0	0.6	0.3
Expo	96.2	94.6	64.9	73.4	0.7	0.7
Lisboa	96.9	97.8	69.0	75.9	0.8	0.8
Tires	95.1	96.0	61.3	85.7	0.6	0.6
Sines	94.3	95.9	73.5	81.5	0.1	0.3
Santarém	96.1	94.3	76.6	69.6	0.7	0.6
Évora	95.8	94.1	58.9	79.7	0.5	0.5
Cmalha	94.6	95.4	69.4	78.8	0.6	0.5
Setúbal	92.8	91.1	74.3	83.6	0.4	0.7
Beja	95.8	95.4	59.5	81.4	0.0	0.5
Global	95.0	94.5	67.6	77.7	0.5	0.6

In terms of global evaluation of the models, both presented a good performance. Regarding temperature, CSUMM obtained a global proximity index of 95% and MEMO 94.5%. The lowest indexes (between 91.1 and 92.8%) occurred, for the two models, at the meteorological stations of Apostiça and Setúbal.

Wind speed performance was not so good. The global index of proximity estimated for CSUMM is lesser than 75% and for MEMO reaches 77.7%. For the inland stations Beja and Évora, CSUMM presents an index near 60%, while MEMO index rounds 80%. The best index estimated for CSUMM corresponds to one of the worst index calculated for MEMO.

In what concerns wind direction, both models have a high global performance with a difference between estimated and measured data lesser or equal a 45°. However, Beja, Setúbal and Sines stations present differences estimated for CSUMM reaching 90°. MEMO results are as well not so good in Sines and Mesquita stations.

A quantitative evaluation also was done taking into account predictions and measurements at four air quality stations, covering 88 hours of observations. Keeping in mind that photochemical models would need several hours to reach an equilibrium state, this analysis was only related to the 2nd simulation day. In the Table II it is possible to find the estimated evaluation parameters, A_u, D and E_d, already described.

In assessing model simulation results for the performance evaluation, there is no rigid criterion for model acceptance or rejection. Based on past photochemical model evaluation, this type of modelling generally produces peak (unpaired) prediction accuracy, overall bias, and gross error statistics in the approximate ranges of ± 15-20%, ± 5-15% and 30-35%, respectively.

Table II. Quantitative air quality evaluation

Evaluation parameters (%)	Numerical Systems	
	MAR	MEMO/MARS
A_u	-1.9	0.7
D	-12.7	-1.9
E_d	26.8	36.6

Concerning the parameter related to the unpaired highest-prediction accuracy, A_u, both systems performed according to the advisable ranges. However, MAR tends to over-predict the highest value and MEMO/MARS tends to under-predict it. The estimated overall bias, D, also falls within the referred ranges. Both systems over-predict O_3 concentrations for the analysed air quality stations. Gross error estimates of all pairs with concentration values higher than 180 $\mu g.m^{-3}$ (public information threshold, 72/92/EEC) were not so reasonable because both systems valuations were out of the 30-35% range, namely MAR's E_d was under and MEMO's E_d over the above range.

In general the two analysed numerical systems could be acceptable due to their performance that falls within the advisable ranges. However, caution is urged in using these ranges as the sole basis for determining the acceptability of model's performance. These ranges should be used in conjunction with graphical procedures to assess overall model performance.

In order to complement the analysis and to study the photochemical pollution that occurs in Lisbon, the concentration fields of O_3 have been plotted with the correspondent wind field. Prognostic airflow and O_3 concentration fields generated by both systems at 10 m above ground level, at 16.00 hours, are presented in figure 3.

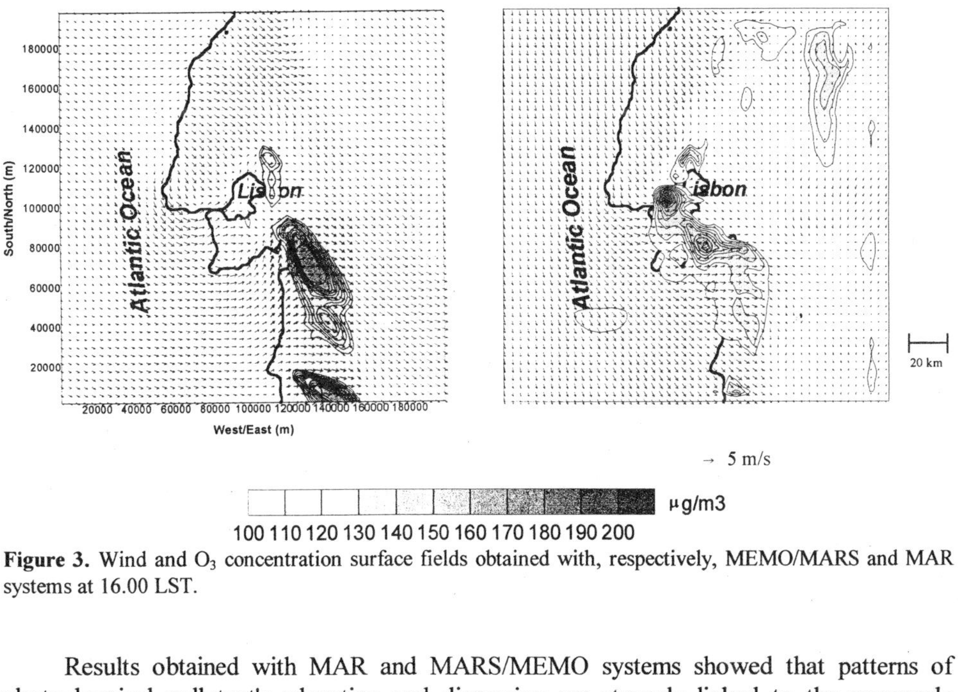

Figure 3. Wind and O_3 concentration surface fields obtained with, respectively, MEMO/MARS and MAR systems at 16.00 LST.

Results obtained with MAR and MARS/MEMO systems showed that patterns of photochemical pollutant's advection and dispersion are strongly linked to the mesoscale circulations that develop in the Lisbon region. In the beginning of the afternoon, a full sea breeze along most of the coastline, is visible in both systems results. This sea breeze front follows the coastline curvature. It is possible to identify the O_3 plume of Lisbon, advected to South in both concentration fields. However, the atmospheric flow simulated by MEMO is responsible for a more pronounced advection of pollutants into land and to South. One peak of O_3 is visible Southeast of Setúbal and another Southeast of Sines. Concentration field estimated with MAR system shows the O_3 plume of Lisbon, advected by the Tejo valley flow, partially retained by a convergence zone formed over the Tejo mouth and Setúbal Península in the late morning. It is also possible to identify the plume transport over Sines.

CONCLUSIONS

The meteorological and air quality campaign realised during the summer of 96 was fundamental to the evaluation of two photochemical numerical systems. Performance of both systems was analysed concerning the meteorological and the air quality results. Regarding meteorological simulations both models results agreed quite well with measured data. Air quality estimates did not accord so reasonably with observations. However, air quality calculations were within the advisable range for the used evaluation parameters.

Nevertheless, the current work purpose, besides the acceptance or rejection of a specific system or model, given the input uncertainties, is also the comparison of the systems among each other for a given problem. Inter-comparison of model's results showed that, in general, simulation calculations were not in a clear accordance.

The LisbEx96 experiment should be more explored in order to deep the evaluation process and to finalise the validation of MEMO/MARS and MAR systems.

ACKNOWLEDGEMENTS

The authors would like to thank to PRAXIS XXI Programme for their helpful support to AMAZOC (PRAXIS/3/3.2/AMB/38/94) and IGLO (PRAXIS/3/3.2/EMG/1949/95) Projects. This work was also supported by DGA (PEAM/P/AMA/603/95).

REFERENCES

Coutinho, M. and Borrego, C., 1995 Application of Transilient Turbulence Theory to a Mesoscale Meteorological Model. Proceedings of the *11th Symposium on Boundary Layers and Turbulence*. American Meteorological Society, Charlotte.

EPA, 1991, Criteria for Assessing the role of transported Ozone/Precursors in Ozone nonattainment areas, EPS-450/4-91-015, Office of Air Quality Planning and Standards, Research Triangle Park, NC (NTIS Nº: PB91-195958.

Kessler, R. and Douglas, S., 1992, User's Guide to the Systems Applications International Mesoscale Model, SYSAPP-92/001. California, San Rafael.

Martins, J., Borrego, C., Hodin, A. and Mejean, P., 1996, Comparison of real scale, wind tunnel and numerical model results of atmospheric dispersion in the vicinity of a building. Proceedings of the 3rd Seminar on Industrial Fires, Riso, Denmark.

Morris, R., Myers, T. and Haney, J., 1990, User's Guide for the Urban Airshed Model, Volume I: User's Manual for UAM (CB-IV), EPA-450/4-90-007A, U.S. Environmental Protection Agency, Research Triangle Park, NC (NTIS Nº: PB91-131227).

Moussiopoulos N., 1989, Mathematische Modellierung mesoskaliger Ausbreitung in der Atmosphare, Fortschr.-Ber. VDI, Reihe 15, Nr. 64, pp. 307.

Moussiopoulos, N., Sahm, P. and Kessler, C., 1995, Numerical Simulation of Photochemical Smog Formation in Athens, Greece a Case Study. *Atmospheric Environment*, **Vol. 29**.

Pielke, R., 1974, A three-dimensional numerical model of the sea breeze over south Florida. *Mon. Wea. Rev.*, **102**, 115-139.

THE USE OF SATELLITE BASED REMOTE SENSING OF POLLUTION TO DEFINE BACKGROUND CONCENTRATIONS FOR ATMOSPHERIC MODELS

Walter Debruyn, Gil Lissens,and Clemens Mensink

VITO, Centre for Remote Sensing and Atmospheric Processes (TAP)
Boeretang 200
B-2400 Mol
Belgium

INTRODUCTION

In this study, we combined two different spaceborne Remote Sensing instruments to deduce daily height resolved ozone concentrations over the European continent. Since this work has been performed as part of the RIFTOZ project (Regional Differences in Tropospheric Ozone in Europe), carried out for the European Community's 4th Framework 'Environment and Climate' Research Programme, our attention focused on ozone retrieval in the troposphere.

Our work to retrieve height resolved ozone concentrations was *statistical in nature*. A *differential approach*, combining measurements from the GOME instrument aboard ERS-2 with measurements from the SBUV-2 instrument aboard NOAA-14, has been developed. The ozone concentrations have been retrieved in vertical ozone "layers" with an as high as possible vertical resolution.

The Global Ozone Monitoring Experiment (GOME) ultraspectral spectrometer aboard the European Remote Sensing Satellite ERS-2 observes solar radiation scattered from the atmosphere; the Differential Optical Absorption Spectroscopy (DOAS) technique is used to derive ozone column densities. The GOME provides a global coverage in three days; its pixel size is approximately 40 x 320 km^2. The Solar Backscattered UltraViolet (SBUV-2) spectrometer aboard the US National Oceanic and Atmospheric Adminstration's NOAA-14 observes solar backscattered radiation in 12 fixed wavelength bands; an optimal estimation retrieval algorithm (Umkehr algorithm) is used to provide ozone concentrations in 12 atmospheric layers with a ground resolution of 183 x 183 km^2 and a local revisit time of 9 days. From this technical description of both instruments, the constraints in geographical and temporal resolution in ozone concentration data are obvious.

RATIONALE

In a sense, the "statistical objective" of this study is to quantify the value added of satellite measurements to the retrieval of tropospheric ozone profiles, in other words, to see if satellite measurements explain part of the variation of actual ozone concentrations versus a standard climatological ozone profile or long-term average ground-based profile.

In this study, the month August 1997 was taken as experimental period; the vertical ozone profiles (i.e. ozone slices or layers) obtained were compared with balloon ozonesoundings at Ukkel (51 N, 4 E), Payerne (47 N, 7 E) and Hohenpeissenberg (48 N, 11 E) measured during the same month. To evaluate the extra information obtained from the satellite based ozone layer retrieval, the ozone layer data were also compared with a standard profile at the geographical zone and time frame considered – the midlatitude summer profile from the MODerate resolution model for LowTRAN-7 (MODTRAN) (Berk et al., 1989) model was used for this purpose – and with the long term mean ozone profile as measured in Ukkel during the month of August for ten years (1987-1997).

PREPARATORY WORK

Selection Of Heigth Resolution Of The Ozone Layers

The vertical resolution of the SBUV-2 instrument is taken as the standard resolution for our analyses, because it is coarser than that of the balloon soundings and standard profiles. Because data from balloon soundings are restricted to heights of around 30 – 35 km (height at which the balloon bursts), 5 layers were withheld, corresponding to layers 1 to 5 (bottom to top) of the SBUV-2 instrument (Table 1).

Table 1. Layer boundaries used by SBUV 'Umkehr' retrievals

Layer number	Boundaries (km)
1	0 – 9.0
2	9.0 – 14.7
3	14.7 – 19.1
4	19.1 – 23.5
5	23.5 – 28.1
6	28.1 – 32.8
7	32.8 – 37.7
8	37.7 – 42.8
9	42.8 – 48.2
10	48.2 – 53.7
11	53.7 – 59.2
12	59.2 - ∝

All ozone concentration data are expressed as Dobson Units (DU) per reference layer.

Spatial Attribution Of Satellite Ozone Profiles To Balloon Launch Sites.

No trajectory tracing of the balloons was performed; the satellite pixels with the shortest distance to a certain balloon launch site were chosen for the purpose of comparison.

RESULTS AND DISCUSSION ON RETRIEVED PROFILES (LAYERS)

The basis for the quantitative evaluation of the value added of satellite data in the estimation of ozone profiles is the reduction in root mean squared error of the satellite ozone profile with the balloon soundings compared to the root mean squared error of the standard profile with the balloon soundings. In essence this amounts to the determination of the success of satellite ozone profiles in estimating the 'natural variability' of daily ozone profiles versus standard summer profiles.

The root mean squared error for the satellite measurements in every layer is defined as:

$$RMSE_l = \sqrt{\frac{1}{n}\sum_{i=1}^{n}\left(O3sat_{i,l} - O3bal_{i,l}\right)^2}$$

with:
l the layer number (from 1 to 5)
n the observation number (from 1 to 93)

Consequently, the root mean squared error for the standard profiles in every layer is defined as:

$$RMSE_l = \sqrt{\frac{1}{n}\sum_{i=1}^{n}\left(O3stp_{i,l} - O3bal_{i,l}\right)^2}$$

Results of the error calculation for each layer and for each method are given in Table 2. The GOME/SBUV satellite data are calculated in the following way. The GOME total ozone column is partitioned in SBUV layers according to the following formula:

$$O3_l = GOME \bullet \frac{O3sbuv_l}{\sum_{l=1}^{l=12} O3subv_l}$$

The idea here was to keep the relative proportions of the ozone concentration in each of the layers as estimated by the SBUV-2 measurements (in 12 layers up to 59.2 km), and use the total ozone of GOME as the total amount of ozone present in these 12 layers. This because of GOME's inherent good performance in measuring total ozone at the geographical sites considered (ESA, 1997).

Standard ozone profiles are derived from the MODTRAN model (ML MODTRAN) and from the August long term profile measured with balloon soundings at Ukkel (Ukkel LT).

Table 2 lists the root mean squared errors relative to the average ozone concentration in each of the layers measured by the balloon soundings.

Table 2. Relative RMSE of satellite ozone profiles and standard profiles (%)

Relative RMSE (%)	Layer1	Layer2	Layer3	Layer4	Layer5
Height (km)	0-9.0	9.0-14.7	14.7-19.1	19.1-23.5	23.5-28.1
GOME/SBUV	41	41	50	10	13
ML MODTRAN	20	91	20	11	23
Ukkel LT	17	40	11	6	7

This table shows that the Ukkel LT profile scores best for all layers. The Ukkel LT profile however does not hold information outside the Ukkel area and it therefore would be 'unfair' to compare satellite ozone profiles to this standard profile. Still, this analysis gives a good idea of the degree of stability of ozone soundings and illustrates the importance of soundings as a complement to satellite observations (or vice versa). If we compare the satellite ozone profile to the ML MODTRAN standard profile, we see that in layer 2 and in the higher altitude layers 4 and 5, the satellites outperform the ML MODTRAN profile. At lower altitudes (layers 1 and 3) the satellite data contain more noise than the MODTRAN standard profile.

Figure 1 below illustrates a typical vertical ozone distribution retrieved from the GOME and SBUV-2 instruments versus an ozone balloon sounding at Ukkel on the same day. The height of the layers is indicated by the dotted lines; the remotely sensed ozone concentration in each layer is given with its root mean squared error.

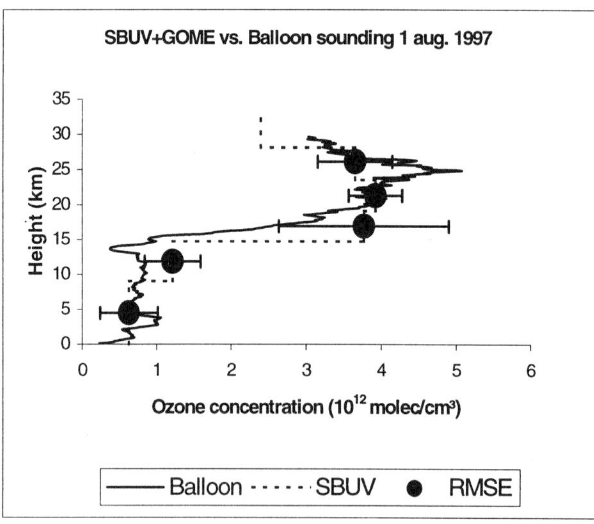

Figure 1. Satellite ozone layers and balloon ozone profiles at Ukkel on August 1, 1997.

This figure illustrates the vertical resolution and the reliability in each layer retrieved with the differential algorithm. It performs best (highest vertical resolution and lowest root mean squared error) from the ozone maximum onwards; below the ozone maximum the error is substantial.

EXTRAPOLATION OVER EUROPE

To suit the needs of the RIFTOZ project, ozone concentration maps have been processed for each day of the month August 1997 for the atmospheric layers 1 to 3 as defined in table 1. To this end a 40 x 40 km^2 grid has been defined over Europe in an area with latitudes between 35 and 75 DEG and longitudes between –15 and 50 DEG. Daily ozone concentrations were calculated at each gridpoint as distance weighted averages of all GOME and SBUV-2 data available within the geographical area considered, using the square root of the distances between the grid point and the pixel center points as distance measures.

Thus the ozone concentration for every pixel $P_{i,j,l}$ with coordinates i,j for every layer l is given by :

$$P_{i,j,l} = \frac{\sum_{n=1}^{N} \frac{1}{\left\| P_{i,j,l} - S_n \right\|^p} \bullet C_{n,l}}{\sum_{n=1}^{N} \frac{1}{\left\| P_{i,j,l} - S_n \right\|^p}}$$

With

p = The power to which the distance from the center point of the satellite pixel to the gridpoint is raised (0.5, 1, 2, 4).

n = 1,…,N = The satellite pixels that are available over Europe.

$\left\| P_{i,j,l} - S_n \right\|$ = Distance "over the globe" from the center point of the satellite pixel to the gridpoint.

$C_{n,l}$ = Ozone concentration in layer l of satellite pixel n

As can be seen from the formula used for extrapolation, the power to which the distance from the pixel center point to the gridpoint is raised (0.5, 1, 2 or 4) determines the extent to which the local ozone concentration is influenced by remote measurements. The quartiles of the data processed within each layer and with each power of the distance measure have been compared to meteorological maps; based on this comparison the power 0.5 was selected as the most realistic parameter. The calculations with distances raised to the power 2 overemphasize the local concentrations at the pixel center points to the extent that the satellite orbits can be clearly traced. This is illustrated in figure 2.

CONCLUSIONS

This study describes the vertical resolution and the reliability (expressed as the root mean squared error) to be expected for tropospheric ozone concentration data obtained via Remote Sensing (satellite retrieval). The apparant low root mean squared error in layer number 2 (9.0-14.7 km) as opposed to the neighbouring layers 1 and 3, is in fact due to the high variablity of the ozone concentrations in the biosphere (layer 1) and in the lower stratosphere (layer 3).

The results obtained in this study are comparable to these in other studies, where height resolved ozone concentrations have been retrieved from satellite based instruments, in casu also the GOME instrument, by means of physical algorithms rather than statistical algorithms. The University of Bremen (Bramstedt et al., 1997) deduced ozone profiles with

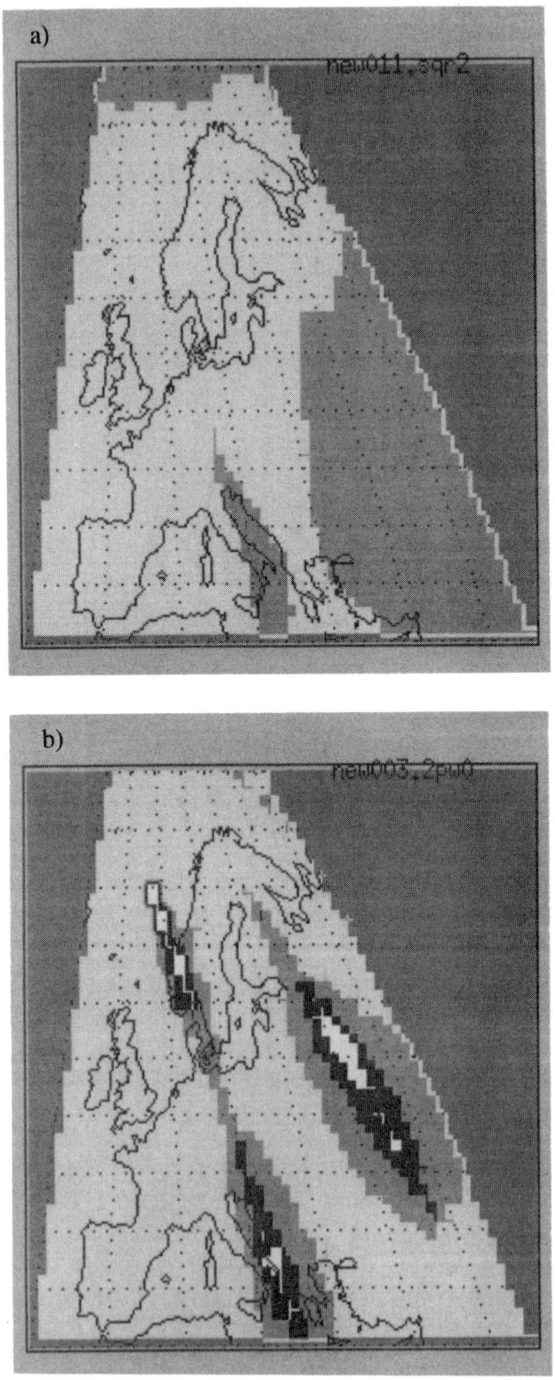

Figure 2. Visualisation of ozone concentration layers by means of quartiles, using the
a) power 0.5 of the distance measure and
b) power 2 of the distance measure.

a vertical resolution of 10 km in the troposphere of the Polar region, using a full retrieval method based on optimal estimation; the Rutherford Appleton Laboratory (Siddans et al., 1997) derived ozone information from the surface to approximately 50 km with a vertical resolution of 5 to 12 km depending on altitude and standard deviations of < 10 % and < 40 % above and below 12 km; the Royal Netherlands Meteorological Institute KNMI (van der A and Kelder, 1998) retrieved ozone concentration values at a step size of 4 km in the stratosphere and one or two values in the troposphere.

The daily ozone concentrations over Europe as obtained by the remote sensing instruments described in this study could serve as *background concentrations* for global and regional scale atmospheric chemistry and transport models. The satellite retrieval techniques could also provide *boundary conditions* for regional (and global) scale models, especially at the top of the model domain.

However, it is obvious that the constraints in spatial and temporal resolution may pose limits to its use for this purpose. A regional scale model would require a horizontal resolution of typically 50 x 50 km^2 and a temporal resolution of one hour or less. This is, for the time being, still beyond the scope of the satellite retrieval method discussed in this study, unless interpolation techniques are applied. Also the vertical resolution in the lower troposphere (9 km in the lowest layer) is too coarse to provide detailed boundary conditions in the form of concentration profiles at the borders of the domain.

ACKNOWLEDGEMENTS

The SBUV/2 measurements from the NOAA-14 platform were kindly given to us by Dr. L. Flynn at the NOAA NESDIS (data obtained with support from the NOAA Climate and Global Change Program Atmospheric Chemistry Element). We also would like to thank Dr. Dirk De Muer at the KMI Ukkel for the kind permission to use the ozone measurements from the KMI balloon soundings, and for giving us the possibility to use similar measurements from Payerne and Hohenpeissenberg. This research has been performed as part of the RIFTOZ project (Regional Differences in Tropospheric Ozone in Europe), filed as ENV4-CT95-0024, under the Environment and Climate chapter of the European Community's 4th Framework Research Programme.

REFERENCES

Berk A., Bernstein L.S. and Rbertson D.C., MODTRAN: A moderate resolution model for Lowtran-7, GL-TR-89-0122, Geophysics Laboratory, Hnascom AFB, 1989.

Bramstedt K., Eichmann K.U., Weber M., Rozanov V., Hoogen R., de Beek R., Buchwitz M., Kurosu T. and Burrows J.P., Ozone profiles from GOME satellite data – Part II: first results from the Arctic winter campaign, in: Proc. 4th Eur. Symp. On Polar Stratospheric Ozone, Schliersee, Germany, 22-26 Sept 1997.

ESA, 6th GOME and SCIAMACHY data and algorithm Scientific Working Sessions, Ukkel, September 29-30 1997.

Siddans R., Reburn W. and Kerridge B.J., Height resolved ozone retrievals in the troposphere and lower stratosphere from GOME, Proc. 3rd ERS Symp., Firenze, Italy, March 1997, ESA SP-414 vol II, 615-620, 1997.

Van der A R.J. and Kelder H., Ozone profile retrieval with GOME, ESA Earth Observation Quarterly, EOQ 58, 14-15, 1998.

DISCUSSION

B. L. WIMAN: When you retrieve ozone data form RS techniques, do you make corrections for aerosols, in particular for layer?

C. MENSINK: Yes, the spectral signature at the sensor is corrected for Broad Band, non structured scattering and absorption phenomena such as those provoked by aerosol. Broad Band correction is done by means of low order polynome fitting. Then, in a second step, the structured spectral signature resulting from trace gas absorption is analysed to retrieve the different trace gas concentrations. In our case we retrieved ozone in this way.

R. YAMARTINO: Your first level was 9 km deep but levels 2 and 3 were 5 km. Could you have adjusted your analysis to give 5 km resolution near the surface?

C. MENSINK: In principle, yes. But, the spectral signature at longer wavelengths, where ozone from below the ozone layer is measured, is weaker than at shorter wavelengths. As solar radiation at these longer wavelengths penetrates to deeper layers of the atmosphere, multiple scattering and surface reflection become important. Thus, to retrieve the ozone concentration near the surface with a reasonable accuracy, the ozone concentration has to be high enough to generate an acceptable signal-to-noise ratio. Therefore we integrate over a thicker atmospheric layer and thus a higher ozone concentration.

A NEW TECHNIQUE TO DERIVE MIXED LAYER DEPTH AND ENTRAINMENT ZONE THICKNESS FROM LIDAR PROFILES

D. G. Steyn[1], M. Baldi[2] and R. M. Hoff[3]

[1] Atmospheric Science Programme, Department of Geography The University of British Columbia, Vancouver, Canada
[2] Institute for Atmospheric Physics, IFA-CNR, Rome, Italy
[3] Atmospheric Environment Service, Egbert, Ontario, Canada

INTRODUCTION

The detection of Mixed Layer Depth (MLD) has been a frequently derived product from both ground-based and airborne lidar (Endlich *et al.*, 1979; Boers *et al.*, 1984 Melfi *et al.*, 1985). The detection of the mixed layer has been based on discontinuities in the aerosol backscatter coefficient (β_a) immediately above the boundary layer and relies on sufficient aerosol production in the surface layer and mixed layer to define the mixed layer top. We will use the lidar backscatter ratio, $b = 1 + \beta_a / \beta_R$, in this paper as a measure of the lidar backscatter (β_R is the backscatter coefficient for pure Rayleigh scattering).

Traditional detection techniques have been based on a measure of the discontinuity in the profile, typically a maximum of -db/dz. Endlich (1979) used a pattern recognition algorithm to match adjacent values of -db/dz which were similar and thus provide horizontal continuity in the boundary layer height even with a noisy signal. Melfi et al. (1984) used an automated detection technique based on a critical absolute backscatter value and also examined the variance of -db/dz which should maximized at the point where convective entrainment is largest.

These techniques are often not reliable. With a noisy signal or low aerosol scattering in the boundary layer, detection of the gradient in backscatter is not always a unique measure of the MLD. In a study from an aircraft mission in the Lower Fraser Valley (Vancouver), British Columbia Hayden *et al.*, 1997 and Hoff *et al.*, 1997, a smoothed set of lidar profiles was used along with a simple derivative technique to determine MLDs. The lidar data were taken as 1 second averages (10 shots) with 12m vertical resolution. Even where the aerosol concentration was uniform in the ML, this technique worked adequately only as long as there was a significant horizontal (10 profiles, or about 2 km of flight leg) and

Air Pollution Modeling and Its Application XIII, edited by S.-E. Gryning and E. Batchvarova, Kluwer Academic/Plenum Publishers, 2000

The idealized profile used is:

$$<b(z)>=\frac{(b_m+b_u)}{2}-\frac{(b_m-b_u)}{2}erf\left(\frac{z-z_m}{s}\right)$$

where z_m is the MLD and s is a measure of the thickness of the entrainment zone. With the definition of the entrainment zone being that region where the aerosol concentration is between the 5th and 95th percentile of the mixed layer value (Nelson *et al.*, 1989), the entrainment zone thickness, EZT, is 2.77s. It should be noted that the z_m and s parameters correspond to translation and dilation factors in formal wavelet analysis. In this technique, we fit only one daughter function (wavelet) to the observed profile. Figure 1 shows the idealized profile with $z_m = 500$ m, $s = 100$ m, $b_u = 2$ and $b_m = 10$.

A process of multidimensional minimization is needed to find "best fit" values of the profile parameters. Initial attempts to use the downhill simplex method (Press *et al.*, 1992) in multidimensions proved unreliable. Implementation of the method of "simulated annealing" achieves very satisfactory results with relatively little computational cost. Routines for implementation of this method are described and presented in Press *et al.* (1992).

RESULTS USING AIRBORNE LIDAR DATA

The lidar data used in this work were obtained from 21 missions carried out over the Lower Fraser Valley of British Columbia during July and August 1993 Hayden *et al.*, 1997 and Hoff *et al.*, 1997. The AES elastic lidar operated at 1.064 um from the National Research Council of Canada's Convair 580 aircraft. The lidar was flown at a nominal altitude of 4300 m and operated at 10 Hz. The data was stored as 10 shot averages giving a horizontal resolution of 1s or approximately 200 m and a vertical resolution of 12 m.

Figures 2 to 6 show five profiles from the August 3 mission. Figures 2 and 3 show typical backscatter profiles having a well defined, polluted (and therefore strongly scattering) mixed layer, underlying a much less polluted (and therefore weakly scattering) overlying inversion layer. Roughly coincident profiles of potential temperature and humidity confirm the MLD in these cases Hayden *et al.* (1997). Figures 4-6 show profiles which are all but intractable using the traditional retrieval algorithms. In Figure 4, the mixed layer has a layer of relatively cleaner air within the boundary layer. This is believed to be due to advection of the overlying more polluted air over a relatively clean region of the valley. Figures 5 and 6 were obtained closer to the colder waters of the Georgia Straight and thus have shallower MLD. Figure 5 shows an overlying plume over the main portion of the mixed layer. Figure 6 shows a profile which is much shallower than those shown in the other retrievals and has an increasing aerosol content right down to the surface.

Table 1 summarizes results of these analyses. It is clear that there is reasonable agreement between the -db/dz technique and the one proposed here only for the first profile (P1). Figures 3 and 4 show profiles for which the gradient retrieval method found very large increases in b near the surface and identified them as very shallow mixed layers, while the new technique gave more correct heights of 425 and 534 m respectively. In Figures 5 and 6, the gradient method identified MLD as 170 m whereas the new method returned the more precise values of 246 and 236 m respectively.

vertical (5 points or 60 m) smoothing of the data. This smoothing not only decreased the value of the high resolution airborne lidar data, but also opened the possibility of aliasing the data. This is especially problematic in cases where horizontal discontinuities existed (near the shoreline where Thermal Internal Boundary Layer, TIBL, development was seen, for example). For this reason, we have developed a new technique to better retrieve the MLD on the unaveraged lidar data.

THE NEW TECHNIQUE

We have examined wavelet techniques to extract the MLD from the lidar profiles. Wavelet transforms have been found not to be well suited to the form of the backscatter profile near the mixed layer. In formal wavelet analysis Meyer (1993), one requires convergence of the integral over all space of the modulus of a wavelet. This requires that the wavelet goes to zero at $\pm\infty$. This can be accomplished by subtracting the Rayleigh value to have the backscatter ratio go to zero at $\pm\infty$. There remains the problem that the wavelet is not defined for negative values and therefore, it would have to be synthetically reflected in the negative domain.

We have chosen a technique which is much simpler and, therefore, easier to automatically process over large volumes of data. This technique fits an idealized backscatter ratio profile $\langle b(z) \rangle$ to the observed data and using a minimization technique to reduce the variance between the two profiles. The idealized profile is one which fits our expectation of the mixed layer, containing a gradient in backscatter from a mixed layer value, b_m, to the value above the ML, b_u.

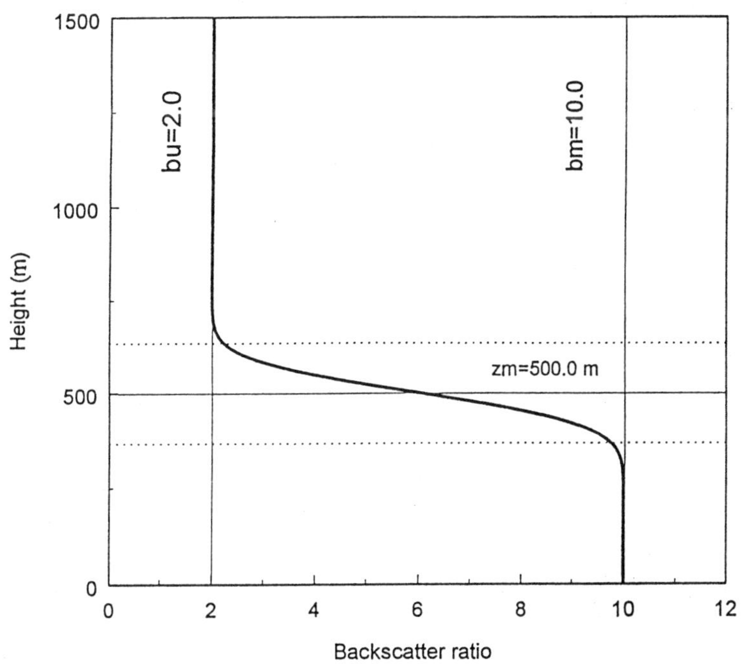

Figure 1. Idealized backscatter profile with b_u=2.0, b_m=10.0, z_m=500.0 m, and s=100.0 m. The dotted lines indicate the EZT, spanning 277 m centered on a MLD of 500 m.

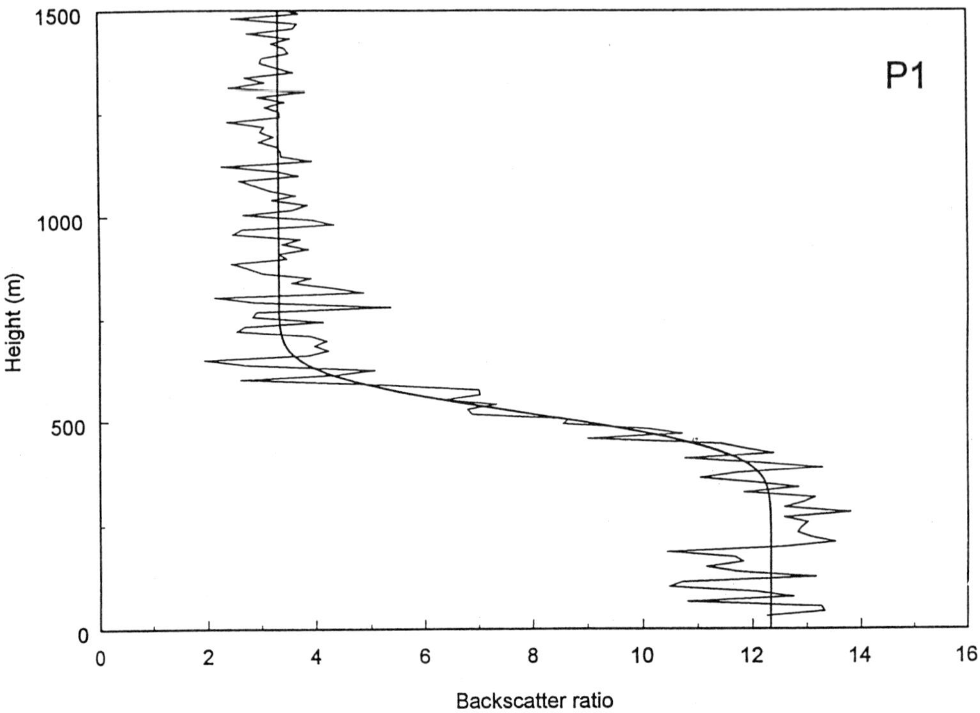

Figure 2. Observed backscatter profile P1 listed in Table 1. Smooth line is fitted idealized profile.

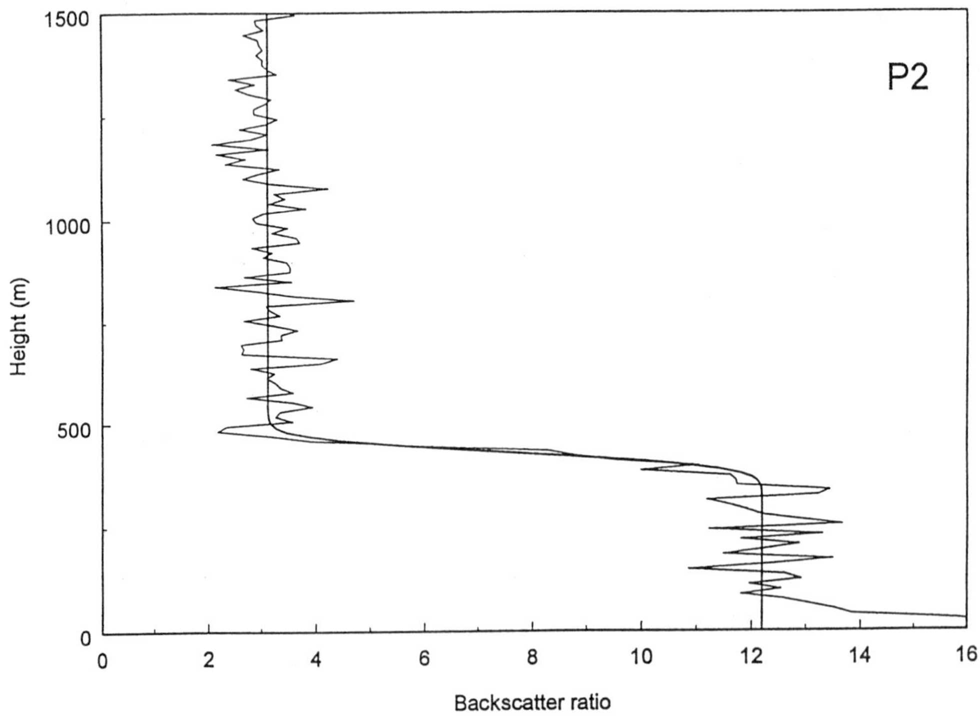

Figure 3. Observed backscatter profile P2 listed in Table 1. Smooth line is fitted idealized profilc.

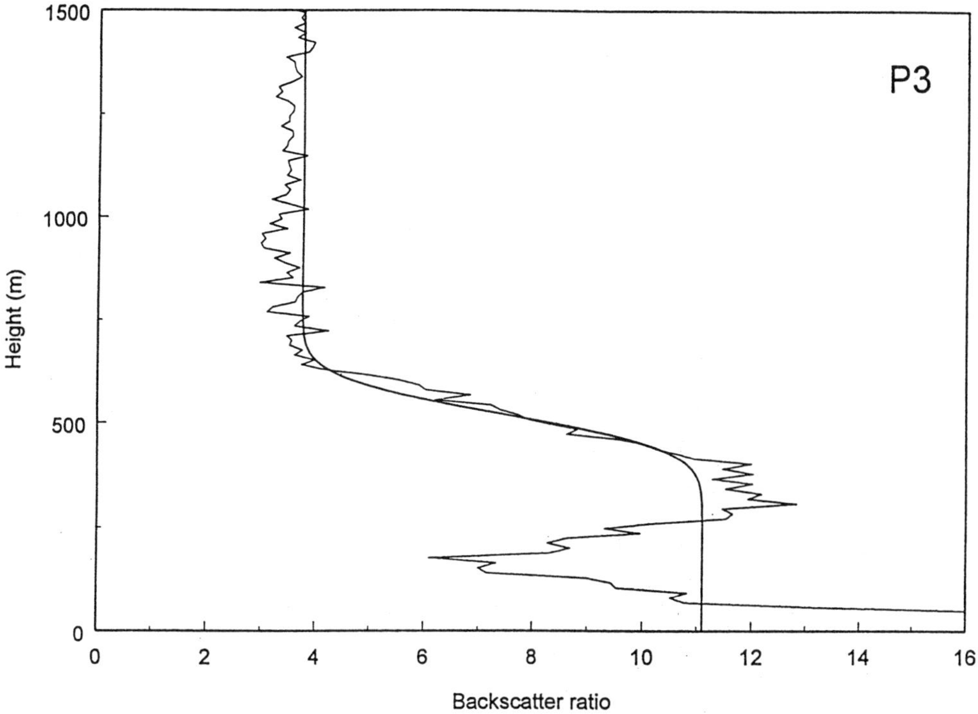

Figure 4. Observed backscatter profile P3 listed in Table 1. Smooth line is fitted idealized profile.

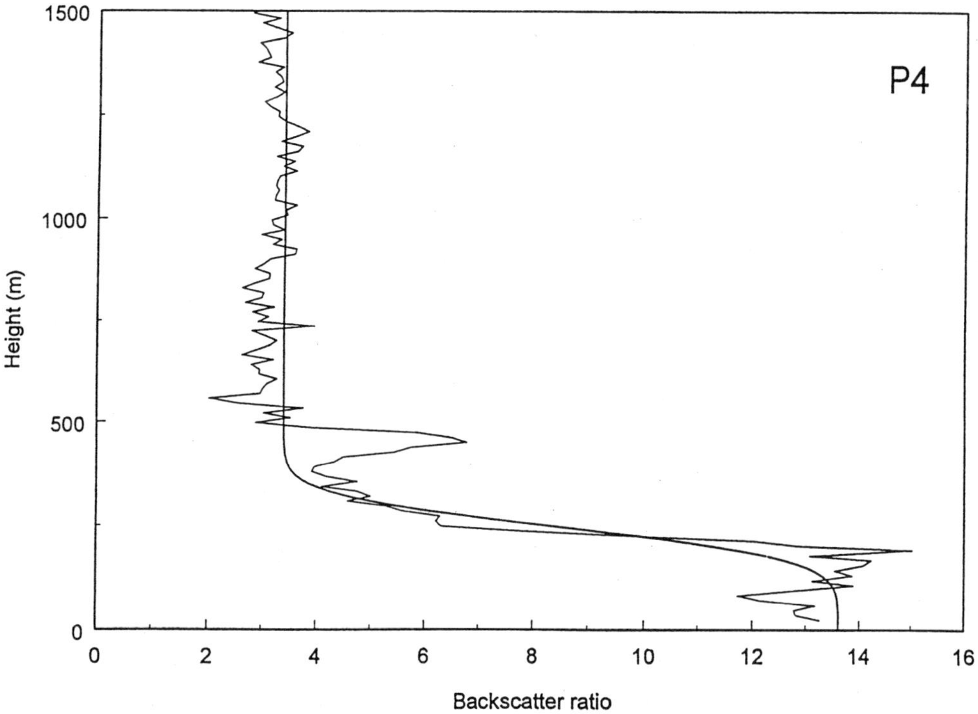

Figure 5. Observed backscatter profile P4 listed in Table 1. Smooth line is fitted idealized profile.

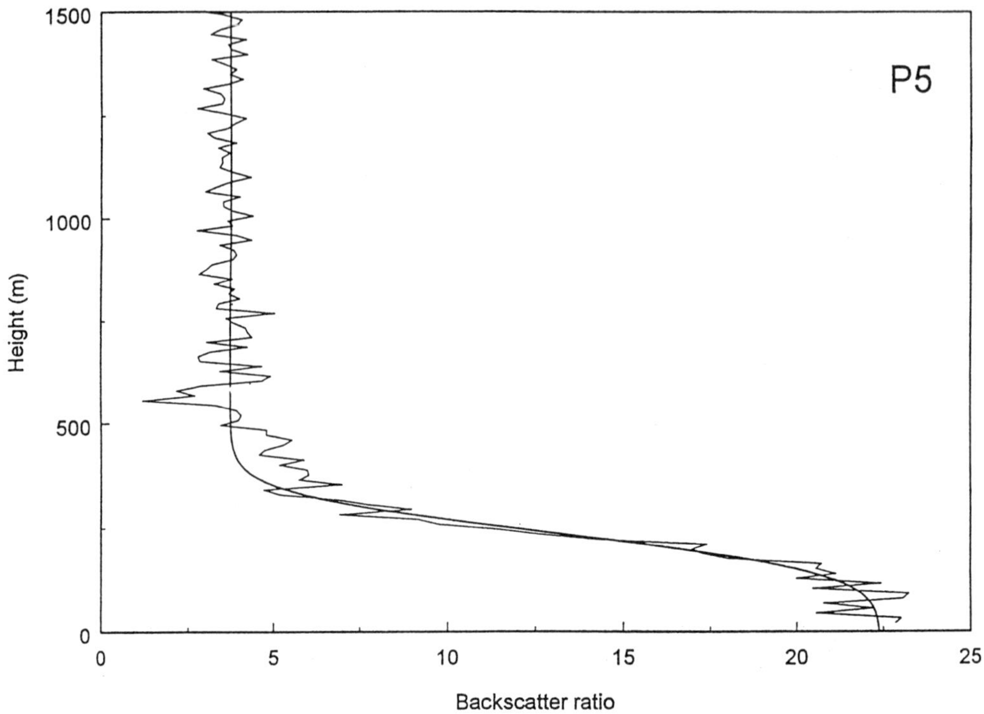

Figure 6. Observed backscatter profile P5 listed in Table 1. Smooth line is fitted idealized profile.

Table 1. Lidar backscatter ratio profiles from 1993.08.03 during Pacific '93 field study. Shaded quantities are MLD detected by other techniques. z_{m1} And z_{m2} are MLD determined by critical backscatter, the upper height is the first height at which backscatter equals the critical value when scanning the profile downwards, the lower height is for upward scanning. z_{md} is MLD detected by the derivative of backscatter. b_u, b_m, z_m And s are defined in the text. f Is root mean square deviation between measured backscatter profile and fitted profile.

Profile	*Leg*	*Lat.*	*Lon.*	z_{m1} (m)	z_{m2} (m)	z_{md} (m)	b_u	b_m	z_m (m)	s (m)
p1	L6N	48.8838	122.515	648	470	552	3.316	12.331	521	105
p2	L8N	48.9815	122.668	482	398	-[1]	3.115	12.682	425	47
p3	L8N	48.7016	122.658	639	402	-[1]	3.578	11.171	534	111
p4	L9S	49.0300	122.746	450	189	169[2]	3.407	13.616	246	87
p5	L9S	49.0435	122.746	341	163	169[2]	3.731	22.623	236	112

Notes:

1) The procedure detected the most significant maximum in -db/dz as that at the surface
2) The procedure was successful in detecting a maximum in -db/dz, but returned a result which appears to be related to features within the mixed layer, rather than at the top.

Acknowledgements

Data were drawn from parts of the Pacific '93 field study funded by Environment Canada. The work was funded by grants awarded by the Atmospheric Environment Service of Environment Canada and the Natural Science and Engineering Research Council of Canada. MB was able to participate through funding from the NATO/CNR senior fellowships programme.

REFERENCES

Boers, R., E. W. Eloranta, and R. L. Coulter, 1984: Lidar observations of mixed layer dynamics: Tests of parameterized entrainment models. *Journal of Climate and Applied Meteorology*, **23**, 247-266.

Endlich, R. M., F. Ludwig and E. E. Uthe, 1979: An automated method for determining the mixing depth from lidar observations. *Atmospheric Environment*, 13, 1051-1056.

Hayden, K.L., K.G. Anlauf, R.M. Hoff, W.J. Strapp, J.W. Bottenheim, H.A. Wiebe, F.A. Froude, J.B. Martin, D.G. Steyn and I.G. McKendry, 1997: The vertical chemical and meteorological structure of the boundary layer in the Lower Fraser Valley during PACIFIC '93. *Atmospheric Environment*, 31 (4), 2089-2106..

Hayden, K.L., K.G. Anlauf, R.M. Hoff and W.J. Strapp, 1994: *PACIFIC '93 Lower Fraser Valley Oxidants Study Aircraft Data Report*, Environment Canada Report, 268p.

Hoff, R.M., M. Harwood, A. Sheppard, F. Froude, B.A. Martin and W. Strapp, 1997: Use of airborne lidar to determine aerosol sources and movement in the Lower Fraser Valley (LFV), British Columbia. *Atmospheric Environment*, 31 (4), 2123-2134.

Melfi, S. H., J. D. Spinhire, S.-H. Chou, and S. P. Palm, 1985: Lidar observation of vertically organized convection in the planetary boundary layer over the ocean., *J. Clim. Appl. Meteorol.*, 24, 806-821.

Meyer, Y., 1993: *Wavelets, Algorithms and Applications* (Translated and revised by R. D. Ryan) SIAM Press, 133p.

Nelson, E., R. Stull and E. Eloranta, 1989: A prognostic relation for entrainment zone thickness. *Journal of Applied Meteorology*, **28**, 885-903.

Press, W. H., S. A. Teukolsky, W. T. Vettering and B. R. Flannery, 1992: Numerical Recipes in FORTRAN: The art of scientific computing, (2nd ed.) Cambridge University Press, 963p.

DISCUSSION

A. LUHAR: With your technique, do you get any (implied) relationship between the entrainment rate and the entrainment zone thickness?

D. STEYN: If we analyze sequences of profiles either in space (from aircraft borne lidar) or time (from surface based, fixed lidar) we find that there are homogeneous stretches in which the ratio of EZT to MLD is consistent with an inversion base to surface heat flux of 0.2. There are however many stretches of data in which the EZT is considerably thicker than this ratio implies. We interpret this as being due to non-stationary entrainment. He implication being that the flux ratio of 0.2 only holds for stationary entrainment.

H. VAN JAARSVELD: Can your technique identify more than one layer? I ask this because what you seem to be measuring is a "bulk" layer in which young and aged pollutants are mixed. Often we want to know the height of the layer in which newly emitted pollutants will be mixed. This may be a height less than the "bulk" mixing layer height.

D. STEYN As we've presented it, the technique can only detect the "bulk" layer you refer to. Of course it would be possible to choose a more complicated fitting function. One obvious candidate would be a sum of two of our profiles with different parameters.

VERIFICATION OF THE OPERATIONAL DACFOS FORECAST OF SURFACE OZONE IN EUROPE

Sissi Kiilsholm[1], Alix Rasmussen[1], and Jens Havskov Sørensen[1]

[1]Danish Meteorological Institute (DMI)
Meteorological Research Division
Lyngbyvej 100
Copenhagen Ø

INTRODUCTION

Verification of ozone forecasting systems has become an even more important subject in connection with the real-time publication of the ozone forecasts to the public. False or no alarms of ozone threshold exeedances should be avoided, so that people who really suffers from smog-episodes will not be unnecessarily disturbed.

The European Environment Agency (EEA) has initiated a work for harmonizing validation of ozone forecasts and ozone data exchange. In a report from a Technical Working Group on Ozone Forecasting and Data Exchange (TWG-DFO)[1] several statistical parameters are taken into account, mostly those concerning the skill to forecast exceedances of the EU public information ozone threshold at 180 $\mu g/m^3$.

This work will only deal with the verification suggested by EEA, except for a modification caused by the few numbers of exceedances of the EU public information threshold in Denmark.

THE OPERATIONAL OZONE FORECASTING MODEL

DMI has developed a operational forecasting system for ozone concentrations in Europe. The system is based on the coupling of a chemical model (EMEP MSC-W's oxidant model),[2] and DMI's 3D lagrangian transport model. The forecast system utilizes analysis and forecast data from the numerical weather prediction model, Danish Meteorological Institute's HIgh Resolution Limited Area Model (DMI-HIRLAM). The whole system is called Danish Atmospheric Chemistry FOrecasting System (DACFOS)[3] .

DACFOS is a photochemical backtrajectory model system. DMI's 3-D transport model can utilize meteorological data from the different versions of DMI-HIRLAM and data from ECMWF's global model.

Air Pollution Modeling and Its Application XIII, edited by S.-E. Gryning and E. Batchvarova, Kluwer Academic/Plenum Publishers, 2000

The system is set up to make 48 hours forecasts, automatically four times a day, for selected receptor points within the EMEP-grid covering all of Europe. For all receptor points, five backward trajectories arriving at equidistantly distributed heights within the atmospheric boundary layer[4] are calculated hourly, meteorological parameters needed for the chemistry model are also provided.

For each trajectory the concentrations of the chemical species are calculated independent of the other trajectories, assuming total mixing in the boundary layer. Calculations are made for 70 chemical species and 130 chemical reactions using a quasi-steady state approximation (QSSA) numerical method.[5]

The forecast will not be able to describe local variations, partly because the emission data are defined on a coarse grid (at present 50 km by 50 km).

THE PUBLIC INFORMATION SYSTEM

In the summer 1998 an information system has been set up at the WWW-page (WWW.dmi.dk) at DMI. Ozone forecasts are shown for areas in the Northern Europe every 6 hour, with a figure for the levels today and one for the levels tomorrow. The levels are provided in four intervals [0-60], [60-120], [120-180], [180- above 180] $\mu g/m^3$ozone, described by low, mean, high and very high.

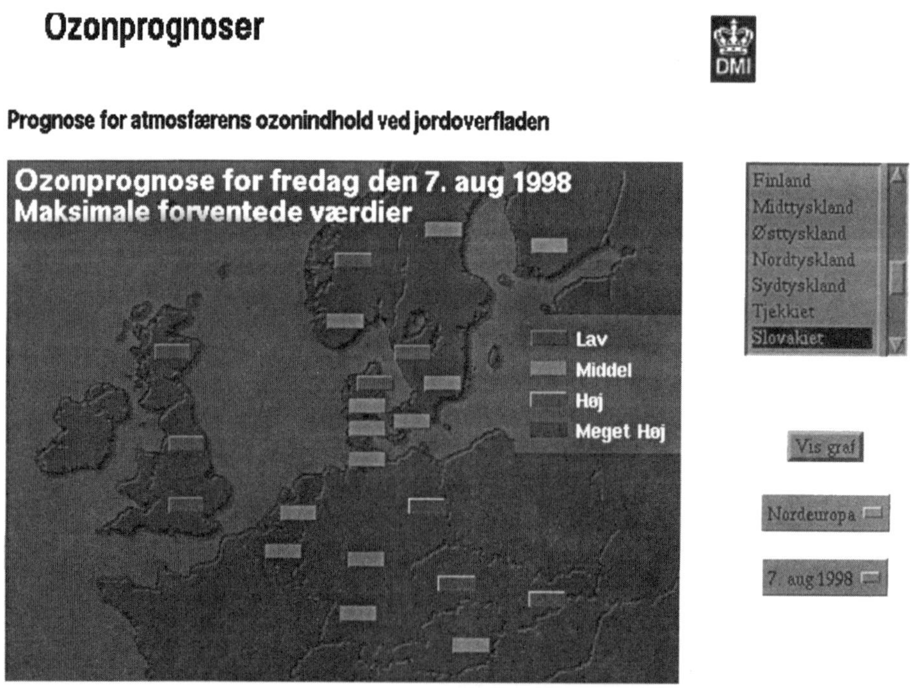

Figure 1. A presentation of the DACFOS forecast from 6 August 1998 for the next day. The danish text on the map says ' Ozone prognoses for Friday 7 Aug 1998. Expected maximum values.'

VERIFICATION OF OZONE FORECASTING MODELS

The EEA has recommended use of 6 statistical parameters.[1] Three parameters are concerning the skill of forecasting using a contingency table, where N is the total number of data points, f = total number of forecast exceedances, m = total number of observed exceedances, a = number of correctly forecast exceedances. The parameters concerning the skill of forecasting exceedances of a threshold are

- the fraction of correct forecast smog events (Probability Of Detection, POD,[6])

$$SP = \frac{a}{m} \cdot 100 \tag{1}$$

- the fraction of realized forecast smog events (Frequency Of Hits, FOH,[6])

$$SR = \frac{a}{f} \cdot 100 \tag{2}$$

- a combination of SP and the 'probability of a null event' PON,[6] in the 'Success Index'

$$SI = \left(\frac{a}{m} + \frac{N + a - m - f}{N - m} - 1 \right) \cdot 100 \tag{3}$$

The next 3 statistical skill parameters are standard statistics, either using persistency as reference model, as in S, or using an uncertainty range, as in H

- The fractional bias, FB, between averaged values for forecast and observation

$$FB = 2 \frac{\left(\overline{P} - \overline{M}\right)}{\left(\overline{P} + \overline{M}\right)} \tag{4}$$

- the skill score, S, where P_i and M_i are forecast, respectively observed values at day i.

$$S = 100 \cdot \left\{ 1 - \frac{\Sigma \left(P_{i+1} - M_{i+1}\right)^2}{\Sigma \left(M_i - M_{i+1}\right)^2} \right\} \tag{5}$$

- the hit score, H, indicates the overlap between the interval $(P \pm r)$ and $(M \pm r)$

$$H = 100 \cdot \frac{1}{N} \sum \frac{2r - \Delta}{2r}, \quad where\, \Delta = min(|P - M|, 2r) \tag{6}$$

Both SP, SR and H have a range between 0 - 100%, SI is ranging from -100 to 100, FB ranges between -2 and +2 and S range from -∞ to 100. The perfect model gives for SR, SP, SI, H and S a value of 100%, and for FB a value of 0. FB and SI has been rescaled in the figures to fit the 0-100% scale, which leads to a perfect model score at 50% for FB and at 100% for SI. In this work H is evaluated with a range $r = 10\ \mu g/m^3$.

If warning of the people, affected to smog events, is the main purpose, the number of unexpected smog events (100-SP) should be low, at the other hand could a high rate of false-alarms reduce the confidence in the forecasts. In cases, when the forecast are used for smog prevention (eg. traffic regulations), the false-alarm rate (100 - SR) should be minimal. The 'Success Index', SI, weights equally the correct forecasting of smog and non-smog events in one index.

The 'event threshold' level suggested by EEA are at 180 $\mu g/m^3$, and only days with ozone concentrations above 120 $\mu g/m^3$ are taken into account. In Denmark, exceedances of the 180 $\mu g/m^3$ threshold occur very seldom. As an example, during the period 1 April - 30 July 1996 only one excess of the population information threshold value were observed in Denmark,[7] and in the summer season 1995, five exceedances were detected[8] . With these low numbers of exceedances, even for long periods, the statistics gives no meaningful results for the suggested 'event threshold'. These conditions are the reason for chosing a lower 'event threshold' in this verification. In the DMI public information system 'high' concentrations are defined to be between 120-180 $\mu g/m^3$. For the verification of DACFOS, 120 $\mu g/m^3$ are used as the 'event threshold', which also is close to the 130 $\mu g/m^3$ level used in US as a level for 'limited health Notice' threshold.[9] The verification is only based on days with higher concentration than 60 $\mu g/m^3$.

OZONE OBSERVATIONS

The stations Harwell and Ladybower, in England, and Lille Valby and Jægersborg, in Denmark, are chosen for a verification in the period June, July, August, September 1997 and June, July 1998. For a longer verification period, ozone measurements from the summer seasons in 1995-1998 are used from Jægersborg.

DMI is measuring surface ozone in Jægersborg, and the other data has kindly been provided at WWW-pages by AEA -technology and NERI.

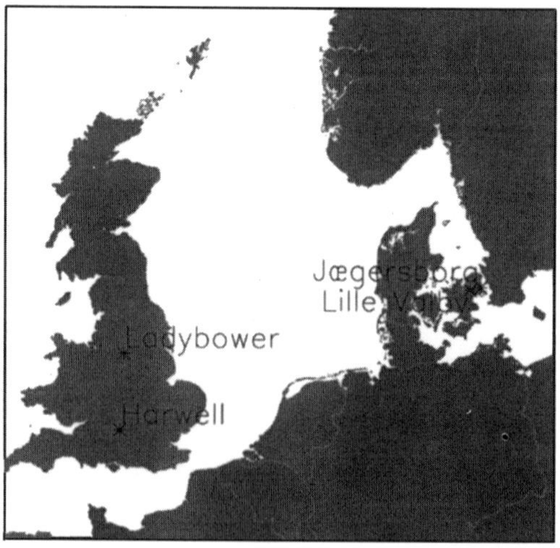

Figure 2. Map of the location of the 4 stations used in the verification.

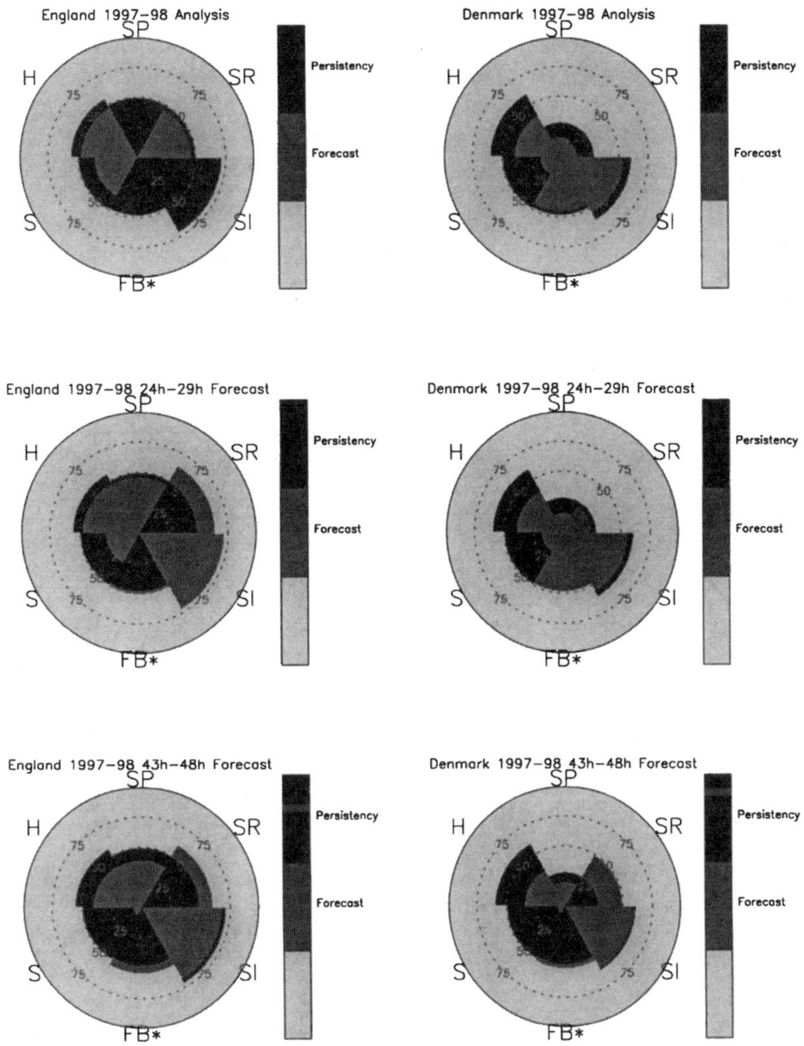

Figure 3. Six statistical parameters for two areas, England and Denmark, shown for three model forecast intervals; analysis, 24h-29h and 43h-48h forecast. Model and persistency values are shown in the same piece, showing the smallest values overlaying the biggest. In case of equal values, only one color is shown. FB* differs from the other parameters by having an ideal value at 50%.

VERIFICATION RESULTS

In figure 3, DACFOS forecast skills are shown for England and Denmark, represented by the four stations described earlier in the text. Model results are calculated for the three intervals: analysis (-6 to -1 hour), 24h-29h forecast and 43h-48h forecast. Intervals of 6 hourly values are used to get values 24 hours a day.

The six calculated statistical parameters are presented in a rose, with the persistency skill value and the forecast skill value overlaying each other, the lowest value on top. A small black piece of cake with a broad grey rim is a presentation of an ideal DACFOS forecast compared to the persistency. In case of equal values for the forecast skill and the persistency skill, the piece of cake has a single color.

DACFOS gives a better forecast performance for England than for Denmark in 1997-1998, especially showed by the skill parameters SR, 'Frequency of hits' eq.(2), and SP, 'Probability of detection' eq.(1). The change of skill with the forecast length are similar for England and Denmark. SP, SI, H and FB has a slightly decreasing skill with forecast length. SR are different from the other parameters by showing an increasing skill, which probably are due to fewer forecasted exceedances. The most drastic changes are detected on S, 'Skill score' eq.(5), in England, where it reduces from around 35% to 5 %. Persistency

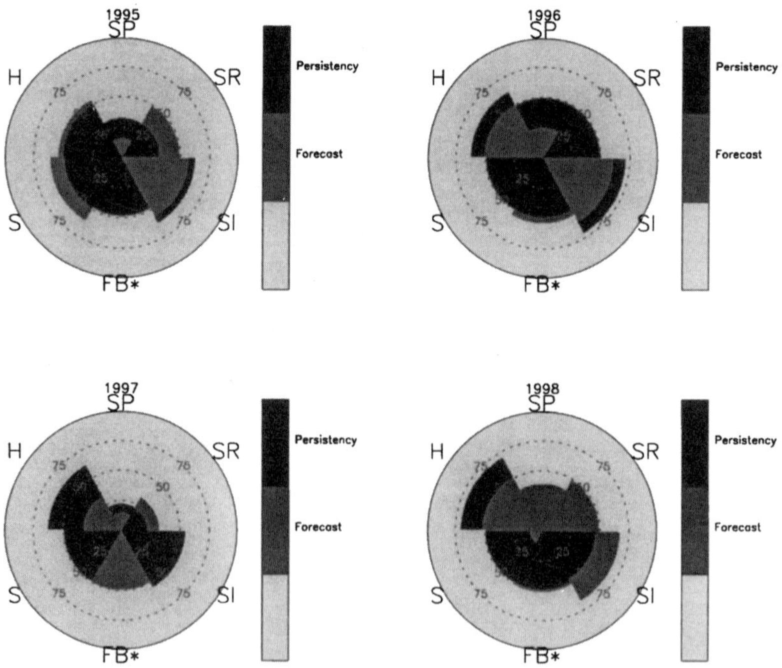

Figure 4. Statistical parameters for Jægersborg (JGB) calculated for the years 1995-1998, based on DACFOS 24-29 hour forecasts. Model and persistency values are shown in the same piece, where the smallest are overlaying the biggest. FB* differs from the other parameters by having an ideal value at 50%.

gives higher SP and SR values for England than for Denmark (50% contra 25%), which probably can be explained by different weather conditions. Single standing, sporadic exceedances of the 'event threshold' occur more often over Denmark, than over England in 1997-98.

A verification for the summer seasons 1995 - 98, in Jægersborg (Denmark), see figure 4, shows varying results from year to year, partly caused by the change of the weather conditions year by year, and partly caused by the few numbers of events in some years.

In general, the numbers of forecasted exceedances are lower than the number of observed exceedances, giving a SP lower than SR for the forecast skills. As seen on figure 4, the observed exceedances occur around twice as often as the forecasted exceedances. The best forecasted year was 1995, with 'skill scores', S, higher than persistency. High statistical scores for the forecast were also seen in 1998, but persistency had a better 'hit and skill score', H and S. It must be mentioned here that the 1998 verification only were based on June and July, with a few exceedances. Persistency are doing well in 1996 for 'hit and skill score', H and S, caused by exceedances on consecutive days .

DISCUSSION

The EEA proposed method for verification of the skill performance of ozone forecast models, gives useful and easy compareable results. The method can be used when the chosen 'event threshold' is low enough to give some events both for measurements and for forecasts, otherwise undefined values for SP and SR will be performed.

Due to very few exceedances of the EEA proposed ozone exceedance threshold in Denmark, an alteration of the exceedance threshold from 180 $\mu g/m^3$ to 120 $\mu g/m^3$ has been made. Unfortunately, this modification makes it impossible to compare with other model verifications, based on the EEA proposed verification principles, but comparisons in time and place are made possible for the DACFOS forecast in Denmark and England.

The comparison of the DACFOS performance skill between England and Denmark shows a better forecast for England than for Denmark in 1997-1998. The DACFOS forecast for Denmark in 1997 had relatively low performance skill compared to the years 1995-1998, so the better forecast for England can not be generalized.

Big differences in the statistical parameters for the years 1995-1998, shows that factors like weather and number of exceedances have a noticeable effect on the parameters. In cases when the statistical parameters should be used for a comparison between different forecast models, long periods of verification must be used in order to avoid influences from irrelevant factors.

REFERENCES

[1] Van Aalst R.M., and De Leeuw F.A.A.M.(ed.). *National Ozone Forecasting Systems and International Data Exchange in Northwest Europe. Draft report of the Technical Working Group on Data Exchange and Forecasting for Ozone episodes in Northwest Europe (TWG-DFO)*. Planned to be published in EEA Topic report series . (1998)

[2] Simpson, D., Andersson-Sköld, Y., and Jenkin,M. E. *Updating the chemical scheme for the EMEP MSC-W oxidantmodel: current status.* EMEP MSC-W Note 2/93 - August 1993.

[3] Jensen M.H., Rasmussen A., Svensmark H. and Sørensen J.H. *Danish Atmospheric Chemistry FOrecasting System (DACFOS). Validation of Semi-operational Ozone Forecasts for Jægersborg, Summer '95.* DMI technical report 96-3 . (1996)

[4] Havskov Sørensen, J., Rasmussen, A., and Svensmark, H. *Forecasting the Atmospheric Boundary Layer Height based on a Bulk Richardson number.* Phys. Chem. Earth, 21 ,No. 5-6,pp. 435-439. (1996)

[5] Simpson, D. *Long-Period Modeling of Photochemical Oxidants in Europe. Model Calculations for July 1985.* Atm. Env., 26A,No. 9,pp. 1609-1636. (1992)

[6] Doswell C.A.III, Davies-Jones R and Keller D.L. *On summary Measures of Skill in Rare Event Forecasting Based on Contingency tables.* Wea. Forecasting, 5, pp.576-585. (1990)

[7] Sluyter R., Van Zantvoort E. *Information Document concerning Air Pollution by Ozone. Overview of the situation in the European Union during the 1996 summer season (April-July).* Report to the Commission by the European Environment Agency European Topic Centre on Air Quality. (1996)

[8] De Leeuw F.A.A.M., Van Zantvoort E. *Exceedance of ozone Threshold Values in the European Community in 1995.. Report to the Commission by the European Environment Agency European Topic Centre on Air Quality.* (1996)

[9] Maine Department of Environmental Protection. *Information Sheet Ground-Level Ozone. Air Quality Index Reports, July 1998.* (1998)

DISCUSSION

H. VAN DOP: Is it your intention to use the model for service to the public or for policy purposes?

S. KIILSHOLM: DACFOS is only used as an informational tool to the smog warning preparedness in Denmark and as a service to the public on the WW, there are no specific plans of using it for policy purposes.

Z. ZLATEV: Comment. It is quite true that the level of 120 ppb is nearly never reached in Scandinavia. However, there are two other critical levels in the last draft of the EU's Ozone Directive (90 ppb and periods in which the ozone concentrations are greater than 60 ppb during 8 successive hours). It is perhaps worthwhile to consider these two levels together with the level of 120 ppb.

S. KIILSHOLM: The proposed level of 90 ppb (~ 180 micro g/m3) gave very few exceedances so a level of 60 ppb for 1-hour means were used. Exceedances of 60 ppb during 8 successive hours would probably be useful for periods of years.

ESTIMATION OF KOLMOGOROV CONSTANT C_0 FROM SONIC ANEMOMETER MEASUREMENTS IN THE ATMOSPHERIC SURFACE LAYER

D. Anfossi [1], G. Degrazia [2], E. Ferrero [3], S.-E.Gryning [4], M.G. Morselli [5], S. Trini Castelli[1]

[1]CNR-Istituto di Cosmogeofisica, Torino, Italy
[2]Universidade Federal de Santa Maria (RS), Brasil
[3]Universita', Dip. Scienze Tecn. Avanz., Alessandria, Italy
[4]RISO National Laboratory, Roskilde, Denmark
[5]ENEL-Polo Ambiente, Milano, Italy

INTRODUCTION

In the inertial subrange the Lagrangian velocity structure function

$$D_i^L(\tau) = \langle [u_i(t+\tau) - u_i(t)]^2 \rangle \qquad i=u,\, v,\, w \tag{1}$$

is defined, according to the Kolmogorov hypothesis (Monin and Yaglom, 1975), as:

$$D_i^L(\tau) = C_0 \; \varepsilon \; \tau, \tag{2}$$

where C_0 is a numerical constant and ε is the ensemble-average rate of dissipation of turbulent kinetic energy.

Hinze (1975) and Tennekes (1982), assuming an exponential form for the Lagrangian autocorrelation function, suggested the following relationship between ε and the Lagrangian integral time scale T_i^L, for times $t < T_i^L$:

$$\varepsilon = \frac{2\;\sigma_i^2}{C_0\;T_i^L} \tag{3}$$

where σ_i^2 is velocity fluctuation variance. Equation (3) is widely used in Lagrangian stochastic models for airborne dispersion studies.

Air Pollution Modeling and Its Application XIII, edited by S.-E. Gryning and E. Batchvarova, Kluwer Academic/Plenum Publishers, 2000

Despite substantial theoretical and experimental work present in the literature, the value of C_0 is not yet established and still represents an open matter (Rodean, 1994). For instance, Hinze (1975) found that C_0 varies between 3 and 10; Anand and Pope (1985) found $C_0 = 2.1$, whereas Sawford (1991) proposed $C_0 = 7$.

In this paper, a theoretical derivation of C_0 proposed by Degrazia and Anfossi (1988) is firstly summarised and then an experimental evaluation of C_0, based on sonic anemometer wind data recorded in the atmospheric surface layer under unstable conditions, is presented.

BRIEF OUTLINE OF THE THEORETICAL C_0 DERIVATION

Starting from the Taylor theorem (1921) and subsequent work by Batchelor (1949), Hinze (1975) and Degrazia et al. (1998) derived the following expression for the Lagrangian time scale T_i^L for large travel times ($t \gg T_i^L$):

$$T_i^L = \frac{\beta_i F_i^E(0)}{4} \tag{4}$$

where $F_i^E(0)$ is the Eulerian energy spectrum evaluated at the origin of the frequency space, normalised by the variance σ_i^2, and β_i is the ratio of the Lagrangian to the Eulerian integral time scales.

It may be observed that equation (4), though derived from Taylor's theory for homogeneous and stationary turbulence, can be applied in non-homogeneous turbulence. Since the frequency representative of more energetic eddies, varies only slightly with the height in the bulk of the PBL (Kaimal et al., 1976, Caughey and Palmer, 1979; Nieuwstadt , 1984; Sorbjan, 1989), it may be assumed that these results can be extended in a first approximation to slightly non-homogeneous flows. Furthermore, previous work (Degrazia et al., 1996), showed that from such an assumption it is possible to derive expressions for the Lagrangian length scales, both in the convective and in the stable boundary layer, that are in good agreement with other derivations.

From equation (4), a C_0 expression may be derived in the following way. Let us recall that

$$\beta_i = T_i^L / T_i^E = 0.55 \; U/\sigma_i \tag{5}$$

in which U is the mean wind speed (Wandel and Kofoed-Hansen (1962), Hanna (1981) and Degrazia and Anfossi, 1988). Then, let us consider the analytical expression of the Eulerian velocity spectrum under unstable conditions (Nieuwstadt, 1984; Sorbjan, 1986; Degrazia et al., 1996)

$$\frac{n\ S_i^E(n)}{w_*^2} = \frac{1.96\ c_i\ f\ (\Psi_\varepsilon)^{\frac{2}{3}}}{\left(f_m^*\right)_i^{\frac{5}{3}}\left[1+\frac{1.5\ f}{\left(f_m^*\right)_i}\right]^{\frac{5}{3}}} \tag{6}$$

where $S_i^E(n) = F_i^E(n)\ \sigma_i^2$ is the Eulerian velocity spectrum, w_* is the convective velocity scale, $\Psi_\varepsilon = (\varepsilon\kappa z)/(w_*^3)$ is the non dimensional molecular dissipation rate function, f is the reduced frequency (nz/U), z is the height above the surface, $\left(f_m^*\right)_i$ is the reduced frequency of the spectral peak regardless the stratification, κ is the von Karman constant and $c_i = \alpha_i\ \alpha_u\ (2\pi k)^{-\left(\frac{2}{3}\right)}$, where $\alpha_u = 0.5 \pm 0.05$ and $\alpha_i = 1,\ 4/3,\ 4/3$ for u, v and w components, respectively (Champagne et al., 1977; Sorbjan, 1989).

Finally, by calculating from equation (6) both σ_i (by integrating it over the whole frequency domain) and $F_i^E(0)$, the following C_0 expression results:

$$C_0 = \frac{14.5\ (a\ c_i)^{(3/2)}\ \kappa}{b} \tag{7}$$

where a and b are numerical constants related to the Eulerian wind spectrum (under unstable conditions, a = 1.96 and b = 1).

A general consideration about equation (7) is that C_0 depends upon numerical constants only, as it should be by definition.

Inserting the numerical values of the various constants in equation (7), the following values for the Kolmogorov constant were obtained (Degrazia and Anfossi, 1988): $C_0 = 3.44 \pm 1.38$ for the crosswind and vertical turbulent velocity components and $C_0 = 2.23 \pm 0.89$ for the longitudinal one.

In conclusion, our theoretical derivation of C_0 is based, from the one hand, on Taylor's statistical diffusion theory, extended to the case of slightly inhomogeneous turbulence and, from the other hand, on observed wind spectral characteristics. This means that C_0 is related to the ratio between Lagrangian and Eulerian statistics, the shape of the Eulerian velocity spectra in different stability conditions and along the three wind velocity components.

TURBULENCE DATA

Turbulence data used in the present analysis, were gathered in the frame of NOPEX experiment. The measurements were collected within agricultural fields of barley and wheat at the Tisby site situated about 45 Km west of Uppsala, Sweden, at 16° 57' E and 59° 46' N. The area around Tisby is almost flat, weakly sloping towards a small river. The typical distance to forested areas is 1.5 km in all directions except south, where the fetch is homogeneous for more than 10 km.

The measurements used in this analysis were carried out in 1995 during a 3 months period starting mid April, and thus represent spring and summer conditions. A very comprehensive measuring program was carried out at this site (Thomsen et al., 1995) including turbulent fluctuations of wind and temperature measured by an ultrasonic anemometer (type Kaijo Denki DAT/TR-61B) at a height of 6.8 meters with a sampling frequency of 20 Hz.

DATA ANALYSIS

19 days of measurements were considered in the present analysis. Hourly raw data of u(t), v(t), w(t), T(t) (72,000 data for each time series) were analysed. In the following, the various steps of the data treatment and computation are resumed.

Wind data were firstly rotated to refer them to a streamline system (Mc Millen, 1988) and the linear trend was removed from each series by the least squares method (Bendat and Piersol, 1971). Then a stationarity test was used to select stationary cases only (Bendat and Piersol, 1971). As a result of this first computational phase, the analysis only continued for those cases in which the three time series - u(t), v(t), w(t) - had simultaneously passed the test.

In the second phase, the wind variances σ_w^2, the surface layer parameters (u_*, $\overline{w'T'}$, L, T_*) and the moments up to the fourth order were computed for each wind component. Only those cases in which the mean horizontal wind speed was greater than 2 ms^{-1} were retained (to exclude the low wind speed cases in which eqs. 3 and 7 are not likely to be valid). Then Eulerian autocorrelation functions $R_i(\tau)$ were computed by way of direct computation and velocity spectra $S_i^E(n)$ were estimated by the standard method, i.e. by $R_i(\tau)$ Fourier transform (Bendat and Piersol, 1971).

In the third phase, Eulerian time scales T_i^E were evaluated from the best fit of $R_i(\tau)$, assumed exponential, from which the Lagrangian time scales T_i^L were obtained from equation (5). Three values of ε were afterwards computed from the three wind component spectra, selecting only those parts of the spectra belonging to the inertial subrange. That is, a best fit between the theoretical spectra, namely:

$$S_i^E(n) = A_i \; u^{2/3} \; \varepsilon^{2/3} \; n^{-5/3}$$

with $A_u = 0.15$ and $A_{v,w} = 0.2$ (Pasquill, 1974) and experimental spectra were performed step by step in the frequency range

$$n_{max} < n < 1/T_i^E,$$

looking for the frequency range in which the exponent was closer to -5/3. Then

$$\bar{\varepsilon} = \left(\varepsilon_u + \varepsilon_v + \varepsilon_w\right)/3$$

was computed and only those cases in which all ε_i satisfied

$$\left|\varepsilon_i - \bar{\varepsilon}\right| / \bar{\varepsilon} < 0.3 \quad (i = u,v,w)$$

were selected.

At the end of this computation and selection procedure 63 cases (hours) remained. Finally estimates of C_0 were derived from equation (3).

RESULTS

To test the general quality of the obtained results and their consistency, we computed the values of the ensemble-average rate of dissipation of turbulent kinetic energy (ε_c), for each time series, according to Kaimal and Finnigan (1994), as follows:

$$\varepsilon_c = \frac{u_*^3}{\kappa z}\left(1.0 + 0.5\left|\frac{z}{L}\right|^{2/3}\right)^{3/2} \tag{8}$$

Fig. 1 shows the comparison among these last values (Y-axis) and those estimated from the spectra and used in the C_0 derivation (X-axis). It can be noticed that there is a fair agreement and this suggests that the overall procedure is reliable.

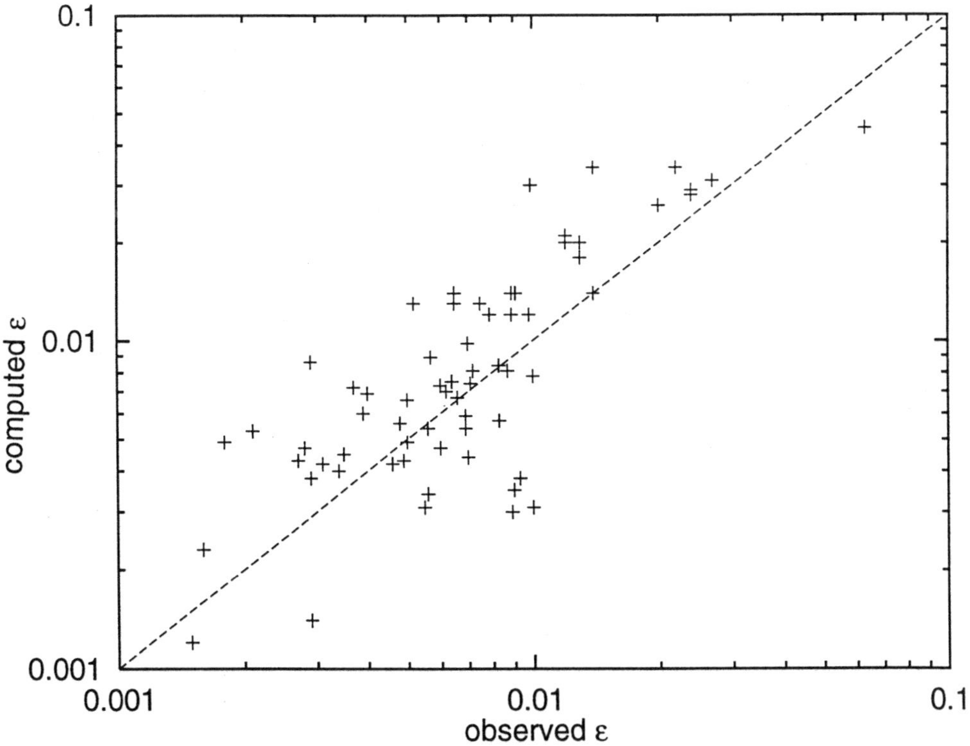

Figure 1. Comparison between turbulent kinetic energy dissipation rate ε estimated in this work (X-axis) and from Kaimal and Finnigan (1994) formula (see equation 8).

Table 1. Results of present C_0 estimations (C_0^{exp}) compared with those predicted by equation 7 (C_0^{th}), under unstable conditions. Standard deviations of these estimates are also reported.

wind component	C_0^{exp}	ΔC_0^{exp}		C_0^{th}	ΔC_0^{th}
u	3.3	1.6		2.2	0.9
v	3.4	1.8		3.4	1.4
w	3.6	0.8		3.4	1.4

Table 1 shows the final results of the present analysis: mean C_0 values obtained for the three wind velocity components (C_0^{exp}) and their standard deviations are indicated in the second and third column, whereas theoretical C_0 values derived by Degrazia and Anfossi (1988) according to equation (7) and their standard deviations are shown in the third and fourth column.

Looking at Table 1, the following considerations can be made:

- the agreement between mean experimental and theoretical C_0 values is generally fair and particularly good as far as the crosswind and vertical components are concerned;
- from the statistical point of view these average values are meaningful, being obtained from a sample of 63 different hourly time series;
- . the experimental standard deviations are of the same order of the theoretic ones. These last values (Degrazia and Anfossi, 1988) were estimated on the basis of the uncertainty of some of the constants appearing in equation (7), such as c_i and the constant appearing in equation (5);
- the theoretical derivation, because of the different values of the constants in the Eulerian spectrum, predicts the same C_0 value for the crosswind and vertical components and a lower value for the longitudinal component. The results of the present analysis appear to yield the same value for the three components;
- the derived C_0 values, though including their possible range of variation indicated by their standard deviations, fit well with the range of values suggested by different authors and resumed in the introduction.

CONCLUSIONS

In this work an experimental evaluation of the Kolmogorov constant C_0, based on surface layer wind data gathered by a sonic anemometer in a flat region in Southern Sweden, is presented. At present, the exact value of C_0 is still an open matter. Obtaining a correct C_0 value, besides from being important in the fundamental atmospheric turbulence studies, is also of great importance in the Ito's type models for the air pollution modelling. For, multiplied by the turbulent kinetic energy dissipation rate ε, it gives the amplitude of the variation of the stochastic component of the velocity fluctuations.

Hourly data were analysed. Wind data were firstly rotated to a streamline system, then a stationarity criterion was used to select stationary cases. Velocity variances and

Eulerian autocorrelation functions were computed and the time scales were estimated. Then the velocity variance spectra were estimated, from which, selecting only the parts belonging to the inertial subrange, the value of ε was derived. Finally C_0 was derived from (equation 3), proposed by Hinze (1975) and Tennekes (1982).

A recent theoretic derivation of C_0 (Degrazia and Anfossi, 1988) was also briefly recalled and its predictions were compared to the present experimental results.

It was found that our present C_0 value is in good agreement with this estimate and with other current theoretical estimations.

REFERENCES

Anand M.S. and Pope S.B. (1985) Diffusion behind a line source in a grid turbulence. *Turbulent Shear Flows*, **4**, edited by L.J.S. Bradbury, F. Durst, B.E. Lauder, F.W. Schmidt and J.H. Whitelaw. Springer-Verlag, Berlin, 46-61

Batchelor G.K.: 1949, 'Diffusion in a field of homogeneous turbulence, I: Eulerian analysis. Aust. J. Sci. Res., **2**, 437-450

Caughey S.J.and Palmer S.G. (1979) Some aspects of turbulence structure throught the depth of the convective boundary layer.Quart. J. Roy. Met. Soc., **105**, 811-827

Champagne F.H., Friehe J.C., La Rue J.C. and Wyngaard J.C. (1977) Flux measurements, flux estimation techniques, and fine-scale turbulence measurements in the unstable surface layer over land. J. Atmos. Sci., **34**, 515-530

Degrazia G.A., Moraes O.L.L., and Oliveira A.P. (1996) An analytical method to evaluate mixing length scales for the Planetary Boundary Layer. J. Appl. Meteor., **35**, 974-977

Degrazia G.A., Anfossi D., Fraga De Campos Velho H. and Ferrero E. (1998) A Lagrangian decorrelation time scale for non-homogeneous turbulence. Boundary-Layer Meteorology, **86**, 525-534

Degrazia G.A. and Anfossi D. (1998) Estimation of the Kolmogorov constant C_0 from classical statistical diffusion theory'. In press on Atmospheric Environment

Du S. (1997) Universality of the Lagrangian velocity structure function constant C_0 across different kinds or turbulence. Boundary-Layer Met., **83**, 207-219

Kaimal J.C. and Finnigan J.J.: 1994, 'Atmospheric boundary layer flows', Oxford University Press, 289 pp

Hanna S.R. (1981) Lagrangian and Eulerian time-scale i the daytime boundary layer. J. Appl. Meteor., **20**, 242-249

Hinze J.O. (1975) *Turbulence*, Mc Graw Hill, 790 pp

Kaimal J.C., Wyngaard J.C., Haugen D.A., Cote' O.R., Izumi Y., Caughey S.J. and Readings C.J. (1976) Turbulence structure in the convective boundary layer. J. Atmos. Sci., **33**, 2152-2169

Monin A.S. and Yaglom A.M. (1975) *Statistical fluid mechanics: Mechanics of turbulence*'

Nieuwstadt F.T.M. (1984) The turbulent structure of the stable nocturnal boundary layer. J. Atmos. Sci., **41**, 2202-2216

Pasquill F. (1974) *Atmospheric diffusion*', Wiley, 429 pp

Rodean H.C. (1994) Notes on the Langevin model for turbulent diffusion of "marked" particles, UCRL-ID-115869 Report of Lawrence Livermore National Laboratory, 122 pp

Sawford B.L. (1991) Reynolds number effects in Lagrangian stochastic models of turbulent dispersion. Phys. Fluids, **A3**, 1577-1566

Sorbjan Z. (1986) Local similarity of spectral and cospectral characteristics in the stable-continuous boundary layer. Boundary-Layer Met., **35**, 257-275

Sorbjan Z. (1989) *Structure of the atmospheric boundary layer*. Prentice Hall, New Jersy

Taylor G.I. (1921) Diffusion by continuous movements. Proc. London Math. Soc., Ser. 2, **20**, 196-211

Tennekes H. (1982) Similarity relations, scaling laws and and spectral dynamics In: Atmospheric turbulence and air pollution modelling (edited by F.T.M. Nieuwstadt and H. van Dop), Reidel, Dordrecht, 37-68

Thomsen A., Barring L., Gryning S.E., Sogaard H and Thorgeirsson H. (1994) Data Report for the NOPEX Tisby site - Concentrated field effort #1, May 27 - June 23, 1994

Wandel C.F. and Kofoed-Hansen O. (1962) On the Eulerian-Lagrangian transform in the statistical theory of turbulence. J. Geophys. Res., **76**, 3089-3093

DISCUSSION

A. LUHAR: How definitive are the values of the constants a and b in your expression (eq. 7) for C_0?

D. ANFOSSI: The values of the constants a and b derive from the values of the numerical constants appearing in the Eulerian wind spectra proposed by the quoted papers. It is difficult to precisely assess how definitive they can be, however they represent the best estimates at present available.

A. LUHAR: Since we always use the factor (C_0 ε) in stochastic models, have you actually tried calculating the uncertainty associated with this factor (and not with C_0 alone)?

D. ANFOSSI: No, we have not calculated the uncertainty of the factor (C_0 ε). However, in doing this work our main interest was to obtain a reliable experimental estimate of C_0 alone since its previous estimates present a large variability of values.

M. ROTACH: What is the uncertainty in the "transformation" of the Eulerian into the Lagrangian integral time scale, respectively, in the method to derive C_0?

D. ANFOSSI: The expression which relates Eulerian to Lagrangian time scale as a function of the turbulence intensity was proposed 4 decades ago and is widely accepted and used in the literature. However the value of the numerical constant shows a certain amount of uncertainty. We estimated the mean value and its standard deviation on the basis of those proposed in the literature.

M. ROTACH: What is the imposed uncertainty on the estimation of C_0?

D. ANFOSSI: The imposed uncertainty is calculated taking into account the variability of the numerical constants coming from the Eulerian wind spectra and from the above discussed Eulerian to Lagrangian time scale relationship.

E. GENIKHOVICH: When deriving the expression for C_0, which was used in your work, did you use the assumption of isotropy?

D. ANFOSSI: No, the estimates of C_0 were derived from eq.3 that contains the characteristics of the energy-containing large eddies. Only small-scale turbulence can be isotropic. As a consequence the value of C_0 changes as a function of the energy-containing eddies for each turbulent flow, according to Landau arguments.

AIR POLLUTANT DIFFUSION IN COMPLEX TERRAIN: AN APPLICATION OF THE CALPUFF MODELING SYSTEM

G. Carizi[(1)]; S. Cinotti[(1)]; F. Gianfelici[(2)], I.Giovannini[(1)], A. Levy[(1)] and L. Presotto[(2)]

[(1)]AQUATER S.p.A. - S. Lorenzo in Campo (PS) - Italy
[(2)]Snamprogetti S.p.A. - S. Donato Milanese (MI) - Italy

ABSTRACT

The CALPUFF modeling system has been applied to estimate the fate of atmospheric pollutants emitted by the stack of a waste disposal incinerator located in an Alpine valley. Specifically designed measurement campaigns have been performed to evaluate either pollutant emission rates and on site meteorological conditions.
The computational results indicate that the model can account for the main features of the thermally and topographically forced valley circulation and thus provide insight into the dispersion pattern and ground level concentrations of atmospheric pollutants in complex topography.

INTRODUCTION

The waste disposal Bolzano incinerator is located at the confluence of two alpine valleys in North East Italy, hence the local wind conditions are determined either by the topographic forcing of mesoscale circulation and the thermally driven valley breeze. As a results the circulation exhibits a complex pattern which is neither homogeneous nor stationary, thus preventing the use of classical Gaussian approach to the evaluation of pollutant dispersion and resulting ground concentrations.
Although rather sophisticated simulation approach are presently available for the analysis of atmospheric circulation and pollutant dispersion on complex topography, they are typically rather demanding as concern availability of input data and computer resources. This usually prevents their use as operational tools and their application is mainly restricted to test cases and calibration of simpler models.
The CALPUFF modelling system, which combines a diagnostic wind model (EPA[3], 1995) with a not stationary Gaussian puff model (EPA[4], 1995) allows to account for main features

of flow field in complex topography and their effect on pollutant dispersion, while keeping the requirements on input data and computational facilities to such a level to be still considered for operational applications.

For these reasons - within the framework of an EIS for the incinerator plant - it has been selected for the evaluation of the valley circulation and dispersion of air pollutant emitted by the plant stack under typical atmospheric scenarios.

While allowing to be operated on rather basic input data, the modeling system can take advantage from the availability of detailed description of site characteristics and local meteorology. In the present application a meteorological campaign, with both ground level and vertical profile monitoring, has been performed with the objectives of obtaining either some insight into local circulation and provide interpolation data for the diagnostic wind model. Data on land use have been obtained from high resolution satellite observation and local topography from a DTM data base with 250 m resolution.

Finally a chemical measurements campaign has been performed to get information on either emitted pollutant and emission rates at plant stack.

THE CALPUFF MODELLING SYSTEM

The CALPUFF modeling, developed by the California Air Quality Board (CARB) is based on a diagnostic wind module - CALMET - for the evaluation of hourly three-dimensional wind fields on complex topography, and a not steady state Gaussian puff model - CALPUFF - for the evaluation of pollutant dispersion.

The diagnostic wind module operates on a square cell Cartesian grid with terrain following vertical layers. The wind field is obtained in two steps. In step one a first guess wind velocity is adjusted for local effects by empirical formulas for:

- **kinematic terrain effects**. Following Liu[8],1980 the magnitude and decay with height above the terrain of the topographically forced vertical wind component is evaluated as a function of atmosphere stability as:

$$w = \left(V_{hor.} \cdot \nabla h_t(x,y)\right)\exp[-kz] \quad \text{with} \qquad k=\frac{N}{|V|}$$

  where V_{hor} is the horizontal wind vector, h_t the terrain height and N the Brunt-Väisäla frequency

- **slope flow.** CALMET assume slope flow direction in the drainage direction, evaluated with the empirical procedure suggested in Allwine[1], 1985. The magnitude of slope flow is evaluated as:

$$V_s = \beta_2\left[1000\left(\frac{|\gamma|}{T_0}\right)h_{\max}s_1\right] \qquad \text{with } s_1 = \left[\left(\frac{\partial h_t}{\partial x}\right)^2 + \left(\frac{\partial h_t}{\partial y}\right)^2\right]$$

where β_2 is a function of time on day, γ the domain scale lapse rate, T_0 the domain scale temperature, h_{max} the maximum terrain height within a user-defined radius of influence of topography and s_1 is a measure of terrain steepness.

- **thermodynamic blocking effects**. The terrain blocking is parameterized in term of the local Froude number:

$$Fr = \frac{V}{N\Delta h_t} \qquad \text{with} \quad \Delta h_t = \left(h_{\max}\right)_{ij} - z_{ijk}$$

where h_{max} is the maximum terrain height within a user specified distance from grid point (i,j) and z_{ijk} is the height above ground of vertical level k. When the Froude number is below a user defined critical values and the local wind has an uphill component, the flow is adjusted to be horizontal and tangent to the terrain, otherwise it is left unchanged.

In step two a weighted averaging of step one wind field and available wind measurements is performed with the formula:

$$(u,v) = \frac{\dfrac{(u,v)_1}{R^2} + \sum_k \dfrac{\left(u_{obs}, v_{obs}\right)_k}{R_k^2}}{\dfrac{1}{R^2} + \sum_k \dfrac{1}{R_k^2}}$$

where summation is extended on the measurement stations available in the domain, and R is a user defined weighting parameters which allow to balance at every grid point the contribution of surrounding stations and of step one wind field.
The vertical profile of horizontal wind velocity is evaluated by similarity laws, while the vertical velocity components are evaluated by mass conservation at each grid point. Zero values of W at model top can be enforced by the method of O'Brien[10],1970.
Finally, overall mass conservation of the three-dimensional wind field is enforced with the Goodin[5],1980 divergence minimization procedure.
In order to evaluate diffusive characteristics of PBL, CALMET include a micro-metorological module for the evaluation of mixing height, sensible heat flux, friction velocity , Monin-Obhukov length and convective velocity scale. The micro-meteorological module adopts an energy balance method (Holtslang[7], 1983) over land while a profile method, based on air sea temperature difference, is used over water.
The diffusive module of the modeling system - CALPUFF - is based on the Gaussian puff approach: each puff is advected by three-dimensional wind field, while its dimensions increase according to the Gaussian formula. The puff dispersion coefficients at each time are evaluated as:

$$\sigma_{y,t_n}^2 = \sigma_{y,T}^2\left(\xi_{t_{n-1}} + \Delta\xi\right) + \sigma_{y,s}^2 + \sigma_{y,b}^2$$
$$\sigma_{z,t_n}^2 = \sigma_{z,T}^2\left(\zeta_{t_{n-1}} + \Delta\zeta\right) + \sigma_{y,b}^2$$

where σ_T represent the turbulent contribution - for which the model implements different formulations, based on the general form (Hanna[6], 1977):

$$\sigma_{y,T} = \sigma_v t f_y\left(\frac{t}{t_{Ly}}\right) \qquad \text{and} \qquad \sigma_{z,T} = \sigma_w t f_z\left(\frac{t}{t_{Lz}}\right)$$

where σ_v and σ_w are the variances of the wind speed components and T_L are the horizontal and vertical Lagrangian time scales; σ_s is the source initial dimension, due to source dimension for area sources or tip stack downwash effects for point sources; σ_b is the plume

buoyancy contribution, parameterized in term of the plume rise. Finally ξ e ζ are parameters defined implicitly by the equation themselves, as indicated in EPA[4],1995.
The model can consider either constant or variable source emission, while both buoyancy and momentum contributions are accounted for in plume rise, with the Briggs[2], 1975 formulations. Partial plume penetration of the inversion layer is evaluated with the Manins[9], 1979 method. Both dry and wet deposition mechanisms are implemented for pollutant removal and a pseudo-first-order chemical reaction mechanism is implemented for the conversions:

$$SO_2 \longrightarrow SO_4^{=} \qquad \text{and} \qquad NO_x\left(NO + NO_2\right) \longrightarrow NO_3^{-}$$

THE MEASUREMENT CAMPAIGN

The numerical modelling of pollutant dispersion have been based on data of two specifically performed measurements campaign:

- the chemical campaign - performed on march 1997 - has provided measured data on the type of pollutants released and their emission rates at the incinerator stack
- the meteorological campaign has been performed between May, 22 and June, 1 1997, with the aim to provide both indication on the local circulation and initialization data for the wind modelling.

Four ground meteorological stations (fig. 1) have been operated along the main Adige river valley. Their position has selected with the aim to get information either on flow channeling along the valley axis and anabatic-katabatic circulation on valley side slopes.
Station *SS1* and *SS3*, located roughly along the main valley axis, were equipped with sensors for the following measurements:

Station *SS1*

- wind speed and direction, air temperature and relative humidity at 2 and 10 m
- global solar radiation, atmospheric pressure and total precipitation at 2 m

Besides at three hourly intervals, the following vertical profile measurements have been performed close to station *SS1*:

radiosounding with tethered balloons up to 1200 m height, with continuous transmission of:

- wind speed and direction
- wet and dry bulb temperature
- atmospheric pressure.

radiosounding with spare balloons, followed with a theodolite up to 2000-2500 m, with acquisition ofd the above parameters

pilot balloons, providing wind speed and directions at vertical intervals of about 50 m, up to 1500 m

Stations *SS3*

- air temperature at 2 and 10 m
- mean wind speed and direction at 10 m

Station SS2 and SS4 were located mid slope at opposite side of the valley, with special aim to get information on valley breeze circulation, and were equipped with the same sensors as station SS3. All stations collected data at half-hourly intervals.
Coherently with local topography, the meteorological campaign has indicated three main circulation patterns:

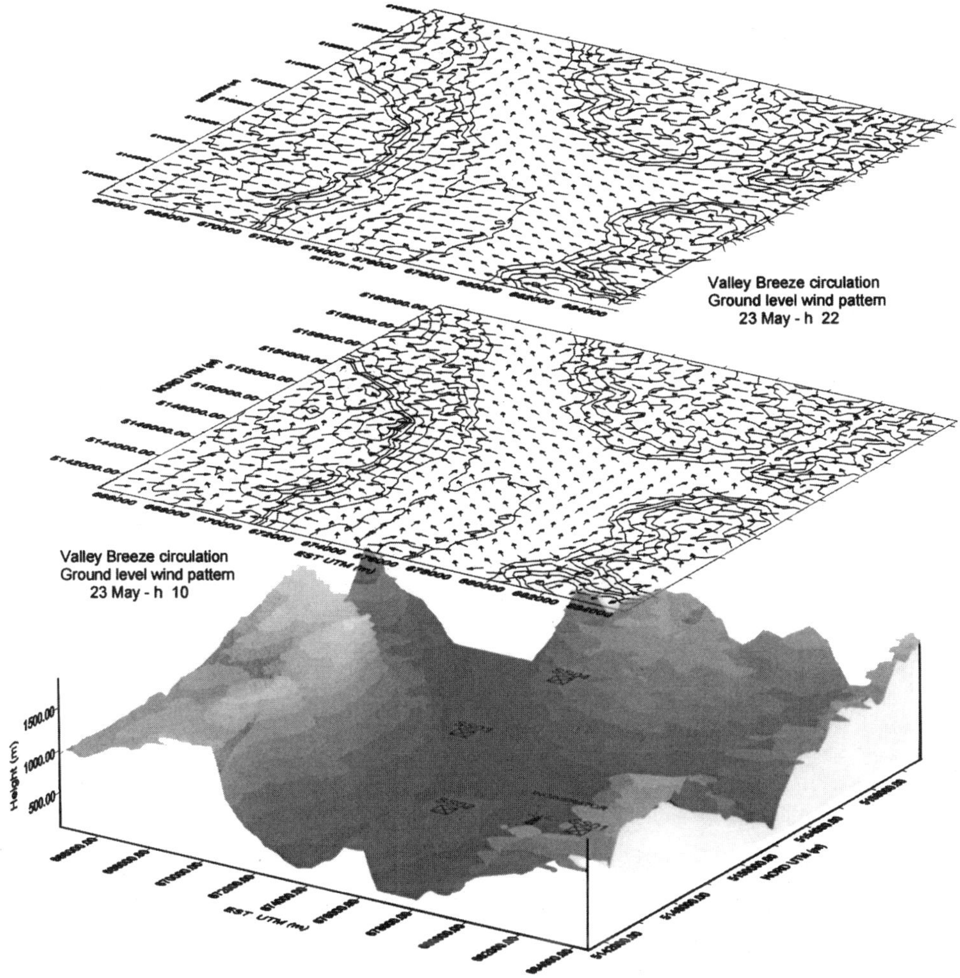

Fig. 1 - *Site topography and computed wind field under valley breeze circulation.*

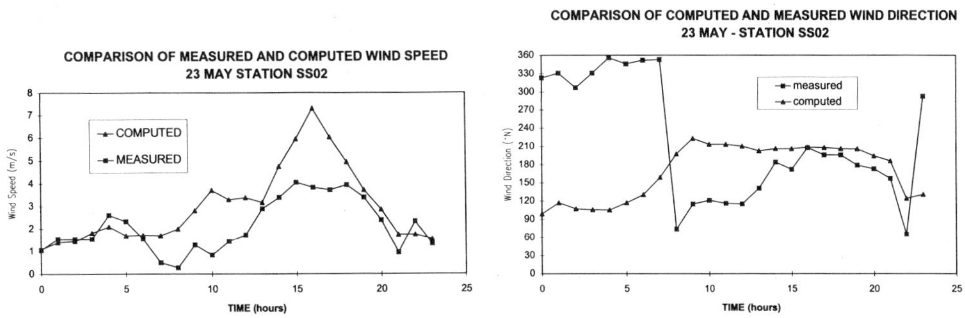

Fig. 2 - *Wind model verification - Comparison of computed and measured winds at SS02 station under valley breeze conditions.*

a) southern or -b) northern wind of appreciable strength, mainly aligned along the valley axis, resulting from the topographic forcing of mesoscale circulation, associated to the transit of depressions
c) valley breeze circulation, with lower wind intensity than in previous case, associated to weak mesoscale circulation
On the basis of the measurements findings, three days - representative of these different scenarios, have been selected for simulation: 23 May - with type c) conditions; 28 May, with type b) circulation;, 29 May, type a) circulation.

NUMERICAL SIMULATIONS

The numerical simulations of the wind field have been performed on a 20kmx20km grid with a 250 m mesh size and 10 unevenly spaced vertical level, ranging from 10 to 2000 m. Local topography has been obtained by a DTM with 250 m resolutions, derived from data of the IDGM (National Geographyc Institute) , while information on land use have been obtained by an in-house database derived from 30 m resolution satellite observations. No reliable information could be obtained concerning other ground characteristics required by CALMET (roughness, albedo, heat flux, Bowen ratio), for which default values have been used.
Input meteorological data have been obtained from the ground level measuring station, and vertical profiles data from the sounding performed at station *SS1*, extrapolated between highest sounding height and top vertical model level.
An attempt at model verification has been made by alternatively excluding one of the four meteorological station from model initialization and then comparing measured and computed wind speed and direction at that position. The verification has resulted in quite satisfactory model performance either under flow channeling conditions and under weaker valley breeze circulation, as exemplified in fig. 2 and 3, relative to SS2 station.
As concern the valley breeze circulation, it is rather interesting to note that, notwithstanding the rather simple schematization adopted in the model, it appears capable of getting main features of the expected overall wind field , with breeze inversion of the flow along valley slopes between morning and night (fig. 1), and mainly aligned along the valley axis at intermediate hours
The pollutant dispersion simulations have been performed for not chemically reactive pollutants. The isolines of maximum ground level concentrations provide some indications on the complex interplay between inhomogeneous wind field and local topography in the definition of pollutant impact: due to varying wind direction, a considerably wide area can be affected by pollutant impact, and topographically forced flow channeling can results in preferential path of pollutant dispersion, as shown in fig. 4 for scenario b) circulation. Finally plume interactions with topography results in preferential pollutant grounding on side slopes of orography, where maximum ground level concentrations are obtained in all three scenarios.

CONCLUSION

The CALPUFF modelling system has been used to investigate flow field and pollutant dispersion in an alpine valley, where the combination of topographic and thermal effects results in rather complex wind patterns.

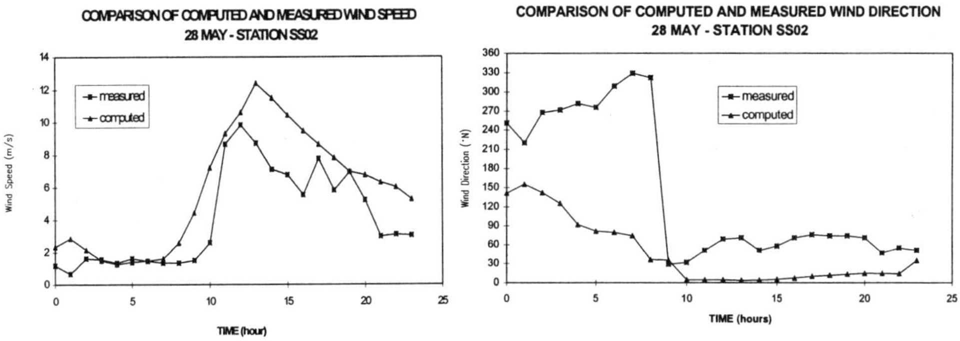

Fig. 3 - Wind model verification - Comparison of computed and measured winds at SS02 station under Northern wind circulation

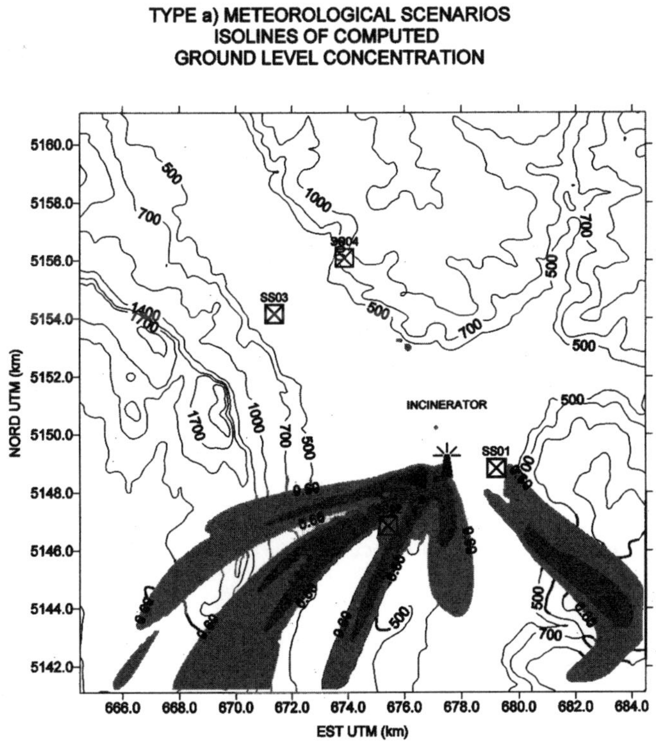

Fig. 4 - CALPUFF model - Isolines of maximum ground level concentration under type b) meteorological conditions

The verification of the meteorological model has shown a rather satisfactory performance, even in an situation where details on site characteristics were quite basic, as commonly occur in operational model applications.

The dispersion simulations has provided some insight into the complex interplay of flow field and topography in determining ground level concentrations of pollutants.

Considering the Italian situation - in which the combination of a complex topography and typically a quite close coastline usually results in the failures of the simplifications of the Gaussian plume approach, on which most regulatory models are based - the satisfactory performance and not very demanding requirements of this modelling system make it a quite appealing choice for EIS study.

REFERENCES

[1] Allwine, K.J.; Whiteman, C.D. - MELSAR: A mesoscale air quality model for complex terrain: Vol. 1 - Overview, technical Description and user's guide - *Pacific Northwest Laboratory, Richland, Washington, 1985*

[2] Briggs, G.A. - Plume rise predictions - in *Lectures on air pollution and environmental impact analyses. American Meteorological Society, Boston MA,* pp. 59 - 111, 1975

[3] EPA (U. S. Environmental Protection Agency) - A User's Guide for the CALMET Meteorological Model - *EPA-45/B-95-002, March 1995*

[4] EPA (U. S. Environmental Protection Agency) - A User's Guide for the CALPUFF Dispersion Model - *EPA-454/B-95-006, July 1995*

[5] Goodin, W.R.; McRae, G.J.; Seinfeld, J.H. - An objective analysis technique for constructing three-dimensional urban scale wind fields - *J. Appl. Meteorol.,* 19, pp. 98-108, 1980

[6] Hanna, S.R. & al. - AMS workshop on stability classification schemes and sigma curves - Summary and recommendations - *Bull. Am. Meteor. Soc.,* 58, pp. 1305-1309, 1977

[7] Holtslag, A.A.M.; Van Ulden, A.P. - A simple scheme for daytime estimates of the surface fluxes from routine weather data - *J. Clim. And Appl. Meteor.,* 22, pp. 517-529, 1983

[8] Liu, M.K.; Yocke, M.A. - Siting of wind turbine generators in complex terrain - *J. Energy,* 4, pp.10-16, 1980

[9] Manins, P.C. - Partial Penetration of an elevated inversion layer by chimney plumes - *Atmospgheric Environ.* , 13, pp. 733-741, 1979

[10] O'Brien, J.J. - A note on the vertical structure of the eddy exchange coefficient in the planetary boundary layer - *J. Atmos. Sci,* 27, pp. 1213-1215, 1970

DISCUSSION

J. LEVITIN: What grid did you use for wind field ?

S. CARIZI: The reconstruction of the wind field is obtained with the same resolution of geophysical maps (Digital terrain Model and LandUse), that is 250 m.

A. STOHL: Have you checked your upper level wind field? These are important for dispersion calculations, but are often very poor in diagnostic wind field models.

S. CARIZI: The cross-validation made on the wind field reconstruction has interested only the first level, that is 10 m. I agree completly with your consideration but at this time no check on upper level wind field has been made.

MASS CONSERVATIVE COUPLING OF NON-HYDROSTATIC METEOROLOGICAL MODELS WITH AIR QUALITY MODELS

M. Talat Odman and Armistead G. Russell

School of Civil and Environmental Engineering
Georgia Institute of Technology
Atlanta, Georgia 30332-0512, USA

INTRODUCTION

High-resolution data produced by recent non-hydrostatic meteorological models (MMs) are expected to significantly improve the characterization of transport in air quality models (AQMs). However, the use of data from non-hydrostatic MMs presents new problems for the air quality modeler. First, there is the *consistency* issue. The wind components together with air density satisfy the continuity equation in MMs. On the other hand, AQMs rely on the species continuity equation to enforce the principle of mass conservation. Though the continuity equation does not appear explicitly in their formulation, AQMs are expected to maintain a uniform mass mixing ratio field for an inert tracer after transport with the winds produced by the MM. This expectation could only occur if the two models used the same *discretization*, i.e., grid, time step, and finite difference forms. However, the models may not share the same grid structure and the forms used for advection in AQMs are usually very different from those in MMs. Also, since the outputs of the MM are stored less frequently than the AQM time step, the input variables cannot be reconstructed exactly at the desired instants. Consequently, the winds and the air density used in AQMs may not be consistent (i.e., they do not satisfy the continuity equation) and the uniform tracer field cannot be maintained. The perturbation of uniform fields is usually more pronounced with data from non-hydrostatic MMs than with data from hydrostatic MMs for the same domains. These perturbations grow in time and may generate instabilities in AQM solutions. A second issue, species or tracer *mass conservation*, presents itself because of the attempts to establish consistency and produce stable results. In some existing AQMs, the conservation of species mass was sacrificed in order to obtain stable results. However, large mass conservation errors are not tolerable in AQMs used to establish source-receptor relationships for the design of emission control strategies. Therefore, it is desirable to establish consistency in a mass-conservative manner.

Air Pollution Modeling and Its Application XIII, edited by S.-E. Gryning
and E. Batchvarova, Kluwer Academic/Plenum Publishers, 2000

METHODOLOGY

The use of data from non-hydrostatic models may lead to unstable solutions due to the consistency problem, as described above. Some widely used AQMs adopted an approach where a uniform tracer is transported along with other species and the concentrations are renormalized using the deviation of the tracer from uniformity as the norm. As a result the uniform mixing ratio field is maintained. However, the approach overlooks that non-linear advection schemes are used in AQMs for high accuracy. These schemes would affect fields of different distributions differently. Most species of interest, especially the emitted ones, have spatial distributions far from being uniform. Therefore, renormalizing the concentration of a species based on the perturbation of a uniform field can increase or decrease its mass artificially. We found that the mass conservation errors introduced by this approach are very large. Here, we will refer to this approach as "the flawed method."

Species mass conservation is not sacrificed if the consistency of meteorological fields is established through the continuity equation. To satisfy the discrete continuity equation in the AQMs, winds (e.g., Flatoy, 1993), or density fields (e.g., Lu et al., 1997), or both fields produced by MMs must be adjusted. Here we will describe three mass conservative methods for assuring consistency: two adjust the vertical wind component and a third adjusts the vertical flux. These methods also differ in the type of vertical advection algorithm used and whether a direct or iterative technique is used to solve for vertical wind component.

Method 1: Inverse Donor Cell

This method requires advection of air in addition to all the species and the use of the donor cell scheme for vertical advection.

$$c_k^{n+1} = c_k^* - \frac{\Delta t}{\Delta z_k}\left(w_{k+1/2}c_k^* - w_{k-1/2}c_{k-1}^*\right) \tag{1}$$

First, we apply horizontal wind components to the air density field, ρ^n, to obtain an intermediate field ρ^*. Any scheme can be used for horizontal advection; we used Bott's scheme (Bott, 1989). Then, we require that the density after the vertical advection operator is applied to ρ^*, be equal to ρ^{n+1}. The latter is calculated from the MM data by using linear interpolation in time. Thus, we need to calculate a new value for vertical velocity that would yield the desired value of density. Substituting concentrations with air densities, Equation (1) is solved for $w_{k+1/2}$ as

$$w_{k+1/2} = \frac{1}{\rho_k^*}\left(\frac{\Delta z_k}{\Delta t}\left(\rho_k^* - \rho_k^{n+1}\right) + w_{k-1/2}\rho_{k-1}^*\right) \tag{2}$$

Since $w_{1/2} = 0$, Equation (2) can be used starting from the surface layer (k=1) and moving to the top, each time using the vertical velocity computed for the previous layer. The winds at the top of the model will usually have a vertical component. The vertical velocities computed from Equation (2) must be used in Equation (1) to advect all other pollutant species. Note that the horizontal advection operator must always be applied before the vertical. Also, since the donor cell scheme is subject to the Courant stability condition, the case when the Courant number exceeds unity requires special treatment. Finally, the linear donor cell scheme is only first-order accurate.

Method 2: Inverse Bott Scheme

This method is very similar to Method 1 except a higher-order vertical advection scheme is used. Higher-order schemes are nonlinear because of the filters or flux limiters they use to eliminate the oscillations in the solution. Their non-linearity makes it difficult to solve for the vertical velocity field that would yield the air density ρ^{n+1} when applied to the horizontally advected air density field ρ^*. While a direct solution is not possible, iterative techniques can be used as suggested by Flatoy (1993). Here we used Bott's scheme with variable grid spacing for vertical advection. This scheme uses second-order area-preserving polynomials to represent the sub-grid distribution of concentration. The vertical velocity field can be calculated using the secant iteration technique.

$$w_{k+1/2}^{i+1} = w_{k+1/2}^{i} - \frac{w_{k+1/2}^{i} - w_{k+1/2}^{i-1}}{\rho_k^i - \rho_k^{i-1}} \left(\rho_k^i - \rho_k^{n+1} \right) \tag{3}$$

where ρ_k^i denotes the density field that the Bott's scheme would produce under the wind field $w_{k+1/2}^i$. To start the iterations two initial guess are needed, $w_{k+1/2}^0$ and $w_{k+1/2}^1$. The wind calculated by inverting the donor-cell scheme serves as $w_{k+1/2}^1$. The other initial guess is taken to be 90% of $w_{k+1/2}^i$. We interpreted convergence as ρ_k^i approaching ρ_k^{n+1} by a certain tolerance. Note that a uniform mixing ratio field advected with this method would be subject to very small perturbations, which can be reduced by decreasing the tolerance. We found that a value of 0.5% is quite effective as a tolerance; smaller values led to unreasonably large number of iterations without too much improvement in the result.

Method 3: Inverse Donor Cell in 3-D and Russell and Lerner Scheme

The third method is based on the approach described by Russell and Lerner (1981) and later used by Easter (1993). First the continuity equation is discretized in the following flux form.

$$\begin{aligned}\rho_{i,j,k}^{n+1} &= \rho_{i,j,k}^{n} - \frac{\Delta t}{\Delta x}\left(F_{i+1/2,j,k}^{n} - F_{i-1/2,j,k}^{n}\right) \\ &- \frac{\Delta t}{\Delta y}\left(G_{i,j+1/2,k}^{n} - G_{i,j-1/2,k}^{n}\right) - \frac{\Delta t}{\Delta z}\left(H_{i,j,k+1/2}^{n} - H_{i,j,k-1/2}^{n}\right)\end{aligned} \tag{4}$$

F, G, H are advective fluxes of air along the *x, y, z* directions, respectively. Then the species continuity equation is also written in the flux form.

$$\begin{aligned}c_{i,j,k}^{n+1} &= c_{i,j,k}^{n} - \frac{\Delta t}{\Delta x}\left(A_{i+1/2,j,k}^{n} - A_{i-1/2,j,k}^{n}\right) \\ &- \frac{\Delta t}{\Delta y}\left(B_{i,j+1/2,k}^{n} - B_{i,j-1/2,k}^{n}\right) - \frac{\Delta t}{\Delta z}\left(C_{i,j,k+1/2}^{n} - C_{i,j,k-1/2}^{n}\right)\end{aligned} \tag{5}$$

For consistency, the fluxes *A, B, C* must be related to their counterparts in Equation (4) by

$$A = Fr_x; B = Gr_y; C = Hr_z \tag{6}$$

where r_x, r_y, r_z are the average mixing ratios (i.e., c/ρ) crossing the grid cell boundaries in the *x, y, z* directions, respectively. We defined F and G as donor cell fluxes and solved Equation (4) for the vertical flux $H^n_{i,j,k+1/2}$ much like the solution for $w^i_{k+1/2}$ in Method 1. For comparison to other methods, we derived the corresponding vertical velocity by interpreting H also as a donor cell flux. The use of first-order accurate donor cell scheme for horizontal advection of air should not introduce large errors since the air density fields do not have the large gradients found in pollutant fields. It is more important to use higher-order schemes for advection of pollutant species. We used Bott's fourth-order area-preserving polynomials in the horizontal and his scheme for variable grid spacing in the vertical to compute the mixing ratios r_x, r_y, r_z.

RESULTS AND DISCUSSION

Using the methods described above in an AQM (Odman and Ingram, 1996), we simulated a tracer experiment conducted during the Central California Air Quality Study (SARMAP). In this experiment inert tracers were released from different locations in the San Francisco Bay area and the Central Valley. All releases started at 6:00 PDT (Pacific Daylight savings Time) on August 3, 1990 and lasted for 3-4 hours. The release rates were approximately uniform. Starting two hours before the releases and using fields generated by the NCAR Meteorological Model (MM5) we ran the model for 48 hours. The AQM grid covered the SARMAP domain with 32x39 cells of 12 km size. In the vertical, the grid extended to the stratosphere with 15 non-uniformly spaced layers starting with a thickness of 60 m at the surface and increasing upward. During the simulation, we did not only follow the tracer concentrations but the adjustments to the vertical wind components as well. When we analyzed these adjustments, we found a strong correlation with the terrain height. The adjustments were largest over the slopes of the coastal range and the Sierra Nevadas.

We found insignificantly small differences among the vertical velocities generated by the three methods. Therefore we will only use the velocity from Method 1 in our comparisons with the vertical wind components provided by MM5. The averages and standard deviations of the MM5 and adjusted vertical wind components as well as their difference are illustrated in Figure 1 for the surface layer. The adjustment reduces the magnitude of the vertical wind by about 25% during the day when there is an upward component with an average of about 0.03 m/s and by 35% at night when the vertical velocity is downward. The standard deviations are also reduced. The standard deviations are smallest early in the morning and largest in the afternoon and remain so up to about level 11 where we reach the free troposphere.

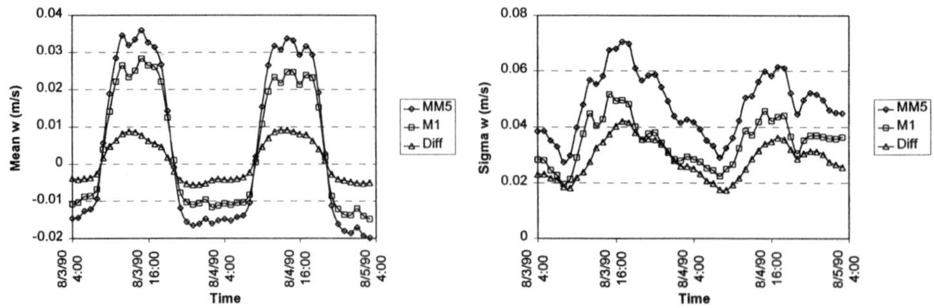

Figure 1. Time series of the mean (left) and standard deviation (right) of the first layer vertical velocity.

Table 1. Mean and standard deviation of vertical wind components in m/s

Time	Layer	MM5 Mean	MM5 σ	Adjusted[1] Mean	Adjusted[1] σ	Difference[2] Mean	Difference[2] σ
700							
	1	-1.22E-02	3.29E-02	-8.55E-03	2.28E-02	-3.65E-03	2.10E-02
	5	1.25E-02	3.65E-02	9.06E-03	3.83E-02	3.42E-03	3.12E-02
	11	1.08E-02	1.01E-01	1.40E-02	1.23E-01	-3.19E-03	8.50E-02
1500							
	1	3.59E-02	6.75E-02	2.83E-02	5.17E-02	7.63E-03	3.74E-02
	5	3.26E-02	1.13E-01	3.00E-02	1.03E-01	2.60E-03	5.88E-02
	11	4.27E-03	1.69E-01	3.39E-02	2.26E-01	-2.96E-02	1.64E-01

[1]Method 1 is used to adjust the vertical velocity.
[2]The mean and standard deviation of the (MM5 - Adjusted) field.

The averages and standard deviations are given in Table 1 for three different layers, the surface (0-60 m), the 5th (386-513 m) and the 11th layers (2155-3570 m). The values are reported at 700 PDT and 1500 PDT on August 3, 1990. In the PBL, relative to the magnitude of vertical velocity, the magnitude of the adjustment decreases with altitude. While the afternoon average adjustment is 21% at the surface it is only 8 % in the 5th layer. The relative difference starts increasing upon reaching the free troposphere. This may be related to the fact that the vertical velocity is small aloft and that the layer thicknesses are increasing with height. Also MM5 sets the vertical velocity to zero at the top of the troposphere while there is no such condition with the adjusted velocity. Data for the second day of the simulation had similar trends. The adjustment to the vertical velocity may be large in certain locations, especially over sloping terrain, but even then the absolute value of the vertical wind is small enough (few tenths of 1 m/s) that we do not expect to see a large impact on tracer or pollutant transport.

The total mass of two tracers, PMCH and PMCP, as predicted by different methods is shown in Figure 2. The results of a simulation with unadjusted wind fields is also shown. The largest conservation error is associated with PMCH mass. This tracer was released from San Jose and has a trajectory over complex coastal terrain. While only 19 kg of tracer was released, the flawed method leads to an estimate of 46 kg of PMCH at 1100 PDT. Other methods are able to maintain total tracer mass at 19 kg as long as the tracer remains within the domain. The decrease in total mass after 2100 PDT is associated with tracer crossing the lateral or top domain boundaries. The mass conservation error of the flawed method is also significant (approximately 25 %) for the PMCP tracer released from Pittsburgh in the San Francisco Bay area. The error is not as significant for other releases from the Central Valley.

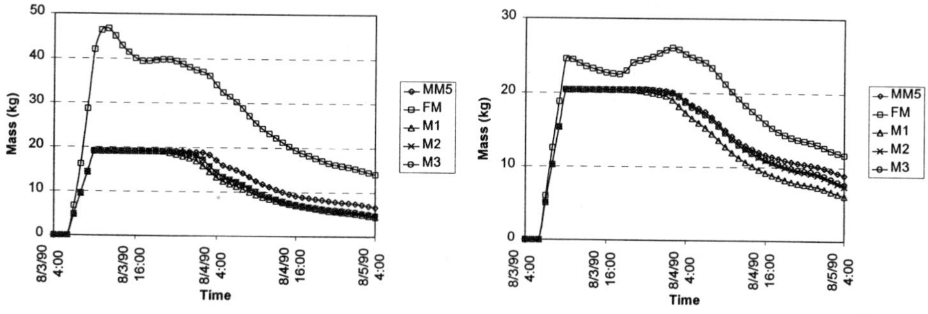

Figure 2. Time series of total mass of PMCH (left) and PMCP (right) as predicted by different methods: non-adjusted wind fields (MM5), flawed method (FM), and mass conservative Methods 1-3 (M1, M2, M3).

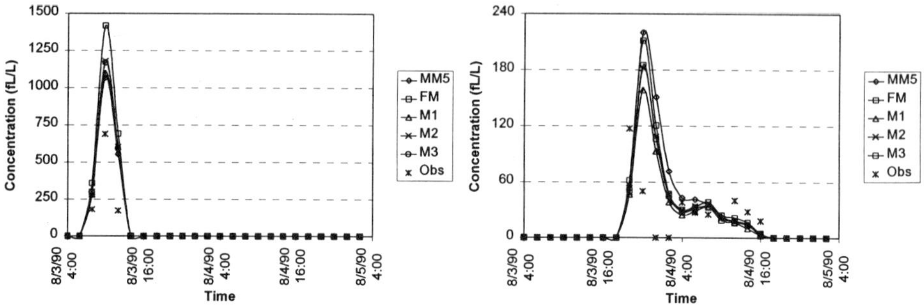

Figure 3. The time series of observed and predicted PMCP concentrations at Brentwood (left) and Friant (right).

In Figure 3 the time series of observed and predicted PMCP concentrations are shown at two sites. The distance from the release point to these sites, Brentwood and Friant, are approximately 30 and 220 km, respectively. The observations are provided as 2-hour averaged concentrations. The predicted concentrations were also averaged to match the averaging period of the observations. The peak concentration predicted by all methods are larger than the observed peak at Brentwood. Compared to the flawed method, all three mass conservative methods yield a peak value closer to the observation, but the overprediction is still of the order of 60-70%. The closest value to the observation is obtained when MM5 winds are used without any adjustment. The predictions of the flawed method are not too far off from those of mass conservative methods. This indicates that the mass conservation errors are small along the trajectory to this site. At Friant the predicted peaks are closer to the observations but they display a time lag. This is not surprising given the relatively larger distance of this site from the release point. In contrast to the former site, the use of unadjusted MM5 winds led to the largest overprediction. The peak predicted by Method 1 is the closest to the observation. At both sites the difference between observed and predicted concentrations are larger than the differences between different methods. Therefore, it is difficult to judge which method performs better in this tracer transport simulation.

Using the same model settings, a five-day ozone episode (August 2-6, 1990) was simulated over the same domain. The ground level ozone concentrations at 1700 PDT on August 5 are shown in Figure 4. The inconsistencies in MM5 wind fields led to instabilities during this simulation and yielded unreasonably high ozone. Methods 2 and 3 produced practically identical results therefore only Method 2 will be discussed. The differences between ozone levels predicted by the flawed method and Method 1 are of the order of 15 ppb near the coastal range, or less inland. Method 1 predicts higher ozone concentrations in major source areas to the south of the San Francisco Bay, such as San Jose. On the other hand, further to the southeast (downwind), ozone concentrations predicted by Method 1 are lower. This may be due to an erroneous increase in the mass of nitrogen oxides in the flawed method. In comparison to Method 1, Method 2 predicted slightly higher (of the order of few ppb) ozone concentrations over the Central Valley and slightly lower concentrations over more complex terrain such as the coastal range and the foot-hills of the Sierra Nevadas. The maximum difference was observed at Fresno where Method 2 predicted 7 ppb higher ozone than Method 1. Recall that the differences between the vertical wind components of Method 1 and Method 2 were negligible. Therefore, the differences in ozone levels characterize the impact of using a more accurate vertical advection scheme in Method 2. Finally, the differences were more pronounced at night and early in the morning. For example at 900 PDT August 5, Method 1 predicted ozone levels 15 ppb lower than the flawed method at Sacramento and 25 ppb lower southwest of the Central Valley.

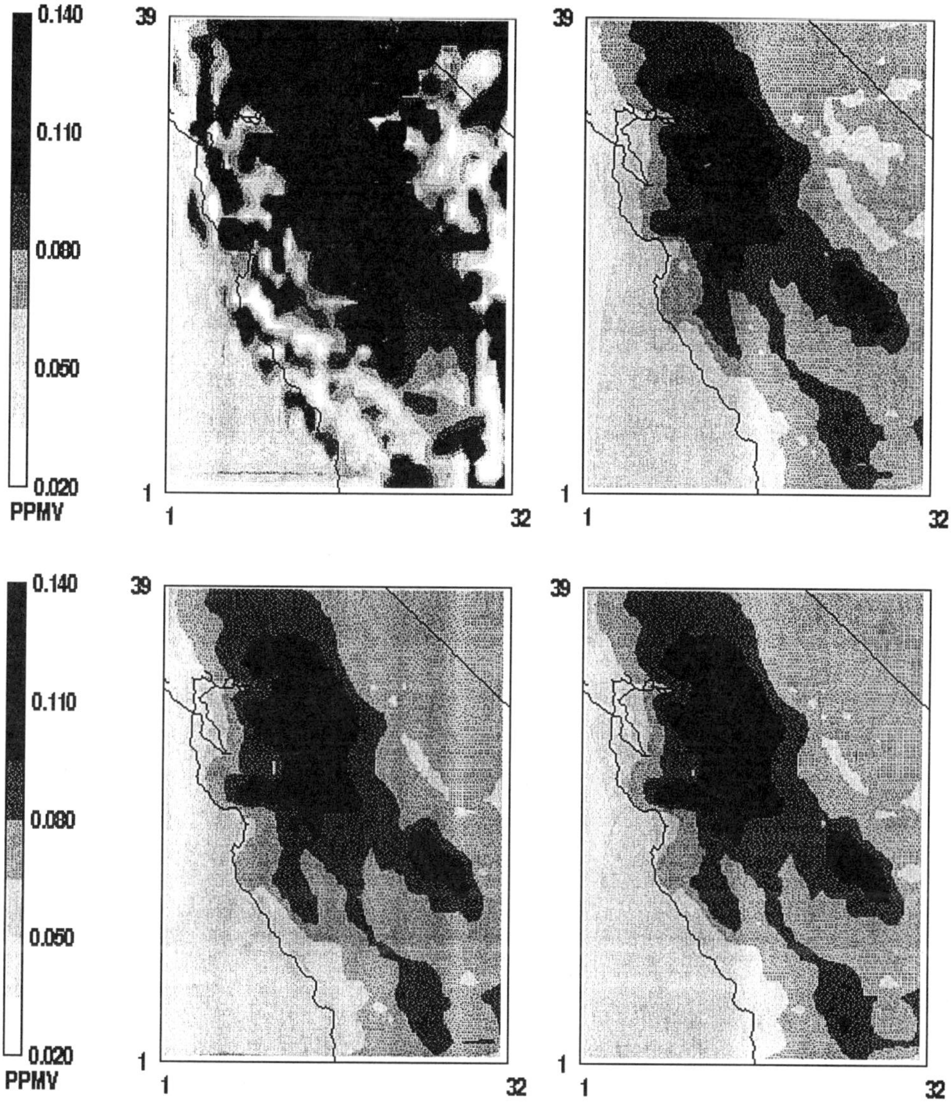

Figure 4. Spatial distribution of surface layer ozone at 1700 PDT August 5, 1990 as predicted by the non-adjusted wind fields (upper-left), flawed method (upper-right), and Methods 1 and 2 (lower left and right).

Figure 5 shows the maximum and the standard deviation of ground level ozone concentrations as predicted by different methods. On the first day, the flawed method predicts about 10 ppb lower peak ozone than the mass conservative methods. It also estimates a maximum concentration of 125 ppb at 900 PDT on August 5. This is about 45 ppb higher than the estimates of other methods. The peak ozone predictions of the flawed method are 10 to 20 ppb higher at night. Method 1 led to the lowest nighttime ozone estimates, with differences of up to 10 ppb from other methods early in the morning. The estimates of Methods 2 and 3 are almost identical. Method 1 resulted in the smallest standard deviation: as much as 1 ppb lower than other methods during the day and 1 to 2 ppb lower than Methods 2 and 3 at night. The nighttime standard deviation of the flawed method was the highest, about 2 to 4 ppb higher than Method 1.

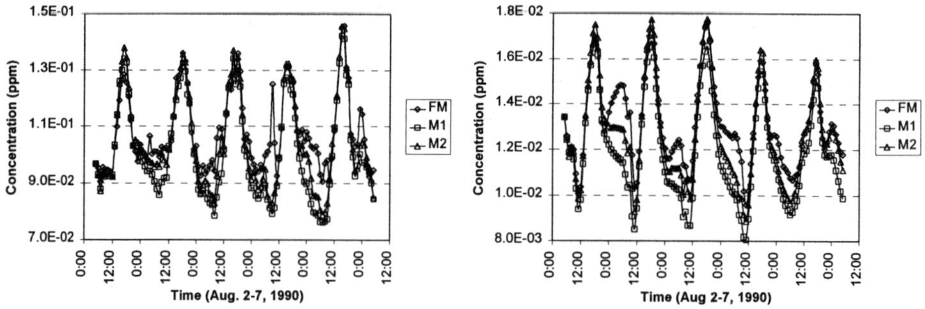

Figure 5. The time series of the maximum ozone concentration (left) and the standard deviation of surface layer ozone (right) as predicted by the flawed method (FM), and mass conservative Methods 1&2 (M1, M2).

CONCLUSIONS

The inconsistency of data from non-hydrostatic MMs may lead to instabilities in AQMs. Meteorological data can be made consistent without sacrificing the mass conservation characteristics of the AQM by adjusting the vertical wind component. This proved to be an effective technique. We implemented and compared three methods for obtaining the vertical wind component from the discrete continuity equation. The methods differ in their vertical advection schemes. We found that the methods produced very similar vertical wind fields but these fields differed from those supplied by the MM. The adjustments reduced the vertical wind speed by an average of about 30% near the surface. The reduction was more pronounced at night when the winds have a weaker downward component. Also, the reduction decreased with altitude in the PBL but started increasing again in the free troposphere where the winds are practically horizontal. Thus, the adjustments may be large relative to the vertical winds but this is generally the case when the wind speeds are small. This implies that overall tracer transport would not be significantly affected by the adjustments. The tracer experiment simulation provided partial support for this hypothesis. Along a limited number of trajectories the original and adjusted wind fields yielded similar results. Unfortunately, the uncertainties in the tracer data and other parameters of the AQM are too large for more definite conclusions. We were unable to determine which one of the three methods best characterizes tracer transport. In an air quality simulation there were small differences in ozone levels predicted by the three methods. These are due to the use of vertical advection schemes of different accuracies. Methods with higher-order schemes seem to perform better but a more complete evaluation is needed at this point.

References

Bott, A., 1989, A positive definite advection scheme obtained by nonlinear renormalization of the advective fluxes, Mon. Wea. Rev. 117:1006.

Easter, R.C., 1993, Two modified versions of Bott's positive-definite numerical advection scheme, Mon. Wea. Rev. 121:297.

Flatoy, F., 1993, Balanced wind in advanced advection schemes when species with long lifetimes are transported, Atmos. Environ. 27A:1809.

Lu, R., Turco, R.P., and Jacobson, M.Z., 1997, An integrated air pollution modeling system for urban and regional scales: 1. Structure and performance, J. Geophys. Res. 102:6063.

Odman, M.T, Ingram, C., 1996, Multiscale Air Quality Simulation Platform (MAQSIP): Source Code Documentation and Validation, MCNC Technical Report ENV-96TR002, Research Triangle Park, North Carolina.

Russell, G.L., and Lerner, J.A., 1981, A new finite-differencing scheme for the tracer equation, J. Appl. Meteor. 20:1483.

DISCUSSION

D. STEYN:

I approached this problem by ensuring that MM and AQM grids for *w* corresponded exactly, then performing a mass consistent interpolation that preserved *w*, and only adjusted horizontal wind components. Can you comment?

T. ODMAN:

Adjusting the horizontal wind components is another alternative. Any of the three wind components, the density, or their combinations may be adjusted to deal with the inconsistency problem. However, there is concern, particularly among the meteorologist, that the adjustments may have a considerable impact on transport. The argument for choosing the horizontal components may be that they are large (compared to *w*) and that they would require relatively small adjustments. The rational for choosing *w* is that advection is not the dominant transport process in the vertical (turbulent diffusion dominates). At first, an average, near-surface adjustment to *w* of about 30% may appear large but the effect on the overall transport is usually small, as I have demonstrated here. On the other hand, since advection is the dominant transport process in the horizontal, the impact of small adjustments to the horizontal winds may not be negligible. I strongly recommend that you analyze the overall transport characteristics (trajectories, concentrations, total mass, etc.) as I performed in this paper.

G. SCHAYES:

You mentioned that the problem of mass conservation arises in non-hydrostatic models. Could you comment why. Could the fact of using a hydrostatic model overcome the problem? This would be justified while using grid sizes of 12 km.

T. ODMAN:

The inconsistency problem arises when the AQM differ from the MM in the grid, time step, and/or finite-difference forms used. Thus, it does not matter whether the model is hydrostatic or non-hydrostatic; the problem can occur in either type of model. In my experience, the inconsistency problem is more pronounced in non-hydrostatic models. However, this may simply be due to the smaller grid sizes and more complex terrain in typical non-hydrostatic model applications. I do not have a direct comparison of the two types of models over the same domain using the same grid sizes.

C. PERSSON: The need for balancing the wind-fields before using them in off-line atmospheric transport models, as you have shown, is very important. For large-scale hydrostatic models the pressure tendency can be included to improve the balancing of the vertical winds. Have you any experience and comments on that?

T. ODMAN: In hydrostatic models the vertical velocity is diagnosed from the horizontal wind components using the pressure tendency. This helps balancing of the winds to some extent, but not to the degree achieved here. For example, uniform tracer fields may be perturbed if the finite-difference forms used in the pressure tendency equation are different than those used in the vertical advection equation. The diagnosing of w requires numerical integration of pressure tendency. Making this integral consistent with the differential forms used for vertical advection is probably more difficult than the method described here for vertical wind adjustment.

MODELING THE DEPOSITION OF ACID FOG TO HIGH-ELEVATION FORESTS: LIQUID WATER CONTENT AND SULFATE CONCENTRATION

John L. Walmsley[1,2], Natty Urquizo[2,3], Jeffrey R. Brook[2]
William R. Burrows[2], and Robert S. Schemenauer[1,2]

[1]Retired
[2]Atmospheric Environment Service
4905 Dufferin Street
Downsview, Ontario M3H 5T4 Canada
[3]Contractor

INTRODUCTION

Objectives and Tools

In this paper, we update the progress report of Urquizo et al. (1998b) on the use of sophisticated statistical analysis programs to estimate the deposition of fogwater and wet acid ions in mountainous terrain. At this stage, our objective is two-fold. First, we estimate liquid water content (LWC) at the Chemistry of High-Elevation Fog (CHEF) site (845 m MSL) on Roundtop Mountain in southern Quebec from routine meteorological measurements at the Sherbrooke weather station (238 m MSL), 78 km to the northeast. Second, we estimate the sulfate concentration in fog using CHEF data, Sherbrooke meteorological data, and upwind emissions and meteorology determined from backward trajectories. In both of these efforts, we used as analysis tools the Classification and Regression Trees (CART) (Breiman et al., 1984; Steinberg and Colla, 1995) and Neuro-Fuzzy Inference Systems (NFIS) (Chiu, 1994).

Data Sources

Between 1985 and 1991, the Chemistry of High Elevation Fog (CHEF) experiment was conducted on three mountains in southern Quebec, Canada. The CHEF project was linked with the Mountain Cloud Chemistry Project (MCCP) in the Appalachian Mountains of the USA. Measurements were made of liquid water content (LWC) as well as standard meteorological parameters (temperature, pressure, relative humidity, wind speed, wind direction and precipitation).

Fogwater and precipitation samples were collected and analyzed for the concentrations of various acid ions. Previous studies (Schemenauer et al., 1995) at two high-elevation (almost 1000 m above sea level) sites showed that the acid deposition from fog was at least as important as that from precipitation.

The fog measuring device [FMD-04R, Advance Electronic Sub-Systems, Inc., Kettleby, ON] was used to measure LWC. Further details appear in Walmsley et al. (1998). Hourly-averaged LWC data obtained from this device were available for various lengths of time between April and October in the period 1988-91. Hourly-averaged data indicating the presence or absence of fog were available for the same period from an optical cloud detector [model MCD-05; now known as the Fog Sentinel, Series DH, Advanced Sensor Technologies, Inc., Abingdon, MD], hereafter referred to as the OCD. When the CHEF site was below cloud base (*i.e.*, not in fog) and precipitation was falling, neither the FMD nor the OCD had a significant response to the precipitation. When the cloud base was below the sampling elevation (*i.e.*, the site was in fog), both instruments responded to the fog. Therefore, no attempt was made to separate out the fog-only cases from those with mixed fog and precipitation.

Data from the Sherbrooke weather station used in this study are available in the Canadian Weather Energy and Engineering Data Sets (CWEEDS) compiled by the Atmospheric Environment Service (AES) in 1993. The hourly values were averaged over periods that matched those of the fogwater samples. From gridded meteorological fields, we computed 72-h back trajectories at 12-h intervals to determine transport direction and average transport distance for a range of transport times. Upwind emissions data and temperatures (Burrows et al., 1997) were also determined from the trajectories.

STATISTICAL ANALYSIS PROGRAMS

Classification and Regression Trees (CART)

The CART program is ideally suited to the problem of analyzing a large amount of data consisting of an observed value to be estimated or predicted (the *predictand*) and a number of possible *predictors*. Both predictand and predictors may assume continuous values or be classified into a finite number of categories. Output from CART is at discrete values and has the form of a decision tree. CART minimizes variance at each "splitting" of the data at internal nodes of the tree. During this process, it selects the relevant predictors from what may be a large number of potential ones. Walmsley et al. (1998) present a conceptual description of how CART operates, from the viewpoint of the user.

Neuro-Fuzzy Inference Systems (NFIS)

NFIS does not eliminate predictors; however, it does provide a continuous-response form of output. NFIS forms cluster centres, using a procedure that is controlled to a certain extent by the user's choice of four input parameters. Membership of data points in the cluster centres is "fuzzy", as a data point may be a member of more than one cluster. Each cluster center is regarded as a basis for rules, which become linear equations in the predictors. Training data are used to determine the coefficients of these equations, yielding a linear least-squares estimation problem.

CART-NFIS Hybrid Method

For LWC prediction, we used CART as a preprocessing step in two ways: to set aside the majority of the observations in which fog was not predicted to occur at the Roundtop CHEF site, and to select the predictors from the remaining data. For acid-ion prediction, CART was used simply for predictor selection. In both applications, NFIS then provided continuous output based on predictors that CART had found to be significant. In this manner, we combined the strengths of the two methods.

PREDICTION OF FOG OCCURRENCE

Of 10 357 hourly records from the 1988-91 (April-October) measurement program, reliable LWC data were available for 7317 hours and OCD data for 4093 hours. Of the latter, 430 observations were in a mid-range, indicating that fog was present for only part of the hour. For simplicity, these were rejected. This left 3252 hours in which fog was not detected and 411 hours in which it was present for the full hour. When fog was not detected, the LWC was set to zero, regardless of its recorded value. The OCD-screening process increased the LWC records from 7317 to 7357. OCD data were available for 3567 of the 7357 records; of these, 3252 were observed to be no-fog cases (Class 0) and the remaining 315 were fog cases (Class 1).

For CART Run 1, the learning set was a randomly selected 80% (N = 2854) of the 1988-91 dataset. The test set was the remaining 20% (N = 713). The results from CART Run 1 are summarized in Table 1, where the learning and test sets have been combined (N = 3567). The *prediction accuracy* or *post agreement* (Stanski et al., 1990) is defined as:

$$A_i = n_{ij}/r_i, \quad i = j, \tag{1}$$

where i is the row number (predictions), j is the column number (observations), n_{ij} is the number of cases in cell (i,j), and r_i is the total of the ith row. The *total accuracy* or *percent correct* (Stanski et al., 1990) appears in the bottom right box. It is computed as the sum of the diagonal elements divided by the total number of cases, N:

$$\text{Total accuracy} = n_{ii}/N, \tag{2}$$

where the tensor summation convention is in effect. Although this number (0.88) appears respectable, it is misleading, as it is dominated by the large number in the (0,0) box (top left). More revealing is the fact that 2856 cases out of 2878, or 0.99, were correctly predicted to fall in Class 0 (no fog). Predictions of Class 1, however, had an accuracy rate of only 0.43. The results clearly demonstrate that the algorithm derived from CART Run 1 can be used with a high degree of confidence to eliminate cases where fog was not observed to occur at Roundtop. The remaining data contain a mixture of fog and no-fog events that can be processed by a subsequent CART application.

Table 1. Contingency table for CART prediction of fog occurrence at Roundtop. Class 0 = no fog; Class 1 = fog. Source: Walmsley et al. (1998)

Predicted Class	Observed Class 0	1	Total	Accuracy
0	2856	22	2878	0.99
1	396	293	689	0.43
Total	3252	315	3567	0.88

The prediction algorithm obtained from Run 1 was used to process the original dataset (N = 7357). As a result, 6022 cases, or 82%, were predicted to fall in Class 0. These can be set aside with high confidence that they are cases in which no fog was observed. The remaining 1335 cases are expected to contain fog and no-fog cases, with the former representing a larger proportion in this reduced dataset than they did in the complete dataset.

PREDICTION OF LWC

The reduced dataset (N = 1335) was processed by CART, with 20% of the data reserved for testing purposes. The resulting algorithm was applied to the reduced data set (the combined learning and test sets). The correlation coefficient is 0.63, suggesting some degree of skill, but the scatter is still rather large. Nevertheless, the predictors selected by CART Run 2 are potentially useful to NFIS, which is unable to eliminate predictors.

When the no-fog results from CART Run 1 are combined with the CART Run 2 results, the correlation coefficient is 0.71, which is a significant improvement over the value obtained (0.60) from the one-step CART application. Table 2 summarizes the combined results of CART Runs 1 and 2. (The total of 6041 cases predicted in the 0 - 0.04 g m^{-3} bin includes 6022 zero values from CART Run 1. From the reduced dataset of 1335, CART Run 2 predicted an additional 19 no-fog cases.) Comparison with the results of the one-step CART application shows that the no-fog and fog prediction accuracies are only slightly improved (0.95 to 0.97 and 0.50 to 0.51, respectively). Although the improvement in the accuracy of fog prediction (*i.e.*, absence or presence) is modest, there has been a very significant improvement in the accuracy of the LWC prediction.

Table 2. Contingency table for the combined CART-CART prediction of LWC (g m^{-3}) at Roundtop. Source: Walmsley et al. (1998)

Predicted LWC	Observed LWC 0-0.04	>0.04	Total	Accuracy
0-0.04	5832	209	6041	0.97
>0.04	645	671	1316	0.51
Total	6477	880	7357	0.88

The predictors from CART Run 2 were used by NFIS with the reduced dataset (N = 1335). The result was a prediction algorithm for calculating values of LWC. The 1335 values were then combined with the 6022 values set aside by application of the CART Run 1 algorithm, all of which had predicted values of 0 g m^{-3}. When the complete prediction of 7357 cases is compared with the observed values, the correlation coefficient is a respectable 0.81, a significant improvement over the results (0.60) obtained from the one-step CART application. The best-fit linear regression line indicates a tendency to underpredict. Nevertheless, about 10 of the 11 highest predicted values lie close to the 1:1 perfect fit line.

Table 3 presents the contingency table resulting from the three-step (CART-CART-NFIS) application. (The total of 6398 cases predicted in the 0 - 0.04 g m^{-3} bin includes 6022 zero values from CART Run 1. From the reduced dataset of 1335, therefore, NFIS predicted an additional 376 no-fog cases.) The accuracy of the no-fog prediction is essentially unchanged from the two-step (CART-CART) process (Table 2). The accuracy of the fog prediction, on the other hand, underwent a significant improvement from 0.51 to 0.66, as did the total accuracy (0.88 to 0.92).

Table 3. Contingency table for the combined CART-CART-NFIS prediction of LWC (g m^{-3}) at Roundtop. Source: Walmsley et al. (1998)

Predicted LWC	Observed LWC 0-0.04	>0.04	Total	Accuracy
0-0.04	6155	243	6398	0.96
>0.04	322	637	959	0.66
Total	6477	880	7357	0.92

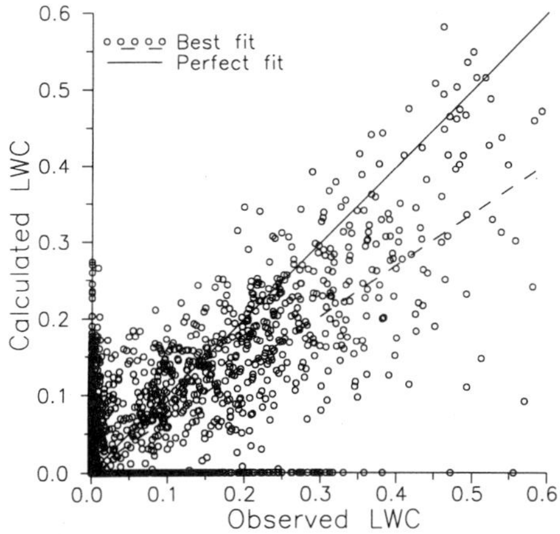

Figure 1. CART-CART-NFIS prediction of LWC (g m^{-3}). N = 7357; best-fit linear regression: y = 0.671x + 0.003 g m^{-3}); correlation coefficient, r = 0.81; rms error = 0.044 g m^{-3}. Source: Walmsley et al. (1998).

PREDICTION OF SULFATE CONCENTRATION

Model development using all fog data led to only one predictor (average wind direction). This was because CART placed too much emphasis on the extreme sulfate-concentration events (> 500 μeq L^{-1}). Of 272 fog samples, the 30 that fell into this category had an average concentration of 884 μeq L^{-1}. Back trajectories for these cases show paths that were consistently over major sources of pollution. Nevertheless, CART had difficulty in identifying predictors when the extreme events were included, so those cases were set aside. Different approaches will be investigated for prediction of extreme events.

CART had more success in identifying predictors for cases less than 500 μeq L^{-1} (N = 242). During the analysis, 16% of the data was reserved for the test set. Four predictors were identified from the total of 129 available. In order of importance, these predictors were: total NO_x emissions (RNNOX) accumulated along the backward trajectory during rain-free hours prior to arrival at the receptor point; total sulfur dioxide emission (TOTSO2) accumulated along the backward trajectory; dew point temperature (TD) at Sherbrooke; and mean wind speed 18 h upwind along the trajectory (WSP4). CART used these predictors to progress through a series of internal nodes and identify six terminal nodes (TNs). A description of how the latter were defined is shown in Figure 2.

```
                              :
             ------------ 1 -------------
             :          RNNOX           :
      ----- 2 -----            -------- 3 --------
      :  TOTSO2   :            :       TD        :
      :           :            :         ------- 4 -------
      :           :            :         :     TOTSO2    :
      :           :            :     --- 5 ---           :
      :           :            :     : WSP4  :           :
   TN 1           2            6     3       4           5
```

Figure 2. Decision tree generated by CART for the prediction of sulfate concentration. The predictor used as a split-variable at each of the five internal nodes is shown. Threshold values at internal nodes are: 1: RNNOX = 785 kg; 2: TOTSO2 = 22 100 kg; 3: TD = 16.3°C; 4: TOTSO2 = 9308 kg; 5: WSP4 = 23.5 m s^{-1}. Terminal nodes (TNs) appear at the bottom. For explanation, see text. Source: Urquizo et al. (1998a).

The tree in Figure 2 is read from top to bottom. All cases are first examined for the magnitude of RNNOX. If RNNOX ≤ 785 kg, then TOTSO2 is examined. If TOTSO2 ≤ 22 100 kg, then the case is classified as TN 1; otherwise it is TN 2. If RNNOX > 785 kg, then TD is examined, etc. Table 4 shows the mean and standard deviation by TN for all the sulfate measurements. The data are broken into the learning set (cases used in developing the tree) and the test set (cases classified using the pre-determined tree).

Table 4. CART Prediction of sulfate concentration (μeq L^{-1}). Observed values > 500 μeq L^{-1} have been eliminated. For explanation, see text. Source: Urquizo et al. (1998a)

	Learning Set			Test Set		
TN	N	AVG	STD	N	AVG	STD
1	98	92	69	18	144	124
2	4	290	144	2	275	42
3	58	140	81	6	138	71
4	2	406	92	1	452	0
5	25	230	109	7	279	123
6	16	289	80	5	313	158
Total	203			39		

Except for TN 1, average values for the test cases correspond reasonably well to the average determined from the learning data. In addition, the distribution of cases is similar between both sets. The correlation coefficient was only 0.54; nevertheless, the CART results are considered to be useful in identifying the relevant predictors.

NFIS was used with the CART predictors in an attempt to improve the results and to produce a continuous distribution of values. The CART-

NFIS results had a correlation coefficient of 0.60, as shown in the scatter plot in Figure 3. This was slightly better than the correlation coefficient obtained with CART (r = 0.54). There is a tendency for the CART-NFIS model to underpredict the higher and overpredict the lower values.

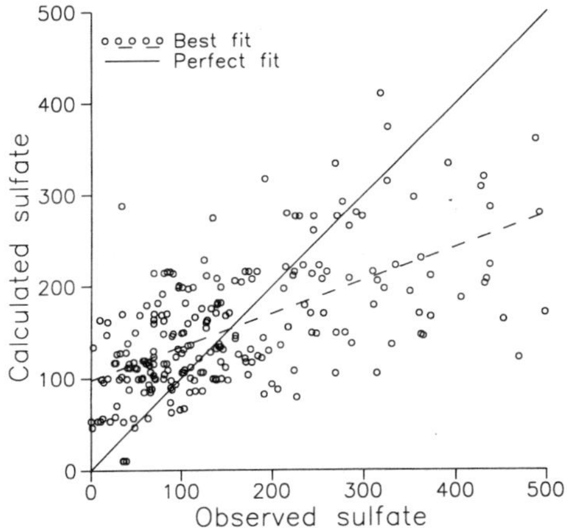

Figure 3. CART-NFIS prediction of wet sulfate concentration (μeq L^{-1}) in fogwater. Observed values ≥ 500 μeq L^{-1} have been excluded. N = 240; best-fit linear regression: y = 0.361x + 97.9 μeq L^{-1}; correlation coefficient, r = 0.60; rms error = 91.5 μeq L^{-1}. Source: Urquizo et al. (1998a).

SUMMARY AND CONCLUSIONS

The classic problem of accurate prediction of an infrequent event was encountered in this project. As shown in Walmsley et al. (1998), fog was observed by the OCD at the Roundtop CHEF site only about 15% of the time amongst 4093 hourly records in the spring, summer and autumn of 1988-91. (The frequency of complete hours of fog was only 10%.) Nevertheless, a methodology was developed that yielded what we consider are acceptable estimates of the LWC in high-elevation fog on Roundtop Mountain from routine meteorological measurements at the Sherbrooke weather station.

Despite some scatter in the result, the correlation coefficient between calculated and observed LWC values (0.81) is greatly superior to earlier prediction attempts (0.32 - 0.53) for the 1988 season made without the aid of CART or NFIS and even the secondary attempts (0.60 - 0.71) made with CART applications to the complete 1988-91 dataset. Likewise, the total accuracy improved from 0.88 to 0.92. When the "infrequent event" problem was fully appreciated, the idea of screening out the predictions of the "frequent event" (the non-occurrence of fog in this project) emerged. We stress that the CART predictions of no-fog, not the observations, were used to perform this screening.

We feel confident that the algorithms derived from the CART and NFIS runs are applicable to LWC prediction at the Roundtop CHEF site for April to September of years outside the 1988-91 period. We reiterate that OCD and LWC data are not needed for applying the CART and NFIS algorithms.

The methodology used in this study provided preliminary estimates of sulfate concentration in fog on high-elevation areas. The hybrid CART-NFIS application yielded a correlation coefficient of 0.60. Efforts to improve these results include searching for a method to screen out and predict the high-concentration cases and investigating the use of predicted LWC as an additional predictor of sulfate concentration.

We plan to examine what steps can be taken to improve the universality of the methods and thereby extend our prediction capability to other parts of southeastern Canada. One aspect could be to apply the methods to other elevations at Roundtop, though this may be difficult without observations for training or verification. A second stage could be to apply the same or similar algorithms at other sites for which verification data are available.

Acknowledgments

Thanks to Richard Tanabe for extracting the hourly observed LWC and FD data from the CHEF database and to Mohammed Wasey for information about the fog measuring device. Dr. Peter Summers drew our attention to the problem of predicting infrequent events. Prof. Douw Steyn asked two very relevant questions at the 22nd ITM, which have served to focus our discussion. The CART software is marketed by Salford Systems, 8880 Rio San Diego Drive, Suite 1045, San Diego, CA 92108. The NFIS software is marketed by The Math Works, Inc., 24 Prime Park Way, Natick, MA 01760-1500.

REFERENCES

Breiman, L., Friedman, J.H., Olshen, R.A., and Stone, C.J., 1984, *Classification and Regression Trees*, Wadsworth & Brooks, Monterey, CA.

Burrows, W.R., Montpetit, J., and Pudykiewicz, J., 1997, CANFIS: A non-linear regression procedure to produce dynamic-statistical air-quality forecast models, in: *Proc., 90th Annual Meeting & Exhibition, Air Waste Management Association, Toronto, Canada, 8-13 June 1997.*

Chiu, S., 1994, Fuzzy model identification based on cluster estimation, *J. Intelligent & Fuzzy Sys.* 2.

Schemenauer, R.S., Banic, C.M., and Urquizo, N., 1995, High elevation fog and precipitation chemistry in southern Quebec, Canada, *Atmos. Environ.* 29;2235.

Stanski, H.S., Wilson, L.J., and Burrows, W.R., 1990, *Survey of Common Verification Methods in Meteorology*, World Meteorological Organization, Geneva. [World Weather Watch Technical Report No. 8, WMO/TD No. 358.]

Steinberg, D. and Colla, P., 1995, *CART: A Supplementary Module for SYSDAT*, Salford Systems, San Diego, CA.

Urquizo, N., Walmsley, J.L., Brook, J.R., and Burrows, W.R., 1998a, The Use of Empirical Models in Estimating Sulfate and Nitrate Deposition from Fog in Mountainous Areas, in: *Proc., First International Conference on Fog and Fog Collection, Vancouver, Canada, 19-24 July 1998*, R.S. Schemenauer and H. Bridgman, eds., Conference on Fog and Fog Collection (Canada), North York, ON.

Urquizo, N., Walmsley, J.L., Burrows, W.R., Schemenauer, R.S., and Brook, J.R., 1998b, Application of the CART and NFIS statistical analyses to fogwater and the deposition of wet sulphate in mountainous terrain, in: *Air Pollution Modeling and Its Application*, **12**, S.-E. Gryning and N. Chaumerliac, eds., Plenum Press, New York. [In press.]

Walmsley, J.L., Burrows, W.R., and Schemenauer, R.S., 1998, The use of routine weather observations to calculate liquid water content in summertime high-elevation fog, *J. Appl. Meteorol.* [In press.]

DISCUSSION

The paper was presented by W. Gong.

M. GARDNER: It is my experience that CART cannot fully explore the nonlinearities in such applications. Have you tried to use artificial neural networks to model the relationship between meteorology and LWC or SO_4 directly? Artificial neural networks have greater flexibility and are capable of modelling very nonlinear systems, whereas CART is simply a data- partitioning process.

J. WALMSLEY: CART is not "simply a data-partitioning process". In doing the partitioning, CART explores the nonlinearities in the process and is capable of building models where Multiple Linear Regression (MLR) cannot. We did, in fact, use the Neural-Fuzzy Inference System (NFIS), as shown in the paper. Aside from the step in which we determined the no-fog events, CART was mainly used to select the predictors for NFIS.

I would like to take this opportunity to thank Dr. Wanmin Gong for presenting the paper when I was unable to attend due to illness.

MODEL VALIDATION, SCIENCE, AND APPLICATION

Peter Builtjes[1] and Andrea Flossmann[2]

[1] TNO-MEP, P.O. Box 342, 7300 AH Apeldoorn, The Netherlands
[2] Université Blaise Pascal, Lab. de Met. Phys. CNRS, 24 Avenue des Landais F-63177 Aubière Cedex, France

INTRODUCTION

Over the last years there is a growing interest to try to establish a proper validation of atmospheric chemistry-transport (ATC) models.
Model validation deals with the comparison of model results with experimental data, and in this way adresses both model uncertainty and uncertainty in, and adequacy of, the experimental data.

Under quality assurance recently a number of reports have been produced in which model validation is adressed in a more formal sense, EU-DG XII (1994), Royal Meteorologocal Society (1995) and Verein Deutscher Ingenieure (1995).
In the more scientific oriented European project EUROTRAC, model validation is especially adressed in the subprojects SATURN and GLOREAM; and model validation is also an important item in the EMEP program under the UN-ECE. For recent papers see Builtjes (1997, 1998) Mohnen (1997), Schlünzen (1997), Berge e.a. (1997).

In the following paragraphs first the difference between diagnostic and integrated model validation will be discussed, followed by an example in which science and validation is adressed, focussing on the processing of atmospheric constituents during a cloud passage. Th next paragraph adresses an example of integrated model validation, a model application study concerning the AOT 40.
Finally, conclusions and recommendations will be presented.

DIAGNOSTIC AND INTEGRATED MODEL VALIDATION

Two different types of model validation, or more modestly stated, model testing,can be distinguished.

Diagnostic model testing is focussed on testing the different processes which are described in ATC-models. Dedicated field studies are the proper way of testing and investigating the different processes incorporated in the overall ATC-models. Processes which can be tested are for example the transport- by means of a tracer experiment-, the

Air Pollution Modeling and Its Application XIII, edited by S.-E. Gryning
and E. Batchvarova, Kluwer Academic/Plenum Publishers, 2000

vertical exchange processes, the dry and wet deposition, the impact of clouds- both physically and chemically-, the photo-dissociation, the emission data, the chemical module, etc.
It should be noted that specific modules, like the chemistry modules can also be tested and developed in a completely controlled environment, in the laboratory, in this way excluding disturbing influences present always in field experiments.
Diagnostic model testing is science oriented, and focusses on the specific scientific question which is under investigation.

Integrated model testing is focussed on the testing of the overall process of atmospheric chemistry-transport. In general, integrated testing will make use of regular monitoring data, and will be less focussed on the use of results of dedicated field campaigns. Integrated testing includes sensitivity analysis and the use of adequate statistical measures. A crucial element is the determination of the spatial representativeness of the measuring data.
Integrated model testing is policy oriented, and focusses on the specific purpose of the model under investigation like episodic ozone, acidification, long term ozone, urban, continental, global scale, etc.

Obviously, diagnostic and integrated model testing have similar elements, and make sometimes use of the same experimental data. Nevertheless, the distinction is usefull in focussing on the different aspects of model validation/testing. (see also Builtjes, 1998, which is in part based on a model validation discussion which was held on the occassion of the 1998 EUROTRAC- symposium)

DIAGNOSTIC MODEL TESTING, science process modelling

Field campaigns are held to adress a specific scientific question, and in this way also to test existing modules which describe the process/question under consideration, or to develop new modules based on the results of the performed field experiment. However, incorporated in a proper analysis of the field experiments use is often made of specific models to try to obtain a coherent picture.
An example will here be given of a field experiment which adresses the problem of the impact of a cloud passage on the processing of atmospheric constituents.
During field campaigns a certain number of meteorological and chemical information is collected. Due to financial and logistic reasons the information is far from being complete.

Here, scientific process modelling is coming into play. The modelling follows several purposes. These purposes, as well as the problems intrinsic to this approach, will be discussed below using as an example the modelling effected for the third GCE (Ground Based Cloud) experiment of EUROTRAC which took place from 22 April to 13 May 1993 on the mountain Great Dun Fell in the north of England. The work is described in a special issue of Atmospheric Environment (see, e.g., Choularton et al, 1997).

The **first** purpose of scientific process modelling is to **complete** information, which could not have been measured. As a very current example we can name here the atmospheric wind fields, at the surface as well as in higher altitudes. These are essential if we want to understand where the air mass under investigation came from and where it went to after being sampled. Normally, during current field campaigns only few measurements of the wind field isolated in space and time are available. They could serve as verification to the wind fields that have been calculated by the model.
As a result of numerous of such modelling studies one fundamental problem regarding this approach has become evident. This is the problem of the initial conditions. To do a correct simulation the model requires **initialisation**. For the GCE experiment the model was initialised with the only complete sounding available, which was taken 90km upwind at the

eastern coastline of north England with a time lag of some hours to the experiment. Thus, during the simulations on the dynamical situation during the experiment (Wobrock et al, 1997) the exact meteorological conditions could not be reproduced. Instead, "only" the variety of possible meteorological conditions and resulting cloud fields could be studied.

A **second** aim of scientific process modelling is to study the processes in the centre of interest of the campaign. For the GCE experiment the goal was to study the processing of atmospheric constituents during a cloud passage. Several measurements were performed which were accompanied by a modelling effort. We will here only briefly address the work of Hallberg et al (1997) considering cloud microphysics. This paper tried to address a recognised problem in cloud modelling, that is, the incapability to reproduce the observed droplet spectrum which would yield the basis for all subsequent interpretation concerning the chemical processes. Several sensitivity studies concerning the mesoscale dynamics and numerous microphysical parameters in Hallberg et al (1997) have elucidated the variability of the modelling results. The conclusions, however, were far from satisfactory as even by varying the unknown parameters the modelled droplet spectrum only approached the measured one but never completely fitted. The closest results were obtained by modifying the residence time of the air volume in cloud, i.e. the dynamics.

These modelling studies, together with numerous others that have been performed until now, suggest two conclusions:

1. The meteorological condition crucially determines transport and is, thus, the prerequisite for all chemical and microphysical studies. If the model badly represents the dynamical situation it is useless to apply it to further investigations on the processes.
2. A model will never be able to reproduce exactly the measured situation due to the incomplete knowledge of initial and boundary conditions. The simulation will, thus, only yield a mean result. Only simulations, varying the unknowns, will allow to investigate the sensitivity of the results. Here, a development of statistical tools is required for the near future to determine the performance of the model and the representativity of its output.

INTEGRATED MODEL TESTING, application 'overall' modelling

Concerning tropospheric ozone, next to research focussing on its impact as greenhouse gas and for human health, attention is growing concerning its impact on the eco-system, forests and crops.

New guidelines have been introduced, the so-called AOT40c for the protection of crops defined as the accumulated exposure over threshold (AOT) with 40 ppb as threshold over may-july, daytime (defined as global radiation above 50 W/m2), with 3000 ppbh as the guideline. A similar guideline is made for forests, AOT40f of 10.000 ppbh over the period april-september. Both guidelines are exceeded over large parts of Europe.

ATC- models focussing on tropospheric ozone have been tested, in a reasonable complete way, concerning their capabilities in describing ozone concentrations, especially during episodes. It should be mentioned here that much more has still to be done in this area, Builtjes, 1998. A possibility might be to create a data-base test-set including only representative and 'validated' stations. The representativeness of stations might also be tested in a field campaign using many stations over a small area.

Concerning longer time averages, like the AOT40, hardly any model testing has been performed uptill now.

The following table gives preliminairy results for the EMEP one-layer trajectory model and the TNO LOTOS (Long term ozone simulation) 3-D eulerian grid model, showing a comparison between model results over the period may-july 1990 for the EMEP-station Kollumerwaard in the Netherlands.

Table 1. AOT-40 in ppbh, Vd is dry deposition velocity, z is height above the surface

Observation		7483	Ratio model/observation
EMEP version 1996	Vd= 0.5 cm/s	23736	3.3
EMEP version 1996	Vd= 0.8 cm/s	18846	2.5
EMEP version 1997	Vd= 0.8 cm/s	11660	1.6
LOTOS	z= 5 m	19564	2.6
LOTOS	z= 1 m	12277	1.6

The results in the table show first that there is a clear overprediction of the model results versus the measurements, a phenomena which also shows at other stations. Second, the model results are very sensitive to the description of the situation close to the surface. This shows in the EMEP results by changing the dry deposition for ozone, and in the LOTOS results with a change in reference height for the ozone concentration. This large sensitivity for AOT 40 to the situation close to the surface is a result of the strong vertical gradient in ozone close to surface and of the threshold value of 40 ppb. This value can be considered as a kind of ozone general background value, so it will happen frequently that the actual ozone level will be either slightly above or slightly below it, which has a large impact on the AOT40 values. Also, because AOT 40 is a dosage value, high ozone levels have also a large impact, they just count up.

This found sensitivity will almost certainly also hold for the measured AOT 40 levels. Consequently, in analysing measured and calculated AOT 40 values, it is of crucial importance to know the exact measuring height (for Kollumerwaard this height is 3.6 m, the standard height for the Dutch network).

In fig.1 and 2 a part of the time series of measured and calculated ozone levels at Kollumerwaard are shown. In general, and especially in fig 1, the agreement between (peak) measured and calculated ozone concentrations is 'reasonable'. However, the comparison for AOT40 is much worse, because the calculated values have a tendency to stay above the 40 ppb threshold, where the measurements often drop below that value. An other element shown in fig. 1 and 2 is the malfunctioning of the measuring equipment from about june 2/3 to june 5/6 (or even further). In case the AOT40 value is calculated automatically from the measured results, this will lead to a too low measured value of AOT40.

A further analysis shows that the calculated ozone values, especially the more averaged, back ground values are too high compared to the measurements in case of northerly winds. Because northerly winds come over the North Sea this might be an indication of a slightly too high (just over 40 ppb!) background ozone value for these summer months. Also the fact that the model sees the Kollumerwaard grid as for about 50 % belonging to the sea, with a consequently low deposition value, and relatively high ozone levels coming from this side, leads to rather high calculated ozone values. In reality the station is about 20 km. inland, which will cause a dry deposition of ozone and a lower ozone value. This subgrid scale, local effect has a large impact on measured and calculated AOT 40 values.

In case measured and calculated ozone values are compared, it is always of great importance to determine the spatial representativeness of the measuring station. In case of AOT 40, which in fact shows details in ozone behaviour much clearer than the ozone concentrations as such, the analysis of this representativeness, including the exact measuring height, is of utmost importance.

Consequently,model validation for AOT40 can only be done in a proper way in case for each station a detailed analysis of the measurements has been carried out including an analysis of the exact situation at the measuring site and its surroundings.

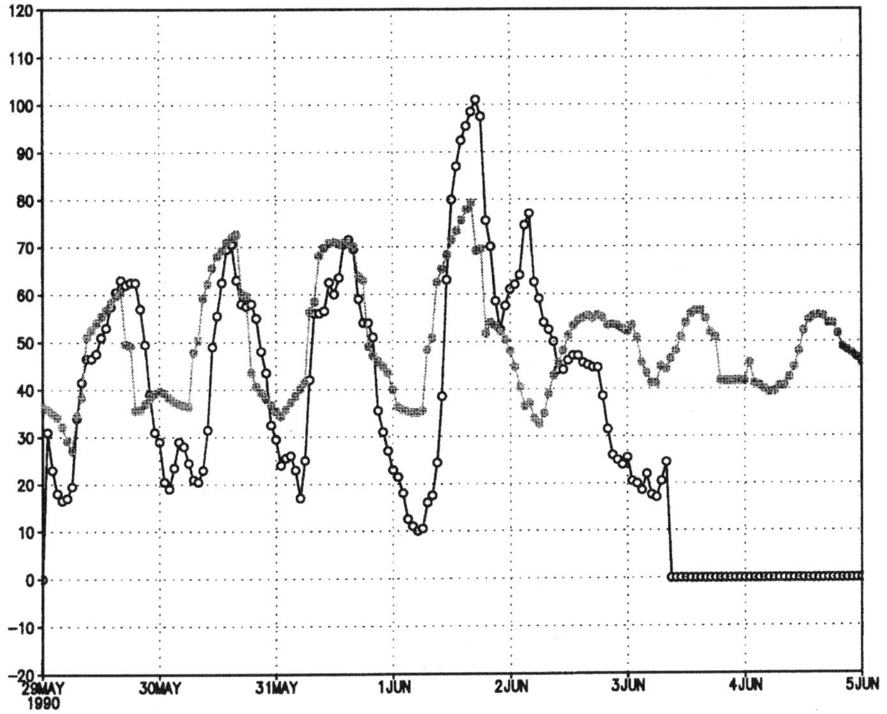

Figure 1. Measured (open circles) and calculated (closed circles) ozone levels at Kolummerwaard, May 29-june 5, 1990.

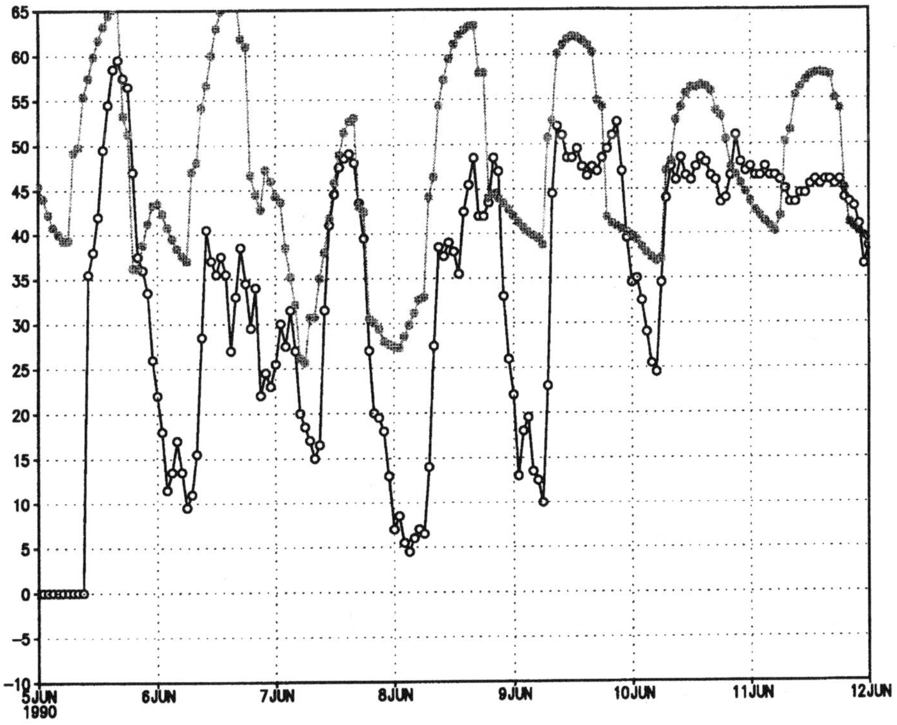

Figure 2. Measured (open circles) and calculated (closed circles) ozone levels at Kolummerwaard, June 5-12, 1990.

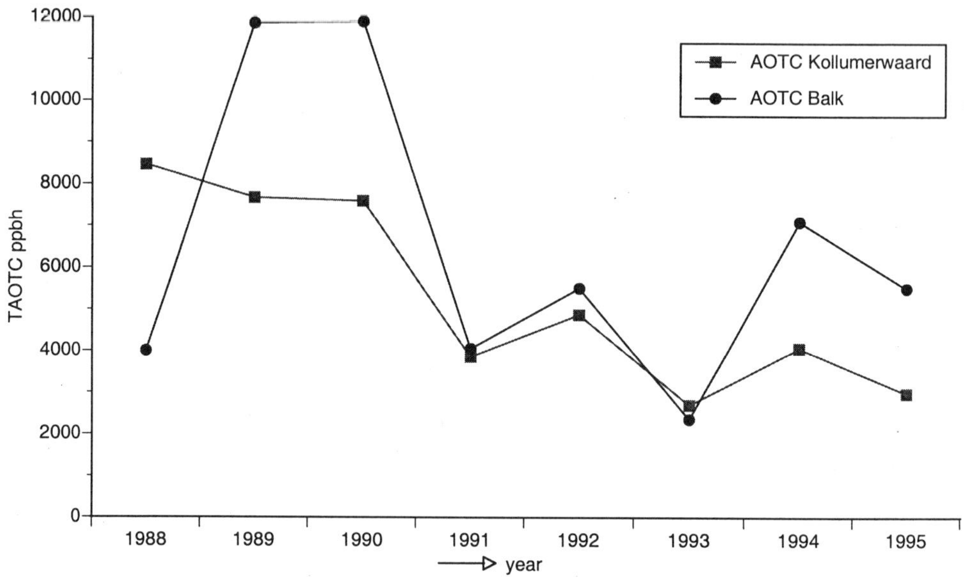

Figure 3. Measured AOT 40 values over 1988-1995 at 2 dutch stations.

Concerning AOT40, a final remark should be made concerning its interannual variability.

In fig 3 measured AOT 40 values are presented for Kolummerwaard and the station Balk, at 20 km. distance, for the years 1988-1995. This figure shows the large variation in AOT 40 over the years, which means that a proper evaluation of AOT40 should cover at least 5 years, and it shows the large spatial variability of measured AOT40 values, showing its limited spatial representativeness in this case.

CONCLUSIONS AND RECOMMENDATIONS

Model validation, model testing can be distinguished in diagnostic and integrated model testing. Diagnostic testing adresses processes and is science oriented, whereas integrated testing is directed to the overall modelling, and is policy oriented.

It is recommanded in model testing to make a distinction between diagnostic and integrated testing, because the demands for the two approaches of testing are to a large extent different.

Diagnostic model testing using field experiments adressing a specific scientific question can only be performed with the help of science process modelling

It is recommended that in field campaigns special attention is given to the modelling of the transport, the dynamics, including an analysis of the sensitivity of the results of the dynamics.

Integrated model testing should be performed by taking the purpose of the model into account.

It is recommendated that in integrated model testing the purpose of the model, either episodic, long term average, local or regional scale etc. should be clearly stated.

Integrated model testing for photo-oxidant models considering longer term averages like AOT 40 is still in its infancy.

It is recommended that photo-oxidant models should be tested for AOT 40 including a full analysis of the field measurements.

REFERENCES

E. Berge, H.A.Jacobson and D.Simpson, 1997, The requirements of data on atmospheric concentrations and depositions for the validation of regional air pollution models Background paper EMEP-WMO workshop Gothenburg

Builtjes, P.J.H., 1996, Modelling and verification of photo-oxidant formation. EUROTRAC Symposium 1996

Builtjes, P.J.H., 1998, Policy development requires verified models: an impossible task? EUROTRAC Symposium 1998

Choularton et al. 1997: The Great Dun Fell Cloud Experiment 1993: an overview. Atm. Env. 31, 2393-2405

EU-DG XII Model evaluation group, 1994 : Guideline for model developers and Model evaluation program, MTH-program, Brussels, Belgium

Hallberg et al. 1997 :Microphysics of clouds: model vs measuremets. Atm. Env 31, 2453-2463

Mohnen,V.A. 1996: Quality assurance in atmospheric measurements and asessments. EUROTRAC Symposium 1996

Royal meteorological Society, 1995: Atmospheric dispersion modelling. guidelines on the justification of choice and use of models, and the communication and reporting of results. RMS-UK

Schlünzen, K.H., 1997: On the validation of high-resolution atmospheric mesoscale models. J. of Wind Eng. and Ind. Aerd. 67&68, 497-492

Verein Deutscher Ingenieure, 1995: Qualitätsschicherung von VDI-Richtlinien. VDI-Kommision Reinhaltung der Luft

Wobrock et al, 1997: Modelling of air flow and cloud field over the northern Pennines. Atm. Env. 31, 2393-2406

DISCUSSION

Z. ZLATEV: You mentioned a lot of issues which should be taken into account in the attempts to validate the model results. An important question is where to start. You mentioned the uncertainties in the input data used in the models (emissions inventories, meteorological data, etc). Is it not a good idea first to try to evaluate these uncertainties?

P. BUILTJES: I fully agree with you. The results of for example the ETEX-experiment learned us a lot about the uncertainties associated with especially the near source dispersion and transport process. However the determination of the accuracy of emission data is more difficult and could involve models themselves in performing inverse modelling and data-assimilation.

E. GENIKHOVICH: The best way to get reasonable answers is to ask reasonable questions. In this case, it means to use reasonable metrics. What metrics should be used when validating the models for prediction of AOT40?

P. BUILTJES: In the first place, we should know what the actual measurement means in terms of spatial representativeness, and height above the surface. Concerning the best statistical measures, a nice overview can be obtained from Elisa Canepa, University of Genova, present at this meeting.

S. RAFAILIDIS: One aspect missing from the work of the scientific community in relation to the validation of numerical models is to follow rigorous quality assurance procedures. Do you think that industrial type quality assurance procedures could be adopted to help solve the uncertainty in the model validation process?

P. BUILTJES: In principle yes, although QA/QC of scientific models is quite different from QA/QC procedures for building aeroplanes. However, it would be useful to define before testing the model the criteria which the model should be able to pass.

COMPARISON OF MODEL RESULTS AND MEASUREMENT DATA ON LONG-RANGE TRANSPORT OF Pb AND Cd POLLUTION

M.Pekar and A.Gusev

Meteorological Synthesizing Centre-East
Str. Kedrova 8, k.1, 117292, Moscow, Russia

1 INTRODUCTION

The EMEP programme is aimed at the investigation and evaluation of transboundary transport of pollutants between European countries. At present in addition to acidifying substances, the programme involves heavy metals in particular lead (Pb) and cadmium (Cd).

Pb and Cd are the most toxic heavy metals in the environment. They are emitted to the atmosphere from industrial activity mainly as fine aerosols and are transported to long distances. Pb and Cd pollution levels were estimated by several European models Baart et al [1], Bartnicki et al. [2]. Mean annual concentrations in air and precipitation were compared with measurement values of the PARCOM/ATMOS network. The verification results demonstrated that models represented relatively accurately (~50%) measured concentrations in air and precipitation for Pb but the difference of model and measured data on Cd was as much as factor 2-3. It was presumed that the main reason for this discrepancy lay in rather uncertain emission data which total value is evaluated within a factor of 2-3 Baart et al.[1]. For western European countries emission uncertainty limits are much less and amount to 20-50% Baart et al [1]. As well it should be mentioned that measurement data are statistically insufficient (8-12 coastal stations) for monitoring of Pb and Cd within Europe and they cannot assure good validation of models.

2 SHORT DESCRIPTION OF THE ASIMD MODEL

The 3D Eulerian model ASIMD has been developed for providing calculations of the deposition and concentration of heavy metals within European scale. Description of the model version with 150-km resolution and results of its application to Pb and Cd on a year by basis are presented in report Pekar [4]. At present the ASIMD version with 50-km resolution and a total of 117x111 grid cells is developed and used for multiannual calculations. In vertical direction nonequdistant grid with four layers is taken. The surface layer is separated. The depths of the following layers are chosen with orientation to the

Air Pollution Modeling and Its Application XIII, edited by S.-E. Gryning
and E. Batchvarova, Kluwer Academic/Plenum Publishers, 2000

levels of meteorological data, 1000, 925, 850 mb. The top of the domain is equal to 2.1 km. Transport, vertical diffusion, dry and wet deposition are solved separately.

Finite - difference approximation of transport equation is realised within the framework of classical "upwind scheme" for which numerical diffusion is well known. Compensation of numerical diffusion is made by correction of advection velocity depending on local gradients. The resulting velocity and real velocity are connected by non-linear equation. For solution of vertical diffusion equation we use the variant of a simple scheme with conservation of 3 moments developed for non-uniform grid, Pekar [5]. A system of parameters required for the description of local conditions of pollution dispersion consists of friction velocity u_*, Monin-Obukhov length scale L, mixing layer height h, vertical diffusion coefficient profiles Kz(z) which were devised on the bases of data on 1000 mb wind and temperature, roughness z0.

The meteorological pre-processor is based on the method of energetic balance evaluation with further application of results of similarity theory, van Ulden and Holtslag [6]. The parameter calculations were made for each 6-hour interval.

A flux of aerosol dry deposition carrying heavy metals is defined by expression:

$$F\left(x,y,z=z_0,\,t\right)=V_d\left(x,y,t\right)\;C(x,y,z_1,t) \tag{1}$$

where z1 - the first calculation level along the vertical, z1 = 50 m, Vd - variable over space and time the deposition velocity.

According to Milford and Davidson [7] mass median diameters (MMD) for Pb and Cd equal to 0.55 μm and 0.84 μm respectively. The parameterisation of dry deposition velocity on a dry surface was made on the basis of Sehmel`s results [8] the following approximations was accepted:

$$V_d^{land}(Pb)=\left(0.02\;u_*^2+0.01\right)\;\left(z_0/10^{-3}\right)^{0.33} \tag{2}$$

$$V_d^{land}(Cd)=\left(0.04\;u_*^2+0.02\right)\;\left(z_0/10^{-3}\right)^{0.30} \tag{3}$$

where: V_d^{land} [cm/s], u_* [m/s], z_0 [m].

Results obtained with the model of Lindfors et al [9] were used for the parameterisation of deposition on the sea surface.

$$V_d^{sea}(Pb)=0.15\;u_*^2+0.013 \tag{4}$$

$$V_d^{sea}(Cd)=0.15\;u_*^2+0.023 \tag{5}$$

where: V_d^{sea} [cm/s], u_* [m/s].

Sink of pollutants due to precipitation scavenging is represented by a linear process (6). The model does not consider the complicated nature of the phenomena.

$$\frac{\partial C}{\partial t}=-\Lambda C,\;\;\Lambda=\frac{WI}{h} \tag{6}$$

where Λ - washout coefficient depends on many parameters of both pollutants and precipitation, *I* - precipitation intensity, *h* - scavenging layer, *W* - scavenging ratio equal to that of concentration in precipitation to concentration in the air. Orders of magnitudes of *W* for heavy metal particles are $\sim 10^5$ that testifies to the effective scavenging. It is set equal to 500000 for Pb and Cd.

3 EMISSION DATA FOR PB AND CD

For modelling two emission data sets were used:

- estimates of ESQUAD project van den Hout (ed.) [10], containing maximum, mean and minimum Pb and Cd emission values. Provisional calculations and a comparison with measurements show that the application of maximum emission values gives better results. Total European emissions of Pb and Cd are 49 kt/year and 1634 t/year correspondingly. The calculations cover 1990-95 during which the same emission was taken.
- estimates of UBA project Berdowski [11], containing emission data on Cd and Pb prepared at TNO. In this case Pb total emission was estimated to be 38 kt/year and Cd total emission - 562 t/year. These fields were also used in calculations for the same period (1990-95) and the results were compared with observed values.

Following the model vertical structure the emission was distributed with two lower layers in proportion: 0.8:0.2 for Pb and 0.3:0.7 for Cd correspondingly with lower 100-m layer and with following 300-m layer. This distribution is conditioned by the fact that Pb emission is mainly resulted from road transport whereas Cd is emitted from high point sources.

4 THE COMPARISON BETWEEN MODEL CALCULATIONS AND MEASUREMENTS

The agreement of calculations and measurement data is determined by the quality of model, adequacy of emission distribution, measurement data quality. Model validation results using PARCOM-ATMOS measurements for 1990 Pekar [4] demonstrated good consistency of concentrations in air for Pb (~50%) and a satisfactory agreement for Cd (~2 times). It is obvious that this validation was limited because of poor statistics. The prolonged period (1990-95) of modelling of Pb and Cd behaviour was intended to provide sufficient statistical data for the comparison of calculations versus measurement data.

For the comparison with calculation results two measurement data set were taken:

- measurement data of EMEP network, Berg et al [3],
- measurement data of PARCOM-ATMOS network.

The comparison was made with measurement data of both sets and separately with PARCOM-ATMOS network data on the assumption that emission and measurements are of relatively high reliability for this region.

4.1 COMPARISON OF CALCULATIONS WITH MEASUREMENT DATA FOR PB

Measurement data of EMEP network along with data of PARCOM/ATMOS network were compared with the results of two types of calculations:

- using emissions according to ESQUAD project (maximum estimates),
- using emissions according to UBA project.

As it follows from Table 1 an overestimation is observed in the first case (PARCOM/ATMOS), and underestimation - in the second case (PARCOM/ATMOS+EMEP). Measurements and calculation results correlate better in case of PARCOM/ATMOS data sample. Air concentrations show considerable model overestimation for the whole measurement data set and rather good consistency of mean values and relatively high correlation for PARCOM/ATMOS data. Calculation results obtained with the UBA project emission show better agreement with measurement data and higher correlation coefficient. At the same time for PARCOM/ATMOS measurements model underestimation of both wet deposition and air concentrations is observed.

Table 1. Results of comparison of calculations with measurement data for lead

Lead	Number of Sites used	Mean Observed	Mean Calculated	Correlation
Wet deposition (mg m^{-2} a^{-1}) (ESQUAD emission)				
1*	33	3.61	3.29	0.63
2**	176	2.19	2.57	0.51
Wet deposition (mg m^{-2} a^{-1}) (UBA emission)				
1	33	3.61	2.46	0.76
2	176	2.19	1.75	0.60
Air concentration (ng m^{-3}) (ESQUAD emission)				
1	22	26.75	25.42	0.68
2	77	22.02	34.99	0.33
Air concentration (ng m^{-3}) (UBA emission)				
1	22	26.75	21.82	0.87
2	77	22.02	25.75	0.65

* - PARCOM/ATMOS data sample

** - PARCOM/ATMOS + EMEP data sample

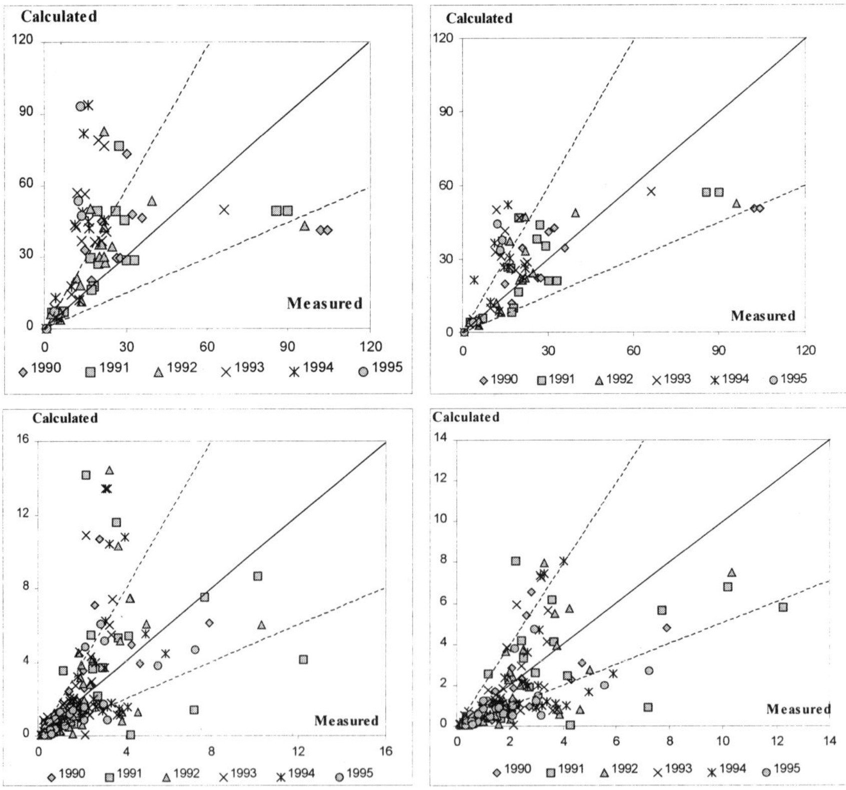

Figure 1. Comparison of calculated air concentrations (upper) and wet deposition (lower) of Pb with measurement data. Calculations were done with ESQUAD project emission (left) and UBA project emission (right). Units for air concentrations: ng m^{-3}, for wet deposition: mg m^{-2}.

4.2 COMPARISON OF CALCULATIONS WITH MEASUREMENT DATA FOR CD

Similar analysis was made for Cd. Comparing measurement data and calculations with ESQUAD project emission an overestimation as much as 2-3 times is observed in case of Cd air concentrations for both sets of measurement data. Calculations with UBA project emission demonstrate significant model underestimation, at the same time the correlation is higher. It may be supposed that the more optimum emission is closer to the ESQUAD project estimates.

Table 2. Results of comparison of calculations with measurement data for Cd

Cadmium	Number of Sites used	Mean Observed	Mean Calculated	Correlation
Wet deposition ($\mu g\ m^{-2}\ a^{-1}$) (ESQUAD emission)				
1*	33	114.83	95.92	0.63
2**	167	74.88	84.97	0.54
Wet deposition ($\mu g\ m^{-2}\ a^{-1}$) (UBA emission)				
1	33	114.83	23.66	0.72
2	167	74.88	18.14	0.62
Air concentration ($ng\ m^{-3}$) (ESQUAD emission)				
1	15	0.43	0.63	0.43
2	55	0.41	0.97	0.34
Air concentration ($ng\ m^{-3}$) (UBA emission)				
1	15	0.43	0.14	0.74
2	55	0.41	0.34	0.50

* - PARCOM/ATMOS data sample

** - PARCOM/ATMOS + EMEP data sample

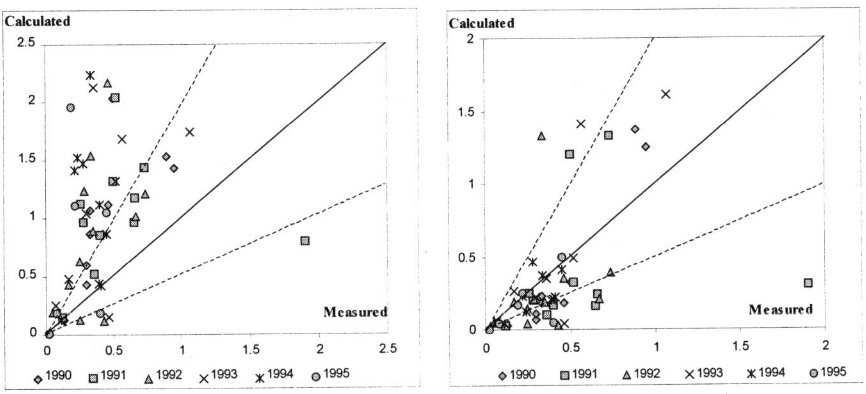

Figure 2. Comparison of calculated air concentrations of Cd with measurement data using ESQUAD project emission (left) and UBA project emission (right). Units: ng m^{-3}.

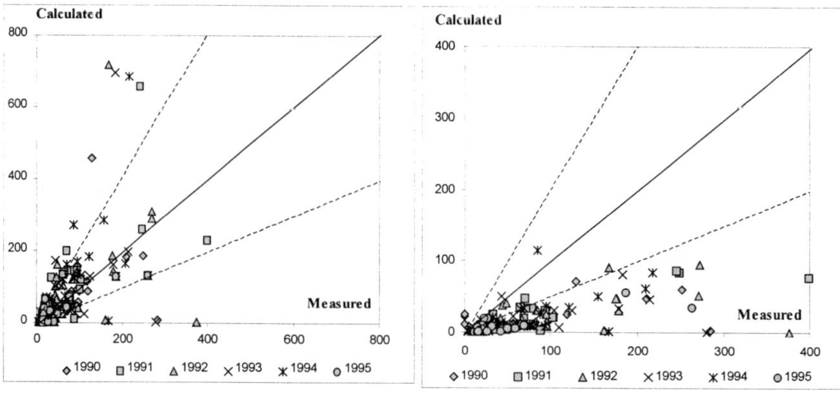

Figure 3. Comparison of calculated air concentrations of Cd with measurement data using ESQUAD project emission (left) and UBA project emission (right). Units: μg m^{-2}.

5 CONCLUSIONS

1. The comparison of measurement data and calculations for Pb using ESQUAD project emission demonstrates the coincidence of mean values within the accuracy of ~15% for wet deposition and ~40% for air concentrations. Similar comparison of measurement data and calculations with the UBA project emission (the total emission is about 25% lower than the ESQUAD project emission) shows considerably better results especially in terms of correlation coefficients. The result of the ASIMD model validation using a considerably large sample of observational data on Pb is in the fact that UBA emission for 1990-95 on the average is the best input for the model.

2. The comparison of measurement data and calculations for Cd using ESQUAD project emission demonstrates a good agreement for both sets of measurement data being used. In case of UBA project emission for Cd (the total emission is about 3 times lower than the ESQUAD project emission) the comparison shows an underestimation of wet deposition and air concentrations. Model results with ESQUAD emission are in a better agreement with measurements. We believe that real emission of Cd is somewhere in-between of estimates provided by the ESQUAD project and the UBA project but closer to the ESQUAD project.

REFERENCES

1. Baart A.C., Berdowski J.J.M. and van Jaarsveld J.A. *Calculation of atmospheric deposition of contaminants on the North Sea* (ed. J.H.Duyzer). TNO-MEP-R95/138, Delft, The Netherlands, 1995.
2. Bartnicki J., Madzelewski H., Pacyna J., Olendrzynski K. *Atmospheric transport and deposition of heavy metals within Europe. Comparison of model result and observation* (Pres. at 10th Int. Conf. on HM in the Environment, Hamburg, Germany, 18-22 Sept), 1995.
3. Berg T., Hjellbrekke A.-G, Skjelmoen J.E. *Heavy metals and POPs within the ECE region.* EMEP/CCC-Report 8/96, NILU, 1996
4. Pekar M. *Regional models LPMOD and ASIMD. Algorithms, parameterisation and results of application to Pb and Cd in Europe scale for 1990.* EMEP/MSC-E Report 9/96, 1996.
5. Pekar M. *Large particle model and its application to sulphur compounds within European scale, in Air Pollution IV* (ed.B.Caussade, H.Power, C.A.Brebbia) pp. 424-434, Computational Mechanics Publications, 1996.

6. van Ulden A.P. and Holtslag A.A.M. *Estimation of atmospheric boundary layer parameters for diffusion applications.* J.Climate Appl.Meteor., 1985, 24, 1196-1207.
7. Milford J.B. and Davidson C.I. *The sizes of particulate trace elements in the atmosphere - a review.* Journal of the Air Pollution Control Association, 1985, 35, *12*, 1249-1260.
8. Sehmel G.A. *Particle and gas deposition: a review.* Atm.Env., 1980, 14, 983-1011.
9. Lindfors V., Joffre S.M. and Damski J. *Determination of the wet and dry deposition of sulphur and nitrogen compounds over the Baltic sea using actual meteorological data.* Finnish Meteorological Institute Contributions N 4, Helsinki, 1991.
10. van den Hout, ed. *The impact of atmospheric deposition of non-acidifying European forestsoils and the North sea,* Main report of the ESQUAD project, RIVM rep.722401003, IMW-TNO rep.R93/329, 1994.
11. Berdowski J.J.M., J. Baas, J.P.J. Bloos, A.J.H. Visschedijk and P.Y.J. Zandveld [1997] *The European Emission Inventory of Heavy Metals and Persistent Organic Pollutants for 1990.* TNO Institute of Environmental Sciences, Energy Research and Process Innovation, Apeldoorn, UBA-FB report 104 02 672/03, p.239

AEROSOLS AT AIR/WATER/LAND INTERFACES: MODELLING AND MEASUREMENTS

Bo L.B. Wiman

Department of Environmental & Energy Systems Studies
Lund University, Institute of Technology
Gerdagatan 13, SE-223 62 Lund, Sweden

INTRODUCTION

Anthropogenic as well as natural aerosol-borne, ecologically active substances are transported and deposited in complex landscapes, such as characterized by forest/open-field, lake/forest, and sea/shore transitions (cf., Wiman et al., 1990). Aerosol mass exchange between the troposphere and natural waters has two important aspects: mass-flow towards (deposition) and from (bubble-bursting) the surface. However, despite substantial progress over the recent two decades uncertainties pertaining to the physico-chemical mechanisms and quantities defining source and, in particular, sink strengths, remain large (e.g., Wiman 1998). Much of the uncertainty is due to the limitations to current theory, to the methodological difficulties involved in field experiments, and to basic problems with generalising laboratory (such as wind-wave tunnel) results. The need for theoretical developments resides mainly in adequate handling of the strongly variable and coupled aerodynamics and hydrodynamics at natural water surfaces, and the resulting effects upon particle size distributions (and thus fluid dynamics behaviour) of aerosols. The methodological/empirical problems originate mainly in the fact that the system is one of multiple phases (ambient gases, hygroscopic as well as non-hygroscopic particles, air film interfacing with water film, in turn interacting with bulk water), wherein rapid processes must be studied, often under harsh experimental conditions implying a compromise between robustness and precision of measuring equipment.

In view of the anthropogenic impacts on the atmosphere, as well as from biogeochemical aspects, this uncertainty is particularly unsatisfactory. Long-range transports from industrialized regions (e.g., Lindfors et al., 1992; Rendell et al., 1993) and from regions subject to wind erosion (e.g., Dulac et al., 1992) often occur across open water. Seaward transports of biogenic organic aerosols (e.g. lipid fragments from terrestrial plants) have also been demonstrated (e.g., Andrews et al., 1997). Paerl (1985; cf. also Rendell et al., 1993) demonstrate that wet and dry deposition can induce long-term effects

Air Pollution Modeling and Its Application XIII, edited by S.-E. Gryning
and E. Batchvarova, Kluwer Academic/Plenum Publishers, 2000

on marine phytoplankton productivity. The case of radioactive aerosols (e.g., Volchok et al., 1970; Walker et al., 1986; Persson et al., 1987) warrants special needs for calculations of particle deposition to water surfaces (such as the Black Sea, and the Baltic). Matschullat (1997) exemplifies how fundamental uncertainties in theory and methods result in needs to re-assess estimates of atmospheric inputs to the sea, bearing upon the possibilities to assess biological effects. Similar to the case for terrestrial systems particle deposition to water bodies may seem negligible relative to reservoirs within the sink (soil, forest biomass, bulk water). However, such a perspective is misleading because of the biological activity and the exchange processes taking place in the often very thin strata interfacing with the atmosphere (water surface film, top soil layer, leaf surfaces, etc.). Hardy et al. (1985) exemplify the importance of coupling deposition rates of particle-borne trace metals to residence times and often strong enrichment of such components within the surface film. Further, the uncertainty with respect to the particle sink efficiency of the oceans, or of complex landscapes, contributes to the uncertainty in calculations of aerosol residence times in the atmosphere, and thus to a poor understanding of biogeochemical cycles and the effects of aerosols on climate (IPCC, 1996). This points to the needs for refining aerosol deposition submodels used in transboundary air pollution modelling (cf., Syrakov et al., 1998).

In the following, examples from empirical studies of aerosols at a lake site and a coastal site are introduced to illustrate some of the questions that modelling has to address; a generalized modelling framework is then introduced; some modelling applications are shown; and further modelling and measurement needs are discussed. Note that model validation is not the issue here; instead, the framework is introduced only with the objective of discussing empirical examples and ways towards stringent modelling-validation procedures.

EXEMPLIFYING FIELD DATA

Based on earlier work on aerosol/canopy interactions (Wiman et al., 1990), a series of field studies and experiments is ongoing at lake and coastal sites in southeast Sweden and, recently, at Black Sea coastal-zone sites (Wiman et al., 1998). Among the objectives are to gain experience with respect to the usefulness and operationality of various measuring systems, and to obtain information on spatially as well as temporally stratified aerosol concentrations and their relations to micrometeorological factors and surface characteristics. Some results for vertical profiles will be introduced below. Measuring equipment includes: a 2-channel Laser Particle Counter (LPC, giving particle number concentrations for optical particle sizes 0.5 to 5 μm, and ≥5 μm); an 8-channel Optical Particle Counter (OPC, providing particle number concentrations for size classes 0.3 to 0.5, 0.5 to 0.7, 0.7 to 1, 1 to 3, 3 to 5, 5 to 7, 7 to 10, and ≥10 μm); Nuclepore 2-stage filter samplers (NPS) equipped with low-volume pumps and filters arranged to make the prefilter capture particles essentially in the coarse particle mode (1st stage filter) and the accumulation mode (2nd stage filter) (cf. Whitby, 1978) with filters taken to Proton Induced X-ray Analysis (PIXE) after exposure; a small, portable Condensation Nuclei Counter (CNC, giving number concentration data for particles ≥0.02 μm); and various micrometeorological sensors, dataloggers and field computers. Among several series of experiments are the following.

(a) Vertical gradient measurements over a lake surface, with the OPC sequentially sampling aerosol particles ≥0.3 μm at two levels (5 and 0.85 m) above the water surface. Concentration differences between these levels are exemplifed in Fig 1. Filter concentration data for S (a typically accumulation-mode oriented aerosol-borne compound), K (typically associated with both accumulation-mode and coarse-mode particles), and Ca (typically coarse-particle oriented) for the same period showed 0±15% (within one standard deviation) decline from 5 to 0.85 m (S), a decline by 11±16% (K), and a decline by 28±14% (Ca). During the period, average day-time (night-time) profiles for relative humidity, rh, showed a decline from close to 100% at the surface to between 70 and 60% (95 and 90%) at the 0.85 and 1 m levels. Day-time temperature stratification was essentially neutral, while night-time temperatures showed a strongly unstable stratification over the first 1.5 m above the surface, thereafter tending towards stable stratification.

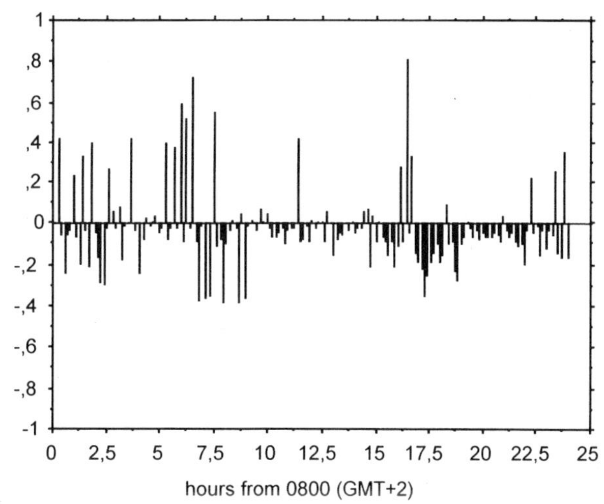

Figure 1. Normalized concentration differences {[c(0.85)-c(5)]/c(5)} between levels 0.85 and 5 m above a lake surface, for particles ≥0.3 μm.

(b) Measuring efforts using a small fishing vessel anchored half a nautic mile from the sea shore, in winds around 2-3 ms^{-1} varying between slightly on-shore and slightly off-shore), with on-board use of an LPC, various NPS combinations, and the CNC. Simultaneously, OPCs plus NPS systems were run on the shore, and at an inland site 10 km north-west of the vessel's position. Concentration profiles obtained with the LPC for particles ≥0.5 and ≥5 μm, respectively, showed a slight concentration decline (by 4±3%) from 2 to 0.5 m for particles in the 0.5-5 μm class, and a tendency (2±6%) towards concentration increase towards the surface for particle ≥5 μm. Fig. 2, which will provide a basis for modelling discucssions below, shows the particle number-vs-size distribution obtained from several hours of sampling with the OPC and the CNC, where CNC data are taken to represent particle size class 0.02 to 0.3 μm.

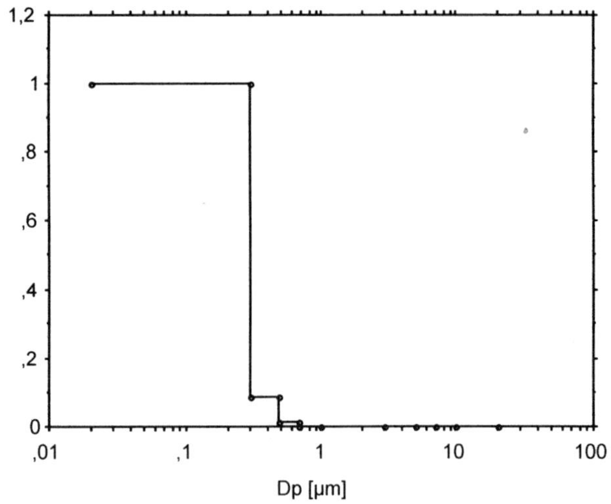

Figure 2. Particle number concentration per $\Delta \ln(Dp)$ (normalized) measured for a coastal zone in SE Sweden. Field data agree well with bi-modal (accumulation and coarse particle mode) log-normal distributions suggested by Whitby (1978).

(c) Several on-shore measurements with a telescopic mast arranged to lean outwards above the water, with LPC, OPC, CNC samplers/inlets and sensors for micrometeorology placed at different levels above the water surface. Fig. 3 portrays a period representative of several other experiments where concentrations of particles ≥0.02 µm were measured at the waterfront at positions 2 and 0.1 m above the surface Wind conditions associated with the data in Fig. 3 changed from gusty along-shore winds and rh-values around 80% to low, stable on-shore winds with rh-values increasing to around 90% or higher. Other on-shore measurements, with both the CNC and LPC, support the picture of downward increasing concentrations of particles in classes from 0.02 µm and larger in strong on-shore, or gusty, winds (with breaking waves), and downward declining concentrations for particles in the 0.02-0.5 µm class in conditions with low winds and generally high rh-values.

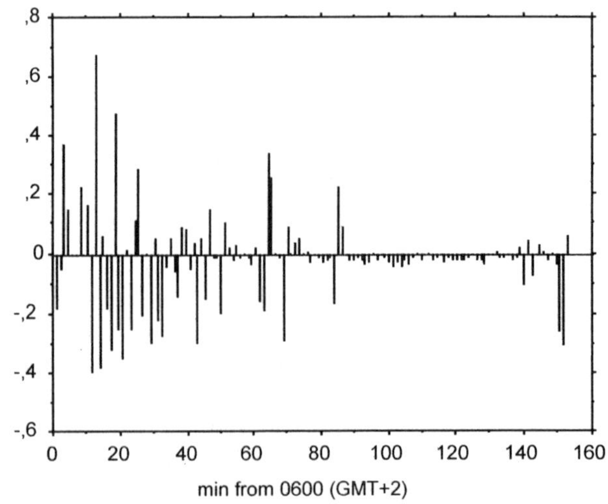

Figure 3. Normalized concentration differences {[c(0.1)-c(2)]/c(2)} between levels 0.1 and 2 m at the sea-shore waterfront, for particles ≥0.02 µm.

In summary, these and other results that cannot be accomodated here suggest that significant aerosol particle depletion as well as enrichment can occur over natural-water surfaces; mechanisms behind this depletion/enrichment need further study for clarification since they involve several sink processes at the water surfaces (e.g., Williams, 1982; Hummelshøj et al., 1992) and also upward Stefan flow (e.g., Slinn et al., 1978) "shielding" the surface from downward particle transfer, as well as source mechanisms (e.g., Blanchard and Woodcock, 1957; Hummelshøj, 1992) in the microlayer. An approach towards investigating field data from some theory and modelling perspectives is introduced below.

THEORY AND MODELLING FRAMEWORK

Depending on whether the analytical approach to modelling aerosol/water surface exchange emanates from aerosol physics (e.g., Fuchs, 1964; Davies, 1966), fluid dynamics engineering (e.g., Blevins, 1984), or meteorology-based atmosphere/surface interaction research (e.g., Krauss, 1972; Slinn and Slinn, 1980; Williams, 1982), a variety of similar, but quantitatively somewhat differing, descriptions obtains for the physical and chemical milieu of the aerosol/water sink and source processes. Detailed discussions of the various air strata close to water surfaces are beyond the scope of this paper. It should be sufficient to note that for the basic transfer mechanisms a simplified description includes: (i), a turbulent layer (some 10 m and upwards above the surface, and determined by gradients in temperature, wind speed, wave height, relative humidity); (ii), an intermediate stratum where turbulence effects decline, developing into (iii) a thin laminary interface layer which encloses the water surface (at low wind speeds) or can be partly disrupted (high wind speeds); (iv), a thin water surface film (a biologically active and sensitive micro-surface layer); and (v), the bulk water. Although a multitude of mechanisms are involved, the focus of the modelling discussions here is on the particle-size distribution of hygroscopic aerosols; this distribution can be expected to change substantially during transfer, thus subjecting the aerosol to changes in the relative importance of deposition mechanisms (*i.a.*, diffusion, direct interception, impaction, gravitational settling). In land/water transitions, such as coastal zones, additional effects are in operation due to changes in aerodynamic roughness, and land/sea air mass circulation.

The stability characteristics of layer (i) vary strongly, seasonally and diurnally and it controls the first stage of the downward transfer of particles with low or negligible gravitational settling velocity (i.e. primarily sub-μm particles, but also influencing coarse mode particles up to a few μm in size). With increasing particle size, sedimentation velocity increases leading to a different physical problem. For hygroscopic particles relative humidity is therefore a non-negligible factor even in this layer, and of increasing importance in the intermediate layer (ii) (in which, over sea surfaces, rh typically takes values around 80-85%). Decreasing tubulent transport in layer (ii) means that for sub-μm particles effects of Brownian diffusion increases in importance as transfer mechanism. A fairly strong gradient in relative humidity can be expected to begin in this layer. At low wind speeds (minor waves only) few transfer mechanisms remain in the final atmospheric layer (iii): diffusion (sub-μm particles) and gravitation (super-μm particles) are the essential ones, unless special phenomena are at hand (such as Stefan flow, thermophoresis, photophoresis, electrical fields; none of these phenomena are well understood in natural deposition environments, however). In this layer rh can be expected to be close to 100% (the lesser the salinity of the water, the closer to saturation; for the viscous layer close to a sea surface relative humidity theoretically would be constrained by Raolt's law to about 98.3 %, although studies by Jenkin (1984) suggest that, in practice, rh-values closer to

100% could occur). The particle size distribution of hygroscopic aerosols transferred to this layer (and also from it) can then be expected to 'slide' and 'deform' substantially along the particle size axis. The time needed for particle growth to equilibrium size is small (cf., e.g., Fairall, 1984); around 10^{-1} s for a 10 μm particle and 10^{-4} s for a 0.1 μm particle.

Among essential problems in describing these transfer processes are the extensions and characteristics of these layers under a variety of wind and wave conditions. For a simplified approach it is helpful to observe that Jenkin (1984) (cf. also Slinn et al., 1978; Blevins, 1984), gives the layer of particular interest for aerosol deposition analyses (the "deposition layer") as about $100\nu/u_*$ (where ν is the kinematic viscosity of air; u_* is friction velocity), so that if u_* is around 12 cms^{-1} (corresponding to about 3 ms^{-1} at 10 m above sea level; cf. Krauss, 1972; Slinn et al., 1978) this gives a "deposition layer" thickness on the order of 1 cm, and some 60% of the change in saturation ratio could occur across this layer. It might be noted that Stefan flow might occur, potentially as rapid as 0.2 cm s^{-1} (Slinn et al., 1978). In order to take some of the above aspects into consideration, the modelling framework used by Wiman and Ågren (1985; cf. also Wiman, 1988) can be used as a basis (cf. Appendix, Eq. 1). In order to exemplify some effects of rh gradients it will be sufficient to investigate steady-state conditions along the vertical coordinate z. This leads to the following version of Eq. 1, where the K_{zz} and D_{zz} terms are henceforth denoted K and D, respectively:

$$(\partial/\partial z)\ \{[K(z)+D(Dp)]\ [\partial c(Dp,z)/\partial z]\} + (\partial/\partial z)[v_S(Dp)c(Dp,z)] - \mathbf{S} + \mathbf{Q} = 0 \qquad (2)$$

One approach to studying the effects of relative humidity would be to incorporate these effects in the **S** and **Q** terms. Another manner is suggested by Piazzola and Despiau (1997), who assume **S**=**Q** and then solve the remaining part of the differential equation, whereafter rh-effects are considered through multiplying the resulting concentration by a function relating particle growth to saturation ratio (relative humidity). However, for situations where field measurements are involved with instruments that measure in fixed particle-size classes, a more practical approach is to perceive the aerosol as occurring in separate classes, and focus on the flux of particles along not only the z-coordinate but also the particle size axis. For the purposes here I assume **S**=**Q**, and also assume balance between formation and removal, so that the following equation obtains:

$$[K(z)+D(Dp)]\ [\partial c(Dp,z)/\partial z] + [v_S(Dp)c(Dp,z)] = 0$$

Following Fairall (1984), Dp (in terms of size equilibrated with ambient rh) can be modelled as a function of z through the use of $Dp = Dp_{dry}\ G(S(z))$, where Dp_{dry} is dry particle diameter (i.e., particle diameter as it would be at saturation ratio S=0). Assuming that particles grow/decline to equilibrium size in negligible time relative to other flux dynamics, neglecting other growth mechanisms (such as coagulation, interactions with other condensing vapours than water), and observing that aerosol particle number-vs-size distributions (and their moments for surface and volume, and often also mass) are best described by log-normal distributions, lead to the following relationships:

$$[K(z)+D(Dp)]\ (\partial c/\partial z) + v_S\ (Dp)\ c(Dp,z) + [K(z)+D(Dp)]\ H(z)\ (\partial c/\partial(\ln Dp)) = 0 \qquad (3)$$

$$H(z) = (1/G)(\partial G/\partial S)(\partial S/\partial z) \qquad (4)$$

Since focus here is on particle concentration data which occur in, or can be transformed into, discrete size intervals j, j+1, etc, each of which can be viewed as $\Delta(\ln Dp_j)$ ($=\ln(Dp_{j+1}) - \ln(Dp_j)$), and since interest here is in the concentration behaviour in each such "size window", a first approximation to the partial differential equation Eq. 3 is, with $D(Dp_j)$ denoted by D_j, the following recursive formulation:

$$[K(z)+D_j]\,(\partial c_j/\partial z) + v_{sj}\,c_j + [K(z)+D_j]\,[1/\Delta\ln(Dp_j)]\,H(z)\,(c_j - c_{j-1}) = 0 \tag{5}$$

In order to explore the general behaviour of this system of ordinary differential equations, and to check numerical algorithms (4th order Runge-Kutta) available for the work with this paper a number of not unrealistic simplifications were made, leading to analytically tractable differential equations (including Bernouilli equations); such studies showed that the model predicts concentration increase in larger particle-size class at the expense of concentration decline in small-particle classes, and also showed negligible discrepancies between numerical and analytical solutions.

EXEMPLIFYING MODEL APPLICATIONS

For a first simplistic application, the particle number distribution obtained (cf. Fig. 2) can be investigated. Neglecting a third but in these measurements not identifiable particle mode below 0.02 μm, Eq. 4 can be used to estimate how the shape of the measured distribution would respond to changes in relative humidity (Fig. 4).

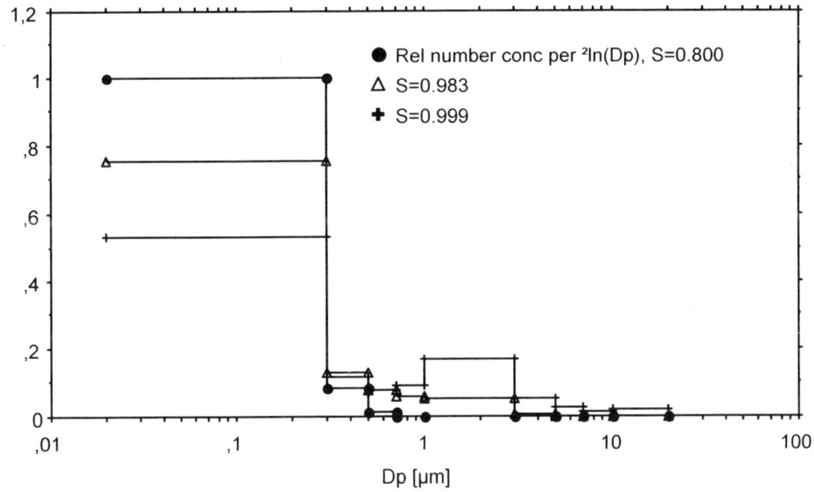

Figure 4. Changes to the measured particle number distribution, assuming it is subjected (in time or space) to changes from the rh prevailing during the measurements (around 80%) to higher values.

If deposition due to gravitation only is considered, and if it is assumed that the amount per particle of a given substance does not change when particles attach more and more water to them, the deposition rate of this substance in the original distribution (the total number concentration of which is conserved in the calculations) will increase by a factor of 2.6, 25, 38, and 240 if the distribution is subject to increased rh from 80 to 90%, 98.3%, 99%, and 99.9% respectively. In addition, it can be seen that under the above assumptions (such as if the distribution is affected by downward movement affecting all particles sizes in the same way) the number concentration recorded by equipment measuring in the respective classes will decline towards the surface for small particles (≥0.02 μm in this example) and increase towards the surface in higher size classes.

Next, a calculation can be performed through the use of Eq. 5 with H(z) assumed negligible (such as would be the case for non-hygroscopic particles) so that eddy diffusivity, diffusion, and gravitational settling are incorporated in the transfer processes, but not particle flux along the size axis. For this modelling example the following submodels and quantities are used for eddy diffusivity K, settling velocity v_S, and temperature T:

K = k u* z F (k is von Karman's constant; F a stability function taking a value around 1 for close-to-neutral conditions which are assumed in this example); v_S and D are calculated from the geometric mean diameters of each class, and according to classical aerosol mechanics; T = Ts - b_l z (Ts water surface tempertaure; b_l taking the value around 0.01 K m^{-1} for neutral conditions). The resulting concentration profiles show decline towards the surface; for the first few classes the gradients are quite small (on the order of 0.1 to 1 %) until a rapid decline towards zero occurs just above the surface; for larger particle sizes, where gravitational settling begins to dominate, similar but more pronounced concentration gradients occur; cf. Fig 5 below where, for simplicity, the essential features are exemplified by the behaviour of classes from 1-3 μm and larger.

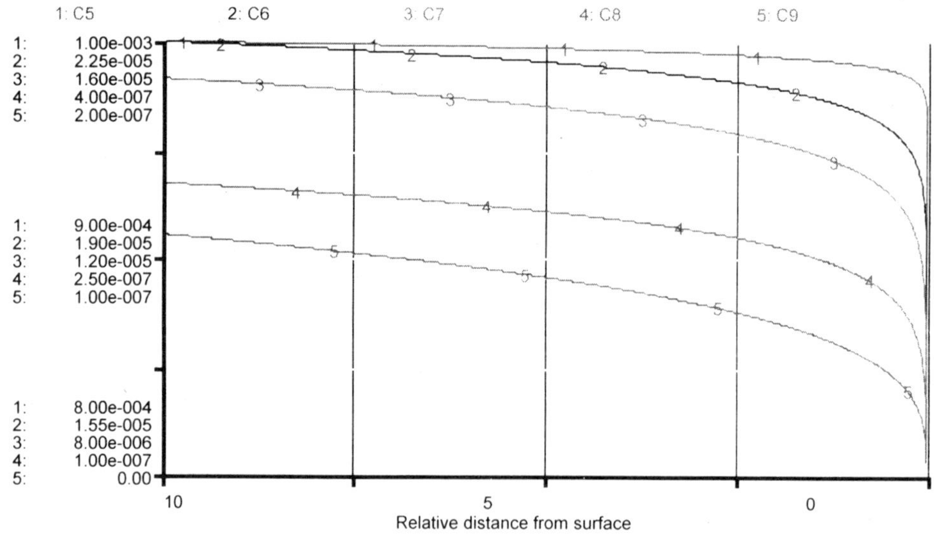

Figure 5. Concentration profiles modelled in the absence of rh effects on particle size, for particle size classes 1-3 (labelled 1:C5), 3-5 (2:C6), 5-7 (3:C7), 7-10 (4:C8), and 10-20 (5:C9) μm, where concentrations at a reference height of 10 units (defined by assumed rate of change in S; cf. Fig. 6) above the surface define the upper boundaries. Scales are chosen to enable optimal comparisons of plotted data.

Finally, a calculation (Fig. 6) is performed with eddy diffusivity, diffusion, and gravitational settling as well as particle growth effects included, with $G(S)=[1+\beta/(1-S)]^{1/3}$ used in the H(S) function, and with ß taking a value in the range 0.1 (continental aerosol) to 1 (maritime aerosol, used here); cf., Fairall (1984).

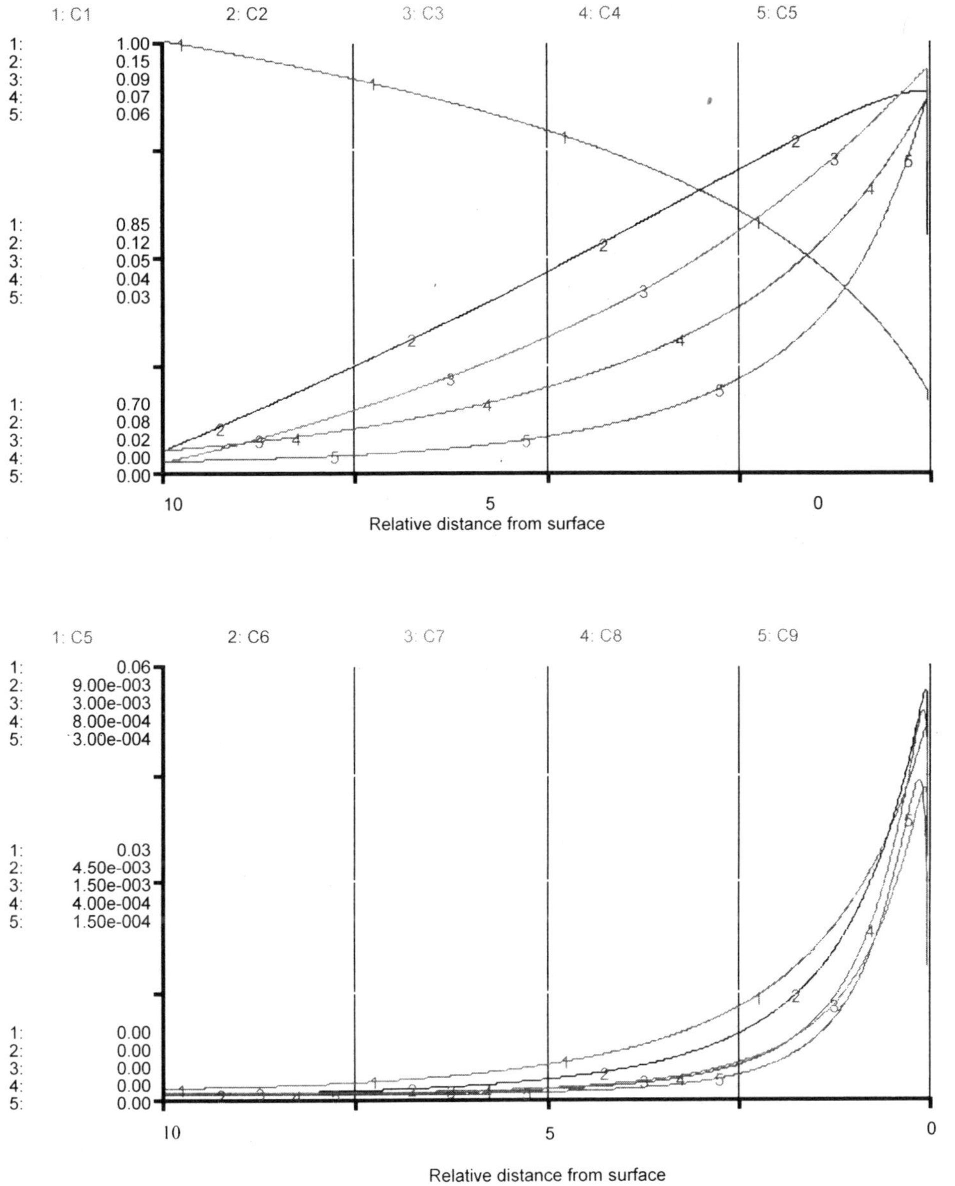

Figure 6 (a) (upper) and (b) (lower). As in Fig. 5, but with effects of changes in rh incorporated so that transfers along the vertical axis as well as he particle size axis affect the build-up of profiles. Note that scales in the diagrams have been chosen to enable optimal comparisons of plotted data. It is assumed that no influx (outflux) of particles due to rh occurs to (from) the lowest (largest) size class (0.02-0.3 (labelled C1), 10-20 (C9) μm, respectively). For S a simple linear decline upwards from S=0.983 to S=0.80 is assumed (with an arbitrary height scaling from 0 to 10 units; in practice, a non-linear decline is likely to occur over a few m, with the major changes over the first few cm).

CONCLUSIONS

A multitude of physico-chemical mechanisms are in operation in the atmosphere/water-surface exchange layers, able to generate declining as well as increasing concentrations downwards towards the air/water interface of particles in various size classes, as exemplified by the measured data. These concentration behaviours are qualitatively consistent with, and thus potentially and partly explained by, the results of a simplified modelling approach that takes into account the simultaneous action by vertical transfer of particles and particle transfer along the size axis due to effect of relative humidity. It is shown that the overall deposition rates of compounds carried by hygroscopic aerosols can be substantially higher than would be suggested by number distributions measured a few m above water surfaces, because a flux of small particles towards larger particle sizes is likely to occur during downward transport, transforming the size distribution. Although not explicitly demonstrated, the rh effects on particle sizes also mean that large particles (such as injected into the air by bubble-bursting) able to move upwards through the high-rh layer close to the interface, and into lower-rh regions, can decline to size ranges subject to long-distance transports. However, several other mechanisms not investigated in this paper, including mixed-mode aerosols (combinations of hygroscopic and non-hygroscopic properties; cf., Svenningsson, 1997), Stefan flow, bubble-bursting, and impaction on waves and jet drops, need to be studied for a more clear understanding of the exchange environment.

In addition to studying vertical stratification of aerosol concentrations in coastal zones, measurements and modelling are needed of the temporal as well as horizontal concentrations variations in these zones. Among important ecological reasons for this are the needs for a fuller understanding of the coupling between the sea and coastal forests (e.g., Eriksson, 1960; Yaalon and Lomas, 1970; Klinger and Erickson III, 1997).

ACKNOWLEDGEMENTS

Field data exemplified in this paper are partly from joint work with Steven E. Lindberg, Oak Ridge National Laboratory, U.S.A. Parts of the research have been based on funding received from the Swedish Natural Science Research Council.

APPENDIX

For an aerosol flowing across a water surface the following pde is a basic description of the rate of change in particle concentration (term I) caused by turbulent exchange (eddy diffusion) and (Brownian) diffusion (II); particle movement due to mean wind, and gravitational and other forces (III); particle collection and other sink mechanisms (IV) and particle generation and other source mechanisms (V):

$$\underset{[I]}{\partial c/\partial t} = \underset{[II]}{\nabla\{(K+D)\nabla c\}} - \underset{[III]}{\nabla(Uc)} - \underset{[IV]}{\mathbf{S}} + \underset{[V]}{\mathbf{Q}} \qquad (1)$$

where

• c (x,y,z,t,Dp) is concentration at time t at coordinate x,y,z of particles (carrying a chemical element or compund of interest) of diameter Dp;
• K (x,y,z,t) is a tensor accounting for turbulent transport of particles (more stringently, turbulence affects particles of different sizes somewhat differently, so that K is Dp-dependent);
• D (x,y,z,t,Dp) is a tensor accounting for diffusional transport of particles ;
• U (x,y,z,t) is the vector ($u_x+v_{x,misc}, u_y+v_{y,mic}, u_z - v_s + v_{z,misc}$) accounting for particle movement with mean wind, settling under gravitation ($v_s(D_p)$), and potentially existing velocities v_{misc} generated by miscellanoeus forces (Stefan flow, thermophoresis, photophoresis, electro-gradient forces, and others)

• $\mathbf{S}(x,y,z,t,D_p)$ and $\mathbf{Q}(x,y,z,t,D_p)$ are sink and source functions, respectively, where $\mathbf{S}$ ($\mathbf{Q}$) relates to the sink (source) efficiency in terms of particle number or mass collected (generated) per unit air volume in unit time by the water surface, and particle number or mass transformed to (from) aerosol phase to gaseous phase.

REFERENCES

Andrews, E., Kreidenweis, S.M., Penner, J.E., and Larson, S.M., 1997, Potential origin of organic cloud condensation nuclei observed at marine site, *Journal of Geophysical Research* 102 D18:21997-22012.

Blanchard, D.C. and Woodcock, A.H., 1957, Bubble formation and modification in the sea and its meteorological significance, *Tellus* 9:145-158.

Blevins, R.D., 1984, *Applied Fluid Dynamics Handbook,* Van Nostrand Reinhold Company, New York.

Davies, C.N., Deposition from moving aerosols, 1966, in: C.N. Davies, ed., *Aerosol Science,* Academic Press, London.

Dulac, F., Bergametti G., Losno, R., Remoudaki, E., Gomes, L., Ezat, U., and Buat-Ménard, P., 1992, Dry deposition of mineral aerosol particles in the marine atmosphere: significance of the large size fraction, in: S.E. Schwartz and W.G.N. Slinn, eds, *Precipitation Scavenging and Atmosphere-Surface Exchange, Volume 2,* Hemisphere Publishing Corporation, Washington.

Eriksson, E., The yearly circulation of chloride and sulphur in Nature; meteorological, geochemical and pedological implications. Part II. *Tellus* 12:63-109.

Fairall, C.W., 1984, Interpretation of eddy-correlation measurements of particle deposition and aerosol flux, *Atmospheric Environment* 18:1329-1337.

Fuchs, N.A., 1964, *The Mechanics of Aerosols,* Pergamon Press, Oxford.

Hardy, J.T., Apts, C.W., Crecelius, E.A. and Fellingham ,G.W., 1985, The sea-surface microlayer: fate and residence times of atmospheric metals, *Limnol. Oceanogr.* 30:93-101.

Hummelshøj, P., Jensen, N.O. and Larsen, S.E., 1992, Particle dry deposition to a sea surface, in: Schwartz S.E. and Slinn W.G.N., eds., *Precipitation Scavenging and Atmosphere-Surface Exchange, Volume 2,* Hemisphere Publishing Corporation, Washington.

IPCC (Intergovernmental Panel of Climate Change), 1996, *Climate Change 1995 - The Science of Climate Change, Contribution of Working Group I to the Second Assessment Report of the Intergovernmental Panel on Climate Change*, Cambridge University Press, Cambridge, U.K.

Jenkin, M.E., 1984, An investigation into the enhancement of deposition of hygroscopic aerosols to wet surfaces in a wind tunnel, *Atmospheric Environment* 16:889-910.

Klinger, L.F. and Erickson III, D.J., 1997, Geophysiological coupling of marine and terrestrial ecosystems, *Journal of Geophysical Research* 102 D21: 25359-25370.

Krauss, E.B., *Atmosphere-Ocean Interactions,* 1972, Clarendon Press, Oxford.

Lindfors, V., Joffre S.M., and Damski, J., 1992, Dry and wet deposition of sulphur and nitrogen compounds over the Baltic Sea, in: S.E. Schwartz and W.G.N. Slinn, eds., *Precipitation Scavenging and Atmosphere-Surface Exchange, Volume 2,* Hemisphere Publishing Corporation, Washington.

Matschullat, J., 1997, Trace element fluxes to the Baltic Sea: problems of input budgets, *Ambio* 26:363-368.

Paerl, H.W., 1985, Enhancement of marine primary production by nitrogen-enriched acid rain, *Nature* 315:747-749.

Persson, C., Rodhe, H. and de Geer, L.-E., 1987, The Chernobyl Accident - a meteorological analysis of how radionuclides reached and were deposited in Sweden, *Ambio* 16:20-31.

Piazzola, J. and Despiau, S., 1997, Vertical distribution of aerosol particles near the air-sea interface in coastal zone, *Journal of Aerosol Sci.* 28:1579-1599.

Rendell, A.R., Ottley, C.J., Jickells, T.D., and Harrison, R.M., 1993, The atmospheric input of nitrogen species to the North Sea, *Tellus* 45B:56-63.

Slinn, W.G.N., Hasse, L., Hicks, B.B., Hogan, A.W., Lal, D., Liss, P.S., Munnich, K.O., Sehmel, G.A. and Vittori, O., 1978, Some aspects of the transfer of atmospheric trace constituents past the air-sea interface, *Atmospheric Environment* 12: 2055-2087.

Slinn, S.A. and Slinn, W.G.N., 1980, Predictions for particle deposition on natural waters, *Atmospheric Environment* 14:1013-1016.

Svenningsson, B., 1997, *Hygroscopic Growth of Atmospheric Aerosol Particles and Its Relation to Nucleation Scavenging in Clouds,* PhD Thesis, Department of Nuclear Physics, Lund University (ISBN 91-628-2764-2)

Syrakov, D., Batchvarova, E., and Wiman, B.L.B., eds, 1998, *Long-Range Air Pollution: from Models to Policies,* Pensoft Publishers, Sofia.

Volchoc, H.L., Feiner, M., Simpson, H.J., Broecker, W.S., Noshkin, V.E., Bowen, V.T. and Willis, E., 1970) Ocean fallout - the Crater Lake experiment, *Journal of Geophysical Research* 75:1084-1091.

Walker, M.I., McKay, W.A., Pattenden, N.J. and Liss, P.S., 1986, Actinide enrichment in marine aerosols, *Nature* 323:141-142.

Whitby, K.T., 1978, The physical characteristics of sulfur aerosols, *Atmospheric Environment* 12:135-159.

Williams, R.M., 1982, A model for the dry deposition of particles to natural water surfaces, *Atmospheric Environment* 16:1933-1938.

Wiman, B.L.B. and Ågren, G.I., 1985, Aerosol depletion and deposition in forests - a model analysis, *Atmospheric Environment* 19: 335-347.

Wiman, B.L.B., 1988, Aerosol capture by complex forest architecture, in: J.T.A Verhoeven, G.W. Heil and M.J.A. Werger, eds, *Vegetation Structure in Relation to Carbon and Nutrient Economy,* SPB Academic Publishing bv, The Hague.

Wiman, B. L. B., Unsworth, M. H., Lindberg, S. E., Bergkvist, B., Jaenicke, R. and Hansson, H.-C., 1990, Perspectives on aerosol deposition to natural surfaces: interactions between aerosol residence times, removal processes, the biosphere and global environmental change, *Journal of Aerosol Sci.* 21: 313-338.

Wiman, B.L.B., 1998, On the interaction between landscape complexity and aerosol sink/source dynamics, in: D. Syrakov, E. Batchvarova and B.L.B. Wiman, eds., *Long-Range Air Pollution: from Models to Policies*, Pensoft Publishers, Sofia.

Wiman, B.L.B., Donev, E., Yurukova, L. & Gaydarova, P.N. , 1998. *Integrated Environmental Diagnosis and Assessment: Montane and Coastal Forest – Structure, Function and Atmospheric Stress Factors.* IEDA Report #1, Sofia, June 1998. Institute of Botany, Bulgarian Academy of Sciences; ISBN 954-9746-01-1.

Yaalon, D.H. and Lomas, J., 1970, Factors controlling the supply and the chemical composition of aerosols in a near-shore and coastal environment, *Agr. Meteorol.* 7:443-454.

PREOPERATIONAL OZONE FORECASTING AT THE DEUTSCHER WETTERDIENST

Stefan Tilmes, Jürgen Rißmann, Ingo Jacobsen, Jörg Zimmermann

Deutscher Wetterdienst
Kaiserleistraße 42
D-63067 Offenbach
stilmes@dwd.d400.de

INTRODUCTION

A great part of the present German research activities concerning tropospheric photochemistry and dynamics is concentrated within the national Tropospheric Research Program (TFS). The first of three principal subjects concerns the development and application of a model hierarchy for the diagnosis and forecast of the distribution of pollutants over Europe (BMBF, 1997). The main emphasis during the first phase of the program is the application of chemistry transport models (*CTMs*) to forecast the ozone distribution based on routinely available input data.
There are different objectives for ozone forecasting. The scientific questions concern the forecasting capabilities of state-of-the-art models and their climatologies. Politics are concerned in case of extreme summer smog events when decisions have to be made e.g. to inform the public. In that case the forecast can also provide initial and boundary values for smaller scale models which can be operated by the local authorities to take further steps.

The role of the Deutscher Wetterdienst (DWD) within this framework is twofold. The DWD provides the routine weather forecast and the computer resources to run the ozone forecasting system. The *LM* (Lokal-Modell, the future operational limited-area model of the DWD; Doms and Schättler, 1997) will be further developed to match the needs of *CTMs*. In the first place this is a two-way interactive nesting procedure.
The second aim is the time critical processing of trace gas measurements and the post-processing of the model output. The representativeness and quality of the available measurements have to be investigated to allow for a proper verification of the forecast. For this purpose a multiannual dataset of about 300 German sites was analysed with sophisticated statistical and algebraic methods. Last not least, the forecast has to be processed for further use.

The following section will summarize the forecasting system and the single modules. Particular needs which arise from the routine application of the system will

Air Pollution Modeling and Its Application XIII, edited by S.-E. Gryning
and E. Batchvarova, Kluwer Academic/Plenum Publishers, 2000

be discussed. In a further section special attention will be paid to the comparability of measured and modelled data and the verification procedures of the forecast. Finally, we will briefly discuss our experiences and give an outlook to our further work.

SETUP OF THE FORECASTING SYSTEM

The ozone forecasting system is a product of a network between research groups at universities and the DWD. To avoid the development of yet another new model approved components are combined to a model system in that way that each participating group can run each submodel. Additionally, the complete chain is installed at the DWD to run the operational forecast. Figure 1 provides an overview of the different groups and their contributions to the forecasting network.

The first link in the chain is the weather forecast with the routine models of the DWD. The *GME* (Global-Modell; Majewski, 1996) delivers the boundary values for the *LM*. Anthropogenic emission rates for the chemical species are predicted by the *EMOD* module (BMBF, 1997) which is developed at the University of Stuttgart. This model uses the data from the weather forecast as input. The EURAD group from the University of Cologne provides the *PPC* (*CTM* preprocessor) and their *CTM* (Ebel et al., 1997). The *PPC* preprocesses the meteorological and emission data and computes biogenic emissions and photolysis rates. Its output is used by the *CTM* which forecasts the fields of the chemical species concentrations and the depositions.

The forecast by the model system is followed by postprocessing also at the DWD. As mentioned above, an option for a further phase of the project is the subsequent simulation of single subregions with a higher resolution in a one-way nesting mode by other institutions. A candidate for such a model is the *DRAIS* model by the University of Karlsruhe (Vogel et al., 1995) which at present is not linked to the operational forecast system and thus not shown in the graph.

The modules of the system

The *GME* and the *LM* together build up the fourth generation of the DWD's numerical weather prediction (NWP) system which will start in the late summer 1998 with pre-operational forecasts four times a day. Some features of the *GME* are listed below:

GME

- hydrostatic
- global hexagonal-triangular grid, local spherical coordinates at each node
- hybrid vertical coordinate
- resolution
  - horizontal: $\approx 55\,\mathrm{km}$ (166 410 grid points/layer)
  - vertical: 31 layers up to 10 hPa
  - time step: $\Delta t = 200\,\mathrm{s}$
- computer resources for a 48 h forecast
  - 1 hour on 120 PEs of a Cray T3E-900
  - 4 GByte output to disk

Due to the high computational cost of the chemical part of the network, the operational runs of the *LM* cannot be used so that a version with a lower spatial resolution is utilized. A two-way interactive nesting option to locally increase the resolution is under developement. Some informative data for the applied version of the *LM*:

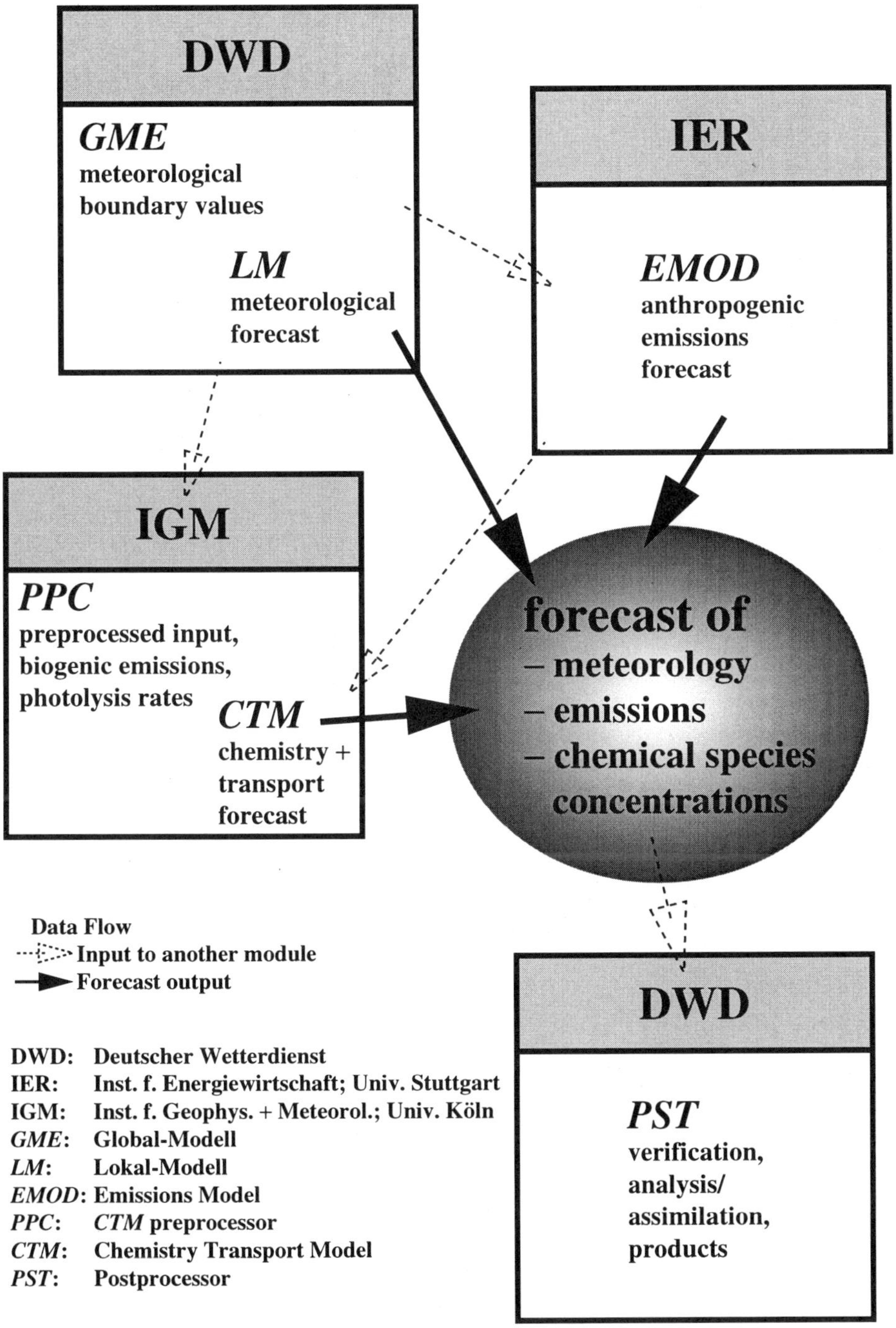

Figure 1. Sketch of the model network for the operational ozone forecast.

LM – air pollution version

- o nonhydrostatic
- o limited area grid, rotated geographic coordinates
- o hybrid vertical coordinate
- o resolution
  - – horizontal: $0.1875^o \approx 21\,\text{km}$ (109×109 grid points/layer)
  - – vertical: 32 layers up to 20 hPa
  - – time step: $\Delta t = 90\,\text{s}$
- o computer resources for a 48 h forecast
  - – 0.5 hours on 64 PEs of a Cray T3E-900
  - – 0.6 GByte output to disk

The *EMOD* module consists of a cache database which holds the necessary basic information used by the emission calculation module. The cache database is generated at the University of Stuttgart and ported to the DWD. The emission calculation module will be operated at the DWD and uses also the meteorological forecast as input. This model computes the emission rates in the desired spatial and temporal resolution for the predicted species.

The *CTM* of the system is the second generation version of the EURAD-*CTM*. Several modifications were made by the Cologne group to bring it into line with the *LM*. These are e.g. the I/O structure, the map projection, and the nonhydrostatic option. This results in the following features of the *CTM* as it is used in the network:

CTM

- o nonhydrostatic
- o same horizontal grid structure as *LM*
- o vertical σ-coordinates
- o resolution
  - – horizontal: same as used by the *LM*
  - – vertical: 16 layers up to 20 hPa
  - – time step: $\Delta t = 600\,\text{s}$
- o RADM2 chemistry (other options are planned for the near future)
- o computer resources for a 48 h forecast (at present)
  - – 3.5 hours on 1 PE of a Cray C98
  - – 1 GByte output to disk

Preparation of the operational forecast

The first version of the model chain with preliminary emission data was set up on a Cray T3E-900 (*LM*) and on a Cray C98 (*PPC*, *CTM*). The performance of the system is listed above. Work is going on to optimize the performance of the *CTM* by rearrangements in the code to meet the requirements of operational models. Also promising is the portation of the model to a single PE of the Fuijitsu VPP700 (a massive parallel system with efficient vector processors) at the ECMWF (European Centre for Medium-Range Weather Forecasts). Another option for the future is a MPI-version of the *CTM* suitable for massive parallel computers like the Cray T3E-900 of the DWD or the VPP700. These efforts are essential because a time limit of about one hour is required for a model run for operational use.

The simulation domain which is the same for the NWP-version of the *LM* covers the whole of Central Europe (see Figure 2). This large area allows for a minimization of the effects of the boundaries on the chemical species concentrations which are a

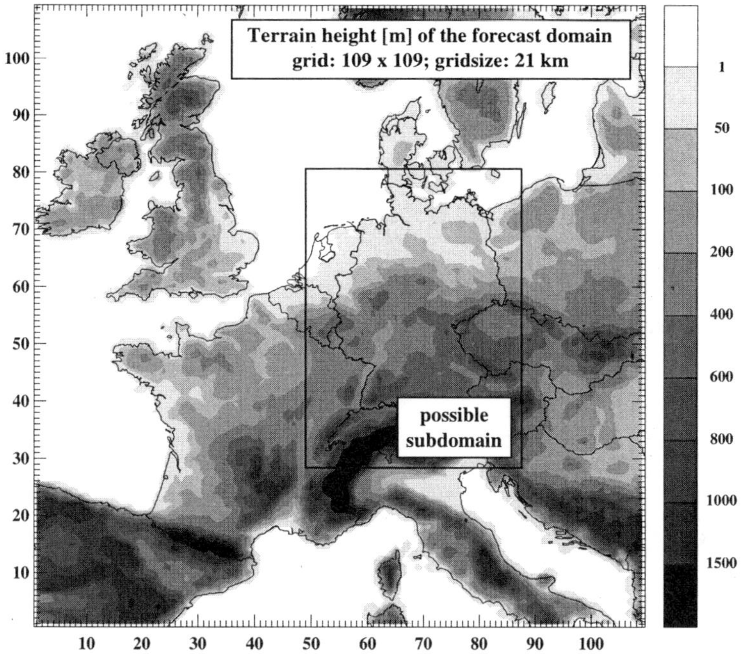

Figure 2. Orography of the forecast domain.

great source of uncertainty within the *CTMs*. The two-way interactive nesting will provide boundary values for subdomains like e.g. Germany to be modelled with a horizontal gridsize of $0.0625^o \approx 7\,\text{km}$ and successively higher resolutions in further subgrids. Test simulations were carried out for episodes in July 1994 and May 1998. First simulations under operational conditions are planned for photosmog episodes in the forthcoming summer.

VERIFICATION STRATEGIES

The operational ozone forecasting requires special strategies for the verification of the results. This section will discuss the measuring data available online which are suitable for this purpose and some considerations concerning the direct and statistical comparison of these measurements with the model results.

Routinely measured station data from the environmental agencies of the countries and states and ozone sonde data from the national meteorological services are available within hours. Because of their limited vertical and temporal resolution satellite data are not yet applicable for the verification of ground-level ozone. The infrastructure necessary for the online reception of the data is under development. For example there are several hundred sites in Germany which provide their data almost in real time. The data from these networks are different from the data available from field campaigns. The QA/QC-level is lower and the surrounding of the sites range from remote to heavily polluted. About 80 % of the sites are polluted by nearby moderate to high traffic. As a consequence, a first step must be the thorough investigation of the data from the sites obtained over several years to get an estimate of the representativeness and quality of each site.

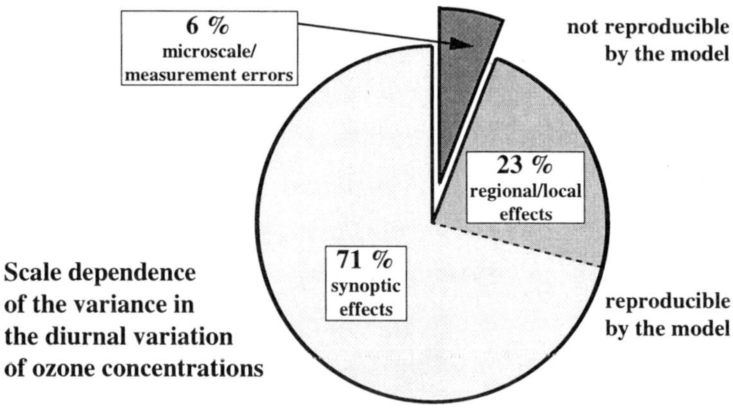

Figure 3. Dependence of the variance in ozone concentrations on the spatial scale as obtained with the principal component analysis of measured diurnal variations.

Analysis of measured time series

We analysed measured time series of more than 300 German sites for the years 1994 to 1996 with the principal component analysis and with the Kolmogorov-Zurbenko filter.

With the Kolmogorov-Zurbenko-Filter (Rao and Zurbenko, 1994) — five-fold application of a 15-day-running mean applied to each hour of the day separately — the time series is divided into a long-term and a short-term component.
The long-term components are strongly correlated between the German stations (82 – 94 %). They rise up to high values in March and start to decline in August. Thus they show an asymmetric seasonal cycle with a spring maximum as compared with the course of solar radiation. This spring maximum of ozone was already described by others (Derwent et al., 1998). This clearly demonstrates that dynamic processes play an important part in ozone production besides photochemistry.
From the long-term components diurnal cycles were extracted in a second step. While the averaged diurnal cycles show large differences depending on season (up to a factor of six) and site of the stations the variances of the short-term components are only weakly correlated with the season. This indicates that they are dominated by dynamics and not by photochemical processes. Furthermore, no obvious correlation was detected between the short-term behaviour of the concentrations of ozone and primary trace gases.

In order to study the observed diurnal variations of ozone at the German sites in more detail a principal component analysis was performed (Tilmes and Zimmermann, 1998). This method allows to separate the variance in the data into independent components. These are the eigenvectors whose contributions to the variance are quantified by the eigenvalues. By computing the correlations between the amplitudes of the eigenvectors of different stations correlation lengths were derived. Thus we get an information on the spatial representativeness of the distinct parts of the variance described by the eigenvectors. More than 70 % of the variance are due to the first eigenvector (see Figure 3). It represents the large-scale synoptic processes with a correlation length above 100 km. The subsequent four eigenvectors which explain 23 % of the variance are connected with regional to local scales of more than 5 km. The remaining 6 % of the variance comprise even smaller scales and mainly measurement errors. So we cannot expect to simulate this part of the time series even with a perfect model. On the other hand a model resolution of about 5 – 10 km will be necessary to explain the information contained in the leading five eigenvectors. Especially in the range between 10 – 20 km the resolution will have a major influence on model accuracy. Similar conclusions were derived by McNair et al. (1996) and Gillani and Pleim (1996).

The principal component analysis is also a powerful tool to characterize individual stations in order to postprocess data (Elbern et al., 1996). Noise and small scale information can be filtered from observations in order to isolate the signal which corresponds to the selected scale. Observations thus filtered can be used to evaluate model output.

Consequences for model verification

It was shown how the ability to reproduce the measurements depends on the resolution of the model and of the input data and on the characteristics of stations as well. This information must be considered whenever modelled and measured data are compared. A separate verification for each class of stations is a good method for this purpose. The verification via difference statistics must be done separately on temporal and spatial datasets. The results of spatial statistics strongly depends — especially in the case of ozone data — on the time of day. Even under comparable meteorological conditions the statistics of time series are very different for different regions. Nevertheless all the calculations reveal important information especially when done separately for the different types of stations. Such procedures will also be useful for the monitoring of measurements. False observations can be detected by conspicuous verification statistics — another feature which can only be realized by an operational forecast which provides long time series of modelled data.

CONCLUSIONS AND OUTLOOK

This presentation gives a short glance on the work going on at the DWD and the cooperating groups within the German TFS program concerning the forecasting of ozone. The setup of an operational system at the DWD is described which is constructed from the contributions of a network of different scientific institutions. This state-of-the-art model system is meant as a tool for policy makers. The possible products range from horizontal maps of predicted ozone concentrations suited for the information of the public to initial and boundary values for regional models operated by the local authorities. Further work has to be done to increase the computational efficiency of the model chain to allow for the routine application of the system. This will provide an modelled dataset as a base for the analysis as it is carried out with the observed data. With this dataset the verification concept can be elaborated further.

The analysis of the observed ozone records shows that it is important to validate models with data from a large number of sites. In this sense the organization of the time critical reception of data on a European scale is an urgent task to enable a thorough postprocessing of the forecast data. Observing the model climate is part of the postprocessing as well as the comparison of long series of forecasts with the climates of the monitoring sites. The experience gained from the verification of the forecast will be the basis for a chemical assimilation cycle. The data assimilation will provide more realistic initial data and physically consistent maps of analysed ground level ozone concentrations.

ACKNOWLEDGEMENTS

We want to thank the groups involved in the network project for the fruitful collaboration. These groups are the EURAD group at the University of Cologne, the IER at the University of Stuttgart, and the IMK at the University of Karlsruhe. We are grateful to the German environmental protection agency (UBA) for providing the ozone data of its stations and the networks of the Federal States. This work was supported by the Bundesministerium für Bildung, Wissenschaft, Forschung und Technologie of the Federal Republic of Germany under grants 07TFS10/LT1-A.2 and 07TFS10/LT1-C.4.

REFERENCES

BMBF (Bundesministerium für Bildung, Wissenschaft, Forschung und Technologie), 1997, *Tropospheric Research Program 1996-2002, Results 1996-1997*, GSF – Forschungszentrum für Umwelt und Gesundheit GmbH, München, http://www.gsf.de/PTUKF/deutsch.html.

Derwent, R.G., Simmonds, P.G., Seuring, S., and Dimmer, C., 1998, Observation and interpretation of the seasonal cycles in the surface concentrations of ozone and carbon monoxide at Mace Head, Ireland from 1990 to 1994, *J. Atmos. Environ.*, 32, 145-157.

Doms, G., and Schättler, U., 1997, *The Nonhydrostatic Limited-Area Model LM (Lokal Modell) of DWD – Part I: Scientific Documentation*, Deutscher Wetterdienst, Offenbach.

Ebel, A., Elbern, H., Feldmann, H, Jakobs, H.J., Kessler, C., Memmesheimer, M., Oberreuter, A., and Piekorz, G., 1997, *Air Pollution Studies with the EURAD Model System (3): EURAD – European Air Pollution Dispersion Model System*, Mitteilungen aus dem Institut für Geophysik und Meteorologie der Universität zu Köln, 120, A. Ebel, M. Kerschgens, F. M. Neubauer, P. Speth, eds., Köln.

Elbern, H., Tilmes, S., and Schmidt, H., 1996, Data assimilation for chemistry-transport-models, *Proceedings of EUROTRAC Symposium '96*, P.M. Borrell, P. Borrel, T. Cvitaš, K. Kelly, and W. Seiler, eds., Computational Mechanics Publications, Southampton, 681-685.

Gillani, N.V., and Pleim, J.E., 1996, Sub-grid-scale features of anthropogenic emissions of NO_x and VOC in the context of regional Eulerian models, *J. Atmos. Environ.*, 30, 2043-2059.

Majewski, D., 1996, *Documentation of the New Global Model GME*, Deutscher Wetterdienst, Offenbach.

McNair, L.A., Harley, R.A., and Russel, A.G., 1996, Spatial inhomogeneity in pollutant concentrations, and their implications for air quality model evaluation, *J. Atmos. Environ.*, 30, 4291-4301.

Rao, S.T., and Zurbenko, I.G., 1994, Detecting and tracking changes in ozone air quality, *J. Air & Waste Manage. Assoc.*, 44, 1089-1092.

Tilmes, S., and Zimmermann, J., 1998, Investigation on the spatial scales of the variability in measured near-ground ozone mixing ratios, submitted to *Geophys. Res. Letters.*

Vogel, B., Fiedler, F., and Vogel, H., 1995, Influence of topography and biogenic volatile organic compounds emission in the state of Baden-Württemberg on ozone concentrations during episodes of high air temperatures, *J. Geophys. Res.*, 100, 22907-22928.

DISCUSSION

Z. ZLATEV: Is it not better to try to select only representative stations and, furthermore, to try to optimize the number of the stations needed?

S. TILMES: In general, all information available should be taken into account. Certainly, a preselection by hand will be helpful to separate some outliers or unphysical data, but it is mainly the task of some objective algorithms to judge on the representativeness of the data in terms of the comparability to modelled data. An optimization of the number of stations is a different task. This is a question of network design and is one potential application of our work on the representativeness of measured and modelled data.

M. SOFIEV: The representativeness of a station highly depends on the averaging period of the data. So, which period have you chosen as the best one for your purposes?

S. TILMES: On the one hand, there are fluctuations in the concentrations that have to be smoothed by a long enough sampling period or the averaging of single measurements. This is usually done when the data are postprocessed before they are published. In the case of ozone the common averaging interval is half an hour to one hour, which is comparable to the usual output frequency of the models. On the other hand, what I was talking about is the spatial and chemical representativeness of the data. First it is necessary to know about the climatological conditions at the observation stations. But second the present regime of the air mass at the site is of great interest, because it largely depends e.g. on the weather situation (winds, temperature, humidity, etc.) and, therefore, can change within hours. Hence, we developed a quantitative measure for the representativeness of ozone concentrations which is valid for observation stations and model grid points as well. With this we can determine the present and climatological regimes of ozone time series and judge the comparability of different time series. This is a strong requirement for the verification of simulation results and also for data assimilation.

LIDAR OBSERVATIONS OF THE PLANETARY BOUNDARY LAYER AEROSOL STRUCTURE AND COMPARISONS WITH IN SITU MEASUREMENTS AND MIXED-LAYER HEIGHT MODEL RESULTS

Orlin Parvanov [1], Ekaterina Batchvarova [2], Alexandar Gizbreht [1]

[1] Institute of Electronics, Bulgarian. Academy of Sciences, 72 Tzarigradsko chausse, Sofia 1784, Bulgaria

[2] National Institute of Meteorology and Hydrology, 66 Tzarigradsko chausse, Sofia 1784, Bulgaria

INTRODUCTION

The development and verification of the mixed-layer height models require experimental data obtained with great dynamics and spatial resolution. Such data can be obtained by lidar aerosol measurements. The interpretation of lidar signals, using the conservative and passive aerosol mixture as a tracer of the turbulence is a basic problem for the lidar measurements. A typical lidar result (Eloranta, 1991) is the visualization of fully developed convection by spatial signal distribution, as capture of the contrast formations is possible only at maximum energy of the vertical currents. In this study an attempt for visualization and objectivization of the macrophysical parameters of the aerosol inhomogeneities in convective turbulence and transition regimes is done. This is a base for the use of aerosol in the measurement of the vertical and the horizontal wind velocity as well as of the turbulence scales in the low troposphere.

EXPERIMENTAL SETUP AND INVESTIGATION METHODS

The lidar observation is taking place in the valley of Sofia. It is situated mainly among the Balkan Mountain (1687 m), Vitosha (2290 m), Liulin (1200 m), Ihtiman Middle Mountain (1300 m). Its size is 30 by 50 kilometers (Figure 1.). The lidar is located on the ground of the Institute of Electronics, Bulgarian Academy of Sciences. On the west the last buildings of an endmost living quarter of Sofia are situated. The average height of the buildings is 8 flours. Industrial zones of Sofia are located eastwards and northwards. 3-5 km far from the probation ground. There are summer houses and field zones on the southwards of the ground, near the foot of Vitosha Mountain, 5 km. The international E9 road passes near the north side of the ground. Tethered and pilot balloons were released in the close proximity of the lidar and gradient measurements were performed. The station for regular

Air Pollution Modeling and Its Application XIII, edited by S.-E. Gryning and E. Batchvarova, Kluwer Academic/Plenum Publishers, 2000

radiosoundings, carried out at 06:00 and 12:00 Universal Time (UT) by the National Institute of Meteorology and Hydrology is situated very close to it.

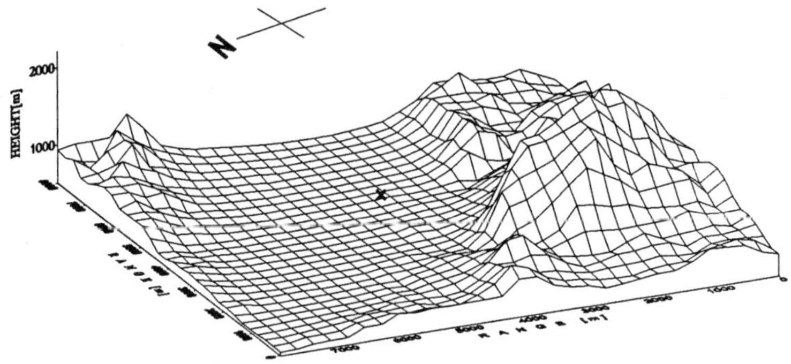

Figure 1. 3-D map of the experimental site. + denotes the location of the lidar.

The meteorological situation during the measuring period, from 10 August till 20 August 1992 is described as a long-lasting period of stable air mass detention over the Balkan peninsula with a high pressure center, located over the East Carpatians. The drought and the lack of cloudiness are specific phenomena during this period of the year.

The experimental setup requires that the measurements with different apparatus and the ground gradient measurements should be done in the course of several days at identical meteorological conditions. The measurements are done continuously and simultaneously in the periods of time from 6 till 12 Standard Local Time (SLT), at sunrise around 6:30 SLT. The obtained information is interpreted on the basis of a juxtaposition of the change of the parameters, registered by different means during the transition from night to day conditions.

A method for Two-dimensional realization of the aerosol field

Lidar method of aerosol sounding consists of registration of echo signals from backscatter laser radiation in atmosphere. Thus can be described by equation

$$P(r) = \frac{\Gamma(r)\tau c P_0 \beta_\pi(r)}{2r^2} T(r), \tag{1}$$

where $P(r)$ is received power from distance r; Γ is the geometric factor of the lidar receiver - telescope type "Cassagrein" 2200X150; β_π is volume backscatter coefficient; P_0 is emitted energy, typically 30 mJ second harmonic of YAG:Nd laser; τ is the time duration of the laser pulse - approximately 15 ns or 5 m in length in space; $T(r)$ is optical transparency to the distance r. The sequence of lidar profiles are collected and two-dimensional data are obtained: $P_i(r_j)$, where i denotes sequential laser shot, j is the number of strobe, corresponding to the height of the sounding through the relation: $r = j * \delta h$, (2) where δh = 7.5 m is the space resolution of the ADC - type waveformrecorder HP8140A (10 bits, 20MHz).

In order to obtain realization of the field of the aerosol volume backscatter coefficient, the following procedures are implemented: First, normalization of the fading according to square of the probation distance. Second, normalization of the radiating energy fluctuations through the input of a “reference channel”. The details on these procedures are described in (Parvanov, 1998). The algorithms for obtaining of the reference channel values are based on 3 different parameters of the lidar data and are used alternatively, according to the noise level in the signals. Third, specifying the activity distance of the lidar. The finding of the signal-noise ratio of the real data is accomplished through algorithms, described in (Parvanov, 1997). Fourth, investigation of statistical characteristics of the time series for assessment of the influence of both the reference channel and the transparency of the sounding traces on the realization quality. The influence of the two factors mentioned above consists in their simultaneous effect on the statistical series of all height, which creates artificial signal correlation and is an obstacle for implementation of analysis of the form, size and drift speed of the aerosol formations in the low troposphere.

EXPERIMENTAL DATA

Lidar Experimental Data

The lidar experimental data are presented on a series of figures as two-dimensional images as follows: the probation heights over the ground are noted on the vertical axis (Figure 3 and 4). The first 150 m do not present in the image because of the lidar geometric factor effect. The hours of beginning and end of every lidar seance are noted on the horizontal axis. The values of the aerosol backscatter coefficients are graded on an eight-degree scale of the gray. For every séance the scale is located linearly between the maximum and the minimum values of the data. For that reason the scale is relative.

The description of the data begins from the earliest hours, from 5:30 SLT. They are laminarly homogeneous images. During the hours after the sunrise till about 8:30 SLT, vertical chaotically located structures are noticed. They are due to both the great homogeneity of the aerosol distribution and the little variation coefficient of its concentration.

On the image of 09:09-09:25 SLT a highly contrasting formation, probably of antropogenic origin can be observed. In the hours after 9:30 SLT an appearance of contrasting formations which base coincides with the lowest heights registered by the lidar can be noticed. These formations increase in size in the course of the day till 11:30 SLT.

Meteorological Experimental data

The meteorological data are presented as a series of temperature profiles obtained through the tethered balloon (Figure 2a). The transformation and the destruction of the ground inversion can be noticed. This process begins at about 7:45 SLT, as the rising velocity of the upper inversion boundary increases. At 9:10 SLT the height has already reached 400 m. The wind profile dynamics shows the specific change in time of the wind velocity, which decreases in the hour immediately before the sunrise. The change of wind velocity and direction at the height of about 700 m, or the base of the elevated inversion, is a specific feature.

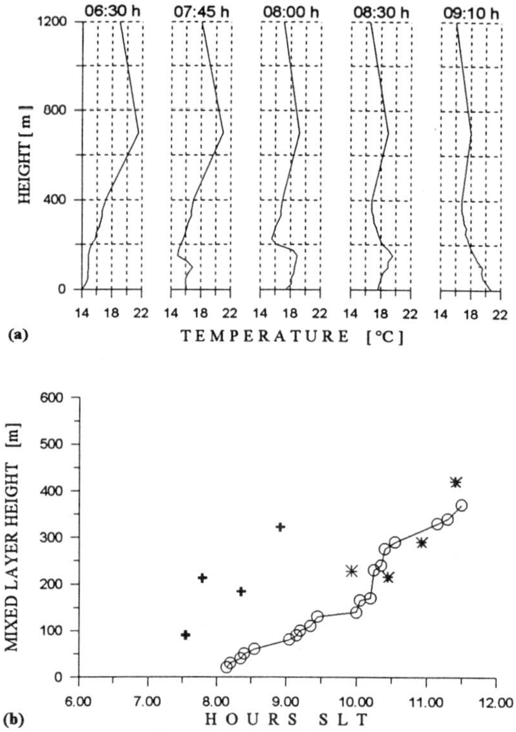

Figure 2a and 2b. A set of aerostat temperature profile measurements on the 19 August 1992. Model calculation of the mixed layer height - “O”; temperature inversion height -”.+ ” ; lidar data for the height of convective cells -” * ”.

Model Data

The mixed-layer model results are obtained through estimates of surface heat and momentum turbulent fluxes obtained from gradient measurements. The model description is presented in (Batchvarova, 1991). The results from the calculations of the mixed-layer growth are presented on Figure 2b.

RESULTS ANALYSIS

On the basis of juxtaposition of the observed changes of the lidar images and the meteorological data behavior, an interpretation of the aerosol stratification change is done.

The aerosol stratification changes are classified as follows: presence of homogeneous layers, random allocation of aerosol eddies and layers with horizontal inhomogeneous.

The homogeneous structure is observed in the hours of stable stratification near the surface and in the residual layer. This structure turns gradually into random allocation of aerosol heterogeneity's, as aerosol signal settling down is observed at the same time. This process is due to the sedimentation of hard particles because of the lack of vertical movements. Antropogenic aerosol horizontal dispersion is observed at the heights of about

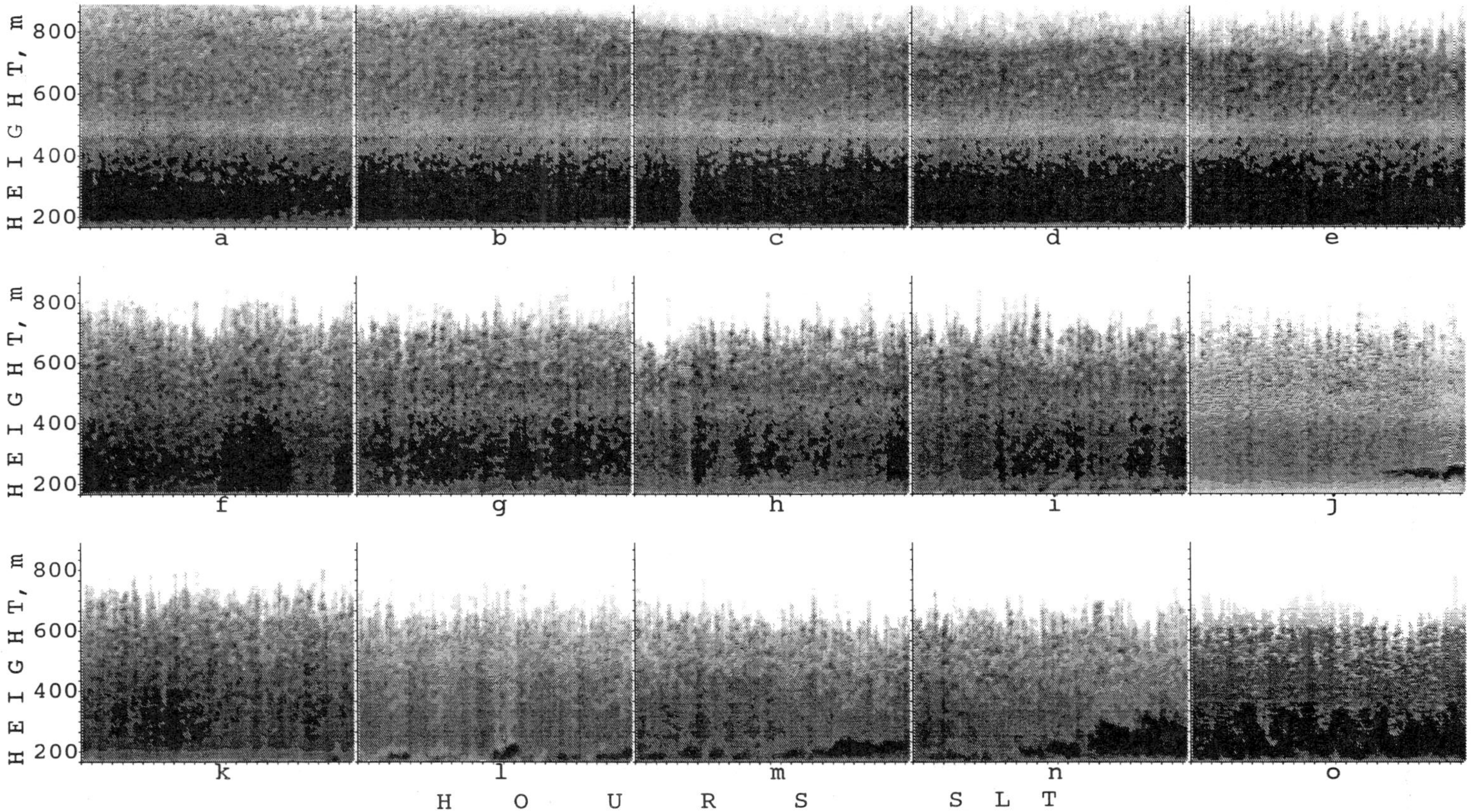

Figure 3. Serial of lidar images obtained on 19 August 1992 at the following hours: a - 5:45; b - 6:08; c - 6:32; d - 6:55; e - 7:24; f - 7:55; g - 8:23; h - 8:47; i - 9:07; j - 9:26; k - 9:44; l - 10:18; m - 10:37; n - 10:58; o - 11:17 SLT.

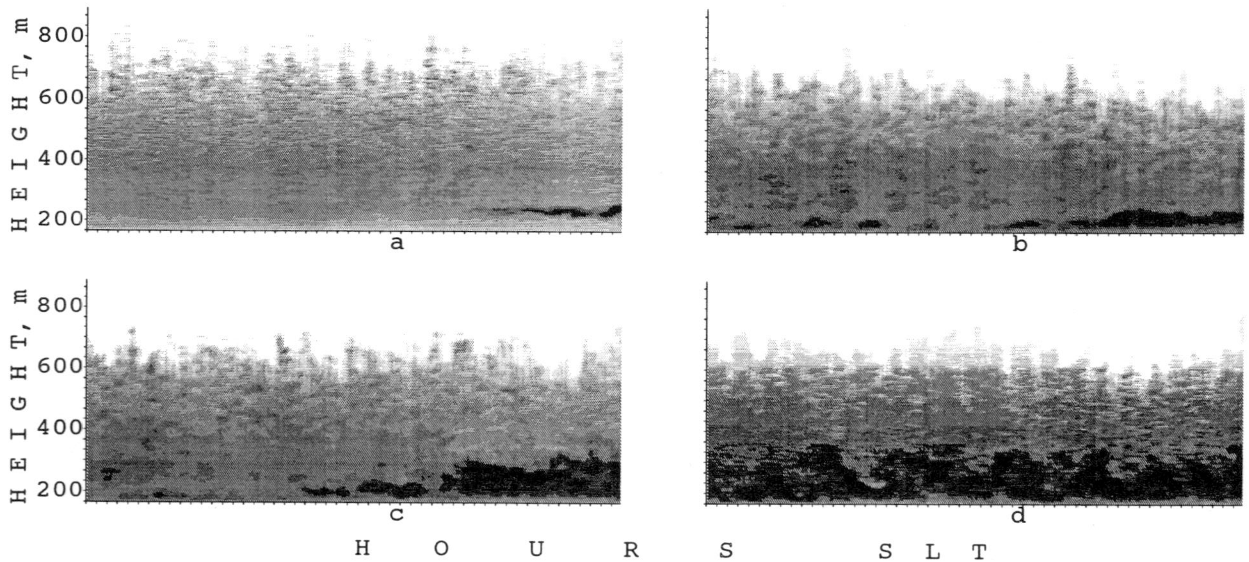

Figure 4. Four lidar images from Figure 3 j, 3m, 3n, 3o, but zoomed in time. At 9:26 SLT it can be seen that an antropogenic aerosol formation (the wind direction is from east i.e. the origin is in industrial zone) is captured within the stable layer up to heights of 350 m. Starting at 10:18 SLT the process of gradually increasing of eddies' scale can be seen. The horizontal advection and the developed convection caused appearance of regular structures like "roll convection".

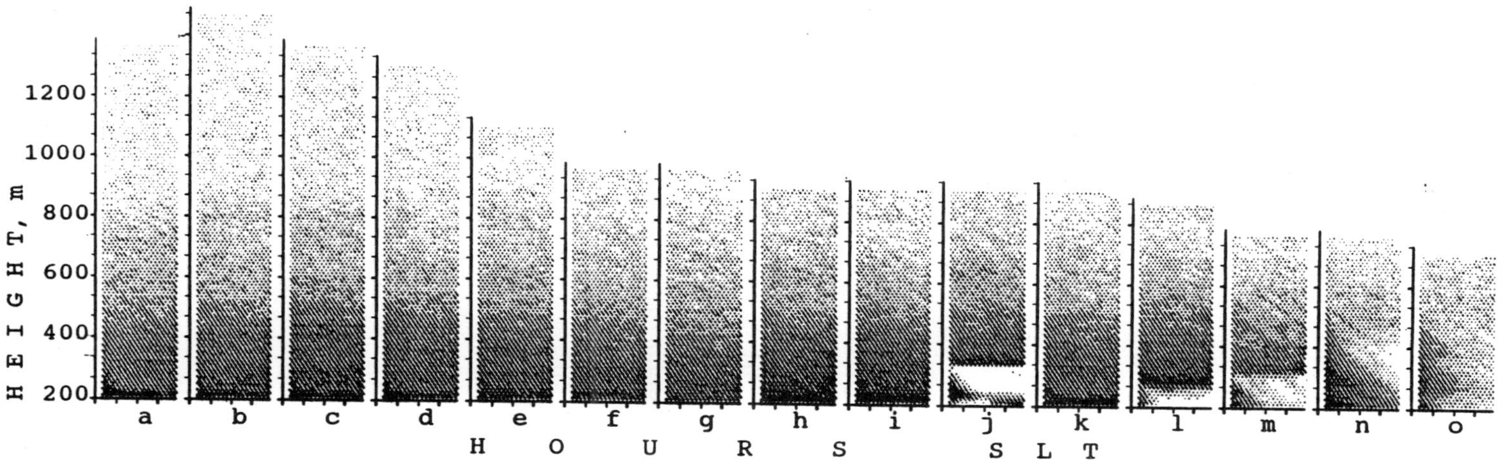

Figure 5. Visualization of the elements of the correlation matrices Mij. A eight grade gray scale is used to note the values from 0 to 1 of the ij element. "i" denotes the height of the sounding / see Equations (1 and 2) / at which a time serial is registered, and "j" denotes next spatial series / see Equation (3) /. The same letter is used to indicate the M and its corresponding two-dimensional lidar realization on Figure 3. The following hours are : a - 5:45; b - 6:08; c - 6:32; d - 6:55; e - 7:24; f - 7:55; g - 8:23; h - 8:47; i - 9:07; j - 9:26; k - 9:44; l - 10:18; m - 10:37; n - 10:58; o - 11:17 SLT.

250 m, which coincide with the upper boundary of the ground inversion. These two stages of development of the aerosol stratification last until the appearance of vertical movements.

After sunrise the ground inversion disintegrates and a convective layer, adjoining to the surface appears at the lidar images. This layer is typical with its horizontal inhomogeneity as a periodically repeated structures. Their growth in height and duration is interpreted as carrying away of the ground antropogenic aerosol by the convective turbulence. At the end of the lidar observation, structures which can be interpreted as "roll convection" are observed at the presence of developed convection and horizontal advection.

Apart from the visual criteria for classification of the processes described above, the behavior of another parameter of the stochastic aerosol field can also be traced.

Following (Stull, 1988), we can describe the turbulence and the aerosol fluctuations as coherent structures in the two-dimensional images. For that purpose we define Mij correlation matrix, with time series correlation coefficients obtained from i and j heights as its elements. The graphic matrix image is presented on Figure 5. Three types of matrix element arrangement by values, coinciding with the three stages of the change of the aerosol stratification type are also observed. This is due to a matrix property, namely, its elements by rows are connected with the values of the spatial correlation function, when the condition:

$$j = i+p, \tag{3}$$

where p is a whole positive number is satisfied.

DISCUSSION AND CONCLUSION

The height of the layer observed in the lidar images as horizontally inhomogeneous with periodic structures are juxtaposed with the model data for the growth of the mixed layer which is shown on Figure 2b. While the destruction or the raising of the ground inversions are an inert process and is registered inoperatively, on the basis of moment values of the heat and humidity fluxes, the moment values of the height which the vertical movement reach can be obtained by model running. The lidar images show the processes of trapping of aerosol by temperature inversion and penetration in stable stratified layers.

The coincidence in the behavior of the model and lidar measurement results show that the aerosol can be used as a tracer for measurement of the scales and the dynamics of the convective turbulence.

REFERENCES

Eloranta E.W., Crum T.D., 1992, Calculation of area-averaged profiles of horizontal wind velocity from volume-imaging lidar data, *Journal of Geoph. Res.*. 97:18395.

Batchvarova E., Gryning S.E., 1991, Applied model for the growth of the day time mixed layer, *Boundary Layer Meteorol*, 56:261.

Parvanov O., Kolev I., Batchvarova E., 1997, Lidar measurements of planetary boundary layer structure and comparison with mixed layer height model results, *Comptes rendus de l'Academie bulgare des Sciences*, 50:37.

Parvanov O., Kolev I., Kaprielov B, 1998, Lidar measurements of the wind velocity profiles in the planetary boundary layer, *Proc of the 19 Int. Laser Radar Conference-10 -16 June, Maryland, USA*, 2:331

Stull R. B., 1988, *An introduction in Boundary Layer Meteorology*, Kluwer Academic Publishers, Dordreht.

POSTER SESSION

REMOTELY SENSED SURFACE HEAT FLUX FOR AIR POLLUTION DISPERSION STUDIES

Koen De Ridder and Clemens Mensink

Flemish Institute for Technological Research (VITO)
Centre for Remote Sensing and Atmospheric Processes
Boeretang 200, B-2400 Mol, Belgium

INTRODUCTION

Model studies of urban to regional scale air pollution dispersion often encounter the difficulty of having to specify surface sensible heat fluxes and atmospheric stability. Since these parameters are not routinely available from meteorological networks, a method is being developed to deduce their spatial distribution from satellite remote sensing. A special effort has been made to use only satellite data that are easily available and to avoid the use of ground-based meteorological data.

METHODOLOGY

The method is based on the surface energy balance equation and on the concept of evaporative fraction. The latter is a parameter that governs the partitioning of available energy (net radiation minus the soil heat flux) between the turbulent fluxes of sensible and latent heat. The evaporative fraction is largely determined by the amount of green, transpiring vegetation, and by the humidity of the upper layer of bare soil patches. The methodology proceeds by the following steps:

- Visible reflectances from a geostationary satellite such as GOES or METEOSAT are used to infer the incoming solar radiation flux, the cloud fraction, and the surface broadband albedo, following a method outlined by Frouin and Gauthier (1990). These parameters in turn allow the computation of the surface radiation balance.
- Red and near-infrared reflectances from NOAA-AVHRR are used to compute the Normalised Difference Vegetation Index (NDVI). This index exploits the high spectral contrast between both wavebands in the presence of vegetation. The fractional vegetation cover is linearly related to the NDVI, as in Wittich and Hansing (1995).
- The microwave Polarisation Difference Temperature (PDT), defined here as the difference between the vertically and horizontally polarised brightness temperatures at

19.35 MHz as measured by DMSP-SSM/I, is converted to moisture content in the upper layers of the soil by a recently developed method (De Ridder and Mensink, 1998).

The land surface is divided in vegetated and bare soil (or urban surface) patches, and the sensible heat flux is separately computed for both surface types. The total sensible heat flux is then obtained as their areal weighted average.

DISCUSSION

The methodology described above was applied to the area of the First International Satellite Land Surface Climatology Project (ISLSCP) Field Experiment (FIFE), located in the Central United States and mainly covered by grassland. The period considered is the Intensive Field Campaign that took place in the first half of August 1987.

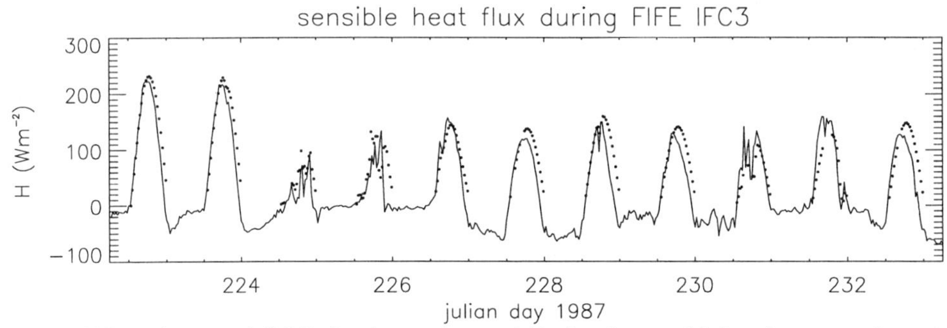

Figure 1. Ground-measured (full line) and remotely sensed (dots) surface sensible heat flux averaged over the FIFE area for the period 10 – 20 August 1987.

Heavy rains on Julian day 226 caused the upper soil layers to change from very dry to wet conditions, which resulted in a decreasing trend of the sensible heat flux. Figure 1 shows that the satellite-based method is capable of reproducing the ground-measured diurnal cycle of the sensible heat flux rather well, most of the time within experimental error. The main problem appears to be that afternoon heat flux values are over-estimated by the present method, which may be due to the fact that differential advection is not represented.

It is stressed that the results presented in Figure 1 (dots) were obtained without using ground-based meteorological data at all. Future developments will include extension of the method to larger areas, including urban surfaces, and to the night-time surface energy balance.

REFERENCES

De Ridder, K., and C. Mensink, 1998, Integration of SSM/I polarised microwave brightness temperature in surface energy balance retrieval schemes. To be submitted to *Remote Sensing of the Environment.*

Frouin, R., and C. Gautier, 1990, Variability of photosynthetically available and total solar irradiance at the surface during FIFE: A satellite description. AMS Symposium on the First ISLSCP Field Experiment (FIFE). Anaheim, CA. February 7-9.

Wittich, K.-P., and O. Hansing, 1995, Area-averaged vegetative cover fraction estimated from satellite data. *Int. J. Biometeorol.*, 38:209.

STUDY OF THE EFFECTS OF THE RILA AND RHODOPE MOUNTAINS ON THE MESOSCALE POLLUTION TRANSPORT

Kostadin Ganev[1] and Reneta Dimitrova[1]

[1]Institute of Geophysics, Bulgarian Academy of Sciences
Acad G. Bonchev str., block 3, Sofia 1113, Bulgaria

INTRODUCTION

There is no doubt that the mountains of the Rila and Rhodope massif cause significant mesoscale deformations of the air flows. On the other hand it is clear that the local flow systems, caused by the horizontal heterogeneity, may have a great influence on the local pollution transport and hence on the detailed pollution pattern. That is why the study (a very preliminary one) of the mesoscale topography effects in the region and their influence on the air pollution transport, which is the objective of the present work, is interesting from both scientific and practical points of view.

NUMERICAL SIMULATIONS AND CONCLUSIONS

A 3D quasi-hydrostatic model of the mesoscale dynamics[1], based on the Businesque approximation (the formulation of Guthman) is used for the purpose. A large number of numerical experiments are carried out under different background conditions. Some of the results are shown in Fig. 1. The simulated flow systems outline the main topography effects, typical for the region - slope winds, channeling of the air flow along the river valleys and blocking effects under stable conditions, etc.

Some pollution transport calculations were carried out together with the dynamic simulations. They were made for an imaginary source, whose location is shown in the figure. The pollution distributions from this source convincingly demonstrate the significance of the topography influences on the local scale pollution in the region. They are obviously different from the conventional "plume" distribution not only in the pollution propagation direction, but also in the pollution field shape.

Air Pollution Modeling and Its Application XIII, edited by S.-E. Gryning and E. Batchvarova, Kluwer Academic/Plenum Publishers, 2000

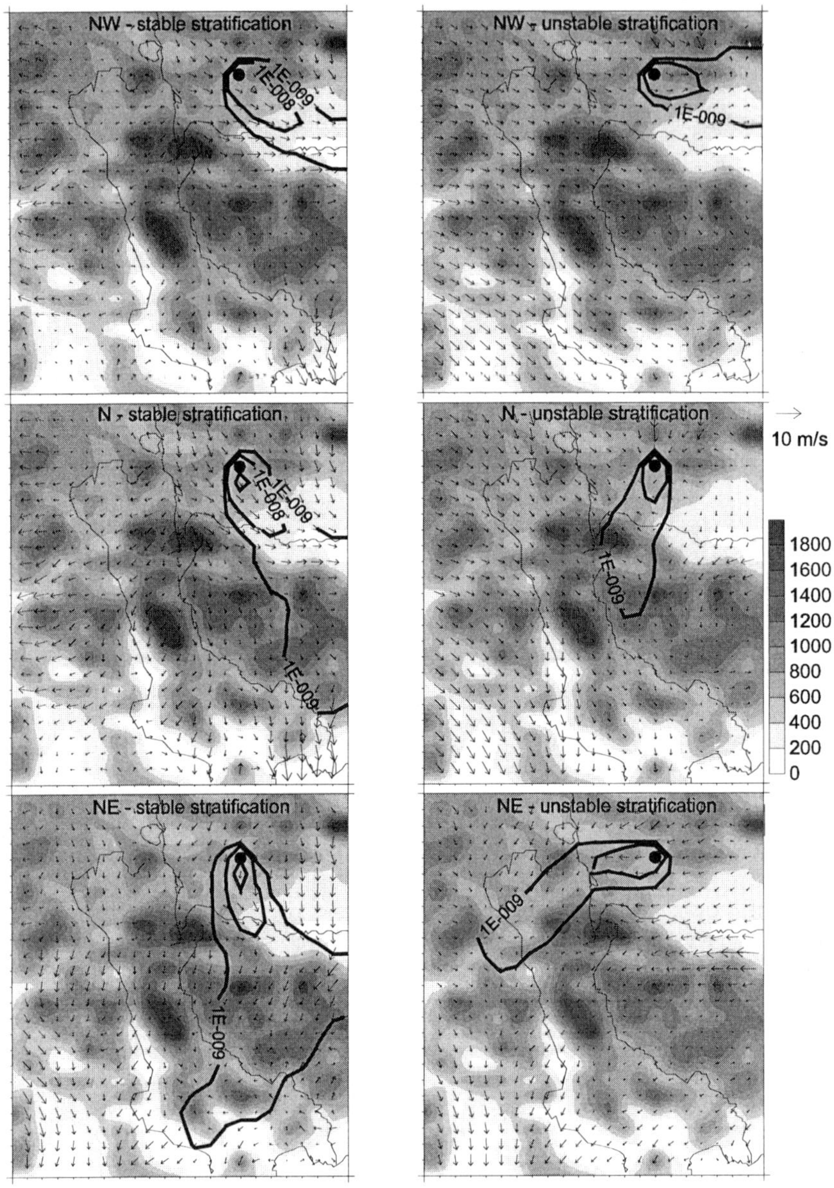

Figure 1. Wind field at $z = 10m$ and corresponding surface concentration fields for different background wind directions and stability conditions. The wind velocity for all the cases is 5m/s. The topography and the main rivers in the region are also shown.

ACKNOWLEDGEMENTS

The present work is sponsored by the National Science Fund under Grant No. H3-628.

REFERENCES

1. Ganev K., Numerical study of the local flow systems in the "Kozloduy" NPP region - some preliminary results. *Bulgarian Geoph. J.*, v. XIX, No 1, 9 - 23 (1993)

COMPARISON OF THE CAPABILITIES OF THE DECPAC AND RAINS MODELS AS TOOLS FOR INTEGRATED ASSESSMENT IN THE ENERGY PLANNING PROCESS

Antoaneta Iotova, Todor Todorov, Dimiter Atanassov

National Institute of Meteorology and Hydrology
Blvd. Tsarigradsko chaussee 66
1784 Sofia, Bulgaria

THE DECPAC AND RAINS MODELS IN BRIEF

DECADES is a joint inter-agency project established to enhance the capabilities for comparative assessment of different electricity generation options and to develop strategies in support of the energy planning and decision making process. A broad and complex set of data for the energy sources, technologies (technical, economic and environmental characteristics) and energy chains is stored in specially developed databases for this purpose. An essential part of the DECADES' software package (DECPAC) is designed to perform power plant, chain and system-level analyses including estimations of the costs and the environmental burdens at all these levels (Vladu, 1997). On the basis of data in the Bulgarian Country Specific Database different analyses and comparative assessments for the Bulgarian electricity sector are realised using the DECPAC. Some realistic scenarios are constructed in order to optimise the possible strategies for the development of this sector in the period 1997 - 2020. As an example, the SO_2 emissions from the Bulgarian thermal power plants under the scenario "coal priority" are presented in Figure 1 where the level "electricity generation" is considered.

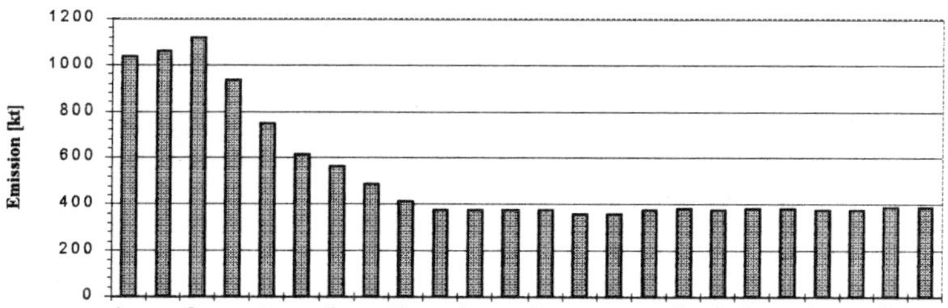

Figure 1. Annual SO_2 emissions [kt] from all Bulgarian thermal power plants.

The Regional Acidification INformation and Simulation (RAINS) model has been developed as a tool for integrated assessment of alternative scenarios for reduction of the emissions of sulphur dioxide

(SO_2), nitrogen oxides (NO_x) and ammonia (Cofala, et al., 1998). As a kind of integrated assessment model, and technology oriented one, the RAINS model can be used for a real environmental aspects analysis of the electricity generation options. The so called meteorological matrixes calculated for 11 years period by corresponding models (developed in the EMEP framework) are incorporated in the RAINS model for this purpose. In the case of Bulgaria, the model is an useful analytical tool for energy planning purposes because of its orientation to calculate SO_2 and NO_x emissions, the cost estimations provided in the model, the dominating role of the fossil fuels (domestic lignite coal especially) for electricity generation in the country, etc. The results from the RAINS model (version 7.2) application for Bulgaria are illustrated in Figure 2.

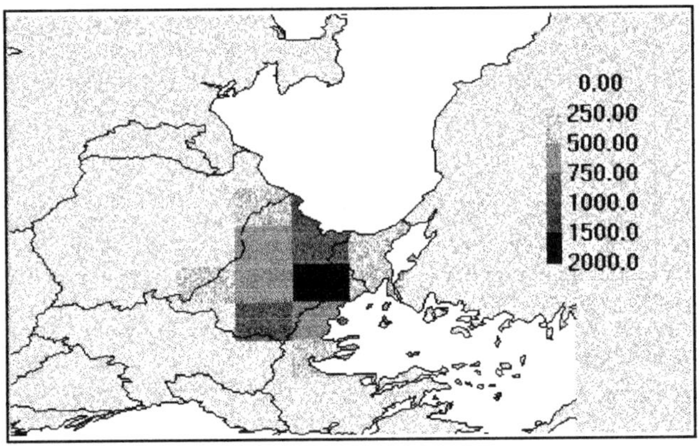

Figure 2. Difference between the SO_2 depositions [mg/m^2.year] in 2010 as a result of two scenarios for the Bulgarian thermal power plants: "maximum technical feasible emission reduction" and "no reduction at all".

COMPARISON BETWEEN THE DECPAC AND RAINS MODELS

Both models contain modules where results from the air pollution modelling are applied. In the DECPAC only the emissions are calculated by simple empirical equations while the RAINS model estimates the total energy-related impact. It has separate modules for the emissions calculation, for their transport and for the deposition estimation over a region (Europe, Asia). The DECPAC model permits more detailed estimations and comparisons of the emissions: by sources (power plants), by energy chains and from the whole system. The RAINS model' estimations are for the whole energy sector or for all the power plants.

In conclusion, the examined models are of use for the energy planning process by providing estimations of the energy related environmental impact and the costs of its reduction as necessary for the development of relevant scenarios. The RAINS model's output is much more complete while the DECPAC provides possibilities to use very detailed input data.

REFERENCES

Cofala, J., Amann, M., Heyes, Ch., Klimont, Z., and Schoepp, W., 1998, Integrated assessment of acidification and tropospheric ozone in Europe: impact of current legislation - In Proceedings of the Swedish-Bulgarian Workshop on Long-Range Air Pollution: From Models to Policies, Pensoft, Sofia

Vladu, I. F., 1997, Comparative assessment of technical, economic and environmental parameters of different generation options - Paper presented during the Technical Committee Meeting/ Workshop, 1 - 12 December 1997, Argonne National Laboratory, Argonne, USA

ESTIMATE OF IMPACT ON REGIONAL OZONE LEVEL IN ONTARIO DUE TO EMISSIONS CHANGE OF FOSSIL POWER GENERATION

Xiude Lin[1] and Wanmin Gong[2]

[1]Ontario Hydro Technologies, Toronto, Ontario, Canada
[2]Atmospheric Environment Service, Downsview, Ontario, Canada

INTRODUCTION

Ontario Hydro's business plan forecasts an increase of its fossil power generation for the next few years. The annual energy production from Ontario Hydro's fossil generating stations is expected to increase from 1997's 25 TWh up to 35 TWh in 1998-2000. This increase will lead to an increase in the stack emissions of NO_x. To assess the impact of the NO_x emissions increase on regional O_3 levels and related environmental exposure, a study was carried out.

In the study, we used a comprehensive air quality modelling system (Lin et al., 1996; Gong et al., 1998), which interfaces MC2 (Mesoscale Compressible Community model) (Tanguay et al., 1990) and ADOM (Acid Deposition and Oxidant Model) (Venkatram et al., 1988). We applied the system to a regional O_3 episodic covering most of eastern North America during the 1988 EMEFS-1 intensive measurement campaign (Hansen et al., 1991). The model simulation was started with a coarse horizontal resolution of 127 km followed by a nesting scheme to zoom into a 21-km fine mesh. Model results were analysed for a broad region covering central and southern Ontario where 80% of Ontario population live and high O_3 episodes occur.

SCENARIO DESIGN, MODELLING RESULTS AND DISCUSSION

Two NO_x emissions scenarios were designed and then applied to the 8-day regional O_3 episode period from July 30 through August 6 of 1988. The first scenario represents a base case including the 1997 stack emissions of NO_x from Ontario Hydro's fossil generating stations. In the second scenario, the Ontario Hydro's emissions are increased by 40% to reflect the generation increase from 25 TWh to 35 TWh in 1998. Two cumulative O_3 exposure indices at grids were introduced for a quantitative analysis of the modelled results. One, SUM60, is defined as the summation of hourly ground level ozone exceeding 60 ppbv over the simulated episode period while the second index, SUM40, uses 40 ppbv as the threshold level.

The results show that the region-averaged indices weighted by the population density from the 1997 emissions scenario to the one with higher stack NO_x emissions. The average of SUM60 decreases by 1.3% while the average of SUM40 by 0.6%. The reason for the decreases in the O_3 exposure lies in the facts that an increase in the stack NO_x emissions induces a near-range depression and a far-range enhancement for O_3 and that the near-range O_3 depression coincides with a more densely populated area rather than the far-range O_3 enhancement.

To further investigate the role of the NO_x emissions change in the context of regional O_3 formation, we calculated O_3 percent changes from the 1997 base case to the 1998 case at all domain grids for afternoon hours from 4 p.m. to 8 p.m. The percent changes were plotted against their corresponding 1997 NO_y (defines as $NO+NO_2+HNO_3+PAN$) levels in Figure 1. It is seen that the O_3 changes are generally correlated with the NO_y levels. An increase in NO_x emissions from 1997 to 1998 generally results in an increase of O_3 formation at locations where

Air Pollution Modeling and Its Application XIII, edited by S.-E. Gryning and E. Batchvarova, Kluwer Academic/Plenum Publishers, 2000

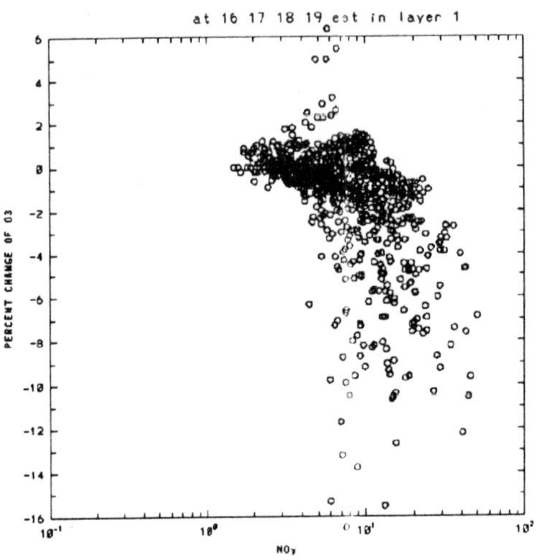

Figure 1. O_3 percent change from 1997 to 1998 vs NOy level.

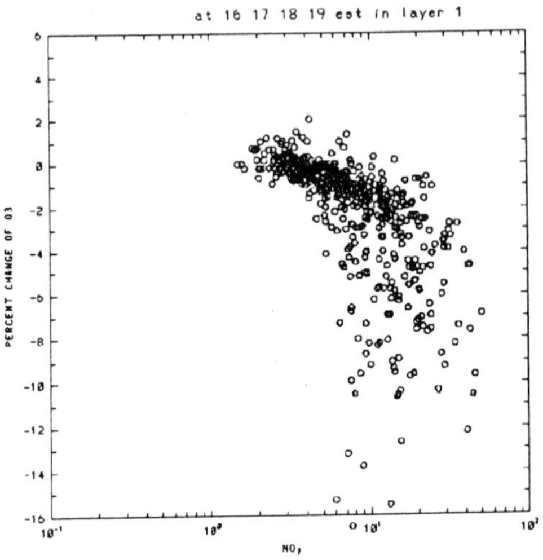

Figure 2. Same as Fig. 1 but for $NO_x/NO_y>0.5$.

NO_y is below 5 ppbv. In the regions with the background NO_y above 10 ppbv, O_3 is suppressed by the NO_x emissions increase. The general feature of the finding is in line with Sillman (1995) and Prevot et al. (1997). The 5-10 ppbv range of NO_y represents a transition zone which separates a NO_x-sensitive O_3 production region from a VOC-sensitive O_3 production region.

When the age of air mass, as defined by the ratio of (NO_y-NO_x) to NO_y, is considered, the correlation between the O_3 changes and the NO_y background could be further refined. By choosing the ratio value below 0.5, we excluded the data pairs associated with older air masses. The remaining data are presented in Figure 2. It is clear that Figure 2 delivers a better correlation between the O_x changes and NO_y levels, and indicates a transition zone at lower NO_y levels than Figure 1. This suggests that the generally held relationships of NO_y to the balance between VOC and NO_x involve a significant photochemical age dependence.

CONCLUSION

In this study, we used a regional air quality modelling system with 21-km horizontal resolution to assess the impact of the NO_x emissions increase of Ontario Hydro. The impact was found minimal. Further analysis indicated that O_3 changes due to the NO_x emissions increase were correlated to NO_y levels and the correlation was affected by the photochemical age of air masses. The finding could have applications in the practice of NO_x emission reduction trading.

REFERENCES

Gong, W., Lin, X., Menard, S., Pellerin, P., and Benoit, R., 1998, A regional modelling study of transport affecting the Canadian southern atlantic region oxidants, The 23rd NATO/CCMS/ITM.

Hansen, D.A., Puckett, K.J., Jansen, J.J., Lusis, M., and Vickery, J.S., 1991, The eulerian model evaluation field study (EMEFS), *Proc. 7th Joint Conference on Applications of Air Pollution Meteorology with AWMA*, AMS, pp 58-62.

Lin, X., Gong, W, and Menard, S., 1996, A new MC2/ADOM modelling system and its application to the study of the oxidants in the Canadian Southern Atlantic Region, *Air Pollution Modelling and Its Application*. XI:275.

Prevot, A.S.H., Staehelin, J., Kok, G.L., Schillawski, R.D., Neininger, B., Staffelbach, T., Neftel, A., Wernli, H., and Dommen, J., 1997, The Milan photooxidant plume, *J. Geophys. Res.* 102:23375.

Sillman, S., 1995, The use of NO_y, H_2O_2, and HNO_3 as indicators for ozone-NO_x-hydrocarbon sensitivity in urban locations, *J. Geophys. Res.* 100:14175.

Tanguay, M., Robert, A., and Laprise, R., 1990, A semi-implicit semi-Lagrangian fully compressible regional forecast model, *Monthly Wea. Rev.* 118:1970.

Venkatram, A., Karamchandani, P.K., and Misra, P.K., 1988, Testing a comprehensive acid deposition model, *Atmos. Environ.* 22:737.

THE USE OF SCAVENGING RATIOS IN AIR POLLUTION TRANSPORT MODELLING, AND CALCULATED RATIO VALUES

Göran Nordlund and Viktor Makara

Finnish Meteorological Institute
Air Quality Research
P.O.Box 503, FIN –00101, Helsinki

INTRODUCTION

The scavenging ratio, i.e. the concentration of a certain pollutant in precipitation divided by the concentration of the pollutant in air may be given arbitrarily good physical relevance by parametrization of the ratio so as to be valid under different dispersion and scavenging conditions. However, in practice the scavenging ratio is a rough instrument, and a certain value of the ratio is assumed to cower a wide range of different dispersion and scavenging cases.

DATA AND CALCULATIONS

For studying the scavenging ratio, daily (24-hour) values were calculated using the measured pollution concentrations in precipitation and in air. The components studied were sulphur (SO2-S and SO4-S) and nitrogen (NO2-N, (HNO3+NO3)-N and (NH3+NH4)-N in air , and NO3-N and NH4-N in precipitation). The data were taken from the Finnish stations in the UN/ECE/EMEP-programme, Utö, Virolahti, Ähtäri and Oulanka over the years 1989-1996.

Average values of the ratios and standard deviations were calculated using all data and separately by excluding the five percent largest values of the ratios. This was done in order to eliminate extreme cases. These extreme cases were, according to separate analyses made, mostly connected with situations, in which there were extremely sharp vertical concentration profiles with very low concentrations in the lowest atmospheric layer.

RESULTS AND CONCLUSION

Due to limited space, it is only possible to present here a selection of the results. Table 1 gives the overall results, i.e. the average values and standard deviations of the scavenging ratios.

Air Pollution Modeling and Its Application XIII, edited by S.-E. Gryning
and E. Batchvarova, Kluwer Academic/Plenum Publishers, 2000

Table 2 shows the relation between the scavenging ratios of different components. In calculating the averages the 95 percentiles were used. These relations between scavenging ratios are physically more interesting than the ratios themselves. Since the monitoring stations used are far from major source areas, the vertical distributions of the different pollutants may be assumed to be relatively similar. The differences in vertical distributions between single cases, which are important for the scavenging ratios, are therefore fairly well eliminated in the relationship between the ratios of different components.

Interesting results were found in studying the scavenging ratios in relation to precipitation amount. According to the results (Figures are shown in the Poster presentation) there is a clear and smooth decrease in the ratio values with increasing precipitation amount. This smooth dependence and the relatively small standard deviations, indicate that the scavenging ratio is a relative robust parameter. Thus, using appropriate parametrization, the scavenging ratio may still be a useful parameter in air pollution dispersion modelling.

Table 1. Scavenging ratios (10^6) of different acidifying components

Utö	SO_4^{2-}/SO_4^{2-}	SO_4^{2-}/SO_2	$SO_4^{2-}/(SO_4^{2-}+SO_2)$	$NO_3^{1-}/(NO_3^{1-}+HNO_3)$	$NH_4^{1+}/(NH_3+NH_4^{1+})$
Average	2.54	3.31	1.20	3.26	1.69
Std. Dev.	9.41	11.66	5.04	9.72	2.68
95% Average	1.78	2.47	0.88	2.51	1.32
Std. Dev.	1.66	2.31	0.73	2.20	1.35
Virolahti					
Average	1.39	1.76	0.62	2.17	0.92
Std. Dev.	1.62	2.63	0.58	1.94	1.52
95% Average	1.17	1.35	0.55	1.89	0.73
Std. Dev.	0.78	1.21	0.37	1.24	0.58
Ähtäri					
Average	1.30	3.60	0.74	3.54	0.76
Std. Dev.	1.79	8.32	0.82	3.60	2.06
95% Average	1.05	2.53	0.63	3.04	0.60
Std. Dev.	0.84	2.91	0.54	1.92	0.52
Oulanka					
Average	1.24	4.60	0.72	4.63	1.21
Std. Dev.	2.43	14.16	1.32	5.40	6.98
95% Average	1.05	2.65	0.57	3.82	0.73
Std. Dev.	1.19	3.91	0.57	2.74	0.76

The notation $SO4^{2-}/SO2$ (here used as an example) denotes values calculated from the daily ratios of $SO4^{2-}$-S in precipitation to the concentration of SO2-S in air. The 95% averages and the corresponding values of standard deviation represent values from which, the 5% largest single scavenging ratios of the respective

Table 2. Relations between scavenging ratios

	R1		R2		R3	
Station	a	b	a	b	a	b
Utö	0.72	1.21	0.71	0.76	1.90	2.91
Virolahti	0.87	1.56	0.62	0.74	2.59	4.22
Ähtäri	0.42	1.11	0.35	0.42	5.07	11.53
Oulanka	0.40	1.40	0.27	0.31	5.23	10.24

$R1= (SO_4^{2-}/SO_4^{2-})/(SO_4^{2-}/SO_2)$
$R2=(SO_4^{2-}/SO_4^{2-})/(NO_3^{1-}/(NO_3^{1-}+HNO_3))$
$R3=(NO_3^{1-}/(NO_3^{1-}+HNO_3))/(NH_4^{1+}/(NH_3+NH_4^{1+}$
In the relationships the 95% percentile values are used. The values a are computed from the averages of Table 1. The values b are based on day-to-day relationship values.

AN INVESTIGATION OF THE DYNAMICAL AND PHYSICO-CHEMICAL FEATURES OF DESERT AEROSOL OVER WEST AFRICA AND EASTERN ATLANTIC

Guy Cautenet and Frédéric Pradelle (Université Blaise Pascal, LaMP), Clermont-Ferrand – France
Michel Legrand and Olivier Chomette (Université de Lille 1, LOA), Lille – France
Gilles Bergametti and Béatrice Marticorena (Université Paris 12, LISA), Paris – France

INTRODUCTION

The possible strong climatic influence of mineral aerosol (almost 1 gigaton of desert dust annually mobilized by wind erosion from the arid or semi-arid areas of Northern Africa is transported towards Mediterranean or Atlantic basins) encourages to develop mesoscale modeling description of all the aspects of the dust life cycle: wind uptake, transport, removal, direct or indirect radiative forcing. This poster deals with a mesoscale description of some of these questions. We have simulated a dust outbreak over Atlantic (at the beginning of May 1997) with the mesoscale model RAMS (Regional Atmospheric Modeling System). The main aspects addressed here are (i) dynamics: checking of dust trajectories using satellite imagery and (ii) an attempt to assess the radiative response using local surface measurements (at Sal Island, Cabo Verde).

THE OBSERVATIONAL DATA

In the beginning of May 1997, a dust cloud was observed by VIS Meteosat imagery while outbreaking over the Mauritanian coast, West Africa (Dulac, 1997, pers. comm.). The Meteosat IRT spectrum (Legrand, 1998, pers. comm.) provided information about aerosol over land. A ground based multichannel radiometer at Sal Island, Cabo Verde (24°W, 16°N) recorded an increase in the optical depth. Finally, meteorological observational records around 1200 UTC in 7 African stations were available from the meteorological African network via the ECMWF. All these data were used to check crudely the accuracy of the modeled dust cloud life cycle.

THE MODELS

We use the RAMS model of the Colorado State University in its non-hydrostatic version (Pielke et al., 1992). The modeled domain area ranges from 40°W to 15°E, and from 5°N to 40°N. The horizontal grid step is 100 km. The vertical is divided into 30 steps, increasing with height. The first 10 levels are located within the lowest 1500 metres. The time step is 60s. The model is initialized, and nudged every 6 hours, using the ECMWF database. The dust source scheme developed by Marticorena and Bergametti (1995) is coupled to RAMS. The aerosol spectrum is derived from Alfaro and Gomes (1998) experimental data: we use a 20 bins spectral scheme from 0.1 to 13 μm. The native radiative model has been replaced by a complex code based upon ECMWF radiative model which accounts for radiative properties of the desert aerosol.

RESULTS

Dynamics: We have checked the capability of the model to account for the meteorological fields: wind, temperature and water. Due to the scarceness of observation stations in Africa, we have limited (i) to control some of the main features of the regional climatology and (ii) to directly compare modeled and measured data. The agreement between model and observed fields was found globally satisfactory.
Dust raise and transport: The total value of mobilized dust mass from May 2 to May 6 is 9.1 Mt, a value which is of the same order than the previous estimates (Marticorena et al, 1997). Over land, the modeled dust source areas and dust cloud trajectory agree generally with IR data. The modeled aerosol outbreak ocurs on May 5. On May 6, the model concentrations over the coastline are decreasing, while the quantities over Cabo Verde islands are increasing as the dust cloud moves westward. From these data, we may conclude that the model results agree qualitatively with satellite data.

Air Pollution Modeling and Its Application XIII, edited by S.-E. Gryning and E. Batchvarova, Kluwer Academic/Plenum Publishers, 2000

Dust removal: We have focussed mainly on dry scavenging (during the episode, limited to a large extent to a continental trajectory north of ITCZ, water clouds may be neglected, in a first approximation). The areas of higher deposition (>5 g/m2) are close to the major source. Near the coastline, the deposition may still reach 1-2 g/m2. The total deposited mass during the 5 days is 5.6 Mt. Wet mechanisms have been briefly simulated: the effect of a stratocumulus cloud in an area where aerosol exists (LWC of 0.4 g/kg, lognormal distribution of droplets, mean radius18 μm) was simulated using an in-cloud capture scheme. All the main processes are considered: brownian processes, thermophoresis, diffusiophoresis, turbulence, gravity impaction. After a 10-h stay within the stratocumulus, about 50% of the mineral matter has been captured by the liquid water phase.

Radiation tests: A first test of the new radiative scheme is presented now. Figure 1 shows a comparison of the measured 669 nm optical depth with the model values of this parameter at Sal Island station (Cabo Verde) at surface level during the dust cloud fly over the island (May 5 and 6). The model value represents the pure effects of the aerosol (excluding any atmospheric gases or water contribution). For this reason, we have subtracted 0.3 from the experimental value, which represents this "background" contribution. If spherical shape is assumed for particles, we note a large underestimation, even though both curves vary rather well in phase. We see that using a surface ratio between 1.5 and 2.5, we can capture the variation of the optical depth (surface ratio is the projected area of the ellipsoid having the same volume than a sphere of radius R, divided by πR^2). The true shape of the aerosol must be carefully checked in view to assess these estimates on experimental basis.

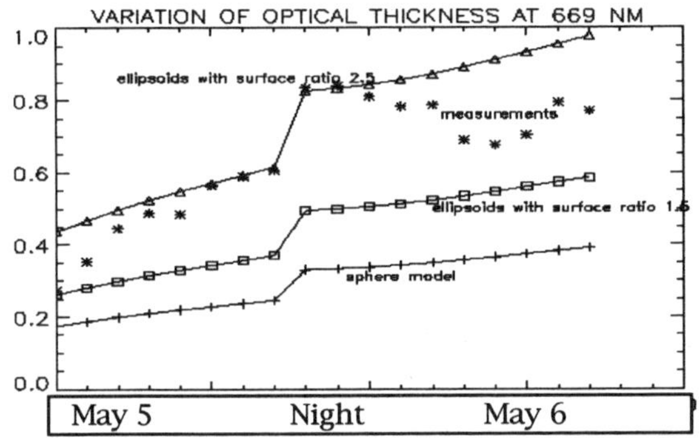

Figure 1. Variation of the 669 nm optical depth at Sal Island on May 5 and 6.

SUMMARY AND CONCLUSIONS

The present poster intends to describe a mesoscale approach of the cycle of a major atmospheric natural contaminant, the desert aerosol. The mesoscale is important because it allows detailed studies of all the involved processes: source, vertical transport, interaction with other components (clouds for instance, or external mixing with another aerosol or gas, ...). It is a necessary step to find parameterizations relevant to global scale. We have presented a simulation of a particular desert dust event in order to check the quality of our modeling with the available observational data (satellite, meteo, radiative). Obviously in the present state, strong constraints such as several vertical soundings in dust concentrations are unfortunately not available. However, it has been possible to retrieve the gross features of the dust cloud trajectory, in particular the place and date of the outbreak towards Cabo Verde islands. The radiative scheme test showed the necessity to consider more realistic shapes for particles.

Acknowledgements

This work is supported by fundings of the French "Centre National de la Recherche Scientifique" (Programme National de Chimie Atmosphérique).

References

Pielke, R. et al (11 names), A comprehensive meteorological modeling system - RAMS. *Meteorol. Atmos. Phys.*, **49**, 69-91, 1992.

Marticorena, B. and G. Bergametti, Modeling the atmospheric dust cycle: 1- Design of a soil-derived dust emission scheme. *J. Geophys. Res.*, **100**, 16415-16430, 1995.

Marticorena, B., G. Bergametti, B. Aumont, Y. Callot, C. N'Doumé and M. Legrand, Modeling the atmospheric dust cycle: 2- Simulation of saharan dust sources. *J. Geophys. Res.*, **102**, 4387-4404, 1997.

Alfaro, S., A. Gaudichet, L. Gomes and S. Maillé , Mineral aerosol production by wind erosion: aerosol particle sizes and binding energies, *Geophys. Res. Letters*, **25**, 991-994, 1998.

COUPLING OF SODAR DATA WITH SIMPLE NUMERICAL MODELS

Dimitrios Melas[1,2], Giulia Abbate[3], and Harry Kambezidis[4]

[1]Laboratory of Atmospheric Physics, University of Thessaloniki, Thessaloniki, Greece
[2]Department of Environmental Studies, University of the Aegean, Mytilene, Greece
[3]ENEA, Casaccia, Italy
[4]National Observatory of Athens, Greece

INTRODUCTION

Contemporary concepts of air quality monitoring and/or assessment focus on the synergistic use of experimental data and numerical models. Along this line, an attempt has been made to combine numerical models and sodar data in order to estimate the relevant meteorological parameters during daytime conditions. The parameters considered include the wind field, the surface fluxes of heat and momentum as well as the mixed-layer depth. The experimental data used in the present study are taken from three doppler sodars that were operating in Rome area.

THE MODELS

The methodology implied in the present paper comprises two numerical models and an analytical model based on similarity relationships for the calculation of the surface fluxes of heat and momentum from sodar data. A full description of the model system is given in Melas et al. (1998b). In brief, the following models are used:

- A three-dimensional diagnostic wind field model that satisfies the continuity equation (Melas et al., 1998a). The input data of the computing program consist of topography height, land use and available measurements of wind speed and direction at various levels.
- A mixed-layer model developed by Batchvarova and Gryning (1996). The model is based on parameterizations of the turbulent kinetic energy budget and the potential temperature jump at the top of the boundary layer. The mixed-layer model can be applied in coastal areas with an irregular coastline and takes into account the spatial and temporal variations of the wind field.
- Similarity methods for the calculation of surface fluxes of heat and momentum from sodar data. The proposed methodology incorporates a framework for calculating the surface fluxes of momentum and heat from sodar measurements of mean wind speed and vertical variance. (Melas, 1993).

RESULTS

The measurements used in the present study were taken during a field campaign that was carried out in Rome and the surrounding area during summer 1992 (Mastrantonio et al., 1994). The equipment consisted of three Doppler sodars which were operating simultaneously at three sites at different

Air Pollution Modeling and Its Application XIII, edited by S.-E. Gryning and E. Batchvarova, Kluwer Academic/Plenum Publishers, 2000

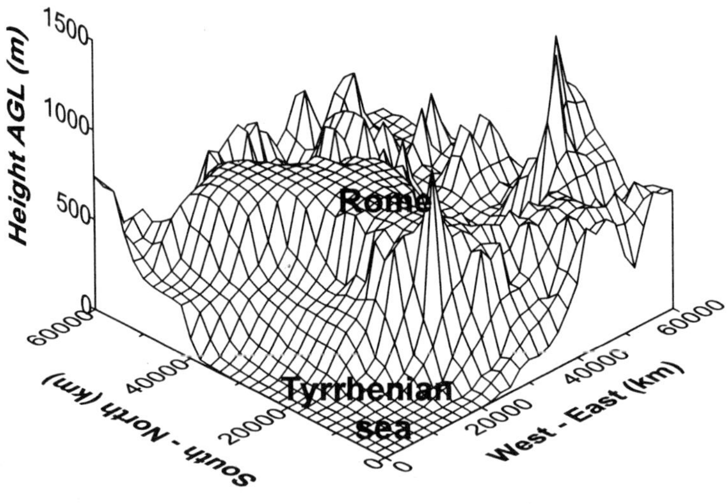

Figure 1. The mixed-layer depth at July 18, 1992, 15.00 LST.

distances from the shoreline. The model system was applied for a domain 60 x 60 km^2 with a grid resolution of 2 km. The results include the diurnal variation of the surface fluxes of heat and momentum, the wind fields and the mixed-layer depth. The results of the diagnostic wind model for the 18th of July, 1992 show that during morning hours a sea breeze system develops over the Tyrrhenian sea and advances rather slowly over the land (not shown here). This has a large impact on both the calculated surface fluxes as well as the mixed-layer depth. Figure 1 shows the horizontal distribution of the mixed-layer depth at 15.00 LST when the sea-breeze system is well developed. As it is revealed in figure 1, the sea breeze has a large impact on the boundary layer development, leading to the development of Internal Boundary Layers (IBLs).

Acknowledgments. The work carried out by the first and the third authors was supported in part by the Greek Ministry of Industry, Research and Technology (contract 5672/30.5.97).

REFERENCES

Gryning, S.-E. and E. Batchvarova, 1996: A Model for the Height of the Internal Boundary Layer over an Area with an Irregular Coastline. Bound.-Layer Meteor., 78, 405-413.

Mastrantonio, G., A. P. Viola, S. Argentini, G. Fiocco, L. Giannini, L. Rossini, G. Abbate, R. Ocone, M. Casonato, 1994: Observations of Sea Breeze Events in Rome and the Surrounding Area by a Network of Doppler Sodars. Bound.-Layer Meteor., 71, 67-80.

Melas, D., 1993: Similarity methods to derive turbulence quantities and mixed layer depth from sodar measurements in the convective boundary layer: A review. Applied Physics B, 57, 11-17.

________, I. Kioutsioukis, and I. C. Ziomas, 1998a: Real time calculation of the dispersion of air pollutants from industrial stacks using a personal computer. Fresenius Env. Bull., 7, 134-140.

________, J. Abbate, D. Haralampopoulos and A. Kelesidis, 1998b: Estimation of Meteorological parameters for air quality measurements: Coupling of sodar data with simple numerical models. Accepted for publication in Journal of Applied Meteorology.

PARAMETERIZATION OF THE VERTICAL TRANSPORTS DUE TO THE CUMULUS CONVECTION USING A MASS-FLUX SCHEME

Orestis Denis Valianatos and Mihaela Caian

National Institute of Meteorology and Hydrology
Bucharest Sos. Bucuresti-Ploiesti 97, Romania

Updraft and downdraft penetrative convective motions are taken into account for a pollutant cloud evolution. The parameterization is based on Tiedtke's (1989) mass-flux scheme. Cumulus convection contribution to large-scale pollutant concentration tendency is function of the net contribution from all clouds to the upward and downward mass-flux, a weighted average of c (pollutant concentration) from all updrafts and downdrafts and large-scale transport contribution (precipitation impact is parameterized following Piedelievre et al., 1990). The amount resulting from evaporation of precipitation in the unsaturated subcloud layer is also included in computation (evaporation due to the downdraft and detrained cloud air is not considered for the concentration amount).

The steady state updraft contribution of the cloud ensemble is given by the entrainment and detrainment rates of concentration and vertical distribution of the updraft mass flux. Entrainment occurs through turbulence exchange (a fraction of the updraft mass flux) and inflow (hypothesised to be directly proportional with the large-scale convergence, following Tiedtke, 1989). Detrainment occurs through turbulent exchange at the edges and outflow at the cloud top, unimodal detrainment being assumed only from the deepest clouds. Cloud base mass-flux is given in the convection parameterization scheme (Tiedtke, 1989) by imposing a moisture balance in the sub-cloud layer.

Downdraft contribution to concentration needs the evaluation of entrainment and detrainment in downdraft mass flux. Constant rates are used as in the original scheme (Tiedtke, 1989) and hence a mass flux independent of height.

These computations have been included in an Eulerian diffusion and transport model, the MEDIA model developed at Meteo-France (Piedelievre et al., 1989). Atmospheric prognostic fields were computed by the ALADIN model, developed at Meteo-France, running operationally on DecAlpha station version in Romania. The fields needed to compute updraft and downdraft impact on the pollutant concentration are provided by the Tiedtke (1989) convection parameterization scheme that was implemented into the ALADIN model code for comparison with the original Bougeault (1985) convection scheme of the model.

The impact of the two convection schemes (Bougeault scheme - Fig. a, b and Tiedtke scheme - Fig. c, d) on the evolution of the pollutant distribution is shown, as the

Air Pollution Modeling and Its Application XIII, edited by S.-E. Gryning
and E. Batchvarova, Kluwer Academic/Plenum Publishers, 2000

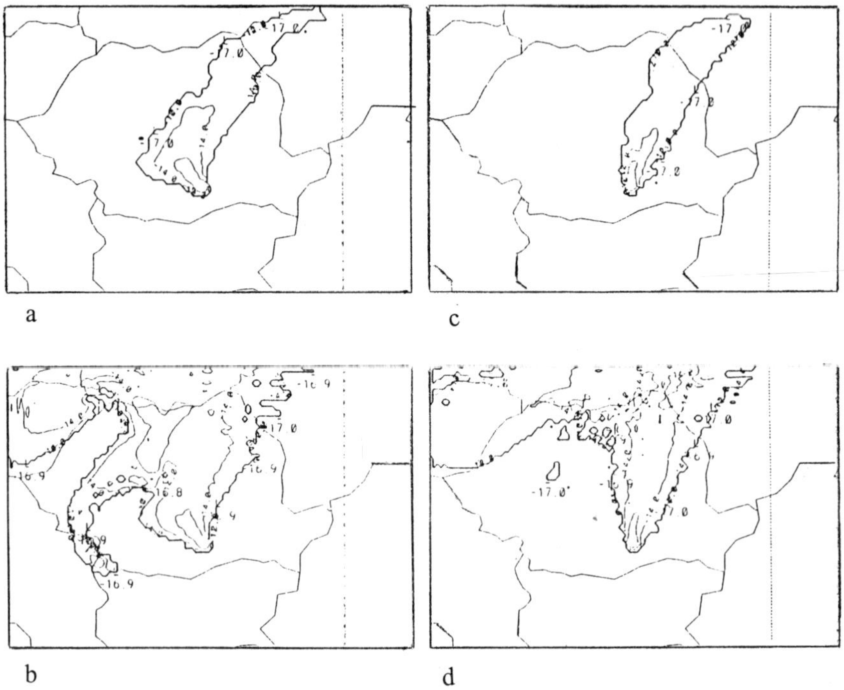

Figure a, b, c, d

effect of precipitation for each scheme for a 30 hours integration case based on 25/06/1996 (Fig. a,c-with precipitation, Fig. b,d-without precipitation). Further studies will compare the results against real data.

REFERENCES

Bougeault, Ph., 1985, A simple Parameterization of Large-scale effects of cumulus convection, Mon. Wea. Rev., 113, 2108-2121.

Piedelievre, J. Ph., Musson-Genon, L., and Bompay, F., 1990, MEDIA-an eulerian model of atmospheric dispersion: first validation on the Chernobyl release, J. Applied Meteor., 29, 1205-1220.

Tiedtke, M., 1989, A comprehensive mass flux scheme for cumulus parameterization in large-scale models, Mon. Wea. Rev., 117, 1779-1800.

Horanyi, A., Batka, M., Bubnova, R., Caian, M., Cordoneanu, E., Ducrocq, V., Geleyn, J., von der Emde, K., Ivanovici, V., Ihasz, I., Janousek, M., Jerczynski, M., Joly, A., Malardel, S., Pescaru, V., Radnoti, G., Seidl, H., Spacek, L., Spiridonov, V., 1993, A version of ARPEGE/IFS for mesoscale dynamical adaptation: ARPEGE/ALADIN. CAS/WGNE Report, Research activities in atmospheric and oceanic modelling, 18, 5.5-5.7.

AN EMERGENCY DECISION SUPPORT SYSTEM WITH INTEGRATED REGIONAL, METEOROLOGICAL NETWORK FOR REAL TIME MODELLING OF ACCIDENTAL ATMOSPHERIC RELEASES IN COMPLEX MOUNTAIN/VALLEY TERRAIN

V. R. D. Herrnberger

Paul Scherrer Institute (PSI)
CH-5232 Villigen PSI, Switzerland

SYSTEM FEATURES

Each of the 4 nuclear sites in Switzerland including the PSI is situated in river valleys surrounded by mountains up to 1000 m of difference in altitude and with slopes of essentially more than 10 %. Particularly our site is in the middle of four connected valleys in the E-W, N -E, SW and SE direction respectively. An operational emergency decision support system (DSS) is being developed for real time simulation of toxic atmospheric releases and their dose impact on this complex environment.

For the estimation of the ground contamination pattern after a short release the system response time lies in between 3' to 30' up to a downwind distance of several 10 km depending on the degree of problem complexity. The temporal and spatial resolutions are half an hour averaging time and 1 km or less in the grid size respectively.

USERS' NEEDS

They are determined by the wide variety of users as scientific, technical, official and political, who differ strongly in their formation, experience, DSS job environment, responsibilities and objectives (Herrnberger et al., 1997). To meet their needs high performance, sufficient reliability, excellent method harmonization and satisfying user-friendliness are necessary but not sufficient system capabilities. They have to be completed by a flexible system structure, which, in our case, consists of two levels of problem complexity and two operational modes. The levels of problem complexity comprise atmospheric flow, dispersion and deposition of pollutants in flat and complex terrain respectively. The operational modes are either interactive and menu-driven or fully automatic, without any interaction of the user. Clearly, the different levels of problem complexity need databases of different amount, resolution, quality and availability. By this structure the user can group different modes with different levels in a way, which is optimum for his objectives and experiences in view of the available key data and modeling resources. In the automatic mode the system operates with standardized release terms to inform immediately the assessors of the actual possible contamination pattern. In emergency cases the term can quickly be adapted to more relevant data without interrupting the automatic mode.

MODEL SELECTION

Originally the automatic operational mode of the DSS was limited to Gaussian plume with site meteorological data. Therefore, its results were valid for flat terrain up to 2 km at the best. They led to uncontrolled errors in the prediction of the contaminated areas in the complex terrain around. In order to achieve more realistic results sophisticated models and relevant meteorological data were needed.

Sufficient know how is available in the world wide scientific community. Consequently, we considered any new in house development in this field as a waste of time and resources. We qualified and selected the suitable models in 4 phases: (1st), to find the state of the art in 38 available dispersion models based on essentially 4 different methods (Eulerian, Langrangian, Particle-in-cell and Gaussian puff); (2nd), to select very few models per method, which are best-referenced with respect to their scientific and technical (software/hardware) performance and user friendliness (Herrnberger et al., 1992/1); (3rd), to gain practical experience and quantitative results by analysis of the relevant tracer experiment SIESTA (Herrnberger et al., 1992/2); (4th), to evaluate the "best" model by qualitative and quantitative performance measures as visualization and statistical analysis (Herrnberger et al., 1994).

The evaluation showed, that the ARAC-II kernel (Sullivan et al., 1993) was one of the most complete and suited model chains: TOPOG, to describe the topography in block geometry; MEDIC, to interpolate the meteorological data; MATHEW, to construct 3-dimensional, mass-consistent (diagnostic) wind fields; ADPIC to transport the contaminant by a Langrangian Monte Carlo simulation of the Hybrid marker particles with radioactive chain decay and buildup.

DATA BASE

The complex terrain around PSI site is taken into account by an extended topographic and meteorological data base- The last is linked on-line to 14 meteorological stations from the existing national network, which give a first guess for the wind above the mountains as well as for the local diurnal flow patterns in the valleys up to distances of 40 km.

The necessary wind fields are synthesized from every half hour in real time. Other relevant data concern atmospheric stratification, particle dispersion and deposition radioactive decay chains and geography.

In emergency situations, as in case of an unexpected ground release due to fire scenarios e.g., the most relevant release data are not available. We proved that a nonnegative least square regression scheme reasonably estimates release rates from field measurements in complex topographical and meteorological situation (Hutter et al., 1997).

DISPERSION IN COMPLEX TOPOGRAPHY

The strong site specific correlation between topography and meteorology and their very importance on atmospheric dispersion and ground contamination by pollutants was demonstrated in most cases by comparing ARAC with Gaussian plume modeling results using the same meteorological data. The annual flow patterns around the PSI site was measured by a rigorous experimental campaign in this area. Up to a down wind distance of 10 km 13 typical flow (and so atmospheric dispersion and ground contamination) situations could be classified (Graber et al., 1996).

Compared to the flat terrain approach (Gaussian plume type) all classes showed good agreement in the very near range of 1 - 2 km and only 2 classes (15 %) for the whole distance.

Stronger effects of the complex topography on the patterns were found for greater distances in most classes (80 %): Either the lateral concentration distributions were asymmetric (30 %) or the concentration patterns differed remarkably in size and extension by wind shear, flow convergence or bifurcation (55 %).

REFERENCES

- Graber, W. K.,Bürki, D., 1996, "Klassifikation von Windfeldern als Grundlage zur Diagnose der aktuellen Ausbreitungssituation (Projekt Windbank unteres Aaretal)", *PSI-Bericht.* 96-11, Paul Scherrer Institute, CH-5232 Villigen PSI.

- Herrnberger V. R. D., Doria, P., Prohaska, G., 1992/1, "Model Evaluation and Selection for Real-time Emergency Applications.“ *Radioprotection*, September, pg. 209-213, Les Ulis, France.

- Herrnberger, V. R. D., Thykier-Nielsen, S., 1992/2, "Comparison of wind-models for real time dispersion simulation of tracer experiments conducted over complex terrain during weak flow conditions." *Air Pollution Modelling and its Application IX*, Edited by H. van Dop and G. Kallos, Plenum Press, New York.

- HERRNBERGER, V. R. D. and DORIA, P., 1994, "Assessment and Verification of Different Types of Dispersion Models in Complex Terrain.". *Air Pollution Modeling and its Application X*, Edited by S. V. Gryning and M. M. Millan, Plenum Press, New York.

- Herrnberger, V. R. D., Graber, W. K., Hutter, M., 1997/1, "The Decision Support System of PSI, and How it Meets the Needs of the Users.", *Radiation Protection Dosimetry*, Vol. 73, Nos 1 - 4, pp, 123 - 126.

- Hutter, M., Herrnberger, V. R. D., 1997/2. "Testing a Source Rate Estimation Method by Application to a Dispersion Experiment.", *Atmospheric Environment*, Vol. 31, No. 20, pp. 3283 - 3290, Pergamon Press, GB.

Sullivan, T. J., Ellis, J. S., Foster, C. S. & K. T., Baskett, J. S., Nasstrom, J. S., Schalk, W. W. III, 1993, "Atmospheric Release Advisory Capability: Real-time Modelling of Airborne Hazardous Materials." *Bull. Amer. Meteor. Soc.*, 74, 2343.

MEASUREMENT AND ANALYSIS OF PHOTOCHEMICAL SMOG OVER İSTANBUL, TURKEY

Mete Tayanç and Gürkan Göçmen

Marmara University, Dept. of Environmental Eng., Göztepe, Istanbul, Turkey

INTRODUCTION

Daily sulfur dioxide and particulate matter concentrations over İstanbul have been monitored since 1985, and the resultant data have been analyzed with statistical models by several studies (Tayanç et al., 1997a; Tayanç et al., 1997b). They found that there has been a decrease in sulfur dioxide concentrations over İstanbul owing to a) the introduction of the cleaner and treated coal, b) switching to natural gas consumption in many residential buildings and industrial activities, and c) better meteorological conditions that favor the dispersion of the air pollutants. Unfortunately, the photochemical pollutants had not been monitored till October 1995. This study focuses on the observations of nitrogen oxide, nitrogen dioxide and ozone conducted by mobile instruments over different regions of İstanbul in one year period (October 1995 - September 1996), and provides information about the temporal characteristics of the photochemical smog over the region.

RESULTS

Developing an accurate and detailed picture of photochemical pollutants in İstanbul is a crucial component of air pollution assessment and serves multiple purposes. It is a result that links the low quality fuel consumption for industrial, heating and transportation purposes with the damage it causes to the environment.

Two mobile stations conducted observations of nitrogen oxides and ozone; one located over the European side and the other one on the Asian side of İstanbul. Observations of the pollutants were not based on a program, thus the measurements have been made randomly in time and place. During the analysis process we partitioned a day into four periods 06:00-10:00, 10:00-15:00, 15:00-22:00 and 22:00-06:00, and the corresponding nitrogen oxide, nitrogen dioxide and ozone data of these periods are studied by statistical techniques. This partitioning scheme is used to overcome the problems of the unorganized, random measurements over different locations. Daily and seasonal analyses revealed that the highest concentrations of photochemical pollutants generally associated with the period of 10:00-15:00 and with the summer months.

Air Pollution Modeling and Its Application XIII, edited by S.-E. Gryning and E. Batchvarova, Kluwer Academic/Plenum Publishers, 2000

Figure 1 illustrates the variation of ozone and nitrogen dioxide in a scatter plot over Saraçhane, where the most frequent observations have been made, in the 06.00-10.00 and 10.00-15.00 periods. Seasonal average concentration levels are depicted with horizontal lines for the period of October 1995 - September 1996 on the figures. Contrary to the expectations, low concentrations of NO_2 and O_3 are found over highly populated regions, and in general, the levels of these pollutants suprisingly do not violate the standards. No any O_3 value exceeds 100 $\mu g/m^3$ in these periods, where the National Ambient Air Quality Standards (NAAQS) of USA and Turkish short-term standards establish 8 hour limit of O_3 as 160 $\mu g/m^3$ and 240 $\mu g/m^3$, respectively. Thus, we cannot talk about an ozone problem over İstanbul. Turkish short-term standard of NO_2 is 300 $\mu g/m^3$. It is obvious that some NO_2 values in 06.00-10.00 period are exceeding the limit. But the majority of the data in this period and all of the data in the other periods are below these limits. Meteorological

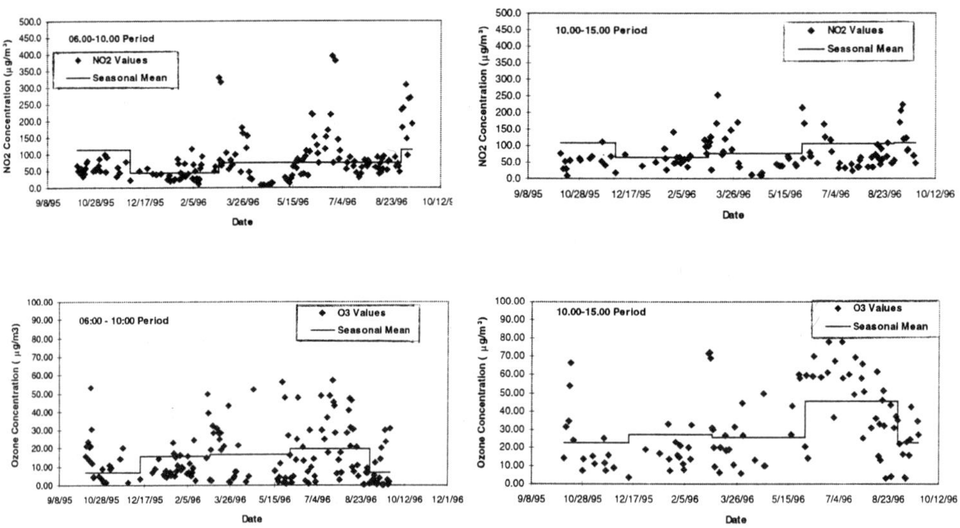

Figure 1. Ozone and nitrogen dioxide concentration levels and seasonal mean values ($\mu g/m^3$).

factors explain this situation to a certain extent. Summer-time persistent moderate northeasterly winds play an important role in the transportation of primary and secondary pollutants towards Marmara and Aegean Sea. Also the Persian-Gulf low-pressure system affecting the region in the summer months prevents the formation of strong inversion conditions, and in turn prevents the building up of air pollutants over İstanbul.

REFERENCES

Tayanç, M., Karaca, M., Saral, A., and Ertürk F., 1997a, Study of the decrease of air pollution concentration levels with the meteorological factors in İstanbul", 22nd *NATO/CCMS International Technical Meeting on Air Pollution Modeling and Its Application*, Clermont-Ferrand, France, June 2-6, 1997.

Tayanç, M., Karaca, M., Saral, A., Ertürk F., and Borhan, Y., 1997b, Investigation of SO_2 dispersion in İstanbul, Turkey", 2nd *European and African Conference on Wind Engineering*, Genova, Italy, June 22-26, 1997.

A REVIEW OF MODELS IN OPERATIONAL USE IN POLAND

Jerzy Burzyński

Institute of Meteorology and Water Management
Division for the Remote Sensing of the Atmosphere
Cracow Branch
Poland, 30-215 Cracow, P. Borowego 14

The paper provides a general overview of the studies on modeling and forecasting of air pollution dispersion in Poland. The dispersion models developed and applied in Poland over the last few years are presented in the table 1, with general description of the type and the way of verification.

The most often used dispersion models types in Poland are schemes based on the Official Polish Regulatory Model. This model is a simple Gaussian model prepared for unqualified users. The concentration's field of PM10 and pollutants in gaseous form (for these kind of gases which are well mixed with air) can be calculated by this model. Deposition of suspended particles can be calculated too. Domain of the calculation should be a flat terrain. Information on orography is taken into account by means of the roughness coefficient. Model is allowed to be used for calculations of the concentrations fields of pollutants emitted from elevated buoyant sources. Concentrations calculated with use of this model are most frequently overestimated.

The dispersion models in Poland are used for the diagnosis of the pollution situations and for the episodes forecasting. The first forecast based on statistical modeling of SO_2 and TSP concentrations was made in Cracow in early 70's. The forecasting system with pollution warnings and emissions control started in Legnica (since 1983) and Głogów (since 1988). The modeling of pollution fields with use of dispersion models and statistical models was experimentally investigated in the Upper Silesia region since 1986 through 1988. This system consisted of three blocks: forecasting of dispersion conditions for this region; statistical forecast of mean diurnal values of pollutant concentrations for 24 hrs.; forecast based on the numerical model for SO_2 for 12 hrs.

The statistical modeling based on the ARMAX type model is use in operational way in Cracow since 1991. The results of pollution concentrations measurements from automatic network are used for the calibration. This statistical model was used in: 1) forecasting of the 24 h. average of SO_2 concentrations; 2) forecasting of the 24 h. average of PM10 concentrations; 3) forecasting of the maximum one hour values of the ozone concentrations; 4) forecasting of the 8 h. average of the ozone concentrations. Forecasts are made for

Air Pollution Modeling and Its Application XIII, edited by S.-E. Gryning
and E. Batchvarova, Kluwer Academic/Plenum Publishers, 2000

Cracow, and forecasts (1, 2) for Katowice district, too. Concentration data are taken from automatic monitoring network, meteorological data were used as independent variables.

The main goal of research in dispersion modeling in Poland now is the problem of traffic pollution. Codes from US and Western Europe are used. In Cracow on the basis of collaboration with „Sendzimir" steel meals a scheme is implemented of operational application of modeling and air pollution forecast in management of this industrial plants.

Table 1. The review of the chosen models in operational use in Poland

Authors; Institution	Name of the code	The model type	Destinations	Verifications
K. Albert, K. Budziński, K. Juda-Rezler; ISIŚ Tech. Un. Warsaw	**SOXNOX**	numerical, eulerian	diagnostic	-
J. Bartnicki; Inst. Meteo Warsaw	**OSMA**	numerical, eulerian spectral	diagnostic, in operational use	comparison with SO_x measurements in 8 European stations
J. Bartnicki, J. Hrehoruk at all; ZMTZA Inst. Meteo Warsaw	**REMOTA**	numerical, eulerian spectral	diagnostic, in operational use	
St. Chruściel, M. Markiewicz; ISIŚ Tech. Un. Warsaw	**URFOR-2**	gaussian, Pasquill type	diagnostic	Data from exp. MONAT-84
St. Chruściel, at all; ISIŚ Tech. Un. Warsaw	**WYTYCZNE**	gaussian, Pasquill type	diagnostic	
P. Holnicki, A. Kaluszko, A. Żochowski; IBS Acad. Sci.	**URFOR3**	numerical, eulerian	for forecast	Data from exp. MONAT-84
P. Holnicki, A. Kaluszko, A. Żochowski; IBS Acad. Sci.	**REGFOR3**	numerical, eulerian	for forecast	Data from exp. MONAT-84
K. Juda-Rezler; ISIŚ Tech. Un. Warsaw, NILU Lillestrom, Norwegian	**MOD**	numerical, eulerian	diagnostic	Data from exp. MONAT-84
K. Twardowska, at all; IM Tech. Un. Warsaw	**STOCH-SO_2**	numerical, eulerian with Kalman filter	for diagnosis and for forecast	Data from exp. MONAT-84, EWAG 87
M. Tomaszewska; ZTA Inst. Meteo Kraków		statistical	for forecast, in operational use in Cracow	with measurement from Cracow automatic Network
M. Tomaszewska; ZTA Inst. Meteo Kraków	**MG**	gaussian, Pasquill type	diagnostic	with correlation spectrometer measurements
M. Uliasz; ISIŚ Tech. Un. Warsaw	**MDMS**	numerical, modes: • Lagrangian, • Eulerian, • Lagrangian-Eulerian	diagnostic	exp Oeresund

REFERENCES

Madany A., Bartochowska M. Review of polish air pollution dispersion models. Research work of the Warsaw Polytechnic (1994)

Walczewski J. System approach to air quality monitoring identification of main problems; Urban Air Pollution edited by I. Allegrini, Springer (1996)

LONG RANGE MODELLING OF POLLUTANT SUBSTANCES IN THE ATMOSPHERE

Degtiarev A.I. and Shtyreva N.V.

Hydrometeocentre of Russia
Bolshoy Predtechenskiy per.
Moscow
Russia

The model for a long range and transboundary transport (MLTT) for harmful emitions is developed for forecasting and monitoring of a large-scale distribution for polluting substances in the atmosphere and their deposition on the surface. The model enables to settle an invoice for distribution and dissipation of industrial pollutants, the forecasting and monitoring of chemical and nuclear emergencies.

Applications of this model are an emergency planning, scientific researches and air quality assessments. The model calculates a gases and aerosols concentrations of pollutants in the atmosphere and their deposition on the surface.

The account of distribution of polluting substances from many sources can be carried out the MLTT. The sources can be persistance, instant or the intensity of sources can vary. The MLTT stimulates an account of individual qualities for each source of the pollution and the given polluting substance. The source can emit one or more polluting substances with the given properties, for example, various pollutant gases or a polydispersive structure of emission. MLTT uses a regional-to-continental or global spatial scale. At present the model works with passive pollutants (a chemical transformation does not count). The model uses complex meteorology in spherical (latitude-longitude) coordinations (2.5°*2.5°) and 8-vertical sigma levels. This model can be realized at PC or workstation.

MLTT is based on a numerical decision of the advective-duffusion equations system of transport pollutants in the atmosphere. The Lagrangian approach for an advective transport of polluting substances is realised in the model. A regular grid is used for decision of the equations system . The significances of wind velocity in the point,where particular particles of pollution are located, are calculated with the helped 3-dimension interpolation. It should note, that the interpolation in the model is the most frequenntly executed procedure, therefore it should satisfy to rigid computing

Air Pollution Modeling and Its Application XIII, edited by S.-E. Gryning
and E. Batchvarova, Kluwer Academic/Plenum Publishers, 2000

requirements. The used procedure is based on the method of construction of a digital-linear function, conterminous in point of a grid with the given function.

The Monte-Carlo method for account of turbulent diffusion in the atmosphere is realized in the model. The Monte-Carlo method is used for the approximation of diffuse members. Parameterization of vertical and horizontal turbulent flows plays the essential role for the correct decision of the given task. Vertical turbulent flows in the surface layer of the atmosphere are determined by a thermal stratification in this layer and the character of the earthly surface. In the planetary boundary layer turbulent flows are calculated with the help of the analytical theory.

The variant of the MLTT in a spherical coordinate system is to the utmost set - up on use of the meteorological information of the operative model of the intermediate term forecast of Hydrometeocentre of Russia. The data of wind velocity components, temperature, humidity, surface pressure and a vertical diffusion coefficient of this model can be directly used by submitted model of polluting substances transport. At absent of total set of the meteorological information a part from these sizes (the coefficient of vertical diffusion, the vertical wind velocity) can be designed directly by in MLTT.

The model used a mecanism of the secondary migration of polluting substances from behind the wind intercepting from a surface. This parameterization also based on a method Monte-Carlo. The probability of secondary migration of a particular particle of pollution on a surface depends on such factors as a module of the surface wind, a type of the surface, roughness, soil moisture.

Using vertical sigma-system it is necessary to take into account, that at including of a new portion of the meteorological informations the communication by probable change of surface pressure needs to be conducted correction vertical coordinate a situation of a particular particle. This operation is carried out for each particle located in the atmosphere by use of the static equation, new and old surface pressure for given point of the integration area.

The distinctive features of the MLTT are: 1) the model of transport composes with the computing model of intermediate term forecasts of the weather of Hydrometeocentre of Russia (or another forecasting model of the similar scale), that it is important for operative accounts, 2) in the model the data about the main meteorological elements on the boundary layer of the atmosphere are taken into account, where the main transport of antropogenic pollution occurs, it is enough in details (at the 4 vertical accounting levels), and, as is known, wind and temperature structures in the low troposphere are rather changed, 3) the model permits to settle an invoice as for the whole Northern hemisphere, as for any given region, that permits essentially to speed up calculations.

Some numerical experiments have produced with the MLTT for the purpose of testing of various algorithms to find out the opportunity to use the model in the forecasting purposes is conducted. The test experiment, conducted under conditions of the constant horizontal wind velocity and the absence of its vertical components has shown, that such algorithms of the model, as horizontal advection, horizontal diffusion and gravitate deposition of aerosoles, adequately the represention of distribution of polluting substances in the atmosphere.

In another experiment the distribution of pollution in the atmosphere and their deposition on the spreading surface was provided with the help of the forecasting meteorological model of the intermediate term forecast weathers SM-15 of Hydrometeocentre of Russia during 17-19.02.93 for a term up to three days. The analysis of the weather situation for 17-19.02.93 in the low troposphere has shown, that received results are quite real.

AEROSOL NITRATE AND NON-SEA-SALT SULFATE OVER THE EASTERN MEDITERRANEAN

Türkan Özsoy[1], Cemal Saydam[2], Nilgün Kubilay[3], Oya B. Nalçacı[1], and İlkay Salihoğlu[3]

[1]Mersin University, Dept. of Environmental Eng., Mersin, Turkey
[2]METU and Turkish Scientific and Research Council, Ankara, Turkey
[3]Institute of Marine Sciences, Middle East Technical University
P.K. 28 Erdemli - İçel 33731 Turkey

INTRODUCTION

Among the biogenic sources of sulfate aerosols, reduced sulfur gases such as DMS constitute and control a significant fraction of the atmospheric nss-sulfate budget (Charlson *et al.*, 1987, e.g.). Certain species of phytoplankton; particularly the dinoflagellate *Phaeocystis pouchetii* and the coccolithophore *Emiliana huxleyi* are known to be important producers of dimethylsulfide (DMS) (Matrai and Keller, 1993, e.g.). Biogenic sulfur, created mainly by seasonal phytoplanktonic activity, could be particularly rich sources in certain regions and dominate the atmospheric sulfur cycle.

Daily samples of aerosol collected at Erdemli, Turkey (36°33'54"N, 34°15'18"E) during October 1991-December 1992 were analyzed by ion chromatography (IC) to determine Cl^-, NO_3^- and $SO_4^=$ concentrations. Elemental compositions of these samples together with air mass trajectory analyses have emphasized the influence of crustal material compared to anthropogenic emissions (Kubilay and Saydam, 1995). The relatively low concentrations of anthropogenic trace elements observed in Erdemli, compared with the western Mediterranean (Bergametti *et al.*, 1989), suggests dominance of marine biogenic DMS contribution to nss-sulfate by phytoplankton blooms.

RESULTS AND DISCUSSION

Aerosol sulfate and nitrate decrease in winter due to scavenging by precipitation. The maximum nss-sulfate occured in June 1992 (21.8±8.5 $\mu g/m^3$) and July 1992 (19.8±6.5 $\mu g/m^3$), whereas nitrate reached 7.7±4.0 $\mu g/m^3$ in April 1992, compared to their means in Table 1.

The episodic transport events are best characterized by case studies, as given for 27-28-29-30 July 1992 below; Air mass back trajectories arriving at 900 hPa level on 27 and 28 July originated from the Black Sea (Fig. 1a,b), where an intense basin-wide coccolithophore (*E. huxleyi*) bloom was observed (Vladimirov *et al.*, 1997) at about the same time, producing the minimum Secchi disk visibility depth of 6.2m. observed since 1923. We consistently measured unusually high nss-sulfate concentrations in aerosols crossing over or originating from the Black Sea in the 25 June-3 August 1992 period.

Table 1. Aerosol nss-sulfate and nitrate concentrations ($\mu g/m^3$) in Erdemli

Sampling date	nss-sulfate (6.57±6.54)	nitrate (2.74±2.73)
27 July 1992	29.29	9.34
28 July 1992	31.44	8.99
29 July 1992	4.24	8.29
30 July 1992	21.79	6.06

The arithmetic means of the entire data set are given in parentheses.

Air Pollution Modeling and Its Application XIII, edited by S.-E. Gryning and E. Batchvarova, Kluwer Academic/Plenum Publishers, 2000

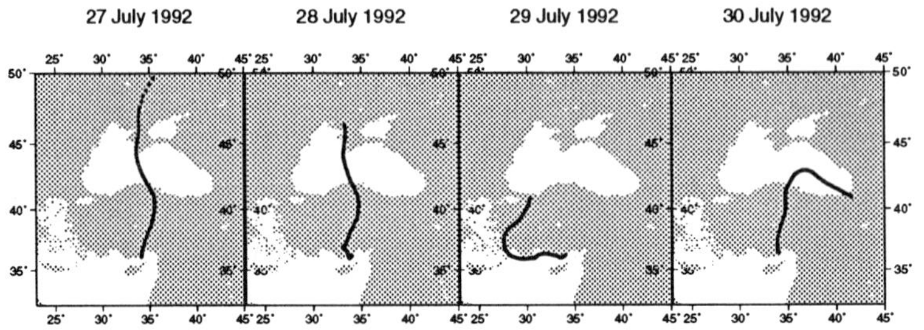

Figure 1. 3-D back trajectories associated with air masses arriving at 900 hPa level (a) 27 July 1992; (b) 28 July 1992; (c) 29 July 1992; (d) 30 July 1992.

Air flows within the boundary layer on 27 and 28 July (Fig. 1a,b) showed similar patterns with maximum nss-sulfate concentrations. On 29 June, the flow at 900 hPa level changed in direction swept the western and south-western coasts of Turkey (see Fig. 1c), resulting in rapid decrease in sulfate concentration. Steady nitrate concentrations in the same period, which are significantly higher than the overall mean (2.74 ($\mu g/m^3$), suggest anthropogenic contributions from land. The rapid decrease on 29 July in contrast to the nitrate suggests a dropout in the biogenic contribution to nss-sulfate level on 29 July, as a result of the temporary diversion from the Black Sea source (see Fig. 1c), which was restored once again when the flow changed direction towards the eastern Black Sea on 30 July 1992 (see Fig. 1d). This increase can be attributed to the anomalous *E. huxleyi* bloom reported by Vladimirov *et al.*, (1997) in the Black Sea.

ACKNOWLEDGEMENTS

We acknowledge the generous help given by Dr. G. Tuncel (Env. Eng. Dept. METU) in making the anion analyses possible. We thank Dr. E. Özsoy for helping to improve the manuscript. Substantial parts of the study were funded by the Turkish Scientific Research Council through the YDABAG-462/G project.

REFERENCES

Bergametti, G., Dutot, A.L., Buat-Menard, P. Losno, R. and Remoudaki, E., 1989, Seasonal variability of the elemental composition of atmospheric aerosol particles over the north-western Mediterranean, *Tellus*, **41**B:353-361.

Charlson, R.J., Lovelock, J.E., Andreae, M.O. and Warren, S.G., 1987, Oceanic phytoplankton, atmospheric sulphur, cloud albedo and climate: A geophysiological feedback, *Nature*, **326**: 655-661.

Kubilay, N. and Saydam, A.C., 1995, Trace elements in atmospheric particulates over the eastern Mediterranean; Concentrations, sources and temporal variability, *Atmospheric Environment*, **29**:2289-2300.

Matrai, P.A. and Keller, M.D., 1993, Dimethylsulfide production in a large scale coccolithophore bloom in the Gulf of Maine, *Continental Shelf Res.*, **13**:831-843.

Vladimirov, V.L., Mankovsky, V.I., Solov'ev, M.V. and Mishonov, A.V., Seasonal and long-term variability of the Black Sea optical parameters. In: *Sensitivity to Change: Black Sea, Baltic Sea and North Sea*, E. Özsoy and A. Mikaelyan, eds., NATO ASI Series 2, Environment-27, Kluwer Academic Publishers, pp. 33-48. (1997).

TURBULENCE PARAMETERIZATION FOR PBL DISPERSION MODELS IN ALL STABILITY CONDITIONS

G. Degrazia[1], D. Anfossi[2], J. Carvalho[2,3], H.F. Campos Velho[4], E. Ferrero[5], C. Mangia[6], U. Rizza[6], and S. Trini Castelli[2]

[1] Universidade Federal de Santa Maria, Departamento de Física - 97119-900, Santa Maria, RS, Brasil
[2] Consiglio Nazionale delle Ricerche, Istituto di Cosmo-Geofísica - 10133, Turin, Italia
[3] Universidade de São Paulo, Instituto Astronômico e Geofísico - 05508-900, São Paulo, SP, Brasil
[4] Instituto Nacional de Pesquisas Espaciais, LAC - 12201-970 São José dos Campos, SP, Brasil
[5] Universita', Dip. Scienze Tecn. Avanz., Alessandria, Italia
[6] Consiglio Nazionale delle Ricerche, Istituto ISIATA, Lecce, Italia

INTRODUCTION

In the atmospheric dispersion models turbulence parameterization is a key parameter. In the present paper, accounting for the current knowledge of the PBL structure and characteristics, a new set of turbulence parameterization to be used is such models has been derived. That is, expressions for the vertical profiles of the velocity standard deviations σ_i, Lagrangian length scale l_{L_i}, time scale T_{L_i} and diffusion coefficient K_i, where $i = 1,2,3$ are proposed. The classical statistical diffusion theory, the observed spectral properties and observed characteristics of energy containing eddies are used to estimate these parameters. These parameterizations give continuous values for the PBL at all elevations $(z_0 \le z \le h)$ and all stability conditions from unstable to stable $(-\infty < L < \infty)$, where h is the neutral or stable PBL height, z_0 the aerodynamic roughness length and L the Monin-Obukhov length.

EXPRESSIONS FOR σ_i, l_{L_i}, T_{L_i} AND K_i

From Taylor's statistical diffusion theory (Degrazia et al.,1998) and spectra of turbulent kinetic energy, where a linear combination of the two turbulent drive mechanisms

(shear + buoyancy) (Frisch,1995) was assumed, the following expressions for σ_i, l_{L_i}, T_{L_i} and K_i result:

$$\sigma_i^2 = \frac{1.06\, c_i \left(\psi_\varepsilon \frac{z}{z_i}\right)^{2/3} w_*^2}{[(f_m^*)_i^c]^{2/3}} + \frac{2.32\, c_i\, \phi_\varepsilon^{2/3}\, u_*^2}{[(f_m^*)^{n+s}]^{2/3}} \tag{1}$$

$$l_{L_i} = 0.14\, z \left\{ \left[\frac{-\overline{L}}{z_i}\frac{z_i}{-L}\right]^{1/2} \frac{1}{(f_m^*)_i^c} + \frac{0.64}{(f_m^*)_i^{n+s}} \right\} \tag{2}$$

$$T_{L_i} = \frac{z}{\sqrt{c_i}} \left\{ \frac{0.14 \left(\frac{-\overline{L}}{z_i}\frac{z_i}{-L}\right)^{1/2}}{[(f_m^*)_i^c]^{2/3}\, w_* \left(\psi_\varepsilon^c \frac{z}{z_i}\right)^{1/3}} + \frac{0.059}{[(f_m^*)_i^{n+s}]^{2/3}\, \phi_\varepsilon^{1/3}\, u_*} \right\} \tag{3}$$

and

$$K_i = 0.14 \sqrt{c_i}\, z \left\{ \frac{\left(\psi_\varepsilon \frac{z}{z_i}\right)^{1/3} \left(\frac{-\overline{L}}{z_i}\frac{z_i}{-L}\right)^{1/2} w_*}{[(f_m^*)_i^c]^{4/3}} + \frac{\phi_\varepsilon^{1/3}\, u_*}{[(f_m^*)_i^{n+s}]^{4/3}} \right\} \tag{4}$$

where w_* is the convective velocity scale, u_* is the local friction velocity, ψ_ε and ϕ_ε are the nondimensional molecular dissipation rate functions, z_i is the CBL height, $(f_m^*)_i^c$ is the reduced frequency of the convective spectral peak, $(f_m^*)_i^{n+s}$ is the reduced frequency of the neutral or stable spectral peak, L is the Monin-Obukohv length, $-\overline{L}/z_i$ is an average stability parameter for the CBL, κ is the Von Karman constant and $c_i = \alpha_i \alpha_u (2\pi\kappa)$ with $\alpha_u = 0.5 \pm 0.05$ and $\alpha_i = 1, 4/3, 4/3$ for u, v and w components, respectively.

In conclusion eqs. (1, 2, 3 and 4) allow the parameterization of σ_i, l_{L_i}, T_{L_i} and K_i for all elevations and all stability conditions.

ACKNOWLEDGEMENTS

We acknowledge financial support provided by CAPES and CNPq.

REFERENCES

Degrazia, G.A., Anfossi, D., Campos Velho, H.F. and Ferrero, E., 1998, A Lagrangian decorrelation time scale for non-homogeneous turbulence, *Boundary-Layer Meteorol.*.86:525.

Frisch, U., 1995, *Turbulence*, Cambridge University Press, Cambridge.

A FAST ADVECTION SCHEME FOR ATMOSPHERIC MODELS WITH LITTLE PSEUDODIFFUSION

Michael V.Galperin

Independent researcher, Studencheskaya str. 31-9, Moscow, 121165 Russia

The important sources of errors in advection computation are considered and a numerical advection scheme developed on this base are presented.

The numerical schemes are distinguished from each other by the methods of restoration of concentration profile with respect to known masses in cells. It seems that accuracy can be improved by application of high approximation order. However it does not take place due to the following reasons:

1.Values of concentration in far points can be used for approximation of one in certain point under existence of high derivative of concentration with respect to space and time. In fact the last condition does not take place and, moreover, atmospheric fronts, precipitation and pollution sources produce the jump discontinuities of concentration. As a result the approximation orders more than 3 or 4 accomplish nothing and even can impair the situation.

2.The approximation of concentration having been using, mass in the cell differs in general from its predetermined value. This circumstance is not taken into account in many schemes at all or forces to introduce rather crude normalisation, limitation, etc.

The first mentioned source of errors can be eliminated, if the mass movement of each cell is computed independently from other ones as it is suggested by Egan and Mahoney (*1972*). The first and second moments of concentration distribution in cell are computed and conserved and the rectangular approximation of concentration profile is used in their model and in the number of other ones. However the accumulation and conservation of second moment result in the part of cell mass initially may locate without the cell and as a consequence a strong distortion of monotony can arise. In addition the second moment calculation requires a large computer time and store.

The presented SCD scheme is based on calculation **the first moment only**. The assumption about rectangular distribution of cell mass accords with conditions, when the first moment (e.g. mass centre) is known only. Therefore the mass is initially evenly distributed inside cell onto maximum possible square at known mass centre (and second moment conservation is not required). The movement and redistribution of individual cell mass within time step does not depend on masses in other cells and the new values of mass and first moment in cells are obtained by superposition.

The presented scheme is stable, positively definite and mass conservative. Its monotony and transportivity are sufficient for most of applications. The boundary effects are absent therefore the boundary conditions can be chosen arbitrarily (in particular they may be zero). The small numerical diffusion makes it possible to calculate the transport of point mass. The mass budget and the export outside the grid limits can be easy calculated. The SCD method is appropriate also in the three-dimensional problems and can be easily adopted for parallel calculation. The scheme can be simply harmonised with any models of vertical distribution, removal and transformation of substances.

There are no the efficient analytical methods for exhaustive estimation of scheme errors. Therefore the set of tests should be applied, each given only part of scheme features. The testing of scheme should be always aimed at disproving of scheme, but its validation. The successful test result

Air Pollution Modeling and Its Application XIII, edited by S.-E. Gryning
and E. Batchvarova, Kluwer Academic/Plenum Publishers, 2000

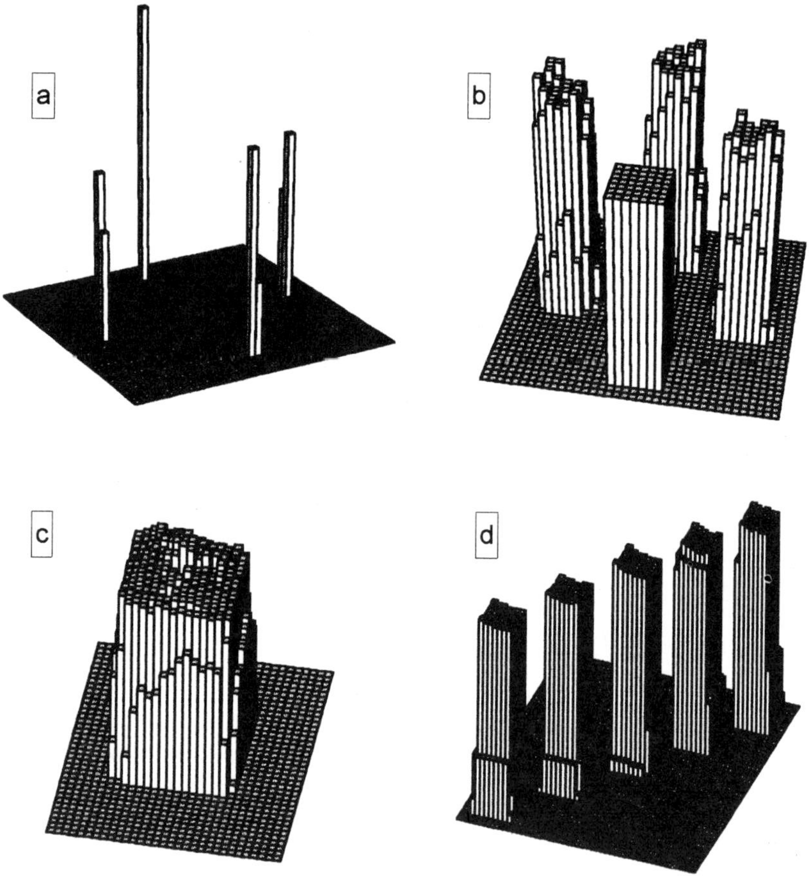

Fig. 1. The test patterns of SCD scheme: *a* and *b* - rotation of one-cell puff («δ-function») and square flat puff by 90°, 180° and 270°; *c* - result of rotation of square puff by 360° around its own centre; *d* - linear movement of square puff along grid diagonal.

permits only to hope that in conditions close to test the scheme can operate satisfactorily and nothing more. The comparison testing was made for presented scheme, original and modifying Egan - Mahoney scheme (modification concerned the method of second moment calculation), original Bott scheme *(Bott,1993)* with Lagrange polynomial of 4-th degree and normalisation, modifying Bott scheme (Bessel polynomial of 3-d degree, flux normalisation), finite - difference MacCormack - Holmgren scheme *(Holmgren,1994)* with approximation of 4-th order with respect to space and of 2-nd one - to time and simplest pseudo-Lagrangian scheme. The testing shows that accuracy of the presented scheme is as good as one of the best of known advection models (Fig. 1). However the presented scheme is many times faster than others with similar accuracy.

REFERENCES

Bott, A.,1993, The monotone area - preserving flux - form advection algorithm: reducing the time - splitting error in two - dimensional flow fields. *Monthly Weather Review* **121,** pp. 2637-2641.

Egan, A.B. and Mahoney, J.R.,1972, Numerical modeling of advection and diffusion of urban area source pollutants. *J. of Applied Meteorology* **11**, pp.312-322.

Holmgren, P.,1994, An advection algorithm and an atmospheric airflow application. *J. of Computational Physics* **115**, pp.27-42.

A MODEL OF PLUME ADVECTION VELOCITY FOR DISPERSION IN URBAN ARRAYS

R.W. Macdonald,[1] R.F. Griffiths,[1] and D.J. Hall[2]

[1]Department of Chemical Engineering, UMIST
PO Box 88, Manchester, M60 1QD, UK
[2]Building Research Establishment
Garston, Watford, WD2 7JR, UK

INTRODUCTION

In the basic Gaussian plume model (GPM) used for predicting concentrations due to continuous plume releases, the mean concentrations in the plume are inversely proportional to the mean plume advection speed U_p. For elevated releases, U_p is evaluated at the height of the release or some other representative height. For ground level releases in a shear layer, U_p varies with distance from the source, since as the plume grows vertically in the boundary layer, it encounters progressively higher wind speeds. From similarity theory, it is possible to show that the correct wind speed to use in this case is $U_p \approx U(0.5\sigma_z)$, that is to say the wind speed evaluated at approximately one half the plume vertical standard deviation.

When a plume is largely within the urban boundary layer, the plume advection speed will be some function of the overall surface geometry, as well as the relative proportions of the plume within and above the urban canopy. The classical similarity theory is not applicable close to the obstacles, and it is not at all obvious what the plume advection speed should be in this case. Knowledge of the plume dispersion parameters σ_y and σ_z alone is insufficient for predicting maximum ground level concentrations. Therefore, in this poster presentation, recent wind tunnel data[1-4] are used to calculate U_p for ground level plumes which have been released in model urban arrays. Using measured concentration and velocity profile data, it has been possible to calculate U_p as a function of distance from the source for various obstacle geometries. Of particular interest is the dispersion at intermediate distance from the source, where the plume dimensions σ_y and σ_z are of the order of the mean width W and height H of the obstacles. This regime is often neglected in dispersion modelling, and most urban GPMs are only applicable when the plume is much bigger than the individual obstacles.

Air Pollution Modeling and Its Application XIII, edited by S.-E. Gryning and E. Batchvarova, Kluwer Academic/Plenum Publishers, 2000

RESULTS

Our wind tunnel[1,2] and field studies[3] have shown that once a plume has passed through one or two rows in an obstacle array, the vertical plume profile can be well modelled by a reflected Gaussian equation with a non-zero effective plume height $z_c \approx H$,

$$C(z) = C_{\max,z}\left[\exp\left(-\frac{(z-z_c)^2}{2\sigma_z^{\,2}}\right) + \exp\left(-\frac{(z+z_c)^2}{2\sigma_z^{\,2}}\right)\right]. \tag{1}$$

By definition, the mean plume velocity in the GPM must satisfy the following integral equation, which is a result of mass flux conservation:

$$U_p(x) = \frac{\iint c(x,y,z)U(x,z)dydz}{\iint c(x,y,z)dydz}. \tag{2}$$

The integration is made over the extent of the plume in the y-z plane. To evaluate the integral, we require the concentration profile (1) as well as an expression for the mean velocity profile, both within and above the urban canopy. These have both been measured in the wind tunnel. However, even for simple parameterisations of the velocity profile, the general integral is not solvable analytically and must be evaluated numerically.[4,5] The results of the present analysis yield values of the ratio U_p/U_H, where U_H is a representative velocity meaured at mean rooftop height upwind of the obstacle arrays. U_p/U_H is found to be a function mainly of distance from the source, and varies from about 0.2-0.3, after one or two rows of obstacles, to 0.4-0.6 at $x/H \approx 20$. If U_H is used directly in a GPM instead of U_p, as is often the case for simple models of obstacle-affected dispersion, then ground level concentrations will be underpredicted by a factor of 2-5. This happens even if the dispersion parameters σ_y and σ_z are accurately known, because $C \propto U_p^{-1}$. Therefore, it is extremely important in this dispersion regime that a correct value of U_p be determined if a GPM is to be used successfully.

Acknowledgements. The work described here was funded by the Hazard Assessment Group of DERA Porton Down. It is part of a larger research and development programme into dispersion in urban areas.

REFERENCES

1. D.J. Hall, R. Macdonald, S. Walker and A.M. Spanton, Measurements of dispersion in simulated urban arrays-a small scale wind tunnel study, BRE Client Report CR244/98, Building Research Establishment Limited (1998).
2. R.W. Macdonald, D.J. Hall and R.F. Griffiths, Scale model study of building effects on dispersion in the urban canopy at intermediate source distances, in Proceedings of the 5th International Conference on Harmonisation within Atmospheric Dispersion Modelling for Regulatory Purposes, Rhodes, Greece, 18-21 May:531 (1998).
3. R.W. Macdonald, R.F. Griffiths and D.J. Hall, A comparison of results from scaled field and wind tunnel modelling of dispersion in arrays of obstacles, *Atm. Env.* (1998). In press.
4. R.W. Macdonald, D.J. Hall and R.F. Griffiths, Measurements of velocity profiles within scale model urban arrays for a gaussian plume model, submitted to *J. Wind Eng. Ind. Aero.* (1998).
5. R.W. Macdonald, D.J. Hall and R.F. Griffiths, Wind tunnel measurements of velocity within simulated urban arrays, BRE Client Report CR243/98, BRE Ltd. (1998).

RESEARCH PROJECT FOR OBSERVATION AND MODELLING OF URBAN AIR POLLUTION IN UK WEST MIDLANDS

R.M. Harrison[1], R.G. Derwent[2], R.H. Maryon[2], M.J. Pilling[3], P.W. Seakins[3], D.E. Heard[3], A.C. Lewis[3], S.A. Penkett[4], R.L. Jones[5], G.R. McGregor[6] and X.-M. Cai[6]

[1]School of Chemistry, University of Birmingham, Edgbaston, Birmingham B15 2TT, UK
[2]Meteorological Office, Room 156, Met Office (APR), London Road Bracknell, Berkshire, RG12 2SY, UK
[3]Department of Chemistry, University of Leeds, Leeds, LS2 9JT, UK
[4]School of Environmental Sciences, University of East Anglia, Norwich NR4 7TJ, UK
[5]Department of Chemistry, University Chemical Laboratory, Lensfield Road Cambridge, CB2 1EW, UK
[6]School of Geography, University of Birmingham, Edgbaston, Birmingham B15 2TT, UK

The project, Pollution of the Urban Midlands Atmosphere (PUMA), is funded by Natural Environment Research Council (NERC) of the UK through a thematic programme on Urban Regeneration and the Environment (URGENT), which aims at a better understanding of natural processes in the shallow sub-surface, urban ecology, hydrological balances and atmospheric dynamics and chemistry in urban environments.

The PUMA project involves the collaboration of the research groups from University of Birmingham, UK Meteorological Office, University of Leeds, University of East Anglia, and Cambridge University. Main objectives of the PUMA project include:

(1) To make measurements of concentrations of a wide range of both long-lived and transient species in order to gain insights into the UK urban atmospheric chemical processes at a very fundamental level and to validate a proposed chemical model. These species include hydrocarbons, carbonyl compounds, oxyacids of nitrogen and the free radical species OH, HO2, RO2 and NO3, which play a key role in atmospheric chemistry.

(2) To apply a mesoscale meteorological model to a major UK conurbation, the West Midlands to obtain the time-dependent and three-dimensional field of relevant meteorological variables as input to chemical models.

Air Pollution Modeling and Its Application XIII, edited by S.-E. Gryning and E. Batchvarova, Kluwer Academic/Plenum Publishers, 2000

(3) To develop a coupled Lagrangian dispersion chemistry model that is capable of predicting both primary and secondary air pollutant concentrations across the conurbation.

(4) To validate the model against high quality measurements of primary pollutants such as carbon monoxide, sulphur dioxide and NOx, as well as secondary pollutants including nitrogen dioxide, and ozone, and also to compare the results with the those derived from an existing model UAM-IV.

Chemical measurement includes five categories as follows: a) continuous fixed site measurements of hourly data for NO, NO_2, O_3, SO_2, CO, PM_{10} and 25 specific hydrocarbons made as part of the UK national and local monitoring networks; b) a mobile monitoring of CO, NO_x and O_3 carried out by a laboratory equipped with GPS technology; c) high density ozone network of more than 50 sensors across the conurbation; d) vertical profiles of O_3, NO_3 and water vapour by tethersonde, LIDAR, and tower; e) campaign measurements of long-lived and transient species. Two campaigns (June-July 1999 and January-February 2000) will make highly comprehensive measurements at the heart of the conurbation. Campaign measurements include: C_2-C_{12} specific hydrocarbon and methane hourly; formaldehyde continuously; formaldehyde, acetaldehyde, acetone and total remaining carbonyls hourly; OH and HO_2 radicals continuously; peroxy radicals, nitrous and nitric acids, photolysis rates of O_3 (to O^1D), NO_2, HONO and HCHO; particulate matters together with sulphate, nitrate, and ammonium (secondary inorganic aerosol) and elemental and organic carbon on a 3-hourly average basis. These measurements allow us to test the importance of HONO and HCHO photolysis in UK urban atmosphere in comparison with ozone photolysis, to estimate the lifetimes of NO_x and VOCs, and to assess the role of NO_3 as an oxidant of hydrocarbons and as a source of nitrate in the nocturnal boundary layer.

Mesoscale meteorological model adopted in the project is the Regional Atmospheric Meteorological System (RAMS). Its validation will be conducted through comparison with the observational data obtained from a mesoscale meteorological measurement network in the West Midlands. The network consists of more than 20 surface stations monitoring hourly meteorological data including wind speed and direction, temperature, humidity, and pressure. These data will also be obtained from one mobile and two fixed 24m meteorological towers, which also provides sensible heat flux, momentum flux, and turbulence variances. The vertical profile of meteorological variables including the mixed layer height will be obtained from a tethersonde. The chemical model will be developed based on the Lagrangian particle model (the UK Met Office dispersion model, NAME) by implementing a very detailed chemical scheme (up to 120 hydrocarbon species) including explicit hydrocarbon chemistry. Available hydrocarbon data from the measurement campaign will be used to identify the most important species and reactions that the model will adopt. The meteorological and chemical data from the above-mentioned two measurement campaigns will be used to validate the model and to gain insights into atmospheric chemical processes in the UK urban environments. This project will also apply the existing UAM-IV to the conurbation as a comparison with the proposed model from two aspects: Eulerian vs. Lagrangian, and the CBM-IV chemistry vs. the new proposed scheme. The PUMA project aims to advance the urban air quality modelling technique appreciably and will be subject to an extensive validation at a level not previously attempted in the UK.

SURVEY ON DISPERSION MODELS FOR DANGEROUS SUBSTANCES

Gerhard Erbes and Ulrike Pechinger

Central Institute for Meteorology and Geodynamics
Hohe Warte 38
A-1190 Vienna / Austria

INTRODUCTION

The Seveso II guideline (96/82/EC) of the European Union on the *Control of major-accident hazards involving dangerous substances* is to be implemented in all European Union member states by February 3, 1999. In Austria this will lead to law adjustments and new regulations on the calculation of safety distances around industrial plants. Such calculations can be done by an approach where worst-case scenarios are defined. Another method is to calculate safety distances by using the probability of a certain accident or damage level. For both of these approaches, a cooperation between experts from government, industry, science and interest groups is necessary.

Within this framework, a survey on dispersion models for dangerous substances was performed to provide an aid to the model selection process (Pechinger and Erbes, 1998). A questionnaire was distributed to 50 model developers, and replies were received for 36 individual models. Model components such as emission scenarios, windfield and dispersion models, cloud processes and presentation of results were compared. A part of the results is shown in Table 1.

Air Pollution Modeling and Its Application XIII, edited by S.-E. Gryning
and E. Batchvarova, Kluwer Academic/Plenum Publishers, 2000

Table 1. Main features of investigated dispersion models for dangerous substances

MODEL	SOURCE TERM					RELEASE PHASE				GAS DENSITY			WIND FIELD			TUR-BULENCE			DISPERSION			CLOUD PROCESSES			OUTPUT			C	R	PRICE (USD)
	P	J	E	F	A	L	V	T	M	D	N	B	C	D	K	S	P	K	B	G	K/S	C	L	M/A	T	H	O			
ADORA 1.2	●	●	●	●	●	●	●	●	●	●	●	●	●			●			●	●		●	●	●/●	●	●	●			7,500 - 20,000
ADREA-HF		●		●	●			●		●	●	●			●			●			●/--		●	--/●	●			●		40,000
ALOHA 5.2.2	●			●	●	●	●	●		●	●	●	●			●			●	●	●/--				●				●	200 - 600
AQPAC 1.2	●	●	●	●	●	●	●			●	●	●	●			●				●	●/--		●	●/●	●				●	1,300
ASAP 1.0	●	●	●	●	●	●	●			●	●	●	●			●				●	●/--		●	●/●	●				●	1,600
CANARY 3.0	●	●	●	●	●	●	●	●	●	●	●	●			●	●			●		●/--		●	●/●	●	●	●			17,500
CHARM 9.0	●	●	●	●	●	●	●	●	●	●	●	●		●		●			●	●	--/●	●	●	●/●	●	●	●		●	20,000
CHARM 10.0	●	●	●	●	●	●	●	●	●	●	●	●			●			●			●/--	●	●	●/●	●	●	●	●	●	20,000
DEGADIS 2.1	●	●				●	●	●	●	●	●	●	●				●		●	●	●/●			●/●						FREE
DRIFT 2.23										●	●		●			●	●		●	●		●		●/●	●			●		3,200 - 8,000
EJECT 1.01		●		●	●	●	●	●					●			●			●	●		●		●/●						3,200 - 8,000
EOLE 3..0	●						●			●	●		●			●			●	●				●/--						ASK VENDOR
ESCAPE 5.02	●	●		●	●	●	●	●		●	●		●			●			●					●/●	●					1,800
FEM3C	●			●	●	●	●			●	●				●			●			●/--			--/●	●			●		FREE
GASP 3.01	●					●																								3,200 - 8,000
GASTAR 3.0		●			●		●	●		●	●		●			●	●		●					●/●	●			●		1,600 / YEAR
HGSYSTEM 3.0	●	●		●	●	●	●	●	●	●	●	●	●			●	●		●	●		●		●/●						FREE
HOTMAC/RAPTAD 4.5	●	●	●				●			●	●	●	●	●	●			●			●/--			●/--	●			●	●	ASK VENDOR
MERCURE 3.2	●	●		●	●			●	●	●	●	●			●			●			●/--		●	--/●	●			●		15,900
MET 2.0	●		●		●		●		●	●	●	●	●			●			●						●				●	800
MIDAS-AT	●	●		●		●	●	●	●	●	●	●	●	●	●	●				●	●/--				●			●	●	ASK VENDOR
OMEGA 3.5		●	●		●				●	●	●	●			●			●			●/--			●/●	●			●	●	100,000
PHAST 5.1	●	●	●	●	●	●	●	●	●	●	●	●	●			●			●	●			●	--/●	●	●	●			19,300
RISKWIT	●	●	●	●	●	●	●	●		●	●	●	●			●	●		●	●			●	●/●	●	●	●			18,000
SAFER 7.8	●		●	●	●	●	●	●	●	●	●	●	●	●		●			●	●		●		●/●		●	●	●	●	18,400
SCIPUFF 1.0	●		●		●	●	●			●	●	●	●	●		●	●		●	●			●	--/●	●			●	●	FREE
SEVEX 1.3	●	●	●		●	●	●	●		●	●				●			●	●	●	●/--			--/●	●	●	●	●		ASK VENDOR
SLAB	●	●		●	●	●	●	●		●	●		●			●	●		?		--/●			--/●					●	FREE
STAR-CD 3.0		●			●		●		●	●	●	●			●			●			●/--	●	●	--/●	●			●		< 24,000 / YEAR
STOER.LAG	●	●	●	●	●	●	●			●	●	●		●	●	●	●				●/--		●	●/●	●	●	●	●	●	6,000
SUPERCHEMS 3.2	●	●	●	●	●	●	●	●	●	●	●	●	●	●		●	●		●	●	--/●	●		●/●	●	●	●			10,000
TRACE 8.0	●	●	●	●	●	●	●	●	●	●	●	●	●			●			●	●		●		●/●	●	●	●			15,000
TRAUMA 3.63		●		●	●	●		●					●						●			●		●/●						3,200
TSCREEN	●	●		●	●	●	●	●		●	●	●	●			●				●				●/--				●		FREE
TWODEE	●						●			●			●			●					--/●									ASK VENDOR
VDI-3783	●	●					●			●			●			EMPIRICAL			EMPIRICAL						●			●		300

Source term: **P** - Pool, **J** - Jet, **E** - Explosion/Fire, **F** - Flash evaporation, **A** -Aerosols
Release phase: **L** - Liquid, **V** - Vapor, **T** - Two-phase, **M** - Multicomponent
Gas density: **D** - Dense, **N** - Neutral, **B** - Buoyant
Wind field: **C** - Constant, **D** - Diagnostic, **K** - Numerical
Turbulence: **S** - Stability classes, **P** - Similarity parameters, **K** - Numerical
Dispersion: **B** - Box/Puff, **G** - Gaussian plume, **K** - Numerical, **S** - Shallow layer
Cloud proc.: **C** - Chemical reactions, **L** - Lift-off, **M** - Moisture, **A** - Aerosols
Output: **T** - Toxicity, **H** - Heat radiation, **O** - Overpressure
C: Treat complex terrain
R: Use real-time meteorological data

Acknowledgements

This study was supported by the Austrian Federal Ministry of the Interior.

REFERENCES

Pechinger, U., and Erbes, G., 1998, Voruntersuchungen zur Anwendung eines Schwergasmodells bei Industrieunfällen. (Summary tables in English). Final report, AE 65.000/142-IV/1/97. Federal Ministry of the Interior, Minoritenplatz 9, A-1010 Vienna, Austria.

MONITORING OF THE ATMOSPHERIC RADIOACTIVITY IN BULGARIA AND ITS POSSIBLE DATA USE IN MODELING

Blagorodka S. Veleva

National Institute of Meteorology and Hydrology - Bulgarian Academy of Sciences
1784 Sofia, BULGARIA

Since early 60's the environmental radioactivity monitoring network of the National Institute of Meteorology and Hydrology has produced daily a set of data on the atmospheric radioactivity. Continuos monitoring of the atmospheric radioactivity - natural and artificial is provided by beta radiometry, gamma spectrometry and recently by radiochemical separation of daily and monthly samples of air particulate material, precipitation and atmospheric fallout. Briefly the main sampling program consists of: atmospheric fallout on the wetted cotton sampler, taken daily at more than 30 stations; air particulate material - at 5 city stations; subsamples from daily precipitation if it is >0.5l/m^2 - at 15 stations; a total monthly deposition on open vessel sampler - at 12 stations.

The response of this monitoring system during the Chernobyl accident allow to obtain a picture of the time and space distribution of the radioactive fallout over the territory of Bulgaria[1], based mainly on the gross beta activity, measured 120 hours after the sampling. The main contamination started to be observed at 1 of May; even 20 to 50 fold increase was detected at the mountain peak stations on April 29. The sum of the Chernobyl fallout for the first two five days' periods for some stations is presented on Table 1, giving an idea about high variability of the deposited activity during the accidental releases. These values do not include the activity of ^{131}I, because of the ashing at 400°C of the cotton material of the deposition samples. The usual values before the accident and today do not exceed 10 Bq/m2 for 5 diurnal period for all of the Bulgarian stations. Air concentration of the beta radionuclides increased also 2 to 5 orders of magnitude. Much more informative maps of daily deposited activity can not be presented here because of the lack of space.

Table 1. Chernobyl fallout of the beta radionuclides for the period May 1 to 5 in Bq/m^2 (D1) and for May 6 to 10 (D2) as a percentage of the D1 at some of the sampling stations

North Bulgarian stations			Sought Bulgarian stations			Mountain peak stations		
Name	D1 kBq/m^2	D2/D1 in %	Name	D1 kBq/m^2	D2/D1 in %	Name	D1 kBq/m^2	D2/D1 in %
Razgrad	28.9	11	Plovdiv	28.9	11	Botev vr.	140.8	12
Ruse	13.6	17	Chirpan	24.4	18	Musala	70.0	11
Kneja	22.6	10	Sliven	24.1	16	Cherni vr.	62.5	28
Varna	11.7	36	Burgas	18.3	18	Snejanka	16.1	102
Pleven	10.2	23	Sofia	13.6	26			

Comparison between the dry deposition velocities obtained for the fresh Chernobyl fallout and the data from the period of atmospheric nuclear test for the days without precipitation show distinguished difference for the main sampling stations in Sofia, Plovdiv, Burgas, Pleven and Varna. For the first half of May 1986 the observed values are in order of magnitude lower compared to those of global fallout from the nuclear tests. This fact might be important when reconstruction of the field of air concentrations is made based on the known deposited activity.

The new development of the monitoring system during the last 6-8 years allows to determine gamma emitting radionuclides by modern low level gamma spectrometry and to start to analyze some alpha radionuclides in addition to the beta radioactive isotopes. Average monthly deposition of ^{137}Cs for 11 Northern Bulgarian stations for the last 6 years varied from place to place and from month to month within the range of 0.01 - 3.5 mBq/m^2 per day. Deposited and air particulate matter activity of ^{137}Cs during May and June 1998 is inside the range of the background fluctuations, so it might be concluded that Bulgaria was not affected by ^{137}Cs releases in Spain. Observed correlation of ^{137}Cs activity to those of cosmogenic ^7Be for some months indicates a similar source and deposition processes. In some cases (windy and dry months) anticorrelation between two isotopes could be explained with more significant resuspension of the ^{137}Cs from the soil.

Plutonium isotopes, released during the Chernobyl accident were in much smaller amount than beta radioactive isotopes. But they are very important to be determined, because Plutonium is one of the most dangerous radionuclides from biological point of view. The information about Plutonium concentrations in the atmosphere is quite scare mainly because of the complicated radiochemical separation and spectrometric technique needed. Results for Plutonium concentrations during the Chernobyl accident in single daily samples of the NIMH monitoring network and examples of the present concentration levels (composed monthly samples) are given in Table 2, where DL means that the value is below detection limit; NA that sample is not analyzed.

Table 2. Deposited activity of Plutonium isotopes and ^{90}Sr during the Chernobyl accident[2] and during the recent years[3]

Sample description	^{90}Sr Bq/m^2 per day	$^{239+240}Pu$ Bq/m^2 per day	^{238}Pu Bq/m^2 per day	$Pu^{239+240}/^{238}Pu$
Chirpan 1-5 May 1986	16.6±0.5	0.055±0.001	0.023 ± 0.001	41.8 %
Vakarel ,6 May 1986	37.2±1.5	0.152 ± 0.003	0.063 ± 0.002	41.4 %
Musala 17 May 1986	2.9±0.1	0.026 ± 0.001	0.0155 ± 0.002	59.6 %
Botev vr., 14 May 1986	NA	0.0125±0.0009	0.0045±0.0005	36%
Varna-Oct.1992	NA	$1\pm0.2x10^{-3}$	DL	
Varna - May 1995	$(4\pm1)x10^{-5}$	$(20\pm4)\ x10^{-5}$	DL	
Pleven - May 1995	$(2.6\pm0.6)\ x10^{-3}$	$(20\pm10)\ x10^{-5}$	DL	

The study of the behavior of the radon progeny in the surface air layer has been started since 1996 in 5 regional laboratories of the NIMH. First results show that background concentrations are related mainly to the mixing processes in the atmospheric boundary layer. The influence of the soil humidity and temperature is not so significant. The introduced parameter "efficient half life time of the isotopic mixture" vary within ±50% of the specific background value for every station if the station is not affected by air mass with different mixture of radionuclides.

REFERENCES

1. Antonov A., B. Veleva, L. Adjarova, M. Kolarova, Time and area distribution of low level radioactivity of technogenic radionuclides in the surface air and fallout over the territory of Bulgaria, Proc. of the 14th Europh. Conf. on Nucl. Physics, Bratislava 1990, World Scient., 1992, Singapore, ed. P. Povinec, 407-419
2. B. Veleva, G. LaRosa, A. Ghods, Simultaneous determination of Sr-90, Pu-239+240 and Am-241 in the atmospheric samples, II Conf. on public prevention in case of emergency and accidents, Sofia, 28-29 Sept, 1993, 71-76 (in Bulgarian)
3. B. Veleva et al., Final Rep.on the Project F336/1.11.1993 "Development of methodology for simultaneous determination of the Pu, Am, Th and Sr-90 in the atmospheric samples", 1996, Bulg. Scientific Foundation.

Concept of a Model System for the Computation of Atmospheric Dispersion Close to Real Situations and the Immediate Assessment of the Radiation Exposure

H. Walter[1], R. Martens[2], H. Thielen[2], T. Sperling[3], K. Maßmeyer[4]

[1] Federal Office for Radiation Protection
Ingolstaedter Landstr. 1
D - 85764 Oberschleissheim
Germany

[2] GRS mbH
Schwertnergasse 1
D - 50667 Koeln
Germany

[3] Institute for Geophysics und Meteorology
University of Cologne
Kerpener Str. 13
D - 50923 Koeln
Germany

[4] University-GH Paderborn
FB Techn. Umweltschutz
An der Wilhelmshoehe 44
D- 37671 Hoexter
Germany

Introduction

The technical development and improvement of computer systems allows to check existing computing procedures related to the judgement of possible consequences after incidents for further development or for the gain of more detailed results. Concerning the mesoscale atmospheric modelling, a significant possibility to improve the results is related to the computation of the transport and dispersion of harmful substances, i.e. radioactive material, with the mean wind and under the effects of the turbulent diffusion, taking into regard the orography of the surroundings of up to 100 km. These atmospheric processes can be computed now in more detail within acceptable computation time, even under emergency circumstances.

Model System

The model system is capable to simulate atmospheric transport and dispersion close to real situations within the meteorological β- and γ-scale, corresponding to a maximum distance of approximately 100 km from the source. Within this model system it is assumed, that the source is a nuclear facility in Germany, using the obligatory Remote Monitoring System for Nuclear Power Plants [KFÜ-System](KTA 1988).

The different parts of the model system take into account the specific requirements of an emergency preparedness system and can be used for emergency centers as well as for regulatory purposes during emergencies.

The model system consists of:

- preprocessors for the computation of surface layer parameters,
- a windfield model with interfaces for local measurements and/or input from operational macroscale models (e.g. numerical weather prediction),
- a Lagrangian dispersion model and
- a model for the assessment of radiation exposures.

The system uses local data of different measuring equipments as well as results from operational flow models of the German Meteorological Service (DM-model).

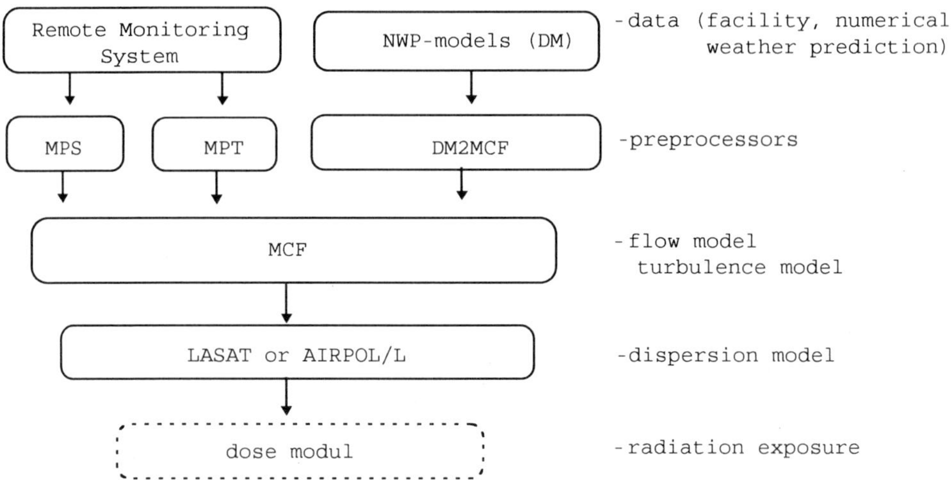

Figure 1. Concept of the model chain with the particular moduls and the data flow.

Within different research programms, several studies for the application and validation of the different models have been accomplished successfully (Thielen et al., 1998; Martens et al.,1998).

References

Kerntechnischer Ausschuß (KTA): Instrumentierung zur Ermittlung der Ausbreitung radioaktiver Stoffe in der Atmosphäre. Sicherheitstechnische Regel des KTA, KTA 1508, Carl Heymanns-Verlag, Köln, 1988.

Martens, R.; Thielen, H.; Sperling, T.; Massmeyer, K.; Walter, H.: Validation of a mass consistent flow model as part of a decision Support System. 10. Fachgespräch zur Umweltradioaktivität im Auftrag des Bundesministeriums für Umwelt, Naturschutz und Reaktorsicherheit (BMU), 28.-30.04.1998, Hamburg, in press.

Thielen, H.; Martens, R.; Schnadt, H.; Maßmeyer, K.; Walter, H.: Further development of the French-German model to calculate atmospheric dispersion following accidental releases from nuclear installations -Meteorological Preprocessor based on SODAR-data. 5th International Conference on Harmonisation within Atmospheric Dispersion Modelling for Regulatory Purposes, 18.-21. May 1998, Rodos, Greece.

A DISPERSION MODEL EVALUATION FOR BUOYANT RELEASES

Ana G. Ulke and María I. Gassmann

Dept. of Atmospheric Sciences - University of Buenos Aires
Buenos Aires - Argentina

The application of a dispersion model of pollutants released in the atmospheric boundary is presented. Based on the bidimensional semiempirical equation, the model incorporates the current understanding of the atmospheric boundary layer dynamics to estimate and parameterize the atmospheric stability, advection, turbulent diffusion and plume rise. The model performance has previously been analyzed against experimental data obtained in non-buoyant releases with a data base covering wide parameters ranges, and showed a correct behavior[1]. The present evaluation involves comparisons of model estimates with measurements from studies with buoyant releases from elevated sources in flat terrain in urban areas (Indianapolis, USA) in different stability conditions[2]. The comparison with tracer data is done for the predicted concentrations with the proposed model (mm) and with an updated Gaussian plume model (gm). All concentrations were normalized by source strength. Only quality 3 data (Q3) have been used. The results here include urban effects on the specification of stability. Predicted normalized crosswind-integrated, maximum arcwise and centerline concentrations were compared with observations. Current quantitative measurements and techniques of model evaluation were obtained and applied[3,4]. The comparison was done with the full dataset and for the data stratified by stability conditions. Confidence intervals on these performance measures were determined by the bootstrap resampling procedure. The obtained statistics for the maximum arcwise concentrations are shown in **Table 1**.

Table 1. Statistics for normalized maximum arcwise concentrations (10^{-9} s/m^3); based on Q3 data

DATA	*MODEL*	*NMSE*	*COR*	*FA2*	*FB*	*FS*
full data (N=363)	obs	0.00	1.000	1.000	0.000	0.000
	mm	0.48	0.504	0.711	0.268	0.152
	gm	0.59	0.385	0.708	-0.154	-0.315
stable (N=32)	obs	0.00	1.000	1.000	0.000	0.000
	mm	0.24	0.784	0.875	0.062	-0.389
	gm	2.47	0.520	0.250	-1.061	-1.168
unstable (N=331)	obs	0.00	1.000	1.000	0.000	0.000
	mm	0.49	0.489	0.695	0.282	0.168
	gm	0.37	0.526	0.752	-0.036	-0.128

The statistics exhibit an acceptable performance with a better behavior of the mm model. For the full dataset it is observed a mean underestimation of the normalized concentrations with the mm model and the opposite relation with the gm model. During stable conditions, the gm model has a poorer performance, while the mm model compares well with the observed concentrations. For atmospheric unstable conditions, the models have similar behavior, with slightly best results for the gm model

Air Pollution Modeling and Its Application XIII, edited by S.-E. Gryning
and E. Batchvarova, Kluwer Academic/Plenum Publishers, 2000

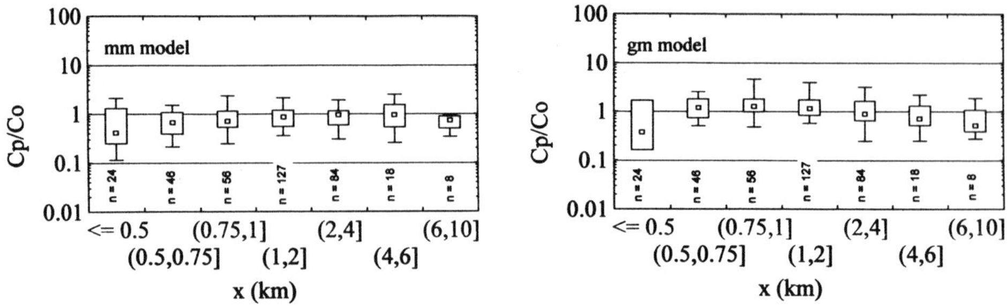

Figure 1. Residual plots of normalized maximum arcwise concentrations analyzed in terms of distance. (The boxes represent percentiles 5, 25, 50, 75 and 95).

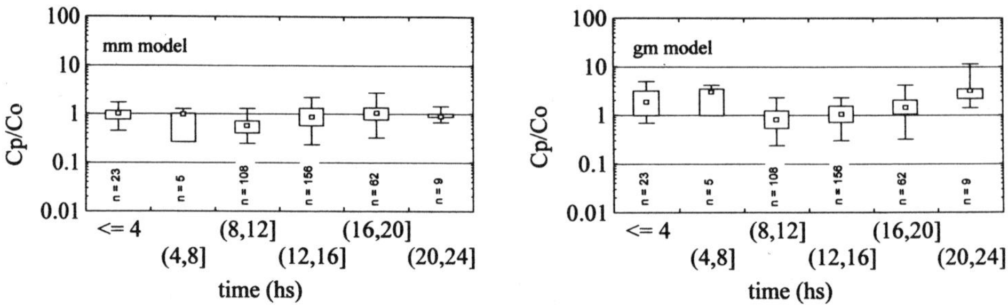

Figure 2. Residual plots of normalized arcwise maximum concentrations analyzed in terms of time.

predictions. Because this is an elevated buoyant source and the observations are at the surface, the obtained large scatter is not unexpected.

The residual plots in **Fig. 1** show that there is some structure to the residuals with distance. From very near the stack to about 750 m downwind the mm model underestimates concentration values and beyond this distance the underprediction is small. The gm model shows underpredictions very near the source, overprediction at intermediate distances and a trend back toward underprediction at the farthest distances. The results' behavior near the source is due to the approximations involved in the treatment of plume rise.

Fig.2 shows the residual as function of time. In general the mm model performs better than the gm model, with less bias and scatter. During nightime hours, the mm model estimates show an adequate behavior. On the contrary, the gm model does significantly overestimate concentrations. During daytime hours, both models underpredict concentrations in the early morning and the trend shifts to unbiased in the following hours for the mm model and the opposite one for the gm model.

REFERENCES

1. A.G. Ulke, A model evaluation exercise for non-buoyant releases from continuous point sources, Preprints of the 5th *International Conference on Harmonisation within Atmospheric Dispersion Modelling for Regulatory Purposes*, 130 (1998).
2. S.R. Hanna and J.C. Chang, Hybrid Plume Dispersion Model (HPDM). Improvements and Testing at Three Field Sites. *Atmos. Env.*, 27A:9,1491 (1993).
3. S. R. Hanna, J. C. Chang and D.G. Strimaitis, Hazardous gas model evaluation with field observations, *Atmos. Env.*, 27A:2265(1993).
4. H. R. Olesen, Toward the establishment of a common framework for model evaluation, in *Air Pollution Modeling and Its Application XI*, S.E. Gryning and F.A. Schiermeier, ed., Plenum Press, New York, 519(1996).

WIND FLOW AND CONCENTRATION FLUCTUATIONS DATA AROUND AN OBSTACLE: A FIELD EXPERIMENTAL WORK

Jane M. Santos[1], Ian D. Roberts[2] and Richard F. Griffiths[3]

[1]Department of Hydraulics, Federal University of Espirito Santo, Cx.P.01-9011 29.060-970 Vitoria ES, Brazil
[2]CBDE, Porton Down, Salisbury, Wiltshire, SP4 0JQ, UK
[3]Environmental Tech. Centre, UMIST, P.O. Box 88 Manchester M60 1QD, UK

INTRODUCTION

Investigations of flow and dispersion around obstacles require experimental data to provide necessary information in order to develop mathematical models and better understanding of the physical processes involved. However, few field trials have been conducted to investigate the influence of stability on the flow and dispersion patterns around an isolated building (e.g. Higson *et al* (1996)); and there is little experimental data available to investigate the dependence of mean values on averaging time. This initial study attempts to examine the influence of atmospheric stability and averaging time on the concentration distribution on the surface of an isolated building.

The experimental data presented here were obtained at Dugway Proving Ground, Utah, USA. A gaseous source was located about 3.5H upwind of the centre of the face of the building (figure 1). Concentration measurements were made using photo ionisation detectors (PID's) at a frequency of 50 Hz. Meteorological data were acquired using three sonic anemometers (y=0.44H, 0.88H and 1.76H) at a frequency of 21Hz. The roughness of the terrain is about 10 mm calculated from the velocity profile in neutral condition. The averaging time was selected based on the time scale of turbulence in the near-wake of the obstacle, which is the order of H/U (where H is the building height and U is the wind speed at one building height). The results presented here were taken from episodes when the central sensor on the windward wall was indicating the highest 5 min mean within a 30 min period when the mean wind direction (averaged over the same period) was normal to the building wall within ±10 degrees and the wind speed was greater than 2 m/s.

RESULTS

Slight changes in the atmosphere stratification do not cause a significant change in the flow around the building. However, strongly unstable or stable stratification tends to yield different flow patterns. For instance, increasing stability in the atmospheric boundary layer causes a decrease in the size of the recirculation region. This effect together with less plume spreading before the impingement on the obstacle produces higher concentration levels on the building surface, as shown in figure 2. This feature can not be observed on the left lateral wall, where the mean concentration values are higher under unstable conditions. This is probably due to misalignment of the wind direction. It should be noted that the concentration fluctuation intensity on the windward wall is not significantly influenced by the atmospheric conditions. On the other hand, on the other walls the concentration fluctuation intensity is higher under unstable condition, which is probably related to the size of the flow separation region and the higher turbulence levels.

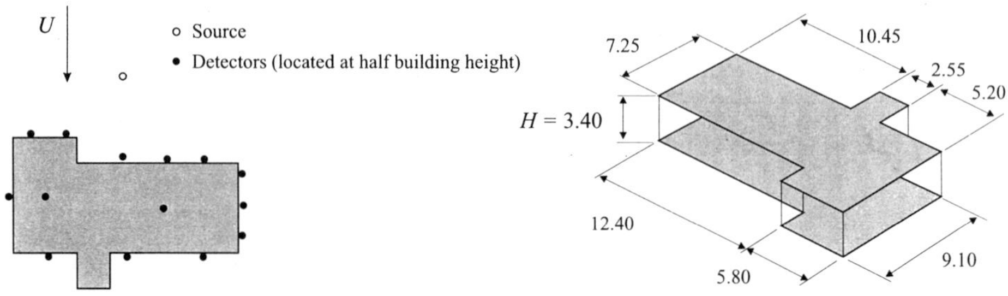

Figure 1 – Schematic representation of the building and detectors (distances in meters).

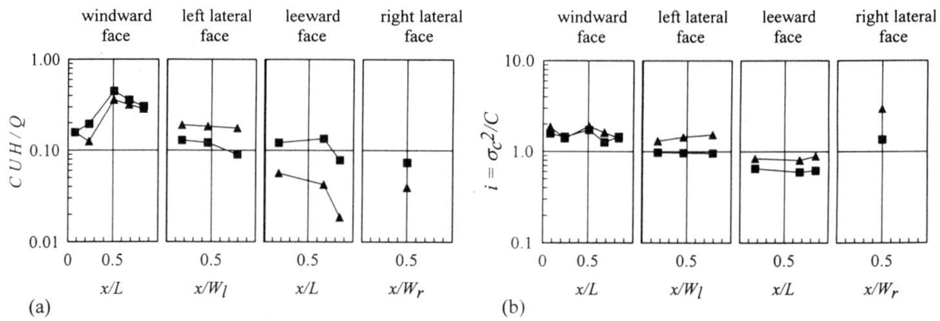

Figure 2 – Normalised mean concentration (a) and concentration fluctuation intensity (b) on the building surface for neutral (■) and unstable (▲) atmospheric conditions.

The analysis of the influence of averaging time was carried out using the averaged values for 30 min and shorter averaging times (5, 10 and 15) selected from the 30 min period when each sensor on the building surface was indicating the highest mean for the averaging time considered. For large averaging time, slow meandering motions have more opportunity to advect the plume back and forth, which means that mean concentration decreases when time averaging increases and fluctuation intensity increases. The results obtained show that the averaging time has stronger influence on the mean concentration than on the concentration fluctuation intensity (i) on the building surface.

The variation of mean concentration values with averaging time is usually expressed as a power law function $(C_1/C_2=(t_1/t_2)^p)$ proposed by Singer (1963), if the averaging times considered are larger than the time scale of the turbulent flow. The mean concentration (C_2) used as a reference value was averaged over the longest period of time (t_2) taken from the selected episode. A value of p equal to –0.2 for open terrain is usually suggested in the literature. The experiments under neutral conditions indicated a mean value of p equal to –0.35 on the windward wall and –0.25 on the leeward wall, while under unstable conditions this value is found to be equal to –0.44 on the windward wall and –0.38 on the leeward wall. This suggests that the averaging time influences more strongly the mean concentration on the windward wall, which indicates that the sizes of the eddies on the windward wall are larger in this region. The values of p found for the measurements on the building surface are clearly higher than the values presented in the literature for open terrain, which clearly indicates that the non-stationary effects of turbulence are even stronger in the flow around an isolated obstacle. The results also show that there is a significant influence of atmospheric conditions on the calculation of the peak-to-mean concentration.

REFERENCES

Higson, H.L., Griffiths, R.F., Jones, C.D. and Hall, D.J., 1996, Flow and dispersion around an isolated building, *Atmos. Environ.*, 30:2859-2870.

Singer, I.A., Imai, K. and Del Campo, R.G., 1963, Peak-to-Mean Pollutant Concentration Ratios for Various Terrain and Vegetation Cover, *J. Air Poll. Contr. Ass.*, 13:40-42

EMISSIONS OF SULPHUR, LEAD,AND BENZO(A)PYRENE IN BULGARIA FOR 1990 AND 1995

Julia Ivancheva[1], Plamen Videnov[1], Stanislav Bogdanov[1], Lidia Asenova[2]

[1] National Institute of Meteorology and Hydrology - Bulgarian Academy of Sciences
[2] Ministry of Environment and Waters

In the frame of activities under EMEP programme emissions of sulphur, lead and benzo(a)pyrene were evaluated and presented in 25x25 km set for the reference years 1990 and 1995. The data on emissions are submitted by the Bulgarian Ministry of Environment and Waters (MoEW), based on economic statistics and CORINAIR methods. At first SNAP-90 was used, but for more recent estimates SNAP-94 was applied. Emissions are presented in two categories - large point sources (LPS) and area sources. According to the definition of LPS in CORINAIR thirtry four LPS of sulphur are identified in the country with yearly emissions in the range 63 - 222500 Mg S/y. For lead the number of LPS is 20 (0.1-19.8 Mg Pb/y) and for B(a)P - 13 LPS (1 - 651 kg B(a)P/y..

The emissions of sulphur are directly related to the sulphur content of the fuel, which for coal normally varies between 0.3 and 1.2 % (up to an extreme value of 4.5 %) and for fuel oil (including heavy fuel oil) from 0.3 up to 3.0 %. LPS are the main sulphur emitters in Bulgaria. The emissions by individual LPS and by month are estimated.

Area sources are formed by many small sources, spread over the area. For sulphur area sources are formed by the traffic, local heating, small industrial boilers and factories. As to estimate Bulgarian area sources for EMEP needs, part of the standard EMEP mesh with resolution 150x150 km is used. Every cell is divided to 36 squares 25x25 km. According to the geographical co-ordinates the LPS emissions are added to the respective cells.

The yearly distribution of emissions of sulphur from area sources for 1990 in the mentioned network is presented. The greater part of these emissions vary between 0.5 and 1.5 Mg/cell/y. Only in the region of Sofia the values are considerably greater - 3 Mg/cell/y.. This may be due to the intensive traffic in the town. The total quantity of sulphur emitted in Bulgaria from the area sources during 1990 is estimated to 182 220 Mg. Both LPS and area sources in 1990 have emitted 986 220 Mg S, i.e. about 1 million of tonnes. The shares of different emitters and the reasons for high sulphur emissions are discussed.

The emission of lead was estimated on the base of average emission factors for production activities, combustion and transport according to CORINAIR. Transport is the main source of lead. From the emissions of LPS these in three cities contribute for 72 % of all LPS. The emission of lead in Sofia as area source is 10 times more than in other regions. Ethylized gasoline with lead content of 0.15 g/l is the main gasoline type used in the country and its combustion represents the major source of atmospheric lead pollution. In next five years the use of ethylized gasoline will be excluded from the road transport sector and consequently the emissions of lead will be sharply cut. Other sources of lead emissions comprise combustion in the industry and process emissions. From the total emissions in the country of 299 Mg Pb/y, 200 Mg are due to road transport and 94 Mg to industrial combustion and technological processes. Because of lack of data for the emissions at ore mining and processing the emissions of Pb are to some extent underestimated.

Emission inventories of B(a)P for the referent years 1990 and 1995 represent a first attempt to asses and express this emission in Bulgaria for the needs of EMEP. Activities connected with B(a)P emissions are presented with production rate (mileage) and emission factors. The emissions themselves are also presented by LPS and area sources. For B(a)P the area sources are the main emitter.

Air Pollution Modeling and Its Application XIII, edited by S.-E. Gryning and E. Batchvarova, Kluwer Academic/Plenum Publishers, 2000

They account for 96% of all emissions in the country. Both emissions from LPS and area sources are split over administrative regions and SNAP categories. The results show that the main B(a)P contamination is related to commercial, industrial and residential combustion. One LPS accounts for 92 % of all emission from LPS.

Comparison of the emissions of lead and B(a)P for 1990 and 1995 using SNAP-90 and SNAP-94 was made.

Table 1. National annual emissions for 1990 and 1995 of lead by source category, Mg/year

SNAP GROUP	SN 90 1990 LPS	SN 90 1990 area	SN 90 1990 total	SN 94 1990 LPS	SN 94 1990 area	SN 94 1990 total	SN 94 1995 LPS	SN 94 1995 area	SN 94 1995 total
01	1.38	0.56	1.94	7.72	0.62	8.34	5.63	0.48	6.11
02		0.49	0.49		3.33	3.33		1.91	1.91
03	70.14	3.52	73.70	135.73	50.63	186.36	82.77	31.51	114.28
04	19.42	0.77	20.2		25.12	25.12		16.28	16.28
07		199.81	199.81		199.21	199.21		153.3	153.3
09		3.09	3.09		13.49	13.49		5.62	5.62
TOTAL	**90.95**	**208.27**	**299.21**	**143.45**	**292.4**	**435.85**	**88.41**	**209.1**	**297.5**

Table 1 shows that the estimates of lead emission for 1990 after SNAP-90 and SNAP-94 differ considerably (e.g. 299 and 436 Mg/y for the totals). While for the LPS the changes are connected mainly with the number and the list of LPS, in the case of area sources the changes are connected with more complex accounting for all emitting activities and processes.

The emissions of B(a)P for all categories in 1995 are bigger than in 1990, except production processes (SNAP 04). The biggest absolute increase of 6 700 kg/y is in the sector of commercial, institutional and residential combustion. This could be attributed to an increase of fuel consumption in this category because of the high electricity and oil prices.

Table 2. Emission of B(a)P from SNAP-94 category 02 in 1990 and 1995

	1990		1995		1990	1995
Fuel	Mg	EF, mg/Mg	Mg	EF, mg/Mg	E, kg	E, kg
BROWN COAL[1]	575416	4.4	38558	4.4	2.5	0.2
BROWN COAL[2]			873191	912	0.0	796.0
HARD COAL[1]	1007588	4.4	13679	7.3	4.4	0.1
HARD COAL[2]			124742	4332	0.0	540.0
BRIQUETTES	565000	26223	725125	26223	14800.0	19000.0
WOOD	665000	2009	1247730	2009	1340.0	2510.0
FUEL OIL	588732	1.1	275789	1.1	0.6	0.3
				TOTAL	**16200.0**	**22900.0**

Note: [1] and [2] indicate two types of combustion for the relevant fuel

Indeed in this category the fuel consumption reported by MoEW increased for all fuels except hard coal and fuel oil as shown on Table 2. The biggest emission increase is for use of briquettes where also the emission factor is the most considerable. The split of brown and hard coal combustion in two part each with different emission factors also leads to emission increase.

References

BC-EMEP, 1993: Bulgarian contribution to EMEP, Annual Report for 1993, NIMH, Sofia, 1994.

BC-EMEP, 1994: Bulgarian contribution to EMEP, Annual Report for 1993, NIMH, Sofia, 1995.

CORINAIR 1992: CORINAIR Inventory Default emission factors handbook (second edition). CEC. CITERA. Jan.1992

EMEP/CORINAIR, 1996: Joint EMEP/CORINAIR Atmospheric Emission Inventory Guidebook, First Edition, Copenhagen: EEA, 1996

CALCULATIONS OF THE AIR POLLUTION AND POLLUTANTS DEPOSITION DUE TO A NON-FERROUS METAL PLANT

Dimiter Syrakov[1], Anna Tzenkova[1], Julia Ivancheva[1], Tatiana Spassova[1], Vasil Andreev[1], and George Djolov[2]

[1] National Institute of Meteorology and Hydrology - Bulgarian Academy of Sciences
[2] University of Venda, Dept. of Physics, South Africa

INTRODUCTION

The Non-ferrous Metal Plant (NMP), located in the Zlatitsa-Pirdop valley, is the major factor exerting anthropogenious influence on environment. This influence is not limited only to the flat part of the valley and it's surrounding slopes but, under certain synoptic conditions, the pollutants emitted from the plant can be transferred to the neighboring regions.

The objectives of the study are:

1.To calculate the climatic fields of air pollution, created by the NMP with respect to gases and aerosols (dust), taking into account dry and wet deposition.

2.To estimate the plant's impact in the air pollution of the neighboring regions.

METHODS

For calculation of the climatic pollution fields, created by a single source or a group of sources, the models GAS_R and AER_R have been used. They are intended for estimating the distribution of gases and aerosols, respectively. The AER_R model includes specific parameterization of aerosol processes (sedimentation, dry and wet deposition), suggested by Galperin. In both models, the calculations are made for a fixed rectangular system of grid points (receptors) with arbitrary prescription of point number and step dimension. The models are PC-oriented and implement a peculiar version of the so-called Gaussean PLUME-model.

The atmospheric stability is determined after the Pasquill-Turner classification. Briggs formulae for sigma's are used. The plume rise is described after Briggs formulae, also. The temperature at chimney level is determined from ground level temperature, using vertical temperature gradients, typical for the different stability classes. The wind velocity at chimney's height is determined from surface wind applying the US-EPA recommended power law, exponent depending on the stability class. With calm weather, plumes, formed at wind speed of 0.5 m/s, are averaged by all directions.

The 3D Eulerian dispersion model EMAP has been used for calculation of the transport to the neighboring regions.

In 1981 and 1987 substantial technological improvements were introduced in NMP. For this reason the modeling calculations were implemented separately for three periods: 1958-1981; 1982-1987 and after 1987. 1996 is used as typical for the last period.

Air Pollution Modeling and Its Application XIII, edited by S.-E. Gryning and E. Batchvarova, Kluwer Academic/Plenum Publishers, 2000

CONCLUSIONS

The main conclusions from model calculations are:

1. The functioning of NMP during the first two periods, from 1958 to 1987, has formed well defined zones of pollution with sulphure dioxide, concentrations exceeding the annual MAC (maximum acceptable concentration) with 50% and more. The technological improvements after 1981 had led to a significant decrease of SO_2 concentrations. A small single zone with exceeding of annual MAC still remains at about 2km northeastwards from the plant.

2. The calculations for 1996 are much more detailed. The following distribution of the pollution in the region, caused by the NMP is found out:

- NMP does not present a danger for the air contamination with copper – according to the modeling calculations the maximum average annual air concentration for 1996 is 0.007$\mu g/m^3$, while the diurnal MAC is 10$\mu g/m^3$. The average seasonal concentrations show the same tendency. The average annual maximum deposition of copper for 24hrs is 671$\mu g/m^2$, i.e. about 187mg/m^2 for the whole year.
- The average annual air pollution with cadmium during 1996 is lower than the average annual standard (0.01$\mu g/m^3$). During the winter, however, the maximum calculated concentration is 0.017$\mu g/m^3$, i.e., in some parts of the region in winter season it can occur an exceeding of daily MACs in case of specific unfavorable meteorological conditions.
- The model calculations of the total deposition of cadmium expressed by its average diurnal values for the year and for the winter and summer seasons show that the whole examined area had been polluted over the acceptable standards. In some zones and for some meteorological conditions high diurnal values of cadmium deposition could be observed. The mean diurnal deposition for 1996 is about 11 times higher than the acceptable standard. The calculated total deposition is 15.3 mg/m^2 for the whole 1996.
- According to the calculations, the air pollution with lead (mean annual and seasonal values) for 1996 is about 10 times lower than MAC, i.e., it probably does not represent danger. The deposition of lead (as an average diurnal value for 1996 and for the seasons), on the other side, is much more higher than the accepted levels. For instance the maximum modeling values of the mean diurnal deposition are about 350$\mu g/m^2$, while the standard is 250$\mu g/m$. During the whole 1996, in the zones of maximum pollution, the deposited quantity of lead is estimated to about 98mg/m^2.

3. The model estimations of the transregional pollution with lead are made on the base of the monthly averaged concentration fields for 1995. The mean annual air concentration takes its maximal values (about 0.82 ng/m^3) in the region of the valley of Zlatitza and Pirdop. This concentrations decrease about 4 times at a distance of 50km from NMP. The mean annual concentration of lead in the precipitations has a maximum also in the same region of Zlatitza-Pirdop valley and is about 3940ng/l . In a radius of 30km around NMP this concentration decrease more than 3 times. The maximum annual dry deposition of lead (52$\mu g/m^2$) is again reached in the same area. In a radius of 50km the deposition diminishes to 20$\mu g/m^2$. The wet deposition of lead obtains its maximum annual value again in the same region equal to 2.7mg/m^2. In a radius of 25km it takes the value of 1mg/m^2. The total annual deposition of lead has a distribution similar to those obtained for the wet deposition. The maximum is about 2.76mg/m^2, in a radius of 20km it reaches 1mg/m^2.

The generalized modeling estimation of the described characteristics is that NMP has a real (significant) influence on the pollution of the environment in a region with a radius of about 200km.

REFERENCES

BC-EMEP, 1993-7: Bulgarian contribution to EMEP, Annual Report for 1993-7, NIMH, Sofia, 1994-7.

PHARE Project BG 9310-03-05-02, Final Report of the Project " Ecological assessment of the impact of the emissions from MDK-Pirdop on the soils in the region of Pirdop-Zlatiza", Section N 6 " The Nature of Climatic Conditions a Prerequisite for Air pollution"

ONLINE COUPLING OF MULTISCALE CHEMISTRY-TRANSPORT MODELS WITH NON-HYDROSTATIC METEOROLOGICAL MODELS

Ralf Wolke, Oswald Knoth,and Annette Münzenberg–St.Denis

Institute for Tropospheric Research
Permoserstr. 15, D–04303 Leipzig, Germany
e–mail: wolke@tropos.de

Air quality models are used to enhance the understanding of pollutant dynamics in the atmosphere, in particular with regard to the relation between emissions and the resulting distribution of chemical species in the troposphere. Such models are extremely expensive in terms of computing time. This is due to the fact that reacting flow calculations are highly stiff in terms of time–stepping. Because explicit ODE solvers require numerous short time steps in order to maintain stability, it seems that especially sparse implicit–explicit solvers have considerable room for efficiency improvements in atmospheric chemistry applications.

In the chemistry–transport code MUSCAT (Knoth and Wolke, 1998a; Wolke and Knoth, 1998) an implicit–explicit (IMEX) time integration scheme is implemented. Explicit second order Runge–Kutta methods for the integration of the horizontal advection are used. The stiff chemistry and all vertical transport processes (turbulent diffusion, advection, deposition) are integrated in an implicit and coupled manner utilizing the second order BDF method. For this part a modification of the code LSODE (Hindmarsh, 1983) is applied with a special linear system solver and a restriction of the BDF order to 2. The nonlinear corrector iteration is performed utilizing a Newton method in which the sparse linear systems are solved by linear Gauss–Seidel iterations. For the integration of the implicit part, smaller time steps are used which are generated by an automatic step size control. The "large" explicit time step is restricted by a modified CFL–criterion. Higher order accuracy and stability conditions for the whole implicit–explicit scheme are investigated in Knoth and Wolke (1998b).

For the spatial discretization of the advection term a third order upwind procedure with additional limiting (Hundsdorfer et al., 1995) is implemented. The fluxes resulting from the horizontal advection are defined as a linear combination of the horizontal fluxes from the current and previous stages of the Runge–Kutta method. These horizontal fluxes are treated as "artificial" sources within the implicit integration. A change of the solution values as in conventional operator splitting is thus avoided.

In Knoth and Wolke (1998a) an online coupling scheme between our chemistry–transport code and the mesoscale non-hydrostatic meteorological model METRAS (Schlünzen et al., 1996) is proposed. The meteorological and the chemistry–transport algorithms have their own separate time step size control. Coupling between meteorology and chemistry–transport takes place at each horizontal advection time step only. The meteorological fields are held constant during the implicit integration cycle. This means that temporal changes of the meteorological values are neglected over the explicit time step of the chemistry–transport model, even though the meteorological solver generally generates these variations with a more accurate time resolution.

For the new version of MUSCAT (Knoth and Wolke, 1998) an improved coupling procedure between the meteorological model and the chemistry–transport code was developed which is adapted to the applied IMEX scheme. Each stage of the IMEX scheme requires a new calculation of the horizontal fluxes and an implicit integration cycle. The new mass conservative coupling provides time–averaged horizontal wind fields for the computation of the horizontal fluxes and time–interpolated meteorological fields (vertical exchange coefficient and wind speed, temperature, humidity, density) during the implicit integration part.

In MUSCAT the horizontal grid is decomposed into subdomains whose sizes depend on available memory and cache. This technique also allows the use of finer or coarser resolutions for individual subdomains. The model simulates each subdomain with its own time step and periodically exchanges information. The step size selection is performed by monitoring the local truncation error. The metorological fields are averaged or interpolated from the equidistant horizontal meteorological grid into the block–structured chemistry–transport grid with different resolutions.

The coupling scheme is applied to meaningful 3D test problems and a ozone scenario in Saxony. For all cases the Euro–RADM chemistry mechanism (Stockwell and Kley, 1994) is used. The results are compared with those of the scheme used previously. The numerical experiments indicate that the proposed coupling procedure between the meteorological model METRAS and the chemistry–transport code MUSCAT works very well. The behavior of the coupling scheme depends considerable on the size of the explicit time step (restricted by the modified CFL–criterion). The coupling procedure can easily be extended to the coupling of other mesoscale meteorological and chemistry–transport codes.

REFERENCES

Hindmarsh, A.C., 1983, ODEPACK, A systematized collection of ODE solvers, in: R.S. Steplman et al., Eds., *Scientific Computing,* (North–Holland, Amsterdam), 55–74.

Hundsdorfer, W., Koren, B., van Loon, M., and Verwer, J.G., 1995, A positive finite–difference advection scheme, *J. Comput. Phys.* **117**, 35–46.

Knoth, O., and Wolke, R., 1998a, An explicit–implicit numerical approach for atmospheric chemistry–transport–modeling, *Atmos. Environ.* **32**, 1785–1797.

Knoth, O., and Wolke, R., 1998b, Implicit–explicit Runge–Kutta methods for computing atmospheric reactive flow, *Appl. Numer. Math.*, in press.

Schlünzen, K.H.,Bigalke, K., Lenz, C.J., Lüpkes, Ch., Niemeier, U., and von Salzen, K., 1996, Concept and realization of the mesoscale transport and fluid model METRAS. METRAS Technical Report 5, Meteorological Institute of the University of Hamburg.

Stockwell, W.R. and Kley, D., 1994, The Euro–RADM mechanism: A gas–phase chemical mechanism for European air quality studies, Berichte des Forschungszentrums Jülich Juel–2868, Jülich.

Wolke, R., and Knoth, O., 1998, Implicit–explicit Runge–Kutta methods applied to atmospheric chemistry–transport modelling, submitted to *Environmental Modelling and Software.*

A MODEL VALIDATION DATA SET RESPECT TO THE AIR POLLUTION DEPOSITION IN NORTHERN WINTER CONDITIONS

Marko Kaasik

Department of Environmental Physics
Tartu University
Tähe 4, Tartu, EE2400 Estonia

The data set is based on field observations carried out in the north-east of Estonia during December 1995 – February 1996. The air pollutants were emitted from stacks of the Estonian and the Baltic thermal power plants (maximum capacity ca. 1.5 GW each), located at 17 km from each other. First of them has two 250 m high stacks, second four 150 m and two 180 m high stacks, producing high momentum and buoyancy plumes. Other sources (Figure 1) are either tens of times weaker (Ahtme, Kohtla- Järve and Sillamäe TPP-s, chemical plants) or have too low stacks for regional-scale transport (Kunda and Slantsy). The kukersite oil shale is fired in both power plants. The chemical composition of fuel and operational regime were rather stable during the season. The flue

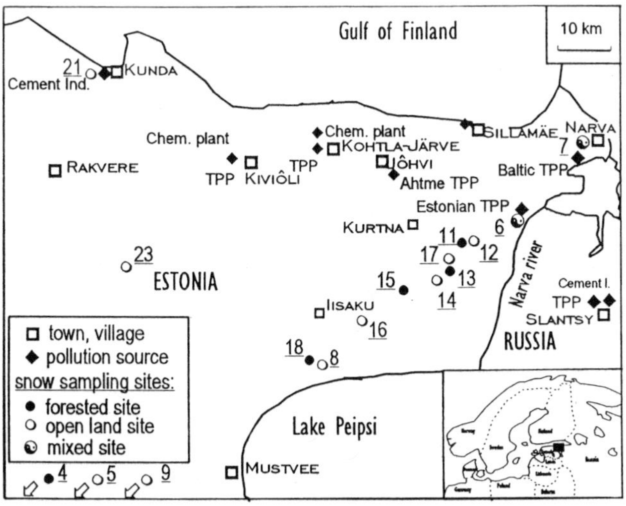

Figure 1. The schematic map of snow sampling sites (numbered) and pollution sources. The allocation of research area (filled rectangle) in northern Europe is shown in lower right corner.

gas was characterised by high content of sulfur dioxide and alkaline fly ash (30 % CaO, other oxides, a number of trace metals).

The landscape near both sources is industrial (open mining, ash plateaus etc.) mixed with forest and shrub. Most of research sites were located at distances 10 – 50 km from sources in sparsely inhabited rural landscape with dominating coniferous forest and open bog. Terrain type is plain or slightly hilly (30 – 90 m above the sea level).

The deposition data are based on 13 snow samples (samples number 9, 21 and 23 excluded, see Figure 1), taken through the entire snow package accumulated during 71 days of permanent snow cover. Three snow layers, accumulated during 20, 26 and 27 days, were separated on the basis of two ice crusts originating from drizzle episodes in the beginning and end of January 1996. Since no positive snow level temperature (in degrees of centigrade) was registered, probably no wash-out or vertical displacement of deposited pollutants occurred. Most of sampling points were located near a straight line from Estonian TPP to SW. Samples were taken either from coniferous forest or open bog. With the exception of two sites near power plants, all sites were chosen not closer than 2 km from main traffic roads and villages. Two samples were taken from Lower-Pedja Nature reserve (approximately 100 km SE from main research area). These were considered as background sample.

The chemical analyses of snow samples were performed at NILU (Norway) and Institute of Ecology (Estonia). Using the factor, cluster and correlation analysis (26 chemical species, all samples and layers), it is proven, that the oil-shale burning products are highly dominating in all samples except of background ones.

Data are partially preprocessed (background deposition fluxes and fraction of fly ash in the total sulfur flux eliminated, fly ash deposition flux derived from fluxes of components), but initial data are also available.

The model validation data set includes following data.

1. meteorological data:

• wind speed and direction at 10 m height, cloud amount and height in each 3 hours, snow level and 2 m temperature – from Narva and Jõhvi stations in the research area;

• precipitaton amount in each 12 hours from same stations;

• radio sounding altitudes, temperatures, wind speeds and directions at ground, 975, 875 and 750 mbar levels from Tallinn (170 - 200 km to west).

2. source data of the Estonian and the Baltic thermal power plants:

• positions, stack heights, opening diameters, flue gas temperatures and velocities;

• average emissions of SO_2 and fly ash from the stacks during snow accumulation periods.

3. geographical data:

• positions of snow sampling sites (geographical and local coordinates);

• vegetation types around sampling sites (open, forested or mixed);

• raster map of land use types in research area.

4. deposition data:

• measured deposition rates (mg/m^3 per day) of SO_2 and fly ash at sampling sites.

Described data set is completed for medium-term average rather than short-term model validation. Its main limitation is absence of concentration measurements in the air. Thus, these data could be used only for validation of bulk process models, not for transport and deposition separately. A combined dispersion-deposition data set is a subject of further research.

The data set is freely available in electronic form (MS Excel workbook or ASCII data files). The contact address is: marko@apollo.aai.ee.

INFLUENCE OF THE NEAR-FIELD GEOMETRY ON FIELD MEASUREMENTS IN URBAN STREET CANYONS

Stilianos Rafailidis

Environmental Research Laboratory, NRCPS "Demokritos,"
GR-153 10 Aghia Paraskevi Attikis, Greece

INTRODUCTION

The potential effect of nearby buildings on air quality observations in the field is inadequately understood. Rafailidis (1998) proposed a method for assessing this influence conservatively and on a long-term basis. It is based on subdividing the wind vector above the roofs into parallel- and normal- to street contributions (Figure 1). Under windy conditions, while U_n exceeds the critical threshold for canyon vortex formation, air circulation and street pollution dispersion are described satisfactorily by the wind tunnel modelling of Rafailidis (1997). Additional flushing by $U_\circ$ mainly advects pollution along the street, without overcoming asymmetries which may develop depending on street canyon configuration (Figure 2). To validate the potential of the above analysis and assess its impact on actual urban canyon situations, this paper examines field observations, focusing on the existence of the suggested persistent vortex circulation and canyon reaeration patterns.

RESULTS AND DISCUSSION

Field street canyon observations from Göttinger Str. in Hannover (NLÖ, 1993) are summarised for 1994 in Figure 3. Only events with $U > 1\text{ms}^{-1}$, at 3m above the roofs, are considered. The street-wind measuring station, at h=1.5m above the street and 2.2m from a side wall, is influenced locally by a nearby pedestrian passage, on the same street side, and two streets intersecting at the opposite side. Nonetheless Figure 3-bottom reveals that, apart from winds oncoming from ±20° from street axis, in all other cases vector summation of the observed V and W components of the street wind (Figure 1) yields a relatively constant result, independent of roof-wind vector rotation. This behaviour is consistent with the action of a persistent vortex rotating inside the canyon, independently from the simultaneous along-the-street advection.

For these oncoming wind conditions then, which constitute a large proportion of the yearly events, the analysis of Rafailidis (1998) may be used to estimate the long-term pollution dispersion in actual street canyons. Furthermore, the field observations suggest that, under windy conditions, vortex motions persist in the street canyon, remaining quite unaffected by disturbances from substantial canyon discontinuities nearby (*e.g.*, a shopping arcade passage in this case). Then, the analysis of Rafailidis (1998) is of practical applicability in actual towns, and may help evaluate systematic errors in field campaigns caused by consistently asymmetric pollution dispersion ensuing from the asymmetric geometry of the measurement canyon.

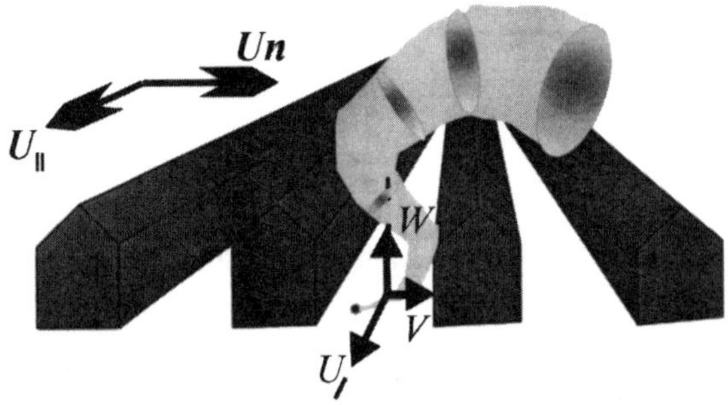

Figure 1. Subdivision of the wind vector above the roofs into two components $U_{\parallel}$ and U_n at that height, and the resulting flow components close to the street (drawn not to scale). The dispersion from a street-level point source is also schematically drawn (the U,V,W vectors and the plume orientations are shown arbitrarily).

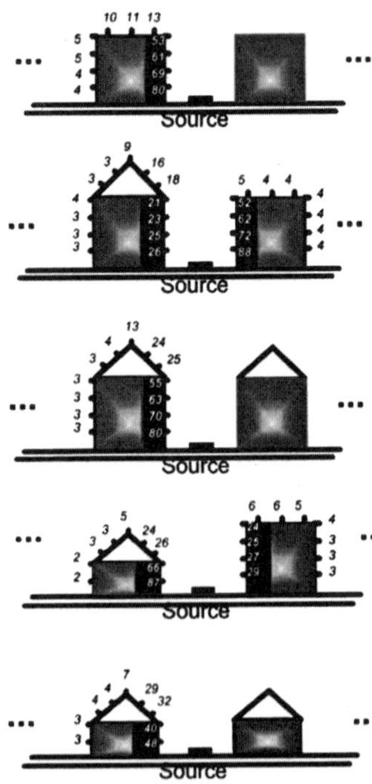

Figure 2. Long-term non-dimensional concentrations $\overline{K}=CUHL/Q$ in different street canyon configurations with equal street-width to building-height. In symmetric geometries the results from one wall are mirrored on the opposite building. For further examples of the analysis, see Rafailidis (1998).

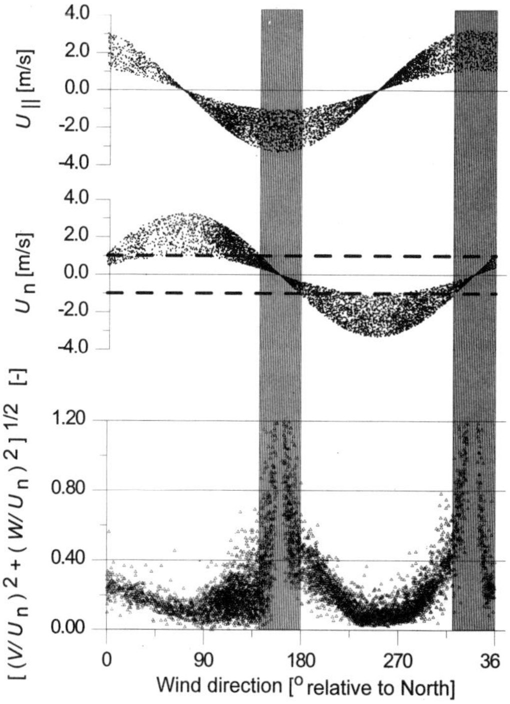

Figure 3. Street-wind components V and W observed during one year Göttinger Str. urban canyon measurements (supplied courtesy of Dipl.-Meteorol. W.J. Müller of the *Niedersächsisches Landesamt für Ökologie*, unpublished). The shaded bands correspond to orientations ≈ ±20° from the street axis. In the middle graph the dashed lines indicate the critical U_n required for stable vortex formation ($U_{n,crit} \approx 1\text{ms}^{-1}$).

REFERENCES

NLÖ, 1993, *Lufthygienische Überwachungssystem Niedersachsen. Luftschadstoffbelastung in Straßenschluchten*. Niedersächsisches Landesamt für Ökologie, Hannover, Germany.

Rafailidis, S., 1997, Urban street canyon pollution dispersion under the influence of coastal breezes, in: *Air Pollution Transport and Diffusion over Coastal Urban Areas*, NATO/CCMS Techn. Rep. No. 214, Kambezidis, H.D., ed., pp. 19-35.

Rafailidis, S., 1998, Using urban site measurements for model validation: Before starting, look at the near-field geometry, in: *Proc. 5th Intl. Conf. on Harmonisation within Atmospheric Dispersion Modelling for Regulatory Purposes*, Rhodos, May, pp. 541-8.

ANALYSIS OF INDUSTRIAL PLUMES SUPERPOSITION EFFECTS IN A MEDITERRANEAN COASTAL REGION

G. Tinarelli, P. Faggian, S. Finardi, G. Brusasca

ENEL/SRI/Polo Ambiente, via Reggio Emilia 39, 20090 Segrate (MI) Italy

INTRODUCTION

The behaviour of pollutant emitted by large industrial sources is a problem of particular interest both at local and regional scales. This is particularly true in the Mediterranean region, where many industrial facilities are located along the coast. In this situation the mesoscale atmospheric circulation generated by the sea/land discontinuities and topography irregularities can originate complex flow structures characterized by periodic reverse of the wind direction. Pollutant dispersed in such conditions can be trapped by the re-circulating flow and give rise to complex plume trajectories dependent on their location, coast shape, terrain and landscape features. When more sources are located in the same region, the different plumes can interact even if they are emitted at a considerable distance. It is therefore relevant for air quality management to know the possible superposition effects at least for the major sources. Such information can be difficult to obtain, particularly in complex terrain, because overlap regions are hardly predictable.

MODEL SIMULATIONS

The atmospheric pollutant dispersion from two thermal power plants, located along the Thyrrenian Coast of Central Italy, has been investigated in order to verify the conditions at which possible superposition effects take place. The prognostic mesoscale (Pielke, 1984) circulation model HERMES (Perdriel, 1990) was used to reproduce the complex structure of both the atmospheric mean flow and turbulence over a 160x160 km^2 region centered on the shoreline. Pollutant dispersion from the two sources was described by the Eulerian K dispersion model implemented into the same code, onto a 120x120 km^2 nested area.

The study has been based on the simulation of five episodes (whose duration ranged from 48 to 72 hours) selected from the data collected during a long term intensive meteorological field campaign that took place during 1995. The periods were chosen in order to represent typical meteorological conditions. Different SO_2 and NO_x emission scenarios have been considered for the thermal power plants. Meteorological simulations have been initialized and driven on the basis the analysis of local measurements and ECMWF meteorological fields. Results have been analyzed in order to identify the circulation

Air Pollution Modeling and Its Application XIII, edited by S.-E. Gryning and E. Batchvarova, Kluwer Academic/Plenum Publishers, 2000

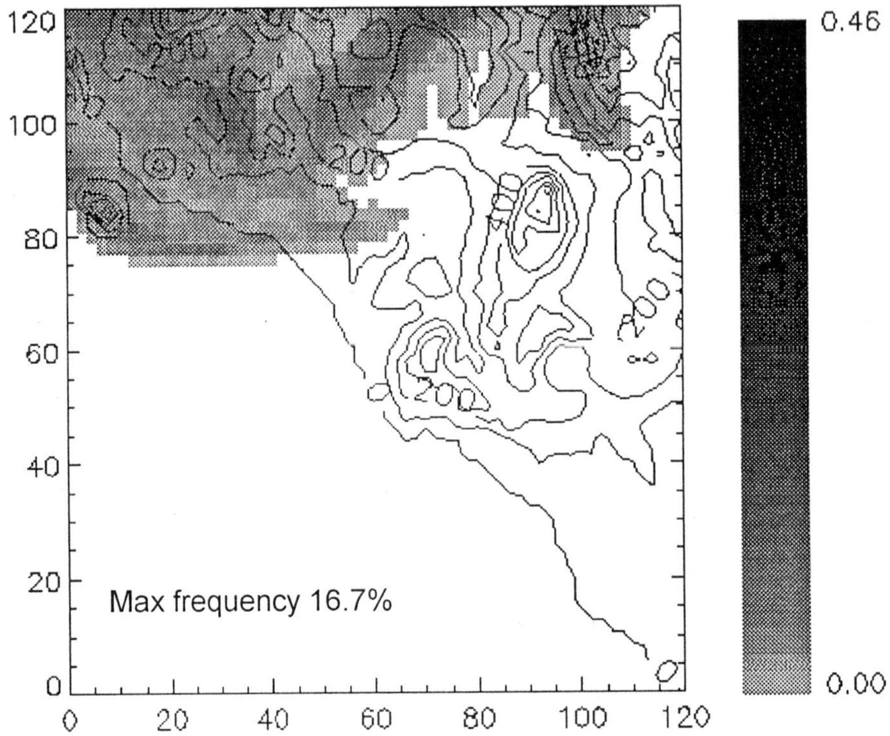

Fig. 1 Spatial distribution of the overlapping frequency over the computational domain. Overlapping events are considered if the contribution coming from one emission is at least 30% of the total g.l.c. level. Only values greater than 20$\mu g/m^3$ have been taken into account. A three days period of November 1995 is considered here.

conditions that can cause interaction between the plumes and overlap of ground level concentrations. The space distribution and statistical relevance of superposition episodes has been evaluated and maps have been produced identifying the presence of possible critical regions. A large number of overlapping episodes was observed into an area located in the Northwestern section of the domain (see Fig. 1). As a general result, the plume interaction effects were to extend the area interested by significant g.l.c. levels (> 20 $\mu g/m^3$) increasing only slightly maximum concentration levels.

Results can help to identify the influence due to the different source characteristics (stack height, plume rise) on the ground level concentration distribution. Some useful suggestions to modulate and balance the emissions of the two power plants can be obtained from the results of the study, in order to avoid critical environmental conditions due to the overlaps or re-circulation of the plumes.

REFERENCES

Pielke R. A. (1984) *Mesoscale Meteorological Modeling*, p.612, Academic Press, New York

Perdriel S. (1990) Note de principe du code Hermes, EDF/DER Report HE-33/90.04 (available from E.D.F., 6 Quai Waitier, 78400, Chatou, France).

AIR POLLUTION STUDIES FOR THE ROMANIAN NUCLEAR POWER PLANT - CERNAVODA BASED ON THE RIMPUFF DISPERSION MODEL

D. Gultureanu[1], B. Gultureanu[2], T. Mikkelsen[3], S. Thykier-Nielsen[3], M. Matenciuc[4]

[1] Department of Physics, “Petroleum-Gas” Univ. of Ploiesti, Ploiesti Prahova, ROMANIA

[2] Department of Computer Sciences, “Petroleum-Gas” Univ. of Ploiesti

[3] Department of Wind Energy and Atmospheric Physics, RISØ National Laboratory, DENMARK

[4] Department of Physics, Economic High School, Mangalia, ROMANIA

The paper presents the results of research work started at RISØ National Laboratory, Denmark and continued at “Petroleum-Gas” University of Ploiesti, Romania regarding application of the real-time episode dispersion model RIMPUFF to Romanian Nuclear Power Plant - Cernavoda with the motivation to understand and quantify severe air pollution scenarios of ground level concentrations and gamma doses.

Records of one hour mean wind and temperature measurements were taken from the nearby to Romanian Nuclear Power Plant - Cernavoda meteorological station and used as input data to the combined flow and diffusion model LINCOM/RIMPUFF for a sequence of severe emissions and wind flow episodes.

The mesoscale dispersion model RIMPUFF (Thykier-Nielsen and Mikkelsen, 1984) is a fast and operational computer code suitable for real time simulation of environmentally hazardous materials and gases released to the atmosphere. RIMPUFF is a puff diffusion code designed for real-time simulation of puff and plume dispersion during time and space changing meteorology.

LINCOM, the flow model driving the dispersion model RIMPUFF, extends the theory for neutral, potential flow over a small hill to mesoscale complex terrain (Troen and de Baas, 1986). LINCOM is a fast diagnostic, non-hydrostatic dynamic flow model based on the solution of linerized versions of the three momentum equations and the continuity equation with a first order spectral turbulent diffusion closure.

The source were considered inside the to Romanian Nuclear Power Plant - Cernavoda. For the simulations we considered a 60km x 60 km domain modelled on a 60 x 90 grid.

The LINCOM/RIMPUFF model has been run on 10 cases including examples of typical daytime convection and also including some less typical, but from an pollution point of view very interesting cases, especially because of the on-shore wind conditions.

The study encounters the modelling of some possible worst case dispersion scenarios in the area.

The area consideredin the study is presented in the next figure:

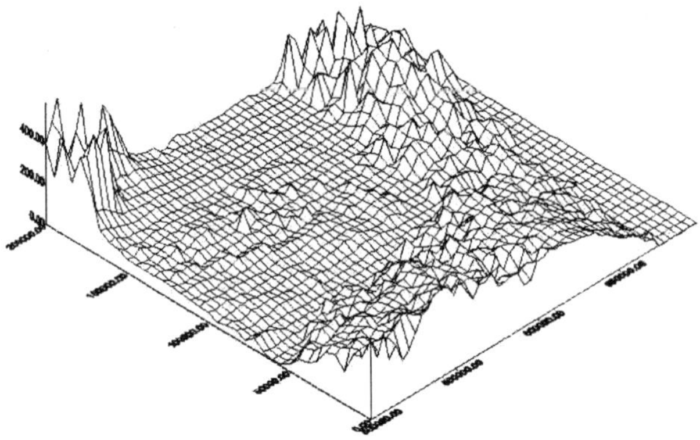

Figure 1 Three dimensional map of the Romanian Nuclear Power Plant - Cernavoda area, considered in the simulation.

Some results of the simulations are presented in the next figures:

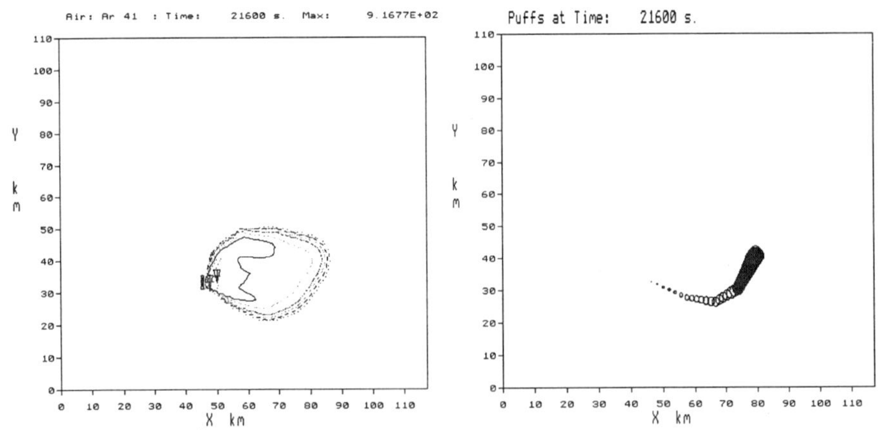

a) b)

Figure 2 a) Concentrations and b) puffs integrated over one hour for a typical daytime convection; Romanian Nuclear Power Plant - Cernavoda area.

REFERENCES

Mikkelsen, T., Larsen, S.E. and Thykier-Nielsen, S.(1984). Description of the Risoe puff diffusion model, Nuclear Technology, 67, 56-65

Mikkelsen, T., Thykier-Nielsen, S., Troen, I., et al. A hard model for complex terrain, Preprints from Eight Symposium on Turbulence and Diffusion, San Diego, CA, April 25-29, 1988, 180-185

Thykier-Nielsen, S, Mikkelsen, T., Kamada, R., and Drake, S.A. (1990). Wind flow model evaluation study for complex terrain, Proceedings of Ninth Symposium on Turbulence and Diffusion, American Meteorological Society, Boston, MA, USA

Troen, I. and de Baas, A. F.(1986). A spectral diagnostic model for wind flow simulation in complex terrain, Proc. Euro. Wind Energy Assoc. Conf., Rome, October 7-9, 1986

AUTOMATED METEOROLOGICAL STATION AND APPROPRIATE SOFTWARE FOR AIR POLLUTION DISTRIBUTION ASSESMENT

Zoran Grsic[1], Predrag Milutinovic[2]

[1]Institute of Nuclear Sciences "Vinca"
P.O.Box 522, 11001 Belgrade, Yugoslavia
[2]Institute of Physics, Zemun
Pregrevica 118, 11080 Belgrade, Yugoslavia

INTRODUCTION

The accident at the research reactor in the Institute "Vinca" in 1958., has initiated forming the meteorological service in the Institute, which has started with operating in 1959. This service has defined it's aims in accordance with regulatory guides of International Atomic Agency, U.S. Nuclear Regulatory Commission and the other sources from literature. The Gaussian atmospheric diffusion models, Pasquill-Giford stability category schemes and classical meteorological measurements were logical choice in that time. Manual and semiautomatic processing of the input meteorological data and nomograms with σ curves, problems in accordance with determineing stability categories are probably in the memory of menu services which were solving problems in atmospheric distribution of some pollutants through atmosphere.

Mass application of the personal computers at the beginning of the eighties, our job made significantly easier. We have developed codes for fast searching through the base of the meteorological data and making pictures of the air pollution distribution in any time covered with the appropriate meteorological measurements.

In order of getting picture of the air pollution plume in practically real time, we had to automate meteorological measurements and to make connection with atmospheric diffusion code. This was important task, especially with the aspect of fast assessment in the accidental situations.

Since April 97', the automated meteorological station is in use in The Institute of Nuclear Science "Vinca". All measured quantities are continuously displayed on tha PC monitor in a digital and graphical form, they are averaging every 10 minutes and sending to the atmospheric diffusion code.

Results of these calculations are used for solving basic and some special atmospheric dispersion problems in the Institute like pollution distribution assessment in practically real time, zoning surrounding of the reactor stack and risk estimation in accordance with accidental releases.

Air Pollution Modeling and Its Application XIII, edited by S.-E. Gryning
and E. Batchvarova, Kluwer Academic/Plenum Publishers, 2000

METEOROLOGICAL MEASUREMENT PROGRAM IN THE INSTITUTE "VINCA"

Meteorological measurement program was outlined in accordance with instruction "...no meteorological measurement program without reference to one or more dispersion models (Sven - Erik Gryning, 1980.)". At meteorological tower 40 meters tall (that's the height of the four stacks placed in the Institute) are placed sensors for the wind direction and speed, temperature at 40 and 2 meters, relative humidity, short and long wave radiation at 2 meters, and precipitation sensor and Geiger counter at 1 meter above surface.

Automated meteorological station

The automatic acquisition system is capable of receiving simultaneously 16 analogue signals with 13 bit resolution. The acquisition system is, in turn, controlled by a microcontroller which packs the received signals into a 42 bytes long message. This message is sent, via RS232 interface to a radio emitter working at 432 MHz. Receiver is located on a laboratory roof which is at distance 300 m from the emitter. The signals that are received are sent, via RS232 interface to a PC.

The operating system used by the PC is a real time and multitasking one. Custom designed software has been designed using C language, as a set of four programs-four separate tasks which communicate via message-passing. User interface is based on graphical windowing system.

Computer program-software for the dispersion prediction

A constitutive part of the automated meteorological station is the Gaussian algorithm of the atmospheric diffusion. This algorithm is translated in the f77 fortran code. For the dispersion prediction it needs as the input meteorological variables all of them listed above. As it was mentioned, type of the model is Gaussian and takes into account the multi source case. Input source parameters are: time of integration, coordinates and height of the sources, it's diameters, source's strength, temperature of pollutants and level above surface of concentration calculation. It can operate in the prognostic mode or in the diagnostic mode. Further, plume rise and effective stack height are calculated using the Brigg's model and wind power low, respectively. Model takes into consideration topography. The calculating domain is max. 160x160 km's.

Results of the dispersion calculations in the XY plane at some level Z, and in the XZ planes for every direction of the wind rose are presented in the tabular and graphical form.

Automated meteorological station together with this software is a fully autonomous system which gives the most usable results at the short distances, up to ~30 km from the sources of air pollution.

REFERENCES

Briggs G.A., 1969, Plume Rise, AEC Critical Review Series, TID-25075.

IAEA Safety Guide No. 50-SG-S3, 1980, Atmospheric Dispersion in Nuclear Power Plant Siting, Vienna

Sven-Erik Gryning,1980, Meteorological Measurement Programs for Air Pollution Studies, Riso-M-2259.

U.S.Nuclear Regulatory Commission, 1976, Regulatory Guide 1.111, Methods for Estimating Atmospheric Transport and Dispersion of Gaseous Effluents in Routine releases from Light-Water-Cooled Reactors. USNRC Office of Standards Development, Washington, D.C.

EVALUATION OF LINEAR WIND FLOW MODELS WITH COMPLEX TERRAIN FIELD DATA

F.N. Dunkerley

Department of Chemical Engineering, UMIST
PO Box 88, Manchester, M60 1QD, UK

INTRODUCTION

In recent years data on wind flow over isolated hills have been obtained in a number of experimental studies and these have been compared with the results from various numerical models. Flow over complex terrain, which encompasses a range of topography from rolling countryside to groups of isolated terrain features, is less well understood and few experimental data are available. The work described in this poster presentation mainly makes use of meteorological data from the MADONA[1] experiment, conducted on the DERA Porton Range, Salisbury, UK in September 1992, to assess the performance of linear wind flow models over the moderately complex terrain encountered on this site. These terrain and meteorological conditions fall well within the limiting criteria suggested for application of the models.

The evaluation study[2] was concerned with the prediction of the mean wind field only and was restricted to linear wind flow models. The three models considered were FLOWSTAR, LINCOM-NEUTRAL and LINCOM-THERMAL. The LINCOM models are complementary but the neutral model was applied to a range of stability conditions. There are a number of important differences between FLOWSTAR and the LINCOM models including the specification of the thermal field, which is still undergoing development in LINCOM_THERMAL. All the models require an input wind field, which strongly influences the predicted result, together with terrain and surface roughness information.

During the MADONA experiment 14 cup-and-vane anemometers were deployed at 10m above ground. The anemometers were located on upwind and lee slopes, peaks or upstream and downstream of terrain features depending on the wind direction. At some locations such straightforward descriptions were not appropriate and a number of terrain features could affect the flow. Thirty minute average model results were compared with the measured data for over 250 periods. The results were classified by stability and wind direction and a quantitative analysis was performed using the BOOT[3] software performance

Air Pollution Modeling and Its Application XIII, edited by S.-E. Gryning
and E. Batchvarova, Kluwer Academic/Plenum Publishers, 2000

measures. Residual scatter plots were also used to qualitatively analyse trends in model results as a function of primary (input) data and these are also presented in this poster.

RESULTS

In general it is expected that the mean flow will speed up over an isolated hill and then decelerate behind it with the development of a wake region downwind. The position of maximum speed-up moves downwind as stability increases and a corresponding deceleration is seen upwind of the terrain. This trend is broadly followed by both the MADONA observations and the model predictions. The linear models mainly underestimated speed-up and this may be attributed to the problem of correctly specifying the input (mesoscale) wind field. As expected they also failed to predict slow-down due to separation in the wake region.

Agreement between observed and predicted mean wind speeds was poor at the more "complex" locations where a number of terrain components may affect the wind field. Model performance improved when locations could be more easily defined (eg peak, upwind slope etc). Significant deviations from the background wind direction are observed where the mean flow is apparently channelled by the shape of the terrain. These directional changes could not be replicated by the linear models.

From the quantitative analysis FLOWSTAR showed statistically good agreement with observations at a number of locations under neutral/stable conditions (classified by h/L_{mo}). LINCOM-NEUTRAL also performed well under neutral conditions classified by z/L_{mo}. The method of stability classification strongly influenced the results of the quantitative assessment. Under the h/L_{mo} system considerably more periods were classed as unstable or stable compared with the z/L_{mo} system where most periods were neutral. The results for LINCOM-THERMAL were highly dependent on the input parameters used to specify the linear potential temperature perturbation. Good results were obtained when arbitrary fixed values were assigned to the input parameters. More sophisticated approaches resulted in some extreme predictions.

Sensitivity analysis was performed for a number of input parameters. A statistically significant change in all the model results was found for a 5% increase in input wind speed.

Acknowledgements. The work described here was funded by the Hazard Assessment Group of DERA Porton Down. It is part of a larger research and development programme into dispersion over complex terrain.

REFERENCES

1. MADONA CDROM Database Version 2.0, CBDE (1995).
2. F.N. Dunkerley, Evaluation of models of wind flow and dispersion in complex terrain, Interim Report No.3 on Agreement 2044/017/CBDE, UMIST ETC (1997).
3. S.R. Hanna, D.G. Strimaitis and J.C. Chang, Hazard response modeling uncertainty volume I: Users guide for evaluating hazardous gas dispersion models, Sigma Research Corporation (1991).

EVALUATION OF SO_2 AND NO_2 CONCENTRATION LEVELS IN VILNIUS (LITHUANIA) USING PASSIVE DIFFUSION SAMPLERS

D.Perkauskas and A.Mikelinskiene

Institute of physics, Savanoriu 231
2028 Vilnius, Lithuania

INTRODUCTION

Sometimes it is useful to use diffusion sampler first introduced by Palmes et al. (1976) in the field of occupational hygiene and used for indoor NO_2 measurement where the sorbed amount must be proportional to the ambient concentration of the gas. The diffusion samplers for the outdoor environment - ambient SO_2 and NO_2 measurements in both urban and rural areas are widely described (Campbell, 1988; Ferm, 1991; Campbell et al., 1994; Carmichael et al., 1995). The aim of this study was to evaluate SO_2 and NO_2 concentration levels in the main district types of Vilnius such as cross roads, dwelling regions, rest places and background level using rather cheap type of diffusion sampler.

EXPERIMENTAL AND RESULTS

The sampler and analytical methods used were similar to those of Palmes et al. (1976) and Ferm (1991). A sampler comprises a polypropylene tube with 70 mm length and 12 mm diameter using only steel net with thread diameter of 0.125 mm with a mesh size of 0.2 mm on the bottom without a Teflon filter. An impregnated filter -(Filtrak or Whatman 40) is placed on top of the sampler. The filter is attached by putting on snap on cap which is made of low density polyethylene.

The impregnation solution for NO_2 consists of trietanolamine. The NO_2 concentration of the extract is determined colorimetrically using Gris reagent. The impregnation solution for SO_2 consists of $NaHCO_3$. The analysis of SO_2 is made by turbidimetric method using BaCl. For a sampling time of one month, the measuring ranges are approximately 0.5-150 μg NO_2/m^3 and 0.6-1000 μg SO_2/m^3.

The method of passive diffusion samplers has been used for evaluation of SO_2 and NO_2 concentration levels in Lithuanian capital Vilnius. SO_2 and NO_2 concentrations have been analysed in 70 samplers twice per 4 seasons during 1995-1996 for each component

Air Pollution Modeling and Its Application XIII, edited by S.-E. Gryning
and E. Batchvarova, Kluwer Academic/Plenum Publishers, 2000

exposed in duration 1 month in main cross-roads, dwelling regions, rest places and in background-suburban layer - so about 600 samplers were exposed and analysed including control parallel sampling. Main results, excluding parallel sampling and some mistakes in choosing sampling place (when samplers were placed too close to low emission sources) are presented in Table 1.

Table 1. SO_2 and NO_2 concentration values in Vilnius during 1995-1996 in μg/m^3

	SO_2					NO_2				
	Mean	**SD**	**min**	**max**	**N**	**Mean**	**SD**	**min**	**Max**	**N**
Cross roads										
Spring	23,1	22,1	2,5	95,6	40	50,7	18,1	8,0	74,8	40
Summer	11,3	13,2	2,5	67,2	40	47,0	14,1	11,6	80,2	40
Autumn	14,3	10,1	2,5	42,6	42	69,0	15,5	46,2	124,2	42
Winter	21,4	15,6	2,5	68,4	42	77,0	19,8	21,1	108,8	42
Dwelling r.										
Spring	29,4	21,7	2,5	69,3	34	26,4	18,5	8,0	84,0	35
Summer	13,2	11,2	1,3	49,1	41	21,5	8,4	7,8	44,6	41
Autumn	12,6	10,4	2,5	44,1	39	34,5	10,3	19,5	57,3	39
Winter	24,5	20,5	2,5	68,4	38	43,6	14,9	17,3	73,6	38
Backgr. Lvl.										
Spring	16,3	9,6	2,5	31,6	13	8,0	4,4	4,6	18,0	13
Summer	6,5	3,3	2,0	13,2	13	8,9	2,9	4,8	14,9	14
Autumn	12,1	12,6	2,5	42,4	16	16,7	6,6	6,1	31,7	16
Winter	16,9	10,7	2,5	38,4	14	15,8	8,9	3,8	35,2	14
Rest places.										
Spring	16,3	15,0	2,5	48,0	8	14,5	7,4	8,4	30,4	8
Summer	9,4	6,6	2,0	20,3	7	15,1	9,5	9,4	36,1	7
Autumn	11,5	7,6	2,5	26,5	9	38,4	17,1	23,3	74,4	9
Winter	22,2	12,9	2,5	38,4	8	37,5	19,1	16,0	70,4	8

The SD in the table 1. represents standard deviation, min. and max. - minimum and maximum values of concentration respectively. As can be seen, the values of SO_2 concentration depend mainly on heating (cold and warm season) and exhibit average values of 7-13 μg/m^3 for warm and 17-24 μg/m^3 for cold seasons. The NO_2 average rates depend strongly on traffic (sampling place) and are highest in crossroads (47 μg/m^3 for warm and 77 μg/m^3 for cold seasons) and lowest at the background-suburban layer (8 μg/m^3 for warm and 17 μg/m^3 for cold seasons). From another point of view the higher SO_2 concentrations in spring than in autumn can be explained by heating season time interval which usually starts in November 1 and deals until the end of April. The higher wind speed values in autumn and winter than in spring and summer causes the more uniform seasonal distribution of air pollution. It should be noted, (Perkauskas et al., 1997) that urban background air pollution by SO_2 and NO_2 is about 30-50% higher than in clean Lithuanian regions.

Also, the measurements of the NO_2 in the 363 urban cities of United Kingdom (Campbell et al., 1994) shows that concentration levels varies from 17.5 until 87.5 μg/m^3. This result is rather close to our measurements. The investigations will be continued.

REFERENCES

Campbell G.W. (1988) Measurements of nitrogen dioxide concentrations in the United Kingdom using diffusion tubes. *Envir. Pollut.* 55: 251-270 (1988).

Campbell G.W., Stedman J.R., and Stevenson K. A survey of nitrogen dioxide concentrations in the United Kingdom using diffusion tubes, July-December 1991, *Atmospheric Environment* 28: 477-486 (1994).

Carmichael G.R., Ferm M., Adikary S., Ahmad J., Mohan M., Hong M-S., Chen L., Fook L., Liu C.M., Soedomo M., Tran G., Suksomsank K., Zhao D., Arndt R., and Chen L.L. Observed regional distribution of sulfur dioxide in Asia, *Water, Air and Soil Pollution*, 85:2289-2294 (1995).

Ferm M. *A sensitive diffusional sampler*. Report. Swedish Environmental Research Institute. Goteborg. (1991).

Palmes E.D., Gunniston A.F., DiMattio J. and Tomczyk C. Personal sampler for nitrogen dioxide, *Am. Ind. Hyg. Ass. J.*, 37:570-577 (1976).

Perkauskas, D., Mikelinskiene, A., Giedraitis, B., Milukaite, A., and Juozefaite, V. Evaluation of SO_2 and NO_2 concentration levels in Lithuania using passive diffusion samplers. *Proceedings of twenty-second NATO/CCMS international technical meeting on air pollution modelling and its application*, Clermont-Ferrand, France, June 2-6, 529-530 (1977).

PARTICIPANTS

The 23rd NATO/CCMS International Technical Meeting on Air Pollution Modelling and Its Application. Riviera Holiday Club, Varna, Bulgaria, September 28 - October 2, 1998

ARGENTINA

Gassmann M. I.

Department of Atmospheric Sciences
University of Buenos Aires
Pabellon II - Piso II, Ciudad Universitaria
1428 Buenos Aires
Tel.: +54 1 576 33 64 ext. 15
Fax: +54 1 576 33 64 ext. 12
E-mail: ulke@at1.fcen.uba.ar

AUSTRALIA

Luhar A.

Division of Atmospheric Research
CSIRO
PMB 1, Aspendale
VIC 3195
Tel.: +61 3 9239 4624
Fax: +61 3 9239 4444
E-mail: ashok.luhar@dar.csiro.au

AUSTRIA

Pechinger U.

Central Institute for Meteorology and Geodynamics
Hohe Warte 38
1190 Vienna
Tel.: +43 1 36 0 26 2402
Fax: +43 1 36 0 26 74
E-mail: Ulrike.Pechinger@zamg.ac.at

BELGIUM

Mensink C.

VITO
Centre for Remote Sensing and Atmospheric Processes
Boeretang 200
2400 Mol
Tel.: +32 14 336841
Fax: +32 14 322795
E-mail: mensinkc@vito.be

Schayes G.

UCL
Institut d'Astronomie et de Geophysique
Universite Catholique de Louvain
2, Chemin du Cyclotron
1348 Louvain-la-Neuve,
Tel: +32 104 73304
Fax: +32 104 74722
E-mail: schayes@astr.ucl.ac.be

BRAZIL

Reis N.

Department of Hydraulics
Federal University of Espirito Santo
Cxp: 01-9011, cep.: 29.060-970
Vitoria, ES
Mailing address:
Environmental Technology Centre
UMIST P.O. box 88
Manchester M60 1QD
U.K.
Tel.: +44 161 200 3984
Fax: +44 161 200 3988
E-mail: mjkpgynr@stud.umist.ac.uk

BULGARIA

Atanassov D.

National Institute of Meteorology and Hydrology
66 Tzarigradsko chaussee
1784 Sofia
Tel.: +359 2 975 3991 ext. 363
Fax: +359 2 880380
E-mail: Dimiter.Atanassov@meteo.bg

Batchvarova E.

National Institute of Meteorology and Hydrology
66 Tzarigradsko chaussee
1784 Sofia
Tel.: +59 2 975 3991 ext. 428
Fax: +359 2 880380
E-mail: Ekaterina.Batchvarova@meteo.bg

Bogdanov S. | National Institute of Meteorology and Hydrology
66 Tzarigradsko chaussee
1784 Sofia
Tel.: +359 2 975 3991 ext. 224
Fax: +359 2 880380
E-mail: Stanislav.Bogdanov@meteo.bg

Dimitrova R. | Institute of Geophysics
Bulgarian Academy of Sciences
Acad. Bonchev Str., Block 3
1113 Sofia
Tel.: +359 2 979 3708
Fax: +359 2 700226
E-mail: rendim@geophys.bas.bg

Ganev K. | Institute of Geophysics
Bulgarian Academy of Sciences
Acad. Bonchev Str., Block 3
1113 Sofia
Tel.: +359 2 979 3708
Fax: +359 2 700226
E-mail: kganev@geophys.bas.bg

Georgiev K. | Central Laboratory for Parallel Processing
Bulgarian Academy of Sciences
Acad. G. Bonchev str. B1 25-A
1113 Sofia
Tel.: +359 2 979 6612
Fax: +359 2 707273
E-mail: georgiev@bas.bg

Georgieva E. | Institute of Geophysics
Bulgarian Academy of Sciences
Acad. Bonchev Str., Block 3
1113 Sofia
Tel.: +359 2 979 3708
Fax: +359 2 700226
E-mail: emilia@geophys.bas.bg

Ivancheva-Kirova J. | National Institute of Meteorology and Hydrology
66 Tzarigradsko chaussee
1784 Sofia
Tel.: +359 2 975 3991 ext.454
Fax: +359 2 880380
E-mail: Ani.Tzenkova@meteo.bg

Kolarova M.	National Institute of Meteorology and Hydrology 66 Tzarigradsko chaussee 1784 Sofia Tel.: +359 2 975 3991 Fax: +359 2 880380 E-mail: Maria.Kolarova@meteo.bg
Miloshev N.	Institute of Geophysics Bulgarian Academy of Sciences Acad. Bonchev Str., Block 3 1113 Sofia Tel.: +359 2 979 3086 Fax: +359 2 700226 E-mail: miloshev@geophys.bas.bg
Panchev S.	University of Sofia, Faculty of Physics 5, James Bourchier, Blvd 1126 Sofia Tel.: +359 2 681394 Fax: +359 2 962 5276 E-mail: spanchev@phys.uni-sofia.bg
Parvanov O.	Institute of Electronics Bulgarian Academy of Sciences 72 Tzarigradsko chaussee Sofia 1784 Tel.: +359 2 750096 Fax: +359 2 975 3201 E-mail: blteam@phys.acad.bg
Prodanova M.	National Institute of Meteorology and Hydrology 66 Tzarigradsko chaussee 1784 Sofia Tel.: +359 2 975 3991 ext.433 Fax: +359 2 880380 E-mail: Maria.Prodanova@meteo.bg
Spassova T.	National Institute of Meteorology and Hydrology 66 Tzarigradsko chaussee 1784 Sofia Tel.: +359 2 975 3991 ext. 363 Fax: +359 2 880380 E-mail: Tatiana.Spassova@meteo.bg
Syrakov D.	National Institute of Meteorology and Hydrology 66 Tzarigradsko chaussee 1784 Sofia Tel.: +359 2 975 3991 ext. 536 Fax: +359 2 880380 E-mail: Dimiter.Syrakov@meteo.bg

Syrakov E.

University of Sofia
Faculty of Physics
5, James Bourchier, Blvd
1126 Sofia
Tel.: +359 2 665 6315
Fax: +359 2 962 5276
E-mail: esyrakov@phys.uni-sofia.bg

Syrakova M.

University of Sofia
Faculty of Physics
5, James Bourchier, Blvd
1126 Sofia
Tel.: +359 2 625 6291
Fax: +359 2 962 5276
E-mail: marsyr@phys.uni-sofia.bg

Todorov T.

National Institute of Meteorology and Hydrology
66 Tzarigradsko chaussee
1784 Sofia
Tel.: +359 2 975 3991 ext.388
Fax: +359 2 880380
E-mail: Todor.Todorov@meteo.bg

Tzenkova-Bratoeva A.

National Institute of Meteorology and Hydrology
66 Tzarigradsko chaussee
1784 Sofia
Tel.: +359 2 975 3991 ext. 454
Fax: +359 2 880380
E-mail: Ani.Tzenkova@meteo.bg

Yordanov D. L.

Institute of Geophysics
Bulgarian Academy of Sciences
Acad. Bonchev Str., Block 3
1113 Sofia
Tel.: +359 2 979 3086
Fax: +359 2 700226
E-mail: emilia@geophys.bas.bg

CANADA

Gong W.

Atmospheric Environment Service
4905 Dufferin Street
Downsview, Ontario M3H 5T4
Tel.: +1 416 739 4883
Fax: +1 416 739 4288
E-mail: wanmin.gong@ec.gc.ca

Lin X.

Environmental Technologies Department
Ontario Hydro Technologies
800, Kipling Ave. KJ135
Toronto, Ontario M8Z 5S4
Tel.: +1 416 207 6403
Fax: +1 416 207 5550
E-mail: lin@rd.hydro.on.ca

Steyn D. G.

Atmospheric Science Programme
Department of Geography
The University of British Columbia
Vancouver, B.C. V6T 1Z2
Tel.: +1 604 822 6407
Fax: +1 604 822 6150
E-mail: douw@geog.ubc.ca

Wilson D. J.

Combustion and Environment Group
Department of Mechanical Engineering
University of Alberta,
Edmonton, Alberta T6G 2G8
Tel: +1 403 492 5467
Fax: +1 403 492 2200
E-mail: david.wilson@ualberta.ca

CZECH REPUBLIC

Bubnik J.

Czech Hydrometeorological Institute
Na Sabatce 17
14306 Praha 4
Tel.: +42 02 4403 2409
Fax: +42 02 4403 2468
E-mail: bubnik@chmi.cz

Macoun J.

Czech Hydrometeorological Institute
Na Sabatce 17
14306 Praha 4
Tel.: +42 02 4403 2410
Fax: +42 02 4403 2468
E-mail: macoun@chmi.cz

DENMARK

Falk A. K. V.

Risø National Laboratory
4000 Roskilde
Tel.: +45 4677 5013
Fax: +45 4677 5970
E-mail: annekatrine.vinther@risoe.dk

Gryning S.-E.	Risø National Laboratory 4000 Roskilde Tel.: +45 4677 5005 Fax: +45 4677 5970 E-mail: sven-erik.gryning@risoe.dk
Jensen A. B.	Danish Toxicology Center Kogle Alle 2 2970 Hørsholm Tel: +45 4576 2055 Fax: +45 4576 2455 E-mail: abj@dtc.dk
Kiilsholm S.	Danish Meteorological Institute Lyngbyvej 100 2100 Copenhagen Ø Tel.: +45 3915 7452 Fax: +45 3915 7460 E-mail: sk@dmi.dk
Sørensen J. H.	Danish Meteorological Institute Lyngbyvej 100 2100 Copenhagen Ø Tel.: +45 3915 7432 Fax: +45 3915 7460 E-mail: jhs@dmi.dk
Zlatev Z.	National Environmental Research Institute Department of Atmospheric Environment Frederiksborgvej 399, 4000 Roskilde Tel.: +45 4630 1149 Fax: +45 4630 1214 E-mail: luzz@sun2.dmu.dk

ESTONIA

Kaasik M.	Department of Environmental Physics University of Tartu Tähe 4 51010 Tartu Tel.: +37 27 410258 Fax: +37 27 410205 E-mail: marko@apollo.aai.ee

FINLAND

Hongisto M.

Finnish Meteorological Institute
Air Quality Research
Vuorikatu 19,
00101 Helsinki
Tel.: +358 9 1929 3150
Fax: +358 9 1929 3146
E-mail: marke.hongisto@fmi.fi

Jylha K.

Finnish Meteorological Institute
Air Quality Research
Vuorikatu 19
00101 Helsinki
Tel.: +358 9 1929 3148
Fax: +358 9 1929 3146
E-mail: kirsti.jylha@fmi.fi

Kukkonen J.

Finnish Meteorological Institute
Air Quality Research
Sahaajankatu 20E,
00810 Helsinki
Tel.: +358 9 1929 5450
Fax: +358 9 1929 5403
E-mail: Jaakko.Kukkonen@fmi.fi

Makara V.

Finnish Meteorological Institute
Air Quality Research
Vuorikatu 19,
00101 Helsinki
Tel.: +358 9 1929 5427
Fax: +358 9 1929 5403
E-mail: viktor.makara@fmi.fi

Pietarila H.

Finnish Meteorological Institute
Sahaajankatu 20 E
00810 Helsinki
Tel: +358 9 1929 5432
Fax: +358 9 1929 5403
E-mail: Harri.Pietarila@fmi.fi

Savunen T.

Finnish Meteorological Institute
Air Quality Research
Sahaajankatu 20E
00810 Helsinki
Tel.: 358 9 1929 5434
Fax: 358 9 1929 5403
E-mail: Tarja.Savunen@fmi.fi

FRANCE

Bompay F.

Meteo France
SCEM/Services Environnement
42, Avenue G. Coriolis
31057 Toulouse Cedex
Tel.: +33 5 6107 8086
Fax: +33 5 6107 8009
E-mail: francois.bompay@meteo.fr

Chaumerliac N.

LAMP/OPGC
Universite Blaise Pascal
63170 Aubiere,
Tel: +33 4 7340 7372
Fax: +33 4 7327 1657
E-mail: Chaumerl@opgc.univ-bpclermont.fr

Isnard O.

Ecole Centrale de Lyon
36, avenue Guy de Collongue
69131 Ecully cedex
Tel.: +33 4 7218 6303
Fax: +33 4 7833 1380
E-mail: isnard@mecaflu.ec-lyon.fr

Poulet D.

Laboratoire de Meteorologie Physique
Universite Blaise Pascal/CNRS/OPGC
24, Avenue des Landais
F-63177 Aubiere Cedex
Tel.: +33 4 7340 7370
Fax: +33 4 7327 1657
E-mail: poulet@opgc.univ-bpclermont.fr

Soulhac L.

Ecole Centrale de Lyon
36, avenue Guy de Collongue
69131 Ecully cedex
Tel.: +33 4 7218 6303
Fax: +33 4 7833 1380
E-mail: soulhac@mecaflu.ec-lyon.fr

GERMANY

Fay B.

Deutscher Wetterdienst
Department Climate and Environment
Frankfurter Str. 135
63067 Offenbach
Tel.: +49 69 8062 2748
Fax: +49 69 8236 1493
E-mail: bfay@dwd.d400.de

Feichter J. | Max-Planck-Institute for Meteorology
Department Theoretical Climate Modelling
Bundesstr. 55,
20146 Hamburg
Phone: +49 40 4117 3317
Fax: +49 40 4117 3298
E-mail: feichter@dkrz.de

Flemming J. | Freie Universität Berlin
Institut für Meteorologie
Carl Heinrich Becher Weg 6-10
12165 Berlin
Tel.: +49 30 8387 1129
Fax: +49 30 8593 9095
E-mail: Flemming@zedat.fu-berlin.de

Graff A. | Federal Environmental Agency
Air Quality Management Section (II 4.1)
Bismarckplatz 1
14193 Berlin
Tel.: +49 30 8903 2323
Fax: +49 30 8903 2285
E-mail: arno.graff@uba.de

Reimer E. | Freie Universität Berlin
Institut für Meteorologie
Carl Heinrich Becher Weg 6-10
12165 Berlin
Tel.: +49 30 8387 1190
Fax: +49 30 8593 9095
E-mail: reimer@zedat.fu-berlin.de

Stohl A. | LMU, Bioklimatologie und Immissionsforschung
Am Hochanger 13
85354 Freising - Weihenstefan
Tel.: +49 81 6171 4742
Fax: +49 81 6171 4753
E-mail: as@atmos1.met.forst.uni-muenchen.de

Thielen H. | GRS Cologne
Schwertnergasse 1
50667 Cologne
Tel: +49 22 1206 8772
Fax: +49 22 1206 8888
E-mail: thi@grs.de

Tilmes S.

Deutscher Wetterdienst
FE 25
Kaiserleistrasse 42
63067 Offenbach
Tel.: +49 69 8062 2746
Fax: +49 69 8236 1493
E-mail: stilmes@dwd.d400.de

Walter H.

Federal Office for Radiation Protection
Ingolstädter Landstr. 1
85764 Oberschleissheim
Tel.: +49 89 3160 3368
Fax: +49 89 3160 3270
E-mail: hwalter@bfs.de

Wolke R.

Institut for Tropospheric Research
Permoserstr. 15,
04303 Leipzig,
Tel.: +49 34 1235 2860
Fax: +49 34 1235 2361
E-mail: wolke@tropos.de

GREECE

Kakaridou M.

Department of Environmental Studies
University of the Aegean
Karantoni 17
81100 Mytilene
Tel: +30 2513 6251
Fax: +30 2512 3783
E-mail: mkak@env.aegean.gr

Kallos G.

University of Athens
Department of Applied Physics
Meteorology Lab.
Panepistimioupolis Bldg. 1
15784 Athens
Tel.: +30 1 7284835/7284923
Fax: +30 1 7295281
E-mail: kallos@skiron.mg.uoa.gr

Melas D.

Department of Environmental Studies
University of the Aegean
Karantoni 17
81100 Mytilene
Tel.: +30 251 21286
Fax: +30 251 23783
E-mail: melas@mailhost.ccf.auth.gr

Rafailidis S.
6 Pavlou Mela str
54621 Thessaloniki
E-mail: rafailidis@hotmail.com

Sahm P.
Aristotle University of Thessaloniki
Department of Mechanical Engineering
Box 483
54006 Thessaloniki
Tel.: +30 31 996011
Fax: +30 31 996012
E-mail: peter@aix.meng.auth.gr

ISRAEL

Levitin J.
Israel Meteorological Service
Research Group
P.O. box 25
50250 Bet Dagan
Tel.: +972 3 9682133
Fax: +972 3 9604854
E-mail: rnd1@meteo-serv.gov.il

ITALY

Anfossi D.
CNR, Istituto di Cosmo-geofisica
Corso Fiume 4
10133 Torino
Tel.: +39 011 6306819
Fax: +39 011 6604056
E-mail: anfossi@to.infn.it

Baldi M.
IFA - CNR
Ist. Atmospheric Physics
Via Fosso del Cavaliere
00133 Roma
Tel.: +39 06 4993 4273
Fax: +39 06 2066 0291
E-mail: baldi@atmos.ifa.rm.cnr.it

Bianconi R.
ENVIROWARE srl
Centro Colleoni, Andromeda 1
Agrate Brianza (MI)
Tel: +39 039 6890073
Fax: +39 039 6881459
E-mail: rbianconi@enviroware.com

Canepa E.

Department of Physiscs
University of Genova
33 Via Dodecaneso
16146 Genova
Tel.: +39 010 3536355
Fax: +39 010 353 6354
E-mail: canepae@fisica.unige.it

Carizi G.

SnamProgetti / Envi
Via Toniolo, 1
61032 Fano (PS)
Phone: +39 721 881772
Fax:: +39 721 881787
E-mail: giorgio.carizi@snamprogetti.eni.it

Golinelli M.

ARA-Servizio Meteorologico
Regionale Emilia-Romagna
Viale Silvani 6
40126 Bologna
Tel: +39 051 284535
Fax: +39 051 284664
E-mail: envi@meteo2.arpamet.regione.emilia-romagna.it

Graziani G.

Environment Institute
Commission of the European Communities, Joint
Research Centre
21020 Ispra (VA)
Tel.: +39 0332 789295
Fax: +39 0332 785466
E-mail: giovanni.graziani@jrc.it

Lanzani G.

Amministrazione Provinciale di Como
Servizio Aria ed Energia
Via Borgovico, 148
22100 Como
Tel: +39 031 230312
Fax: +39 031 230345
E-mail: provcoar@tin.it

Longoni M. G.

Regione Lombardia
Via Fara 26
Milano
Tel.: +39 02 6765 2460
Fax: +39 02 6706222
E-mail: giu_maf_model@hotmail.com

Martini A. D.

Regione Lombardia
Via Fara 26
Milano
Tel.: +39 02 6765 2440
Fax: +39 02 6706222
E-mail: giu_maf_model@hotmail.com

Mosca S.

ENVIROWARE srl
Centro Colleoni, Andromeda 1
Agrate Brianza (MI)
Tel.: +39 039 6890073
Fax: +39 039 6881459
E-mail: smosca@enviroware.com

Tinarelli G.

ENEL/SRI Area Ambiente
Via Rubattino 54
20134 Milano
Tel.: +39 02 7224 3936
Fax: +39 02 7224 3915
E-mail: tinarelli@cram.enel.it

Tirabassi T.

Institute Fisbat of CNR
Via Gobetti 101
40129 Bologna
Tel.: +39 051 6399601
Fax: +39 051 6399654
E-mail: tirabassi@area.bo.cnr.it

Trini Castelli S.

Istituto di Cosmogeofisica del C.N.R.
Corso Fiume 4
10133 Torino
Tel.: +39 11 6306816
Fax: + 39 11 6604056
E-mail: trini@to.infn.it

JAPAN

Kitada T.

Department of Ecological Engineering
Toyohashi University of Technology
Tempaku-cho
441-8580 Toyahashi
Tel.: +81 5324 46902
Fax: +81 5324 46929
E-mail: kitada@earth.eco.tut.ac.jp

Kouchi A.

Nagasagi Research & Development Center
Mitsubishi Heavy Industries
Technical Headquarter
5-717-1 Hukahori-Machi
Nagasaki 850-91
Tel.: +81-9583 42832
Fax: +81 9583 42385
E-mail: kouchi@ngsrdc.mhi.co.jp

LITHUANIA

Perkauskas D.

Institute of Physics
Savanoriu 231
2028 Vilnius
Tel: +370 2641527
Fax: +370 2321975
E-mail: donatasp@ktl.mii.lt

NORWAY

Iversen T.

Department of Geophysics
University of Oslo
P.O.Box. 1022 - Blindern
0315 Oslo
Tel.: +47 2285 5821
Fax: +47 2285 5269
E-mail: trond.iversen@geofysikk.uio.no

Mylona S.

DNMI
P.O.Box 43 - Blindern
0313 Oslo
Tel.: +47 2296 3325
Fax: +47 2296 3050
E-mail: sophia.mylona@dnmi.no

Slørdal L. H.

Norwegian Institute for Air Research
Instituttvägen 18,
2007 Kjeller
Tel.: +47 6389 8165
Fax: +47 6389 8050
E-mail: leiv.haavard.slordal@nilu.no

PORTUGAL

Borrego C.

Department of Environment and Planning
University of Aveiro
3810 Aveiro
Tel.: +351 34 370832
Fax: +351 34 28600
E-mail: borrego@ua.pt

RUSSIA

Galperin M. V.

Independent Researcher
Studencheskaya str. 38-31
Moscow 121165
Tel.: +7 095 249 1280,
E-mail: galperin@edunet.ru

Genikhovich E.

Main Geophysical Observatory
7, Karbysheva Street
194021 St. Petersburg
Fax: +7 812 247 8661
E-mail: ego@main.mgo.rssi.ru

Gusev A.

Meteorological Synthesizing Centre-East, EMEP
Kedrova Str. 8, k.1
117292 Moscow
Tel.: +7 095 124 4758
Fax: +7 095 310 7093
E-mail: msce@glasnet.ru

Rivin G. S.

Institute of Computational Technologies SB,
RAS, 6 Ac. Lavrentjev Ave
630090 Novosibirsk
Tel: +7 383 232 4257
Fax: +7 383 234 1342
E-mail: rivin@adm.ict.nsc.ru

Sofiev M. A.

Institute of Program Systems
Russian Academy of Sciences
Stroiteley str. 4, bldg. 1 app 18
117311 Moscow
Tel.: +7 095 930 0961
Fax: +7 095 930 0961
E-mail: mas@glasnet.ru

SERBIA

Grsic Z.

Institute of Nuclear Sciences "Vinca"
P.O.Box 522
11001 Belgrade
Tel: +381 11 4440871, Ext.621
Fax: +381 11455943
E-mail: grsa@rt270.vin.bg.ac.yu

Milutinovic P.

Institute of Physics
Pregrevica 118
11080 Belgrade
Phone: +381 11 3160260, Ext 110
E-mail: presa@exp.phy.bg.ac.yu

SOUTH AFRICA

Djolov G.

University of Venda, Department of Physics
P.Bag X5050
Thohoyandou 0920
Northern Province
Tel.: +27 159 824757 ext.2317
Fax: +27 159 824749
E-mail: djolov@caddy.univen.ac.za

SPAIN

Calbó J.

Department of Physics
Universitat de Girona
Escola Politecnica Superior, EPS II
17071 Girona
Tel.: +34 972 418491
Fax: +34 972 418098
E-mail: jcalbo@speedy.udg.es

Salvador R.

CEAM, Parque Tecnologico,
Calle 4, Sector Oeste
46980 Paterna (Valencia)
Tel: +34 93 4011745
Fax: +34 93 3340255
E-mail: salvador@ceam.es

SWEDEN

Persson C.

Swedish Meteorological and Hydrological Institute
601 76 Norrköping
Tel.: +46 11 495 8196
Fax: +46 11 495 8001
E-mail: christer.persson@smhi.se

Robertson L.

Swedish Meteorological and Hydrological Institute
601 76 Norrköping
Tel.: +46 11 495 8281
Fax: +46 11 495 8001
E-mail: lroberts@smhi.se

Wiman B. L. B.
School of Nat. Resources Man. & Agenda-21 Research
Faculty of Natural Sciences
Kalmar University
P.O. Box 905
391 29 Kalmar
Tel: +46 480 446200
Fax: +46 480 446262
Email: milj-bwi@hermes.lu.se

Zilitinkevich S.
Uppsala University
Department of Earth Sciences
Meteorology (MIUU)
Villavägen 16
752 36 Uppsala
Tel: +46 18 4717193
Fax: +46 18 551124
E-mail: sergej@big.met.uu.se

SWITZERLAND

de Haan P.
Swiss Federal Institute of Technology, GIETH,
Winterthurerstrasse 190
8057 Zürich
Tel.: +41 1 635 5211
Fax: +41 1 362 5197
E-mail: dehaan@geo.umnw.ethz.ch

Herrnberger V. R. D.
Paul Scherrer Institute
5232 Villigen
Tel: +41 56 310 2111
Fax: +41 56 310 2309
E-mail: "PSW225::METEO"@psiclb.psi.ch

Rotach M.
Swiss Federal Institute of Technology, GIETH,
Winterthurerstrasse 190
8057 Zürich
Tel.: +41 1 635 5222
Fax: +41 1 362 5197
E-mail: rotach@geo.umnw.ethz.ch

THE NETHERLANDS

Builtjes P.
TNO-MEP
P.O. Box 342
7300 AH Apeldoorn
Tel.: +31 55 549 3038
Fax: +31 55 549 3252
E-mail: builtjes@mep.tno.nl

van Dop H.

Institute for Marine and Atmospheric Research
Princetonplein
3584 CC Utrecht
Tel.: +31 30 2533154
Fax: +31 30 2543163
E-mail: H.vanDop@fys.ruu.nl

van Jaarsveld J. A.

RIVM
P.O. Box 1
3720 BA Bilthoven
Tel.: +31 30 2742818
Fax: +31 30 2287531
E-mail: hans.van.jaarsveld@rivm.nl

TURKEY

Özsoy E.

METU - Institute of Marine Sciences
P.O. Box 28
Erdemli 33731 Icel
Tel.: +90 324 521 2406
Fax: +90 324 521 2327
E-mail: ozsoy@ims.metu.edu.tr

Tayanc M.

Marmara University
Department of Environmental Engineering
Kuyubasi 81040, Kadikoy
Istanbul
Tel.: +90 216 349 2711-272
Fax: +90 216 348 0293
E-mail: mtayanc@marun.edu.tr

UNITED KINGDOM

Cai X.

School of Geography
University of Birmingham
Edgbaston
Birmingham B15 2TT
Tel: +44 121 414 5533
Fax: +44 121 414 5528
Email: x.cai@bham.ac.uk

Dunkerley F.

Environmental Technology Centre
UMIST P.O. Box 88
Manchester M60 1QD
Tel.: +44 161 200 3984
Fax: +44 161 200 3988
E-mail: fay.dunkerley@umist.ac.uk

Fisher B.

University of Greenwich
School of Earth and Environmental Sciences
Medway Campus
Chatham Maritime, Kent ME4 4TB
Tel.: +44 181 331 9911
Fax: +44 181 331 9805
E-mail: fb04@gre.ac.uk

Griffiths R. F.

Environmental Technology Centre
Department of Chemical Engineering
UMIST P.O Box 88
Manchester M60 1QD
Tel.: +44 161 200 3980
Fax: +44 161 200 4399
E-mail: richard.f.griffiths@umist.ac.uk

Gardner M. W.

School of Environmental Sciences
University of East Anglia
Norwich NR4 7TJ
Tel.: +44 1603 593115
Fax: +44 1603 507719
E-mail: m.w.gardner@uea.ac.uk

Kapias T.

Environmental Technology Centre
Department of Chemical Engineering
UMIST P.O. Box 88
Manchester M60 1QD
Tel: +44 161 200 3855
Fax: +44 161 200 3988
E-mail: mjkmeztk@fs1.ce.umist.ac.uk

U.S.A.

Odman M. T.

School of Civil and Environmental Engineering
Georgia Institute of Technology
Atlanta Georgia 30332-0512
Tel.: +1 404 894 2783
Fax: +1 404 894 8266
E-mail: talat.odman@ce.gatech.edu

Rao S. T.

New York State Dep. of Environmental Conservation
Office of Science and Technology
50 Wolf Road - Room 198
Albany, NY 12233-3259
Tel.: +1 518 457 3200
Fax: +1 518 485 8410
E-mail: strao@air.dec.state.ny.us

Schiermeier F. A.

Atmospheric Sciences Modeling Division
U.S.EPA/NOAA
Research Triangle Park, NC 27711
Tel.: +1 919 541 4542
Fax: +1 919 541 1379
E-mail: francis.schiermeier@noaa.gov

Wilkinson J. G.

School of Civil and Environmental Engineering
Georgia Institute of Technology
Atlanta Georgia 30332-0512
Rel: +1 404 894 3089
Fax: +1 404 894 9724
E-mail: jwilkins@tset.ce.gatech.edu

Yamartino R. J.

Sigma Research Division of Earth Tech
196 Baker Avenue, Concord
MA 01742
Tel.: +1 978 371 4265
Fax: +1 978 371 2468
E-mail: rjy@src.com

AUTHOR INDEX

SUBJECT INDEX